Defensive Mutualism in Microbial Symbiosis

MYCOLOGY SERIES

Editor
J. W. Bennett
Professor
Department of Plant Biology and Pathology
Rutgers University
New Brunswick, New Jersey

Founding Editor
Paul A. Lemke

1. *Viruses and Plasmids in Fungi*, edited by Paul A. Lemke
2. *The Fungal Community: Its Organization and Role in the Ecosystem*, edited by Donald T. Wicklow and George C. Carroll
3. *Fungi Pathogenic for Humans and Animals (in three parts)*, edited by Dexter H. Howard
4. *Fungal Differentiation: A Contemporary Synthesis*, edited by John E. Smith
5. *Secondary Metabolism and Differentiation in Fungi*, edited by Joan W. Bennett and Alex Ciegler
6. *Fungal Protoplasts*, edited by John F. Peberdy and Lajos Ferenczy
7. *Viruses of Fungi and Simple Eukaryotes*, edited by Yigal Koltin and Michael J. Leibowitz
8. *Molecular Industrial Mycology: Systems and Applications for Filamentous Fungi*, edited by Sally A. Leong and Randy M. Berka
9. *The Fungal Community: Its Organization and Role in the Ecosystem, Second Edition*, edited by George C. Carroll and Donald T. Wicklow
10. *Stress Tolerance of Fungi*, edited by D. H. Jennings
11. *Metal Ions in Fungi*, edited by Güfcnther Winkelmann and Dennis R. Winge
12. *Anaerobic Fungi: Biology, Ecology, and Function*, edited by Douglas O. Mountfort and Colin G. Orpin
13. *Fungal Genetics: Principles and Practice*, edited by Cees J. Bos
14. *Fungal Pathogenesis: Principles and Clinical Applications*, edited by Richard A. Calderone and Ronald L. Cihlar
15. *Molecular Biology of Fungal Development*, edited by Heinz D. Osiewacz
16. *Pathogenic Fungi in Humans and Animals: Second Edition*, edited by Dexter H. Howard
17. *Fungi in Ecosystem Processes*, John Dighton
18. *Genomics of Plants and Fungi*, edited by Rolf A. Prade and Hans J. Bohnert
19. *Clavicipitalean Fungi: Evolutionary Biology, Chemistry, Biocontrol, and Cultural Impacts*, edited by James F. White Jr., Charles W. Bacon, Nigel L. Hywel-Jones, and Joseph W. Spatafora
20. *Handbook of Fungal Biotechnology, Second Edition*, edited by Dilip K. Arora
21. *Fungal Biotechnology in Agricultural, Food, and Environmental Applications*, edited by Dilip K. Arora
22. *Handbook of Industrial Mycology*, edited by Zhiqiang An
23. *The Fungal Community: Its Organization and Role in the Ecosystem, Third Edition*, edited by John Dighton, James F. White, and Peter Oudemans
24. *F ungi: Experimental Methods in Biology*, Ramesh Maheshwari
25. *Food Mycology: A Multifaceted Approach to Fungi and Food*, edited by Jan Dijksterhuis and Robert A. Samson
26. *The Aspergilli: Genomics, Medical Aspects, Biotechnology, and Research Methods*, edited by Gustavo H. Goldman and Stephen A. Osmani
27. *Defensive Mutualism in Microbial Symbiosis*, edited by James F. White, Jr. and Mónica S. Torres

Defensive Mutualism in Microbial Symbiosis

Edited by

James F. White, Jr.
Rutgers University
New Brunswick, New Jersey, U. S. A.

Mónica S. Torres
Rutgers University
New Brunswick, New Jersey, U. S. A.

CRC Press
Taylor & Francis Group
Boca Raton London New York

CRC Press is an imprint of the
Taylor & Francis Group, an **informa** business

CRC Press
Taylor & Francis Group
6000 Broken Sound Parkway NW, Suite 300
Boca Raton, FL 33487-2742

Printed in the United States of America on acid-free paper
10 9 8 7 6 5 4 3 2 1

International Standard Book Number-13: 978-1-4200-6931-0 (Hardcover)

Library of Congress Cataloging-in-Publication Data

Defensive mutualism in microbial symbiosis / editors: James F. White and Monica S. Torres.
p. cm. -- (Mycology series ; v. 27)
Includes bibliographical references and index.
ISBN 978-1-4200-6931-0 (alk. paper)
1. Endophytic fungi. 2. Mutualism (Biology) 3. Symbiosis. I. White, James F. (James Francis), 1953- II. Torres, Monica S. III. Title. IV. Series.

QK604.2.E53D44 2009
579.5'17852--dc22 2008054550

Visit the Taylor & Francis Web site at
http://www.taylorandfrancis.com

and the CRC Press Web site at
http://www.crcpress.com

Dedication

This book is dedicated to Bianka N. Ruska

(December 28, 2006–May 24, 2008).

She brought sunshine and joy into this world.

Contents

Part III Eukaryotic Defensive Symbionts

Part IV Fungal Endophyte as Model System to Understand Defensive Mutualism

Preface

Protective effects on eukaryotic hosts stemming from microbial colonization are relatively common in nature (Janzen, 1977) but their study escalated after Keith Clay (1988) proposed the defensive mutualism hypothesis to explain the widespread distribution of endophytic fungi in grasses. Since then there has been a deluge of research evaluating defensive mutualisms in several kingdoms of organisms; and much has been learned about the phenomenon of microbial-based defensive mutualisms. We believe that the time is right for a work that unifies microbial-based defensive mutualisms in diverse systems into a single text for students, teachers, and investigators. We have not attempted to produce the definitive treatment of defensive mutualisms and fully expect that more comprehensive treatments will follow this book. We have also not tried to guide the authors to write chapters that fit any central theses or hypotheses. Instead, we requested only that they cover well the biological systems that are the topics of their chapters with respect to the defensive mutualism theme.

To organize this book, we initially intended to classify mutualisms based on defensive properties, including anti-herbivory disease protection, abiotic stress tolerance, etc.; however, it soon became clear that it is difficult to find simple defensive mutualisms with single defensive effects. The majority of the defensive mutualisms that have been studied closely have multiple defensive effects on hosts. For example, clavicipitaceous grass endophytes may show protection from herbivory, diseases, and abiotic stresses. We also evaluated the idea of classifying mutualisms based on the habitat of the host species, a system that was particularly artificial. Finally, we organized this book by having an introductory chapter and a set of chapters providing an overview of the subject, followed by chapters that group the defensive mutualisms based simply on whether the microbial partner is prokaryotic or eukaryotic. A final part contains chapters on plant endophytes from the perspective of their use as a model for the study of microbial-based defensive mutualisms.

This book is an attempt to unify writings of diverse defensive mutualistic systems into a single work in order to permit the reader to assimilate and acquire a perspective on defensive mutualisms, particularly those involving microbial partners. It is our hope that this book will make a contribution by facilitating discussions of defensive mutualisms on a higher level or across disciplines, and will provide the information necessary to resolve some of the questions and uncertainties surrounding the concept of defensive mutualism.

James F. White, Jr.
Mónica S. Torres

REFERENCES

Clay, K. 1988. Fungal endophytes of grasses: A defensive mutualism between plants and fungi, *Ecology* 69:10–16.

Janzen, D.H. 1977. Why fruits rots, seeds mold, and meat spoils. *The American Naturalists* 111:691–713.

Editors

James F. White, Jr. is professor and chair of the Plant Biology and Pathology Department in the School of Environmental and Biological Science at Rutgers University, New Brunswick, New Jersey. He conducts research on the biology of fungal endophytes, is author of more than 150 articles, and is editor of several reference books on the biology, taxonomy, and phylogeny of fungi including the following: *Biotechnology of Acremonium Endophytes of Grasses* (1994), *Microbial Endophytes* (2000), *The Clavicipitalean Fungi* (2004), and *The Fungal Community: Its Organization and Role in the Ecosystem* (2005). Dr. White received his MS in mycology and plant pathology from Auburn University, Alabama and his PhD in mycology from the University of Texas, Austin, Texas. He teaches both undergraduate and graduate students in fungal biology. He served as the founding secretary of the International Symbiosis Society. He is currently the associate editor of the journal *Mycologia* and the *Encyclopoedia of Microbiology* published by Elsevier.

Mónica S. Torres received her BS in biology from the National University of La Plata (Argentina), her MS in agricultural nematology at Ghent University (Belgium), and her PhD in mycology and plant pathology from Rutgers University. After completing her PhD she joined the National University of Mar del Plata (Argentina) as assistant professor where she was involved in fungal endophyte research, teaching and seed testing for farmers and industry. She is currently a postdoctoral associate in the Department of Plant Biology and Pathology at Rutgers University. Her scientific interests are in the area of taxonomy, phylogeny, and evolution of the *Clavicipitaceae* and biology of fungal endophytes in natural and agricultural ecosystems. She is currently a member of the Seed Health Technical Committee for the International Seed Testing Association where she is reviewing and developing protocols for detection of fungal endophytes by seed testing labs. At Rutgers University, she mentors and teaches undergraduate students in fungal biology.

Acknowledgments

We are grateful to our families (Csaba Ruska, Shari White, Nate White, and April White) and many friends for supporting us during this endeavor. We thank all the authors for their contributions and the staff at Taylor & Francis for guidance and support during the planning, writing, and editing process.

Contributors

Janet L. Andersen Mathematics Department, Hope College, Holland, Michigan (deceased)

Charles W. Bacon United States Department of Agriculture, Agriculture Research Service, Russell Research Center, Toxicology and Mycotoxin Research Unit, Athens, Georgia

Gerald Bills Centro de Investigación Básica, Merck Sharp & Dohme de España, Madrid, Spain

Thomas L. Bultman Biology Department, Hope College, Holland, Michigan

Gregory P. Cheplick Department of Biology, College of Staten Island, City University of New York, Staten Island, New York

Keith Clay Department of Biology, Indiana University, Bloomington, Indiana

Michael H. Cortez Center for Applied Mathematics, Cornell University, Ithaca, New York

Cameron R. Currie Department of Bacteriology, University of Wisconsin-Madison, Madison, Wisconsin

John Dighton Rutgers Pinelands Field Station, New Lisbon, New Jersey

Steven Forst Department of Biological Sciences, University of Wisconsin, Milwaukee, Wisconsin

José Manuel García Garrido Department of Soil Microbiology and Symbiosis Systems, Experimental Station of Zaidín, Spanish National Research Council, Granada, Spain

Randy Gaugler Center for Vector Biology, Rutgers University, New Brunswick, New Jersey

Olga Genilloud Centro de Investigación Básica, Merck Sharp & Dohme de España, Madrid, Spain

Anthony E. Glenn United States Department of Agriculture, Agriculture Research Service, Russell Research Center, Toxicology and Mycotoxin Research Unit, Athens, Georgia

Hans-Dieter Görtz Biological Institute, University of Stuttgart, Stuttgart, Germany

Elizabeth A. O'Grady Department of Biological Sciences, University of Wisconsin, Milwaukee, Wisconsin

Edward Allen Herre Smithsonian Tropical Research Institute, Balboa, Ancon, Panama

Dorothy M. Hinton United States Department of Agriculture, Agriculture Research Service, Russell Research Center, Toxicology and Mycotoxin Research Unit, Athens, Georgia

Yoram Kapulnik Department of Agronomy and Natural Resources, Institute of Plant Sciences, Agricultural Research Organization, the Volcani Center, Bet Dagan, Israel

Yong-Ok Kim Department of Biology, University of Washington, Seattle, Washington and Montana State University, Bozeman, Montana

Hinanit Koltai Department of Ornamental Horticulture, Institute of Plant Sciences, Agricultural Research Organization, the Volcani Center, Bet Dagan, Israel

Heather S. Koppenhöfer Center for Vector Biology, Rutgers University, New Brunswick, New Jersey

James D. Lawrey Department of Environmental Science and Policy, George Mason University, Fairfax, Virginia

Ainslie E. F. Little Department of Bacteriology, University of Wisconsin-Madison, Madison, Wisconsin

Luis C. Mejía Department of Plant Biology and Pathology, School of Environmental and Biological Sciences, Rutgers University, New Brunswick, New Jersey

Deborah S. Millikan Department of Microbiology, University of Georgia, Athens, Georgia

Nydia Morales-Soto Department of Biological Sciences, University of Wisconsin, Milwaukee, Wisconsin

Nancy A. Moran Department of Ecology and Evolutionary Biology, University of Arizona, Tucson, Arizona

Kerry M. Oliver Department of Ecology and Evolutionary Biology, Center for Insect Science, University of Arizona, Tucson, Arizona

Erika L. Olson Department of Biological Sciences, University of Wisconsin, Milwaukee, Wisconsin

Gen Omura Department of Biology, Faculty of Science, Kobe University, Kobe, Japan

David Overy Centro de Investigación Básica, Merck Sharp & Dohme de España, Madrid, Spain

Fernando Peláez Centro de Investigación Básica, Merck Sharp & Dohme de España, Madrid, Spain

Timothy J. Pennings Mathematics Department, Hope College, Holland, Michigan

Anna Maria Pirttilä Department of Biology, University of Oulu, Oulu, Finland

Alison J. Popay Raukura Research Centre, Hamilton, New Zealand

Michael Poulsen Department of Bacteriology, University of Wisconsin-Madison, Madison, Wisconsin

Regina S. Redman Department of Biology, University of Washington, Seattle, Washington

Rusty J. Rodriguez U.S. Geological Survey, Seattle, Washington and Department of Biology, University of Washington, Seattle, Washington

Giovanna Rosati Department of Biology, University of Pisa, Pisa, Italy

Eric W. Schmidt Department of Medicinal Chemistry, University of Utah, Salt Lake City, Utah

Martina Schrallhammer Department of Zoology, Biological Institute, University of Stuttgart, Stuttgart, Germany, and Department of Biology, University of Pisa, Pisa, Italy

Michael Schweikert Department of Zoology, Biological Institute, University of Stuttgart, Stuttgart, Germany

Ajay P. Singh Department of Plant Biology and Pathology, Rutgers University, New Brunswick, New Jersey

Vartika Singh Department of Plant Biology and Pathology, Rutgers University, New Brunswick, New Jersey

Holly Snyder Department of Biological Sciences, University of Wisconsin, Milwaukee, Wisconsin

Eric V. Stabb Department of Microbiology, University of Georgia, Athens, Georgia

Terrence J. Sullivan Biology Department, Hope College, Holland, Michigan

Toshinobu Suzaki Department of Biology, Faculty of Science, Kobe University, Kobe, Japan

Mariusz Tadych Department of Plant Biology and Pathology, Rutgers University, New Brunswick, New Jersey

Mónica S. Torres Department of Plant Biology and Pathology, Rutgers University, New Brunswick, New Jersey

Nicholi Vorsa Department of Plant Biology and Pathology, Rutgers University, New Brunswick, New Jersey, and Philip E. Marucci Center for Blueberry and Cranberry Research and Extension, Chatsworth, New Jersey

Piippa R. Wäli Department of Biology, University of Oulu, Oulu, Finland

James F. White, Jr. Department of Plant Biology and Pathology, Rutgers University, New Brunswick, New Jersey

Chares F. Wimpee Department of Biological Sciences, University of Wisconsin, Milwaukee, Wisconsin

Claire Woodward U.S. Geological Survey, Seattle, Washington, and Department of Biology, University of Washington, Seattle, Washington

Part I

Overview of Mutualistic Associations and Defense

1

Introduction: Symbiosis, Defensive Mutualism, and Variations on the Theme

James F. White, Jr. and Mónica S. Torres

CONTENTS

1.1 Symbiosis

Symbiosis is a term that is widely used in multiple fields. In the social sciences, the term "symbiosis" may apply to any beneficial association between distinct units (e.g., parts of the self in Jungian philosophy; see Jung, 1959); or symbiosis may refer to an unhealthy dependence between two people that detracts from individuals realizing their full independent potentials (Horner, 1985). In recent years, the term has been used to refer to working or living in harmony with the biotic and abiotic environment (see Chertow, 2000). In popular biology, the term is sometimes used to denote any association between two living organisms of different species (Margulis and Sagan, 1986). But this popular view is an over simplification of a more complex and variable relationship (Margulis and Fester, 1992). Most organisms that engage in symbiotic interactions may also exist independently and a range of outcomes of the interaction is possible depending on many variables. Further, in an evolutionary sense, symbiosis is not a phenomenon of individual organisms but rather of populations. In this sense the term symbiosis refers to any short- or long-term association between populations of different species where the survival or "evolutionary fitness" of one or more population partners is enhanced by the association. Symbioses may be diffuse associations as in some generalist pathogen–host associations, where the pathogen population may associate with multiple host species. Symbioses may also be very specific as is seen in the case of many biotrophic fungal plant pathogens, e.g., powdery mildews, where the pathogens are closely adapted for growth on a single host species or a very narrow range of closely related host species. Symbioses may be transitory as in some facultative plant pathogens like *Pythium* spp., where the pathogen may exist for part of its life as a saprotroph and may irregularly exist as a plant pathogen; or they may be continuous as in obligate endosymbionts of plants like *Epichloë/Neotyphodium* species that cannot exist outside the host plant.

Evolutionarily, symbioses exist because of enhanced evolutionary fitness in at least one of the participant populations. In the case of microbial symbionts, if fitness in the host population is reduced, the association may be characterized as "parasitism" or "pathogenicity"; if fitness in the host population is not affected, the symbiosis is "commensalism"; and if fitness of the host population is enhanced, the association is "mutualism" (Douglas, 1994).

1.2 Types of Mutualisms

The scientific and popular literature abound with articles, books, and chapters about mutualism, so much that it is difficult for anyone to grasp the breadth of the field. Mutualisms are very common in nature—organisms of different species cooperate in some way to increase fitness of the mutualistic unit. In the chapter "The natural history of mutualisms," Dan Janzen (1985) articulated the problem with the statement "all terrestrial higher plants, vertebrates, and arthropods are involved in one diffuse mutualism and many are involved in several." Mutualisms of one sort or another are more the rule in nature than the exception. Finding meaningful ways to classify these mutualisms is difficult because they often have several diverse effects on hosts.

Dan Janzen (1985) categorized all mutualisms into four broad categories: (1) seed dispersal mutualisms, (2) pollination mutualisms, (3) resource harvest mutualisms, and (4) protection mutualisms. A diffuse seed dispersal mutualism is evident between birds and berry producing plants. Many examples can be found where plants rely on animals of one kind or another to disperse their seeds. Pollination mutualisms are widespread where plants rely on insects, birds, bats, or other animals to effect the transfer of gametes between plant individuals. An interesting variant of pollination mutualism involves the fungal endophytes of genus *Epichloë* (Ascomycetes) where spermatia of the fungus are vectored between compatible mating types of the fungus by a symbiotic fly which feeds on the postfertilized mycelium of the fungus (Bultman and White, 1987; see Chapter 18). The fungus depends on the fly to complete its mating cycle and the fly depends on the fungus for its nourishment. Resource harvest mutualisms are numerous; an example is the fungus gardens of attine ants where ants cultivate fungi in their underground nests and use the fungus culture to process nondigestible plant materials and as a food source (Chapela et al., 1994; see Chapter 10). Similarly, many animal ruminants maintain a complex assemblage of microbes in their rumens to degrade nondigestible plant materials and produce vitamins and other nutrients that the host cannot otherwise produce. The mycorrhizae of many land plants may be classified as resource harvest mutualisms where fungi absorb soil nutrients and pass them on to the plant host. The mycorrhizal fungi benefit in obtaining much of their carbon and nitrogen from the host autotroph (Jeffries and Barea, 2001; see Chapters 12 through 14).

The protection mutualisms are those mutualisms where the host derives some defensive benefit from the mutualistic interaction. The classic example is the case of the "ant plants" where ants live on plants, feeding on nutrients from plant nectarines and attack other insects that may visit the plant host to feed on plant leaves or nectarines (Boucher, 1985; Davidson and McKey, 1993). Another classic example of protection mutualism is the *Septobasidium* model system. *Septobasidium* grows epiphytically as flattened patches on the bark and young stems of woody plants associated with colonies of scale insects (Couch, 1931). The scale insects feed on medullary ray cells in the woody tissues of their host plants. The fungus depends on the scale insect for food supply since nutrients are assimilated from the scale insects by fungal hyphae that penetrate the insect body. The fungus depends not only on the insect for food but also on the young larval stages for dispersal when the larvae crawl out to other insect colonies or to a new tree. In turn, the fungal thallus protects the insect from environmental extremes and potential predators such as birds and hymenopterous wasps (Couch, 1931). Under the fungal thallus the remains of the scale insects consumed by the fungus can be commonly found, but some part of the colony remains uninfected, receiving the benefit of protection.

In 1988, Keith Clay began to explore fungus–plant protection mutualisms and began to use the term "defensive mutualism" to refer to this category of mutualisms (see Chapter 2). In the years that followed Keith Clay's initial publication, numerous investigators have examined potential defensive mutualistic effects in diverse symbiotic systems. Numerous investigators have contributed factual details and ideas to a collective understanding of defensive mutualism.

1.3 Stress Protection: An Expanded Concept of Defensive Mutualism

Defensive mutualism is generally thought of as pertaining to protection from herbivores, predators, or pathogens (see Chapter 2). However, it is increasingly clear that mutualistic microbes may protect hosts from stresses, whether biotic or abiotic in origin (see Chapters 12, 14, 20, and 21). This seems logical since organisms may perceive and react to environmental stressors regardless of their biotic or abiotic origins. For example, plants may respond to stress from drought, heavy metals in soils, or infection by disease fungi in similar ways (Gechev et al., 2006; see Chapter 20). Stress often results in production of reactive oxygen species (ROS) in tissues of plants. In the absence of ROS protective compounds or "ROS quenching" compounds, ROS would destroy cell components such as membranes, proteins, and nucleic acids. Plant endosymbionts that secrete or stimulate production of compounds that counteract or "quench" ROS, such as the amino acid proline, fungal sugar trehalose, or the sugar alcohol mannitol, may enhance tolerance of hosts to biotic and abiotic stressors alike. In some of the grass endophyte mutualisms, endosymbionts are thought to defend grass hosts from desiccation, heavy metals, and disease organisms through mechanisms involving ROS quenching compounds.

[The lichen symbiosis is a mutualism where the basis of the mutualism is probably enhanced abiotic and biotic stress tolerance (Lange, 1992; see Chapter 11).] The fungal partner in many lichens produces a wide array of lichen compounds. Some lichen compounds have been shown to render lichens resistant to solar radiation, a defense against abiotic stress factor. Lichen compounds have also been shown to defend lichen thalli from herbivory of invertebrates and parasitism by microbes. Lange (1992) expressed the hypothesis that lichen compounds are "generalized adaptations to life in extreme environments," essentially defending the symbiotic pair from abiotic and biotic stressors.

A more complete view of defensive mutualism thus includes any modification or augmentation in host physiology that enables hosts to better tolerate stress of any origin and survive and reproduce as a result of it.

1.4 The Concept of "Symbiotic Relativism"

It is clear that the effects of any symbiosis are relative to the host's environment and developmental stage. Such symbiotic relativism appears to be more the rule than the exception in symbioses. Which type of mutualism is more important to host fitness in a particular symbiosis thus depends on the circumstances of the organisms and may vary depending on the stage of the life cycle and/or environmental conditions of the host. Insect predatory nematodes and their bacterial symbionts may provide a useful example of symbiotic relativism (Forst and Clarke, 2002; see also Chapters 7 and 8). In the adult nematodes "offensive mutualism" may be very important. Here, the nematodes use toxin and enzyme-producing endosymbiotic bacteria to infect, then immobilize and kill their insect prey. However, the same bacterial endosymbionts grow within the body of the insect filling it with metabolites that prevent consumption by other insects and colonization by microbes. This mutualistic effect is a "food resource defense" where the food resource is protected in the symbiosis. Thus, the nematode rapidly kills the insect prey using the endosymbiont (offensive mutualism), then protects its carcass from competitors and scavengers (see Chapter 7). The nematode then deposits eggs within the body of the insect and the larvae are nourished for a period of time on the insect's remains. The food resource defensive aspect is critical for survival of the progeny of the nematode. Because the progeny live and grow for a period of time within the toxic carcass of the insect, the progeny are protected from predation and perhaps infections by nematode pathogenic fungi and bacteria that are abundant in soils. In this latter phase, the mutualistic function is "progeny defensive mutualism" where the progeny are defended by the endosymbiont. Thus, from the perspective of host fitness, there are at least three potential components to defensive mutualism between the predatory nematode and its bacterial endosymbiont and the predominant component depends on the life cycle stage. The predatory nematode symbioses are also "resource harvest mutualisms" following the classification of mutualisms outlined by Janzen (1985) since mycorrhizae and the bacterial endosymbionts of predatory nematodes function in aiding the host to acquire nutrients (see Chapter 7). Here, the endosymbiotic bacteria, once in the carcass of the insect degrade components of the insect carcass, creating a "nutrient soup" on which the nematode and its progeny feed.

Another example of symbiotic relativism is seen in the "parasitism–mutualism continuum." For example, some endophytes generally considered mutualists may reduce plant host fitness under extreme low light or low nutrient conditions (Saikkonen et al., 1988; Cheplick, 2007; see also Chapters 14, 15, 18, and 19). The cost of the symbiosis to the host plant in terms of nutrients required to maintain the symbiosis may exceed the benefits from the symbiosis. In this case, the symbiosis is more parasitism with a net negative impact on host fitness. On the other hand changes in light and soil nutrients may move the symbiotic function toward mutualism on the parasitism–mutualism continuum.

A recently discovered example of symbiotic function relativism is the symbiosis between the new world palm *Ireartea deltoidea* and its fungal endophyte *Diplodia mutila* (Alvarez et al., 2008). The fungus grows asymptomatically within plants of the palm, deterring feeding by insects under certain conditions or causing disease and mortality under other conditions. Direct sunlight converts the fungus from a mutualist into a pathogen that causes a mortal disease in seedlings. Seedlings that grow in the shade are defended from insect herbivores while those in the tree gaps are destroyed by the endosymbiont. Thus, whether the fungal association results in a defensive mutualism or a disease is in part directly determined by or relative to external abiotic conditions. The symbiosis may function mutualistically to defend the host from herbivores or, in the case of disease, it may function to enable the fungus to proliferate on the host, reproduce, and disseminate to new hosts. Through responsiveness to the abiotic environment symbiotic relativism may thus have an important impact on host ecologies.

1.5 Symbiosis Complexity

Defining mutualism in terms of categories based on benefits or presumed benefits suggest a static relationship in which the currency and outcomes of the symbiosis are relatively definable within simple, stable, and constant categories. It must be recognized, however, that symbioses between organisms may not fit perfectly into any single category with multiple effects on hosts (positive, negative, and neutral); and that the predominant outcome of a symbiotic interaction impacting fitness of the organisms at any one point in time may be relative to the conditions that the organisms are experiencing at that time. For example, mycorrhizal symbioses are generally thought of as resource harvest mutualisms (or nutritional mutualisms) where plants benefit by increased absorption of soil nutrients by the mycorrhizal fungi, but the interactions between mycorrhizae and plant hosts may also be defensive mutualisms under the right conditions. Mycorrhizae may protect plants from biotic (root-feeding nematodes, insects, root pathogenic fungi, etc.) (Dehne, 1982; Newsham et al. 1995; Smith, 1988) and abiotic (heavy metals and drought) stresses (Michelsen and Rosendahl, 1990; García-Garrido and Ocampo, 2002; see Chapters 12 through 14).

It is evident that this broad category of mutualism, like any category, has the limitation of potential grouping of several mutualistic functions of the symbioses into a single category that when dissected may be seen to be more complicated. Such as the example provided by the predatory nematodes where multiple mutualistic functions are evident: "offensive mutualism," "food resource defensive mutualism" (see Chapters 7 and 8), "progeny defensive mutualism," and "nutritional mutualism." Enhanced fitness of the host may be considered to be the result of all components of the mutualism.

Some symbiotic systems involve more than two partners such as the case of the fungus-gardening ants that use antibiotic-producing bacteria to control fungal garden parasites (Currie, 2001; see Chapter 10). Fungus-growing ants (Attini, Formicidae) and their fungi (Lepiotaceae, Basidiomycotina) is a well-studied symbiotic system (Weber, 1966). The ants forage on substrates that they use for the cultivation of fungal mycelium. The fungal gardens of attine ants are host to a parasitic fungus of genus *Escovopsis* (Ascomycotina). Attine ant cuticles are coated with masses of *Streptomyces,* a filamentous bacterium (actinomycete) which produces antibiotics that suppress the growth of the specialized garden-parasite *Escovopsis.* Because symbiotic interactions between organisms are complex and the benefits to hosts depend on specific circumstances, it seems important to dissect and examine symbioses closely to evaluate all possible ecological functions of the symbioses.

Microbe propensity for production of secondary metabolites resulting from interactions with hosts provide the raw materials for many of the defensive interactions with hosts (see Chapters 4, 5, and 16). The various defensive roles of microbial-produced molecules may provide clues regarding potential applications in medicine and agriculture (see Chapters 22 and 23).

1.6 Conclusions

In this book, we have included chapters on modeling microbial-based defensive mutualisms, evaluating their bases, and exploring variations of the defensive mutualism theme in several kingdoms of hosts (see Chapters 4, 6, and 9). In these defensive mutualisms, the microbe generally plays some role in defending the host or its food resources from consumption or biotic and/or abiotic stresses. In many cases, the mechanisms of defense are not understood; while in other symbioses the defensive properties are well characterized chemically or biologically. Beyond assuming the broadest possible definition of defensive mutualism and providing some general observations on defensive mutualisms, we have not made an attempt to synthesize new mutualism principles or extrapolate from the collective studies on defensive mutualism as expressed by the authors of the individual chapters. We also do not intend this work to be a definitive treatment of defensive mutualism. Instead it is our hope that this work represents enough of the diversity for scientists and students to gain a good perspective on the phenomenon of microbial-based defensive mutualism.

REFERENCES

Alvarez, P., White, J. F., Jr. Gil, N., Svenning, J. C., Balslev, H., and T. Kristiansen. 2008. Light converts endosymbiotic fungus to pathogen, influencing seedling survival and host tree recruitment. Available from *Nature Proceedings* (http://hdl.handle.net/10101/npre.2008.1908.1).

Boucher, D. H. 1985. *The Biology of Mutualism. Ecology and Evolution.* Oxford University Press: New York.

Bultman, T. L. and J. F. White, Jr. 1987. Pollination of a fungus by a fly. *Oecologia* 75:317–319.

Chapela, I. H., Rehner, S. A., Schultz, T. R., and U. G. Mueller. 1994. The evolutionary symbiosis between fungus growing ants and their fungi. *Science* 266:1691–1694.

Cheplick, G. P. 2007. Costs of fungal endophyte infection in Lolium perenne genotypes from Eurasia and North Africa under extreme resource limitation. *Environmental and Experimental Botany* 60:202–210.

Chertow, M. R. 2000. Industrial symbiosis: Literature and taxonomy. *Annual Review of Energy and Environment* 25:313–337.

Clay, K. 1988. Fungal endophytes of grasses: A defensive mutualism between plants and fungi. *Ecology* 69:10–16.

Couch, J. N. 1931. Memoirs: The biological relationship between Septobasidium retiforme (B.&C.) Pat. and Aspidotus osborni New. and Ckll. *Quarterly Journal of Microscopical Science* 2–74:383–438.

Currie, C. R. 2001. Prevalence and impact of a virulent parasite on a tripartite mutualism. *Oecologia* 128:99–106.

Davidson, D. W. and D. McKey. 1993. The evolutionary ecology of symbiotic ant relationships. *Journal of Hymenoptera Research* 2:13–83.

Dehne, H. W. 1982. Interaction between vesicular-arbuscular mycorrhizal fungi and plant pathogens. *Phytopathology* 72:1115–1119.

Douglas, A. E. 1994. *Symbiotic Interactions.* Oxford University Press, Oxford.

Forst, S. and D. Clarke. 2002. Bacteria-nematode symbiosis. In *Entomopathogens Nematology*, ed. R. Gaugler, pp. 57–77. CABI Publishing: Wallingford, UK.

García-Garrido, J. M. and J. A. Ocampo. 2002. Regulation of the plant defense response in arbuscular mycorrhizal symbiosis. *Journal of Experimental Botany* 53:1377–1386.

Gechev, T. S., Breusegem, F. V., Stone, J. M., Denev, I., and C. Laloi. 2006. Reactive oxygen species as signals that modulate plant stress responses and programmed cell death. *BioEssays* 28:1091–1101.

Horner, T. M. 1985. The psychic life of the young infant: Review and critique of the psychoanalytic concepts of symbiosis and infantile omnipotence. *American Journal of Orthopsychiatry* 55(3):324–344.

Janzen, D. H. 1985. The natural history of mutualisms. In *The Biology of Mutualism*, ed. D. H. Boucher, pp. 40–99. Oxford University Press, New York.

Jeffries, P. and J. M. Barea. 2001. Arbuscular mycorrhiza a key component of sustainable plant soil ecosystems. In *The Mycota. Fungal Associations*, ed. B. Hock, pp. 95–113. Springer Verlag: Berlin.

Jung, C. 1959. *The Archetypes and the Collective Unconscious.* The collected works of C. G. Jung, vol. 9(1). Princeton University Press: Princeton.

Lange, O. L. 1992. Pflanzenleben unter streß. Flechten als Pioniere der Vegetation an Extremstandorten der Erde. University of Würzburg, Germany.

Margulis, L. and R. Fester. (eds.). 1992. *Symbiosis as a Source of Evolutionary Innovation: Speciation and Morphogenesis.* MIT Press, Cambridge, MA.

Margulis, L. and D. Sagan. 1986. *Microcosms.* Summit Books, New York.

Michelsen, A. and S. Rosendahl. 1990. The effect of VA mycorrhizae fungi, phosphorus, and drought stress on growth of Acacia nilotica and Leucana leucocephala seedlings. *Plant and Soil* 124:1247–1300.

Newsham, K. K., Fitter, A. H., and A. R. Watkinson. 1995. Arbuscular mycorrhiza protect an annual grass from root pathogenic fungi in the field. *Journal of Ecology* 83:991–1000.

Saikkonen, K., Faeth, S. H., Helander, M., and T. J. Sullivan. 1988. Fungal endophytes: A continuum of interactions with host plants. *Annual Review of Ecology and Systematics* 29:319–343.

Smith, G. S. 1988. The role of phosphorous nutrition in interactions of vesiculararbuscular mycorrhizal fungi with soilborne nematodes and fungi. *Phytopathology* 78:371–374.

Weber, N. 1966. The fungus growing ants. *Science* 121:587–604.

2

Defensive Mutualism and Grass Endophytes: Still Valid after All These Years?

Keith Clay

CONTENTS

2.1 Introduction

Dan Janzen (Janzen, 1985) suggested that all mutualisms could be classified into four basic functional groups: seed dispersal, pollination, resource harvest (including food processing), and protection. In 1988 I published the paper "Fungal endophytes of grasses: A defensive mutualism between plants and fungi" (*Ecology* 69: 10–16). The primary hypothesis was that many grass–endophyte associations represent a defensive mutualism where endophytes produce physiologically active alkaloid compounds that help to protect their host plants against herbivory. I was honored to learn that this is the most highly cited paper ever written on endophytes (as of February 2008). An analysis using Web of Science (search using defens* mutual*) suggested that there were a few publications using the phrase "defensive mutualism" before 1988 but none of this work was cited. After 1988 there was a rapid increase in citations and new publications referring to "defensive mutualism" (see Figure 2.1).

While the 1988 *Ecology* paper may have helped to launch this particular buzz phrase into the scientific lexicon, the idea of defensive (or protective) mutualism goes back much further. Thomas Belt, in his 1874 book *The Naturalist in Nicaragua*, described the protection of acacia trees from herbivores by ants (Belt, 1874). He wrote, "These ants form a most efficient standing army for the plant, which prevents not only the mammalia from browsing on the leaves, but delivers it from the attacks of a much more dangerous enemy—the leaf-cutting ants". For these services the ants are not only securely housed by the plant, but are provided with a bountiful supply of food…." Now, 120 years later, a burgeoning literature suggests that symbiotic microbes may protect hosts from a variety of biological enemies. Hosts obtaining protection include insects and other arthropods (Haine, 2008; Oliver et al., 2008), nongrass plants

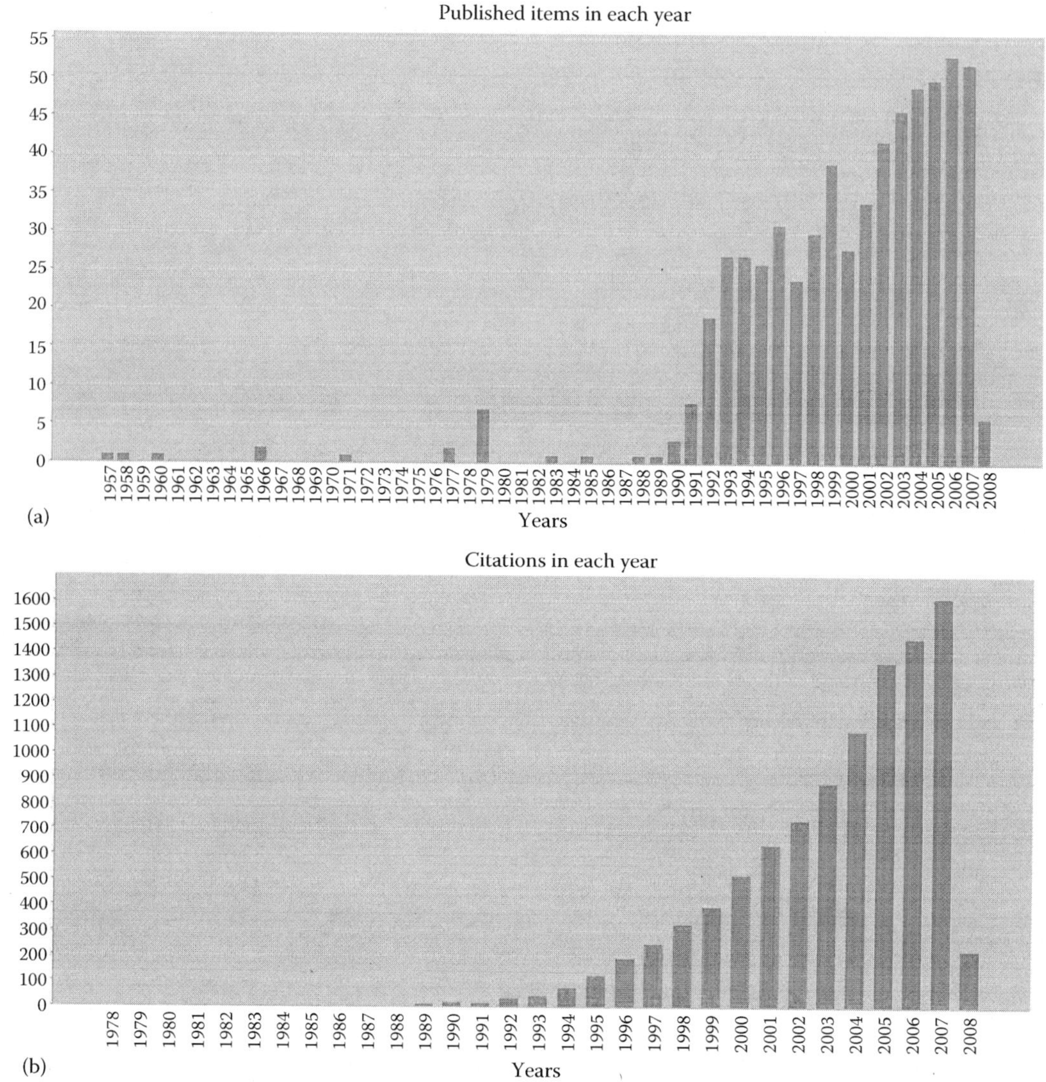

FIGURE 2.1 Results from Web of Science analysis of number of papers published using defensive mutualism in the title or keywords (a) and citations of those papers (b).

(Kucht et al., 2004; Ralphs et al., 2008), and never before examined grass species (Koh and Hik, 2007), providing more general support for the defensive mutualism hypothesis.

The best framework for understanding grass–endophyte interactions may still remain defensive mutualism, especially for seed-transmitted, alkaloid-producing *Neotyphodium* endophytes. A recent phylogenetic analysis of the fungal family Clavicipitaceae (Spatafora et al., 2007) suggests that the common ancestor to the grass endophytes was derived from insect pathogens, which also produced biologically active secondary metabolites involved in insect pathogenicity. Thus, biologically active secondary compounds represent a conserved feature of this fungal clade. With the host shift to plants, the biochemical machinery of endophytes was well adapted for defense against herbivores (see also Torres et al., 2007).

The purpose of this chapter is to consider whether this framework for viewing grass–endophyte interactions is still valid after 20 years, and whether there are more tenable alternatives. It is not intended as a review-type paper; many reviews have already been published. Rather it is intended as a personal consideration of the interaction between grasses, endophytes, alkaloids, and herbivores. Perusal of the literature since 1988 provides several perspectives. A major idea is that grass–endophyte interactions are often mutualistic and a dominant mechanism is the protection of host plants from consumers by endophyte-produced alkaloids. At the opposite end is the idea that toxic endophytes are the product of plant breeding and are found only in a small number of agriculturally important grasses. It may be that neither view is correct nor reality is somewhere in between. The operative question is where? I do not purport to answer this question here. More research and data are required. Instead, I highlight a number of specific concepts and processes in light of grass–endophyte interactions with the goal of gaining a better sense of what we know and what we do not know, and where to go from here.

2.2 It Must Be Mutualism

Theoretical models demonstrate that endosymbionts transmitted vertically through maternal lineages will not persist unless they provide some fitness benefit to the host. Vertical transmission will select reduced virulence and ultimately beneficial symbionts because symbiont reproductive success is completely dependent on host reproductive success (Ewald, 1987; Lipsitch et al., 1996). Purely seed- (or egg-) transmitted symbionts are de facto evidence of mutualism where infected hosts have higher fitness than uninfected hosts. Empirical evidence from many systems strongly supports the predicted correlation between transmission mode of parasites and fitness effect on hosts (Bull et al., 1991; Ferdy and Godelle, 2005; Stewart et al., 2005).

This theory suggests that seed-transmitted endophytes of grasses, such as those found in tall fescue (*Lolium arundinaceum*), should have a positive fitness effect on their hosts. Otherwise, they should eventually be lost from the host population. There are several mechanisms by which uninfected plants can be produced by infected plants (e.g., imperfect transmission to seeds, loss from dormant seeds, and lack of colonization of new tillers). If uninfected plants have higher fitness, they will leave more uninfected offspring and eventually displace infected plants from the population. Many endophytes produce a sexual stage that is well adapted for contagious spread by spores. With contagious spread, all bets are off and there is no necessary reason to presume that the relationship should be mutualistic. Virulent pathogens can have high prevalence in host populations with efficient contagious spread. Nevertheless, many endophytes that completely or partially sterilize host plants also exhibit mutualistic characteristics. Where no known mechanism of contagious spread exists, endophyte associations should be presumed to be mutualistic based on theoretical and empirical evidence unless demonstrated otherwise.

2.3 A Question of Timing

Mutualistic seed-transmitted endophytes should increase in frequency within host populations over time whereas seed-transmitted endophytes with detrimental effects on hosts should decrease in frequency. Samples taken at a single point in time are of limited value because they provide no information about the trajectory. In particular, studies showing variable infection frequencies in host populations are interesting but provide no direct information on the costs or benefits of infection and the nature of the interaction. Variable and/or low levels of endophyte infection (e.g., Saikkonen et al., 2000; Spyreas et al., 2001; Bazely et al., 2007) do not indicate whether infection frequency is declining from a higher level, increasing from a lower level, changing in a nondirectional manner, or remaining stable. Several studies, mostly with tall fescue, have shown that endophyte infection frequency increases over time (Clay et al., 2005 and references therein), consistent with a mutualistic interaction. For example, in a controlled field experiment where replicate 5 m × 5 m plots were all initially sown with 50:50 mixture of infected and uninfected tall fescue seed, infection frequency increased to 80% over 5 years in plots subjected to the greatest herbivore pressure (Figure 2.2). Infection increased in all plots, but to a lesser extent in plots with reduced herbivory. These results demonstrate a positive effect of endophyte infection on host fitness that is herbivore-dependent.

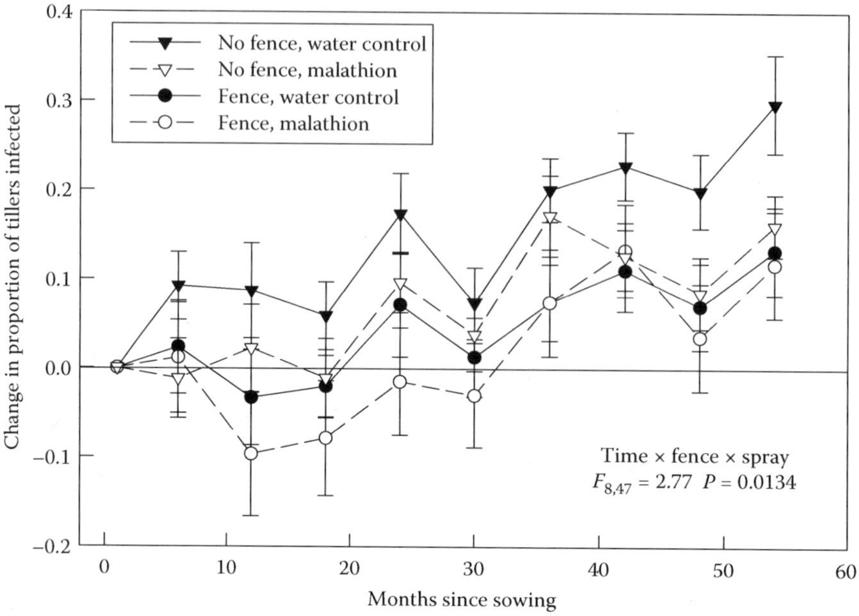

FIGURE 2.2 The change in endophyte frequency in experimental tall fescue plots among herbivory treatments over time. The change in frequency was determined by subtracting the initial proportion of tillers infected in that plot from the proportion of tillers infected on each date. The change in proportion is bounded by −0.5 and 0.5 (0% and 100% infected, respectively). Over time, infection increased in all plots (time, $F_{8,47} = 29.6$, $P < 0.0001$). Treatments diverged over time (fence × insecticide × time interaction, $F_{8,47} = 2.8$, $P < 0.01$). Symbols show means ± SE. (Reprinted from Clay, K., Holah, J., and Rudgers, J.A., *Proc. Natl. Acad. Sci. USA*, 102, 12465, 2005. With permission).

In a recent study, Koh and Hik (2007) found that endophyte infection frequency in *Festuca altaica* in alpine meadows (Yukon Territory, Canada) was a positive function of grazing pressure by collared pikas (*Ochotona collaris*) and hoary marmots (*Marmota caligata*). Rather than manipulating herbivore pressure experimentally as above, they took advantage of the fact that grazing by these animals is concentrated near boulder fields, which provide a refuge against predation. Fescue grasses further from boulder fields receive less grazing pressure as a result. Given that boulder fields do not move easily, this spatial pattern was presumably long-standing. They found a highly significant decrease in endophyte infection frequency with distance from boulder fields, consistent with the idea of defensive mutualism.

More studies are needed where dynamics of endophyte infection are tracked over time, preferably in response to particular environmental conditions. Stable or declining intermediate infection frequencies are not in conflict with the hypothesis of defensive mutualism, but may instead indicate that there is little or no herbivore pressure to provide a selective advantage for infected plants. The establishment of fences to prevent large grazers could reduce the advantage of endophyte infection, possibly leading to its decrease over time. For example, Palmer et al. (2008) recently demonstrated that 10 years of large mammal exclusion in an African savanna led to the breakdown of a protective ant–plant mutualism. However, small mammals (e.g., voles) and insects might be unaffected by fences and still consume plant material. Alternatively, increase in grazing pressure (e.g., from domestic livestock, from native herbivores released from predation) could lead to rapid increase in both infection frequency within populations and dominance of endophyte-infected grasses within the community. This may reflect the situation in China where endophyte-infected *Achnatherum* species may be increasing with grazing pressure by livestock (Wei et al., 2006; Li et al., 2007). Similarly, insect outbreaks could provide an occasional, but strong, selective advantage to infected hosts that might not be evident most of the time. Clay (1997) presented a graphical model of how grazing pressure could affect endophyte infection frequency and the dominance of endophyte-infected hosts in communities. Observing these processes may take a long time—possibly many decades. In dense swards there may be little seedling establishment but instead only vegetative growth and reproduction of existing genotypes. The fundamental idea remains that endophytes that are only vertically transmitted through seed must be mutualistic in order to persist over time (see also Lipsitch, 1996). While there may be fluctuations in endophyte host fitness, and there may be particular environmental conditions that are detrimental to endophyte-infected hosts, the mean fitness of endophyte-infected hosts must be higher than that of uninfected plants on average over time in field populations.

2.4 It's the Alkaloids, Stupid!

A vast body of research supports the hypothesis that the primary role of secondary compounds like alkaloids, terpenoids, glucosides, and tannins is to deter herbivores from feeding on plant tissues (Fraenkel, 1959; Williams et al., 1989; Rosenthal and Berenbaum, 1992). However, insects and other herbivores are not just passive participants in this interaction but can evolve and coevolve mechanisms to avoid or detoxify deterrent secondary compounds. It has been suggested that much of the plant and insect diversity has been generated through the process of chemical coevolution (Thompson, 1982; Berenbaum et al., 1986). Much human agriculture has arisen from breeding to eliminate secondary compounds from fruits, seeds, and leaves.

The widespread distribution of alkaloid compounds in grasses associated with fungal endophyte infection suggests, by analogy with plant secondary compounds, that they play a role in defense against herbivores. Past research on endophytes has revealed that, as a group, they are capable of producing a variety of alkaloid compounds. However, one endophyte strain or species may produce only one or more of the various alkaloid types while others produce no alkaloids whatsoever (Siegel et al., 1990; Bush et al., 1997; Leuchtmann et al., 2000; Faeth et al., 2006). Different types of alkaloids (peramine, lolitrem, loline, and ergot alkaloids) have different modes of action (e.g., peramine-feeding deterrent, lolitrem—neurotoxin). Some affect insects, others affect mammals, and others affect both (Clay and Cheplick, 1989). Interestingly, there is no evidence for mammalian resistance to ergot alkaloids, which are the most widespread endophyte alkaloids.

Siegel et al. (1990) reported that the large majority of seed-transmitted endophytes examined produced at least one type of alkaloid and many produced several types. The presence of alkaloids and the specific type of alkaloid had a large effect on aphid deterrence in experimental trials, as did the particular aphid species. The fact that one aphid species was unaffected by certain alkaloids does not negate the potential selective advantage of resistance to the other species. There was no evidence of deterrence in the absence of alkaloids. This suggests that chemical assays for alkaloid types and concentrations can substitute for animal feeding trials. If there are no alkaloids in endophyte-infected grasses then there is likely to be no effect against herbivores. Interestingly, Leuchtmann et al. (2000), in a comparison of 18 European grasses infected by *Epichloë* and *Neotyphodium* endophytes, found that stroma-producing *Epichloë* endophytes were typically free of alkaloids while seed-transmitted *Neotyphodium* endophytes often contained high concentrations of lolines. This pattern is consistent with the prediction of seed-transmitted endophytes being mutualistic.

There is evidence that endophytes testing negative for standard alkaloid compounds can still exhibit herbivore-deterrent effects. For example, Brem and Leuchtmann (2001) found enhanced herbivore resistance of *Epichloë*-infected *Brachypodium sylvaticum* compared to uninfected plants even though previous studies had never detected alkaloids using standard assays. Similarly, Jones et al. (2000) found that cattle avoided endophyte-infected robust needlegrass (*Achnatherum robustum*, formerly *Stipa robusta*), even though no alkaloids were detected (see also Faeth et al., 2006). These results suggest that there may be unknown herbivore-deterrent endophyte compounds not detected by standard assays.

2.5 Cattle Eat More than Caterpillars

The grass family is second to none in terms of its importance as a food plant to humans and many other herbivores, both vertebrates and invertebrates. However, it is somewhat anomalous in its relative lack of secondary compounds. While inherently poisonous grasses do exist (i.e., reed canarygrass, *Phalaris arundinacea*), they are the exception and not the rule. Instead, grasses are characterized by the occurrence of silica in leaves and stems, which may deter animals from feeding on grasses (Gali-Muhtasib et al., 1992). Grazing mammals often have teeth that are well-adapted for dealing with silica-containing grass blades but highly specialized chemical detoxification mechanisms are less frequent in mammalian herbivores than in insects. Few mammals are specialists on a particular plant family or genus, although many grazers rely heavily on the grass family. To a large extent, grasslands would not exist without large mammalian grazers (McNaughton, 1984).

In grasses and grasslands, mammalian herbivores consume vastly more plant biomass than insects. As a result, they will provide stronger selection favoring traits that reduce herbivory. Even small mammals like voles can have major effects on grassland vegetation. For example, Clay et al. (2005) found that vole manipulation (via fencing treatments) had a far larger effect on grassland vegetation and dominance of endophyte-infected tall fescue than did insect manipulation (via insecticide spraying). Similarly, Koh and Hik (2007) found that grazing rodents had a highly significant effect on endophyte infection frequency of *Festuca altaica* in alpine meadow communities. Historical accounts from Africa, Asia, and North America also describe increases of particular grasses, now known to be endophyte-infected, with increasing grazing pressure by livestock (Shaw, 1873; Bailey, 1903; Bor, 1960). This suggests that natural selection provided by heavy grazing favors more toxic endophyte associations within a grass species and/or more toxic endophyte-infected grasses within a community.

2.6 Alternative Hypotheses or Variations on a Theme?

My reading of the literature suggests that there are four primary alternative hypotheses to "defensive mutualism." However, all of them implicitly accept defensive mutualism as a major feature of grass–endophyte interactions. Three of them are specific elaborations about when and where defensive mutualism occurs and the fourth suggests that there are other, nondefense related benefits of endophyte infection. The conclusion therefore is that defensive mutualism remains the primary conceptual framework for understanding grass–endophyte interactions. The various commentaries and alternatives presented since the 1988 *Ecology* paper represent variations on an old theme. Nevertheless, each alternative is worth considering.

2.7 In Defense of Drought

This hypothesis suggests that there are other benefits to endophyte-infection apart from defense against biological pests. In particular, endophyte infection results in biochemical and physiological changes in the host plant that improve its ability to tolerate abiotic stresses or take up limiting nutrients from its environment. Most data come from *Lolium* spp. Enhanced drought tolerance by endophyte-infected grasses is well documented. For example, West et al. (1993) showed that endophyte-infected tall fescue is able to grow better and regrow than uninfected tall fescue following drought (see also Bouton et al., 1993). Lewis et al. (1997) demonstrated that water supply deficit explained 43% of the variance in endophyte-infection rate of *Lolium* spp. in France with higher levels of infection found in more Mediterranean regions where summer drought stress is common. Apart from drought tolerance, reduced carbon mineralization rates, and percentage soil nitrate (Franzluebbers and Hill, 2005), reduced microbial biomass and respiration, altered carbon and nitrogen pools (Franzluebbers and Studemann, 2005), suppression of archael and high G + C gram-positive bacterial communities, and greater soil carbon sequestration

(Jenkins et al., 2006) have been observed in endophyte-infected tall fescue vs. uninfected agricultural soil. None of these potential effects preclude resistance against pests and may operate in tandem with defensive mutualism.

2.8 Plant Breeding Beats Mother Nature

This alternative (e.g., Saikonnen et al., 1998, 2006) suggests that the toxic effects of endophyte infection in tall fescue and perennial ryegrass are the consequences of domestication, plant breeding, and artificial selection, and are rarely, if ever, observed in native grasses. I find that this is a straw man argument based on specious comparisons with Arizona fescue (*Festuca arizonica*). The basic argument is (1) toxic, endophyte-infected tall fescue is an agronomic species, (2) nontoxic, endophyte-infected Arizona fescue is a wild species, and (3) therefore, agronomic species are toxic and native species are not. Another plausible argument is (1) tall fescue grows at lower elevations, (2) Arizona fescue is a montane species, and (3) therefore, lowland species are toxic while montane species are nontoxic.

Besides faulty logic, I find several problems with the agronomic grass hypothesis. One is that it ignores the many examples of herbivore deterrence from wild endophyte-infected grasses. Additional data would be welcome, but should come from a wider variety of wild grasses rather than detailed studies of one species. A second problem is that it ignores the fact that agronomic pasture and turf grasses are little removed from their wild ancestors and exist commonly as feral or wild populations in many areas. Every pasture and turf grass exists in the wild and almost all widespread grasses have agronomic uses. Wild vs. agronomic is a blurry distinction. A third problem is confusion about plant and endophyte breeding given that *Neotyphodium* endophytes are asexual. While particular preexisting endophyte strains may be favored during plant breeding, new endophyte genotypes are not created. The argument that tall fescue and perennial ryegrass are Frankensteins of the plant world does not hold up to careful scrutiny, even though it does make for lively debate and a steady stream of review papers. But even if tall fescue alone exhibits herbivore deterrence with endophyte infection, it is by far the most common and widespread endophyte interaction in many regions of the world and therefore represents the status quo. In any case, the agronomic hypothesis does not negate the defensive mutualism hypothesis but rather suggests that it is limited to a particular subset of endophyte hosts.

2.9 Genes and Genotypes

Another common hypothesis is that many grasses benefit from endophyte infection but others do not, and the interaction depends on genotypic variation within host populations (Cheplick and Cho, 2003; Tintjer and Rudgers, 2006). This is demonstrably true but the significance of this effect in the larger framework of grass–endophyte interactions is questionable. There are two reasons to question the genotype effect. One is that host grass genotypes are ephemeral while endophyte genotypes are constant, at least for seed-transmitted *Neotyphodium* endophytes. For obligately outcrossed species like tall fescue, every individual plant represents a distinct genotype and all their progeny are themselves unique. In contrast, the vertically transmitted endophyte is a clonal lineage that infects thousands or millions of different host genotypes. Thus, while variation in the benefit of endophyte infection to individual host genotypes is interesting, it is the average effect of a given endophyte genotype across all host genotypes that is most significant from an ecological and evolutionary perspective.

There is a second, technical problem with the primacy of plant genotype in determining the outcome of grass–endophyte interactions. Many of the studies demonstrating significant variation in endophyte effects among host genotypes are based on rearing seeds in controlled, benign environments. In natural systems subjected to the rigors of climate, competitors, pathogens, etc., perhaps only 1 of 100 or 1 of 100,000 seedlings survives to reproduction. What determines which one survives this selective sieve? The endophyte-infection status, and the benefits that host genotype gains from infection, could be an overriding factor. It is likely that cohorts of endophyte-infected hosts become progressively more mutualistic as self-thinning and mortality of genotypes occur. This graphical model is presented in

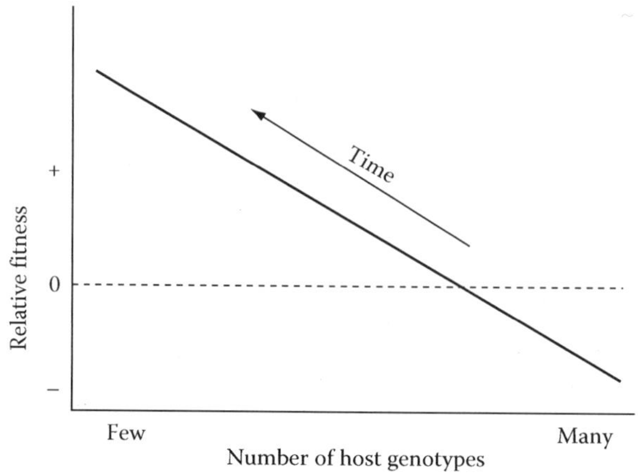

FIGURE 2.3 Conceptual model showing how the average degree of mutualism increases over time within a single grass generation as the number of grass genotypes declines with self-thinning and mortality.

Figure 2.3. The conclusion about the nature of the grass–endophyte interaction may depend on the point in time you look. Many of the empirical results showing significant genotypic variation in endophyte interactions may represent experimental artifacts of growing seed under idealized environmental conditions, and are of less significance in nature. I predict that if one examines established populations, a stronger and less variable measure of mutualism would be obtained compared to examination of unselected seed progeny. Long-term field experiments subjected to ambient levels of climatic variation, pest pressure, interspecific competition, etc. will be more informative than greenhouse experiments conducted under optimal conditions.

Support for the idea of selection favoring more mutualistic combinations comes from systems where host plant genotype is propagated indefinitely by plant clonal growth. In these systems, the life span of the host genotype is equivalent to the life span of the endophyte genotype. For example, Harberd (1961) described individual endophyte-infected clones of red fescue (*Festuca rubra*) in England that spanned many hectares and were estimated to be many centuries old. Similarly, *Cyperus virens* sedges infected by *Balansia cyperi* produce viviparous plantlets, and clustering of infected plants in coastal marshes, suggesting that highly mutualistic plant–fungal genotype combinations persisted and were propagated over many generations (Clay, 1986; see also Clay and Kover, 1996). More recently, Pan and Clay (2003) demonstrated that *Epichloë*-infected clones of *Glyceria striata* exhibited significantly more clonal growth (via stolons) than uninfected plants from the same population. This result is also consistent with the idea that the most mutualistic plant–fungal combinations persist and propagate in plant populations. Sampling ramets from established populations, propagating them, disinfecting a subset with fungicides, and then comparing performance of infected vs. uninfected clones of the same genotype under realistic field conditions could lead to different conclusions than similar trials with seeds.

2.10 Geographical Mosaic

A fourth alternative hypothesis is the geographical mosaic of coevolution hypothesis of Thompson (2006). This theory was not developed for any particular system but can be applied to grass–endophyte interactions. It suggests that the interaction between host and endophyte varies with local biotic and abiotic conditions, and is not uniform across their range. Infection rates may vary with drought stress, soil fertility, or altitude (Lewis et al., 1997; McCormick et al., 2001; Bazely et al., 2007) as the relative costs and benefits of infection change. For example, the relative performance of endophyte-infected vs. uninfected tall fescue varied along a nutrient gradient with the greatest benefit of infection accruing in high fertility conditions (Cheplick et al., 1989). Similarly, particular sites or populations might be subjected to high levels of grazing pressure or insect herbivory while other areas might have less (Clay and Brown, 1997; Clay et al., 2005; Koh and Hik, 2007). Moreover, in small or isolated populations, there could be genetic bottlenecks in plant and/or endophyte populations that might lead to fixation of less than optimal genotypic combinations. Kover et al. (1997) found that essentially every host plant of *Danthonia spicata* was infected by a distinct genotype of *Atkinsonella hypoxylon*, a sexual species. In contrast, an allozyme survey of tall fescue populations in the eastern United States suggested that virtually all host plants were infected by a single endophyte genotype (Leuchtmann and Clay, 1990). Thus, the genetic diversity of plant and endophyte populations could limit potential responses to changing environmental conditions.

The geographic mosaic hypothesis will be best tested by field experiments replicated across large areas of the host species' range where the relative fitness of endophyte-infected and uninfected plants is monitored in relation to local agricultural conditions (e.g., pest pressure, soil fertility, climate, etc.). To my knowledge, these experiments have never been done outside agricultural field trials (e.g., Bouton et al., 1993). Surrogate measures such as infection frequency or alkaloid content (e.g., Saikkonen et al., 2000; Spyreas et al., 2001; Faeth et al., 2006; Bazely et al., 2007) can provide some interesting patterns but they provide no information on time series or underlying mechanisms.

I concur that there is variation in the costs and benefits of endophyte infection and variation in the genetic makeup of interacting populations. This has been explicit from the earliest discussions of grass–endophyte interactions (Clay, 1988a,b). But this does not negate the defensive mutualism hypothesis. The question is by which mechanism or mechanisms does endophyte infection increase host fitness. This question occurs within the context of theoretical predictions that purely seed-borne endophytes must be mutualistic to persist. Defense against pests still appears to be the major mechanism of mutualism for host populations in the field. There is no reason that other mechanisms like drought tolerance could not be operating simultaneously. It is important to note that defensive mutualism will not be observed in greenhouse or growth chamber experiments where natural enemies are eliminated, or in garden-type experiments where pests are controlled.

2.11 Final Thoughts

Since I first began working on endophytes during my PhD work at Duke University in the late 1970s (Clay, 1984), the field has changed tremendously. Much of the research understandably has been driven by economic concerns and has been focused on cool-season pooid grasses. However, they represent just a tiny fraction of plant–endophyte interactions or indeed of macroorganism–microorganism interactions. For example, active host defense by microbial symbionts is apparently widespread in arthropods (Currie et al., 1999; Lively et al., 2005; Haine, 2008; Oliver et al., 2008). Fungal endophyte relationships are recognized as being ubiquitous in all plants (Arnold et al., 2000) where they may also play a defensive role (Arnold et al., 2003). Seed-transmitted, alkaloid-producing endophytes are being found in other plant families (Ralphs et al., 2008) and seed-transmitted clavicipitaceous endophytes are now being reported from dicotyledonous plants (Kucht et al., 2004).

A great diversity of systems and opportunities are opening up that are all predicated on the defensive mutualism hypothesis. And perhaps now is the time to examine more critically a wider range of grass systems beyond the C-3 pooid grasses. I think that we need less talk and more action, less rhetoric and more research. While there is debate about the relevance of defensive mutualism for cool-season grass–endophyte interactions, the concept is being applied to an ever-expanding diversity of biological systems. The diversity of molecular mechanisms and natural products produced by symbiotic bacteria and fungi make them ideal partners in defensive partnerships, may be even better than ants.

ACKNOWLEDGMENTS

My thanks to the many graduate students, undergraduates, post-docs, and collaborators who have contributed to my research on grass–endophyte interactions and to the NSF for their continued financial support.

REFERENCES

Arnold, A. E., Maynard, Z., Gilbert, G., Coley, P. D., and T. A. Kursar. 2000. Are tropical fungal endophytes hyperdiverse? *Ecology Letters* 3:267–274.

Arnold, A. E., Mejía, L., Kyllo, D., Rojas, E., Maynard, Z., and E. A. Herre. 2003. Fungal endophytes limit pathogen damage in a tropical tree. *Proceedings of the National Academy of Sciences USA* 100:15649–15654.

Bailey, V. 1903. Sleepy grass and its effect on horses. *Science* 17:392–393.

Bazely, D. R., Ball, J. P., et al. 2007. Broad-scale geographic patterns in the distribution of vertically-transmitted, asexual endophytes in four naturally-occurring grasses in Sweden. *Ecography* 30:367–374.

Belt, T. 1874. *The Naturalist in Nicaragua*. The University of Chicago Press, Chicago (reprinted 1985).

Berenbaum, M. R., Zangerl, A. R., and J. K. Nitao. 1986. Constraints on chemical coevolution: Wild parsnips and the parsnip webworm. *Evolution* 40:1215–1228.

Bor, N. 1960. *The Grasses of Burma, Ceylon, India, and Pakistan*. Pergamon Press, New York.

Bouton, J. H., Gates, R. N., Belesky, D. P., and M. Owsley. 1993. Yield and persistence of tall fescue in the southeastern coastal plain after removal of its endophyte. *Agronomy Journal* 85:52–55.

Brem, D. and A. Leuchtmann. 2001. *Epichloë* grass endophytes increase herbivore resistance in the woodland grass *Brachypodium sylvaticum. Oecologia* 126:522–530.

Bull, J. J., Molineux, I. J., and W. R. Rice. 1991. Selection of benevolence in host-parasite system. *Evolution* 45:875–882.

Bush, L. P., Wilkinson, H. H., and C. L. Schardl. 1997. Bioprotective alkaloids of grass-fungal endophyte symbioses. *Plant Physiology* 114:1–7.

Cheplick, G. P. and R. Cho. 2003. Interactive effects of fungal endophyte infection and host genotype on growth and storage in *Lolium perenne. New Phytologist* 158:183–191.

Cheplick, G. P., Clay, K., and S. Marks. 1989. Interactions between infection by endophytic fungi and nutrient limitation in the grasses *Lolium perenne* and *Festuca arundinacea. New Phytologist* 111:89–98.

Clay, K. 1984. The effect of the fungus *Atkinsonella hypoxylon* (Clavicipitaceae) on the reproductive system and demography of the grass *Danthonia spicata. New Phytologist* 98:165–175.

Clay, K. 1986. Induced vivipary in the sedge *Cyperus virens* and the transmission of the fungus *Balansia cyperi* (Clavicipitaceae). *Canadian Journal of Botany* 64:2984–2988.

Clay, K. 1988a. Fungal endophytes of grasses: A defensive mutualism between plants and fungi. *Ecology* 69:10–16.

Clay, K. 1988b. Clavicipitaceous fungal endophytes of grasses: Coevolution and the change from parasitism to mutualism. In *Co-evolution of Fungi with Plants and Animals*, eds. D. L. Hawksworth and K. Pirozynski, pp. 79–105. Academic Press, London.

Clay, K. 1997. Fungal endophytes, herbivores and the structure of grassland communities. In *Multitrophic Interactions in Terrestrial Systems,* eds. A. C. Gange and V. K. Brown, pp. 151–169. Blackwell Scientific Publishers, Oxford.

Clay, K. and V. K. Brown. 1997. Infection of *Holcus lanatus* and *H. mollis* by *Epichloë* in experimental grasslands. *Oikos* 79:363–370.

Clay, K. and G. P. Cheplick. 1989. Effect of ergot alkaloids from fungal endophyte-infected grasses on the fall armyworm (*Spodoptera frugiperda*). *Journal of Chemical Ecology* 15:169–182.

Clay, K. and P. Kover. 1996. Evolution and stasis in plant/pathogen associations. *Ecology* 77:997–1003.

Clay, K., Holah, J., and J. A. Rudgers. 2005. Herbivores cause a rapid increase in hereditary symbiosis and alter plant community composition. *Proceedings of the National Academy of Science USA* 102:12465–12470.

Currie, C. R., Scott, J. A., Summerbell, R. C., and D. Malloch. 1999. Fungus-growing ants use antibiotic producing bacteria to control garden parasites. *Nature* 398:701–704.

Ewald, P. 1987. Transmission modes and evolution of the parasitism-mutualism continuum. *Annals of the New York Academy of Sciences* 503:295–306.

Faeth, S. H., Gardner, D. R., et al. 2006. Temporal and spatial variation in alkaloid levels in *Achnatherum robustum*, a native grass infected with the endophyte *Neotyphodium. Journal of Chemical Ecology* 32:307–324.

Ferdy, J. B. and B. Godelle. 2005. Diversification of transmission modes and the evolution of mutualism. *American Naturalist* 166:613–627.

Fraenkel, G. S. 1959. The raison d'etre of secondary plant substances. *Science* 129:1466–1470.

Franzluebbers, A. J. and N. S. Hill. 2005. Soil carbon, nitrogen, and ergot alkaloids with short- and long-term exposure to endophyte-infected and endophyte-free tall fescue. *Soil Science Society of America* 69:404–412.

Franzluebbers, A. J. and J. Studemann. 2005. Soil carbon and nitrogen pools in response to tall fescue endophyte infection, fertilization, and cultivar. *Soil Science Society of America Journal* 69:396–403.

Gali-Muhtasib, H. U., Smith, C. C., and J. J. Higgins. 1992. The effect of silica in grasses on the feeding behavior of the prairie vole, *Microtus ochrogaster. Ecology* 73:1724–1729.

Haine, E. R. 2008. Symbiont-mediated protection. *Proceedings of the Royal Society B-Biological Sciences* 275:353–361.

Harberd, D. J. 1961. Observations on population structure and longevity of *Festuca rubra* L. *New Phytologist* 60:184–206.

Janzen, D. H. 1985. The natural history of mutualisms. In *The Biology of Mutualism*, ed. D. H. Boucher, pp. 40–99. Oxford University Press, New York.

Jenkins, M. B., Franzluebbers, A. J., and S. B. Humayoun. 2006. Assessing short-term responses of prokaryotic communities in bulk and rhizosphere soils to tall fescue endophyte infection. *Plant Soil* 289:309–320.

Jones, T. A., Ralphs, M. H., Gardner, D. R., and N. J. Chatterton. 2000. Cattle prefer endophyte-free robust needlegrass. *Journal of Range Management* 53:427–431.

Koh, S. and D. S. Hik. 2007. Herbivory mediates grass-endophyte relationships. *Ecology* 88:2752–2757.

Kover, P. X., Dolan, T. E., and K. Clay. 1997. Potential versus realized transmission rates of a vertically and horizontally transmitted plant pathogen. *Proceedings of the Royal Society of London B* 264:903–909.

Kucht, S., Gross, J., et al. 2004. Elimination of ergoline alkaloids following treatment of *Ipomoea asarifolia* (Convolvulaceae) with fungicides. *Planta* 219:619–625.

Leuchtmann, A. and K. Clay. 1990. Isozyme variation in the *Acremonium/Epichloë* fungal endophyte complex. *Phytopathology* 80:1133–1139.

Leuchtmann, A., Schmidt, D., and L. P. Bush. 2000. Different levels of protective alkaloids in grasses with stroma-forming and seed-transmitted *Epichloë/Neotyphodium* endophytes. *Journal of Chemical Ecology* 26:1025–1036.

Lewis, G. C., Ravel, C., Naffaa, W., Astier, C., and G. Charmet. 1997. Occurrence of *Acremonium* endophytes in wild populations of *Lolium* spp. in European countries and a relationship between level of infection and climate in France. *Annals of Applied Biology* 130:227–238.

Li, C. J., Gao, J. H., and Z. B. Nan. 2007. Interactions of *Neotyphodium gansuense*, *Achnatherum inebrians*, and plant-pathogenic fungi. *Mycological Research* 111:1220–1227.

Lipsitch, M., Siller, S., and M. A. Nowak. 1996. The evolution of virulence in pathogens with vertical and horizontal transmission. *Evolution* 50:1729–1741.

Lively, C. M., Clay, K., Wade, M. J., and C. Fuqua. 2005. Competitive coexistence of vertically and horizontally transmitted parasites. *Evolutionary Ecology Research* 7:1183–1190.

McCormick, M. K., Gross, K. L., and R. A. Smith. 2001. *Danthonia spicata* (Poaceae) and *Atkinsonella hypoxylon* (Balansiae): Environmental dependence of a symbiosis. *American Journal of Botany* 88:903–909.

McNaughton, S. J. 1984. Grazing lawns: Animals in herds, plant form, and coevolution. *American Naturalist* 124:863–886.

Oliver, K. M., Campos, J., et al. 2008. Population dynamics of defensive symbionts in aphids. *Proceedings of the Royal Society B-Biological Sciences* 275:293–299.

Palmer, T. M., Stanton, M. L., Young, T. P., Goheen, J. R., Pringle, R. M., and R. Karban. 2008. Breakdown of an ant-plant mutualism follows the loss of large herbivores from an African savanna. *Science* 319:192–195.

Pan, J. J. and K. Clay. 2003. Infection by the systemic fungus, *Epichloë glyceriae*, alters clonal growth of its grass host, *Glyceria striata. Proceedings of the Royal Society of London B* 270:1585–1591.

Ralphs, M. H., Creamer, R., et al. 2008. Relationship between the endophyte *Embellisia* spp. and the toxic alkaloid swainsonine in major locoweed species (*Astragalus* and *Oxytropis*). *Journal of Chemical Ecology* 34:32–38.

Rosenthal, G. and M. Berenbaum. 1992. *Herbivores Their Interaction with Secondary Plant Metabolites.* Academic Press, New York.

Saikkonen, K., Faeth, S. H., Helander, M., and T. J. Sullivan. 1998. Fungal endophytes: A continuum of interactions with host plants. *Annual Review of Ecology and Systematics* 29:319–343.

Saikkonen, K., Ahlholm, J., Helander, M., Lehtimaki, S., and O. Niemelainen. 2000. Endophytic fungi in wild and cultivated grasses in Finland. *Ecography* 23:360–366.

Saikkonen, K., Lehtonen, P., et al. 2006. Model systems in ecology: Dissecting the endophyte-grass literature. *Trends in Plant Science* 11:428–433.

Shaw, J. 1873. On the changes going on in the vegetation of South Africa. *Botanical Journal of the Linnean Society* 14:202–208.

Siegel, M. R., Latch, G. C. M., Bush, L. P., Fannin, N. F., Rowan, D. D., Tapper, B. A., Bacon, C. W., and M. C. Johnson. 1990. Fungal endophyte-infected grasses: Alkaloid accumulation and aphid response. *Journal of Chemical Ecology* 16:3301–3315.

Spatafora, J. W., Sung, G. H., et al. 2007. Phylogenetic evidence for an animal pathogen origin of ergot and the grass endophytes. *Molecular Ecology* 16:1701–1711.

Spyreas, G., Gibson, D. J., and M. Basinger. 2001. Endophyte infection levels of native and naturalized fescues in Illinois and England. *Journal of the Torrey Botanical Society* 128:25–34.

Stewart, A. D., Logsdon, J. M., and S. E. Kelley. 2005. An empirical study of the evolution of virulence under both horizontal and vertical transmission. *Evolution* 59:730–739.

Thompson, J. N. 1982. *Interaction and Coevolution*. John Wiley & Sons, New York.

Thompson, J. N. 2006. *The Geographic Mosaic of Coevolution*. The University of Chicago Press, Chicago.

Tintjer, T. and J. A. Rudgers. 2006. Grass–herbivore interactions altered by strains of a native endophyte. *New Phytologist* 170:513–521.

Torres, M. S., Singh, A. P., Vorsa, N., Gianfagna, T., and J. F. White, Jr. 2007. Were endophytes pre-adapted for defensive mutualism? *Proceedings of the 6th International Symposium on Fungal Endophytes of Grasses*, pp. 63–67.

Wei, Y. K., Gao, Y. B., Xu, H., Su, D., Zhang, X., Wang, Y. H., Lin, F., Chen, L., Nie, L. Y., and A. Z. Ren. 2006. Occurrence of endophytes in grasses native to northern China. *Grass and Forage Science* 61:422–429.

West, C. P., Izekor, E., Turner, K. E., and A. A. Elmi. 1993. Endophyte effects on growth and persistence of tall fescue along a water-supply gradient. *Agronomy Journal* 85:264–270.

Williams, D. H., Stone, M. J., et al. 1989. Why are secondary metabolites (natural-products) biosynthesized. *Journal of Natural Products* 52:1189–1208.

3

Overview of Defensive Mutualism in the Marine Environment

Chares F. Wimpee, Elizabeth A. O'Grady, and Erika L. Olson

CONTENTS

3.1 Introduction

The marine environment abounds with microbial symbiosis. Eukaryotic, bacterial, and archaeal microbes form an enormous variety of associations with marine animals, with seaweeds, and with one another. Although those marine microbial symbioses that have received the most widespread attention are of a nutritional nature (e.g., photosynthetic symbiosis, thiotrophic symbiosis, and methanotrophic symbiosis), a variety of microbial symbioses are defensive, while others probably serve mixed functions of nutrition and defense. It is probably unnecessary to state that we have barely begun to study these associations, and that countless symbioses have yet to reveal themselves. What is covered in this chapter is only a small number of defensive microbial symbiosis in the marine environment. Certain topics will be dealt with in greater detail in other chapters. So, although the coverage here will be broader than in other chapters, it might (unfortunately) be shallower, because of the diversity of topics, and with admittedly disproportionate detail given to the best-studied systems.

This chapter addresses defenses against abiotic and biotic factors in the marine realm, with the caveat that some of these defenses are as yet implied, and not yet experimentally supported. However, certain defenses have excellent experimental support and lend significant credibility to hypothesized defenses of similar nature. The expanse of coverage will include (either demonstrated or implied) defense against UV light, toxic abiotic chemicals, toxic byproducts of metabolism, predation, infection, recovery from stress, and settling of epibiotic organisms. On the host side, the organisms involved include protists, sponges, corals, tunicates, bryozoans, worms, fish, squid, and seaweeds. The symbionts include both eukaryotic and prokaryotic microbes. Furthermore, the associations range from highly complex microbial assemblages (e.g., those associated with sponges, tunicates, and corals) to associations involving only one or a few types of symbionts (e.g., light organs). Certain associations appear obligate (at least ecologically, if not physiologically), while others are facultative. Host/symbiont associations are highly specific in some associations, but apparently not in others, and transmission of symbionts is vertical in some cases, horizontal in others, and undetermined in most. Thus, the chapter has little thematic cohesiveness other than a geographic one, in that these symbioses occur in the largest and most complex ecosystem on earth, the oceans.

3.2 Defense against Abiotic Factors

3.2.1 UV Protection

In symbiotic associations that have been studied primarily from the perspective of nutrient transfer or exchange, the possible defensive/protective function of the microbes might have been overlooked. An example that has received significant attention, however, is that of protection against UV light. In shallow water marine organisms that engage in photosynthetic symbiosis, there is a trade-off: depths that permit photosynthetically active radiation (PAR) to reach the organisms also allow damaging UV light to penetrate. Shibata (1969) demonstrated the existence of a UV-absorbing compound in shallow-dwelling corals. It was later demonstrated by Siebeck (1988) that there is a correlation between depth and UV sensitivity, suggesting that organisms in the shallower depths possess UV-absorbing compounds. Subsequently, the presence of UV-screening mycosporine-like amino acids (MAAs) has been demonstrated in many of these symbiotic associations. In certain systems, however, there is some confusion over who is protecting whom. A broad assumption has been that MAAs are produced by the microbial symbionts, since animals were not thought capable of producing them for lack of genes for the shikimic acid pathway. This assumption would leave open the possibility of either symbiotic or dietary acquisition of MAAs. Dionisio-Sese et al. (1997) found that photosynthesis is severely inhibited when cyanobacterial symbionts (*Prochloron* sp.) isolated from their host ascidian, *Lissoclinum patella*, are irradiated with UV light. However, when intact ascidian colonies are irradiated with UV, *Prochloron* cells subsequently show no inhibition of photosynthesis, suggesting that the transparent host tissue acts as a UV shield. Consistent with this hypothesis, several MAAs were identified in the surface tunic of the host, leading those authors to propose that the tunicate is protecting its symbionts, not vice versa. This does not resolve the issue of the source of the MAAs, however. Although there is a higher concentration of MAAs in the host cells than in the symbiont cells, this evidence does not rule out the possibility that symbiont-synthesized MAAs are transported to and concentrated by the host. Maruyama et al. (2003) have evidence in two ascidians harboring *Prochloron* symbionts that MAAs are concentrated in bladder cells of the host tunic. However, the same group (Hirose et al., 2004) found that nonsymbiotic ascidians had tunics transparent to both, visible and UV light, while those harboring symbiotic *Prochloron* had UV-absorbing MAAs. Taken together, one interpretation is that in symbiotic ascidians, microbially produced MAAs are transferred to and concentrated by host tunic cells. More recently, Hirose et al. (2006) have found in the symbiotic ascidian *Didemnum molle* that concentrations of MAAs as well as pigments and spicules were higher in shallow waters where UV penetration is the highest, and suggest that *D. molle* might use these to adjust light conditions for its photosynthetic symbionts.

In the cnidarians, the primary photosynthetic symbionts are dinoflagellates in the genus *Symbiodinium.* Apart from the fixation of carbon by *Symbiodinium* and its well-documented transfer to the host, these algae are implicated in the production of UV-blocking MAAs. Banaszak and Trench (1995) showed that cultured isolates of *Symbiodinium microadriaticum* synthesize more MAAs in a combination of PAR + UV light than in PAR alone, which suggests an inducible response to UV exposure. However, the ability to synthesize MAAs is not universal among dinoflagellate symbionts, at least in culture. In a study by Banaszak et al. (2000), it was shown that *Symbiodinium* phylotype A produces MAAs in culture, although phylotypes B and C do not. Furthermore, certain aposymbiotic and nonsymbiotic cnidarians have been found to have MAAs (Banaszak and Trench, 1995). In those cases in which the symbionts do not synthesize MAAs, it has been suggested that the compounds are obtained in the diet (Banaszak and Trench, 1995). In contrast to the earlier study, Banaszak et al. (2006) screened 54 species of symbiotic cnidarians and found that *Symbiodinium* extracts in all clades, A, B, C, D, and E, contain MAAs. (In a related context, Stat et al. (2008) have presented evidence that Clade A *Symbiodinium* borders on being parasitic in that its release of photosynthate in the symbiotic state is substantially less than that of the more typically symbiotic Clade C. This illustrates the difficulty in assigning labels like "mutualism" to these very complex associations.) To add to the confusion, Starcevic et al. (2008) have obtained evidence that the genes encoding enzymes in the shikimic acid pathway have been laterally transferred to the sea anemone *Nematostella vectensis* from both bacterial and dinoflagellate donors. While these findings do not preclude the possibility of symbiont production of sunscreens in some taxa, it also indicates that it is not necessarily so universally.

Sponges in the photic zone face similar dangers from UV irradiation, and it is therefore not surprising that MAAs have been found in sponge extracts (e.g., Karentz et al., 1991; Bandaranayke et al., 1996; McClintock and Karentz, 1997; Shick and Dunlap, 2002). Many of these sponges have microalgal or cyanobacterial symbionts and it is speculated that these symbionts might be the source of the sunscreens (Steindler et al., 2002). Protection from UV is not restricted to animal/algal symbiosis. UV-screening MAAs have also been identified in a marine ciliate, *Maristentor dinoferus*, which harbors dinoflagellate symbionts in the genus *Symbiodinium* (Sommaruga et al., 2006). It is unclear whether the ciliate host or the algal symbiont is producing the MAA compounds.

3.2.2 Defense against Toxic Chemicals

Other abiotic perils exist in the marine environment. Few environments are as brutally toxic as hydrothermal vents at which superheated water laden with hydrogen sulfide and heavy metals boils out of the crust. Predicted by plate tectonic theory, the existence of these vents was verified in 1977 during dives by a Woods Hole Oceanographic team in the deep sea submersible vessel Alvin (Ballard, 1977; Lonsdale, 1977). No one was prepared for the finding that these vents support communities as lush and productive as rain forests and coral reefs, populated most prominently by giant tube worms and large bivalves. Since the vents are located far below the photic zone, microbiologists immediately recognized that the communities must be driven by chemoautotrophic bacteria, especially sulfur oxidizers. It was Colleen Cavanaugh who extended this idea by hypothesizing that animals at the vent sites are likely to be supported by symbiosis with chemoautotrophs (Cavanaugh et al., 1981). This was subsequently shown to be the case (e.g., Cavanaugh, 1983). It has also been hypothesized that the sulfur-oxidizing microbes associated with these animals might serve not only a nutritive function (by supplying the host with reduced carbon obtained through chemoautotrophy), but also a detoxification function (discussed by Nelson and Fisher, 1995 and Stewart and Cavanaugh, 2006). Indeed, it seems possible that the earliest advantage in associating with sulfur oxidizers might have been that these microbes protect the host by detoxifying H_2S, and this later became a more intimate nutritional relationship. Associations between invertebrates and H_2S-oxidizing bacteria have since been discovered in a variety of sulfur-rich habitats, many of them more accessible than deep-sea vents. Taylor et al. (2007) has suggested that sulfur-oxidizing symbionts of sponges might serve a similar detoxification function. A freshwater example is the swamp worm *Alma emini*, whose sulfur-oxidizing symbionts are thought to serve a sulfide-detoxifying function (Maina and Maloiy, 1998).

Hydrogen sulfide is not the only toxic material to which vent inhabitants are exposed. There are also toxic concentrations of heavy metals. The Pompeii worm, *Alvinella*, is noteworthy for two reasons: (1) it arguably has the highest temperature tolerance of any animal and (2) it has a dense coating of episymbiotic microbes associated with its integument (Gaill et al., 1987). In addition to the apparent chemosynthetic function of the epibionts (Campbell et al., 2001, 2003), it has been suggested that these microbes also serve a defensive role for their hosts by detoxifying metals. In support of this hypothesis, Jeanthon and Prieur (1990) showed that bacteria isolated from the integument of *Alvinella* were resistant to cadmium, zinc, arsenate, and silver, and showed tolerance to high levels of copper. Subsequent studies (discussed in Hardivillier et al., 2004) also implicated a metal detoxification role for microbial symbionts of vent animals. In addition, the finding (Vetriani et al., 2005) of mercury resistance in free-living moderate thermophiles from hydrothermal vents suggests that heavy metal resistance might be widespread in the vent microbial community, so it will not be surprising if additional symbionts are discovered, which might play such a defensive role. A skeptic would (and should) be quick to point out that what is lacking is the evidence that these metal-resistant symbionts reduce the metal load for their hosts as well as for themselves. As is the case in essentially all these microbial symbioses, future studies are needed to test these hypotheses.

3.3 Defense against Predation and Infection

What follows is only a brief discussion of an area of research that is burgeoning into a major effort. Secondary chemical compounds produced by invertebrates (especially sponges and tunicates) and their microbial symbionts represent what might be among our last untapped natural reservoirs of clinically important compounds. The potential practical application of these compounds tends to take center stage, while their ecological importance has received less attention. Clearly, however, there has been strong selective pressure in favor of the production of these compounds, and a rational hypothesis is that they function in a defensive role. Much of what is summarized here is extensively reviewed by Piel (2004, 2006) and Simmons et al. (2008).

3.3.1 Secondary Compounds in Sponges

Much of the recent research on sponges has been driven by the fact that they are prolific producers of secondary compounds with potential or demonstrated antitumor and antibacterial properties. It is generally assumed that a primary role of such secondary compounds is defense against predation and microbial invasion. Chemical defense against predation is well documented in sponges (e.g., Pawlik et al., 1995; Chanas et al., 1996; Haygood et al., 1999; Kubanek et al., 2000; Waddell and Pawlik, 2000a,b; Amsler et al., 2001; Becerro et al., 2003; Burns et al., 2003; McClintock et al., 2005; Paul et al., 2007). Furthermore, there is a substantial and growing body of evidence indicating that the sources of many such secondary compounds are microbial symbionts, of which there is an enormous variety in sponges (Newbold et al., 1999; Hentschel et al., 2001; Li and Liu, 2002; Proksch et al., 2002, 2003; Piel, 2004). Amazingly, microbial symbionts can make up to 40% of the sponge tissue volume (Friedrich et al., 2001) and represent a wide phylogenetic diversity (Hentschel et al., 2002). The symbionts include archaea, numerous heterotrophic and photosynthetic bacteria, and a variety of eukaryotic microbes, including algae and fungi (Lee et al., 2001; Taylor et al., 2007). These symbionts are thought to play a variety of roles, including provision of nutrients for the host (including photosynthetic products), translocation of metabolites, digestion of various compounds, nitrogen fixation, stabilization of the sponge skeleton, and (in some cases) defense against predation, infection, and biofouling (Lee et al., 2001).

Efforts to find direct evidence for a microbial origin of biologically active secondary compounds have in many cases been frustrated by the difficulty in culturing the symbionts and rigorously demonstrating synthesis of the compound in isolation. The evidence is more often correlative. Initial evidence for microbial synthesis took the form of chemical analysis of specific cell fractions of symbiotic sponges. For example, Unson and Faulkner (1993) used chemical analysis of flow-cytometrically separated cells from the sponge *Dysidea herbacea* to show that a unique group of polychlorinated compounds isolated from the whole sponge tissue is concentrated in symbiotic filamentous cyanobacterial cells. The same group (Bewley et al., 1996) used chemical analysis of symbiont cell populations in the marine sponge *Theonella swinhoei* to show that the macrolide swinholide A is concentrated in a unicellular bacterial population, while an antifungal cyclic peptide was localized in a filamentous bacterial population.

The utilization of microbial symbionts for defense implies some degree of coevolution, and raises the question of whether the sponge microbial community is specific to the host. The filter-feeding lifestyle of sponges might suggest that their microbial communities are reflective of the planktonic microbial community. Instead, however, it has been demonstrated that the microbial symbionts of sponges differ significantly from the microbes found in the surrounding seawater, arguing strongly for host/symbiont specificity (Hentschel et al., 2002). Furthermore, the same study demonstrated that different sponge species from geographically nonoverlapping regions harbor similar symbiont communities, supplying additional support for the hypothesis of a sponge-specific symbiont community. A new candidate phylum, "Poribacteria" has been proposed, based on the finding of deeply branching 16S sequences amplified from the total DNA isolated from three different sponge genera (Fieseler et al., 2004). The proposed phylum forms a monophyletic cluster with less that 75% 16S sequence identity to other bacterial groups.

Genomic analysis of Poribacteria is in progress (Fieseler et al., 2006), and should ultimately provide valuable information on the metabolic capabilities of this group. It is tempting to speculate that these and other possibly sponge-specific microbes have coevolved with their sponge hosts. In support of coevolution, Schmitt et al. (2007) have evidence of vertical transmission of a complex assemblage of sponge symbionts.

3.3.2 Secondary Compounds in Tunicates

Although somewhat eclipsed by the highly prolific sponges, tunicates are also a rich source of secondary compounds (e.g., Rinehart, 1987; Vervoortl et al., 1998; Proksch et al., 2002, 2003; Davis et al., 2003; Lehrer et al., 2003; Sung et al., 2007). In most cases, the sources of these compounds have not been determined, and some are likely to be of host origin. In those cases in which a microbial origin has been identified, some are dietary (for example, see Proksch et al., 2002; Simmons et al., 2008). Thus, a microbial origin does not guarantee a symbiotic origin. However, in certain cases, a strong case has been made for a microbial origin for secondary compounds in tunicates, and this is likely to be the case in a great many more (reviewed by Piel, 2004, 2006; Simmons et al., 2008). In some cases, the evidence is correlative; i.e., the compounds isolated are essentially identical to those from known microbial sources. In other cases, microbes isolated from tunicates have been shown to synthesize the compounds under study (reviewed by Piel, 2004). The progress in characterization of the microbial communities of tunicates has lagged behind that of the sponges, and undoubtedly greater insight into defensive role of these symbionts will come from further study of these assemblages.

3.3.3 Microbial Symbionts in Corals

Like the similarly sessile sponges and tunicates, corals also harbor complex microbial communities (apart from their well-studied photosynthetic symbionts), and early indications are that members of these symbiont communities might also perform a presumably defensive role analogous to the associations found in sponges and tunicates. The effort to characterize the microbial community in corals has been driven in large part by the increasing prevalence of bleaching (loss of symbiotic microalgae) and bacterial diseases of coral that threaten the very existence of reefs in the future (see reviews by Richardson, 1998 and Rosenberg et al., 2007a,b). Knowlton and Rohwer (2003) suggest that bleaching and disease might be a response to destabilization of mutualistic microbial symbiosis.

In corals, there has not (yet) been as much effort directed toward the search for secondary compounds of clinical importance as there has been in sponges and tunicates. This might change, as it has recently become more clear that the coral microbial community is enormously diverse, and is unlike that found in the surrounding seawater (Rosenberg et al., 2007a and references therein), and unlike that found in sponges (Taylor et al., 2007). This implies that bacteria from the surrounding seawater are excluded, possibly by antimicrobial activity effective only on nonsymbionts. There have been early indications of this.

For example, Kelman et al. (1998) showed that extracts of the soft coral *Parerythropodium fulvum fulvum* have antimicrobial activity against several pathogenic bacteria but not against coral-symbiotic microbes. Although the authors found that the small number of cultured symbiotic strains could not account for the antimicrobial activity, they suggested that it is possible that other, uncultured symbionts (which greatly outnumber the cultured isolates) might be responsible for synthesis of the defensive compound(s). Kim et al. (2000a,b) showed that crude extracts of gorgonian corals show antifungal activity. The study was motivated by the finding that *Aspergillus sydowii* is a pathogen of sea fans (Smith et al., 1996; Geiser et al., 1998). Those authors did not speculate on the cellular source of the antifungal activity, however. Since then, significant progress has been made in characterizing coral microbial communities, and certain trends have emerged. The diversity is vast, perhaps on the order of that found in sponges, and includes many taxa not previously encountered (Rohwer et al., 2001, 2002). We are a long way from knowing the full extent of coral microbial diversity. However, based on the finding that the majority of the over 400 sequences encountered in a culture-free study of multiple coral species were encountered only once. Rohwer et al. (2002) estimated the number of recognizably different taxa to exceed 6000. More recently, Koren and Rosenberg (2006) similarly found that the majority of the 400 sequences found by culture-independent methods in *Oculina patagonica* were encountered only once, again indicating that we are not yet close to estimating the diversity of these coral symbiont communities. Similar communities are found in geographically distant colonies of the same species, while different communities are found in separate species that co-occur geographically (Rosenberg et al., 2007a). Furthermore, the microbial communities are compartmentalized within the coral and are distinctly different in mucus, skeleton, and tissue (Koren and Rosenberg, 2006; Rosenberg et al., 2007a). Microbial communities from different coral species are quite different from one another, and differ from the microbial community in the water column (Frias-Lopez et al., 2002). An additional layer of complexity was revealed when Klaus et al. (2007) showed that the microbial communities associated with the same species of coral, *Montastraea annularis*, is significantly different at different depths. Recently, Kellogg (2004) and Wegley et al. (2004) have shown that corals also harbor archaea as members of their microbial communities. The roles of these archaea have not yet been determined. It has been determined, however, that the archaea do not appear to form species-specific associations with corals in the way bacteria do.

3.3.3.1 Coral Probiotic Hypothesis

The complexity of the coral symbiont community, its changeability, the relatively rapid recovery from bleaching, and the apparent acquisition of immunity to microbial disease have led Reshef et al. (2006) to propose the Coral Probiotic Hypothesis. The rationale for this hypothesis is that corals adapt more rapidly than expected to changing conditions, including bleaching and microbial challenges, and that this adaptation is intimately linked to their microbial symbiont community. This hypothesis grew in part from the unexpected development of apparent immunity to *Vibrio shiloi* infection in the coral *O. patagonica*. In their 2006 paper, Reshef et al. report that *V. shiloi*, previously shown to cause bleaching disease in *O. patagonica* (Kushmaro et al., 1996, 1997), could no longer be detected in the corals, and that previously infective *V. shiloi* could no longer infect those corals. The main thrust of the hypothesis is that there is a dynamic relationship between the holobiont (the coral host plus its microbial symbionts) and the environment. Because the makeup of the symbiont community can change quickly, selection for the optimal holobiont allows the corals to rapidly adapt to changing environmental conditions. Stated in NeoDarwinian terms, the coral holobiont is able to draw from a much greater genetic diversity than the coral host alone, which allows adaptation that is much more rapid than could be accounted for by mutation and selection in the host coral. Said another way, the hypothesis argues that the bacterial symbionts of corals perform an essential defensive role by enhancing recovery from stress and disease.

Related to the Coral Probiotic Hypothesis, recovery of corals from bleaching stress might be enhanced by another population of microalgal symbionts referred to as endolithic algae. These algae are neither zooxanthellae nor endosymbionts in the endodermal cells as the zooxanthellae are. Instead, these are ectosymbionts that inhabit the coral skeleton (Highsmith, 1981). Shashar et al. (1997) describe the coral skeleton as a refuge for endolithic algae, and suggest that the algae might provide UV-blocking MAAs (see Section 3.2.1), and that the coral tissue and skeletal material might be protecting the endolithic algae from the inhibitory effects of high irradiance. It might have been more accurate to add that the symbiotic zooxanthellae provide a major source of protection for endolithic algae, since it is the zooxanthellae that absorb the majority of light. However, another proposed protective mechanism might be at work here; it has been suggested that the observed bloom of endolithic algae following bleaching of zooxanthellae might help the coral weather the stress and help in the recovery of the coral host until reestablishing its population of zooxanthellae (Fine and Loya, 2002; Fine et al., 2005). Support of this hypothesis comes not only from the work by those investigators, but also from earlier measurements of transfer of photosynthate from endolithic algae to host corals (Schlichter et al., 1995). Though less efficient than transfer from zooxanthellae to the host, the increased population density of endolithic algae following bleaching (presumably a consequence of higher irradiance) could compensate, and provide the coral host sufficient carbon and energy to weather the stress. It should be noted that the best evidence for this hypothesis comes from one species of coral, *O. patagonica*, which shows an unusual ability to successfully recover after bleaching stress. Whether this mechanism operates in other corals is not known. Thus, the protective role of endolithic algae might not be universal.

3.3.4 Microbial Communities in Sea Anemones

Sea anemones, which are close cousins of the corals, also harbor bacterial symbionts. A preliminary study by Har and Thompson (2008) found 74 16S sequences which had <97% identity to known taxa. (For readers less familiar with bacterial sequence diversity, 16S rRNA sequences with <99% identity are generally agreed to represent separate species.) Almost no work has been done to test the possibility of a mutualistic role for these symbionts, but by analogy with the closely related corals, such a role is entirely plausible. Work on this anemone system continues in the Thompson laboratory, and is likely to reveal not only an interesting cache of novel species, but also potentially a plethora of new functions.

3.3.5 Genomic and Metagenomic Approaches

For all the circumstantial evidence for microbial production of defensive secondary compounds in invertebrates, direct evidence is often elusive, owing in large part to the difficulty in culturing the symbionts. Piel et al. (2004) have used a metagenomic approach to circumvent this problem. They cloned the genes for an antitumor compound from the symbiont metagenome of the sponge *T. swinhoei*, a sponge that has been shown to produce polyketides. The sequence and organization of the polyketide synthesis gene cluster isolated from the *Theonella* symbiont metagenome showed that the genes are clearly bacterial and phylogenetically related to the *ped* gene cluster cloned by the same group (Piel, 2002) from the symbiont metagenome of the beetle *Paederus fuscipes*. In the beetle symbiont, the genes encode the defensive antitumor polyketide pederin. This metagenomic approach was challenging, due to the unusual complexity of the sponge symbiont community. In striking contrast to the sponge, the beetle symbionts are virtually a monoculture of bacteria closely related to *Pseudomonas aeruginosa*.

In a similar strategy (Schmidt et al., 2005), the genome of *Prochloron didemni*, the uncultured symbiont of the ascidian *L. patella* was cloned, and the gene cluster encoding biosynthesis of Patellamide A and C was identified and confirmed by heterologous expression of the genes in *E. coli*. Patellamides are cytotoxic peptides with potential clinical applications. It had been previously suggested that their source was the *Prochloron* symbionts of *L. patella*, but until the work of Schmidt and collaborators was complete, this had remained speculative.

Schirmer et al. (2005) used PCR to identify polyketide synthase (PKS) genes from the microbial community associated with the marine sponge *Discodermia dissolute*. They followed this with a screening of metagenomic library, revealing several PKS gene clusters. An additional example of a genomic cloning strategy is provided by Lopanik et al. (2006), who cloned the PKS gene fragments responsible for synthesis of bryostatin from the uncultured bacterial symbiont of the marine bryozoan *Bugula neritina*. A strategy involving PCR amplification revealed PKS genes of unequivocal bacterial origin in 20 different sponge species (Fieseler et al., 2007). In the same study, three bacterial PKS gene clusters were characterized in large insert metagenomic libraries from two of the sponges. A PCR strategy was also employed by Kennedy et al. (2008) to identify PKS genes from the *Haliclona simulans* symbiont metagenome. Thus, the indirect approach of genomic and metagenomic cloning will undoubtedly prove to be a useful route for identification of secondary product sources, as well as potential clinical exploitation of the compounds.

3.3.6 Direct Evidence of Microbial Protection against Predation

The above discussion provides alluring (if sometimes circumstantial) evidence for a role of microbial symbionts in protection against predation and infection in invertebrates by virtue of secondary compound production. It is gratifying that, in some cases, direct evidence exists for microbial protection against predation. Davidson et al. (2001) showed evidence that *Candidatus Endobugula sertula*, a bacterial symbiont of the bryozoan *Bugula neritina* (Haygood and Davidson, 1997), was responsible for the synthesis of the bryostatins, a class of polyketides. Lopanik et al. (2004) followed this with a study that showed that bryostatins produced by *Ca. Endobugula sertula* deterred fish predation on larvae of *Bugula meritina*. Furthermore, Lim and Haygood (2004) showed that a bacterial symbiont (*Candidatus Endobugula glebosa*) of the sister bryozoan species *Bugula simplex* also makes bryostatins, providing evidence that such a strategy apparently also exists in a second bryozoan. Surprisingly, however, the same group (Lim-Fong et al., 2008) found that although related bacterial symbionts are present in four sibling species of *Bugula*, only two have evidence of symbiont-derived defensive compounds. It is interesting to note that the seemingly more vulnerable sessile adult bryozoans do not appear to use bryostatins as defensive compounds (Lopanik et al., 2004).

Another fascinating case in which microbial defense against predation has been directly demonstrated experimentally involves certain marine isopods and their episymbiotic cyanobacteria (Lindquist et al., 2005). The isopods, which inhabit coral reefs in Papua, New Guinea, are coated with cyanobacteria, giving them a distinctive red color. Experiments in which the cyanobacteria were removed from the isopods showed a far higher rate of predation of isopods by fishes. Furthermore, addition of a crude extract from the cyanobacteria to control food pellets reduced consumption of the pellets by 70%. Together, these results provide a strong indication that the episymbionts confer a valuable protective function on their hosts. There are at least two morphotypes of cyanobacteria associated with the isopods, and the authors of the study suspect that it is the larger of the two types which produce the defensive compounds. Phylogenetic analysis indicates that the smaller morphotype is apparently related to free-living *Synechococcus*, and is probably horizontally derived. It is not thought that these smaller cyanobacteria produce the protective compounds, since *Synechococcus* is not known to produce any unusual secondary metabolites that could be recruited for a defensive role. The larger episymbiotic cyanobacteria are likely to be the ones producing the defensive compounds. These cells are apparently vertically transmitted, since newly emerged juveniles appear to be inoculated with symbiotic cyanobacteria by direct contact with the mother. The isopods have also been seen to consume their episymbionts, indicating that they are farming the bacteria. Thus, the relationship apparently has a dual role of protection and nutrition.

Protection from predation apparently extends to microbe–microbe symbiosis, as well. Marine ciliates in the genus *Euplotidium* have ectosymbiotic bacteria that protect the host from predation by another ciliate, *Litonotus lamella* (Rosati et al., 1999). These ectosymbionts have been described as "epixenosomes," or "external foreign bodies" (Verni and Rosati, 1990), and have been found to be related to the Verrucomicrobia (Petroni et al., 2000). Epixenosomes are the only known examples of bacterial ectosymbionts in the Verrucomicrobia group. Among the protists, the majority of bacterial symbionts (both ecto- and endo-) are in the Proteobacteria (Amann et al., 1995). Although the host does not require its epixenosomes in laboratory culture, the symbionts clearly have survival value for the host in nature. The mechanism of defense is via an extrusion apparatus, which has led to speculation that the defensive trichocysts of ciliates might have arisen from bacterial symbionts (Petroni et al., 2000).

3.4 Defensive Light Organ Symbiosis

Of the myriad bioluminescent organisms in the oceans, the vast majority are intrinsically luminous (i.e., their capacity to emit light is a property of their own metabolism), and are therefore not relevant to this book. In certain animals, however, light emission is a consequence of microbial symbiosis.

This is true of several species of fish and squid. In terms of structural, physiological, and genetic details, it is safe to say that light organ symbiosis (particularly that of sepiolid squid) represents the best characterized marine bacterial symbiosis of any type. Thus, considerably more detail can be provided here than for other microbial symbiosis.

The fish and squid known to have symbiotic bioluminescent bacteria undoubtedly represent a small percentage of the total yet to be discovered. They include various species of angler fish, flashlight fish, hatchet fish, pony fish, and at least 35 species of squid (Ruby and McFall-Ngai, 1992; Haygood, 1993). Although some of these animals use bioluminescence for offensive purposes (e.g., as a lure in angler fish) or for signaling (e.g., flashlight fish), the majority of symbiotic luminous animals apparently use light for predator evasion.

Perhaps the most common use of bioluminescence (by virtue of the number of different organisms that apparently do this) is predator evasion by counterillumination, a camouflage strategy in which the ventral side of the animal is illuminated in order to break up its silhouette, thus hiding the animal from "sit and wait" predators below (Harvey, 1952; Clarke, 1963). Even in the dim midwater zone, downwelling skylight is sufficient for animals to cast a shadow, putting them at risk. Hastings (1971) showed experimentally that the pony fish *Leiognathus equulus* (which possesses a symbiotic light organ) exhibits behavior consistent with defensive counterillumination. Significantly, the fish displays a ventral glow only in the presence of light. In darkness, the fish shuts off the light. This makes sense, if counterillumination hides a silhouette resulting from downwelling light. Furthermore, McFall-Ngai and Morin (1991) showed that, within limits, three species of pony fish, *Leiognathus nuchalis*, *L. splendens*, and *Gazza minuta*, increase ventral illumination with increased ambient light. Counterillumination also appears to be the strategy in the Hawaiian bobtail squid *Euprymna scolopes*. Long assumed to be the case because of the ventral location of the light organ, this hypothesis received support from the work of Jones and Nishiguchi (2003), who showed that the squid is capable of modulating the intensity of ventral light emission depending on the intensity of the downwelling light. Unlike the pony fish, which emit, its ventral bioluminescence during daylight hours, *Euprymna* is active at night, and adjusts its counterillumination intensity to match the downwelling moonlight and starlight.

Morin et al. (1975) proposed a different predator evasion strategy in the flashlight fish *Photoblepharon palpebratus*, based on observations of captured individuals as well as schools of fish. In flashlight fish, the symbiotic light organ is located beneath each eye, and the on–off light emission is accomplished by means of a shutter. One strategy of predator evasion is presumed to occur through a "blink and run" behavior, in which the fish blinks the light and abruptly changes direction. An additional proposed evasion strategy occurs in schools of flashlight fish, in which the blinking of the aggregation is thought to visually confuse predators.

Unlike a variety of other bioluminescent systems in which light emission is tightly regulated and light is emitted on demand, luminous bacteria emit a constant glow, once induced to do so. In symbiotic systems, this presents a problem for the host, since constant light emission would put the host at risk. To be most advantageous as a defensive mechanism, light emission must be modulated. This is accomplished through various shutters, chromatophores, lenses, and light guides (Herring, 1978, 1990). The structural sophistication of light organs illustrates the intense positive selective pressure for this defensive symbiosis.

3.4.1 Hosts

Of the animal hosts harboring bioluminescent bacterial symbionts, those that are best characterized are pony fish (leiognathid fishes) and sepiolid squid. The pony fish have a circumesophageal light organ. Light emission from the luminous bacterial symbionts is reflected from the mirrored inner dorsal surface of the gas bladder, thus directing it downward (McFall-Ngai, 1983a,b). The reflective surface is composed of guanine layers acting as a 1/4-wave reflector, a common structure that has apparently evolved more than once in marine animals. Among the sepiolid squid, the species that has received (by far) the most attention is *E. scolopes*. *Euprymna* has a ventral ink sac-associated symbiotic light organ equipped with a lens, a reflector, and a mechanism for modulating light intensity (McFall Ngai and Montgomery, 1990; Montgomery and McFall-Ngai, 1994).

3.4.2 Symbionts

Light organ symbionts include several strains that have been successfully cultured and an uncertain (and undoubtedly large) number that have not been cultured. Cultured species include *Vibrio fischeri*, *Vibrio logei*, *Photobacterium leiognathi*, *Photobacterium mandampamensis*, *Photobacterium phosphoreum*, and *Photobacterium kishitanii*. *V. fischeri* has a range of hosts that (so far) hold the record for spanning the widest evolutionary distance, inhabiting the light organs of Moncentrid fish (Ruby and Nealson, 1976) as well as squid of the genera *Euprymna* (Ruby and McFall-Ngai, 1992) and *Sepiola* (Fidopiastis et al., 1998). *V. logei* has been found to be an inhabitant of the light organ of the squid genus *Sepiola* (Fidopiastis et al., 1998). The four *Photobacterium* species inhabit the light organs of a variety of fish (Ast and Dunlap, 2004, 2005; Dunlap and Ast, 2005; Ast et al., 2007; Kaeding et al., 2007). Attempts have been made to culture additional light organ symbionts (e.g., those of flashlight fish), but thus far with no success. All known light organ symbionts are in the order Vibrionales. The symbiotic habit appears curiously prevalent in two groups: the genus *Photobacterium* and the *V. fischeri* group (the main players in the latter being *V. fischeri* and *V. logei*). The *V. fischeri* group occupies an ambiguous taxonomic position, consistently branching outside the main clade of Vibrios, and there has been a recent proposal (Urbanczyk et al., 2008) to change the name of the genus to *Aliivibrio* to reflect its distance from other *Vibrio* species. There are a number of other luminous species within the main clade of Vibrios, but none thus far is a known light organ symbiont. The possible exception is the uncultured symbiont of *Kryptophanaron alfredi*, the *luxA* sequence of which is more similar to that of *Vibrio harveyi* than to *V. fischeri* or *Photobacterium* (Haygood, 1990).

3.4.3 Establishment of Light Organ Symbiosis

Because of its accessibility, both physically and experimentally, the symbiotic association between *E. scolopes* and *V. fischeri* is by far the best characterized light organ symbiosis, and the only one in which the stages of symbiont establishment have been described. This discussion will be brief, since a vast and expanding literature exists on the subject (e.g., see reviews by McFall-Ngai, 1999; Visick and McFall-Ngai, 2000; Nyholm and McFall-Ngai, 2004; Visick and Ruby, 2006). Acquisition of bacterial symbionts is horizontal, occurring within the first 12 h after hatching (McFall-Ngai and Montgomery, 1990; McFall-Ngai and Ruby, 1991; Montgomery and McFall-Ngai, 1994). In a process that has been extensively characterized, *V. fischeri* enter the juvenile squid through three pores on each side of the undeveloped organ, facilitated by ciliated appendages. Upon acquisition of the symbionts, the ciliated appendages regress, the three pores coalesce into one pore on each side, and the light organ undergoes a series of developmental changes culminating in the adult structure within days (Nyholm and McFall-Ngai, 2004). Although it has been clearly demonstrated that the presence of *V. fischeri* induces morphogenesis of the juvenile light organ, Claes and Dunlap (2000) have shown that much of light organ development is "hardwired," and will occur later in aposymbiotic squid, given sufficient time. A possible interpretation, therefore, is that *V. fischeri* induces earlier development of certain structural features of the light organ that would otherwise develop on their own at a later stage.

Whether symbiont acquisition is horizontal in all light organ symbiosis is unknown. However, the assumption is that it is common, if not universal. Even in host species in which colonization has not been experimentally determined, the diversity of symbiont genotypes within one host argues for horizontal acquisition. As is the case in *Sepiola*, it has been shown that the light organ symbionts of fish do not necessarily comprise a monoculture of just one genotype (Dunlap et al., 2004). Furthermore, in at least three fish species, *Acropoma japonicum*, *Photopectoralis panayensis*, and *Photopectoralis bindus*, two different species of luminous bacteria, *Photobacterium mandapamensis* and *Photoacterium leiognathi*, have been shown to coexist in the same light organ (Kaeding et al., 2007).

3.4.4 Host–Symbiont Specificity in Light Organ Symbiosis

The success of different strains of *V. fischeri* to colonize *Euprymna* is variable, implying the evolution of some degree of strain specificity (Lee and Ruby, 1994; Nishiguchi et al., 1998). Furthermore, the squid are able to distinguish light-producing strains from dark strains (Visick et al., 2000). Also, as mentioned above (Fidopiastis et al., 1998), the squid genus *Sepiola* can contain mixed symbiont community comprising both *V. fischeri* and *V. logei*. The mix is variable among individuals, demonstrating that host specificity within *Sepiola* is perhaps somewhat relaxed. *V. fischeri* and *V. logei* are very closely related, and it is likely that symbiotic strains of the two species share certain surface features or other characteristics that permit colonization of *Sepiola*. It has recently been shown that an important symbiosis factor in *V. fischeri* is the RscS protein (Yip et al., 2005, 2006), prompting the hypothesis that this could be a major factor in determining the colonization competence of different strains.

3.4.5 Maintenance of the Symbiosis

In those cultured luminous bacterial species in which the regulation has been closely studied, light emission is under quorum-sensing control (reviewed by Miller and Bassler, 2001), and thus occurs only after the population reaches sufficient density. This is likely to be the case in most (if not all) light organ symbionts, although only *V. fischeri* has been investigated extensively in this context (reviewed by Visick, 2005). Furthermore, light is an expensive commodity to produce (Harvey, 1952), and is thus very sensitive to the physiological state of the bacteria. Light emission is most intense in the mid to late exponential growth phase of the bacteria, and ceases in stationary phase. This presents a problem in terms of maintenance of bacteria in a bright state. The light organ symbionts cannot be allowed to reach stationary phase, or the light turns off. The solution is a strategy well-known to microbiologists: semicontinuous culture. Semicontinuous culture approximates a chemostat, which is a system in which growth of microbes is balanced by dilution of the culture with fresh medium, such that the microbes remain at a constant growth rate and population density. In semicontinuous culture, the process is stepwise, wherein the culture is periodically diluted, rather than continuously diluted. In the light organ system of *E. scolopes*, bioluminescent *V. fischeri* are periodically ejected from the organ, leaving a small inoculum to again grow up to the requisite density to emit light (Boettcher et al., 1996). This ejection occurs every morning, at which time 90%–95% of the *V. fischeri* symbionts are released into the seawater. The remaining bacteria have reached sufficient density by nightfall to provide the light required for defensive camouflage. Such a semicontinuous culture apparently operates in various fish (Haygood and Distel, 1993; Wada et al., 2008), although the details have not been quantified as they have been in *Euprymna*. The nature of luminous bacterial physiology makes it likely that such a system would be common, and probably necessary, in all light organ symbiosis.

3.5 Protective Gut Symbiosis in the Aquatic Environment

It has become quite clear that the human microbiota, especially that of the gut, is not merely a collection of benign commensals, rather they are active participants in the biology of a complex host/microbe ecosystem (e.g., Xu and Gordon, 2003; Ley et al., 2006; Gill et al., 2006). Among the vital roles of these microbial symbionts is to serve a defensive function by influencing development of the immune system (Mazmanian et al., 2005; Pamer, 2007) and by keeping potential pathogens (either resident or invasive) in check (e.g., Mazmanian et al., 2008).

These findings have prompted the initiation of the NIH-funded Human Microbiome Project and the European Commission's Metagenomics of the Human Intestinal Tract project (for an overview, see Mullard, 2008). Furthermore, these findings vindicate (at least to some extent) proponents of probiotics (primarily in the form of microbial dietary supplements) as enhancers of human health (Marteau et al., 2001).

In light of the increasing awareness of the importance of the human gut microbiota, it is probably not surprising to find a probiotic role of gut symbiontic microorganisms in the marine environment. In the aquatic world, research on probiotics has been driven primarily by the aquaculture industry. Certain microbes, notoriously the Vibrios, plague the farming of both shellfish and finfish. High density monoculture of animals, whether vertebrates or invertebrates, is intrinsically artificial and unstable, and the predictable outbreaks of devastating microbial disease has resulted in skyrocketing use of antibiotics, which brings with it the unavoidable selection for antibiotic-resistant microbes (for review, see Cabello, 2006). A different and more natural approach has been to investigate the possibility of microbial defense against pathogens.

The organisms popularly farmed (e.g., shrimp, oysters, and fish) harbor complex microbial communities, some members of which might confer natural protection against pathogens. The goal of these efforts would be to provide protective microbes in the feed for the farmed organisms, thus populating their intestinal tracts with defensive allies, rather than administering antibiotics. As early as 1990, Onarheim and Raa suggested that resident gut bacteria might protect host fish from pathogenic invasion. Part of the effort, therefore, has been to identify naturally occurring symbiotic gut microbes that confer such a defensive function. The study of the gut microbiota of marine animals is very much in its infancy. It is not known, for example, how many members of the microbial community of marine animals are resident and how many are transient. At present, the only aquatic gut model receiving large-scale attention is the freshwater zebrafish (Rawls et al., 2004), and that work has a primarily developmental perspective.

Probiotic defensive effects of resident or artificially added microbes have been found for invertebrates as well as for fish. A few examples in invertebrates: in *Artemia* (brine shrimp), bacterial strains that confer protection against pathogenic *Vibrio proteolyticus* are also capable of colonizing the crustacean host (Verschuere et al., 2000). In that study, it was shown that live cells, not a cell extract, were necessary to exert the protective action, supporting the hypothesis that this is a defensive role of natural microbial symbionts. Probiotic treatment with *Bacillus subtilis* BT23 reduced mortality of black tiger shrimp due to *V. harveyi* by 90% (Vaseeharan and Ramasamy, 2003). *B. subtilis* UTM 126, isolated from the intestine of pacific white shrimp, protects juvenile shrimp not only from infection by *V. harveyi*, but also from *V. alginolyticus* and *V. parahaemolyticus* (Balcazar and Rojas-Luna, 2007). Additional naturally occurring symbionts of pacific white shrimp that show probiotic activity are *Vibrio alginolyticus* UTM 102, *Roseobacter gallaeciensis* SLV03, and *Pseudomonas aestumarina* SLV22 (Balcazar et al., 2007). In another example, the predominance of *Vibrio gazogenes* in the gut of the prawn *Penaeus merguiensis* (Oxley et al., 2002) hints at a probiotic role. It is possible that the *V. gazogenes* outcompetes a variety of other Vibrios for resources in the prawn gut, including those with pathogenic potential.

The following are some examples of protective fish-associated microbes. As early as 1991, Westerdahl and collaborators found that 28% of bacteria isolated from the intestine of *Scophtalmus maximus* (turbot) inhibited the fish pathogen *Vibrio anguillarum* (Westerdahl et al., 1991). The same group (Olsson et al., 1992) showed that intestinal isolates from both turbot and dab (*Limanda limanda*) had an inhibitory effect on *V. anguillarum*. The bacterial isolate *Weissella hellenica* DS-12 from flounder intestine showed antibacterial activity against a variety of fish pathogens in the genera *Edwardsiella*, *Pasteurella*, *Aeromonas*, and *Vibrio* (Cai et al., 1998). Similarly, Hjelm et al. (2004) showed that intestinal isolates in the genera *Vibrio* and *Roseobacter* were antagonistic toward *V. anguillarum*, and Chabrillón et al. (2005) found that pathogenic *V. harveyi* were inhibited by isolates from *Sparus aurata* (gilthead sea bream). The probiotic role of marine microbes might be inducible, at least in certain cases. For example, Ruiz-Ponte et al. (1999) showed that the protective antibacterial role of *Roseobacter* strain BS107 is enhanced by the presence of *V. anguillarum* strain 408. Trypsin sensitivity implicates a proteinaceous signal molecule released by Vibrio.

In freshwater, naturally occurring symbionts of rainbow trout, primarily in the genus *Pseudomonas,* showed significant antagonistic activity toward the fish pathogen *V. anguillarum* (Gram et al., 1999; Spanggaard et al., 2001). Also in rainbow trout, naturally occurring symbionts (*Bacillus* and *Aeromonas* species) protect the host not only against pathogenic vibrios (*V. anguillarum* and *V. ordalii*) but also against *Aeromonas salmonicida*, *Lactococcus garvieae*, *Streptococcus iniae*, and *Yersinia ruckeri*, reducing mortality by 83%–100% (Brunt et al., 2007).

3.6 Defenses against Fungal Infection

Although, for economic reasons, much attention has been focused on the probiotic potential of gut symbiotic microbes in aquaculture, other examples are known. For example, Gil-Turnes et al. (1989) showed that the episymbiotic bacteria (*Aeromonas* sp.) that coat the exterior of the embryos of the shrimp *Palaemon macrodactylus* protect the embryos from infection by the pathogenic fungus *Lagenidium callinectes*. Similarly, Gil-Turnes and Fenical (1992) showed that larvae of the lobster *Homarus americanus* are protected from fungal infection by epibiotic bacteria. It is possible that such a strategy might be found in other crustaceans as well. Schmidt et al. (2000) found that a bacterial symbiont (*Candidatus* Entotheonella palauensis) of the sponge *T. swinhoei* contains theopalauamide, an antifungal agent. Outside the animal kingdom, the seaweed *Lobophora variegate* produces a potent fungicide (Lobophorolide) that defends the alga from pathogenic invasion (Kubanek et al., 2003). The similarity of the compound to microbially synthesized polyketides has led the authors to hypothesize a possible symbiotic origin of this fungicide. Although there are only a handful of cases that have been revealed, such defensive relationships are probably widespread in the aquatic environment.

3.7 Detoxification of Host Metabolites

Although not rigorously demonstrated yet, a potential protective role of microbial symbionts in some systems is the detoxification of host metabolic products. It has been suggested, based on metagenomic analysis of the symbiotic microbial community associated with the marine oligochaete worm *Olavius algarvensis*, that one function carried out by certain symbionts is detoxification of ammonium and urea (Woyke et al., 2006). This suggestion is based on finding, among symbiont metagenomic sequences, homologues of known ammonium and urea transporters as well as a urease homologue. Such a detoxifying function would be consistent with the highly reduced nephridia (which normally operate in elimination of nitrogen waste) in these worms. In addition, the same authors suggest that microbial symbionts might also utilize certain host waste fermentation products, since genes for mono- and dicarboxylate transport and utilization were found in the symbiont metagenome.

Similarly, Taylor et al. (2007) suggests that nitrifying symbiotic bacteria of sponges might be responsible for lowering the levels of ammonia and nitrite, which are metabolic products of both the host and its symbionts. Another need for detoxification involves algal symbiosis. Harboring high concentrations of photosynthetic microsymbionts presents an interesting paradox: the obvious nutritional advantage of the association also carries the risk of hyperoxic stress resulting from the production of reactive oxygen species (ROS) that can damage both host and symbiont (Furla et al., 2005). Richier et al. (2005) found a greater diversity of the antioxidant enzyme superoxide dismutase in symbiotic anemones as compared to nonsymbiotic anemones. Furthermore, the same study found a decrease in antioxidant activity in aposymbiotic anemones compared to their symbiotic counterparts. These authors conclude that the symbiotic state leads to higher tolerance of hyperoxia. However, since both host and symbiont make antioxidant enzymes, it is unclear whether there is any mutual protection involved. The authors of the study suggest that it might be the animal host protecting its algal symbionts from ROS, not vice versa. But in an analogous photosynthetic symbiosis, the freshwater *Paramecium bursaria*, which carries green algal symbionts (*Chlorella* sp.), Hortnagl and Sommaruga (2007) suggest that the algal symbionts are responsible for minimizing oxidative stress in the host. Furthermore, since UV light enhances production of ROS, it could also be that MAAs (Section 3.2.1) provide an indirect defense against ROS.

3.8 Inhibition of Epibionts

In addition to avoiding predation and infection, sessile marine organisms must prevent excessive settling of epibionts on their body surfaces. Since this area has been extensively reviewed by Krug (2006), only a brief overview is provided here. Analogous to allelopathy in terrestrial plants, there are mechanisms by which marine organisms, both animal and algal, inhibit the settling of larvae and spores of competitor species and epibionts. It is not surprising to encounter mechanisms in which microbial symbionts have been recruited for this purpose. Surface-associated bacteria in animals and seaweeds have been implicated in defending the host against the settling of both macro- and microepibionts. For example, it was shown by Holmström et al. (1992) that surface-associated bacteria from the tunicate *Ciona intestinalis* inhibit the settling of larvae from barnacles and tunicates. The same group (Holmström and Kjelleberg, 1999; Holmström et al., 2002) found that surface-associated *Pseudoalteromonas* species inhibit settling of fouling organisms. Thakur et al. (2004) found that epiphytic bacteria in the genus *Bacillus* found associated with the sponge *Ircinia fusca* have antibacterial activity that might prevent the growth of other bacteria on the sponge surface. Similarly, Kanagasabhapathy et al. (2004) found that surface bacteria isolated from various sponges inhibit fouling bacteria. In corals, Harder et al. (2003) and Dobretsov and Qien (2004) have isolated surface-associated bacteria that inhibit growth of fouling bacteria. Other examples are found in seaweeds. Boyd et al. (1999) showed that 21% of microbes associated with marine macroalgae inhibit the growth of biofouling bacteria. Armstrong et al. (2001) describe the apparently protective role of epiphytic bacteria on seaweed surfaces, protecting the host from the settling of other bacteria. Another example is in the seaweed genus *Ulva*. Epiphytic bacteria (*Pseudoalteromonas tunicata* and *Phaeobacter* sp.) on *Ulva* produce compounds inhibitory to the settling of algal and fungal spores, invertebrate larvae, and other bacteria (Rao et al., 2007). It was shown that surprisingly low cell densities of *Pseudoalteromonas tunicata* and *Phaeobacter* were required to elicit the inhibitory response, lending support to the hypothesis that this mechanism operates in the natural setting.

3.9 Conclusions

The examples discussed so briefly in this chapter are by no means exhaustive. Hopefully, they serve to illustrate the extraordinarily rich diversity of protective/defensive microbial symbiosis in the marine environment. Almost no system has been described in detail, and even in those in which significant progress has been made, there are still many burning questions. In nearly every case described, we still seek the "holy grails" of host/symbiont specificity, symbiont transmission, and of course physiological and ecological cost/benefit analysis. This field is very much in its larval stages, and progress will come from a combination of ecological, physiological, structural, genetic, and genomic/metagenomic approaches, as well as choosing tractable model systems. Provided we still have a marine environment to study in the future, there are countless microbial symbioses yet to be described.

REFERENCES

Amann, R., Ludwig, W., and K. H. Schleifer. 1995. Identification and in situ detection of individual microbial cells without cultivation. *Microbiology Reviews* 59:143–169.

Amsler, C. D., McClintock, J. B., and B. J. Baker. 2001. Secondary metabolites as mediators of trophic interactions among Antarctic marine organisms. *American Zoologist* 41:17–26.

Armstrong, E., Yan, L., Boyd, K. G., Wright, P. C., and J. G. Burgess. 2001. The symbiotic role of marine microbes on living surfaces. *Hydrobiologia* 461:37–40.

Ast, J. C., Cleenwerck, I., Engelbeen, K., Urbanczyk, H., Thompson, F. L., DeVos, P., and P. V. Dunlap. 2007. *Photobacterium kishitanii* sp. nov., A luminous marine bacterium symbiotic with deep-sea fishes. *International Journal of Systematic and Evolutionary Microbiology* 57:2073–2078.

Ast, J. C. and P. V. Dunlap. 2004. Phylogenetic analysis of the lux operon distinguishes two evolutionarily distinct clades of *Photobacterium leiognathi. Archives of Microbiology* 181:352–361.

Ast, J. C. and P. V. Dunlap. 2005. Phylogenetic resolution and habitat specificity of members of the Photobacterium phosphoreum species group. *Environmental Microbiology* 7:1641–1654.

Balcazar, J. L. and T. Rojas-Luna. 2007. Inhibitory activity of probiotic *Bacillus subtilis* UTM 126 against vibrio species confers protection against vibriosis in juvenile shrimp (*Litopenaeus vannamei*). *Current Microbiology* 55:409–412.

Balcazar, J. L., Rojas-Luna, T., and D. P. Cunningham. 2007. Effect of the addition of four potential probiotic strains on the survival of pacific white shrimp (*Litopenaeus vannamei*) following immersion challenge with *Vibrio parahaemolyticus. Journal of Invertebrate Pathology* 96:147–150.

Ballard, R. D. 1977. Notes on a major oceanographic find. *Oceanus* 20:25–44.

Banaszak, A. T., LaJeunesse, T. C., and R. K. Trench. 2000. The synthesis of mycosporine-like amino acids (MAAs) by cultured, symbiotic dinoflagellates. *Journal of Experimental Marine Biology and Ecology* 249:219–233.

Banaszak, A. T., Santos, M. G. B., Lajeunesse, T. C., and M. P. Lesser. 2006. The distribution of mycosporine-like amino acids (MAAs) and the phylogenetic identity of symbiotic dinoflagellates in cnidarian hosts from the Mexican Caribbean. *Journal of Experimental Marine Biology and Ecology* 337:131–146.

Banaszak, A. T. and R. K. Trench. 1995. Effects of ultraviolet (UV) radiation on microalgal-invertebrate symbiosis. II. The synthesis of mycosporine-like amino acids in response to UV in *Anthopleura elegantissima* and *Cassiopeia xamachana*. *Journal of Experimental Marine Biology and Ecology* 194:233–250.

Bandaranayke, W. M., Bemis, J. E., and D. J. Bourne. 1996. Ultraviolet absorbing pigments from the marine sponge Dysidea herbacea: Isolation and structure of a new mycosporine. *Comparative Biochemistry and Physiology* 115C:281–286.

Becerro, M. A., Thacker, R. W., Turon, X., Uriz, M. J., and V. J. Paul. 2003. Biogeography of sponge chemical ecology: Comparisons of tropical and temperate defenses. *Oecologia* 135:91–101.

Bewley, C. A., Holland, N. D., and D. J. Faulkner. 1996. Two classes of metabolites from *Theonella swinhoei* are localized in distinct populations of bacterial symbionts. *Experientia* 52:716–722.

Boettcher, K. J., Ruby E. G., and M. J. McFall-Ngai. 1996. Bioluminescence in the symbiotic squid *Euprymna scolopes* is controlled by a daily biological rhythm. *Journal of Comparative Physiology* 179:65–73.

Boyd, K. G., Adams, D. R., and J. G. Burgess. 1999. Antibacterial and repellent activities of marine bacteria associated with algal surfaces. *Biofouling* 14:227–236.

Burns, E., Ifrach, I., Carmeli, S., Pawlik, J. R., and M. Ilan. 2003. Comparison of anti-predatory defenses of Red Sea and Caribbean sponges. I. Chemical defense. *Marine Ecology Progress Series* 252:105–114.

Brunt, J., Newaj-Fyzul, A., and B. Austin. 2007. The development of probiotics for the control of multiple bacterial diseases of rainbow trout, Oncorhynchus mykiss (Walbaum). *Journal of Fish Diseases* 30:573–579.

Cabello, F. C. 2006. Heavy use of prophylactic antibiotics in aquaculture: A growing problem for human and animal health and for the environment. *Environmental Microbiology* 8:1137–1144.

Cai, Y., Benno, Y., Nakase, T., and T.-K. Oh. 1998. Specific probiotic characterization of *Weissella hellenica* DS-12 isolated from flounder intestine. *Journal of General and Applied Microbiology* 44:311–316.

Campbell, B. J., Jeanthon, C., Kostka, J. E., Luther, G. W., and S. C. Cary. 2001. Growth and phylogenetic properties of novel bacteria belonging to the epsilon subdivision of the proteobacteria enriched from *Alvinella pompejana* and deep-sea hydrothermal vents. *Applied Environmental Microbiology* 67:4566–4572.

Campbell, B. J., Stein, J. L., and S. C. Caryl. 2003. Evidence of chemolithoautotrophy in the bacterial community associated with *Alvinella pompejana*, a hydrothermal vent polychaete. *Applied Environmental Microbiology* 69:5070–5078.

Cavanaugh, C. 1983. Symbiotic chemoautotrophic bacteria in marine invertebrates from sulphide-rich habitats. *Nature* 302:58–61.

Cavanaugh, C., Gardiner, S. L., Jones, M. L., Jannasch, H. W., and J. B. Waterbury. 1981. Prokaryotic cells in the hydrothermal vent tube worm *Riftia pachyptila* Jones: Possible chemoautotrophic symbionts. *Science* 213:340–342.

Chabrillón, M., Rico, R. M., Arijo, S., Díaz-Rosales, P., Balebona, M. C., and M. A. Moriñigo. 2005. Interactions of microorganisms isolated from gilthead sea bream, *Sparus aurata* L., on *Vibrio harveyi*, a pathogen of farmed Senegalese sole, *Solea senegalensis* (Kaup). *Journal of Fish Diseases* 28:531–537.

Chanas, B., Pawlik, J. R., Lindel, T., and W. Fenical. 1996. Chemical defense of the Caribbean sponge *Agelas clathrode* (Schmidt). *Journal of Experimental Marine Biology and Ecology* 208:185–196.

Claes, M. and P. V. Dunlap. 2000. Aposymbiotic culture of the sepiolid squid *Euprymna scolopes*: Role of the symbiotic bacterium *Vibrio fischeri* in host animal growth, development, and light organ morphogenesis. *Journal of Experimental Zoology* 286:280–296.

Clarke, W. D. 1963. Function of bioluminescence in mesopelagic organisms. *Nature* 198:1244–1246.

Davidson, S., Allen, S. W., Lim, G. E., Anderson, C. M., and M. G. Haygood. 2001. Evidence for the biosynthesis of bryostatins by the bacterial symbiont "*Candidatus Endobugula sertula*" of the bryozoan Bugula neritina. *Applied Environmental Microbiology* 67:4531–4537.

Davis, R. A., Christensen, L. V., Richardson, A. D., da Rocha, R. M., and C. M. Ireland. 2003. Rigidin E, a new pyrrolopyrimidine alkaloid from a Papua New Guinea tunicate *Eudistoma* species. *Marine Drugs* 1:27–33.

Dionisio-Sese, M. L., Ishikura, M., Maruyama, T., and S. Miyachi. 1997. UV-absorbing substances in the tunic of a colonial ascidian protect its symbiont, *Prochloron* sp., from damage by UV-B radiation. *Marine Biology* 128:455–461.

Dobretsov, S. and P. Y. Qien. 2004. The role of epibiotic bacteria from the surface of the soft coral *Dendronephthya* sp. in the inhibition of larval settlement. *Journal of Experimental Marine Biology and Ecology* 299:35–50.

Dunlap, P. V. and J. C. Ast. 2005. Genomic and phylogenetic characterization of luminous bacteria symbiotic with the deep-sea fish *Chlorophthalmus albatrossis* (Aulopiformes: Chlorophthalmidae). *Applied Environmental Microbiology* 71:930–939.

Dunlap, P. V., Jiemjit, A., Ast, J. C., Pearce, M. M., Marques, R. R., and C. R. Lavilla-Pitogo. 2004. Genomic polymorphism in symbiotic populations of *Photobacterium leiognathi. Environmental Microbiology* 6:145–158.

Fidopiastis, P. M., von Boletzky, S., and E. G. Ruby. 1998. A new niche for *Vibrio logei*, the predominant light organ symbiont of squids in the genus *Sepiola. Journal of Bacteriology* 180:59–64.

Fieseler, L., Hentschel, U., Grozdanov, L., Schirmer, A., Wen, G., Platzer, M., Hrvatin., et al. 2007. Widespread occurrence and genomic context of unusually small polyketide synthase genes in microbial consortia associated with marine sponges. *Applied Environmental Microbiology* 73:2144–2155.

Fieseler, L., Horn, M., Wagner, M., and U. Hentschel. 2004. Discovery of the novel candidate phylum "Poribacteria" in marine sponges. *Applied and Environmental Microbiology* 70:3724–3732.

Fieseler, L., Quaiser, A., Schleper, C., and U. Hentschel. 2006. Analysis of the first genome fragment from the marine sponge-associated, novel candidate phylum Poribacteria by environmental genomics. *Environmental Microbiology* 8:612–624.

Fine, M. and Y. Loya. 2002. Endolithic algae: An alternative source of photoassimilates during coral bleaching. *Proceedings of the Royal Society London B* 269:1205–1210.

Fine, M., Meroz-Fine, E., and O. Hoegh-Guldberg. 2005. Tolerance of endolithic algae to elevated temperature and light in the coral *Montipora monasteriata* from the southern Great Barrier Reef. *The Journal of Experimental Biology* 208:75–81.

Frias-Lopez, J., Zerkle, A. L., Bonheyo, G. T., and B. W. Fouke. 2002. Partitioning of bacterial communities between seawater and healthy, black band diseased, and dead coral surface. *Applied Environmental Microbiology* 68:2214–2228.

Friedrich, A. B., Fischer, I., Proksch, P., Hacker, J., and U. Hentschel. 2001.Temporal variation of the microbial community associated with the Mediterranean sponge *Aplysina aerophoba. FEMS Microbial Ecology* 38:105–113.

Furla, P., Allemand, D., Shick, J. M., Ferrier-Pages, C., Richier, S., Plantivaux, A., Merel, P.-L., and S. Tambutte. 2005. The symbiotic anthozoan: A physiological chimera between alga and animal. *Integrative and Comparative Biology* 45:595–604.

Gaill, F., Desbruyeres, D., and D. Prieur. 1987. Bacterial communities associated with "Pompei worms" from the East Pacific Rise hydrothermal vents: SEM, TEM obšervations. *Microbial Ecology* 13:129–139.

Geiser, D. M., Taylor, J. W., Ritchie, K. B., and G. W. Smith. 1998. Cause of sea fan death in the West Indies. *Nature* 394:137–138.

Gil-Turnes, M. S. and W. Fenical. 1992. Embryos *of Homarus americanus* are protected by epibiotic bacteria. *The Biological Bulletin* 182:105–108.

Gil-Turnes, M. S., Hay, M. E., and W. Fenical. 1989. Symbiotic marine bacteria chemically defend crustacean embryos from a pathogenic fungus. *Science* 246:116–118.

Gill, S. R., Pop, M., DeBoy, R. T., Eckburg, P. B., Turnbaugh, P. J., Samuel, B. S., Gordon, et al. 2006. Metagenomic analysis of the human distal gutmicrobiome. *Science* 312:1355–1359.

Gram, L., Melchiorsen, J., Spanggaard, B., Huber, I., and T. Nielsen. 1999. Inhibition of *Vibrio anguillarum* by *Pseudomonas fluorescens* strain AH2—a possible probiotic treatment of fish. *Applied Environmental Microbiology* 65:969–973.

Harder, T., Lau, S. C. K., Dobretsov, S., Fang, T. K., and P. Y. Qian. 2003. A distinctive epibiotic bacterial community on the soft coral *Dendronephthya* sp. and antibacterial activity of coral tissue extracts suggest a chemical mechanism against bacterial epibiosis. *FEMS Microbial Ecology* 43:337–347.

Har, J. and J. Thompson. 2008. Characterizing the microbiome of the estuarine cnidarian *Nematostella vectensis.* Abstract N-265, American Society for Microbiology General Meeting, Boston MA.

Hardivillier, Y., Leignel, V., Denis, F., Uguen, G., Cosson, R., and M. Laulier. 2004. Do organisms living around hydrothermal vent sites contain specific metallothioneins? The case of the genus *Bathymodiolus* (Bivalvia, Mytilidae). *Comparative Biochemistry and Physiology, Part C* 139:111–118.

Harvey, E. N. 1952. *Bioluminescence*. Academic Press, Inc., New York. 649 pp.

Hastings, J. W. 1971. Light to hide by: Ventral luminescence to camouflage the silhouette. *Science* 173:1016–1017.

Haygood, M. G. 1990. Relationship of the luminous bacterial symbiont of the Caribbean flashlight fish, *Kryptophanaron alfredi* (family Anomalopidae) to other luminous bacteria based on bacterial luciferase (luxA) genes. *Archives of Microbiology*. 154:496–503.

Haygood, M. 1993. Light organ symbiosis in fishes. *Critical Reviews in Microbiology* 19:191–216.

Haygood, M. G. and S. K. Davidson. 1997. Small-subunit rRNA genes and in situ hybridization with oligonucleotides specific for the bacterial symbionts in the larvae of the bryozoan *Bugula neritina* and proposal of "*Candidatus Endobugula sertula*." *Applied Environmental Microbiology* 63:4612–4616.

Haygood, M. G. and D. L. Distel. 1993. Bioluminescent symbionts of flashlight fishes and deep-sea anglerfishes form unique lineages related to the genus *Vibrio*. *Nature* 363:154–156.

Haygood, M. G., Schmidt, E. W., Davidson, S. K., and D. J. Faulkner. 1999. Microbial symbionts of marine invertebrates: Opportunities for microbial biotechnology. *Journal of Molecular Microbiology and Biotechnology* 1:33–43.

Hentschel, U., Hopke, J., Horn, M., Friedrich, A. B., Wagner, M., and B. S. Moore. 2002. Molecular evidence for a uniform microbial community in sponges from different oceans. *Applied Environmental Microbiology* 68:4431–4440.

Hentschel, U., Schmid, M., Wagner, M., Fieseler, L., Gernert, C., and J. Hacker. 2001. Isolation and phylogenetic analysis of bacteria with antimicrobial activities from the Mediterranean sponges *Aplysina aerophoba* and *Aplysina cavernicola*. *FEMS Microbial Ecology* 35:305–312.

Herring, P. J. (Ed.). 1978. *Bioluminescence in Action*. Academic Press, London, pp. 569.

Herring, P. J. 1990. Bioluminescent communication in the sea. In *Light and Life in the Sea*, (Eds.) P. J. Henning, A. K. Campbell, M. Whitfield, and L. Maddock, pp. 357. C.U.P., Cambridge.

Highsmith, R. C. 1981. Lime-boring algae in coral skeletons. *Journal of Experimental Marine Biology and Ecology* 55:267–281.

Hirose, E., Hirabayashi, S., Hori, K., Kasai, F., and M. Watanabe. 2006. UV protection in the photosymbiotic ascidian *Didemnum molle* inhabiting different depths. *Zoological Science* 2:57–63.

Hirose, E., Ohtsuka, K., Ishikura, M., and T. Maruyama. 2004. Ultraviolet absorption in ascidian tunic and ascidian-*Prochloron* symbiosis. *Journal of the Marine Biological Association of the UK* 84:789–794.

Hjelm, M., Bergh, O., Riaza, A., Nielsen, J., Melchiorsen, J., Jensen, S., Duncan, H., et al. 2004. Selection and identification of autochthonous potential probiotic bacteria from turbot larvae (*Scophthalmus maximus*) rearing units. *Systematic and Applied Microbiology* 27:36–371.

Holmström, C., Egan, S., Franks, A., McCloy, S., and S. Kjelleberg. 2002. Antifouling activities expressed by marine surface associated Pseudoalteromonas species. *FEMS Microbial Ecology* 41:47–58.

Holmström, C. and S. Kjelleberg. 1999. Marine *Pseudoalteromonas* species are associated with higher organisms and produce biologically active extracellular agents. *FEMS Microbial Ecology* 30:285–293.

Holmström, C., Rittschof, D., and S. Kjelleberg. 1992. Inhibition of settlement by larvae of Balanus amphitrite and Ciona intestinalis by a surface-colonizing marine bacterium. *Applied Environmental Microbiology* 58:2111–2115.

Hortnagl, P. and R. Sommaruga. 2007. Photo-oxidative stress in symbiotic and aposymbiotic strais of the ciliate *Paramecium bursaria*. *Photochemical and Photobiological Sciences* 6:842–847.

Jeanthon, C. and D. Prieur. 1990. Susceptibility to heavy metals and characterization of heterotrophic bacteria isolated from two hydrothermal vent polychaete annelids, *Alvinella pompejana* and *Alvinella caudata*. *Applied and Environmental Microbiology* 56:3308–3314.

Jones, B. W. and M. K. Nishiguchi. 2003. Counterillumination in the Hawaiian bobtail squid, *Euprymna scolopes* Berry (Mollusca: Cephalopoda). *Marine Biology* 144:1151–1155.

Kaeding, A., Ast, J. C., Pearce, M. M., Urbanczyk, H., Kimura, S., Endo, H., Nakamura, M., and P. V. Dunlap. 2007. Phylogenetic diversity and cosymbiosis in the bioluminescent symbioses of "*Photobacterium mandapamensis*". *Applied and Environmental Microbiology* 73:3173–3182.

Kanagasabhapathy, M., Nagata, K., Fujita, Y., Tamura, T., Okamura, H., and S. Nagata. 2004. Antibacterial activity of the marine sponge *Psammaplysilla purpurea*: Importance of its surface-associated bacteria. *Oceans'04* 3:1323–1329.

Karentz, D., McEuen, F. S., Land, M. C., and W. C. Dunlap. 1991. Survey of mycosporine-like amino acid compounds in Antarctic marine organisms: Potential protection from ultraviolet exposure. *Marine Biology* 108:157–166.

Kellogg, C. A. 2004. Tropical Archaea: Diversity associated with the surface microlayer of corals. *Marine Ecology Progress Series* 273:81–88.

Kelman, D., Kushmaro, A., Loya, Y., Kashman, Y., and Y. Benayahu. 1998. Antimicrobial activity of a Red Sea soft coral, *Parerythropodium fulvum fulvum*: Reproductive and developmental considerations. *Marine Ecology Progress Series* 169:87–95.

Kennedy, J., Coding, C. E., Jones, B. V., and Dobson, A. D. W. 2008. Diversity of microbes associated with the marine sponge, *Haliclona simulans*, isolated from Irish waters and identification of polyketide synthase genes from the sponge metagenome. *Environmental Microbiology* 10:1888–1902.

Kim, K., Harvell, C. D., Kim, P. D., Smith, G. W., and S. M. Merkel. 2000a. Fungal disease resistance of Caribbean sea fan corals (*Gorgonia* sp.). *Marine Biology* 136:259–267.

Kim, K., Kim, P. D., Alker, A. P., and C. D. Harvell. 2000b. Chemical resistance of gorgonian corals against infections. *Marine Biology* 137:393–401.

Klaus, J., Janse, I., Sandford, R., and B. W. Fouke. 2007. Coral microbial communities, zooxanthellae, and mucus along gradients of seawater depth and coastal pollution. *Environmental Microbiology* 9:1291–1305.

Knowlton, N. and F. Rohwer. 2003. Multispecies microbial mutualisms on coral reefs: The host as a habitat. *American Naturalist* 162(4 Suppl.):S51–S62.

Koren, O. and E. Rosenberg. 2006. Bacteria associated with mucus and tissues of the coral *Oculina patagonica* in summer and winter. *Applied and Environmental Microbiology* 72:5254–5259.

Krug, J. 2006. Defense of benthic invertebrates against surface colonization by larvae: A chemical arms race. *Progress in Molecular and Subcellular Biology Subseries Marine Molecular Biotechnology*. In *Antifouling Compounds*, (Eds.) N. Fusetani and A. S. Clare. Springer-Verlag, Berlin, Heidelberg.

Kubanek, J., Jensen, P. R., Keifer, P. A., Sullards, M. C., Collins, D. O., and W. Fenical. 2003. Seaweed resistance to microbial attack: A targeted chemical defense against marine fungi. *Proceedings Natural Academy of Sciences USA* 100:6916–6921.

Kubanek, J., Pawlik, J. R., Eve, T. M., and Fenical. 2000. Triterpene glycosides defend the Caribbean reef sponge *Erylus formosus* from predatory fishes. *Marine Ecology Progress Series* 207:69–77.

Kushmaro, A., Loya, Y., Fine, M., and E. Rosenberg. 1996. Bacterial infection and coral bleaching. *Nature* 380:396–403.

Kushmaro, A., Rosenberg, E., Fine, M., and Y. Loya. 1997. Bleaching of the coral Oculina patagonica by Vibrio AK-1. *Marine Ecology Progress Series* 147:159–165.

Lee, Y. K., Lee, J.-H., and H. K. Lee. 2001. Microbial symbiosis in marine sponges. *The Journal of Microbiology* 39:254–264.

Lee, K.-H. and E. G. Ruby. 1994. Competition between *Vibrio fischeri* strains during initiation and maintenance of light organ symbiosis. *Journal of Bacteriology* 176:1985–1991.

Lehrer, R. I., Tincu, J. A., Taylor, S. W., Menzel, L. P., and A. J. Warin. 2003. Natural peptide antibiotics from tunicates: Structures, functions and potential use. *Integrative and Comparative Biology* 43:313–322.

Ley, R. E., Peterson, D. A., and J. I. Gordon. 2006. Ecological and evolutionary forces shaping microbial diversity in the human intestine. *Cell* 124:837–848.

Li, Z.-Y. and Y. Liu. 2002. Marine sponge *Craniella austrialiensis*-associated bacterial diversity revelation based on 16S rDNA library and biologically active Actinomycetes screening, phylogenetic analysis. *Letters in Applied Microbiology* 43:410–416.

Lim, G. E. and M. G. Haygood. 2004. "*Candidatus* Endobugula glebosa," a specific bacterial symbiont of the marine bryozoan *Bugula simplex. Applied and Environmental Microbiology* 70:4921–4929.

Lim-Fong, G. E., Regali, L. A., and M. G. Haygood. 2008. Evolutionary relationships of "*Candidatus* Endobugula" bacterial symbionts and their *Bugula* bryozoan hosts. *Applied and Environmental Microbiology* 74:3605–3609.

Lindquist, N., Barber, P. H., and J. B. Weisz. 2005. Episymbiotic microbes as food and defence for marine isopods: Unique symbioses in a hostile environment. *Proceedings of the Royal Society B* 272:1209–1216.

Lonsdale, P. 1977. Clustering of suspension-feeding macrobenthos near abyssal hydrothermal vents at oceanic spreading centers. *Deep-Sea Research* 24:857–863.

Lopanik, N., Lindquist, N., and N. Targett. 2004. Potent cytotoxins produced by a microbial symbiont protect host larvae from predation. *Oecologia* 139:131–139.

Lopanik, N. B., Targett, N. M., and N. Lindquist. 2006. Isolation of two polyketide synthase gene fragments from the uncultured microbial symbiont of the marine bryozoan *Bugula neritina. Applied and Environmental Microbiology* 72:7941–7944.

Maina, J. N. and G. M. O. Maloiy. 1998. Adaptations of a tropical swamp worm, *Alma emini,* for subsistence in a H_2S-rich habitat: Evolution of endosymbiotic bacteria, sulfide metabolizing bodies, and novel processes of elimination of neutralized sulfide complexes. *Journal of Structural Biology* 122:257–266.

Marteau, P. R., deVrese, M., Cellier, C. J., and J. Schrezenmeir. 2001. Protection from gastrointestinal diseases with the use of probiotics. *American Journal of Clinical Nutrition* 73(Suppl.):430S–436S.

Maruyama, T., Hirose, E., and M. Ishikura. 2003. Ultraviolet-light-absorbing tunic cells in didemnid ascidians hosting a symbiotic photooxygenic prokaryote, Prochloron. *The Biological Bulletin* 204:109–113.

Mazmanian, S. K., Liu, C. H., Tzianabos, A. O., and D. L. Kasper. 2005. An immunomodulatory molecule of symbiotic bacteria directs maturation of the host immune system. *Cell* 122:107–118.

Mazmanian, S. K., Round, J. L., and D. L. Kasper. 2008. A microbial symbiosis factor prevents intestinal inflammatory disease. *Nature* 453:620–625.

McClintock, J. B., Amsler, C. D., Baker, B. J., and R. W. M. Van Soest. 2005. Ecology of Antarctic marine sponges: An overview. *Integrative and Comparative Biology* 45:359–368.

McClintock, J. B. and D. Karentz. 1997. Mycosporine-like amino acids in 38 species of subtidal marine organisms from McMurdo Sound, Antarctica. *Antarctic Science* 9:392–398.

McFall-Ngai, M. J. 1983a. Adaptations for reflection of bioluminescent light in the gas bladder of *Leiognathus equulus* (Perciformes Leiognathidae). *Journal of Experimental Zoology* 227:23–33.

McFall-Ngai, M. J. 1983b. The gas bladder as a central component of the leiognathid bacterial light organ symbiosis. *American Zoologist* 23:907.

McFall-Ngai, M. J. 1999. Consequences of evolving with bacterial symbionts: Insights from the squid–vibrio associations. *Annual Review of Ecological System* 30:235–256.

McFall-Ngai, M. J. and M. K. Montgomery. 1990. The anatomy and morphology of the adult bacterial light organ of *Euprymna scolopes* Berry (Cephalopoda:Sepiolidae). *The Biological Bulletin* 179:332–339.

McFall-Ngai, M. J. and J. G. Morin. 1991. Camouflage by disruptive illumination in leiognathids, a family of shallow water, bioluminescent fishes. *Journal of Experimental Biology* 156:119–137.

McFall-Ngai, M. J. and E. G. Ruby. 1991. Symbiont recognition and subsequent morphogenesis as early events in an animal-bacterial mutualism. *Science* 254:1491–1494.

Miller, M. B. and B. L. Bassler. 2001. Quorum sensing in bacteria. *Annual Review of Microbiology* 55:165–199.

Montgomery, M. K. and M. J. McFall-Ngai. 1994. Bacterial symbionts induce host organ morphogenesis during early postembryonic development of the squid *Euprymna scolopes. Development* 120:1719–1729.

Morin, J. G., Harrington, A., Nealson, K., Krieger, N., Baldwin, T. O., and J. W. Hastings. 1975. Light for all reasons: Versatility in the behavioral repertoire of the flashlight fish. *Science* 190:74–76.

Mullard, A. 2008. The inside story. *Nature* 453:578–580.

Nelson, D. C. and C. R. Fisher. 1995. Chemoautotrophic and methanotrophic endosymbiotic bacteria at deep-sea vents and seeps. In *Deep-Sea Hydrothermal Vents*, Ed. D. M. Karl, pp. 125–167. Boca Raton, FL: CRC Press.

Newbold, R. W., Jensen, P. R., Fenical, W., and J. R. Pawlik. 1999. Antimicrobial activity of Caribbean sponge extracts. *Aquatic Microbial Ecology* 19:279–284.

Nishiguchi, M. K., Ruby, E. G., and M. J. McFall-Ngai. 1998. Competitive dominance among strains of luminous bacteria provides an unusual form of evidence for parallel evolution in sepiolid squid-vibrio symbioses. *Applied and Environmental Microbiology* 64:3209–3213.

Nyholm, S. and M. J. McFall-Ngai. 2004. The winnowing: Establishing the squid-vibrio symbiosis. *Natural Reviews Microbiology* 8:632–642.

Olsson, J. C., Westerdahl, A., Conway, P. L., and S. Kjelleberg. 1992. Intestinal colonization potential of turbot (*Scophthalmus maximus*)- and dab (*Limanda limanda*)-associated bacteria with inhibitory effects against *Vibrio anguillarum. Applied and Environmental Microbiology* 58:551–556.

Onarheim, A. M. and J. Raa. 1990. Characteristics and possible biological significance of an autochthon flora in the intestinal mucus of sea-water fish. In *Micro-Biology in Poicilotherms*, Ed. R. Lesel, pp. 197–201. Amsterdam: Elsevier Science Publishers.

Oxley, A. P. A., Shipton, W., Owens, L., and D. McKay. 2002. Bacterial flora from the gut of the wild and cultured banana prawn, *Penaeus merguiensis. Journal of Applied Microbiology* 93:214–223.

Pamer, E. G. 2007. Immune responses to commensal and environmental microbes. *Nature Immunology* 8:1173–1178.

Paul, V., Arthur, K. E., Ritson-Williams, R., Ross, C., and K. Sharp. 2007. Chemical defenses: From compounds to communities. *The Biological Bulletin* 213:226–251.

Pawlik, J. R., Chanas, B., Toonen, R., and W. Fenical. 1995. Defenses of Caribbean sponges against predatory reef fish. I. Chemical deterrency. *Marine Ecology Progress Series* 127:183–194.

Petroni, G., Spring, S., Schleifer, K.-H., Verni, F., and G. Rosati. 2000. Defensive extrusive ectosymbionts of *Euplotidium* (Ciliophora) that contain microtubule-like structures are bacteria related to Verrucomicrobia. *Proceedings Natural Academy of Sciences USA* 97:1813–1817.

Piel, J. 2002. A polyketide synthase-peptide synthetase gene cluster from an uncultured bacterial symbiont of Paederus beetles. *Proceedings of the Natural Academy of Sciences USA* 99:14002–14007.

Piel, J. 2004. Metabolites from symbiotic bacteria. *Natural Product Reports* 21:519–538.

Piel, J. 2006. Bacterial symbionts: Prospects for the sustainable production of invertebrate-derived pharmaceuticals. *Current Medical Chemistry* 13:39–50.

Piel, J., Hui, D., Wen, G., Butzke, D., Platzer, M., Fusetani, N., and S. Matsunaga. 2004. Antitumor polyketide biosynthesis by an uncultivated bacterial symbiont of the marine sponge *Theonella swinhoei. Proceedings of the Natural Academy of Sciences USA* 101:16222–16227.

Proksch, P., Ebel, R., Edrada1, R. A., Schupp, P., Lin, W. H., Sudarsono, Wray, V., and K. Steube. 2003. Detection of pharmacologically active natural products using ecology. Selected examples from Indopacific marine invertebrates and sponge-derived fungi. *Pure and Applied Chemistry* 75:343–352.

Proksch, P., Edrada, R. A., and R. Ebel. 2002. Drugs from the seas—current status and microbiological implications. *Applied Microbiology and Biotechnology* 59:125–134.

Rao, D., Webb, J. S., Holmström, C., Case, R., Low, A., Steinberg, P., and S. Kjelleberg. 2007. Low densities of epiphytic bacteria from the marine alga *Ulva australis* inhibit settlement of fouling organisms. *Applied Environmental Microbiology* 73:7844–7852.

Rawls, J. F., Samuel, B. S., and J. I. Gordon. 2004. Gnotobiotic zebrafish reveal evolutionarily conserved responses to the gut microbiota. *Proceedings of the Natural Academy of Sciences USA* 101:4596–4601.

Reshef, L., Koren, O., Loya, Y., Zilber-Rosenberg, I., and E. Rosenberg. 2006. The coral probiotic hypothesis. *Environmental Microbiology* 8:2067–2073.

Richier, S., Furla, P., Plantivaux, A., Merle, P.-L., and D. Allemand. 2005. Symbiosis-induced adaptation to oxidative stress. *Journal of Experimental Biology* 208:277–285.

Rinehart, K. L. 2000. Antitumor compounds from tunicates. *Medical Research Review* 20:1–27.

Rinehart, K., Gloer, J., Cook, J., Carter, J., Mizsak, S., and T. Scahill. 1987. Structures of the didemnins, antiviral and cytotoxic depsipeptides from a Caribbean tunicate. *Journal of the American Chemical Society* 103:1857–1859.

Richardson, L. L. 1998. Coral diseases: What is really known? *Trends in Ecology and Evolution* 13:438–443.

Rohwer, F., Breitbart, M., Jara, J., Azam, F., and N. Nowlton. 2001. Diversity of bacteria associated with the Caribbean coral Montastrea franksi. *Coral Reefs* 20:85–95.

Rohwer, F., Seguritan, V., Azam, F., and N. Knowlton. 2002. Diversity and distribution of coral-associated bacteria. *Marine Ecology Progress Series* 243:1–10.

Rosati, G., Petroni, G., Quochi, S., Modeo, L., and F. Verni. 1999. Epixenosomes: Peculiar epibionts of the hypotrich ciliate *Euplotidium itoi* defend their host against predators. *The Journal of Eukaryotic Microbiology* 46:278–282.

Rosenberg, E., Kellogg, C. A., and F. Rohwer. 2007a. Coral microbiology. *Oceanography* 20:114–122.

Rosenberg, E. O., Koren, L., Reshef, R., Efrony, and I. Zilber-Rosenberg. 2007b. The role of microorganisms in coral health, disease and evolution. *Nature Reviews Microbiology* 5:355–362.

Ruby, E. G. and M. J. McFall-Ngai. 1992. A squid that glows in the night: Development of an animal–bacterial mutualism. *Journal of Bacteriology* 174:4865–4870.

Ruby, E. G. and K. H. Nealson. 1976. Symbiotic association of *Photobacterium fischeri* with the marine luminous fish *Monocentris japonicua*: A model of symbiosis based on bacterial studies. *The Biological Bulletin* 151:574–586.

Ruiz-Ponte, C., Samain, J. F., Sanchez, J. L., and J. L. Nicolas. 1999. The benefit of a roseobacter species on the survival of scallop larvae. *Marine Biotechnology (NY)* 1:52–59.

Schirmer, A., Gadkari, R., Reeves, C. D., Ibrahim, F., DeLong, E. F., and C. R. Hutchinson. 2005. Metagenomic analysis reveals diverse polyketide synthase gene clusters in microorganisms associated with the marine sponge *Discodermia dissoluta. Applied Environmental Microbiology* 71:4840–4849.

Schlichter, D., Zscharnack, B., and H. Kerisch. 1995. Transfer of photoassimilates from endolithic algae to coral tissue. *Naturwissenschaften* 82:561–564.

Schmidt, E. W., Nelson, J. T., Rasko, D. A., Sudek, S., Eisen, J. A., Haygood, M. G., and J. Ravel. 2005. Patellamide A and C biosynthesis by a microcin-like pathway in *Prochloron didemni*, the cyanobacterial symbiont of *Lissoclinum patella. Proceedings of the Natural Academy of Sciences USA* 102:7315–7320.

Schmidt, E. W., Obraztsova, A. Y., Davidson, S. K., Faulkner, D. J., and M. G. Haygood. 2000. Identification of the antifungal peptide-containing symbiont of the marine sponge *Theonella swinhoei* as a novel proteobacterium, "*Candidatus* Entotheonella palauensis." *Marine Biology* 136:969–977.

Schmitt, S., Weisz, J. B., Lindquist, N., and U. Hentschel. 2007. Vertical transmission of a phylogenetically complex microbial consortium in the viviparous sponge *Ircinia felix*. *Applied Environmental Microbiology* 73:2067–2078.

Shashar, N., Banaszak, A. T., Lesser, M. P., and D. Amrami. 1997. Coral endolithic algae: Life in a protected environment. *Pacific Sciences* 51:167–173.

Shibata, K., 1969. Pigments and a UV-absorbing substance in corals and a blue-green alga living in the Great Barrier Reef. *Plant and Cell Physiology* 10:325–335.

Shick, J. M. and W. C. Dunlap. 2002. Mycosporine-like amino acids and related gadusols: Biosynthesis, accumulation, and UV-protective functions in aquatic organisms. *Annual Review of Physiology* 64:223–262.

Siebeck, O. 1988. Experimental investigation of UV tolerance in hermatypic corals (Scleractinia). *Marine Ecology Progress Series* 3:95–103.

Simmons, T. L., Coates, R.C., Clark, B. R., Engene, N., Gonzalez, D., Esquenazi, E., Dorrestein, P. C., and W. H. Gerwick. 2008. Biosynthetic origin of natural products isolated from marine microorganism–invertebrate assemblages. *Proceedings Natural Academy of Sciences USA* 105:4587–4594.

Smith, G. W., Ives, L. D., Nagelkerken, I. A., and K. B. Ritchie. 1996. Caribbean sea fan mortalities. *Nature* 383:487.

Sommaruga, R., Whitehead, K., Shick, J. M., and C. S. Lobban. 2006. Mycosporine-like amino acids in the zooxanthella–ciliate symbiosis *Maristentor dinoferus*. *Protist* 157:185–191.

Spanggaard, B., Huber, I., Nielsen, J., Sick, E. B., Pipper, C. B., Martinussen, T., Slierendrecht W. J., and L. Gram. 2001. The probiotic potential against vibriosis of the indigenous microflora of rainbow trout. *Environmental Microbiology* 3:755–765.

Starcevic, A., Akthar, S., Dunlap, W. C., Shick, J. M., Hranueli, D., Cullum, J., and P. F. Long. 2008. Enzymes of the shikimic acid pathway encoded in the genome of a basal metazoan, *Nematostella vectensis*, have microbial origins. *Proceedings Natural Academy of Sciences USA* 105:2533–2537.

Stat, M., Morris, E., and R. Gates. 2008. Functional diversity in coral–dinoflagellate symbiosis. *Proceedings Natural Academy of Sciences USA* 105:9256–9261.

Steindler, L., S. Beer, and M. Ilan. 2002. Photosymbiosis in intertidal and subtidal tropical sponges. *Symbiosis* 33:263–273.

Stewart, F. J. and C. M. Cavanaugh. 2006. Bacterial endosymbioses in Solemya (Mollusca: Bivalvia)—model systems for studies of symbiont–host adaptation. *Antonie Van Leeuwenhoek* 90:343–360.

Sung, P.-J., Lin, M.-R., Chen, J.-J., In, S.-F., Wu, Y.-C., Hwang, T.-L., and L. S. Fang. 2007. Hydroperoxysterols from the Tunicate *Eudistoma* sp. *Chem. Pharm. Bull.* 55:666–668.

Taylor, M. W., Radax, R., Steger, D., and M. Wagner. 2007. Sponge-associated microorganisms: Evolution, ecology, and biotechnological potential. *Microbiology and Molecular Biology Reviews* 71:295–347.

Thakur, N. L., Anil, A. C., and W. E. G. Müller. 2004. Culturable epibacteria of the marine sponge *Ircinia fusca*: Temporal variations and their possible role in the epibacterial defense of the host. *Aquatic Microbial Ecology* 37:295–304.

Unson, M. D. and D. J. Faulkner. 1993. Cyanobacterial symbiont synthesis of chlorinated metabolites from *Dysidea herbacea* (Porifera). *Experientia* 44:1021–1022.

Urbanczyk, H., Ast, J. C., Higgins, M. J., Carson, J., and P. V. Dunlap. 2008. Reclassification of *Vibrio fischeri, Vibrio logei, Vibrio salmonicida* and *Vibrio wodanis* as *Aliivibrio fischeri* gen. nov., comb. nov., *Aliivibrio logei* comb. nov., *Aliivibrio salmonicida* comb. nov. and *Aliivibrio wodanis* comb. nov. *International Journal of Systematic and Evolutionary Microbiology* 57:2823–2829.

Vaseeharan, B. and P. Ramasamy. 2003. Control of pathogenic Vibrio spp. by *Bacillus subtilis* BT23, a possible probiotic treatment for black tiger shrimp *Penaeus monodon*. *Letters Applied Microbiology* 36:83–87.

Verni, F. and G. Rosati. 1990. Peculiar Epibionts in *Euplotidium itoi* (Ciliata, Hypotrichida). *The Journal of Eukaryotic Microbiology* 37:337–343.

Verschuere, L., Heang, H., Criel, G. Sorgeloos, P., and W. Verstraete. 2000. Selected bacterial strains protect *Artemia* spp. from the pathogenic effects of *Vibrio proteolyticus* CW8T2. *Applied Environmental Microbiology* 66:1139–1146.

Vervoortl, H. C., Pawlik, J. R., and W. Fenical. 1998. Chemical defense of the Caribbean ascidian *Didemnum conchyliatum*. *Marine Ecology Progress Series* 164:221–228.

Vetriani, C., Chew, Y. S., Miller, S. M., Yagi, J., Coombs, J., Lutz, R. A., and T. Barkayl. 2005. Mercury adaptation among bacteria from a deep-sea hydrothermal vent. *Applied Environmental Microbiology* 71:220–226.

Visick, K. 2005. Layers of signaling in a bacterium–host association. *The Journal of Bacteriology* 187:3603–3606.

Visick, K., Foster, J., Doino, J., McFa-Ngai, M. J., and E. G. Ruby. 2000. *Vibrio fischeri lux* genes play an important role in colonization and development of the host light organ. *Journal of Bacteriology* 182:4578–4586.

Visick, K. L. and M. J. McFall-Ngai. 2000. An exclusive contract: Specificity in the *Vibrio fischeri–Euprymna scolopes* partnership. *Journal of Bacteriology* 182:1779–1787.

Visick, K. L. and E. G. Ruby. 2006. *Vibrio fischeri* and its host: It takes two to tango. *Current Opinion in Microbiology* 9:632–638.

Wada, M., Barbara, G., Mizuno, N., Azuma, N., Kogure, K., and Y. Suzuki. 2008. Expulsion of symbiotic luminous bacteria from pony fish, *Leiognathus nuchalis*. In *Bioluminescence and Chemiluminescence: Progress and Perspectives*, pp. 99–102. Singapore: World Scientific Publishing Co. Pte. Ltd.

Waddell, B. and J. R. Pawlik. 2000a. Defenses of Caribbean sponges against invertebrate predators: I. Assays with hermit crabs. *Marine Ecology Progress Series* 195:125–132.

Waddell, B. and J. R. Pawlik. 2000b. Defenses of Caribbean sponges against invertebrate predators: II. Assays with sea stars. *Marine Ecology Progress Series* 195:133–144.

Wegley, L., Yu, Y., Breitbart, M., Casas, V., Kline, D. I., and F. Rohwer. 2004. Coral-associated Archea. *Marine Ecology Progress Series* 273:89–96.

Westerdahl, A., Olsson, J. C., Kjelleberg, S., and P. L. Conway. 1991. Isolation and characterization of turbot (Scophtalmus [sic] maximus)-associated bacteria with inhibitory effects against *Vibrio anguillarum. Applied Environmental Microbiology* 57:2223–2228.

Woyke, T., Teeling, H., Ivanova, N. N., Huntemann, M., Richter, M., Gloeckner, F. O., Boffelli, D., et al. 2006. Symbiosis insights through metagenomic analysis of a microbial consortium. *Nature* 443:950–955.

Xu, J. and J. I. Gordon. 2003. Honor thy symbionts. *Proceedings of the Natural Academy of Sciences USA* 100:10452–10459.

Yip, E. S., Geszvain, K., DeLoney-Marino, C. R., and K. L. Visick. 2006. The symbiosis regulator RscS controls the syp gene locus, biofilm formation and symbiotic aggregation by *Vibrio fischeri. Molecular Microbiology* 62:1586–160.

Yip, E. S., Grublesky, B. T., Hussa, E. A., and K. L. Visick. 2005. A novel, conserved cluster of genes promotes symbiotic colonization and sigma 54-dependent biofilm formation by *Vibrio fischeri. Molecular Microbiology* 57:1485–1498.

Part II

Prokaryotic Defensive Symbionts

4

Microbial Symbionts for Defense and Competition among Ciliate Hosts

Hans-Dieter Görtz, Giovanna Rosati, Michael Schweikert, Martina Schrallhammer, Gen Omura, and Toshinobu Suzaki

CONTENTS

4.1 Introduction

Like multicellular organisms, ciliates are attacked by predators and also affected by harmful microorganisms. Ciliates have only limited means to escape predators; some are able to enhance their swimming speed when getting into contact with a predator or turn backward rapidly. Some ciliates are able to release nasty cell organelles called extrusomes, attacking predators with volleys of tiny arrows. The extrusomes best studied are the trichocysts of *Paramecium*. Trichocysts are effective in defending *Paramecium* against predators such as the ciliate *Didinium* (Harumoto and Miyake, 1990). Extrusomes of other ciliates may contain microbicidal substances (Masayo et al., 1999; Miyake et al., 2001); predators tend to avoid these ciliates. A further possible method of defense is the morphological deformation of ciliates, e.g., *Euplotes* may alter its shape in response to the presence of kairomones released by certain predators of this protozoon (Kusch, 1993). Most ciliates, however, lack means of defense against predators. In the ciliate *Euplotidium*, this deficiency is overcome by its epixenosomes, epibiotic bacteria that play a significant role in defending its host against predators.

Most free-living protozoa feed on microorganisms. Thus, they should be prepared to face microbial attacks. However, while ingested microorganisms are rapidly killed and digested, microbial infections almost exclusively happen via ingestion. It appears that some microbes are able to resist digestion and escape the phagosome. Depending on the metabolic features of the invaders, they may either adapt to intracellular life or overgrow the host cell within hours to a few days. Once a bacterium has adapted to intracellular life, it may become a favorable symbiont. It may even become essential for the host. The symbiont may then provide the host with specific metabolic traits or functions.

Görtz (1983) speculated previously that intracellular bacteria provide their hosts with some resistance against further infections. In recent years, evidence has been found that certain symbionts may support protozoan hosts in defense against predation and/or harmful microbial infections. Some examples will be discussed in this chapter. In many cases, however, the significance of symbiotic bacteria for their hosts has not yet been elucidated. Even intracellular symbionts may have a function for their hosts in intraspecific competition. The significance of symbionts in this case may be regarded ambivalent, the symbiosis being favorable or harmful.

4.2 Defensive Symbiosis: The Ciliate *Euplotidium* and Its Epixenosomes

Ciliates of the genus *Euplotidium* are sand-dwelling marine hypotrichs. Since Noland (1937) erected the genus, six species have been determined. At least three of these species, namely, *Euplotidium itoi* (Ito, 1958), *E. arenarium* (Magagnini and Nobili, 1964), and *E. prosaltans* (Tuffrau, 1985), together with a species of the genus *Gastrocirrhus*, share a peculiar feature: a broad band of particles running along the right and left borders of the cell body and forming a sort of "scarf" at the dorsal anterior end. On the basis of only optical microscopic observation, these particles had been interpreted as extrusomes. Later, electron microscopical analyses of specimens of *E. itoi* and *E. arenarium* collected from the Mediterranean Sea demonstrated that the particles are always extracellular and they are episymbionts rather than extrusomes (Verni and Rosati, 1990). As their real nature was unknown, they were referred to as "epixenosomes," from the ancient Greek: external alien bodies. In all likelihood, all the above mentioned structures are actually of the same entity and epixenosomes are present in different *Euplotidium* and *Gastroiyrrhus* species.

Epixenosomes possess many unique characteristics. During the reproductive stage (form I), they are spherical (0.5 μm in diameter), bacteria-like in morphology, and divide by binary fission. In the mature stage (form II), they are unable to divide and have a far more complicated internal structure to which a functional compartimentalization corresponds (Rosati et al., 1996). The most prominent structure is a ribbonlike extrusive apparatus (EA) coiled up around a central core, surrounded by a basket of regularly arranged microtubule-like tubules (Figure 4.1).

The nature and phylogenetic affiliation of epixenosomes were resolved by comparative sequence analysis of 16S rRNA gene sequences. They belong to *Verrucomicrobia,* a distinct lineage within phylogenetic bacterial trees containing a small number of cultured species and a number of members only recognized by their 16S rRNA gene sequences (Petroni et al., 2000). Morphological and molecular data concur with the idea that epixenosomes of *E. itoi* and *E. arenarium* (the only two analyzed up to now) are representatives of the same new species of a new genus (Rosati et al. in preparation).

The association between *Euplotidium* and epixenosomes is constant in the natural environment (epixenosomes have been found in every specimen examined soon after collection over more than 15 years) and can be indefinitely maintained in well-fed lab cultures. The host cell cycle is perfectly coordinated with the multiplication and the differentiation of epixenosomes from form I to form II. Already during the very early stages of *Euplotidium* morphogenesis, the number of dividing epixenosomes begins to increase and the cortical region in which they lie begins to widen. Then, the gradual widening of this cortical region, its increase in length, and the recovery of the typical nondividing features proceed along precise steps together with the ciliate morphogenesis (Giambelluca and Rosati, 1996). All these observations suggest a positive, strict relationship between these two organisms.

The association is, very likely, vital for epixenosomes since attempts to grow the episymbionts in culture in artificial media for different bacteria failed. Although this failure may be attributed either to the indispensability of the association or the peculiarity of the organism and its metabolism, preliminary results from molecular studies revealed that the number and the organization of the ribosomal operons of epixenosomes have characteristics typical of obligate symbiotic prokaryotes (G. Petroni, personal communication).

On the other hand, epixenosomes are not vital for their host. When *Euplotidium* stops dividing (for example, during starvation), epixenosomes stop dividing too; then all form I epixenosomes change into form II and are gradually lost. Once conditions are normalized, a certain percentage of epixenosome-free euplotidia recover their reproductive capacity; in this way, strains without epixenosomes were obtained for both *E. itoi* and *E. arenarium.* These strains behave similar to the epixenosome-bearing strains and reproduce even faster. In some way, their morphogenesis is simplified; when they divide, their cortical region corresponding to the epixenosomal band does not widen.

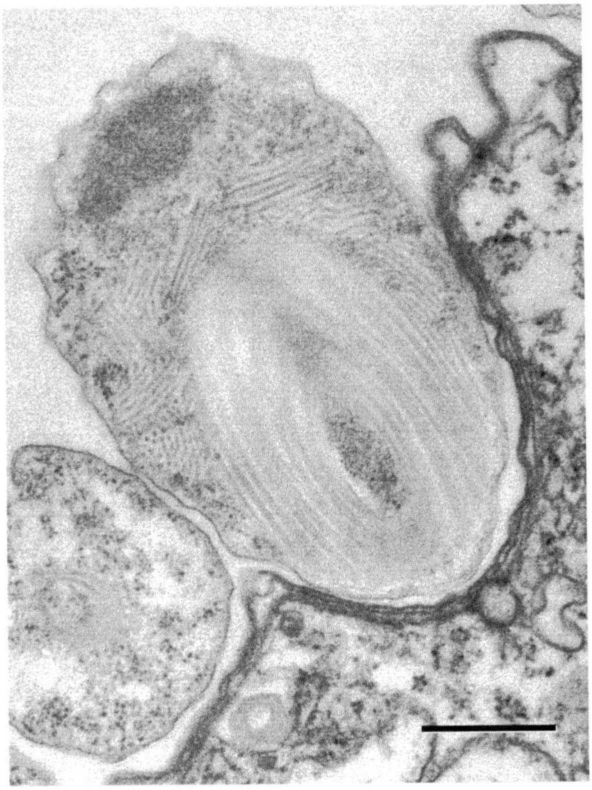

FIGURE 4.1 Epixenosome. Longitudinal section of an epixenosome form II and (partially, lower left) form I. Mind the central EA. Bar = 0.5 μm.

The question arises: why does *Euplotidium* keep its epixenosomes in natural environment? The answer is: epixenosomes provide the host with an important ecological advantage, i.e., defense against predators. This has been experimentally demonstrated through the contemporaneous availability of stocks with and without epixenosomes. Both types of stocks were exposed to a ciliate predator, previously reported as *Litonotus lamella* by Ricci and Verni (1988); on the basis of both a deeper morphological analysis (performed with light microscope as well as SEM and TEM) and an original molecular analysis (Rosati et al., 2008), this ciliate is now recognized as *Amphileptus marina*. *A. marina* shares its habitat with *E. itoi* and its feeding behavior is well known (Ricci and Verni, 1988; Ricci et al., 1996). It involves several basic steps: detection of the prey from some distance, discharge of toxicysts upon direct cell-to-cell contact, search for the stricken prey, and ingestion of the prey.

It has been observed that, upon direct cell-to-cell contact, *A. marina* discharges its toxicysts and paralyses *E. itoi* both without and with the epixenosomes. However, while it is able to ingest the former (Figure 4.2), it never ingests the latter. The only difference between the two potential prey types being the absence or the presence of epixenosomes, it can be inferred that the presence of epixenosomes prevents the engulfment of euplotidia stricken by the predator. A number of stricken prey, not ingested by the predator, are able to recover normal behavior in a short time once transferred to pure sea water. It may be assumed that in the natural environment, where the toxic substance of the toxicysts can be rapidly dispersed, *E. itoi* with epixenosomes enjoy a high probability of survival, certainly higher than that of *E. itoi* without epixenosomes, part of which are eaten (Rosati et al., 1999). So epixenosomes serve a function for their host that in other ciliates is served by extrusomes (Harumoto and Miyake, 1990; Miyake and Harumoto, 1996).

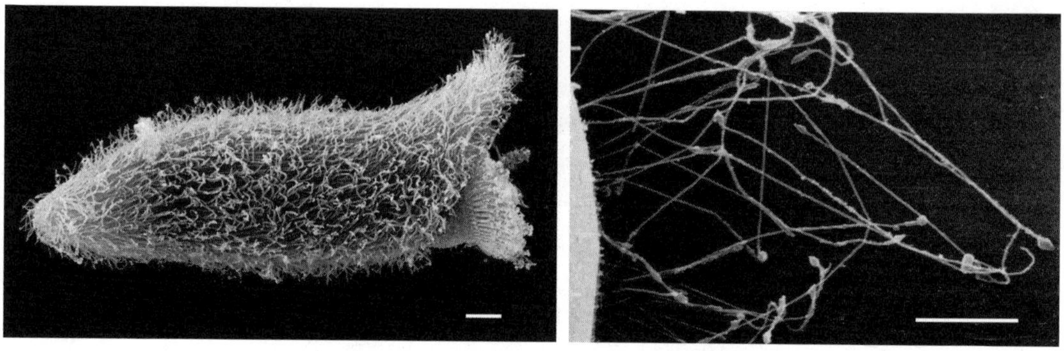

FIGURE 4.2 Left—*A. marina* engulfing *E. itoi* having no epixenosomes. Right—Portion of *E. itoi* with ejected epixenosomal tubes. Bars = 10 μm. (Adapted from Rosati, G., Petroni, G., Quochi, S., Modeo, L., and Verni, F., *J. Eukaryot. Microbiol.*, 46, 278, 1999.)

What is the mechanism of the defensive function? We obtained experimental evidences that it is based upon the ejection of the EA. It can be hypothesized that the predator toxicyst discharge triggers ejection and ejection of the EA disturbs the predator in the recognition steps necessary for finding and engulfing the stricken prey (Figure 4.2).

The defensive role of epixenosomes is, at present, the only one reported for epibiotic or intracellular bacterial symbionts in ciliates. The significance in defense against predators could account for the apparent absence of *euplotidium* without epixenosomes in the tide pools from which all our experimental organisms have been collected, and the defensive function is very likely an important factor in stabilizing and maintaining such a specialized symbiotic relationship. Apparently, form II epixenosomes "sacrifice" themselves so that the rest of the bacterial population (epixenosomes form I) and their ciliate host survive. Anyway, the heroism of epixenosomes would not be so extreme. Indeed, as the total number of symbionts per cell is roughly constant during generations, in all likelihood, a spontaneous loss (detachment or ejection) of the old form II epixenosomes takes place during the division of *Euplotidium*.

4.3 Protection of *Paramecium bursaria* from Microbial Infections by Symbiotic *Chlorella*

The green paramecium, *Paramecium bursaria*, harbors about 300 algal symbionts (Figure 4.3). The green *Chlorella* symbionts are favorable for the host in providing it with considerable amounts of carbohydrates, chiefly maltose (Muscatine et al., 1967; Brown and Nielsen, 1974). Moreover, *Chlorella*-bearing *P. bursaria* gain the capability of photoaccumulation (Iwatsuki and Naito, 1981; Niess et al., 1981). Another advantage of *Chlorella*-bearing paramecia appears to be a resistance against microbial infections provided by *Chlorella* (Görtz, 1982).

Once it has lost *Chlorella*, *P. bursaria* can nevertheless survive and multiply. Symbiont-free cells are called aposymbionts. Aposymbiont, i.e., white *P. bursaria*, may be infected by yeast or bacteria (Bomford, 1965; Görtz, 1982; Abendschein et al., in preparation). These observations led to the suggestion that aposymbiotic *P. bursaria* may generally bear the risk of microbial infections and, on the other hand, endocytobiotic *Chlorella* may confer some resistance upon its hosts. Resistance against microbial infections should be important for protozoa feeding on microbial prey. In natural habitats, white *P. bursaria* are rare. Nevertheless, green paramecia can be freed from the algae by various methods. This opens the possibility of experimental infections by incubating white paramecia with microorganisms. In an investigation with various yeast species, all of them were taken up as endocytobionts by aposymbiotic *P. bursaria* (Figure 4.4) (Abendschein et al., in preparation).

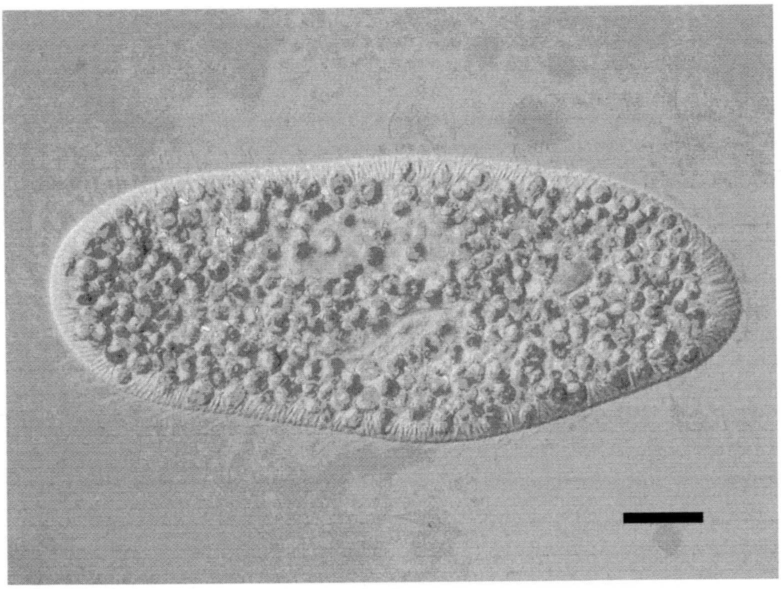

FIGURE 4.3 (See color insert following page 206.) *P. bursaria* with *Chlorella* symbionts. Phase contrast. Bar = 10 μm.

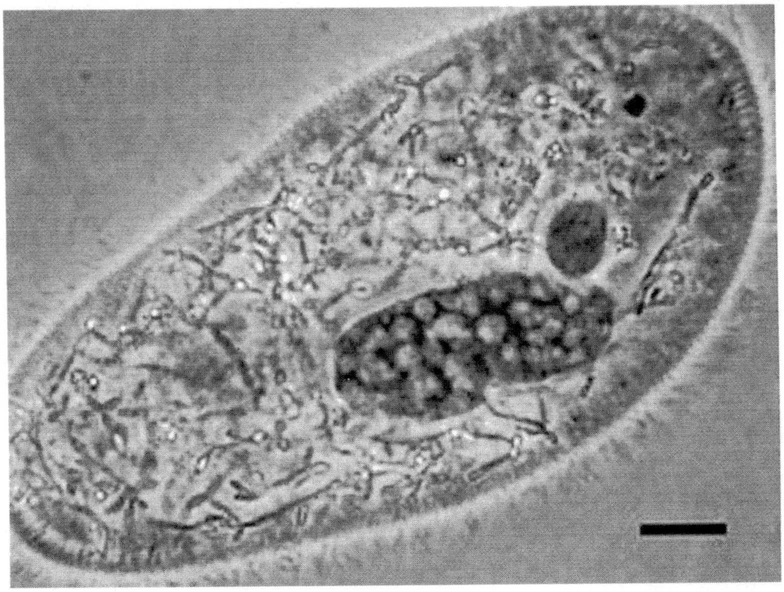

FIGURE 4.4 *P. bursaria* without *Chlorella* symbionts. The cell is colonized by *Yarrowia lipolytica*. Many budding yeasts are spread all over the paramecium. Stained with lacto-aceto-orcein, phase contrast. Bar = 10 μm.

Among paramecia, *P. bursaria* has a unique life history. With its green symbionts, *P. bursaria* may survive long periods of food shortage and thus does not depend on a continuous supply of bacteria. The symbiosis with *Chlorella* enables this *Paramecium* to be the only pelagic *Paramecium* (Nyberg, 1988). Other species of *Paramecium* are benthic and often found in local aggregations in littoral habitats.

Another consequence of its feeding strategy may be that *P. bursaria* is an extreme outbreeder with a clonal life expectancy of 6,000 divisions (Jennings, 1939; Sonneborn, 1957). The symbiosis with *Chlorella* may, on the other hand, be regarded as a burden in competition with other ciliates under certain conditions, e.g., green *P. bursaria* may not start multiplying as rapidly as aposymbiotic paramecia in case of sudden availability of food (Görtz, 1982) because of the algal endocytobionts. The reason is that the algae participate in the resources of the host cell.

A preadaptation of *P. bursaria* for hosting *Chlorella* may be the reason for the readiness of *P. bursaria* to take up and maintain microbial symbionts. Though the molecular details and mode of *Chlorella*-uptake by *P. bursaria* are not clear (Kodama and Fujishima, 2007), this preadaptation may imply the possibility, we may even call it the risk, of being infected by pathogenic microorganisms too. Infections by other microorganisms had been observed by Pringsheim (1928), Bomford (1965), and Görtz (1982). The observation of infections by various free-living yeast species that remain viable and multiply within the host cells and thus are maintained as endocytobionts. These observations show that there may be little selectivity not only during infection but also concerning maintenance of the endocytobionts. Only in case of extreme starvation, host cells are killed by the endocytobiotic yeasts after two weeks. Possibly, host cells are being exhausted by the endocytobionts under unfavorable conditions, the same would also hold true for *Chlorella*-bearing *P. bursaria*.

P. bursaria may be taken as an example that eukaryotic cells together with their endocytobionts must be regarded as evolutionary units. The same may hold true for white *P. bursaria* infected with microorganisms. Even in these associations, host paramecia could not be secondarily infected by other microorgansims. *P. bursaria*-bearing yeasts or bacteria appear to be defended by its symbiotic microorganisms (Görtz, 1982) too. So, other microorganisms may take the part of *Chlorella* in defense of the symbiotic system *P. bursaria*.

In *P. bursaria* colonized by yeasts or bacterial symbionts, *Chlorella* may displace these microorganisms. When *Chlorella* reinfects the paramecia, the number of nonalgal endocytobionts decreases while the number of chlorellas taken up increases due to their multiplication. This could indicate that *Chlorella* endocytobionts have better mechanisms for the uptake of certain metabolites of the host cell, metabolites that may be of limited access for symbionts. However, the mechanism by which endocytobiotic microorganisms are displaced after reinfecting host paramecia with chorellas is yet to be elucidated.

4.4 Protozoa Infected by Killer-Symbionts Outcompete Symbiont-Free Cells

Several genera of ciliates, e.g., *Paramecium*, are known to contain species occupying ecological niches overlapping each other, namely, concerning their prey organisms (Gause, 1934; Landis, 1981, 1988). Overlapping niches limit propagation and cause interspecific competition. When the killer trait in *Paramecium* was found by Sonneborn (1938, 1943), it was immediately recognized as a competitive advantage for killer cells. Killer paramecia bear bacterial symbionts that produce toxins. Whereas the killer cells themselves are resistant against the toxins, other cells of the same or related *Paramecium* species are killed. Neither the toxins have been identified nor the mechanism is known by which resistance of the host cells is brought about (Preer et al., 1974; Kusch and Görtz, 2006). After the first detection of the phenomenon, a number of different bacterial killer-symbionts were found not only in *Paramecium* (Preer et al., 1974; Fokin and Görtz, 1993) but also in other protists such as *Acanthamoeba* (Fritsche et al., 1993). The best investigated symbionts are the species of the genus *Caedibacter* (see for reviews Preer et al., 1974; Pond et al., 1989; Kusch and Görtz, 2006).

All *Caedibacter* bacteria present an obligatory symbiotic lifestyle. When the genus was created, the capability of repression of a peculiar proteinaceous body, a so-called R body (refractile body, refractile in phase contrast microscopy), was emphasized to be a unique feature of the genus. R bodies could be used as an easily recognizable morphological trait (Preer et al., 1974; Preer and Preer, 1982; Görtz and Schmidt, 2004). R bodies are protein sheets rolled up to form a body that is about 0.5 µm wide and long (Figures 4.5 and 4.6). Phylogenetic analyses revealed that the genus *Caedibacter* is a polyphyletic assemblage with *Caedibacter taeniospiralis* belonging to the *Gammaproteobacteria* (Beier et al., 2002) and *Caedibacter caryophilus* belonging to the *Alphaproteobacteria* (Springer et al., 1993; Schrallhammer et al., 2006).

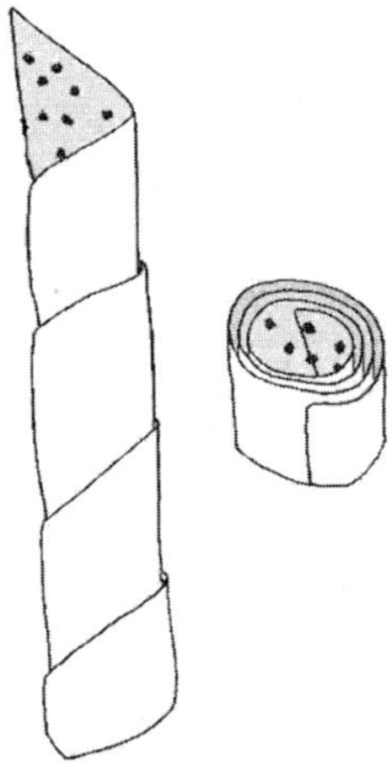

FIGURE 4.5 R bodies, coiled and stretched. R bodies of *Caedibacter* are coiled proteinaceous ribbons that may be unrolled by appropriate triggers such as acid pH.

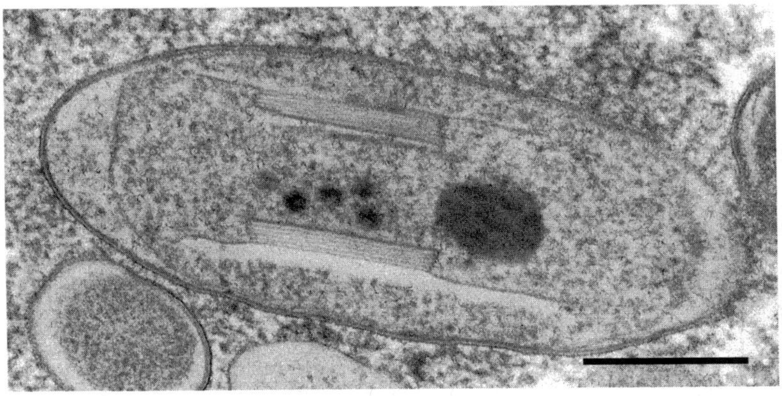

FIGURE 4.6 *C. caryophilus* in the macronucleus of *P. caudatum*. Some of the bacteria contain R bodies with phages capsids (arrows). Bar = 0.5 μm.

Toxicity in *Caedibacter* was found to be connected with the occurrence of the R body (Dilts and Quackenbush, 1986). Paramecia-bearing *Caedibacter* may eventually release a few of their symbionts that may then be swallowed by other paramecia. When bacteria containing such R bodies are ingested by a *Paramecium*, R bodies have been shown to stretch and by this, destroy the phagosome (Jurand et al., 1971). The stretched R body is thought to introduce the toxin into the cell. R body proteins are encoded by phage or plasmid genes (Quackenbush and Burbach, 1983; Quackenbush et al., 1986a,b; Heruth et al., 1994). There are good indications that the toxins are also encoded by phage or plasmid genes (Quackenbush, 1988; Pond et al., 1989; Jeblick and Kusch, 2005).

Of special interest is the nature of the toxins. They cause different prelethal symptoms in sensitive paramecia and four different types of killing have been observed up to now: hump killing, spin killing, vacuolization, and paralysis (Sonneborn, 1959). Jurand et al. (1971) compared the prelethal effects of hump killing, spin killing, and vacuolization and suggested that the toxins are related in that they affect the osmoregulatory properties of biological membranes.

Jeblick and Kusch (2005) analyzed the sequence of the plasmid pKAP298 of *C. taeniospiralis* and identified as putative toxin a protein that has a homology with ATPases of the Soj-/ParA-family and also to a membrane-associated ATPase that is involved in eukaryotic ATPase-dependent ion carriers. This protein presumably has effects on the *Paramecium* cell membrane, where it might disturb the osmoregulative abilities of the ciliate and lead to hump-killing. First analyses in *C. caryophilus*, which renders its host a paralysis killer, indicate that in this case also, the toxin interacts with the *Paramecium* membranes (M. Schrallhammer, unpublished).

Lethal effects on competitors increase the relative abundance of infected paramecia in their natural habitats. In coculture, infected strains outcompete uninfected strains in both *P. tetraurelia* and *P. novaurelia* (Kusch et al., 2000, 2002). Sonneborn (1938, 1943) described the killer trait in *Paramecium* as a possible advantage in intraspecific competition, as shown in culture. The ecological significance of the killer trait in *Paramecium* was later demonstrated by Landis (1981, 1987) and confirmed by Kusch et al. (2002). In their study, the protozoan predators *Amoeba proteus* (Amoebozoa, Gymnamoebia) and *Didinium nasutum* (Ciliophora) were cultivated with either killer paramecia as a food source or paramecia that did not harbor *Caedibacter*. However, a toxic effect of killer paramecia on the predators was not observed.

From these observations, it might be concluded that a killer symbiosis is of mutual advantage for both the symbiont and its host. However, a closer view elucidates the parasitic nature of the bacteria. *Caedibacter*-free cells need less food and grow faster than infected cells. Linka et al. (2003) have shown that *Caedibacter* species are energy parasites and may therefore limit the reproduction of their host. Like other parasitic bacteria, *Caedibacter* has a specific ATP/ADP antiporter that recruits energy from the host cell. Under unfavorable conditions, *Caedibacter* may even overgrow its host cell, finally killing it (Schmidt et al., 1987). While the killer toxins are apparently encoded by phages or plasmids, resistance may be brought about by the bacteria. Schmidt et al. (1987) observed that *C. caryophilus* that had lost the ability to produce R bodies, so presumably had lost its phages, still protected its host paramecia against killer toxins. Nevertheless, it is the action of the phages or plasmids in the killer symbionts that outcompetes paramecia without killer symbionts.

Caedibacter is poorly infectious, if at all, and it seems that transmission is chiefly vertical. Therefore, it was most surprising to find these killer symbionts in a nucleus of a *Paramecium*. It was tempting to assume that the killer symbionts had been introduced by coinfection with infectious bacteria. Fokin et al. (2004) observed that noninfective bacteria could coinfect the macronucleus of *Paramecium caudatum* together with the infectious bacterium *Holospora obtusa*. Thus, *Caedibacter* could have invaded the macronucleus of *P. caudatum* by coinfection with an infectious bacterium too. In fact, double infections with *Caedibacter* and infectious *Holospora* bacteria have been observed. In this context, it should also be mentioned that not all strains of a *paramecium* species may support the growth of killer symbionts. Sonneborn (1943) found out that the genetic condition of a potential host is crucial for the maintenance of *C. taeniospiralis* in a cell. He showed that for a special gene, an allele K is needed for the bacteria to exist while K in the homozygous state does not support the symbionts (Schneller et al., 1959; Balsley, 1967).

In conclusion, though being rather of a parasitic nature, killer symbionts may provide their host paramecia with competitive advantage over noninfected cells. The killer trait depends, however, on the presence and genetic activity of phages or plasmids of the bacterial symbionts. Though the toxins have not been identified in any of the many killer symbionts known, there are good indications that these toxins interact with biological membranes and disturb the osmoregulatory mechanisms of toxin-sensitive competitors.

4.5 Significance of Bacteria in Dinoflagellates

Dinoflagellates are a sister phylum of ciliates and apicomplexans within the alveolate protists. Dinoflagellates are able to establish endocytobioses with a huge diversity of eukaryote and prokaryote microorganisms. On the other hand, some dinoflagellate species can be found as endoctyobionts in other protozoans (Lobban et al., 2002) and metazoans, then called zooxanthellae (e.g., Trench, 1993; Van den Heok et al., 1995; Rowan, 1998; Baker, 2003).

Interactions and the evolutionary significance of the members of such associations were rarely investigated, but associations of dinoflagellates with phototrophic partners obviously open up new ecological niches. While eukaryotes as endocytobionts in dinoflagellates were investigated intensively in the context of the endosymbiosis theory of the origin of cell organelles (e.g., Schnepf, 1992a,b; Delwiche and Palmer, 1997; Schnepf and Elbrächter, 1999), prokaryotes as endocytobionts in dinoflagellates were neglected in the past but became of scientific and practical interest in the context of harmful algal blooms. Toxic algal blooms can have a strong impact on fishery and aquaculture industries worldwide by killing or contaminating stocks, leading to a wide variety of diseases of sea birds, marine mammals, and humans ingesting marine organisms (Shumway, 1990; Anderson, 1995, 1997; Sellner et al., 2003). Some harmful alga, when blooming, can cause amnesia (amnesic shellfish poisoning, ASP), neurotoxic effects (paralytic shellfish poisoning, PSP), and diarrhea (diarrhetic shellfish poisoning, DSP) in humans and sometimes even lead to death after consumption of seafood (e.g., Landsberg, 2002).

First report of an intracellular bacterium in dinoflagellates was published by Silva in 1952 who later proposed the hypothesis of toxin production in dinoflagellates by their prokaryote endocytobionts (Silva, 1962, 1990). Toxicity was linked directly to blooms and can be correlated to the relative abundances of dinoflagellates. However, since the beginning of research on dinoflagellate toxicity in harmful blooms, the producer of the toxins has been discussed controversially. Moreover, the occurrence of toxic events was even reported in the apparent absence of dinoflagellate blooms (Cembella et al., 1987).

Interactions between bacteria and dinoflagellates were rarely examined at the cellular level. This is partly due to difficulties in cultivating dinoflagellates axenically and because endocytobionts cannot be cultivated separately. Results concerning the production of toxins are contradictory in nearly all cases investigated. Whereas toxicity was related to algal blooms, and especially dinoflagellates, more and more bacteria were found to be able to produce the same toxins (e.g., Silva, 1982; Sasner et al., 1984; Kodama et al., 1988; Kodama, 1990; Doucette and Trick, 1995; Franca et al., 1995; Gallacher et al., 1996; Kopp et al., 1997; Kirchner et al., 1999; Hold et al., 2001a,b). Toxicity was reported during the absence of toxic bacteria too (Sasner et al., 1984; Kodama, 1990). Axenically cultured dinoflagellates retained the ability to produce high amounts of toxins (Boczar et al., 1988; Bomber et al., 1989; Kim et al., 1993a). On the other hand, Abott and Ballantine (1957) and Ray and Wilson (1957) showed that the toxic dinoflagellate *Gymnodinium breve* looses its toxicity when cultured axenically.

Production of ciguatoxin by the dinoflagellate *Ostreopsis lenticularis* was correlated with the presence of bacteria of the genus *Nocardia* while the bacteria themselves were shown to be nontoxic (Tosteson et al., 1986, 1989). Contrary to this, Abbott and Ballantine (1957) reported a reduction of toxin level while the number of associated bacteria increased. The transfer of toxicity from toxic dinoflagellates of *Alexandrium lusitanicum* (=*Gonyaulax tamarensis*) to a previously nontoxic *A. lusitanicum* by the transfer of bacteria was successful (Silva and Sousa, 1981; Silva, 1990) but remains unconfirmed.

The considerably high variance of intraspecific toxin concentrations (Ogata et al., 1987; Tosteson et al., 1989; Kim et al., 1993b) can be used to argue both, in favor or against the symbiosis theory of toxin production. Moreover, bacteria were reported to produce related or even identical toxic substances in the absence of dinoflagellates (Doucette, 1995; Gallacher and Smith, 1999).

Toxin production of bacteria, as in dinoflagellates, seems to be correlated to the available amount of nutrients. Toxin production was highest in starved bacteria (Kodama, 1990; Doucette and Trick, 1995) and starved dinoflagellates (Anderson et al., 1990). Toxin production was also found to be high during the growth phase (Silva, 1982; Tosteson et al., 1989; Lewis et al., 2001). However, the significance of toxin production in the environment is not clear (e.g., Cembella, 2003). Besides its known effects on higher trophic levels, toxin production may have some negative effect on grazers and have therefore been suggested as chemical protectants against grazing (Cembella, 2003; Selander et al., 2006). Results studying the effects of PSP toxins on grazers are contradictory and vary from nondetectable (Colin and Dam, 2003; Bricelj et al., 2005; Prince et al., 2006; Kubanek et al., 2007) to reduced feeding (Teegarden, 1999), delayed development, decreased fecundity (Guisande et al., 2002), and lethality (Sykes and Huntley, 1987; Colin and Dam, 2003; Bricelj et al., 2005), sometimes within the same study.

Intracellular bacteria were found in toxic and nontoxic dinoflagellates. Toxic dinoflagellate *P. lima* and *P. maculosum* without intracellular bacteria were reported by Zhou and Fritz (1993) and in

P. lima (Rausch de Traubenberg et al., 1995a). Okadaic acid and intracellular bacteria did not colocalize as revealed by immunocytochemistry (Rausch de Traubenberg et al., 1995b), but localization of toxins with microscopical methods seems to be difficult (Anderson and Cheng, 1988; Doucette and Anderson, 1993). Attempts to localize derivatives of STX resulted in specific localization close to the periphery of the dinoflagellates (Figure 4.7). Intense signals correlate with the sutures embracing the thecal plates, suggesting that the toxins are associated with alveolar membranes (Figure 4.7) rather than cytoplasmic or intranuclear bacteria. Intracellular bacteria were not detected by electron microscope in cells of different *Alexandrium* species producing saxitoxin (Schweikert et al., 2000).

Cyanobacteria are potent producers of neurotoxins and were found in close association with dinoflagellates, but in few genera only. They were called phagosomes by Geitler (1959) and identified as cyanobacteria by Norris (1967). Some ultrastructural investigations were made but no clear results were given about their intracellular or extracellular locations. However, in most literature it is mentioned to be extracellular, sometimes outside the cytoplasm only (Schnepf and Elbrächter, 1999). From morphological observations, they were said to be related to *Synechocysis* and *Synechococcus* genera. Associated cyanobacteria were found in different species of *Ornithocercus* (Hallegraeff and Jeffrey, 1984; Schnepf and Elbrächter, 1992; Janson et al., 1995), *Citharistes* (Figure 4.8) (Gordon et al., 1994), and *Histioneis* (Gordon et al., 1994), but were also reported intracytoplasmic by Lucas (1991) in *Amphisolenia globifera*. Recent phylogenetic investigations resulted in a huge diversity of cyanobacteria associated with dinoflagellates related to different phylogenetic groups (Foster et al., 2006a).

Observations of marine dinoflagellates in environmental samples with epifluorescence microscopy indicate a much higher number of associations between cyanobacteria and dinoflagellates than published so far (Elbrächter and Schweikert, 1997). Although such associations can be found frequently in oligotrophic waters, toxin producing cyanobacteria as symbionts were not found to be associated

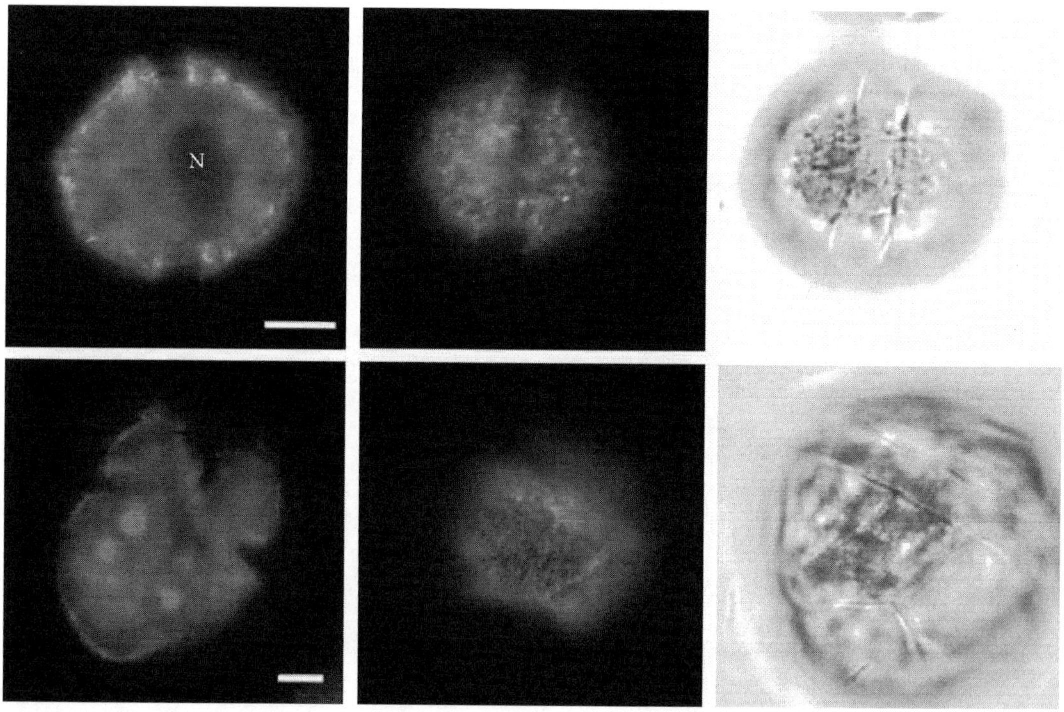

FIGURE 4.7 (See color insert following page 206.) Immunolocalization of saxitoxin (STX/neoSTX) in *A. lusitanicum* (upper row) and *Alexandrium fundyense* (bottom row). Right micrographs are differential interference contrast images. Bars = 10 μm.

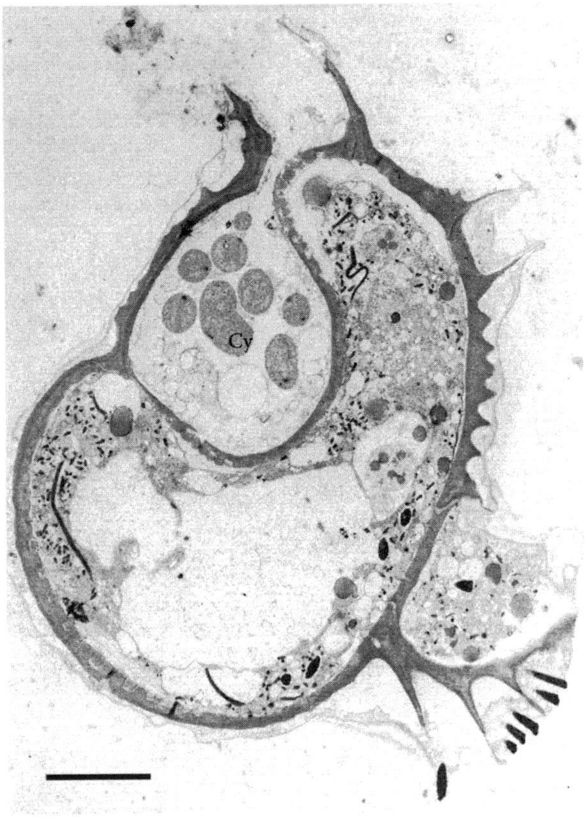

FIGURE 4.8 *Citharistes* sp. with associated cyanobacteria. Bar = 10 μm.

with dinoflagellates so far. Symbionts seem to be restricted to the polyphyletic cyanobacterial taxons *Synechococcus* and *Synechocystis* (Foster et al., 2006b). Besides some investigations on endocytobiotic cyanobacteria, phylogenetic investigations of intracellular bacteria in dinoflagellates were neglected chiefly due to the difficulty of discriminating between symbiotic and phagocytized bacteria. Fluorescence in situ hybridization together with the design of oligonucleotides specific for different phylogenetic taxa may now help in confirming the symbiotic nature of endocytobionts. Actual results display a huge diversity of different phylogenetic groups of endocytobionts (Table 4.1). Surprisingly, *C. caryophilus*, known as an intranuclear endocytobiont of *Paramecium*, was identified in the cytoplasm of a freshwater dinoflagellate.

Like toxin production, bioluminescence in dinoflagellates may have originated from symbiotic bacteria. Dinoflagellates are the most common organisms generating bioluminescence when reaching high cell concentrations. They can highlight moving objects like ships and ship waves (Rohr et al., 2002), swimming animals (Mensinger and Case, 1992; Rohr et al., 1998), and braking waves at coastlines (Stokes et al., 2004). Bioluminescence of dinoflagellates is thought to have antipredation functions by reducing predator grazing through discontinued feeding behavior of the predator because of the light effects (Esaias and Curl, 1972; White, 1979; Buskey and Swift, 1983; Buskey et al., 1985). Additionally, "burglar alarm" effects of bioluminescence in dinoflagelles are proposed to be resulting in an attraction of secondary predators to primary predators reducing grazing pressure on phytoplankton (Young, 1983; Mensinger and Case, 1992; Abrahams and Townsend, 1993; Fleisher and Case, 1995).

Repeatedly and in symbioses among various taxa, a transfer of numerous genes from endocytobiotic bacteria into the host nuclei has been discussed. The acquisition of new physiological properties via lateral

TABLE 4.1

Examples of Dinoflagellates with Intracellular Bacteria (in Alphabetical Order)

Name	Location of Bacteria	Microscopy	References
Adenoides eludens	Cytoplasm	TEM	Hoppenrath et al. (2003)
Alexandrium lusitanicum	Cytoplasm	TEM	Silva (1962, 1982, 1990) and Franca et al. (1995)
Alexandrium tamarense/ fundyense complex	Cytoplasm	TEM	Lewis et al. (2001)
Alexandrium tamarense	Cytoplasm of hypnocysts	TEM	Schweikert (2003)
Amphidinium carteri	Intracellular	LM	Gold and Pollingher (1971)
Amphidinium herdmanii	Cytoplasm	TEM	Dodge (1973)
Amphisolenia globifera	Cytoplasm	TEM	Lucas (1991)
Amphisolenia thrinax	Cytoplasm	TEM	Lucas (1991)
Ceratium tripos	Intracellular	LM	Gold and Pollingher (1971)
Cochlodinium heterolobatum	Cytoplasm	LM	Silva (1962, 1967, 1982)
Crepidoodinium autrale	Cytoplasm	TEM	Lom et al. (1993)
Crypthecodinium cohnii	Intracellular	LM	Gold and Pollinger (1971)
Dinophysis acuminata	Cytoplasm	TEM	Lucas and Vesk (1990)
Exuviella baltica	Intracellular	LM	Silva (1959)
Durkinskia capesis	Cytoplasm	TEM	Pienaar et al. (2007)
Glenodinium foliaceum	Cytoplasm, nucleus	LM, TEM	Silva (1978, 1982), Silva and Franca (1985), Doucette et al. (1998), and Gold and Pollingher (1971)
Gonyaulax spinifera	Cytoplasm	LM	Silva (1962, 1982)
Gymnodinium acidotum	Vacuoles	TEM	Wilcox and Wedemayer (1984) and Farmer and Roberts (1990)
Gymnodinium catenatum	Bacteria-like structures	TEM	Franca et al. (1996)
Gymnodinium lei	Cytoplasm, food vacuoles		Gaines and Elbrächter (1987)
Gymnodinium splendens (syn. *G. sanguineum*)	Cytoplasm, nucleus, vacuoles	TEM, LM	Gold and Pollingher (1971) and Silva (1978, 1982)
Gyrodinium instriatum	Cytoplasm, nucleus	LM/TEM/FISH	Silva (1967, 1982), Silva and Franca (1985), Doucette et al. (1998), and Alverca et al. (2002)
Gyrodinium lebouriae	Cytoplasm, food vacuoles	TEM	Lee (1977)
Heterocapsa circularisquama	Cytoplasm, food vacuoles	TEM, FM	Maki and Imai (2001)
Katodinium glandulum	Cytoplasm	TEM	Dodge (1973)
Noctiluca scintillans	Cytoplasm	TEM	Lucas (1982) and Kirchner et al. (1999)
Peridinium balticum	Cytoplasm	TEM	Pienaar (1980) and Chesnick and Cox (1986)
Peridinium cinctum	Cytoplasm	TEM, FISH	Calado et al. (1999) and Schweikert and Meyer (2001)
Prorocentrum lima	Cytoplasm	TEM, FISH	Rausch de Traubenberg et al. (1995a)
Prorocentrum micans	Intracellular, in vacuoles	LM	Silva (1959), Gold and Pollingher (1971), and Rausch de Traubenberg and Soyer-Gobillard, (1990)
Prorocentrum scutellum	Cytoplasm	LM	Silva (1982)
Prorocntrum minimum	Cytoplasm	LM	Silva (1982)
Scippsiella trochoidea	Cytoplasm	LM	Silva (1962, 1982)
Woloszynskia pascheri	Chloroplast stroma	TEM	Wilcox (1986)

Note: FISH, fluorescence in situ hybridization; LM, light microscopy; TEM, transmission electron microscopy.

gene transfer from endocytobionts could also explain the scattered distribution of toxin production in dinoflagellates or even the ability to show bioluminescence. Lateral gene transfer from bacteria to some strains of dinoflagellates was proven for Rubisco Form II (Morse, 1995; Palmer, 1995, 1996; Whitney et al., 1995; Delwiche and Palmer, 1997; Jenks and Gibbs, 2000).

4.6 Conclusions

Ciliates and other protists form a variety of symbioses with microorganisms. The selective significance of a symbiosis depends upon the adaptive values of the contributions of each partner to the association, and it may be questioned as to what are the valuable contributions by microbial symbionts. Certainly, the risk of predation, on the one hand, and infection by harmful microorganisms, on the other hand, is high for protists. Even though there are costs, any advantage in defense brought about by symbionts, whether epibiotic or intracellular, is worth paying. With advanced abilities in defense, the unit of host and symbiont may stabilize in a given niche or even colonize new ecological niches. Improving abilities in defense against predators as well as harmful pathogens may therefore be the main reason for the establishment of symbioses. Another possible reason for the establishment of a firm symbiosis, certainly, is any advantage in intraspecific competition due to the contributions of a microbial symbiont such as the one observed in the killer symbioses of *Paramecium*.

The significance of intracellular bacteria in defense and competition is largely based on the evolution of sophisticated adaptations to an intracellular mode of life. Relevant features are partly encoded on plasmid or phage genomes. It is, therefore, not surprising that many microorganisms are strictly host specific; others may shuttle between different hosts. We must be aware that new human pathogens may emerge from protists reservoirs such as the one that has occurred with *Legionella pneumophila* (Horn et al., 1999; Görtz, 2002; Görtz and Michel, 2003; Horn and Wagner, 2004).

REFERENCES

Abbott, B. C. and D. Ballantine. 1957. The toxin from *Gymnodinium veneficum* Ballantine. *Journal of Marine Biology Association UK* 36:169–189.

Abrahams, M. V. and L. D. Townsend. 1993. Bioluminescence in dinoflagellates: A test of burglar alarm hypothesis. *Ecology* 7:258–260.

Alverca, E., Biegala, I. C., Kennaway, G. M., Lewis, J., and S. Franca. 2002. In situ identification and localization of bacteria associated with *Gyrodinium instriatum* (Gymnodiniales, Dinophyceae) by electron and confocal microscopy. *European Journal of Phycology* 37:523–530.

Anderson, D. M. 1997. Bloom dynamics of toxic *Alexandrium* species in the northeastern U.S. *Limnology and Oceanography* 42:1009.

Anderson, D. M. and T. P.-O. Cheng. 1988. Intracellular localization of saxitoxins in the dinoflagellate *Gonyaulax tamarensis*. *Journal of Phycology* 24:17–22.

Anderson, D. M., Kulis, C. M., Sullivan, J. J., and C. Lee. 1990. Dynamics and physiology of saxitoxin production by the dinoflagellates *Alexandrium* spp. *Marine Biology* 104:511–524.

Baker, A. C. 2003. Flexibility and specificity of coral-algal symbiosis: Diversity, ecology, and biogeography of *Symbiodinium*. *Annual Review of Ecology, Evoution, and Systematics* 34:661–689.

Balsley, M. 1967. Dependence of the kappa particles of stock 7 of *Paramecium aurelia* on a single gene. *Genetics* 56:125–131.

Beier, C. L., Horn, M., Michel, R., Schweikert, M., Görtz, H.-D., and M. Wagner. 2002. The genus *Caedibacter* comprises endobionts of *Paramecium* spp. related to the *Rickettsiales* (*Alphaproteobacteria*) and to *Francisella tularensis* (Gammaproteobacteria). *Applied and Environmental Microbiology* 68:6043–6050.

Boczar, B. A., Beitler, K., Liston, J., Sullivan, J. J., and R. A. Cattolico. 1988. Paralytic shellfish toxins in *Protogonyaulax tamarensis* and *Protogonyaulax catenella* in axenic culture. *Plant Physiology* 88:1285–1290.

Bomber, J. W., Rubio, M. G., and D. R. Norris. 1989. Epiphytism of dinoflagellates associated with the disease ciguatera: Substrate specificity and nutrition. *Phycologia* 28:360–368.

Bomford, R. 1965. Infection of alga-free *Paramecium bursaria* with strains of *Chlorella, Scenedesmus* and a yeast. *Journal of Protozoology* 12:221–224.

Bricelj, V. M., Connell, L., Konoki, K., MacQuarrie, S. P., Scheuer, T., Catterall, W. A., and V. L. Trainer. 2005. Sodium channel mutation leading to saxitoxin resistance in clams increase risk of PSP. *Nature* 434:763–767.

Brown, J. A. and P. J. Nielsen. 1974. Transfer of photosynthetically produced carbohydrate from endosymbiotic chlorellas to *Paramecium bursaria. Journal of Protozoology* 21:569–570.

Buskey, E. J. and E. Swift. 1983. Behavioural responses of the coastal copepod *Acartia hudsonica* (Pinhey) to stimulated dinoflagellate bioluminescence. *Journal of Experimental Marine Biology and Ecology* 72:43–58.

Buskey, E. J., Reynolds, G. T., Swift, E., and A. J. Walton. 1985. Interactions between copepods and bioluminescent dinoflagellates: Direct observations using image intensification. *Biological Bulletin* 169:530.

Calado, A. J., Hansen, G., and O. Moestrup. 1999. Architecture of the flagellar apparatus and related structures in the type species of *Peridinium, P. cinctum* (Dinophyceae). *European Journal of Phycology* 34:179–191.

Cembella, A. D. 2003. Chemical ecology of eukaryaotic microalgae in marine ecosystems. *Phycologia* 42:420–447.

Cembella, A. D., Sulivan, J. J., Boyer, G. L., Taylor, F. J. R., and R. J. Andersen. 1987. Variation in paralytic shellfish toxin composition within the protogonyaulax tamarensis/catarella species complex: Red tide dinoflagellates. *Biochemical Systematics and Ecology* 15:137.

Chesnick, J. M. and E. R. Cox. 1986. Specialization of endoplasmatic reticulum architecture in response to a bacterial symbiosis in *Peridinium balticum* (Pyrrhophyta). *Journal of Phycology* 22:291–298.

Colin, S. P. and H. G. Dam. 2003. Effects of the toxic dinoflagellate, *Alexandrium fundyense,* on the copepod, *Acartia hudsonica*: A test of the mechanisms that reduce ingestion rates. *Marine Ecology Progress Series* 248:56–66.

Delwiche, C. F. and J. D. Palmer. 1997. The origin of plastids and their spread via secondary symbiosis. *Plant Systematics and Evolution* 11(Suppl.):53–86.

Dilts, J. A. and R. L. Quackenbush. 1986. A mutation in the R-body-coding sequences destroys expression of the killer trait in *P. tetraurelia. Science* 232:641–643.

Dodge, J. D. 1973. *The Fine Structure of Algal Cells.* Academic Press, London.

Doucette, G. J. 1995. Interactions between bacteria and harmful algae. *Natural Toxins* 3:65–74.

Doucette, G. J. and D. M. Anderson. 1993. Intracellular distribution of saxitoxin in *Alexandrium fundyense*. In: *Toxic Phytoplankton Blooms in the Sea*, Smayda, T. J. and Shimizu, Y. (eds.). Elsevier Science Publishers B.V., Amsterdam, p. 863 ff.

Doucette, G. J., Kodama, M., Franca, S., and S. Gallacher. 1998. Bacterial interactions with harmful algal bloom species: Bloom ecology, toxigenesis, and cytology. In: *Physiological Ecology of Harmful Algal Blooms*, vol. G41, Anderson, D. M, Cembella, A. D, and Hallegraeff, G. M. (eds.). NATO ASI Series. Springer-Verlag, Berlin, Heidelberg, pp. 619–647.

Doucette, G. J. and C. G. Trick. 1995. Characterization of bacteria associated with different isolates of *Alexandrium tamarense*. In: *Harmful Algal Blooms*, Lessus, G., Azul, C., Erard, E., Gentien, P., and Marcaillou, C. (eds.). Lavoisier, Paris, pp. 33–38.

Elbrächter, M. and M. Schweikert. 1997. Untersuchungen zu den tiefen Wassermassen und plankto logische Beobachtungen im tropischen Westpazifik während der SONNE-Fahrt Nr. 113 (TROPAC), 6.7 Biologische Probennahmen. *Berichte aus dem Institut fuer Meereskunde* 288:124–128.

Esaias, W. E. and H. C. Curl. 1972. Effect of dinoflagellates bioluminescence on copepod ingestion rates. *Limnology and Oceanography* 17:901–905.

Farmer, M. A. and K. R. Roberts. 1990. Organelle loss in the endosymbiont of *Gymnodinium acidotum* (Dinophyceae). *Protoplasma* 153:178–185.

Fleisher, K. J. and J. F. Case. 1995. Cephalopod predation facilitated by dinoflagellate bioluminescence. *Biological Bulletin* 189:263–271.

Fokin, S. I. and H.-D. Görtz. 1993. *Caedibacter macronucleorum* sp. nov., a bacterium inhabiting the macronucleus of *Paramecium duboscqui. Archiv für Protistenkunde* 143:319–324.

Fokin, S. I., Skovorodkin, I. N., Schweikert, M., and H.-D. Görtz. 2004. Co-infection of the macronucleus of *Paramecium caudatum* by free-living bacteria together with the infectious *Holospora obtusa. Journal of Eukaryotic Microbiology* 51:417–424.

Foster, R. A., Carpenter, E. J., and B. Bergman. 2006a. Unicellular cyanobionts in open ocean dinoflagellates, radiolarians and Tintinnids: Ultrastructural characterization and immuno-localization of phycoerythrin and nitrogenase. *Journal of Phycology* 42:453–463.

Foster, R. A., Coller, J. L., and E. J. Carpenter. 2006b. Reverse transcription PCR amplification of cyanobacterial symbiont 16S rDNA sequences from single non-photosynthetic eukaryotic marine planctonic host cells. *Journal of Phycology* 42:243–250.

Franca, S., Pinto, L., Alvito, P., Sousa, I., Vasconcelos, V., and G. J. Doucette. 1996. Studies on prokaryotes associated with PSP producing dinoflagellates. In: *Harmful and Toxic Algal Blooms*, Yasumoto, T., Oshima, T., and Fukuyo, Y. (eds.). IOC of UNESCO, Paris, pp. 347–350.

Franca, S., Viega, S., Mascarenhas, V., Pinto, L., and G. J. Doucette. 1995. Prokaryotes in association with a toxic *Alexandrium lusitanicum* in culture. In: *Harmful Marine Algal Blooms*, Lassus, P., Arzul, G., Erard-le Denn, E., Gentien, P., and Marcaillou-le-Baut, C. (eds.). Lavoisier, Paris, pp. 45–51.

Fritsche, T. R., Horn, M., Seyedirashti, S., Gautom, R. K., Schleifer, K.-H., and M. Wagner. 1993. In situ detection of novel bacterial endosymbionts of *Acanthamoeba* spp. phylogenetically related to members of the order *Rickettsiales. Applied and Environmental Microbiology* 65:206–212.

Gaines, G. and M. Elbrächter. 1987. Heterotrophic nutrition. In: *The Biology of Dinoflagellates*, Taylor, J. F. R. (ed.). Blackwell, Oxford, pp. 224–268.

Gallacher, S., Flynn, K. J., Lewis, J., Munro, P. D. Birkbeck, and T. H. 1996. Bacterial production of sodium channel blocking toxins. In: *Harmful Algal Blooms*, Yasumoto, T., Oshima, Y., and Fukuyo, Y. (eds.). IOC of UNESCO, Paris, pp. 355–358.

Gallacher, S. and E. A. Smith. 1999. Bacteria and paralytic shellfish toxins. *Protist* 150:245–255.

Gause, G. F. 1934. *The Struggle for Existence*. Williams & Wilkins, Baltimore, MD.

Geitler, L. 1959. Syncyanosen. In: *Handbuch der Pflanzenphysiologie,* vol. IX, Ruland, W. (ed.). Springer, Berlin, pp. 530–545.

Giambelluca, M. A. and G. Rosati. 1996. Behavior of epixenosomal band during morphogenesis of *Euplotidium itoi. European Journal of Protistology* 31:77–80.

Gold, K. and U. Pollingher. 1971. Occurrence of endosymbiotic bacteria in marine dinoflagellates. *Journal of Phycology* 7:264–265.

Gordon, N., Angel, D. L., Neori, A., Kress, N., and B. Kimor. 1994. Heterotrophic dinoflagellates with symbiotic cyanobacteria and nitrogen limitation in the Gulf of Aquaba. *Marine Ecology Progress Series* 107:83–88.

Görtz, H.-D. 1982. Infection of *Paramecium bursaria* with bacteria and a yeast. *Journal of Cell Science* 58:445–453.

Görtz, H.-D. 1983. Endonuclear symbionts in ciliates. In: *Intracellular Symbiosis*, Jeon, K. W. (ed.). Supplement 14. Int. Rev. Cytol., Academic Press, New York, pp. 145–176.

Görtz, H.-D. 2002. Bacterial symbionts of protozoa in aqueous environments—potential pathogens? In: *Emerging Pathogens*, Greenblatt, C. and Spigelman, M. (eds). Oxford University Press, Oxford, pp. 25–37.

Görtz, H.-D. and R. Michel. 2003. Bacterial symbionts in protozoa in aqueous environments—potential pathogens? In: *Emerging Pathogens*, Greenblatt, C. and Spigelman, M. (eds.). Oxford University Press, Oxford, pp. 25–37.

Görtz, H.-D. and H. J. Schmidt. 2004. *Caedibacter, Holosporaceae, Lyticum, Paracaedibacter, Pseudocaedibacter, Pseudolyticum, Tectibacter* and *Polynucleobacter*. In: *Bergey's Manual of Systematic Bacteriology*, vol. 2, Garrity, G. M. (ed.). Springer Verlag, New York.

Guisande, C., Frangopupous, M., Maneiro, I., Vergara, A. R., and I. Riveiro. 2002. Ecological advantages of toxin production by the dinoflagellate *Alexandrium minutum* under phosphorus limitation. *Marine Ecology Progress Series* 225:169–176.

Hallegraeff, G. M. and S. W. Jeffrey. 1984. Tropical phytoplankton species and pigments of continental shelf waters of North-West Australia. *Marine Ecology Progress Series* 20:59–74.

Harumoto, T. and A. Miyake. 1990. Defensive function of trichocysts in Paramecium. *Journal of Experimental Zoology* 260:84–92.

Heruth, D. C., Pond, F. R., Dilts, J. A., and R. L. Quackenbush. 1994. Characterization of genetic determinants for R body synthesis and assembly in *Caedibacter taeniospiralis* 47 and 166. *Journal of Bacteriology* 176:3559–3567.

Hold, G. L., Smith, E. A., Rappe, M. S., Maas, E. W., Moore, E. R. B., Stroempl, C., Stephen, J. R., Prosser, J. I., Birkbeck, T. H., and S. Gallacher. 2001a. Characterisation of bacterial communities associated with toxic and non-toxic dinoflagellates: *Alexandrium* spp. and *Scrippsiella trochoidea. FEMS Microbiology Ecology* 37:161–173.

Hold, G. L., Smith, E. A., Birkbeck, T. H., and S. Gallacher. 2001b. Comparison of paralytic shellfish toxin (PST) production by the dinoflagellates *Alexandrium lusitanicum* NEPCC 253 and *Alexandrium tamarense* NEPCC 407 in the presence and absence of bacteria. *FEMS Microbiology Ecology* 36:223–234.

Hoppenrath, M., Schweikert, M., and M. Elbrächter. 2003. Morphological reinvestigation and epitype characterization of the sand dwelling dinoflagellate *Adenoides eludens* (Dinophyceae). *European Journal of Phycology* 38:385–394.

Horn, M., Fritsche, T. R., Gautom, R. K., Schleifer, K.-H., and M. Wagner. 1999. Novel bacterial endosymbionts of *Acanthamoeba* spp. Related to the *Paramecium caudatum* symbiont *Caedibacter caryophilus. Environmental Microbiology* 1:357–367.

Horn, M. and M. Wagner. 2004. Bacterial endosymbionts of free-living amoebae. *Journal of Eukaryotic Microbiology* 51:509–514.

Ito, S. 1958. Two new species of marine ciliate. *Euplotidium itoi* sp. nov. and *Gastrocirrhus trichocystus* sp. Nov. *Zoological Magazine* (*Tokyo*) 67:184–187.

Iwatsuki, K. and Y. Naito. 1981. The role of symbiotic *Chlorella* in photoresponse of *Paramecium bursaria. Proceedings of the Japan Academy* 57 B:318–323.

Janson, S., Rai, A. N., and B. Bergman. 1995. Intracellular cyanobiont *Richelia intracellularis*: Ultrastructure and immunolocalization of phycoerythrin, nitrogenase, Rubisco and glutamine synthetase. *Marine Biology* 124:1–8.

Jeblick, J. and J. Kusch. 2005. Sequence, transcription activity and evolutionary origin of the R-body coding plasmid pKAP298 from the intracellular parasitic bacterium *Caedibacter taeniospiralis. Journal of Molecular Evolution* 60:164–173.

Jenks, A. and S. P. Gibbs. 2000. Immunolocalization and distribution of form II Rubisco in the pyrenoid and chloroplast stroma of *Amphidinium carterae* and form I Rubisco in the symbiont-derived plastids of *Peridinium foliaceum* (Dinophyceae). *Journal of Phycology* 36:127–138.

Jennings, H. S. 1939. Genetics of *Paramecium busaria*. II. Age and death of clones in relation to the result of conjugation. *Journal of Experimental Zoology* 96:17–52.

Jurand, A., Rudman, B. M., and J. R. Jr. Preer. 1971. Prelethal effects of killing action by stock 7 of *Paramecium aurelia. Journal of Experimental Zoology* 177:365–388.

Kim, C. H., Sako, Y., and Y. Ishida. 1993a. Comparison of toxin composition between populations of *Alexandrium* spp., from geographically distant areas. *Nippon Suisan Gakkaishi* 59:641–646.

Kim, C. H., Sako, Y., and Y. Ishida. 1993b. Variation of toxin production and composition in axenic cultures of Alexandrium catenella and *A. tamerense. Nippon Suisan Gakkaishi* 59:633–639.

Kirchner, M., Sahling, G., Schütt, C., Döpke, H., and G. Uhlig. 1999. Intracellular bacteria in the red tide-forming heterotrophic dinoflagellate *Noctiluca scintillans. Archiv fur hydrobiologie* 54:297–310.

Kodama, M. 1990. Possible links between bacteria and toxin production in algal blooms. In: *Toxic Marine Phytoplankton*, Graneli, E (ed.). Elsevier Science Publishing, Amsterdam.

Kodama, Y. and M. Fujishima. 2007. Infectivity of *Chlorella* species for the ciliate *Paramecium bursaria* is not based on sugar residues of their cell wall components, but on their ability to localize beneath the host cell membrane after escaping from the host digestive vacuole in the early infection process. *Protoplasma* 231:55–63.

Kodama, M., Ogata, T., and Sato, S. (1988). Bacterial production of saxitoxin. *Agricultural and Biological Chemistry* 52:1075–1077.

Kopp, M., Doucette, G. J., Kodama, M., Gerdts, G., Schütt, C., and L. K. Medlin. 1997. Phylogenetic analysis of selected toxic and non-toxic bacterial strains isolated from the toxic dinoflagellate *Alexandrium tamarense. FEMS Microbiology Ecology* 24:252–257.

Kubanek, J., Snell, T. W., and C. Pirkle. 2007. Chemical defense of the red tide dinoflagellates *Karenia brevis* against rotifer grazing. *Limnology and Oceanography* 52:1026–1035.

Kusch, J. 1993. Predator-induced morphological chances in *Euplotes* (Ciliata): Isolation of the inducing substance released from *Stenostomum sphagnetorum* (Turbellaria). *Journal of Experimental Zoology* 265:613–618.

Kusch, J., Czubatinski, L., Wegmann, S., Hübner, M., Alter, M., and P. Albrecht. 2002. Competitive advantages of *Caedibacter*-infected Paramecia. *Protist* 153:47–58.

Kusch, J. and H.-D. Görtz. 2006. Towards an understanding of the killer trait: *Caedibacter* endocytobionts in *Paramecium. Progress in Molecular and Subcellular Biology* 41:61–76.

Kusch, J., Stremmel, M., Schweikert, M., Adams, V., and H. J. Schmidt. 2000. The toxic symbiont *Caedibacter caryophila* in the cytoplasm of *Paramecium novaurelia. Microbiology and Ecology* 40:330–335.

Landis, W. G. 1981. The ecology, role of the killer trait, and interactions of five species of the *Paramecium aurelia* complex inhabiting the littoral zone. *Canadian Journal of Zoology* 59:1734–1743.

Landis, W. G. 1987. Factors determining the frequency of the killer trait within populations of the *Paramecium aurelia* complex. *Genetics* 115:197–205.

Landis, W. G. 1988. Ecology. In: *Paramecium*, Görtz, H.-D. (ed.). Springer Verlag, Berlin, pp. 419–436.

Landsberg, J. L. 2002. The effects of harmful algal blooms on aquatic organisms. *Reviews in Fisheries Science* 10:113–389.

Lee, R. E. 1977 Saprophytic and phagocytic isolates of the colourless heterotrophic dinoflagellate *Gyrodinium lebouriae* Herdman. *Journal of Marine Biology Association UK*. 57:303–315.

Lewis, J., Kennaway, G., Franca, S., and E. Alverca. 2001. Bacterium–dinoflagellate interactions: Investigative microscopy of *Alexandrium spp.* (Gonyaulacales, Dinophyceae). *Phycologia* 40:280–285.

Linka, N., Hurka, H., Lang, B. F., Burger, G., Winkler, H. H., Stamme, C., Urbany, C., Seil, I., Kusch, J., and H. E. Neuhaus. 2003. Phylogenetic relationships of non-mitochondrial nucleotide transport proteins in bacteria and eukaryotes. *Gene* 306:27–35.

Lobban, C. S., Schefter, M., Simpson, A. G. B., Pochon, X., Pawlowski, J., and W. Foissner. 2002. *Maristentor dinoferus* n. gen., n. sp., a giant heterotrich ciliate (Spirotrichea: Heterotrichida) with zooxanthellae, from coral reefs on Guam, Mariana Islands. *Marine Biology* 140:411–423.

Lom, J., Rohde, K., and I. Dykova. 1993. *Crepidoodinium australe*, new species, an ectocommensal dinoflagellate from the gills of *Sillago ciliata*, an estuarine fish from the New South Wales coast of Australia. *Diseases of Aquatic Organism* 15:63–72.

Lucas, I. A. N. 1982. Observations on *Noctiluca scintillans* Macartney (Ehrenb.) (Dinophyceae) with notes on an intracellular bacterium. *Journal of Plankton Research* 4:401–409.

Lucas, I. A. N. 1991. Symbionts of the tropical Dinophysales (Dinophyceae). *Ophelia* 33:213–224.

Lucas, I. A. N. and M. Vesk. 1990. The fine structure of two photosynthetic species of Dinophysis (Dinophysiales, Dinophyceae). *Journal of Phycology* 26:345–357.

Magagnini, G. and R. Nobili. 1964. Su *Euplotes woodruffi* Gaw e su *Euplotidium arenarium* n. sp. (Ciliata, Hipotrichida). *Monitore Zoologia Italia* 72:178–202.

Maki, T. and I. Imai. 2001. Relationships between intracellular bacteria and the bivalve killer dinoflagellate *Heterocapsa circularisquama* (Dinophyceae). *Fisheries Science* 67:794–803.

Masayo, N. T., Hideo, I., and T. Harumoto. 1999. Toxic and phototoxic properties of the protozoan pigments Blepharismin and Oxyblepharismin. *Photochemistry and Photobiology* 69:47–54.

Mensinger, A. F. and J. F. Case. 1992. Dinoflagellate luminescence increases susceptibility of zooplankton to teleost predation. *Marine Biology* 112:207–210.

Miyake, A. and T. Harumoto. 1996. Defensive function of trichocysts in *Paramecium* against the predatory ciliate *Monodinium balbiani. European Journal of Protistology* 32:128–133.

Miyake, A., Harumoto, T., and I. Hideo. 2001. Defence function of pigment granules in *Stentor coeruleus. European Journal of Protistology* 37:77–88.

Morse, D. 1995. A nuclear encoded form II RUBISCO in Dinoflagellates. *Science* 269:17.

Muscatine, L., Karakashian, S., and M. Karakashian. 1967. Soluble extracellular products of algae symbiotic with a ciliate, a sponge and a mutant hydra. *Comparative Biochemistry and Physiology* 20:1–12.

Niess, D., Reisser, W., and W Wiessner. 1981. The role of endosymbiotic algae in photoaccumulation of *Paramecium bursaria. Planta* 152:268–271.

Noland, L. E. 1937. Observations on marine ciliates of the gulf coast of Florida. *Transactions of the American Microscopic Society* 56:160–171.

Norris, R. E. 1967. Algal consortisms in marine plankton. In: *Proceedings of the Seminar on Sea, Salt and Plants*, Krishnamurthy, V. (ed.). Catholic Press, Bhavnagar, India, pp. 178–189.

Nyberg, D. 1988. The species concept and breeding systems. In: *Paramecium*, Görtz, H.-D. (ed). Springer Verlag, Berlin, pp. 41–58.

Ogata, T., Ishimaru, T., and M. Kodama. 1987. Effect of water temperature and light intensity on growth rate and toxicity change in *Protogonyaulax tamarense. Marine Biology* 95:217–220.

Palmer, J. D. 1995. Rubisco rules fall; gene transfer triumphs. *Bioessays* 17:1005–1008.

Palmer, J. D. 1996. RUBISCO surprises in dinoflagellates. *Plant Cell* 8:343–345.

Petroni, G., Spring, S., Schleifer, K.-H., Verni, F., and G. Rosati. 2000. Defensive extrusive ectosymbionts of *Euplotidium* (Ciliophora) that contain microtubule-like structures are bacteria related to *Verrucomicrobia. Proceedings of the National Academy of Science USA* 97:1813–1817.

Pienaar, R. N. 1980. The ultrastructure of *Peridinium balticum* (Dinophyceae) with particular reference to its endosymbionts. *Elektronenmikroskopie-vereniging van Suidelike Afrika Verrigtings* 10:75–76.

Pienaar, R. N., Sakai, H., and T. Horiguchi. 2007. Description of a new dinoflagellate with a diatom endosymbiont, *Durinskia capensis* sp. nov. (Peridiniales, Dinophyceae) from South Africa. *Journal of Plant Research* 120:247–258.

Pond, F., Gibson, I., Lalucat, J., and R. L. Quackenbush. 1989. R-body-producing bacteria. *Bacteriological Reviews* 53:25–67.

Preer, J. R. Jr. and L. B. Preer. 1982. Revival of names of protozoan endosymbionts and proposal of *Holospora caryophila* nom. nov. *International Journal of Systematic Bacteriology* 32:140–141.

Preer, J. R. Jr., Preer, L. B., and A. Jurand. 1974. Kappa and other endobionts in *Paramecium aurelia. Bacteriological Reviews* 38:113–163.

Prince, K., Lettieri, L., McCurdy, K. J., and J. Kubanek. 2006. Fitness consequences for copepods feeding on a red tide dinoflagellate: Deciphering the effects of nutritional value, toxicity, and feeding behaviour. *Oecologia* 147:479–488.

Pringsheim, E. G. 1928. Physiologische Untersuchungen an *Paramecium bursaria*, II. *Archiv für Protistenkunde* 64:361–418.

Quackenbush, R. L. 1988. Endosymbionts of killer paramecia. In: *Paramecium*, Görtz, H.-D. (ed.). Springer, Berlin.

Quackenbush, R. L. and J. A. Burbach. 1983. Cloning and expression of DNA sequences associated with the killer trait of *Paramecium tetraurelia* stock 47. *Proceedings of the National Academy of Science USA* 80:250–254.

Quackenbush, R. L., Cox, B. J., and J. A. Kanabrocki. 1986a. Extrachromosomal elements of extrachromosomal elements of paramecia, and their extrachromosomal elements. In: *Extrachromosomal Elements in Lower Eukaryontes*, Wickner, R. B., Hinnebusch, A., Lambowitz, A. M., Gunsalus, I. C., and Hollaender, A. (eds). Plenum Press, New York, pp. 265–278.

Quackenbush, R. L., Dilts, J. A., and B. J. Cox. 1986b. Transposonlike elements in *Caedibacter taeniospiralis. Journal of Bacteriology* 166:349–352.

Rausch de Traubenberg, C., Géraud, M.-L., Soyer-Gobillard, M.-O., and D. Emdadi. 1995a. The toxic Dinoflagellate *Prorocentrum lima* and its associated bacteria. I. An ultrastructural study. *European Journal of Protistology* 31:318–326.

Rausch de Traubenberg, C. and M.-O. Soyer-Gobillard. 1990. Bacteria associated with a photosynthetic dinoflagellate in culture. *Symbiosis* 8:117–133.

Rausch de Traubenberg, C., Soyer-Gobillard, M.-O., Géraud, M.-L., and M. Albert. 1995b. The toxic Dinoflagellate *Prorocentrum lima* and its associated bacteria. II. Immunolocalisation of okadaic acid in axenic and non-axenic cultures. *European Journal of Protistology* 31:383–388.

Ray, S. M. and W. B. Wilson. 1957. Effect of unialgal and bacteria-free cultures of *Gymnodinium breve* on fish. U.S. *Fish and Wildlife Service Special Report*, Fisheries. No. 211.

Ricci, N., Morelli, A., and F. Verni. 1996. The predation of *Litonotus* on *Euplotes,* a two step cell–cell recognition process. *Acta Protozoologica* 35:234–268.

Ricci, N. and F. Verni. 1988. Motor and predatory behavior of *Litonotus lamella. Canadian Journal of Zoology* 66:1973–1981.

Rohr, J., Hyman, M., Fallon, S., and M. I. Latz. 2002. Bioluminescence flow visualization in the ocean: An initial strategy based on laboratory experiments. *Deep-Sea Research* 49:2009–2033.

Rohr, J., Latz, M. I., Fallon, S., Nauen, J. C., and E. Hendricks. 1998. Experimental approaches towards interpreting dolphin-stimulated bioluminescence. *Journal of Experimental Biology* 201:1447–1460.

Rowan, R. 1998. Diversity and ecology of zooxanthellae on coral reefs. *Journal of Phycology* 34:407–417.

Rosati, G., Giambelluca, M. A., and E. Taiti. 1996. Epixenosomes peculiar epibionts of the protozoon ciliate *Euplotidium itoi*: Morphological and functional cell compartmentalization. *Tissue and Cell* 28:313–320.

Rosati, G., Modeo, L., and F. Verni. 2008. Micro-game hunting: Predatory behaviour and defensive strategies in ciliates. In: *Microbial Ecology Research Trends*, Dijk, Thijis Van (ed.). Nova Science Publishers, Inc., pp. 65–86.

Rosati, G., Petroni, G., Quochi, S., Modeo, L., and F. Verni. 1999. Epixenosomes, peculiar epibionts of the hypotrich Ciliate *Euplotidium itoi*, defend their host against predators. *Journal of Eukaryotic Microbiology* 46:278–282.

Sasner, J. J., Ikawa, M., and T. L. Foxall. 1984. In: *Seafood Toxins*, Ragelis, E. P. (ed). American Chemical Society Symposium Series 262, Washington, DC, pp. 391–406.

Schmidt, H. J., Görtz, H.-D., Pond, F. R., and R. L. Quackenbush. 1987. Characterization of *Caedibacter* endonucleosymbionts from the macronucleus of *Paramecium caudatum* and the identification of a mutant with blocked R body synthesis. *Experimental Cell Research* 174:49–57.

Schneller, M. V., Sonneborn, T. M., and J. A. Mueller. 1959. The genetic control of kappa-like particles in *Paramecium aurelia. Genetics* 44:533–534.

Schnepf, E. 1992a. From Parasitism to symbiosis: The dinoflagellate sample. In: *Alage and Symbioses: Plants, Animals, Fungi, Viruses, Interactions Explored*, Reisser, W. (ed.). Biopress Limited, Bristol, England, pp. 699–710.

Schnepf, E. 1992b. From Prey via endosymbiont to plastid: Comparative studies in Dinoflagellates. In: *Origins of Plastids*. Lewin, R. E. (ed.). Chapman & Hall, New York and London, pp. 53–76.

Schnepf, E. and M. Elbrächter. 1992. Nutritional strategies in Dinoflagellates. A review with emphasis on cell biological aspects. *European Journal of Protistology* 28:3–24.

Schnepf, E. and M. Elbrächter. 1999. Dinophyte chloroplasts and phylogeny—a review. *Grana* 38:81–97.

Schrallhammer, M., Fokin, S. I., Schleifer, K.-H., and G. Petroni. 2006. Molecular characterization of the obligate endosymbiont "Caedibacter macronucleorum" Fokin and Görtz 1993 and of its host *Paramecium duboscqui* strain Ku4–8. *Journal of Eukaryotic Microbiology* 53:499–506.

Schweikert, M. 2003. Cell wall ultrastructure and intracytoplasmic bacteria in hypnocysts of toxic *Alexandrium tamarense* (Dinophyceae). *Protistology* 3:138–144.

Schweikert, M., Bürk, C., Dietrich, R., Hanke, P., Hummert, C., and F. Brümmer. 2000. Localization of PSP-toxins in dinoflagellates of the genus *Alexandrium. 9th International Congress of Harmful Algal Blooms (HAB 2000)*, Hobart, Tasmania, Australia, February 7–11.

Schweikert, M. and B. Meyer. 2001. Characterization of intracellular bacteria in the freshwater dinoflagellate *Peridinium cintum. Protoplasma* 217:177–184.

Selander, E., Thor, P., Toth, G., and H. Pavia. 2006. Copepods induce paralytic shellfish toxin production in marine dinoflagellates. *Proceedings of the Royal Society London Series B Biological Sciences* 273:1673–1680.

Sellner, K. G., Doucette, G. J., and G. J. Kirkpatrick. 2003. Harmful algal blooms: Causes, impacts, and detection. *Journal of Industrial Microbiology and Biotechnology* 30:383–406.

Shumway, S. E. 1990. A review of the effects of algal blooms on shellfish and aquaculture. *Journal World Aquaculture Society* 21:65–104.

Silva, E. S. E. 1959. Some observations on marine dinoflagellate cultures. I. Prorocentrum micans Ehr. and Gyrodinium sp. *Notas e Estudos do Instituto de Biologia Maritima* 12:1–15.

Silva, E. S. E. 1962. Some observations on marine dinoflagellate cultures. III. *Gonyaulax spinifera, G. tamarenis* and *Peridinium trochoideum. Notas e Estudos do Instituto Biologia Maritima* 26:1–21.

Silva, E. S. E. 1967. *Cochlodinium heterolobatum* n. sp.: Structure and some cytophysiological aspects. *Journal of Protozoology* 14:745–754.

Silva, E. S. E. 1978. Endonuclear bacteria in two species of dinoflagellates. *Protistologica* 14:113–119.

Silva, E. S. E. 1982. Relationship between dinoflagellates and intracellular bacteria. In: *Marine Algae in Pharmaceutical Science*, Hoppe, H. A. and Levring, T. (eds.). Walter de Gruyter, Berlin, New York., vol. 2, pp. 269–288.

Silva, E. S. E. 1990. Intracellular bacteria: The origin of dinoflagellate toxicity. *Journal of Environmental Pathology, Toxicology, and Oncology* 10:124–128.

Silva, E. S. E. and S. Franca. 1985. The association dinoflagellate-bacteria: Their ultrastructural relationship in two species of dinoflagellates. *Protistologica* 21:429–446.

Silva, E. S. E. and I. Sousa. 1981. Experimental work on the dinoflagellate toxin production. *Arquivos do Instituto Nacional de Saude* 6:381–387.

Sonneborn, T. M. 1938. Mating types in *P. aurelia*: Diverse conditions for mating in different stocks; occurrence, number and interrelations of the type. *Proceedings of the American Philosophical Society* 79:411–434.

Sonneborn, T. M. 1943. Gene and cytoplasm. I. The determination and inheritance of the killer character in variety 4 of *P. aurelia*. II. The bearing of determination and inheritance of characters in *P. aurelia* on problems of cytoplasmic inheritance, pneumococcus transformations, mutations and development. *Proceedings of the National Academy of Sciences USA* 29:329–343.

Sonneborn, T. M. 1957. Breeding systems, reproductive methods, and species problems in protozoa. In: *The Species Problem*, Mayr, E. (ed.). American Association of Advanced Sciences, Washington, DC, pp. 155–324.

Sonneborn, T. M. 1959. Kappa and related particles in *Paramecium. Advances in Virus Research* 6:229–356.

Springer, N., Ludwig, W., Amann, R., Schmidt, H. J., Görtz, H.-D., and K.-H. Schleifer. 1993. Occurrence of fragmented 16S rRNA in an obligate bacterial endosymbiont of *Paramecium caudatum. Proceedings of the National Academy of Sciences USA* 90:9892–9895.

Stokes, M. D., Grant, G. B., Latz, M. I., and J. Rohr. 2004. Bioluminescence imaging of wave-induced turbulence. *Journal of Geophysical Research* 109:1–8.

Sykes, P. F. and M. E. Huntley. 1987. Acute physiological reaction of *Calanus pacificus* to selected dinoflagellates: Direct observations. *Marine Biology* 94:19–24.

Teegarden, G. J. 1999. Copepod grazing selection and particle discrimination on the basis of PSP toxin content. *Marine Ecology Progress Series* 181:163–176.

Tosteson, T. R., Ballantine, D. L., Tosteson, C. G., Bardales, A., Durst, H. D., and T. B. Higert. 1986. Comparative toxicity of *Gambierdiscus toxicus, Ostreopsis cf. lenticularis*, and associated microflora. *Marine Fisheries Review* 48:57–59.

Tosteson, T. R., Ballantine, D. L., Tosteson, C. G., Hensley, V., and A. T. Bardales. 1989. Associated bacterial flora, growth, and toxicity of cultured benthic dinoflagellates *Ostreopsis lenticularis* and *Gambierdiscus toxicus. Applied and Environmental Microbiology* 55:137–141.

Trench, R. K. 1993. Microalgal-invertebrate symbioses: A review. *Endocytobiosis and Cell Research* 9:135–176.

Tuffrau, M. 1985. Une nouvelle espèce du genre *Euplotidium* Noland 1937: *Euplotidium prosaltans* n. sp. (Cilié Hypotriche). *Cahiers de Biologie Marine* 26:53–62.

Ukeles, R. and B. M. Sweeny. 1969. Influence of Dinoflagellate trichocysts and other factors on the feeding of *Crassostrea virginica* larvae on *Monochrysis lutheri. Limnology and Oceanography* 14:403–410.

Van den Heok, C., Mann, D. G., and H. M. Jahns. 1995. *Algae: An Introduction of Phycology*. Cambridge University Press, UK.

Verni, F. and G. Rosati. 1990. Peculiar epibionts in *Euplotidium itoi* (Ciliata, Hypotrichida). *Journal of Eukaryotic Microbiology* 37:337–343.

White, H. H. 1979. Effects of dinoflagellates bioluminescence on the ingestion rates of herbivorous zooplankton. *Journal of Experimental Marine Biology and Ecology* 36:217–224.

Whitney, S. M., Shaw, D. C., and D. Yellowlees. 1995. Evidence that some dinoflagellates contain a ribulose-1,5-bisphosphate carboxylase/oxygenase related to that of the alpha-proteobacteria. *Proceedings of the Royal Society London Series B Biological Sciences* 259:271–275.

Wilcox, L. W. 1986. Prokaryotic endosymbionts in the chloroplast stroma of the dinoflagellate *Woloszynskia pascheri. Protoplasma* 135:71–79.

Wilcox, L. W. and G. J. Wedemayer. 1984. *Gymnodinium acidotum* Nygaard (Pyrrhophyta), a dinoflagellate with an endosymbiotic cryptomonad. *Journal of Phycology* 20:236–242.

Young, R. E. 1983. Oceanic bioluminescence: An overview of general functions. *Bulletin of Marine Science* 33:829–845.

Zhou, J. and L. Fritz. 1993. Utrastructure of two toxic marine dinoflagellates, Prorocentrum lima and Protocentrum maculosum. *Phycologia* 32:444–450.

5

Bacterial Chemical Defenses of Marine Animal Hosts

Eric W. Schmidt

CONTENTS

5.1 Introduction

In the intense predation environment of the oceans—especially tropical oceans—many animals have obvious, visible means of self-defense. Molluscs are usually shelled; corals are encased in calcium carbonate and are capable of stinging attackers, and so on. Standing out as glaring exceptions to the defensive rule, the soft-bodied marine animals such as sponges and sea slugs manage to survive and thrive often as major contributors to biomass in a given environment. It has long been postulated that marine animals lacking physical defenses would instead be chemically defended, relying on small molecules to deter predation via numerous possible allelochemical pathways. Based upon this hypothesis, well over 10,000 marine natural products have been isolated from invertebrate animals. Many of these compounds are potent toxins, leading to clinical use in cancer therapy (Miljanich, 2004; Newman and Cragg, 2004). Many more have been shown to provide chemical defenses by various mechanisms, including feeding deterrence, toxicity, warning signaling, and prevention of surface colonization (fouling) (McClintock and Baker, 2001). The striking thing about these chemicals is that many, if not most, are actually synthesized by bacteria living within their animal hosts. Thus, bacterial symbionts allow many of the soft-bodied animals to exist in predation-intense environments, such as coral reefs.

Of course, the participation of microbes in chemical interactions with eukaryotes is well known from land organisms, as documented in other sections of this book. There are several factors that make marine animal–bacteria symbioses distinctly different, especially biological factors. For example, many more phyla of eukaryotes live in the ocean than on land, and animal lifestyles are relatively much more diverse (Brusca et al., 2003). Perhaps the most important difference is the relative amount of knowledge concerning the chemistry of marine animals. Of the many marine "animal" molecules that likely originate in bacterial symbionts, there is an immense body of knowledge concerning the presence and variation of chemical motifs by time, space, and species (Blunt et al., 2007). Because of this information, it is possible to readily form and test hypotheses about bacteria–animal symbioses in the ocean. Such hypotheses are more than a scientific curiosity: they are very useful for biotechnology and improve our understanding of diversity in the ocean. Since many invertebrate marine natural products are potential pharmaceuticals but suffer from a problem of supply, symbiosis studies also offer one potential avenue to supply promising new chemical entities (Haygood et al., 1999; Schmidt, 2005).

Small molecule marine natural products from animals form many different structural "families." Often, a single family will have dozens of representatives, all of which are chemically slightly different from each other and could be envisioned to occupy phylogenetic trees depicting chemical relationships (Schmidt et al., 2000). A large amount of data are available concerning the biological, geographical, and temporal occurrence of these small molecules. By overlaying the chemical phylogeny with these details, key insights into symbiosis and biosynthesis can be readily obtained. These molecules are synthesized to hit specific biochemical targets and thus reflect an ancient coevolution of small chemical entities with large proteins, nucleic acids, and other cellular structures. In particular, symbiotic interactions are highly informative of the evolution of small molecules and can even be used for the rational genetic synthesis of new small molecules (Donia et al., 2006). This is because specific symbionts are closely related to each other, producing chemistry that falls within a relatively narrow range of the chemical phylogenetic tree. These natural products often feature hypervariable regions, sampling a large amount of chemical space with a small amount of underlying biosynthetic diversity. In addition to the structure–activity relationships discernible from hypervariability, the feature is useful for biosynthetic engineering. Biosynthetic pathways within symbiotic bacteria are rapidly evolving, possibly in response to mutating biochemical targets or shifting ecological needs. By studying rapid changes, it is possible to replicate them in the laboratory to produce new small molecules at will. These advantages stand in contrast to studying small molecules from free-living (environmental) bacteria, for which there is no clear ecological relationship. It is difficult to purposefully cultivate close relatives of bacteria that synthesize specific small molecules, and even if close relatives are found, the underlying biosynthetic pathways are often quite divergent.

It should be emphasized that experimental evidence for bacterial synthesis of natural products in symbioses has been obtained in only a few cases. Even in these cases, it remains possible that the story is more complicated. For example, both plants and their endosymbiotic fungi are sometimes capable of producing identical defensive compounds, but using apparently convergent enzymes. In another complex case, "fungal" metabolites are synthesized by endosymbiotic bacteria living within filamentous fungi (Partida-Martinez and Hertweck, 2005). Recent data from insect symbionts demonstrate widespread lateral gene transfer from bacteria to host, leading to the possibility that both animals and bacteria may produce identical compounds using nearly identical genes (Hotopp et al., 2007). Finally, it is commonly suggested that two coevolving organisms could cooperate to produce a natural product. This scenario is highly unlikely if one considers the complex enzymology involved. However, it is reasonable to expect that host animals could contribute precursors for bacterial metabolism. For example, genetically modified erythromycin biosynthetic proteins in bacteria can be fed otherwise unavailable subunits that are incorporated into the final compounds (Stassi et al., 1998). In addition, animals commonly modify metabolites to detoxify them; this process can often accidentally increase toxicity, for example, in aflatoxin modification by human liver P450s (Guengerich et al., 1996). It is highly likely that animals will modify "symbiotic" metabolites via common detoxifying mechanisms. Other bacteria living within animals could similarly impact bacterial metabolism (Hamann, 2003). Thus, although there is excellent experimental and circumstantial evidence for bacterial production of allelochemicals in marine animals, readers are cautioned to keep an open mind.

5.2 Symbiotic Chemical Interactions

Traditionally, metabolites were assigned to be bacterial or animal in origin based upon the physical separation of cells by one of several methods, followed by chemical analysis of the purified cell types. These methods provided the key data that allowed the symbiotic hypothesis to be explored, but location does not indicate biosynthetic source and may in fact be misleading (discussed in more detail below). More recently, molecular and microbiological methods have provided much more direct evidence, revealing bacterial sources for several animal metabolites. Examples will be described from sponges, ascidians, bryozoans, and molluscs.

5.2.1 Ascidians

Nearly 1000 natural products have been isolated from these invertebrate chordates, which survive by filter feeding. Of these, strong evidence exists that two major classes, comprising over 70 compounds, are synthesized by bacterial symbionts. These groups include the patellamides/patellins and the tambjamines. Many other "ascidian" metabolites are also likely to be bacterial products.

5.2.1.1 Patellamides

These compounds and their relatives are cyclic peptides that are linked in an amide bond via their N and C termini (Figure 5.1) (Ireland and Scheuer, 1980). The compounds are further modified by heterocyclization of cysteine, serine, or threonine residues, yielding thiazole, thiazoline, and oxazoline moieties. An additional group within this family contains serine and threonine residues that are prenylated, probably by dimethylallylpyrophosphate rather than being heterocyclized (Carroll et al., 1996). The compounds have been isolated from a variety of ascidians from the Family Didemnidae in varying yields, often in great abundance (gram-per-kilogram). The compounds have been screened for biological activity in many different assays, but for the most part they are only toxic to higher eukaryotes such as human cells. The compounds are often quite potently cytotoxic, with IC_{50} values down to therapeutically relevant nM levels. In fact, one compound, trunkamide (Figure 5.1), was a preclinical anticancer candidate (Carroll et al., 1996; Wipf and Uto, 2000). Based upon this toxicity level and the relative abundance within ascidians, it is reasonable to presume that the compounds may deter predation or offer some other allelochemical advantage to the animals.

Didemnid ascidians living in tropical oceans also often harbor symbiotic cyanobacteria, *Prochloron* spp., which have thus far eluded numerous cultivation attempts (Lewin and Cheng, 1989). The symbiosis is important to the nutritional state of the host, with demonstrated exchange of nutrients between animal

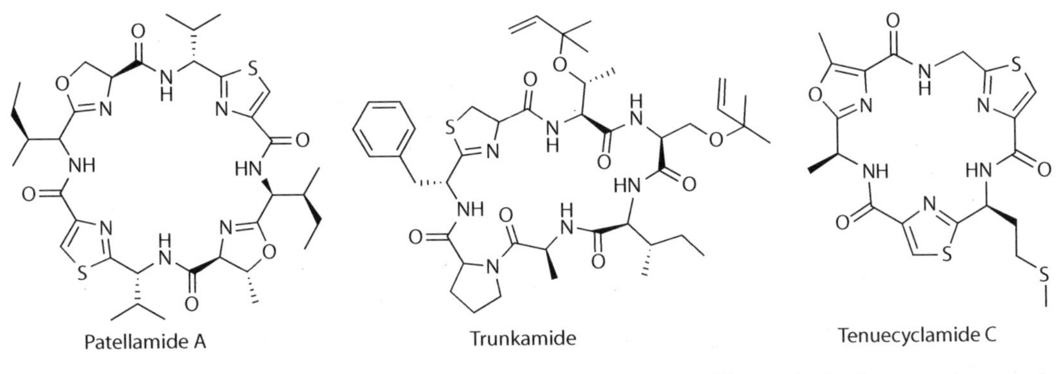

FIGURE 5.1 Patellamide A and trunkamide from uncultivated symbionts of ascidians and related compound tenuecyclamide C from free-living cyanobacteria.

and bacteria. In some cases, nearly all the reduced carbon required by the host ascidian could be provided by bacterial photosynthesis (Koike et al., 1993). The chemists who isolated the first patellamide relatives noted the presence of *Prochloron*, leading to the idea that symbiotic bacteria could be responsible for synthesizing these cytotoxic small molecules (Ireland and Scheuer, 1980). In later studies, ascidian cells were physically separated from *Prochloron*, and patellamide-like compounds were found solely in the cyanobacterial cell population (Degnan et al., 1989). However, experiments by another research group found that patellamides could be found throughout the animal–*Prochloron* association and could not be localized to single cell types (Salomon and Faulkner, 2002). A molecular biological solution was required to determine the true source of these metabolites.

To determine the source of compounds, the didemnid ascidian *Lissoclinum patella* was collected in Palau (Micronesia), and *Prochloron didemni* was physically separated from the host and preserved. In a *P. didemni* genome sequencing project, it was found that patellamides A and C were directly encoded on a precursor peptide, PatE (Schmidt et al., 2005). In addition, the *pat* locus to the patellamides also encoded four other essential enzymes. In *Escherichia coli* cell culture, these four enzymes and the PatE precursor led to production of patellamides, representing a solution to the supply problem (Donia et al., 2006). Interestingly, in another study DNA was directly cloned from enriched *Prochloron*, leading to detection of patellamides in *E. coli* culture without the requirement of sequencing (Long et al., 2005). Based upon these experiments, didemnid ascidians represented the first marine group for which the source question had been definitively answered, from whole animals to genome sequencing to production in *E. coli*.

There are about 60 relatives of patellamides, including peptides with six to eight amino acids, variable oxidation patterns, and replacement of the oxazoline moiety with the isoprene motif. To determine the origin of this variability, a comparative analysis was performed using *Prochloron*-containing ascidians from across the tropical Pacific (Donia et al., 2006). First, the patellamide (oxazoline-containing) group was examined. Strikingly, 29 different pathways were cloned that were >99% identical to each other at the DNA and protein levels over ~11 kbp. The only major difference was focused to the exact region encoding product structures; everything else was universally conserved. This indicated that directed mutations in a small sequence region could yield a large library of natural products. Indeed, PCR-based mutation of this region led to the synthesis in *E. coli* of wholly unnatural patellamide derivatives. More recently, the pathway to prenylated compounds was isolated, revealing yet another unexpected evolutionary story (Donia et al., 2008). In these pathways, genes and proteins were >99% identical to the patellamide-like pathways, except that there was a difference in the middle of the cluster in which genes were as low as 40% identical to the patellamide pathway. The prenylating pathways could be functionally expressed in *E. coli*. Since there were no new gene types compared to the patellamide pathway, it was apparent that these genes at ~40% identity are responsible for prenylation in one case and heterocyclization in another. Finally, within the prenylating pathways, the mutational story was similar to that found for patellamides, in which pathways were >99% identical except in the exact region encoding diverse prenylated products (Table 5.1).

Because of strong similarities to microcin and lantibiotic biosynthetic pathways and the presence of related compounds in numerous cyanobacteria, the group of compounds exemplified by patellamides were recently given the family name "cyanobactins" (Donia et al., 2007). The symbiotic cyanobactins, because of their unique evolutionary story that was clearly revealed in symbiosis, provide a series of biotechnological tools. For example, large libraries of modified, cyclic peptides may be synthesized using genetic engineering. Such compounds would be of exceptional use in drug discovery and genetic–organic synthesis.

5.2.1.2 Tambjamines

These small, pyrrole-containing alkaloids (Figure 5.2) were first isolated as toxic metabolites from the bryozoan *Sessibugula translucens* and its nudibranch predator, the nembrothid *Tambja abdere* (Carté and Faulkner, 1983). Nudibranchs are known to concentrate toxic chemicals from their foods, presumably for defense against predation (Faulkner, 1992). In fact, tambjamines were demonstrated to provide protection from predation, even to some degree against the predatory nudibranch *Roboastra tigris* (Carté

TABLE 5.1
If Compounds of the Patellamide Family Are Linearized and Aligned, It Can Be Seen that There Are Overlapping Families with Varying Modifications

Peptide	AA#1	AA#2	AA#3	AA#4	AA#5	AA#6	AA#7	AA#8
1	Ccy[a]	Val	Ccy	Phe	Tcy	Val	Pro	
2	Tcy[b]	Val	Ccy	Phe	Ccy	Phe	Pro	
3	Tcy	Val	Ccy	Ile	Tcy	Val	Ccy	Ile
4	Scy[c]	Val	Ccy	Ile	Tcy	Val	Ccy	Ile
5	Ccy	Ile	Ccy	Met	Ccy			
6	Ccy	Val	Ccy	Met	Ccy	Leu	Pro	
7	Ccy	Phe	Ccy	Ile	Ccy	Val	Pro	Val
8	Ccy	Leu	Ccy	Ile	Ccy	Val	Pro	Val
9	Tcy	Ala	Ccy	Ile	Tpren	Phe		
10	Thr	Ala	Ccy	Ile	Tpren	Phe		
11		Phe	Ccy	Tpren	Spren	Ile	Ala	Pro
12	Tpren[d]	Leu	Ccy	Tpren	Val	Pip[f]		
13	Tpren	Leu	Ccy	Tpren	Val	Pro		
14	Tpren	Leu	Ccy	Tpren	Ile	Pro	Val	Pro
15	Tpren	Val	Ccy	Tpren	Ile	Pro	Val	Pro
16	Spren[e]	Phe	Ccy	Tpren	Val	Pro	Val	Pro
17	Tpren	Val	Ccy	Tpren	Phe	Pro	Val	Pro
18	Spren	Phe	Pro	Ccy	Ile	Pro	Ile	

Note: By using symbiosis to observe natural mutations underlying these differences, it is possible to replicate them in the laboratory and generate new compounds. Amino acid numbering system for comparison only: all peptides represented are cyclic. a, Ccy = Cys thiazole or thiazoline; b, Tcy = Thr oxazole or oxazoline; c, Scy = Ser oxazole or oxazoline; d, Tpren = β-*O*-prenyl Thr; e, Spren = β-*O*-prenyl Ser; f, Pip = 3-amino-2-piperidone derived from Glu; Yellow: heterocyclization; Blue: prenylation.

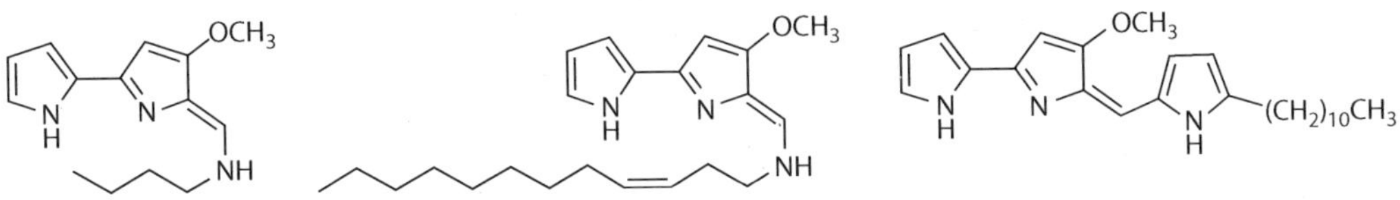

FIGURE 5.2 Tambjamine C from marine animals compared to tambjamine YP1 and undecylprodigiosin from bacteria.

and Faulkner, 1986). Subsequently, the compounds were isolated from other bryozoans and from the ascidian, *Atapozoa* sp. (Paul et al., 1990). The ascidian contained abundant amounts of tambjamines, which were shown to prevent predation by many different fishes and invertebrates. As with bryozoans, *Atapozoa* sp. is preyed upon by nembrothid nudibranchs, *Nembrotha* spp., which also concentrated tambjamines to levels that could deter predation. Nembrothid nudibranchs may be attracted to tambjamine-containing organisms, be they bryozoans or the taxonomically distant ascidians.

Because of their wide occurrence in diverse organisms, it was postulated that tambjamines may be bacterial products (Carté and Faulkner, 1986; Matsunaga et al., 1986). Indeed, they are structurally related to prodiginines, which are well-known bright red pigments from bacteria. Red bacteria were visualized within *S. translucens*, lending some support to the notion that bacteria could be the source of tambjamines. However, tambjamines themselves are deeply pigmented (often deep green), and pigmentation in *Atapozoa* was associated with ascidian cells (Lindquist and Fenical, 1991). This situation is reminiscent of that of the patellamides, in which "localization" studies gave conflicting results.

Tambjamine-like alkaloids have been isolated from a *Streptomyces* sp. and also from *Pseudoalteromonas tunicata* (Kojiri et al., 1993; Franks et al., 2005). As the name suggests, *P. tunicata* was first isolated from the surface of an ascidian (aka tunicate), and thus, there is a clear link from bacteria to compound production in ascidians. Unfortunately, the source ascidians were not known to contain tambjamines, and therefore some contention remains in this field. Recently, a gene cluster was identified encoding the biosynthesis of tambjamines using a method that was nearly identical to those reported in the identification of the patellamide gene cluster (Burke et al., 2007). These genes provide tools for more thorough investigation of the biosynthetic source of tambjamines within ascidians (and bryozoans) and the potential for *de novo* biosynthesis of tambjamines in nembrothids by dietary bacteria.

5.2.2 Bryozoans

Like ascidians, these animals are also sedentary filter feeders, which have provided the source for about 200 new small molecules. One group, the bryostatins, has been extensively studied and is synthesized by specific bacterial symbionts. In fact, this group was the first to be characterized at the molecular level and provided a model for all other studies in the field.

5.2.2.1 Bryostatins

These polyketides from the bryozoan *Bugula neritina* are extremely potent cytotoxic agents, variants of which have been in clinical trials against cancers for more than a decade (Pettit et al., 1982; Kortmansky and Schwartz, 2003). As complex polyketides of this type are strongly associated with bacterial metabolism, it was suspected that the bryostatins might have a bacterial origin. This origin would be technologically important, since bryostatins are isolated in extremely low yields from the producing bryozoans; about 1–2 g of bryostatin 1 (Figure 5.3) is isolated per metric ton of bryozoan (Mann, 2002). By finding a bacterial source, important pharmaceutical products could be provided by bacterial culture or cloning of biosynthetic genes.

In the early 1980s, specific symbiotic bacteria were described lining the pallial sinus of *B. neritina* larvae (Woollacott, 1981). In the 1990s, these bacteria were identified as gamma-proteobacteria, *Endobugula sertula*, using 16S rRNA gene sequence analysis (Haygood and Davidson, 1997; Davidson and Haygood, 1999). Since *E. sertula* could not be cultured (and still eludes cultivation), experiments on this symbiosis led to the development of tools for metagenomic analysis that are widely used in biosynthetic studies. This study also marked the first molecular investigation into putative chemical source symbionts in the marine environment and heralded the arrival of a new field of inquiry. Although *E. sertula* was shown to be the only major bacterial strain associated with larvae, and treatment of *B. neritina* with antibiotics diminished bryostatin production, direct evidence of bryostatin biosynthesis was lacking.

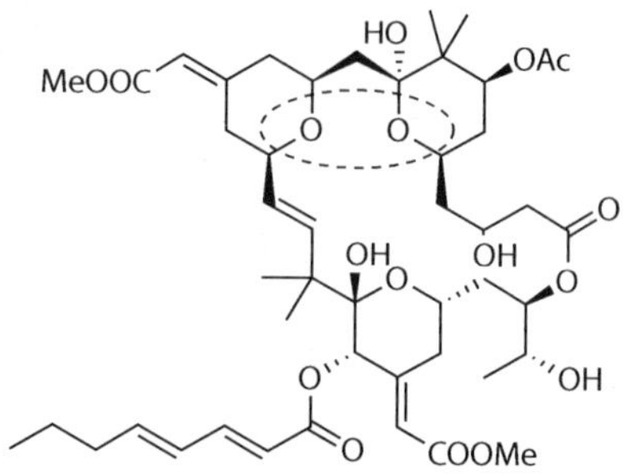

Bryostatin 1

FIGURE 5.3 Bryostatin 1 from a bryozoan. The circle indicates ether linkages that helped to confirm the gene cluster.

The discovery of a bryostatin biosynthetic gene cluster in *E. sertula* put to rest concerns about the origin of the bryostatins (Davidson et al., 2001). The *bry* cluster contained genes for several proteins classified as modular type I polyketide synthase (PKS) and was the only PKS within *E. sertula* (Hildebrand et al., 2004; Sudek et al., 2007). PKS proteins link acetate and malonate derivatives via decarboxylative Claisen reactions to form carbon chains. In the modular PKSs, each added enzyme active site is used only once, and thus gene sequence gives a fairly accurate picture of the final polyketide structures. In the bryostatin case, the order of modules within the PKS genes reasonably correlated with the final structure, with a few modest exceptions. Particularly, compelling evidence is provided by the correlation of tetrahydropyran rings in bryostatin with tetrahydropyran biosynthetic machinery in the correct modules of the *bry* cluster. Transcriptional analysis further validates the active synthesis of *bry*-derived proteins in the organism.

The bryostatin symbiosis has been studied from the ecological perspective. Bryostatins have been shown to protect *B. neritina* larvae from predation (Lindquist and Hay, 1996; Lopanik et al., 2004). Indeed, bryostatins and *E. sertula* are extremely abundant within larvae in comparison to adult bryozoans. A recent microscopic analysis revealed that bryostatins are concentrated on external surfaces of the larvae, while *E. sertula* lines the pallial sinus (Sharp et al., 2007). Thus, bryostatins do not colocalize with the producing cells. Both *E. sertula* and bryostatins were followed through an entire bryozoan life cycle, providing key insights into the maintenance and horizontal transfer of *E. sertula* as well as the ecological role of bryostatins. For example, bryostatins accumulate on the surface of developing adult tissues, but they appear to be "shed" once the chitinous exoskeleton is fully developed. This result is consistent with a protective effect of bryostatins on soft tissues and a physical defense of fully developed tissues.

5.2.3 Cnidarians

Over 3000 small molecules have been isolated from soft corals, sea whips, and related animals. Unlike hard corals, which possess imposing physical defenses, soft corals are vulnerable to predation. Several studies implicate small molecules in coral chemical defense; these are mostly terpenes and derivatives thereof. Corals often harbor symbiotic eukaryotes, the dinoflagellate zooxanthellae. A published report details biosynthesis of a coral metabolite group, the pseudopterosins, by symbiotic dinoflagellates.

5.2.3.1 Pseudopterosins

Pseudopterosins (Figure 5.4) are well known as the first commercialized marine natural products, which found use as anti-inflammatory agents in wrinkle cream (Look et al., 1986; Mayer et al., 1998). Because relatively large amounts of pseudopterosins are present within the gorgonians, *Pseudopterogorgia elisabethae*, the animals are harvested from shallow Caribbean waters to supply the compound. Nonetheless, access to symbiotic organisms or enzymes could provide a better source or could be used

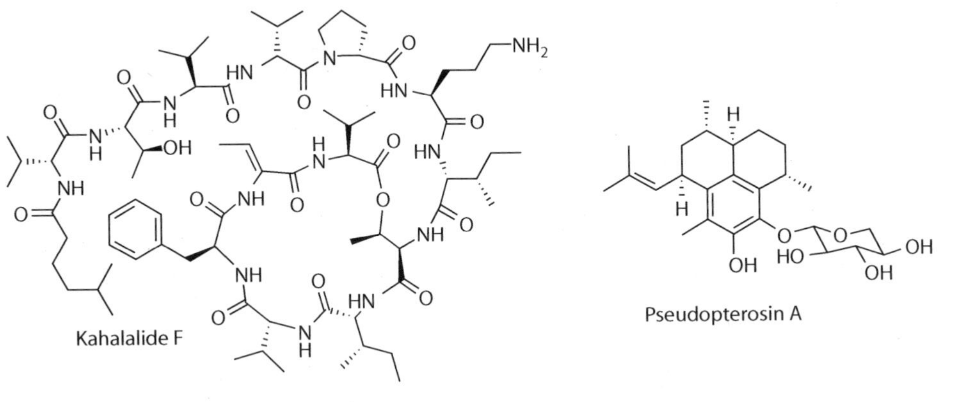

FIGURE 5.4 Some mollusc and soft coral compounds known to be of symbiotic origin

for analog synthesis. Dinoflagellates were removed from *P. elisabethae* and incubated with radioactive precursors. Label was incorporated into pseudopterosins, indicating *de novo* biosynthesis in dinoflagellates, and in addition, a terpene cyclase was isolated (Mydlarz et al., 2003; Kohl and Kerr, 2004). Wounding of the coral initiates production of pseudopterosins in some circumstances, further implicating a chemical defense role for the compounds (Newberger et al., 2006). Dinoflagellate symbionts isolated from flatworms also synthesize natural products—in this case polyketides (Kobayashi and Ishibashi, 1993). Although this chapter is intended to focus on bacterial symbiosis with animals, the dinoflagellate examples are worth keeping in mind, reflecting the unique and complex environment of the sea.

5.2.4 Molluscs

Most small-molecule research on this group has focused on the soft-bodied molluscs, such as nudibranchs and sea hares, which lack physical defenses. Over 1000 compounds are documented from the phylum. It has long been thought that these animals usually obtain toxic chemical defense compounds from their diets, which include diverse organisms such as sponges, ascidians, and algae (Faulkner, 1992). Another intriguing possibility is that instead of just compounds, natural product-synthesizing bacteria may also be transferred, enabling *de novo* synthesis in both molluscs and their prey (Hill et al., 2005). In some cases, a sea hare obtains toxins from a cyanobacterial diet, providing a clear-cut example of "bacterial" synthesis of an animal natural product (Harrigan et al., 1998; Luesch et al., 2002). Others have been shown to synthesize their own terpene or polyketide feeding deterrents (Ireland and Scheuer, 1979; Cimino et al., 1983). Often, these terpenes are similar or identical to toxins isolated from whole sponges (Cavagnin et al., 2001). It is unclear whether bacteria are transferred from sponge to mollusc in these cases or whether some convergent biosynthetic process is at work. It is tempting to think that a cell transfer process may be required, since several soft-bodied molluscs are capable of transferring intact, functional chloroplasts or stinging nematocysts into distal structures in their bodies. Molluscs seem programmed to recognize cells or organelles to sequester them from their diet and to move them through their bodies (Brusca et al., 2003). However, there is only one well-studied example of a symbiotic source, and much work remains to determine whether this phenomenon is more general.

5.2.4.1 Kahalalides

Kahalalides (Figure 5.4) are cytotoxic cyclic peptides that are currently in clinical trials against cancer (Hamann et al., 1996; Lopez-Macia et al., 2001; Rademaker-Lakhai et al., 2005). They exhibit hypervariability of structure, similar to the patellamides, but unlike the patellamides, they appear more likely to be synthesized via the nonribosomal mechanism. The compounds were isolated from the sacoglossan mollusc, *Elysia rufescens*, and its algal prey, *Bryopsis* sp., and shown to chemically defend both organisms from predation (Becerro et al., 2001). *Vibrio* spp. bacteria were isolated from both the algae and the surface of *E. rufescens* and found to produce kahalalide F (Figure 5.4) in culture (Hill et al., 2005). Thus, it appears that *E. rufescens* is able to synthesize kahalalides *de novo* using bacteria that it obtains from its algal diet. This compelling story may be widespread; it is possible that numerous classes of soft-bodied molluscs are attracted to and concentrate bacteria rather than small molecules. If so, the current dogma needs to be revised.

5.2.5 Sponges

Phylum Porifera has been the source for, by far, the majority of marine invertebrate natural products, with nearly 10,000 compounds reported. Sponges also have an ancient geological record of association with bacteria going back 600 million years (Brunton and Dixon, 1994). It comes as no surprise that sponges and their metabolites have been a major subject for symbiosis studies, and a growing literature covers symbiotic bacteria from a variety of perspectives (Taylor et al., 2007). Below, only chemical symbioses with underlying molecular or cultivation evidence will be described. Most studies using molecular biological approaches tackle sponges from the polyphyletic order Lithistida or from the genus *Dysidea*. Both groups are extremely common in tropical seas and are replete with diverse natural products, some of which are potential drug leads. In addition, both have key features that make them ideal for the study of symbiosis. *Dysidea* spp. sponges are pantropical, grow in shallow waters, and have obligate nutritional symbioses with cyanobacteria that constitute up to 40% of their biomass (Flatt et al., 2005). Often, the cyanobacteria are filamentous and are thus readily manipulated in field experiments. Lithistid sponges are such prolific metabolite producers that they have been called chemical "star performers or host to the stars"—the real stars being symbiotic bacteria (Bewley and Faulkner, 1998). In fact, they are full of bacteria, with estimates of 40% body weight being composed of bacteria. In addition, they often contain small molecules resembling those isolated from bacteria. While lithistids tend to grow in deeper water and are a bit less convenient than *Dysidea* to work with, they make up for it with their extremely diverse bacterial and chemical repertoire. These sponges are just among the first to be examined among many hundreds or thousands of potential species with interesting natural products.

5.2.5.1 Theopederin

Sponges of the genus *Theonella* are large bacteria fermentors and harbor a very diverse bacterial population (Schmidt et al., 2000; Hentschel et al., 2002). They often contain potently toxic polyketide and nonribosomal peptide natural products, which are likely to have a feeding deterrent role, although this remains to be tested (Bewley and Faulkner, 1998). The theopederin/onnamide group is one such toxic polyketide class (Figure 5.5), which can be active at picomolar levels against some human cell lines (Sakemi et al., 1988; Fusetani et al., 1992). Interestingly, the carbon skeletons of these compounds are nearly identical to pederin (Figure 5.5), the toxic product of terrestrial beetles of the genus *Paederus* (Pavan and Bo, 1953). These animals are otherwise known as blister beetles because they eject a noxious defensive secretion that causes intense, painful blisters in humans. Pederin has been shown to deter predation by spiders, and feeding studies clearly delineated its polyketide origins (Cardani et al., 1973; Kellner and Dettner, 1996). The presence of complex polyketides in a range of different animals on land and in the sea strongly suggested a bacterial biosynthesis. Indeed, *Paederus* beetles are often infected with symbiotic *Pseudomonas* spp., which long eluded cultivation. When these pseudomonads were transferred from toxic to nontoxic beetles, toxicity was also transferred, further implicating the bacteria in toxin production (Kellner, 2001).

Using a metagenomic approach, a putative pederin biosynthetic gene cluster was cloned from a metagenomic sample of *Paederus fuscipes* and sequenced (Piel, 2002; Piel et al., 2005). This cluster encoded a series of modular type I PKS-nonribosomal peptide synthetase (NRPS) proteins. The order of the modules reflected the structure to a large degree. Probably the most convincing evidence that this cluster encoded pederin production was observed in the form of a surprising evolutionary artifact. Most of the sponge theopederin/onnamide series are much larger than pederin: these compounds contain an extended polyketide chain terminating in the amino acid arginine, while pederin lacks these features. The pederin gene cluster encoded modules that should make the sponge compound, including the extended polyketide chain and a clear NRPS module encoding arginine. However, the cluster was

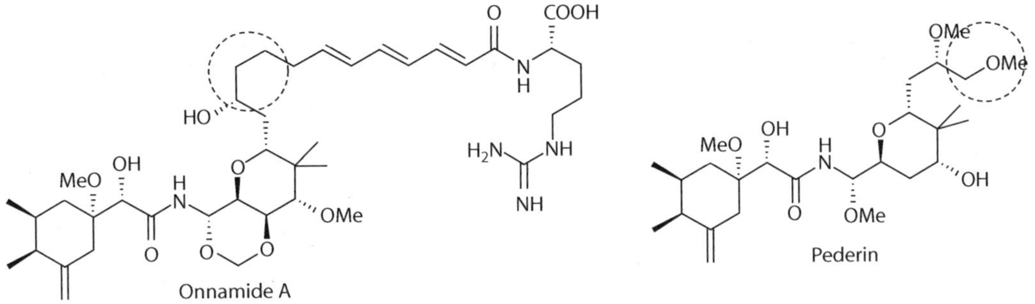

FIGURE 5.5 Symbiotically derived molecules from lithistid sponges. The circle indicates the region in which the pederin gene cluster is interrupted by an oxidase.

interrupted by an inserted, in-frame oxidase domain that could oxidatively cleave a biosynthetic intermediate, leading to the correct pederin small molecule. This insertion greatly increased the probability had that the correct genes had been identified and had provided tools for cloning theopederin/onnamide genes from *Theonella swinhoei* sponges. A fragment of the putative onnamide biosynthetic cluster was found in one out of 60,000 metagenomic cosmid clones from the sponge. Domain and module order was relatively conserved in comparison to the pederin cluster and were clearly bacterial in nature (Piel et al., 2004a,b). These results strongly suggest a bacterial source for these toxic compounds within sponges.

5.2.5.2 Theopalauamide and Swinholide

Theopalauamide and swinholide compounds are high- and low-nanomolar cytotoxins, respectively (Figure 5.6), to human cell lines (Kitagawa et al., 1990; Schmidt et al., 1998). They and their relatives are commonly isolated from *T. swinhoei*, as is true in the onnamide example above. Because of the incredible abundance (~1 mg compound per g wet weight) of compounds and their potent activity, they are

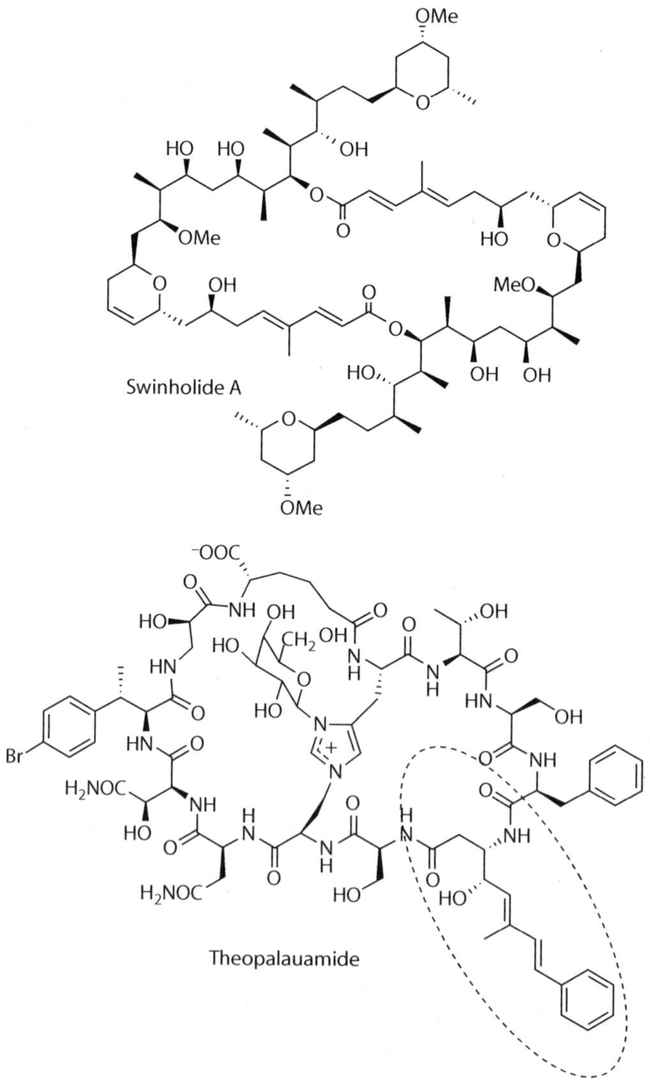

FIGURE 5.6 Symbiotically derived molecules from lithistid sponges. Circled is the portion of theopalauamide that is polyketide derived and is similar to a polyketide portion of microcystins.

likely to be candidates for feeding deterrence as well. In one study, extracts of *T. swinhoei* from the Red Sea, which reportedly contains theopalauamide and swinholide relatives, were potent feeding deterrents against wrasse and sea urchin predators (Burns et al., 2003). Based upon structural considerations, swinholide appears to be a polyketide, while theopalauamide seems like a hybrid polyketide-nonribosomal peptide. In fact, the polyketide motif in theopalauamide is very similar to the polyketide portion of microcystins, which are cyanobacterial products (Neilan et al., 1999). Swinholide bears similarity to bacterial polyketides as well, and more recently, swinholide A itself was isolated from a cyanobacterial mat (Andrianasolo et al., 2005).

In cell separation experiments using *T. swinhoei* in Palau, theopalauamide was isolated only from a filamentous bacterial population, while swinholide was found solely in a unicellular bacterial population (Bewley et al., 1996). No compound was found in a surface layer of cyanobacteria inhabiting *T. swinhoei*, and in addition, the compounds were not detectable in sponge cells. Molecular analysis of the filamentous bacteria revealed that they represented a new, deeply branching subgroup of delta-proteobacteria aligned most closely with the myxobacteria although also near the sulfate reducers. These filamentous, nonphotosynthetic bacteria, *Entotheonella palauensis*, were shown to be present in numerous theonellid sponges (Schmidt et al., 2000). They were also detected in other types of lithistids later (Schirmer et al., 2005). Numerous derivatives of theopalauamide are known, and it was shown that 16S rRNA gene sequence varied with changes in the resulting chemical structures. In the resulting manuscript, the idea that symbionts could be useful for the study of biosynthetic mutations was first elaborated (Schmidt et al., 2000). This idea was fully explored in the example of patellamide-like molecules, as described above.

Many other lithistid metabolites are likely to be bacterial in origin, although firm evidence is lacking. One intriguing example is that of the microsclerodermins, discovered in deep-water samples of *Microscleroderma* sp. (Schmidt and Faulkner, 1998) exactly identical structures were identified from cultivated, terrestrial myxobacteria. Although it is common to find similar structures in marine animals and bacteria, exact identity between structures has been found only in a few cases.

5.2.5.3 Halogenated Metabolites

Several halogenated natural products have been isolated from *Dysidea* (*Lamellodysidea*) *herbacea* (Ridley et al., 2005a). *Dysidea* and other sponges commonly contain cyanobacterial symbionts, *Oscillatoria spongeliae*, which have eluded cultivation, meaning that research has required cell separation or metagenomic techniques (Ridley et al., 2005b).

Some of the *Dysidea* compounds are brominated diphenyl ethers, a motif that is extremely rare except for synthetic compounds (Unson et al., 1994). In addition, a number of chlorinated peptide metabolites have been isolated from *D. herbacea* (Unson and Faulkner, 1993). Closely related compounds including one with an identical planar structure to sponge compounds have been isolated from cyanobacteria. Cell separation studies demonstrated that these compounds were localized in *O. spongeliae*, not in sponge cells, while terpenoid metabolites were found in sponge cells only (Unson and Faulkner, 1993). In fact, this was the first study to demonstrate a bacterial origin for a sponge natural product. It was thus a groundbreaking study that initiated a new era in marine natural products research. Later, cell separation studies demonstrated that brominated diphenyl ethers were present in *O. spongeliae* and as crystalline material within the sponge, constituting up to 12% of the total sponge weight (Unson et al., 1994). However, the compounds were not found in sponge cells or in other bacterial cells.

Barbamide (Figure 5.7), from the free-living cyanobacterium *Lyngbya majuscula*, exemplifies the halogenated peptide group (Ramaswamy et al., 2006). In particular, the extremely rare amino acid trichloroleucine is present in both barbamide and *D. herbacea* compounds, such as dysidinin. The barbamide biosynthetic pathway was cloned, leading to the discovery and characterization of a novel flavin-dependent halogenase (Chang et al., 2002). Using PCR primers for this halogenase in two independent studies, candidate halogenating genes were identified in *O. spongeliae* and confirmed to be from the cyanobacteria using *in situ* hybridization (Flatt et al., 2005; Ridley et al., 2005b). The halogenase was only found in *Dysidea* spp. that contained trichloroleucine compounds and not in *Dysidea* with different chemistry. These results strongly suggest that *O. spongeliae* are the true producers of dysidenin and related compounds (Figure 5.7). By 16S rRNA gene sequence analysis, *O. spongeliae* appears to be closely related to *Lyngbya* spp.

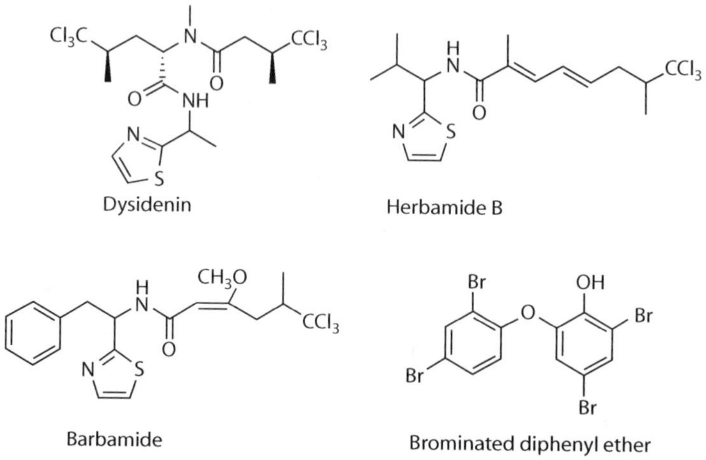

FIGURE 5.7 Symbiotically derived molecules from *Dysidea* sponges. Dysidenin from sponges, barbamide from free-living cyanobacteria, and herbamide B from both.

5.2.5.4 Others

Manzamines (Figure 5.8) are complex, potently active metabolites of unusual biosynthetic origin and have been isolated from a phylogenetically diverse group of sponges including *Xestospongia* sp. (Sakai et al., 1986; Baldwin and Whitehead, 1992). From this sponge, an actinomycete of the genus *Micromonospora* was isolated and shown to produce manzamines in culture (Hill et al., 2004; Peraud, 2006). This study was the first in which a complex marine invertebrate natural product was identified in cultivable symbiotic bacteria using a target-directed approach. The strain produced a few milligrams in culture while the sponge contains gram per kilogram of the compounds.

5.2.5.5 Okadaic Acid

The dinoflagellate *Prorocentrum lima*, separated from the host sponge *Halichondria okadai*, was shown to produce okadaic acid. Previously, this complex polyketide was isolated from the whole sponge (Sugiyama et al., 2007).

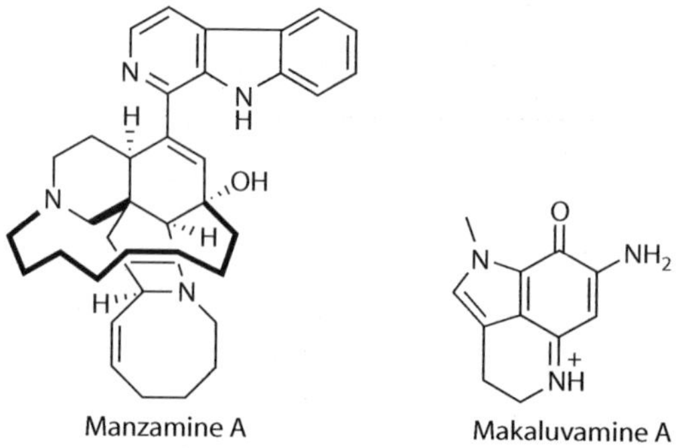

FIGURE 5.8 Other sponge compounds of symbiotic origin.

5.2.5.6 Makaluamine A

Makaluamine A is a complex alkaloid that was first isolated from sponges of the genus *Zyzzya* and later from eukaryotic myxomycetes (Ishibashi et al., 2001).

5.3 Discussion

As shown by the above examples, the last 10 years have seen the blossoming of molecular and microbiological evidence supporting the symbiotic hypothesis of natural product synthesis in marine invertebrates. Major evidence falls along two lines: some studies focus on metagenomic analysis on difficult-to-culture symbiotic bacteria, while others rely on cultivation and the chance that the cultivated strains will produce a molecule of interest.

Cultivation studies have shown that manzamines are produced by bacteria in sponges, while tambjamines are probably bacterial products in ascidians and bryozoans. The algal/mollusc products, kahalalides, are also found in cultivated bacteria. A number of compounds have been found in cultivated organisms not associated with marine animals. For example, identical patellamide relatives have been found in cyanobacteria, ascidians, and sponges. A chlorinated peptide has been found in both sponges and free-living cyanobacteria.

Molecular biological studies have been highly revealing, providing conclusive proof of symbiotic synthesis in some cases and strong evidence in others. The complete gene clusters for patellamides in ascidians and bryostatins in bryozoans have been cloned and sequenced, and an extremely large fragment of the theopederin/onnamide pathway has been sequenced from a sponge symbiosis. Smaller fragments have been sequenced in studies of chlorinated metabolites in *Dysidea*. In the case of patellamides and relatives, heterologous expression provides a potential solution to the supply problem.

In addition to solving the supply problem or providing a new level to the diversity of marine invertebrates in the ocean, the bacterial synthesis of animal natural products has profound benefits for the study of pathway evolution (and hence genetic engineering). Instead of seeing just single natural products from marine invertebrate symbionts, these bacteria synthesize families of related compounds. In the patellamide case, isolated islands of hypervariability within an absolutely conserved background were observed to directly encode new natural products. Since these compounds usually have very well defined natural targets, the changes to these cassettes probably reflect evolution to hit new targets or a chemical "arms race" in which both agent and target are rapidly evolving. Notably, since these mutations defy statistical analysis (0% change over 11 kbp, up to 60% change over 48 base pairs), these directed changes are clearly and directly functional. Patellamides are secreted, and thus their products are intended to target exogenous organisms. Recently, genes targeting other organisms were dubbed "exogenes" (Olivera, 2006). It is unclear whether other symbiotic natural products exhibit localized hypervariability. Such variability has already been demonstrated to be useful for drug discovery and biotechnology.

Finally, it should not be forgotten that several important natural products from sponges, soft corals, and flatworms are produced by eukaryotic organisms such as myxomycetes and dinoflagellates. It remains unclear whether these eukaryotes are the true sources of metabolites or whether they have other bacterial symbionts themselves. The application of molecular methods to these systems will be immensely powerful to untangle this complex scenario.

There are two fundamental assumptions about marine natural products and symbiosis that were useful in the development of the field but have since been disproved conclusively. Because these ideas are widely quoted and are leading researchers astray, it is worth describing the evidence against them here. These ideas are (1) localization of metabolites in a cell type is indicative of biosynthetic source and (2) producing cell biomass should be proportional to the amount of natural product synthesized.

The first idea, that metabolites can be found in producing cells, was necessary to design and make sense of early cell separation studies. In addition, since many free-living bacteria in the ocean do not secrete small molecules (which would then diffuse and become essentially useless to the producer), this idea seemed to make sense. However, systems that have been intensively studied and for which the most rigorous molecular data are available concretely disprove this notion. For example, bryostatins are clearly secreted and found throughout the bryozoan host; they are not localized to *E. sertula*, although they are proximal to *E. sertula*. The patellamides have been found both inside and outside *Prochloron*, depending upon experimental design. In *E. coli* expression systems ~100% of patellamides and their chemical relatives are secreted into the media. Thus, location is *not* correlated with production in the best-studied systems. Studies that identify molecules within certain animal cell types should be interpreted cautiously, especially when those molecules are highly lipophilic.

The second idea was that the amount of bacteria proposed to produce a metabolite should be proportional to the amount of compound found in a whole organism. Bacteria present in very small amounts should not be capable of producing large amounts of small molecule. This idea could conceivably be wrong on several accounts. First, some marine invertebrates are very long-lived and could easily accumulate bioactive metabolites over a long period of time. Second, bacteria often efficiently produce enormous amounts of small molecules when selected to do so. Third, it is possible due to lateral transfer that more than one organism produces a small molecule, or that only some strains within an identical 16S background produce the molecule of interest. Available data support a lack of correlation between production and producer biomass. For example, manzamine derivatives are found in the gram-per-kilogram level in whole sponges, yet elegant molecular work demonstrates that the producing *Micromonospora* sp. is present in very low abundance. By contrast, *Prochloron* is usually extremely abundant within ascidians, yet patellamide abundance is highly variable (from none up to gram-per-kilogram levels). Moreover, rigorous molecular data clearly reveal the complexity of biosynthetic pathways within seemingly identical strains. The patellamide example is described above. In addition, studies of the discodermolide-containing sponge indicated that filamentous *Entotheonella* spp. bacteria contain hundreds of nonribosomal peptide biosynthetic pathways in a single sponge. As there are no known single strains of bacteria containing more than ~20 such pathways, it is likely that there is immense diversity of *Enotheonella*, resulting in diverse biosynthetic pathways within the sponge. Because of the complexity of bacterial associations within marine animals, the amount of isolated natural product is *not* proportional to the producing strain biomass.

Despite the above caveats, it is clear from recent data that many if not most defensive natural products in marine animals are ultimately derived from bacteria. The evolutionary and physiological mechanisms by which animals acquire and exploit these bacteria will fuel biological research into the symbiosis question. The pharmaceutical and biotechnological reasons for this research are also compelling. Finally, the symbiotic variability that has been revealed by chemistry adds another dimension to the already amazing biodiversity of marine habitats.

ACKNOWLEDGMENT

Our work in symbiosis is funded by NSF and NIH (GM).

REFERENCES

Andrianasolo, E. H., H. Gross, D. Goeger, M. Musafija-Girt, K. McPhail, R. M. Leal, S. L. Mooberry, and W. H. Gerwick. 2005. Isolation of swinholide A and related glycosylated derivatives from two field collections of marine cyanobacteria. *Organic Letters* 7:1375–1378.

Baldwin, J. E. and R. C. Whitehead. 1992. On the biosynthesis of manzamines. *Tetrahedron Letters* 33:2059–2062.

Becerro, M. A., G. Goetz, V. J. Paul, and P. J. Scheuer. 2001. Chemical defenses of the sacoglossan mollusk *Elysia rufescens* and its host Alga *Bryopsis* sp. *Journal of Chemical Ecology* 27:2287–2299.

Bewley, C. A. and D. J. Faulkner. 1998. Lithistid sponges: Star performers or hosts to the stars. *Angewandte Chemie International Edition* 37:2162–2178.

Bewley, C. A., N. D. Holland, and D. J. Faulkner. 1996. Two classes of metabolites from *Theonella swinhoei* are localized in distinct populations of bacterial symbionts. *Experientia* 52:716–722.

Blunt, J. W., B. R. Copp, W. P. Hu, M. H. Munro, P. T. Northcote, and M. R. Prinsep. 2007. Marine natural products. *Natural Product Reports* 24:31–86.

Brunton, F. R. and O. A. Dixon. 1994. Siliceous sponge-microbe biotic associations and their recurrence through the Phanerozoic as reef mound constructors. *Palaios* 9:370–387.

Brusca, R. C., G. J. Brusca, and N. J. Haver. 2003. *Invertebrates*. Sinauer Associates, Sunderland, MA.

Burke, C., T. Thomas, S. Egan, and S. Kjellberg. 2007. The use of functional genomics for the identification of a gene cluster encoding for the biosynthesis of an antifungal tambjamine in the marine bacterium *Pseudoalteromonas tunicata. Environmental Microbiology* 9:841–814.

Burns, E., I. Ifrach, S. Carmeli, J. R. Pawlik, and M. Ilan. 2003. Comparison of anti-predatory defenses of Red Sea and Caribbean sponges. I. Chemical defense. *Marine Ecology Progress Series* 252:105–114.

Cardani, C., C. Fuganti, D. Ghiringhelli, P. Grasselli, and M. Pavan. 1973. The biosynthesis of pederin. *Tetrahedron Letters* 30:2815–2818.

Carroll, A. R., J. C. Coll, D. J. Bourne, J. K. MacLeod, T. M. Zabriskie, C. M. Ireland, and B. F. Bowden. 1996. Patellins 1–6 and trunkamide A: Novel cyclic hexa-, hepta- and octa-peptides from colonial ascidians, *Lissoclinum* sp. *Australian Journal of Chemistry* 49:659–667.

Carté, B. and D. J. Faulkner. 1983. Defensive metabolites from three nembrothid nudibranchs. *The Journal of Organic Chemistry* 48:2314–2318.

Carté, B. and D. J. Faulkner. 1986. Role of secondary metabolites in feeding associations between a predatory nudibranch, two grazing nudibranchs, and a bryozoan. *Journal of Chemical Ecology* 12:795–804.

Cavagnin, M., E. Mollo, F. Castelluccio, M. T. Ghiselin, G. Calado, and G. Cimino. 2001. Can molluscs biosynthesize typical sponge metabolites? The case of the nudibranch *Doriopsilla areolata. Tetrahedron* 57:8913–8916.

Chang, Z., P. Flatt, W. H. Gerwick, V. A. Nguyen, and D. H. Sherman. 2002. The barbamide biosynthetic gene cluster: A novel marine cyanobacterial system of mixed polyketide synthase (PKS)-nonribosomal peptide synthetase (NRPS) origin involving an unusual trichloroleucyl starting unit. *Gene* 296:235–247.

Cimino, G., S. De Rosa, S. De Stefano, G. Sodano, and G. Villani. 1983. Dorid nudibranch elaborates its own chemical defense. *Science* 219:1237–1238.

Davidson, S. K. and M. G. Haygood. 1999. Identification of sibling species of the bryozoan *Bugula neritina* that produce different anticancer bryostatins and harbor distinct strains of the bacterial symbiont *Candidatus* endobugula sertula. *Biological Bulletin* 196:273–280.

Davidson, S. K., S. W. Allen, G. E. Lim, C. M. Anderson, and M. G. Haygood. 2001. Evidence for the biosynthesis of bryostatins by the bacterial symbiont "*Candidatus* Endobugula sertula" of the bryozoan *Bugula neritina. Applied and Environmental Microbiology* 67:4531–4537.

Degnan, B. M., C. J. Hawkins, M. F. Lavin, E. J. McCaffrey, D. L. Parry, A. L. Vandenbrenk, and D. J. Watters. 1989. New cyclic peptides with cytotoxic activity from the ascidian *Lissoclinum-patella. Journal of Medicinal Chemistry* 32:1349–1354.

Donia, M., B. J. Hathaway, S. Sudek, M. G. Haygood, M. J. Rosovitz, J. Ravel, and E. W. Schmidt. 2006. Natural combinatorial peptide libraries in cyanobacterial symbionts of marine ascidians. *Nature Chemical Biology* 2:729–735.

Donia, M. S., J. Ravel, and E. W. Schmidt. 2008. A global assembly line to cyanobactins from symbiotic and free-living cyanobacteria. *Nature Chemical Biology* 4:341–343.

Faulkner, D. J. 1992. Chemical defenses of marine molluscs, pp. 119–163 in V. J. Paul, ed., *Ecological Roles of Marine Natural Products.* Cornell University Press, Ithaca, NY.

Flatt, P., J. T. Gautschi, R. W. Thacker, M. Mustafija-Girt, P. Crews, and W. H. Gerwick. 2005. Identification of the cellular site of polychlorinated peptide biosynthesis in the marine sponge *Dysidea* (Lammelodysidea) *herbacea* and symbiotic cyanobacterium *Oscillatoriá spongeliae* by CARD-FISH analysis. *Marine Biology* 147:761–774.

Franks, A., P. Haywood, C. Holmström, S. Egan, S. Kjellberg, and N. Kumar. 2005. Isolation and structure elucidation of a novel yellow pigment from the marine bacterium *Pseudoalteromonas tunicata. Molecules* 10:1286–1291.

Fusetani, N., T. Sugawara, and S. Matsunaga. 1992. Bioactive marine metabolites. 41. Theopederins A–E, potent antitumor metabolites from a marine sponge, *Theonella* sp. *The Journal of Organic Chemistry* 57:3828–3832.

Guengerich, F. P., Y. F. Ueng, B. R. Kim, S. Langouet, B. Coles, R. S. Iyer, R. Thier, T. M. Harris, T. Shimada, H. Yamakzaki, B. Ketterer, and A. Guillouzo. 1996. Activation of toxic chemicals by cytochrome P450 enzymes: Regio- and stereoselective oxidation of aflatoxin B1. *Advances in Experimental Medicine and Biology* 387:7–15.

Hamann, M. T. 2003. Enhancing marine natural product structural diversity and bioactivity through semisynthesis and biocatalysis. *Current Pharmaceutical Design* 9:879–889.

Hamann, M. T., C. S. Otto, P. J. Scheuer, and D. C. Dunbar. 1996. Kahalalides: Bioactive peptides from a marine mollusk *Elysia rufescens* and its algal diet *Bryopsis* sp. *The Journal of Organic Chemistry* 61:6594–6600.

Harrigan, G. G., H. Luesch, W. Y. Yoshida, R. E. Moore, D. G. Nagle, V. J. Paul, S. L. Mooberry, T. H. Corbett, and F. A. Valeriote. 1998. Symplostatin 1: A dolastatin 10 analogue from the marine cyanobacterium *Symploca hydnoides*. *Journal of Natural Products* 61:1075–1077.

Haygood, M. G. and S. K. Davidson. 1997. Small-subunit rRNA genes and in situ hybridization with oligonucleotides specific for the bacterial symbionts in the larvae of the bryozoan *Bugula neritina* and proposal of "*Candidatus* endobugula sertula". *Applied and Environmental Microbiology* 63:4612–4616.

Haygood, M. G., E. W. Schmidt, S. K. Davidson, and D. J. Faulkner. 1999. Microbial symbionts of marine invertebrates: Opportunities for microbial biotechnology. *Journal of Molecular and Microbiological Biotechnology* 1:33–34.

Hentschel, U., J. Hopke, M. Horn, A. B. Friedrich, M. Wagner, J. Hacker, and B. S. Moore. 2002. Molecular evidence for a uniform microbial community in sponges from different oceans. *Applied and Environmental Microbiology* 68:4431–4440.

Hildebrand, M., L. E. Waggoner, H. Liu, S. Sudek, S. Allen, C. Anderson, D. H. Sherman, and M. Haygood. 2004. bryA: An unusual modular polyketide synthase gene from the uncultivated bacterial symbiont of the marine bryozoan *Bugula neritina*. *Chemistry & Biology* 11:1543–1552.

Hill, R. T., M. T. Hamann, O. Peraud, and N. Kasanah. 2004. Manzamine-producing actinomycetes. WO 2004 013297.

Hill, R. T., M. T. Hamann, J. Enticknap, and K. V. Rao. 2005. Kahalalide-producing bacteria in W. I. P. Organization, ed. WO 2005042720 A2.

Hotopp, J. C., M. E. Clark, D. C. Oliveira, J. M. Foster, P. Fischer, M. C. Torres, J. D. Giebel, N. Kumar, N. Ishmael, S. Wang, J. Ingram, R. V. Nene, J. Shepard, J. Tomkins, S. Richards, D. J. Spiro, E. Ghedin, B. E. Slatko, H. Tettelin, and J. H. Werren. 2007. Widespread lateral gene transfer from intracellular bacteria to multicellular eukaryotes. *Science* 317:1753–1756.

Ireland, C. M. and P. J. Scheuer. 1979. Photosynthetic marine mollusks: *In vivo* ^{14}C incorporation into metabolites of the sacoglossan *Placobranchus ocellatus*. *Science* 205:922–923.

Ireland, C. M. and P. J. Scheuer. 1980. Ulicyclamide and ulithiacyclamide, 2 new small peptides from a marine tunicate. *Journal of the American Chemical Society* 102:5688–5691.

Ishibashi, M., T. Iwasaki, S. Imai, S. Sakamoto, K. Yamaguchi, and A. Ito. 2001. Laboratory culture of the myxomycetes: Formation of fruiting bodies of *Didymium bahiense* and its plasmodial production of Makaluvamine A. *Journal of Natural Products* 64:108–110.

Kellner, R. L. L. 2001. Horizontal transmission of biosynthetic capabilities for pederin in *Paederus melanurus* (Coleoptera: Staphylinidae). *Chemoecology* 11:127–130.

Kellner, R. L. L. and K. Dettner. 1996. Differential efficacy of toxic pederin in deterring potential arthropod predators of *Paederus* (Coleoptera: Staphylinidae) offspring. *Oecologia* 107:293–300.

Kitagawa, I., M. Kobayashi, T. Katori, M. Yamashita, J. Tanaka, M. Doi, and T. Ishida. 1990. Absolute stereostructure of swinholide A, a potent cytotoxic macrolide from the Okinawan marine sponge *Theonella swinhoei*. *Journal of the American Chemical Society* 112:3710–3712.

Kobayashi, J. and M. Ishibashi. 1993. Bioactive metabolites of symbiotic marine organisms. *Chemical Reviews* 93:1753–1770.

Kohl, A. C. and R. G. Kerr. 2004. Identification and characterization of the pseudopterosin diterpene cyclase, elisabethatriene synthase, from the marine gorgonian, *Pseudopterogorgia elisabethae*. *Archives of Biochemistry and Biophysics* 424:97–104.

Koike, I., M. Yamamuro, and P. C. Pollard. 1993. Carbon and nitrogen budgets of 2 ascidians and their symbiont, *Prochloron*, in a tropical seagrass meadow. *Australian Journal of Marine and Freshwater Research* 44:173–182.

Kojiri, K., S. Nakajima, H. Suzuki, A. Okura, and H. Suda. 1993. A new antitumor substance, BE-18591, produced by a streptomycete: Fermentation, isolation, physicochemical and biological properties. *The Journal of Antibiotics* 46:1799–1803.

Kortmansky, J. and G. K. Schwartz. 2003. Bryostatin-1: A novel PKC inhibitor in clinical development. *Cancer Investigation* 21:924–936.

Lewin, R. A. and L. Cheng, eds. 1989. *Prochloron—A Microbial Enigma*. Chapman & Hall, New York.

Lindquist, N. and W. Fenical. 1991. New tambjamine class alkaloids from the marine ascidian *Atapozoa* sp. and its nudibranch predators: Origin of the tambjamines in *Atapazoa*. *Experientia* 47:504–506.

Lindquist, N. and M. E. Hay. 1996. Palatability and chemical defense of marine invertebrate larvae. *Ecological Monographs* 66:431–450.

Long, P. F., W. C. Dunlap, C. N. Battershill, and M. Jaspars. 2005. Shotgun cloning and heterologous expression of the patellamide gene cluster as a strategy to achieve sustained metabolite production. *ChemBioChem* 6:1760–1765.

Look, S. A., W. Fenical, G. K. Matsumoto, and J. Clardy. 1986. The pseudopterosins: A new class of antiinflammatory and analgesic diterpene pentosides from the marine sea whip *Pseudopterogorgia elisabethae* (Octocorallia). *The Journal of Organic Chemistry* 51:5140–5145.

Lopanik, N., N. Lindquist, and N. Targett. 2004. Potent cytotoxins produced by a microbial symbiont protect host larvae from predation. *Oecologia* 139:131–139.

Lopez-Macia, A., J. C. Jimenez, M. Royo, E. Giralt, and F. Albericio. 2001. Synthesis and structure determination of kahalalide F. *Journal of the American Chemical Society* 123:11398–11401.

Luesch, H., G. G. Harrigan, G. Goetz, and F. D. Horgen. 2002. The cyanobacterial origin of potent anticancer agents originally isolated from sea hares. *Current Medicinal Chemistry* 9:1791–1806.

Mann, J. 2002. Timeline: Natural products in cancer chemotherapy. *Nature Reviews Cancer* 2:143–148.

Matsunaga, S., N. Fusetani, and K. Hashimoto. 1986. Bioactive marine metabolites. VIII. Isolation of an antimicrobial blue pigment from the bryozoan *Bugula dentata*. *Experientia* 42:84.

Mayer, A. M., P. B. Jacobson, W. Fenical, R. S. Jacobs, and K. B. Glaser. 1998. Pharmacological characterization of the pseudopterosins: Novel anti-inflammatory natural products isolated from the Caribbean soft coral, *Pseudopterogorgia elisabethae*. *Life Sciences* 62:PL401–PL407.

McClintock, J. B. and B. J. Baker, eds. 2001. *Marine Chemical Ecology*. CRC Press LLC, Boca Raton, FL.

Miljanich, G. P. 2004. Ziconotide: Neuronal calcium channel blocker for treating severe chronic pain. *Current Medicinal Chemistry* 11:3029–3040.

Mydlarz, L. D., R. S. Jacobs, J. Boehnlein, and R. G. Kerr. 2003. Pseudopterosin biosynthesis in *Symbiodinium* sp., the dinoflagellate symbiont of *Pseudopterogorgia elisabethae*. *Chemistry & Biology* 10:1051–1056.

Neilan, B., E. Dittman, L. Rouhainen, R. A. Bass, V. Schaub, K. Sivonen, and T. Börner. 1999. Nonribosomal peptide synthesis and toxigenicity of cyanobacteria. *Journal of Bacteriology* 181:4089–4097.

Newberger, N. C., L. K. Ranzer, J. M. Boehnlein, and R. G. Kerr. 2006. Induction of terpene biosynthesis in dinoflagellate symbionts of Caribbean gorgonians. *Phytochemistry* 67:2133–2139.

Newman, D. J. and G. M. Cragg. 2004. Marine natural products and related compounds in clinical and advanced preclinical trials. *Journal of Natural Products* 67:1216–1238.

Olivera, B. M. 2006. Conus peptides: Biodiversity-based discovery and exogenomics. *Journal of Biological Chemistry* 281:31173–31177.

Partida-Martinez, L. P. and C. Hertweck. 2005. Pathogenic fungus harbours endosymbiotic bacteria for toxin production. *Nature* 437:884–888.

Paul, V. J., N. Lindquist, and W. Fenical. 1990. Chemical defenses of the tropical ascidian *Atapozoa* sp. and its nudibranch predators *Nembrotha* spp. *Marine Ecology Progress Series* 59:109–118.

Pavan, M. and G. Bo. 1953. Pederin, toxic principle obtained in the crystalline state from the beetle *Paederus fuscipes* Curt. *Physiologia Comparata et Oecologia* (*International Journal of Comparative Physiology and Ecology*) 3:307–312.

Peraud, O. 2006. Isolation and characterization of a sponge-associated actinomycete that produces manzamines. Pages 231. University of Maryland, College Park, Baltimore, MD.

Pettit, G. R., C. L. Herald, D. L. Doubek, D. L. Herald, E. Arnold, and J. Clardy. 1982. Isolation and structure of bryostatin 1. *Journal of the American Chemical Society* 104:6846–6848.

Piel, J. 2002. A polyketide synthase-peptide synthetase gene cluster from an uncultured bacterial symbiont of *Paederus* beetles. *Proceedings of the National Academy of Sciences of the United States of America* 99:14002–14007.

Piel, J., D. Hui, N. Fusetani, and S. Matsunaga. 2004a. Targeting modular polyketide synthases with iteratively acting acyltransferases from metagenomes of uncultured bacterial consortia. *Environmental Microbiology* 6:921–927.

Piel, J., D. Hui, G. Wen, D. Butzke, P. M., N. Fusetani, and S. Matsunaga. 2004b. Antitumor polyketide biosynthesis by an uncultivated bacterial symbiont of the marine sponge *Theonella swinhoei*. *Proceedings of the National Academy of Sciences of the United States of America* 101:16222–16227.

Piel, J., D. Butzke, N. Fusetani, D. Hui, M. Platzer, G. Wen, and S. Matsunaga. 2005. Exploring the chemistry of uncultivated bacterial symbionts: Antitumor polyketides of the pederin family. *Journal of Natural Products* 68:472–479.

Rademaker-Lakhai, J. M., S. Horenblas, W. Meinhardt, E. Stokvis, T. M. de Reijke, J. M. Jimeno, L. Lopez-Lazaro, J. A. Lopez Martin, J. H. Beijnen, and J. H. Schellens. 2005. Phase I clinical and pharmacokinetic study of kahalalide F in patients with advanced androgen refractory prostate cancer. *Clinical Cancer Research* 11:1854–1862.

Ramaswamy, A. V., P. M. Flatt, D. J. Edwards, L. T. Simmons, B. Han, and W. H. Gerwick. 2006. The secondary metabolites and biosynthetic gene clusters of marine cyanobacteria. Applications in biotechnology in P. Proksch and W. E. G. Müller, eds., *Frontiers in Marine Biotechnology*, Taylor & Francis, Boca Raton, FL.

Ridley, C. P., P. R. Bergquist, M. K. Harper, D. J. Faulkner, J. N. Hooper, and M. G. Haygood. 2005a. Speciation and biosynthetic variation in four dictyoceratid sponges and their cyanobacterial symbiont, *Oscillatoria spongeliae. Chemistry & Biology* 12:397–406.

Ridley, C. P., D. John Faulkner, and M. G. Haygood. 2005b. Investigation of Oscillatoria spongeliae-dominated bacterial communities in four dictyoceratid sponges. *Applied and Environmental Microbiology* 71:7366–7375.

Sakai, R., T. Higa, C. W. Jefford, and G. Bernardinelli. 1986. Manzamine A, an antitumor alkaloid from a sponge. *Journal of the American Chemical Society* 108:6404–6405.

Sakemi, S., T. Ichiba, S. Kohmoto, G. Saucy, and T. Higa. 1988. Isolation and structure elucidation of onnamide A, a new bioactive metabolite of the marine sponge, *Theonella* sp. *Journal of the American Chemical Society* 110:4851–4853.

Salomon, C. E. and D. J. Faulkner. 2002. Localization studies of bioactive cyclic peptides in the ascidian *Lissoclinum patella. Journal of Natural Products* 65:689–692.

Schirmer, A., R. Gadkari, C. D. Reeves, F. Ibrahim, E. F. DeLong, and C. R. Hutchinson. 2005. Metagenomic analysis reveals diverse polyketide synthase gene clusters in microorganisms associated with the marine sponge *Discodermia dissoluta. Applied Environmental Microbiology* 71:4840–4849.

Schmidt, E. W. 2005. From chemical structure to environmental biosynthetic pathways: Navigating marine invertebrate-bacteria associations. *Trends in Biotechnology* 23:437–440.

Schmidt, E. W., C. A. Bewley, and D. J. Faulkner. 1998. Theopalauamide, a bicyclic glycopeptide from filamentous bacterial symbionts of the lithistid sponge *Theonella swinhoei* from Palau and Mozambique. *Journal of Organic Chemistry* 63:1254–1258.

Schmidt, E. W. and D. J. Faulkner. 1998. Microsclerodermins C-E, antifungal-cyclic peptides from the lithistid sponges *Theonella* sp. and *Microscleroderma* sp. *Tetrahedron*, 54:3403–3056.

Schmidt, E. W., A. Y. Obraztsova, S. K. Davidson, D. J. Faulkner, and M. G. Haygood. 2000. Identification of the antifungal peptide-containing symbiont of the marine sponge *Theonella swinhoei* as a novel delta-proteobacterium, *Candidatus* Entotheonella palauensis. *Marine Biology* 136:969–977.

Schmidt, E. W., J. T. Nelson, D. A. Rasko, S. Sudek, J. A. Eisen, M. G. Haygood, and J. Ravel. 2005. Patellamide A and C biosynthesis by a microcin-like pathway in *Prochloron didemni*, the cyanobacterial symbiont of *Lissoclinum patella. Proceedings of the National Academy of Sciences of the United States of America* 102:7315–7320.

Sharp, K. H., S. K. Davidson, and M. G. Haygood. 2007. Localization of *Candidatus* Endobugula sertula and the bryostatins throughout the life cycle of the bryozoan *Bugula neritina. The ISME Journal* 1:693–702.

Stassi, D. L., S. J. Kakavas, K. A. Reynolds, G. Gunawardana, S. Swanson, D. Zeidner, M. Jackson, H. Liu, A. Buko, and L. Katz. 1998. Ethyl-substituted erythromycin derivatives produced by directed metabolic engineering. *Proceedings of the National Academy of Sciences of the United States of America* 95:7305–7309.

Sudek, S., N. B. Lopanik, L. E. Waggoner, M. Hildebrand, C. Anderson, H. Liu, A. Patel, D. H. Sherman, and M. G. Haygood. 2007. Identification of the putative bryostatin polyketide synthase gene cluster from *Candidatus* Endobugula sertula, the uncultivated microbial symbiont of the marine bryozoan *Bugula neritina. Journal of Natural Products* 70:67–74.

Sugiyama, N., K. Konoki, and K. Tachibana. 2007. Isolation and characterization of okadaic acid binding proteins from the marine sponge *Halichondria okadai. Biochemistry* 46:11410–114120.

Taylor, M. W., R. Radax, D. Steger, and M. Wagner. 2007. Sponge-associated microorganisms: Evolution, ecology, and biotechnological potential. *Microbiology and Molecular Biology Reviews* 71:295–347.

Unson, M. D. and D. J. Faulkner. 1993. Cyanobacterial symbiont biosynthesis of chlorinated metabolited from *Dysidea herbacea* (Porifera). *Experientia* 49:349–353.

Unson, M. D., N. D. Holland, and D. J. Faulkner. 1994. A brominated secondary metabolite synthesized by the cyanobacterial symbiont of a marine sponge and accumulation of the crystalline metabolite in the sponge tissue. *Marine Biology* 119:1–11.

Wipf, P. and Y. Uto. 2000. Total synthesis and revision of stereochemistry of the marine metabolite trunkamide A. *Journal of Organic Chemistry* 65:1037–1049.

Woollacott, R. M. 1981. Association of bacteria with bryozoan larvae. *Marine Biology* (*New York*) 65:155–158.

6

Is the Vibrio fischeri–Euprymna scolopes *Symbiosis a Defensive Mutualism?*

Eric V. Stabb and Deborah S. Millikan

CONTENTS

6.1 Introduction

The light-organ symbiosis between the bioluminescent bacterium *Vibrio fischeri* and the Hawaiian bobtail squid *Euprymna scolopes* is a fascinating association rife with intriguing biology. This sepiolid squid is a small nocturnal inhabitant of shallow sandy reefs in the Hawaiian archipelago that allows *V. fischeri*, and only this bacterium, to colonize epithelium-lined crypts of a specialized light-emitting organ. This "light organ" is located just ventral to the squid's ink sac in the mantle cavity. Once they have colonized the light organ, the bacterial symbionts emit a bluish light, and adaptations by the host allow it to direct and modulate the emitted light. Below, we will discuss how the light organ's architecture suggests that it functions in the camouflaging behavior referred to as counterillumination (Figure 6.1), wherein marine animals emit light downward, roughly matching the downwelling light from above to obscure their silhouette from animals beneath them in the water column (Clarke, 1963; Dahlgren, 1916; Harper and Case, 1999; Latz, 1996; McFall-Ngai and Morin, 1991; Warner et al., 1979).

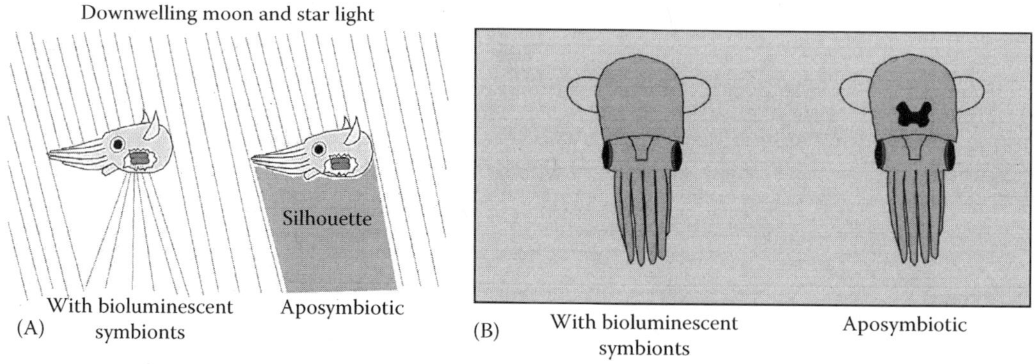

FIGURE 6.1 Proposed counterillumination mechanism of *E. scolopes*. Panel A: It has been proposed that *E. scolopes* with bioluminescent symbiotic *V. fischeri* can emit a controllable, ventrally directed luminescence thereby obscuring their silhouette from a predator beneath them in the water column. Panel B: A cartoon of *E. scolopes* viewed ventrally and back-lit illustrates in an idealized way how the light organ and integral ink sac might be obscured by bioluminescence and how the rest of the animal may have different degrees of translucence.

Research on the *V. fischeri–E. scolopes* symbiosis has gathered momentum, but many big-picture questions remain unanswered. *E. scolopes* was first described by Berry nearly a century ago (Berry, 1912), and detailed studies of bacterial infections in *Euprymna morsei* light organs were provided by Kishitani in the 1920s. Although the original reports are now difficult to find, they are nicely reviewed in E. Newton Harvey's book *Bioluminescence* (Harvey, 1952). While early studies focused on the behavior, ecology, and embryology of *E. scolopes*, in the last 20 years research into the *V. fischeri–E. scolopes* association has increasingly focused on using this symbiosis as a model for mechanistic studies of mutualistic host–bacteria interactions (Stabb, 2006). This focus stems in large measure from experimental tractability. Most notably, this symbiosis can be reconstituted in the laboratory, allowing detailed analyses of how a very specific host–bacterium relationship is initiated. Many exciting discoveries have been made, yet fundamental questions about the ecology of these symbiotic partners and the significance of the bioluminescence resulting from their association remain relatively obscure.

The purpose of this chapter is first to orient the reader to key background information in the field, encompassing some mechanistic studies as well as what is known about the ecology of the partners and the evidence for mutualism in the symbiosis. Then we will describe the premise and basic model of counterillumination. We will next present the evidence for and against counterillumination being the selective advantage of this symbiosis for the host squid, and weigh this evidence along with other hypotheses that seek to explain the ecological significance of light-organ bioluminescence. Finally, we will close by highlighting gaps in our current knowledge and future experimental strategies that promise to yield important new insights.

6.2 Background

6.2.1 Reconstitution of the Symbiosis in the Laboratory Enables Experimental Studies of Early Infection

V. fischeri symbionts are acquired by *E. scolopes* through horizontal transmission, with each new generation of hatchling squid obtaining *V. fischeri* from the environment. Thus, *V. fischeri* and *E. scolopes* exist for some period of time outside the symbiosis, although at least for the host this aposymbiotic stage is brief. The ability to reconstitute this symbiosis experimentally stems from the fact that both partners can be maintained without great difficulty in the laboratory. *V. fischeri* can be cultured readily, and we have observed doubling times of less than 30 min in standard rich media.

When samples from *E. scolopes* light organs are plated onto solid media, *V. fischeri* colonies are recovered with high efficiency, and the light-organ infections are monospecific, meaning that they contain only *V. fischeri* and no other bacteria (Boettcher and Ruby, 1990, 1994; Ruby and McFall-Ngai, 1991). *E. scolopes* can also be kept in the laboratory, where it readily produces eggs and hatchlings. *E. scolopes* was initially bred in aquaria for embryological studies (Arnold et al., 1972), but later the posthatching initiation and development of their light-organ symbiosis was also studied (Montgomery and McFall-Ngai, 1995; Wei and Young, 1989). Eventually, *E. scolopes* was raised through a full life cycle (Claes and Dunlap, 2000; Hanlon et al., 1997).

E. scolopes eggs are symbiont free, and exposing the hatchlings to *V. fischeri* inocula in seawater results in symbiotic infection (Ruby and McFall-Ngai, 1991; Wei and Young, 1989). Arguably, this infection in the laboratory approximates natural horizontal transmission of symbiont. Populations of *V. fischeri* have been examined in *E. scolopes* habitats, so inoculum levels that approximate environmental populations of *V. fischeri* can be used. When fresh hatchlings are placed in nearshore Hawaiian seawater, the onset of luminescence is detectable within 8 h (Ruby and McFall-Ngai, 1991), a time frame that parallels infections in the laboratory. Colonization by small numbers of *V. fischeri* that have not yet begun to bioluminesce, occurs even earlier (Asato, 1993; Ruby and Asato, 1993).

Numerous articles have reviewed the advances in our mechanistic understanding of this symbiotic infection (Geszvain and Visick, 2006; McFall-Ngai, 2000; Nyholm and McFall-Ngai, 2004; Ruby, 1999; Ruby and McFall-Ngai, 1999; Stabb, 2006; Visick, 2005; Visick and McFall-Ngai, 2000; Visick and Ruby, 2006), and we will provide only a brief overview of the infection process. The bilobed *E. scolopes* light organ resides in the mantle cavity, where it is exposed to environmental seawater. Upon hatching, two fields of cilia on the light-organ surface begin beating and shedding mucus (Nyholm et al., 2002). Bacteria in the seawater get stuck in this mucus, thereby concentrating them from the dilute environment (Nyholm et al., 2000). Motility by the bacteria is then required as they swim from aggregates in the mucus to pores on the surface of the light organ (Graf et al., 1994; Millikan and Ruby, 2002, 2004; Nyholm et al., 2000). Symbionts enter the pores and travel through ducts and eventually into epithelium-lined pockets referred to as antechambers and crypts (Sycuro et al., 2006). Although many *V. fischeri* cells can be concentrated in aggregates outside the light organ, relatively few cells appear to survive this journey and initiate colonization (Dunn et al., 2006; McCann et al., 2003; Ruby and Asato, 1993).

The initial *V. fischeri* colonists proliferate, and both the increase in their cell density and some environmental cue in the host trigger them to begin producing bioluminescence (Boettcher and Ruby, 1990; Bose et al., 2007). *V. fischeri* mutants that do not produce bioluminescence are attenuated in colonization (Bose et al., 2008; Visick et al., 2000), suggesting that the host either detects dark infections and sanctions them or generates an environment in the light organ in which bioluminescence is physiologically advantageous for the symbionts. This underscores the importance of bioluminescence in the symbiosis and can be interpreted as an indication that the host limits the success of dark "cheaters" among the symbionts.

6.2.2 Ecology and Behavior of *V. fischeri*

V. fischeri has been isolated from a variety of marine environments, and is perhaps best known for colonizing light-emitting organs of specific fish and squid. The "bobtail" squid that serve as hosts for *V. fischeri* are in the genera *Sepiola* or *Euprymna*, which are found in the Mediterranean Sea and Pacific Ocean, respectively (Fidopiastis et al., 1998; Nishiguchi, 2002; Nishiguchi et al., 1998). *V. fischeri* also colonizes the light organs of monocentrid "pinecone" fish (Fitzgerald, 1977; Ruby and Nealson, 1976). However, *V. fischeri* is not restricted to monospecific associations and has been isolated from the gut consortia of fish in nonspecific symbioses (Makemson and Hermosa, 1999; Ruby and Morin, 1979; Sugita and Ito, 2006). Moreover, *V. fischeri* is found free living in the water column and in sediments (Lee and Ruby, 1994b; Ruby et al., 1980). Numerically, it is not particularly abundant in these environs, but *V. fischeri* is widespread and occurs in habitats where it has no known light-organ hosts. The metabolic and genomic flexibility of *V. fischeri* suggest that in contrast to obligate symbionts this bacterium probably has an important environmental lifestyle (Ochman and Moran, 2001; Ruby et al., 2005).

To adjust to the host and other environments, *V. fischeri* possesses an impressive array of regulatory systems (Geszvain and Visick, 2006; Hastings and Greenberg, 1999). *V. fischeri* is well known for a group behavior known as "quorum sensing," wherein a diffusible pheromone mediates changes in gene expression when populations reach high cell density (Hastings and Greenberg, 1999). Quorum sensing and the redox-responsive ArcA/ArcB regulatory system tune the luminescence of *V. fischeri* to express maximally when cells are crowded, yet in relatively oxidative conditions (Bose et al., 2007). Because regulation of luminescence is governed in part by a diffusible pheromone, the potential exists for a subpopulation of *V. fischeri* to incite a group decision to luminescence in the light organ. Understanding the basis for the behavior of *V. fischeri* in the light organ should help elucidate the role of luminescence for the bacterium (Stabb, 2005) and may reveal mechanisms by which the host manipulates the symbionts' luminescence.

6.2.3 Ecology and Behavior of *E. scolopes*

E. scolopes is a solitary nocturnal predator that feeds mainly on shrimps and polychaetes, and grows from hatchlings just a few millimeter in length to thumb-sized adults (Moynihan, 1983; Shears, 1988). The adults probably live less than a year (Hanlon et al., 1997; Singley, 1983). Females lay clutches of tens or hundreds of eggs, and as mentioned above, each new generation must acquire *V. fischeri* from the surrounding marine environment (Wei and Young, 1989), although infection is so rapid that no uninfected *E. scolopes* individual has ever been found.

Most observations of *E. scolopes* are made in shallow sandy reef areas, but it is unclear how confined it is to such habitats. We, and others, typically use flashlights and nets to collect animals at night by wading in knee-deep water. It is not uncommon to find animals less than a meter from shore, sometimes in water no more than 10 cm deep. However, there are reports of *E. scolopes* being collected offshore near the surface (R. Young, personal communication) and even at depths around 200 m (Berry, 1912). So while most observations of *E. scolopes* are in nearshore shallow water, it is not clear the extent to which this reflects the fact that this is where researchers tend to look for them. Similarly, we tend to find *E. scolopes* adults on the sandy bottom, not up in the water column; however, almost certainly the animals are aware of our presence before we spot them, so their behavior may be perturbed. In aquaria, *E. scolopes* spend most of their time on the bottom (Moynihan, 1983), and we tend to assume that this is their habit; however, it would be most accurate to say that little is known about their natural behavior.

Camouflaging seems to be a general strategy for *E. scolopes* (Anderson and Mather, 1996; Shears, 1988), which routinely buries or coats itself with sand and uses chromatophores to change colors from a palette that is similar to the background (Figure 6.2). The squid can even be observed swimming with a sand coat, which they can apparently discard with remarkable speed and control (Shears, 1988). Even without the sand coat, their coloration blends in well with the sandy reef bottoms, as anyone who has collected these animals can attest. The animals also sometimes emit ink blobs and then jet away, when threatened. However, in our observations and those of Anderson and Mather (1996), the animals rarely jet very far from the site of an encounter.

Little is known about which predators *E. scolopes* may be hiding from. Hawaiian monk seals occasionally feed on *E. scolopes* (Goodman-Lowe, 1998), and we have observed lizard fish struggling with a catch of *E. scolopes*; however, we have a poor understanding of the threats it faces. Also lacking are observations of luminescence by these squid in their natural habitat. Every wild-caught *E. scolopes* tested has been found to emit bioluminescence; however, this was determined upon placing the animals in a sensitive luminometer and not by observations in the wild. There are reports that the animals "flash" their luminescence when disturbed (Moynihan, 1983; A. Wier and M.J. McFall-Ngai personal communication; R. Young, personal communication), but this has not been documented in a natural setting.

6.2.4 Evidence of Mutualism in this Symbiosis

The selective advantage of the symbiosis for *V. fischeri* appears clear. *V. fischeri* is provided a privileged niche in the *E. scolopes* light organ, where it grows rapidly (Ruby and Asato, 1993). Not only are they provided nutrients to support this growth, but because the squid can prevent infection by other

FIGURE 6.2 An adult *E. scolopes*. An adult Hawaiian bobtail squid sits on the coral sand bottom of an aquarium. The mottled appearance is due to controllable chromatophores that are expanded to produce a relatively dark pattern of reddish and yellowish brown pigments in this picture. Chromatophores can also be retracted to small points, rendering the mantle largely translucent. The mantle length of this typical adult specimen is ~2.5–3 cm. (Photo credit of Jeffrey L. Bose).

microorganisms, *V. fischeri* cells would appear to have the host immune system effectively protecting them from predatory grazing or competitive antibiosis by other microbes. Moreover, colonization of the *E. scolopes* light organ is not a dead end. Each morning these squid settle to the bottom, cover themselves with sand, and the light of dawn triggers the expulsion of most *V. fischeri* cells into the environment (Boettcher et al., 1996a). The *V. fischeri* left behind grow to repopulate the light organ by dusk. Presumably, because of this diurnal "venting" process, *V. fischeri* populations are relatively high in habitats occupied by *E. scolopes* (Lee and Ruby, 1994b). Further ecological studies (Lee and Ruby, 1994a) support the theory that in the Hawaiian reefs occupied by *E. scolopes*, the ability to colonize this host is a strong selective force on *V. fischeri*. So, while *V. fischeri* can be found in many marine environments, its populations appear to be enhanced by *E. scolopes*.

Presumably, the host also benefits from the association, and the *V. fischeri–E. scolopes* symbiosis is a mutualism. The *E. scolopes* light organ grows and maintains a monospecific culture of *V. fischeri* and has the means to control the release of its bioluminescence. Moreover, we (and others) have observed that stressed animals appear to clear their light organs of *V. fischeri* symbionts altogether, suggesting that the host has ultimate control and is not suffering an unavoidable infection. The symbiosis probably provides a benefit to *E. scolopes* by giving it a controllable source of bioluminescence for some behavior, and this is usually attributed to a camouflaging "counterillumination" behavior mentioned above and discussed at greater length below. Although an additional nutritional or other benefit of the symbionts cannot be ruled out, *E. scolopes* raised through a complete life cycle in the absence of *V. fischeri* did not appear to suffer due to lack of symbionts (Claes and Dunlap, 2000).

6.3 Counterillumination Model

Strategies to avoid predation are numerous and varied in the marine environment and can involve highly specialized adaptations. Familiar examples include animals with the ability to camouflage themselves through pigmentation or shape that mimics the surrounding substrate, such as rock or kelp beds, or the ability to bury into sand or hide among rocks. However, in the open ocean, the ability to "hide" from predators is difficult, given the lack of substrate. In spite of this difficulty, many mesopelagic organisms including luminous fishes, crustaceans, and cephalopods have found a way to camouflage themselves by producing light in discrete regions of their body (Clarke, 1963). This use of light is called counterillumination, which is most often used to describe the ability to match downwelling light using ventrally directed luminescence (Dahlgren, 1916; Harper and Case, 1999; Latz, 1996; McFall-Ngai and Morin, 1991; Warner et al., 1979; Young, 1977). In this way, light production along the ventral surface of an organism can be used to hide the animal's silhouette if viewed from below in the water column (e.g., Figure 6.1).

Downwelling light in the mesopelagic zone is unidirectional and predictable in its wavelength and low intensity, but in shallow water light is more variable both in quantity and quality. Moreover, light in shallow water is multidirectional owing to reflection, shadows, and surface effects. This potentially renders counterillumination a more difficult strategy for shallow-water animals to use effectively. However, McFall-Ngai and Morin provided convincing evidence that counterillumination behavior does occur in shallow-water leiognathid fishes, although they suggest that "disruptive illumination" might be a better description of the mottled appearance that the ventral bioluminescence imparts on the animals (McFall-Ngai and Morin, 1991).

The concept of "disruptive illumination" is an important distinction from most discussions of counterillumination, and it may be especially relevant to the role of the *E. scolopes* light organ. However, for most part we will use "counterillumination" to encompass partial or complete illumination, disruptive illumination, or intermittent countershading, with the commonality to all of these being the ability of the animal to produce light that mimics their surroundings to whatever extent, in order to camouflage themselves. One exception is that of the cookie cutter shark discussed next, where camouflage is not the apparent role of counterillumination. In general though we will use "counterillumination" in reference to hiding behaviors, and we will sometimes use "camouflaging" in conjunction with "counterillumination" to reinforce this distinction.

Several factors may contribute to the effectiveness of counterillumination. It has been argued that animals should have many evenly distributed light sources, because those with fewer or less-distributed sources of bioluminescence would be less able to produce an even field of light, minimizing or negating any camouflaging effect of counterillumination (Johnsen et al., 2004). Consistent with this argument, counterillumination is usually attributed to an array of ventrally directed photophores. In *Histioteuthis* squid, for example, many small light organs are scattered over the ventral surface of the body, head, and arms. However, the majority of shallow-water luminous animals, including *E. scolopes*, have a single light-emitting organ. It has also been pointed out that effective counterillumination should mimic both the quantity and the quality (e.g., wavelength) of downwelling light (Johnsen et al., 2004; McFall-Ngai and Morin, 1991). Otherwise, luminescence could have the opposite effect and become an attractant for predators. Finally, the habits and visual acuity of the predator being hidden from are important.

6.4 Weighing the Evidence for Counterillumination and Other Models

6.4.1 Does Light-Organ Anatomy Suggest Counterillumination?

The idea that the *E. scolopes* light-organ functions in counterillumination gathered momentum from close examination of its architecture (McFall-Ngai and Montgomery, 1990). The light organ is situated in the mantle cavity just ventral to the bulk of the ink sac, and it is oriented such that light is emitted ventrally. The light-organ crypts occupied by bioluminescent symbionts are situated between a reflective layer of reflectin protein (Crookes et al., 2004) and a muscle-derived lens (Montgomery and McFall-Ngai, 1992), which together appear to direct and control ventral light emission. In the adult animals, the ink sac apparently can be drawn around the light organ, perhaps in the manner of a shutter, and anatomical observations of several fixed specimens suggest that controlled movement of reflective tissue and ink sac diverticula could modulate the emission of light (McFall-Ngai and Montgomery, 1990). The animals may also be able to control luminescence by modulating the oxygenation of the light organ (Boettcher et al., 1996b). This could effectively control luminescence because oxygen is one of the reactants

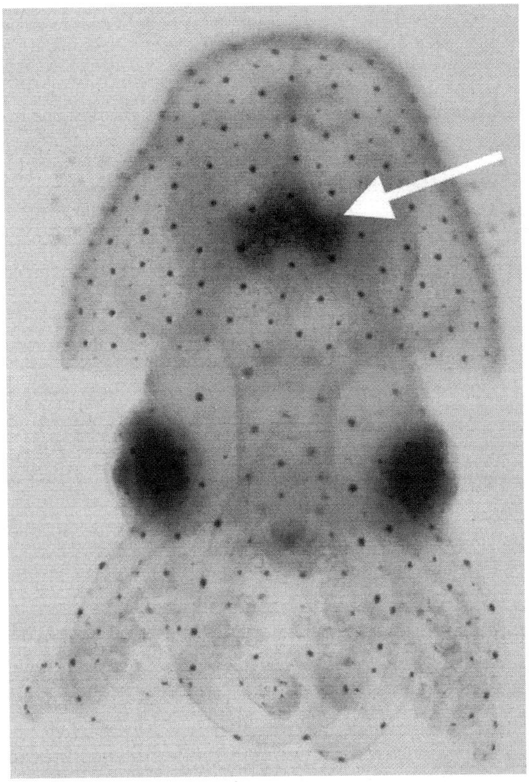

FIGURE 6.3 Backlit juvenile *E. scolopes*. This animal has retracted its chromatophores to small points rendering its mantle largely translucent. A white arrow points to the light organ and ink sac. The other prominent dark (opaque) shaded structures are the eyes. The mantle length of this typical hatchling juvenile is ~1–1.5 mm. (Photo credit of Dawn M. Adin).

of the *V. fischeri* luciferase reaction. Thus, the *E. scolopes* light organ has elements consistent with counterillumination, specifically a controllable ventrally emitted light.

In at least one respect, the architecture of the light organ seems imperfectly adapted for camouflaging by counterillumination. The *E. scolopes* light organ does not cover the entire underside of the animal the way multiple photophores of some mesopelagic squid do. This would arguably represent a poor or incomplete counterillumination apparatus. On the other hand, symbiotic luminescence would provide counterilluminating cover for a large opaque structure in the squid, the ink sac integral to the light organ (this is illustrated with a juvenile in Figure 6.3), which might provide a selective advantage.

6.4.2 Does Host Behavior Fit the Counterillumination Hypothesis?

It is difficult to say whether *E. scolopes* behavior supports the counterillumination hypothesis simply because very little is known about this animal's behavior. As discussed in the final section, answering many questions about the behavior and ecology of *E. scolopes* would help put light-organ function in a natural context and inform our discussion of its potential role(s). At least it seems clear that *E. scolopes* is nocturnal and spends at least some time in the water column, which is consistent with counterillumination.

One mysterious behavior of *E. scolopes* deserves special mention. Multiple investigators have seen *E. scolopes* "flash" bioluminescence (Moynihan, 1983; A. Wier and M.J. McFall-Ngai personal communication; R. Young, personal communication). Typically, in situations when these animals were severely disturbed or roughly handled, they often released ink. This suggests that flashes might be an attempt to distract or confuse predators that get too close, perhaps as part of an escape response. Obviously, visible flashes are inconsistent with camouflaging, and it may be that flashes reflect the real role of the light organ or that the light organ has two very distinct roles.

6.4.3 How Conclusive Are Attempts to Measure Counterillumination in *E. scolopes*?

Jones and Nishiguchi (2004) have provided the most important and convincing data for evaluating counterillumination in the *V. fischeri–E. scolopes* symbiosis. They placed adult *E. scolopes* in small chambers and plotted the amount of light emitted from the animals, as a function of the intensity of the downwelling light. To differentiate the room lights from animal bioluminescence, the overhead lights were turned off and animal luminescence was measured immediately thereafter. Controls that included chambers without animals ensured that phosphorescence of the seawater, which might be expected to correlate with overhead light intensity, was not mistaken for bioluminescence. As predicted by the counterillumination hypothesis, bioluminescence released from animals was more intense as the intensity of the overhead illumination was increased (Jones and Nishiguchi, 2004). An exception to this was observed at the highest light intensities, when animals began to dim the luminescence they emitted. Although not predicted by the counterillumination hypothesis, it seems plausible that at uncommonly perhaps artificially high light intensities the animals would not be able to match the downwelling light and would abandon counterillumination behavior. More importantly, this observation helps to validate the experimental setup, because it is unclear how any nonbiological explanation for the data could yield this result at high light intensity.

Two results of the Jones and Nishiguchi study hint at either the imperfection of counterillumination or the technical difficulty of such experiments. First, although luminescence emitted by the animals correlated with the intensity of downwelling light, the animal luminescence appeared dimmer than that of the downwelling light (Jones and Nishiguchi, 2004). However, even imperfect matching of the downwelling light (imperfect counterillumination) might have a selective advantage over no counterillumination at all. Also, it is important to remember that animal luminescence was measured after the lights were turned off, but presumably before the squid could react and alter the luminescence. Therefore, an unexpectedly quick reaction by the squid could explain this aspect of the data. This brings up a second result not predicted by the counterillumination hypothesis; sometimes animals were remarkably slow in dimming their luminescence after the lights were turned out. In such animals, luminescence continued for a minute or more after the lights were out. This seems to be a counterproductive behavior for an animal trying to be stealthy, but then again the animals may never (or rarely) encounter such a rapid and absolute change in light intensity, in nature.

Overall, the work of Jones and Nishiguchi strongly supports the counterillumination hypothesis; however, skeptics can still point to an obvious gap in our observations. Specifically, to our knowledge nobody has photodocumented counterillumination by *E. scolopes*, either in the laboratory or in nature. At least in the laboratory this might seem easy, by simply putting *E. scolopes* in a clear-bottom container and viewing it from underneath with dim diffuse light overhead. In our experience, however, animals put in this situation are prone to simply sitting on the bottom, with their arms curled under them and their chromatophores darkened. Presumably, counterillumination would be most useful when the animals are up in the water column, and as discussed below future work aimed at viewing animals behaving in this way could provide important information.

It is also worth noting that to our knowledge the wavelength of light emitted from *E. scolopes* has not been compared to the background light in its habitat(s). This is important, because for counterillumination to be an effective camouflage it must presumably match the background both in intensity (as Jones and Nishiguchi measured) and in quality. Moynihan (1983) reported that flashes seen from disturbed squid appeared "green," although we have observed the expected bluish light emitted by bacteria in *E. scolopes* juveniles (Stabb, 2005). It seems unlikely that a green light would effectively match the backgrounds we are familiar with in the shallow sandy reefs of Hawaii; however, measurements of the emitted light and environmental light are needed to resolve the issue.

6.4.4 Counterillumination Is Not Necessarily a "Defensive" Camouflage

If counterillumination is used as camouflage by *E. scolopes*, this might reflect an offensive, rather than defensive, strategy. *E. scolopes* might use counterillumination to hide from potential prey. However, at least in aquaria we have not observed *E. scolopes* approach the prey from above. Rather they seem to attack horizontally with their tentacles to capture shrimps. It is also uncertain whether the prey that *E. scolopes* feeds on would be easier to catch if the squid possesses a ventral counterillumination mechanism; however, with so much unknown about *E. scolopes* behavior (e.g., its feeding habits in deeper water), it is impossible to rule out camouflaging counterillumination as an offensive strategy.

6.4.5 Alternatives (or Amendments) to the Counterillumination Camouflage Hypothesis

If *E. scolopes* does not use bioluminescence in a camouflaging counterillumination behavior, or if this is not the primary selective advantage of symbiotic bioluminescence, then what might be the main role of the light organ? Proposing alternate answers to this question may prove to be a useful exercise, both by preventing tunnel-visioned interpretation of existing data and by pointing future research in new directions. Below, we discuss a few possible uses for the light organ other than as a silhouette-obscuring camouflage.

E. scolopes might use counterillumination as a prey attractant. Widder presented evidence that the cookie cutter shark, *Isistius brasiliensis*, uses counterillumination not to obscure its entire ventral surface, but rather to obscure all but an image shaped like a smaller fish (Widder, 1998). She convincingly argues that this might be used as a lure to attract predators of small fish, including the swordfish, tunas, and porpoises that *I. brasiliensis* attacks. It is not clear to us whether a distinct luring image would be produced by the shape of the *E. scolopes* light organ, but we cannot rule this out. Notably, this model predicts the same sort of correlation between ambient light and emitted luminescence that was observed by Jones and Nishiguchi (2004).

Alternatively, luminescence *per se*, and not the counterillumination of a particular alluring shape, might be used to attract prey. This would seem to be a risky proposition for *E. scolopes* hunting in the water column. *E. scolopes* may have defensive strategies other than avoiding detection, but these are not immediately apparent (e.g., it lacks spines or known predator-deterring toxins), making stealth an important asset for *E. scolopes*. On the other hand, it seems plausible that the shrimp or polychaetes that *E. scolopes* feeds on might display phototaxis, and with so little knowledge about the behavior and ecology of *E. scolopes* we cannot rule out the possibility that *E. scolopes* exploits such a behavior of its prey.

Perhaps, the most plausible alternative model for the role of the light organ center is its potential to flash brightly. As mentioned above, such flashes have been observed when animals were collected or held in captivity (Moynihan, 1983; A. Wier and M.J. McFall-Ngai personal communication; R. Young personal communication). The ability to flash visible luminescence could suggest a role in intraspecies communication; however, the fact that it has been observed when the animals are disturbed (e.g., physically grabbed) suggests a role as a startling tactic (Herring, 1977). Another sepiolid squid in Hawaii, *Heteroteuthis hawaiiensis*, found in open ocean waters at depths of up to 600 m, can use its light organ to create a glowing cloud of bacteria mixed with mucus (Dilly and Herring, 1978; Young, 1995; Young and Roper, 1976). If *E. scolopes* can emit a brief flash of light with or without expelling bacterial symbionts, this might be used to startle or confuse the predators. Like the expulsion of ink blobs by *E. scolopes* (Anderson and Mather, 1996), this behavior could be used once camouflage has failed.

The bioluminescent defense response of *H. hawaiiensis* may be used in addition to the likely use of its light organ (and multiple light organs on its arms and body) to camouflage its silhouette during the low-level light of dusk and dawn when it migrates to shallower waters (150–200 m) to feed (Dilly and Herring, 1978; Young, 1995; Young and Roper, 1976). This underscores the point that *E. scolopes* may use its light organ for dual roles, and both counterillumination and another behavior may have selective advantages. Or from another perspective, a flashing behavior might be the selective force driving *E. scolopes* light-organ evolution, but this might come at the cost of a relatively large opaque structure that would make the squid more visible to predators, but by emitting a dim counterilluminating light when the flashing behavior is not in use *E. scolopes* might ameliorate the disadvantage of having a light organ. Again, such a scenario would be consistent with the data. Jones and Nishiguchi (2004) reported data supporting counterillumination, and yet counterillumination may not be the primary purpose for the *E. scolopes* light organ.

6.5 Future Directions

Many big-picture questions regarding the *V. fischeri–E. scolopes* symbiosis remain unanswered. Notably, as discussed in this chapter, the functional significance of the light organ for the host is uncertain, although evidence points to a role in camouflaging by counterillumination. Future behavioral and ecological research on *E. scolopes* would be welcome and could shed light on one of the most central issues of the symbiosis.

6.5.1 The Life and Times of *E. scolopes*

Many fundamental questions about the ecology and behavior of *E. scolopes* remain opaque. How much of their life is spent in shallow water? Do they often venture deeper? Can reports of *E. scolopes* in the mesopelagic zone be substantiated? What are they feeding on in such habitats? How much time do they spend in the water column or on the bottom? What is the quality of the light in their environs? Does this match the light they emit? Do they hunt exclusively in the water column? What are their main predators? What are the hunting techniques and visual acuity of these predators? Answering these questions may not be easy. Observing the natural habits of these nocturnal well-camouflaged animals has obvious barriers. For example, it will be difficult for divers to follow the nighttime behavior of these animals without disrupting their routine. Remote tagging and tracking devices are probably not feasible, although the advent of nanotechnologies and miniaturization could facilitate such approaches.

One technology that could provide a powerful tool for studying *E. scolopes* in the environment is the use of autonomous underwater vehicles (AUVs). Fidopiastis and Clark (personal communication) have proposed that a robot with the capacity to swim and crawl might be trained to follow *E. scolopes* individuals, monitoring and recording their behavior. This promises to be much less disturbing to the animals than observations by humans, and could yield unparalleled insights into the behavior of *E. scolopes* over a range of environments. Remotely operated vehicles have been used to monitor marine animals such as jellyfish, and AUV technology has been applied to tracking bass (Zhou et al., 2007). Moreover, an AUV was used to measure patterns of marine bioluminescence (Blackwell et al., 2002). If an AUV can be trained to track *E. scolopes* individuals and potentially to monitor their luminescence, this has great potential for answering many unresolved questions about their behavior.

6.5.2 What about the Flashes?

The fact that *E. scolopes* occasionally emits bright flashes may belie the true function of the light organ; however, observations of this behavior are rare, and to our knowledge it has only been observed in disturbed captive animals. A squid-tracking AUV might be able to document flashes in the wild, and to determine the behavior of the animal before and after the flash. Is it used when the animals are threatened by a predator? Is it used in conjunction with the release of ink? Do the animals change their swimming direction in an evasive manner following a flash? Do the flashes have a temporary blinding or distracting effect on a would-be predator? Or are flashes used as intraspecies signals perhaps eliciting some altogether unknown behavior? Any observation of light-organ flashing in the wild would be tremendously important.

Observations of flashing in the laboratory might also provide insight into the function of the light organ for *E. scolopes*. Because the flashes seem to be rare, it may be useful to set up a digital image recording system that monitors the animals constantly, detects flashes, and then saves the data before and after a flash event. A clear-bottom tank could be used with a camera (or array of cameras) so that observations could be made of the animal's ventral surface, although flashes should be detectable regardless of how the animals direct them. To test whether flashes are used in defensive responses or communication, it will probably be necessary to place multiple squid or predators in the same tank, which could require a relatively large setup. Although such an experiment is not without obstacles, it could yield important data and might be easier than observations, in nature.

6.5.3 More Tests of Counterillumination

One conspicuously missing piece of evidence for counterillumination by *E. scolopes* is the lack of an image documenting this effect. A squid-tracking AUV might capture such an image in the environment, which would provide strong support for this theory. However, carefully crafted laboratory studies might provide similar and compelling evidence. As noted above, animals observed in clear-bottomed containers tend to simply sit on the bottom. A clear-bottomed raceway with water flowing through it (e.g., a clear pipe) might force the animals off the bottom, allowing visualization from underneath the swimming *E. scolopes*. Alternatively, a useful setup might include a camera underneath a clear-bottomed container set to capture several hours worth of images, or to be triggered by the motion of the animals. In the absence of a human investigator, the animals may eventually leave the bottom (e.g., to capture prey or mate) and could then be visualized. Sensitive digital cameras capable of capturing low-light images could document the squid's silhouette and whether or not it is obscured by counterillumination.

A more advanced apparatus for measuring counterillumination would also be useful, if counterillumination can be elicited by animals going about relatively normal behaviors in the laboratory. One such methodology that has been employed productively is to use an overhead light that provides illumination that is seen as constant the animals, by that is actually "chopped" into short bursts of light, such that a photomulitiplier detector set "out of phase" with the incident light will measure only the counterillumination (Latz and Case, 1992). In this way, quantitative measurements of counterillumination can be made on animals that are not forced into cramped confines as they were in the apparatus used by Jones and Nishiguchi (2004), and it may provide a telling picture of natural counterillumination behavior. For example, the apparent under-illumination by animals observed by Jones and Nishiguchi may simply reflect a perturbation of the animals associated with putting them in the experimental apparatus.

The ultimate proof that camouflaging provides an antipredatory selective advantage would be an ecological prey study in which *E. scolopes* with luminescent symbiotic bacteria and animals raised with nonluminescent symbionts or no symbionts are presented to an appropriate predator and in an environment that mimics their natural one. Such experiments are obviously difficult to set up; however, such an approach was used to demonstrate an antipredatory effect of counterillumination in the fish *Porichthys notatus* (Harper and Case, 1999). A potential predator in these experiments could be the lizardfish of the Synodontidae family, which we suspect may be an important natural predator of *E. scolopes*, although as noted above more research needs to be done in this regard. Such experiments would need to be done with pools of animals isolated from one another, so that cross contamination of luminescent bacteria to nonluminous animals cannot occur. The animals would also have to be monitored closely for potential flashing behavior, to distinguish their possible antipredatory value from that of counterillumination. If done convincingly, such experiments could settle the issue of the selective advantage of symbiotic bioluminescence in the *V. fischeri–E. scolopes* mutualism.

ACKNOWLEDGMENTS

The authors are especially indebted to Margaret McFall-Ngai and Richard Young for reading drafts of this chapter and providing helpful suggestions. We also thank Dawn Adin and Jeffrey Bose for the photographs of *E. scolopes*, Bryan W. Jones, Michele Nishiguchi, and Pat Fidopiastis for helpful conversations, and numerous colleagues who have provided healthy skepticism of counterillumination in the *V. fischeri–E. scolopes* symbiosis. EVS was supported in this work by a grant from the National Science Foundation (CAREER MCB-0347317).

REFERENCES

Anderson, R. C. and J. A. Mather. 1996. Escape responses of *Euprymna scolopes* Berry, 1911 (Cephalopoda: Sepiolidae). *Journal of Molluscan Studies* 62:543–545.

Arnold, J., Singly, C., and L. Williams-Arnold. 1972. Embryonic development and post-hatching survival of the sepiolid squid *Euprymna scolopes* under laboratory conditions. *Verliger* 14:361–364.

Asato, L. M. 1993. Morphological and physiological changes in *Vibrio fischeri* during the initiation and release from the mutualistic association with its sepiolid host, *Euprymna scolopes*. Masters thesis, University of Southern California.

Berry, S. S. 1912. The Cephalopoda of the Hawaiian islands. In *Bulletin of the United States Bureau of Fisheries*. Government Printing Office, Washington, pp. 255–362.

Blackwell, S., Case, J., Glenn, S., Kohut, J., Moline, M. A., Purcell, M., Schofield, O., and C. VonAlt. 2002. A new AUV platform for studying near shore bioluminescence structure. In *Proceedings of the 12th International Symposium on Bioluminescence and Chemiluminescence*, eds. P. Herring, L. J. Kricka, and Stanley, P. E., pp. 197–200. London: World Scientific.

Boettcher, K. J. and E. G. Ruby. 1990. Depressed light emission by symbiotic *Vibrio fischeri* of the sepiolid squid *Euprymna scolopes. Journal of Bacteriology* 172:3701–3706.

Boettcher, K. J. and E. G. Ruby. 1994. Occurrence of plasmid DNA in the sepiolid squid symbiont *Vibrio fischeri. Current Microbiology* 29:279–286.

Boettcher, K. J., Ruby, E. G., and M. J. McFall-Ngai. 1996a. Bioluminescence in the symbiotic squid *Euprymna scolopes* is controlled by a daily biological rhythm. *Journal of Comparative Physiology* 179:65–73.

Boettcher, K. J., Ruby, E. G., and M. J. McFall-Ngai. 1996b. Bioluminescence in the symbiotic squid *Euprymna scolopes* is controlled by a daily biological rhythm. *Journal of Comparative Physiology A* 179:65–73.

Bose, J. L., Kim, U., Bartkowski, W., Gunsalus, R. P., Overley, A. M., Lyell, N. L., Visick, K. L., and E. V. Stabb. 2007. Bioluminescence in *Vibrio fischeri* is controlled by the redox-responsive regulator ArcA. *Molecular Microbiology* 65:538–553.

Bose, J. L., Rosenberg, C. S., and E. V. Stabb. 2008. Effects of *luxCDABEG* induction in *Vibrio fischeri*: Enhancement of symbiotic colonization and conditional attenuation of growth in culture. *Archives of Microbiology* 190:169–183.

Claes, M. F. and P. V. Dunlap. 2000. Aposymbiotic culture of the sepiolid squid *Euprymna scolopes*: Role of the symbiotic bacterium *Vibrio fischeri* in host animal growth, development, and light organ morphogenesis. *Journal of Experimental Botany* 286:280–296.

Clarke, W. D. 1963. Function of bioluminescence in mesopelagic organisms. *Nature* 198:1244–1246.

Crookes, W. J., Ding, L. L., Huang, Q. L., Kimbell, J. R., Horwitz, J., and M. J. McFall-Ngai. 2004. Reflectins: The unusual proteins of squid reflective tissues. *Science* 303:235–238.

Dahlgren, U. 1916. Production of light by animals. *Journal of the Franklin Institute* 181:525–556.

Dilly, P. N. and P. Herring. 1978. The light organ and ink sac of *Heteroteuthis dispar. Journal of Zoology, London* 186:47–59.

Dunn, A. K., Millikan, D. S., Adin, D. M., Bose, J. L., and E. V. Stabb. 2006. New *rfp*- and pES213-derived tools for analyzing symbiotic *Vibrio fischeri* reveal patterns of infection and *lux* expression *in situ. Applied Environmental Microbiology* 72:802–810.

Fidopiastis, P. M., von Boletzky, S., and E. G. Ruby. 1998. A new niche for *Vibrio logei*, the predominant light organ symbiont of squids in the genus *Sepiola. Journal of Bacteriology* 180:59–64.

Fitzgerald, J. M. 1977. Classification of luminous bacteria from the light organ of the Australian pinecone fish, *Cleidopus gloriamaris. Archives of Microbiology* 112:153–156.

Geszvain, K. and K. L. Visick. 2006. Roles of bacterial regulators in the symbiosis between *Vibrio fischeri* and *Euprymna scolopes. Progress in Molecular and Subcellular Biology* 41:277–290.

Goodman-Lowe, G. D. 1998. Diet of the Hawaiian monk seal (*Monachus schauinslandi*) from the northwester Hawaiian islands during 1991 to 1994. *Marine Biology* 132:535–546.

Graf, J., Dunlap, P. V., and E. G. Ruby. 1994. Effect of transposon-induced motility mutations on colonization of the host light organ by *Vibrio fischeri. Journal of Bacteriology* 176:6986–6991.

Hanlon, R. T., Claes, M. F., Ashcraft, S. E., and P. V. Dunlap. 1997. Laboratory culture of the sepiolid squid *Euprymna scolopes*: A model system for bacteria-animal symbiosis. *Biological Bulletin* 192:364–374.

Harper, R. D. and J. F. Case. 1999. Counterillumination and its antipredatory value in the plainfin midshipman fish *Porichthys notatus. Marine Biology* 134:529–540.

Harvey, E. N. 1952. *Bioluminescence*. Academic Press, New York.

Hastings, J. W. and E. P. Greenberg. 1999. Quorum sensing: The explanation of a curious phenomenon reveals a common characteristic of bacteria. *Journal of Bacteriology* 181:2667–2668.

Herring, P. 1977. Luminescence in cephalopods and fish. *Symposia of the Zoological Society of London* 38:127–159.

Johnsen, S., Widder, E. A., and C. D. Mobley. 2004. Propagation and perception of bioluminescence: Factors affecting counterillumination as a cryptic strategy. *Biological Bulletin* 207:1–16.

Jones, B. W. and M. K. Nishiguchi. 2004. Counterillumination in the Hawaiian bobtail squid, *Euprymna scolopes* Berry (Mollusca: Cephalopoda). *Marine Biology* 144:1151–1155.

Latz, M. I. 1996. Physiological mechanisms in the control of bioluminescent countershading in a midwater shrimp. In *Zooplankton: Sensory Ecology and Physiology*, eds. P. H. Lenz, D. K. Hartline, J. Purcell, and MacMillan, D. L., pp. 163–174. Amsterdam: Gordon and Breach.

Latz, M. I. and J. F. Case. 1992. Slow photonic and chemical induction of bioluminescence in the midwater shrimp, *Sergestes similis* Hansen. *Biological Bulletin* 182:391–400.

Lee, K-H. and E. G. Ruby. 1994a. Competition between *Vibrio fischeri* strains during initiation and maintenance of a light organ symbiosis. *Journal of Bacteriology* 176:1985–1991.

Lee, K-H. and E. G. Ruby. 1994b. Effect of the squid host on the abundance and distribution of symbiotic *Vibrio fischeri* in nature. *Applied Environmental Microbiology* 60:1565–1571.

Makemson, J. C. and G. V. Hermosa. Jr. 1999. Luminous bacteria cultured from fish guts in the Gulf of Oman. *Luminescence* 14:161–168.

McCann, J., Stabb, E. V., Millikan, D. S., and E. G. Ruby. 2003. Population dynamics of *Vibrio fischeri* during infection of *Euprymna scolopes. Applied Environmental Microbiology* 69:5928–5934.

McFall-Ngai, M. and J. G. Morin. 1991. Camouflage by disruptive illumination in Leiognathids, a family of shallow-water, bioluminescencent fishes. *Journal of Experimental Biology* 156:119–137.

McFall-Ngai, M. J. 2000. Negotiations between animals and bacteria: The "diplomacy" of the squid-*Vibrio* symbiosis. *Comparative Biochemistry and Physiology. Part A, Molecular and Integrative Physiology* 126:471–480.

McFall-Ngai, M. J. and M. K. Montgomery. 1990. The anatomy and morphology of the adult bacterial light organ of *Euprymna scolopes* Berry (Cephalopoda: Sepiolidae). *Biological Bulletin* 179:332–339.

Millikan, D. S. and E. G. Ruby. 2002. Alterations in *Vibrio fischeri* motility correlate with a delay in symbiosis initiation and are associated with additional symbiotic colonization defects. *Applied Environmental Microbiology* 68:2519–2528.

Millikan, D. S. and E. G. Ruby. 2004. *Vibrio fischeri* flagellin A is essential for normal motility and for symbiotic competence during initial squid light organ colonization. *Journal of Bacteriology* 186:4315–4325.

Montgomery, M. K. and M. J. McFall-Ngai. 1992. The muscle-derived lens of a squid bioluminescent organ is biochemically convergent with the ocular lens. *The Journal of Biological Chemistry* 267:20999–21003.

Montgomery, M. K. and M. J. McFall-Ngai. 1995. The inductive role of bacterial symbionts in the morphogenesis of a squid light organ. *American Zoologist* 35:372–380.

Moynihan, M. 1983. Notes on the behavior of *Euprymna scolopes* (Cephalopoda: Sepiolidae). *Behavior* 85:25–41.

Nishiguchi, M. K. 2002. Host-symbiont recognition in the environmentally transmitted sepiolid squid-*Vibrio* mutualism. *Microbial Ecology* 44:10–18.

Nishiguchi, M. K., Ruby, E. G., and M. J. McFall-Ngai. 1998. Competitive dominance among strains of luminous bacteria provides an unusual form of evidence for parallel evolution in Sepiolid squid-*Vibrio* symbioses. *Applied Environmental Microbiology* 64:3209–3213.

Nyholm, S. V., Deplancke, B., Gaskins, H. R., Apicella, M. A., and M. J. McFall-Ngai. 2002. Roles of *Vibrio fischeri* and nonsymbiotic bacteria in the dynamics of mucus secretion during symbiont colonization of the *Euprymna scolopes* light organ. *Applied Environmental Microbiology* 68:5113–5122.

Nyholm, S. V. and M. J. McFall-Ngai. 2004. The winnowing: Establishing the squid-*Vibrio* symbiosis. *Nature Reviews Microbiology* 2:632–642.

Nyholm, S. V., Stabb, E. V., Ruby, E. G., and M. J. McFall-Ngai. 2000. Establishment of an animal-bacterial association: Recruiting symbiotic vibrios from the environment. *Proceedings of Natural Academy of Sciences USA* 97:10231–10235.

Ochman, H. and N. A. Moran. 2001. Genes lost and genes found: Evolution of bacterial pathogenesis and symbiosis. *Science* 292:1096–1099.

Ruby, E. G. 1999. The *Euprymna scolopes–Vibrio fischeri* symbiosis: A biomedical model for the study of bacterial colonization of animal tissue. *Journal of Molecular Microbiology and Biotechnology* 1:13–21.

Ruby, E. G. and L. M. Asato. 1993. Growth and flagellation of *Vibrio fischeri* during initiation of the sepiolid squid light organ symbiosis. *Archives of Microbiology* 159:160–167.

Ruby, E. G., Greenberg, E. P., and J. W. Hastings. 1980. Planktonic marine luminous bacteria: Species distribution in the water column. *Applied Environmental Microbiology* 39:302–306.

Ruby, E. G. and M. J. McFall-Ngai. 1991. Symbiont recognition and subsequent morphogenesis as early events in an animal-bacterial mutualism. *Science* 254:1491–1494.

Ruby, E. G. and M. J. McFall-Ngai. 1999. Oxygen-utilizing reactions and symbiotic colonization of the squid light organ by *Vibrio fischeri. Trends in Microbiology* 7:414–420.

Ruby, E. G. and J. G. Morin. 1979. Luminous enteric bacteria of marine fishes: A study of their distribution, densities, and dispersion. *Applied Environmental Microbiology* 38:406–411.

Ruby, E. G. and K. H. Nealson. 1976. Symbiotic association of *Photobacterium fischeri* with the marine luminous fish *Monocentris japonica*: A model of symbiosis based on bacterial studies. *Biological Bulletin* 151:574–586.

Ruby, E. G., Urbanowski, M., Campbell, J., Dunn, A., Faini, M., Gunsalus, R., Lohstroh, P., Lupp. C., McCann, J., Millikan, D., Schaefer, A., Stabb, E., Stevens, A., Visick, K., Whistler, C., and E. P. Greenberg. 2005. Complete genome sequence of *Vibrio fischeri*: A symbiotic bacterium with pathogenic congeners. *Proceedings of Natural Academy of Sciences USA* 102:3004–3009.

Shears, J. S. 1988. The use of a sand-coat in relation to feeding and diel activity in the sepiolid squid *Euprymna scolopes. Malacologia* 29:121–133.

Singley, C. T. 1983. *Euprymna scolopes*. In *Cephalopod Life Cycles*, ed. P. R. Boyle. London: Academic Press.

Stabb, E. V. 2005. Shedding light on the bioluminescence "paradox." *ASM News* 71:223–229.

Stabb, E. V. 2006. The *Vibrio fischeri-Euprymna scolopes* light organ symbiosis. In *The Biology of Vibrios*, eds. F. L. Thompson, B. Austin, and Swings, J., pp. 204–218. Washington: ASM Press.

Sugita, H. and Y. Ito. 2006. Identification of intestinal bacteria from Japanese flounder (*Paralichthys olivaceus*) and their ability to digest chitin. *Letters Applied Microbiology* 43:336–342.

Sycuro, L. K., Ruby, E. G., and M. McFall-Ngai. 2006. Confocal microscopy of the light organ crypts in juvenile *Euprymna scolopes* reveals their morphological complexity and dynamic function in symbiosis. *Journal of Morphology* 267:555–568.

Visick, K. L. 2005. Layers of signaling in a bacterium-host association. *Journal of Bacteriology* 187:3603–3606.

Visick, K. L., Foster, J., Doino, J., McFall-Ngai, M., and E. G. Ruby. 2000. *Vibrio fischeri lux* genes play an important role in colonization and development of the host light organ. *Journal of Bacteriology* 182:4578–4586.

Visick, K. L. and M. J. McFall-Ngai. 2000. An exclusive contract: Specificity in the *Vibrio fischeri–Euprymna scolopes* partnership. *Journal of Bacteriology* 182:1779–1787.

Visick, K. L. and E. G. Ruby. 2006. *Vibrio fischeri* and its host: It takes two to tango. *Current Opinion in Microbiology* 9:632–638.

Warner, J. A., Latz, M. I., and J. F. Case. 1979. Cryptic bioluminescence in a midwater shrimp. *Science* 203:1109–1110.

Wei, S. L. and R. E. Young. 1989. Development of symbiotic bacterial bioluminescence in a nearshore cephalopod, *Euprymna scolopes. Marine Biology* 103:541–546.

Widder, E. A. 1998. A predatory use of counterillumination by the squaloid shark, *Isistius brasiliensis. Environmental Biology of Fishes* 53:267–273.

Young, R. E. 1977. Ventral bioluminescent countershading in midwater cephalopods. *Symposia of the Zoological Society of London* 38:127–159.

Young, R. E. 1995. Aspects of the natural history of pelagic cephalopods of the Hawaiian mesopelagic-boundary region. *Pacific Science* 49:143–155.

Young, R. E. and C. F. Roper. 1976. Bioluminescent countershading in midwater animals: Evidence from living squid. *Science* 191:1046–1048.

Zhou, J., Clark, C., and J. Huissoon. 2007. SIFT approach used in fish tracking for autonomous underwater vehicle. *Proceedings of the 2007 International Symposium on Unmanned Untethered Submersible Technology (UUST)*. http://www.ausi.org/events/uust/proceedingsForm.pdf

7

Entomopathogenic Nematode and Bacteria Mutualism

Heather S. Koppenhöfer and Randy Gaugler

CONTENTS

7.1 Introduction

Entomopathogenic nematodes of the families Heterorhabditidae and Steinernematidae are lethal insect endoparasites characterized by their association with bacteria in the genera *Photorhabdus* and *Xenorhabdus*, respectively. This association is obligate mutualism in nature, with each partner requiring the other to complete its life cycle. Nematode growth and reproduction depend upon conditions established in the insect host by the bacterium. On the one hand, the bacteria contribute anti-immune proteins to assist the nematode in overcoming host defenses and antimicrobials that suppress competitors (Forst and Clarke, 2002). On the other, the bacteria lack invasive powers and are dependent upon the nematode to locate and penetrate suitable hosts. The nematode–bacterium complex is an important natural enemy of soil insects and plays a significant role in the regulation of soil food webs. Interest in their potential for use as biological insecticides has fueled decades of research.

Entomopathogenic nematodes show a broad, worldwide geographical range and are common in the soil environment. Seventy-five (61 steinernematid and 14 heterorhabditid) nematode species have been described. The two families share the same general life history but belong to different clades within the order Rhabditida. Heterorhabditids were derived from free-living bacteriovorous ancestors, whereas steinernematid origins are ambiguous. Twenty-three bacterial species (20 *Xenorhabdus*, 3 *Photorhabdus*) have been described, but most bacterial associates are yet to be examined. The bacteria are Gram-negative, nonfermentative rods, belonging to the family *Enterobacteriaceae* (Thomas and Poinar, 1979). *Photorhabdus* and *Xenorhabdus* are a unique group phenotypically and genotypically distinct from other genera within the family. They are insect pathogens although one species, *Photorhabdus asymbiotica*, has been occasionally reported as an opportunistic human pathogen (Peel et al., 1999). Although similar, *Photorhabdus* are bioluminescent and secrete anthraquinone pigments and catalase, which are traits absent in *Xenorhabdus*. Similarities in the mutualism and parasitism of the bacteria–nematode complexes resulted from convergent evolution (Poinar, 1993). That is, each complex independently evolved their pathogenic and symbiotic relationships.

Other bacteria may show conditional associations with an entomopathogenic nematode, but the natural symbiont is the most efficient partner for pathogenicity, reproduction, and development of the nematode

(Bonifassi et al., 1999; Sicard et al., 2004b). Each steinernematid species is mutualistically associated with only one species of *Xenorhabdus*; however, a few *Xenorhabdus* species are symbionts of multiple steinernematid species (Table 7.1). In contrast to the *Steinernema–Xenorhabdus* complex, different strains of one heterorhabditid species, *Heterorhabditis bacteriophora*, are associated with two different species and subspecies of *Photorhabdus*. One of these bacterial species, *Photorhabdus temperata*, is also associated with several other heterorhabditid hosts. Other multiple symbiotic associations are likely to exist.

TABLE 7.1

Bacteria/Nematode Associations

Photorhabdus/Heterorhabditis		
P. luminescens subsp. *luminescens*	*H. bacteriophora* Brecon	Thomas and Poinar (1979); Boemare et al. (1993); Fischer-Le Saux et al. (1999)
P. luminescens subsp. *akhurstii*	*H. indica*	Fischer-Le Saux et al. (1999)
P. luminescens subsp. *kayaii*	*H. bacteriophora*, Turkey	Hazir et al. (2004)
P. luminescens subsp. *laumondii*	*H. bacteriophora* HP88	Fischer-Le Saux et al. (1999)
P. luminescens subsp. *thracensis*	*H. bacteriophora*, Turkey	Hazir et al. (2004)
P. temperata	*H. zealandica*, *H. megidis*, *H. bacteriophora* NC1	Fischer-Le Saux et al. (1999)
P. asymbiotica subsp. *asymbiotica*	Unknown host	Fischer-Le Saux et al. (1999) Akhurst et al. (2004)
P. asymbiotica subsp. *australis*	*Heterorhabditis* sp. (similar to *H. indica*)	Akhurst et al. (2004) Gerrard et al. (2006)
Xenorhabdus/Steinernema		
X. nematophila	*S. carpocapsae*	Poinar and Thomas (1965) Thomas and Poinar (1979)
X. beddingii	*S. longicaudum*	Akhurst and Boemare (1988)
X. bovienii	*S. affine*, *S. feltiae*, *S. intermedium*, *S. kraussei*	Akhurst and Boemare (1988)
X. budapestensis	*S. bicornutum*	Lengyel et al. (2005)
X. cabanillasii	*S. riobrave*	Tailliez et al. (2006)
X. doucetiae	*S. diaprepesi*	Tailliez et al. (2006)
X. ehlersii	*S. serratum (nomen nudum)*	Lengyel et al. (2005)
X. griffiniae	*S. hermaphroditum*	Tailliez et al. (2006)
X. hominickii	*S. karii*, *S. monticolum*	Tailliez et al. (2006)
X. innexi	*S. scapterisci*	Lengyel et al. (2005)
X. indica[a]	*S. thermophilum (S. abbasi)*	Somvanshi et al. (2006) Tailliez et al. (2006)
X. japonica	*S. kushidai*	Nishimura et al. (1994)
X. koppenhoeferi	*S. scarabaei*	Tailliez et al. (2006)
X. kozodoii	*S. arenarium*, *S. apuliae*	Tailliez et al. (2006)
X. mauleonii	S. sp. isolated from island of St. Vincent located in Caribbean Sea	Tailliez et al. (2006)
X. miraniensis	Steinernematid nematode isolated from Mirani, Queensland Australia	Tailliez et al. (2006)
X. poinarii	*S. glaseri*, *S. cubanum*	Akhurst and Boemare (1988)
X. romanii	*S. puertoricense*	Tailliez et al. (2006)
X. stockiae	*S. siamkayai*	Tailliez et al. (2006)
X. szentirmaii	*S. rarum*	Lengyel et al. (2005)

[a] At the time of publication species name not effectively published.

The complex is extraordinary not only in having a pathogenic relationship with one phylum (Arthropoda) and a mutualistic one with a second (Nematoda or Proteobacteria), but also in displaying three distinct modes of symbiosis: offensive, resource harvest (i.e., nutritional), and defensive mutualism. We submit that the tripartite nematode-bacteria-insect interaction is a powerful, easily manipulated, and seriously underutilized model system to study mechanisms governing symbiotic as well as pathogenic interactions.

7.2 Life Cycle

The life cycle of the entomopathogenic nematode has two distinct phases: free-living phase within the soil and parasitic phase within a host insect. The third-stage infective juvenile (Figure 7.1A) is a nonfeeding, developmentally arrested, nonaging, and free-living stage responsible for finding and penetrating suitable soil insect hosts. This environmentally tolerant stage is equivalent to the dauer juvenile of *Caenorhabditis elegans*, but differs in that steinernematid and heterorhabditid infective juveniles carry bacterial symbionts monxenically in the intestinal tract. The bacteria are in a nearly quiescent state, protected within their nematode host from unfavorable conditions in the soil. The bacterial symbionts are not free-living and cannot persist in the soil. In *Heterorhabditis*, *Photorhabdus* primarily colonizes the anterior region of the intestine just posterior to the basal bulb and, to varying degrees, is also located throughout the remainder of the intestine (Endo and Nickle, 1991; Ciche and Ensign, 2003). *Steinernema* species have a specialized bilobed intestinal vesicle that is colonized by *Xenorhabdus* (Figure 7.1B) (Bird and Akhurst, 1983).

Insect seeking entomopathogenic nematodes deploy two broad categories of foraging strategies: ambush and cruise foraging. Ambush foragers (e.g., *Steinernema carpocapsae*, *S. scapterisci*) tend to nictate, standing stationary on their tail in a straight position elevating the anterior 95% of the body from the substrate (Campbell and Gaugler, 1993). Nictation enables the nematode to attach to passing insects. Some ambush species can leap through the air toward host cues. Ambushers respond poorly to volatile cues released by hosts, ignoring insects placed only millimeters distant (Lewis et al., 1992), and are found in the uppermost soil-litter layer. Cruise foragers (e.g., *S. glaseri*, *Heterorhabditis bacteriophora*) neither nictate nor attach well to passing hosts, but are highly mobile and responsive to long-range host volatiles. Thus, ambushers tend to parasitize highly mobile, surface-adapted insect species, whereas cruisers tend to parasitize sedentary insect species inhabiting the soil profile.

An entomopathogenic nematode is capable of parasitizing only a single host, so infective juveniles must carefully assess the suitability of a potential host before committing irreversibly to infection. Lewis et al. (1996) demonstrated that contact with the host cuticle, presumably via cuticular hydrocarbons, provides important host recognition cues that excite a response to locate portals of entry. High recognition responses were recorded for hosts supporting high levels of reproduction, whereas noninsect arthropods such as isopods were not recognized. Natural body openings are key portals of entry, including the spiracles which open into the insect respiratory system and the mouth and anus which lead to the soft midgut. In addition, infective stage heterorhabditids are armed with a dorsal tooth which can be wielded to penetrate through the host cuticle at thin intersegmental regions (Bedding and Molyneux, 1982).

After the infective stage penetrates to the hemocoel of a suitable insect host, the protective cuticle retained from the second-stage juvenile is shed and the bacterial symbiont is released. Steinernematids expel the bacteria into the host hemocoel via the anus, whereas heterorhabditids regurgitate their bacterial partners through the mouth (Figure 7.1C) (Ciche and Ensign, 2003). The bacteria quickly replicate in the nutrient-rich insect blood and in the connective tissues surrounding the insect midgut (Silva et al., 2002). Most nematode species recover from the infective juvenile's developmentally arrested state in the hemocoel and initiate development again. *Steinernema carpocapsae*, however, is capable of recovering in the insect gut prior to entering the hemocoel (Sicard et al., 2004a).

The host immune response is an important obstacle to entomopathogenic nematodes and bacteria in making the transition from a free-living state to parasitism. Here, the mutualist partners cooperate in evading or suppressing the host immune system to avoid phagocytosis and encapsulation. They also cooperate in killing the insect hosts, usually in 24–48 h. The bacteria alone, however, are extraordinarily

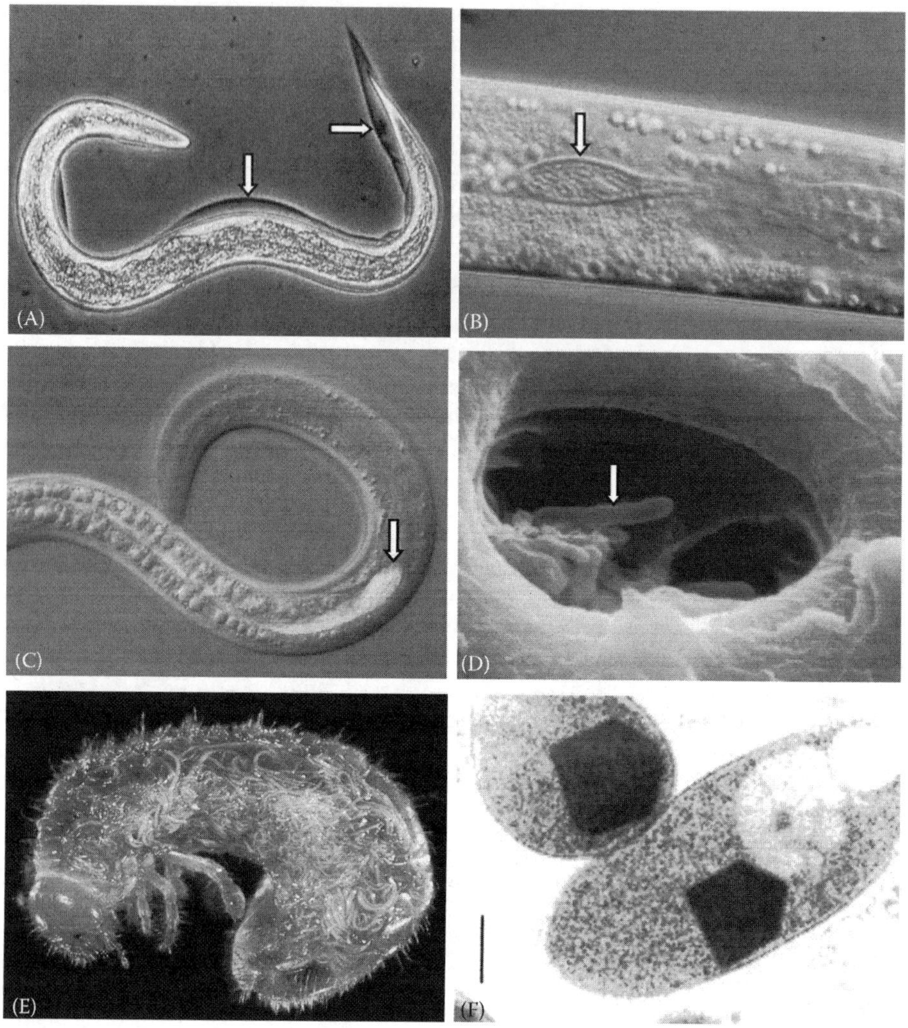

FIGURE 7.1 (See color insert following page 206.) (A) Third-stage infective juvenile of *Steinernema carpocapsae* nematode. Note protective sheath (arrows) encasing the nematode. (Photo by Gaugler, R. *Entomopathogenic Nematology*, Gaugler, R. (Ed.), CABI Publishing, Wallingford, U.K., 35–36. With permission.) (B) Anterior region of the intestine of infective juvenile of *Steinernema carpocapsae* showing the vesicle (arrow) which is colonized by cells of the nematode's bacterial symbiont, *X. nematophila*. (Courtesy of Christian Laumond and Noël Boemare INRA, France.) (C) Bolus of GFP-labeled *Photorhabdus luminescens* in the intestine of infective juvenile of *Heterorhabditis bacteriophora* nematode immediately prior to regurgitation. (From Ciche, T.A., Darby, C., Ehlers, R.-U., Forst, S., and Goodrich-Blair, H., *Biological Control*, 38, 22, 2006. With permission; Courtesy of T. Ciche.) (D) Scanning electron micrograph of the oral cavity of *Steinernema carpocapsae* showing cells of its associated symbiont, *X. nematophila* (arrow). (Courtesy of L. LeBeck.) (E) Larvae of the Japanese beetle (*Popillia japonica*) with the cuticle excised to reveal the developing adult nematodes of *Heterorhabditis bacteriophora* within. (Photos by Wang, Y. and Gaugler, R., *Entomopathogenic Nematology*, Gaugler, R. (Ed.), CABI Publishing, Wallingford, U.K., 35–36. With permission.) (F) Crystalline inclusion bodies (black areas) in cells of *Photorhabdus temperata*. (From Boemare, N., in *Entomopathogenic Nematology*, Gaugler, R. (Ed.), CABI Publishing, Wallingford, U.K., 2002, 35–36. With permission; Courtesy of EMIP, INRA, Montpellier, France.)

lethal to most insects, with LD50 values as low as 1 cell per insect (Clarke and Dowds, 1995). This quick kill may reduce the necessity to evolve a high degree of adaptation to a specific host as many nematode–bacterial complexes, most notably *S. carpocapsae–X. nematophila* and *H. bacteriophora–P. luminescens*, have host ranges that encompass hundreds of insect species. Conversely, other complexes, notably *S. scapterisci–X. innexi*, show a much narrower degree of host specificity.

Symptoms of infection are highly characteristic. Bacterial pigments cause insects killed by steinernematid-*Xenorhabdus* infections to become beige, tan, or grayish-, yellowish-, or walnut-brown, whereas insects killed by heterorhabditid-*Photorhabdus* infections become orange, red to reddish-brown, or greenish and the host tissues become ropey and highly viscous. These pigments and viscosity are also evident in artificial growth media. Moreover, *Photorhabdus* produces enzymes that cause the insect cadaver to glow with a faint but visible yellow–green luminescence. A property of many marine species, *Photorhabdus* are the only terrestrial bacteria known to be bioluminescent, a property of many marine species (Peel et al., 1999). The enzyme catalyzing light emission is a bacterial luciferase. The function of this bioluminescence is not yet understood.

Xenorhabdus and *Photorhabdus* produce phenotypically variant cell types (Forst and Clarke, 2002). The primary variant is associated with the gut of infective juvenile nematodes, whereas the secondary variant spontaneously arises when the bacteria are in culture or in the insect cadaver during the late stages of nematode reproduction (Akhurst, 1980). The two cell types differ morphologically and physiologically. Primary cells are hemolytic and highly motile; they adsorb certain dyes and produce crystalline inclusions, antibiotics, lipases, proteases, and in *Photorhabdus*, bioluminescence (Forst et al., 1997)—traits that are greatly reduced or lost in the secondary cells. The variant cell types are equally pathogenic against insects, but the secondary variant of *Photorhabdus* does not support growth and reproduction of *Heterorhabditis* (Gerritsen and Smits, 1993, 1997).

The key step in bacterial transition from insect pathogens to nematode symbionts is multiplying and breaking down insect host tissues, thereby provisioning the cadaver with nutrients essential for nematode growth. The bacterial symbionts further produce broad-spectrum antibiotic compounds, which suppress secondary invaders from the soil or insect gut that otherwise compete for nutrients and oxygen (Webster et al., 2002). In addition to preserving the insect cadaver until the nematode life cycle is complete, antibiotic production is a key mechanism to assure nematode–bacterial specificity. In short, the bacterial partner converts the host into a self-contained microhabitat ideal for nematode development and reproduction.

The nematodes feed directly upon the cells of their symbiont (Figure 7.1D) as well as on liquefying host tissues, molting twice to become adults (Poinar, 1990). Steinernematids are amphimictic, so infective juveniles may develop into either males or females, so reproduction requires host invasion by both sexes. By contrast, heterorhabditid infective juveniles mature into self-fertilizing hermaphrodites with ovotestes, permitting host colonization when even a single individual finds and infects an insect. Subsequent heterorhabditid generations are a mix of amphimictic and hermaphroditic stages. Mating behaviors differ sharply in the two genera, with *Steinernema* males coiling around the female at the vulva, and *Heterorhabditis* males align parallel to the female or hermaphrodite.

Both genera are oviparous initially with eggs laid within the hemocoel which hatch in 2 days. If food is abundant, the developing nematodes proceed through four juvenile molts to become adults. Several generations may be produced in a single host until the cadaver is completely colonized (Figure 7.1E). When nematode density is high and nutrients become limited within the insect cadaver, infective juvenile formation is promoted. Here, second-stage juveniles do not develop into normal third-stage juveniles but take an alternative developmental pathway to become third-stage infective juveniles encased within the cuticle of the second-stage juvenile. The retained cuticle provides protection from desiccation (Campbell and Gaugler, 1991) and nematophagous fungi (Timper and Kaya, 1989). Infective juveniles are formed during *endotokia matricida* in which intrauterine development results in the death and consumption of the parent hermaphrodite or female by the developing infectives (Wang and Bedding, 1996; Johnigk and Ehlers, 1999). Thus, the parent provides a nutrient supply critical for juvenile development when insect nutrients are depleted.

An essential feature of infective juvenile formation is reassociation of the bacterial symbionts with their nematode partners. Colonization of the nematode intestine is initiated by a few ingested cells which, rather than being digested, adhere to the nematode vesicle lining (Martens et al., 2003). Fimbriae are bacterial cell surface appendages proposed to mediate adherence to gut epithelial cells (Binnington and Brooks, 1993; Forst and Nealson, 1996). After infective juvenile formation is complete, these bacteria proliferate to complete colonization of the gut before becoming quiescent. Nonsymbiotic species do not adhere and are unable to multiply and colonize the gut. Thus, the specificity of the nematode–bacterium relationship is tightly controlled by exclusion of bacterial competitors and recruitment by the infective juvenile of its symbiont.

When insect resources have been exhausted, infective juveniles armed with a fresh supply of bacteria emerge from the empty shell of the insect. Steinernematid infective juveniles produced via *endotokia matricida* emerge mainly from natural openings of the maternal cadavers, while heterorhabditids emerge mostly transcuticularly from the parent (Baliadi et al., 2001). Emergence from the insect cadaver may be via natural body openings or transcuticularly at intersegmental areas where the cuticle is thin. Encased within the retained cuticle of the second-stage juvenile, infective juveniles move into the soil to become free-living again. Emergence is closely related to soil moisture, with reduced emergence into dry soil (Koppenhöfer et al., 1997). The infective stage is armed with enlarged amphids (chemosensory organs) and somatic muscles, morphological adaptations related to this stage's central function: host search. Under ideal conditions, steinernematids emerge 6–11 days and heterorhabditids 12–14 days after infection (Kaya and Koppenhöfer, 1999). Each infected insect cadaver produces hundreds of thousands of new host-seeking symbiont-equipped infective juveniles.

7.3 Offensive Mutualism: Tag-Team Killing of an Insect

The nematodes and bacteria combine forces to defeat the insect immune response elicited by hemocoel invasion. The bacteria are not released until several hours after penetration (Wang et al., 1995), so the nematode plays the essential role in immune suppression in the earliest stage of infection. Consider the example provided by the entomopathogenic nematode *S. glaseri* which naturally parasitizes scarab larvae. The nematode is recognized and rapidly entrapped by host blood cells after penetration, but the immune response soon weakens and the nematodes break free by vigorous movements (Wang and Gaugler, 1999). Live axenic infective juveniles introduced to scarab larvae by intrahemocoelic injection not only avoid encapsulation but even protect subsequently injected freeze-killed nematodes, demonstrating that the bacteria are not central to this phase of immune suppression. The mechanism involves surface coat protein secreted by *S. glaseri* which destroys host hemocytes. When the SCP3a surface coat protein is isolated and then coinjected with latex beads into the larval hemocoel, the beads are protected from phagocytosis. This finding suggests that the nematode's symbiotic bacteria would be similarly protected when they are most vulnerable: immediately after release. In short, the nematode secretes anti-immune proteins which ensure its own survival as well as preparing the way for its partner.

Once recognized by the host immune system, the bacteria are subjected to hemocyte aggregation and ensuing nodulation, which are activated in part by eicosanoids (Stanley and Miller, 2006). *Photorhabdus* and *Xenorhabdus* hamper hemocyte aggregation and nodule formation by inhibiting the activity of phospholipase A2, the enzyme responsible for induction of the insect eicosanoid pathway (Park et al., 2003; Kim et al., 2005). Furthermore, like many other gram-negative pathogenic bacteria, *Photorhabdus* encodes a type III secretion system (TTSS), which translocates effector proteins into eukaryotic host cells. One of the effectors, LopT, protects the cells by suppressing phagocytosis and reducing nodulation (Brugirard-Ricaud et al., 2004, 2005). A dedicated TTSS has not been found in *Xenorhabdus* (Goodrich-Blair and Clarke, 2007), but a flagellar TTSS is required for the secretion of lipase (Park and Forst, 2006).

Lipopolysaccharide (LPS) is a major component of the cell outer membrane in both *Photorhabdus* and *Xenorhabdus* (Dunphy and Thurston, 1990). In *Xenorhabdus*, the LPS is cytotoxic by way of the lipid A moiety, which binds to and destroys insect hemocytes (Dunphy and Webster, 1988, 1991). Furthermore, LPS suppresses melanization by inhibiting phenoloxidase activity (Dunphy and Webster, 1991). In *Photorhabdus*, the role of LPS is not clearly understood, but may be important in counteracting antimicrobial peptides produced by the insect immune system (Goodrich-Blair and Clarke, 2007). By contrast, *Xenorhabdus* suppresses the transcription of insect genes encoding antimicrobial peptides (Park et al., 2007). Early in infection, *Photorhabdus* produces a trans stilbene antibiotic, (E)-1,3-dihydroxy-2-(isopropyl)-5-(2-phenylethenyl) benzene, that inhibits phenoloxidase activity (Eleftherianos et al., 2007). This compound serves a dual role by countering the host immune response and protection from microbial competitors.

After overcoming the immune system, the bacteria produce a number of toxins to kill their insect hosts. Among the most effective virulence factors in the bacterial arsenal are the Tc (toxin complex) protein toxins. Tc's are high molecular weight insecticidal toxins that are lethal to insects even when administered orally (Bowen et al., 1998). In *Photorhabdus*, loci for four different complexes (Tca, Tcb, Tcc, Tcd) have been described, of which Tca is known to disrupt the insect midgut epithelium in a manner similar to the δ-endotoxin of *Bacillus thuringiensis* (Blackburn et al., 1998). The counterparts of the Tc's in *Xenorhabdus* are termed Xpt, *Xenorhabdus* protein toxin (XptA1, XptA2, XptB1, XptC1) of which XptA1 is central for the expression of insecticidal activity and XptB1 and XptC1 are required for full virulence (Morgan et al., 2001; Sergeant et al., 2003). XptA1 binds to the host gut and its "hollow box" structure permits the protein to act as a receptacle for the XptB1 and XptC1. (Lee et al., 2007), making the XptA1 300 times more toxic to lepidopterous larvae than by itself. Unlike *Photorhabdus* Tca, the Xpt mode of action is different from that of the *B. thuringiensis* toxin system.

Both bacterial genera secrete extracellular cytotoxic proteins known as hemolysins (Brillard et al., 2001, 2002). In *Xenorhabdus*, the xenorhabdolysin (C1) hemolysin is extremely virulent and triggers apoptosis in insect as well as mammalian cells (Vigneux et al., 2007). This cytotoxin is an essential toxin required for full *Xenorhabdus* virulence (Cowles and Goodrich-Blair, 2005). The role of hemolysins in *Photorhabdus* is less clear as these cells retain their virulence even without hemolysin secretion (Brillard et al., 2002). However, other toxins secreted by *Photorhabdus* serve as important virulence factors. The *Photorhabdus* Mcf1 and Mcf2 (makes caterpillars floppy) toxins result in rapid loss of insect body turgor and death (Daborn et al., 2002; Waterfield et al., 2003). Mcf1 destroys hemocytes and the insect midgut by triggering massive apoptosis, whereas the site and mode of action of Mcf2 remains unknown. These toxins show strong homology to each other but diverge at their N-termini encoding different effector domains (Waterfield et al., 2003). The different effector domains may allow for distinct sites of actions within the insect. Additional *Photorhabdus* toxins act to destroy the insect gut. *Photorhabdus* specifically colonizes the area between the basal membrane and midgut epithelium, where it expresses the gut active Tca along with PrtA, an RTX-like metalloprotease; together they induce a massive programmed cell death of the midgut epithelium (Silva et al., 2002).

In addition to killing the insect, these toxins and others that have not been presented here (Table 7.2) may facilitate bioconversion of the insect tissue to provide nutrition for the developing nematodes. Many virulence factors are engaged for the nematode/bacteria infection process. However, it is not clear which mechanisms are universal or nematode, bacterium, or insect host specific. As new nematode species are being isolated, each having a mutualistic association with a specific bacterium, more opportunities arise to uncover the specificities involved in bacterial virulence.

7.4 Resource Harvest Mutualism: What Is for Dinner?

In resource harvest mutualism, one symbiont gathers nutrients that are converted into a form the other partner can use, as in the well-studied case of termites that depend upon protozoan gut inhabitants to secrete cellulases that breakdown cellulose. The nematode–bacterial relationship is analogous to that of termite–protozoan symbiosis with regard to this digestive or nutritional mutualism. In both examples, the microbial symbionts ingest and hydrolyze food particles to provision a nutrient base for themselves and their partners. *Xenorhabdus* and *Photorhabdus* species achieve high cell densities in the host hemocoel, producing numerous lytic exoenzymes during the growth phase that are involved in insect cadaver bioconversion including lipases, phospholipases, chitinase, protease, and DNases. Late in infection, a metalloprotease is also produced by *Photorhabdus* that may facilitate tissue bioconversion (Bowen et al., 1998). The production of these enzymes is greatly reduced in the secondary variant, explaining in part why nematode reproduction tends to be limited when secondary cells proliferate. The bacterial symbionts are involved in iron acquisition, LPS synthesis, and lipid synthesis, all activities beneficial to nematode nutrition. For example, Watson et al. (2005) demonstrated that a *P. temperata* mutant defective in scavenging iron did not support nematode growth and development in vitro, but the addition of exogenous iron to the media restored nematode growth and development. Bacterial pathogenicity was

TABLE 7.2 Factors Produced by *Photorhabdus* and *Xenorhabdus* to Increase Virulence and Survival of the Nematode/Bacterium Complex

	Genus	Virulence Factor	Reference
Offensive mutualism		**Secretion System**	
	Photorhabdus	Type III secretion system (TTSS)	Brugirard-Ricaud et al. (2004, 2005)
		LopT effector protein	
	Xenorhabdus	Flagellar TTSS, lipase secretion	Park and Forst (2006)
		Proteases	
	Photorhabdus	PrtS metalloprotease	Held et al. (2007)
		PrtA zinc metalloprotease	Bowen et al. (2003)
	Xenorhabdus	XrtA (Protease II)	Caldas et al. (2002); Park and Forst (2006)
		Hemolysins	
	Photorhabdus	PhlA hemolysin	Brillard et al. (2002)
	Xenorhabdus	XhlA hemolysin	Cowles and Goodrich-Blair (2005)
		XaxAB (α-Xenorhabdolysin)	Ribeiro et al. (2003); Vigneux et al. (2007)
		Toxins	
	Photorhabdus	Toxin complexes (Tc's)	Bowen et al. (1998)
		Makes caterpillars floppy (Mcf1 and Mcf2)	Daborn et al. (2002); Waterfield et al. (2003)
		Photorhabdus virulence cassettes (PVCs)	Yang et al. (2006)
		Photorhabdus insect related binary proteins, PirAB	Waterfield et al. (2002)
	Photorhabdus/ Xenorhabdus	Txp40 ubiquitous cytotoxic protein	Brown et al. (2006)
	Xenorhabdus	*Xenorhabdus* protein toxins (Xpt's)	Sergeant et al. (2003)
		MrxA pore-forming toxin	Banerjee et al. (2006)
		Other Virulence Factors	
	Photorhabdus/ Xenorhabdus	Lipase	Clarke and Dowds (1995); Dunphy (1995); Givaudan and Lanois (2000)
		LPS endotoxin	
	Xenorhabdus	Lecithinase	Thaler et al. (1995)
Defensive mutualism		**Antibiotics**	
	Photorhabdus	Carbapenem	Derzelle et al. (2002)
		Genistine	Sztaricskai et al. (1992)
		Hydroxystilbenes	Paul et al. (1981); Richardson et al. (1988)
		Anthra-quinones	Richardson et al. (1988); Li et al. (1995)
	Xenorhabdus	Xenorhabdins	McInerney et al. (1991a)
		Xenorxides	Li et al. (1998)
		Xenocoumacins	McInerney et al. (1991b)
		Benzylideneacetone	Ji et al. (2004)
		Indole derivatives	Paul et al. (1981)
		Bacteriocins	
	Photorhabdus	Lumicins	Sharma et al. (2002)
		Photorhabdicins	Ffrench-Constant et al. (2003)
	Xenorhabdus	Xenorhabdicins	Boemare et al. (1992); Thaler et al. (1995)
		Ant-Deterrent Factor (ADF)	
	Photorhabdus/ Xenorhabdus	Small, extracellular, possibly nonproteinacious compound(s)	Baur et al. (1998); Zhou et al. (2002)

also reduced in this mutant relative to the wild-type bacteria, suggesting that iron acquisition plays a role beyond nutrition.

Entomopathogenic nematodes grow and reproduce at best suboptimally in the absence of their symbiotic partner. There is growing evidence that proteinaceous crystalline inclusions found in *Xenorhabdus* and *Photorhabdus* (Figure 7.1F) play a nutritional role in this aspect of the partnership (Bintrim and Ensign, 1998; Goodrich-Blair and Clarke, 2007). The crystal proteins make up 40% of bacterial total protein and are high in essential amino acids (Couche et al., 1987). When the two crystal proteins from *Photorhabdus*, cipA and cipB, were expressed in *Escherichia coli*, the cells supported nematode development. Although resembling *B. thuringiensis* crystals, the symbiont crystals are not insecticidal, but rather appear to provide an important protein source needed for nematode nutrition (You et al., 2006). That is, the crystal inclusions are essential to maintain nematode–bacterial mutualism.

7.5 Defensive Mutualism: Protection of the Insect Cadaver

The bacterial symbionts produce a diverse array of antimicrobial compounds during late log growth that protect the insect cadaver from colonization by microbes from the gut or the soil environment. The cadaver is maintained for 1–2 weeks in a state much like a test tube holding a pure bacterial culture and the developing nematodes. By protecting the cadaver, the bacteria ensure full use of the insect resources for the developing nematodes and maintain the integrity of the symbiotic relationship.

Bacteriocins associated with *Photorhabdus* and *Xenorhabdus* are proteinaceous, large molecule compounds considered to be narrow spectrum antibiotics. Specificity between the nematode and its bacterial symbiont is a result of the exclusion of bacterial competitors and recruitment by the infective-stage juvenile of its specific symbiont. Thus, bacteriocins assist the natural symbiont in outcompeting congeneric bacteria and preserving specificity in *Steinernema–Xenorhabdus* and *Heterorhabditis–Photorhabdus* relationships (Boemare et al., 1997).

R-type and F-type pyocins are phage tail-like bacteriocins. The R-type pyocins are similar to contractile but nonflexible structures of bacteriophage tails and are thought to be derived from the common ancestor of phage P2, whereas the F-type pyocins resemble flexible but noncontractile tail structures and are thought to be derived from phage λ(Nakayama et al., 2000). Xenorhabdicin is an R-type pyocin produced by *X. nematophila* (Thaler et al., 1995) with antibacterial activity against other *Xenorhabdus* strains and species, *P. luminescens*, and species of the sister taxon *Proteus*. *Photorhabdus* species produce phage-related photorhabdicins that resemble both R-type and F-type pyocins (Ffrench-Constant et al., 2003). Photorhabdicins are active against other *Photorhabdus* strains as well as more distantly related bacteria such as *E. coli*.

Lumicins from *Photorhabdus* are a novel class of bacteriocins (Sharma et al., 2002). The lumicin loci encode killer proteins and multiple immunity proteins that protect the native cell from the lethal effects of the lumicin. Four loci have been described, lum1–4, of which lum3 is conserved among species of *Photorhabdus* (Marokhazi et al., 2003). Divergence in sequences encoding killer and immunity proteins are likely to occur between closely related bacterial strains to aid the symbiont during interstrain competition within the insect host. Like photorhabdicins lumicins are active against other *Photorhabdus* spp. as well as *E. coli*. This suggests that *Photorhabdus* bacteriocins may not be exclusively active against closely related symbionts but may also be involved in clearing of the insect gut microflora and invading soil bacteria.

Photorhabdus and *Xenorhabdus* secrete a variety of nonproteinaceous, small molecule compounds that are generally active against distantly related bacteria, yeasts, and fungi (Akhurst, 1982), with some also showing activity against insects and nematodes (Li et al., 1998). The bioactive compounds are genus specific, and most species produce more than one group of secondary metabolites. *Photorhabdus* species produce derivatives of anthraquinones, stilbenes, carbapenem, genistein, and photobactin (Li et al., 1998; Derzelle et al., 2002; Ciche et al., 2003) whereas *Xenorhabdus* produce xenorhabdins, xenocoumacins, indole derivatives, and benzylideneacetone (Li et al., 1998; Ji et al., 2004).

Anthraquinones are pigments responsible for the red color expressed by insect cadavers infected by *Heterorhabditis* species (Richardson et al., 1988). Pigment function remains unclear, but some of the derivatives show activity against microorganisms (Li et al., 1995; Brachmann et al., 2007). Hydroxystilbene antibiotics suppress gram-positive and gram-negative bacteria by inhibiting RNA synthesis (Sundar and Chang, 1992). One stilbene derivative, 3,5-dihydroxy-4-isopropylstilbene, is not only active against bacteria, but also strongly fungicidal, nematicidal, and insecticidal (Li et al., 1995; Eleftherianos et al., 2007). This compound shows a unique triple function in inhibiting the insect immune system, killing the insect, and protecting the cadaver from secondary invaders (Eleftherianos et al., 2007). A carbapenem-like molecule is secreted with broad-spectrum activity against gram-negative bacteria. Due to its early secretion and specific activity, the likely purpose is to prevent the insect gut microflora from invading the cadaver after insect death (Derzelle et al., 2002).

The largest group of *Xenorhabdus*-produced antibiotics are the xenorhabdins, members of the pyrothine family of antibiotics (McInerney et al., 1991a). All xenorhabdins show significant activity against gram-positive bacteria but little against gram-negative species. Some are highly antifungal, and one compound, xenorhabdin II, is insecticidal (Li et al., 1998). Xenorxides, oxidized products of xenorhabdins, are particularly broad spectrum, showing activity against gram-positive bacteria, yeast, and fungi (Li et al., 1998). Xenocoumacins represent another class of antibiotics, and like the xenorhabdins, are mainly effective against gram-positive bacteria (McInerney et al., 1991b). *Xenorhabdus* antibiotics that are active against both gram-positive and gram-negative species are indole derivatives (Paul et al., 1981). Although structurally different from the hydroxystilbenes of *Photorhabdus*, indole derivatives appear to share the same mode of antibiosis by inhibiting RNA and protein synthesis (Sundar and Chang, 1993). Benzylideneacetone is a monoterpenoid in a new class of antibiotics secreted by *X. nematophila* with specific activity against gram-negative bacteria (Ji et al., 2004).

The bacteria further provide insect deterrent factors that offer protection from foraging insects (Baur et al., 1998). Dead insects are an attractive food resource for scavengers which may consume the cadaver or compromise the cadaver cuticle which is a barrier against microbial invasion. The bacterial associates of some entomopathogenic nematode species produce an ant-deterrent factor, ADF, which discourages arthropod scavengers such as ants from feeding on the cadaver. ADF is small (<10 kDa), extracellular, acid sensitive, and heat stable (Zhou et al., 2002), characteristics shared with the insect-feeding deterrent thuringiensin, the exotoxin of *B. thuringiensis* (Mohd-Salleh and Lewis, 1982). The ability of ADF to deter ant feeding is dependent upon the ant species, bacterial strain, and variant of the ADF-producing symbiont. ADF is not produced in sufficient titer in early infections to repel ant scavengers; peak production and repellency does not occur until 4–5 days after infection (Zhou et al., 2002). Thus ADF is another beneficial compound produced by symbiotic bacteria that increases nematode fitness.

Antibiotic and ADF production requires a large investment from bacterial metabolism, and therefore uses resources. Without this cost, bacteria could invest more in nutrient uptake and proliferation and therefore better adapt to conditions outside of the tripartite relationship (Sicard et al., 2005). Not being able to do so keeps them tightly bound to their nematode partner.

7.6 Applications

Efforts to use entomopathogenic nematodes to control insect pests were initiated soon after the first species was described, and even before it was understood that a bacterial partner was present. In 1929, Glaser and Fox (1930) found *S. glaseri* infected Japanese beetle larvae, an introduced soil insect causing devastating damage to New Jersey agriculture, with larval densities estimated at more than 500 million per square mile. A survey determined that the nematode parasite had an extremely limited distribution. Development of in vitro mass rearing methodology (McCoy and Glaser, 1936) led to billions of nematodes being produced and released from 1939 to 1942 throughout New Jersey (563 release sites) in an effort to colonize *S. glaseri* for biological control of the Japanese beetle. However, these early workers were unaware of the nematode's associated bacterium, *X. poinarii*. Gaugler et al. (1992) demonstrated that the antimicrobial agents used to inhibit artificial media contamination during mass rearing resulted in loss of the bacterial partner, and only the nematode half of the partnership was released. Thus, one of the most ambitious and least known experiments in biological control ended in failure.

Not until the efforts of Poinar and Thomas (1966) 40 years later was the role of the associated bacteria as mutualist partner and key etiological agent responsible for insect virulence recognized. Commercial development of the nematode–bacterium complex as biological insecticides subsequently proceeded rapidly. A seminal step occurred in 1983 with the formation of Biosys, a California-based company that pioneered large-scale in vitro production and formulation technologies (Georgis, 2002). The availability of large quantities of gratis nematodes stimulated a burst of field testing and widespread awareness of their potential. Unable to compete effectively against chemical insecticides on the basis of cost, stability or efficacy, nematodes and their symbiotic bacteria nevertheless occupy a niche market where chemicals are restricted or ineffective. Nearly 20 small companies currently market a dozen species of insecticidal nematode products in the United States, Europe, Japan, India, and Australia. They are used against pests of cranberries (black vine weevil, cranberry girdler), turfgrass (mole cricket, billbugs, white grubs), artichokes (artichoke plume moth), mushrooms (sciariid flies), citrus (root weevils), ornamentals (black vine weevil, wood borers), and many other insect pests in horticulture, agriculture, home, and garden.

Commercial interest in the bacterial partners has received equally intense attention. *Xenorhabdus* and *Photorhabdus* secrete more than 30 bioactive secondary metabolites displaying a remarkably rich diversity of activities including insecticidal, antimicrobial, insect repellant, nematicidal, antiulcer, and even anticancer (Webster et al., 2002). Most research emphasis has been placed on the plethora of insecticidal toxins isolated from bacterial cultures (the nematode partners also produce toxins but remain nearly unstudied). Despite the large size of the Tc proteins, TcdA has been expressed in plants, although at a lower oral toxicity than the native toxin (Liu et al., 2003). Considering the need for alternatives to *B. thuringiensis* for deployment in transgenic plants, this area of study seems certain to receive continued attention.

The bacterial secondary metabolites show antibiotic activity against microbes of agricultural and medical importance. For example, xenorxides produced by *X. bovienii* suppress the mammalian pathogenic fungus *Aspergillus fumigatus* and the mammalian pathogenic bacterium *Staphylococcus aureus* (Li et al., 1998). The indole derivative nematophin is effective against methicillin-resistant *St. aureus*, MRSA (Paik et al., 2003), and metabolites produced by different *Xenorhabdus* species are active against the primary mastitis pathogens, *St. aureus*, *E. coli*, and *Klebsiella pneumoniae* (Furgani et al., 2008). Xenocoumacins show potent antiulcer activity when dosed orally to rats with stress-induced gastric ulcers (McInerney et al., 1991b), presumably due to activity against the *Helicobacter pylori* bacterium associated with most gastric ulcers. Despite the impressive range of antibiotics demonstrated by the bacterial metabolites across five decades of investigation, in addition to numerous patents, no significant commercial effort to apply these discoveries has yet been reported. Similarly, the anticancer activity shown by xenorxides in cell cultures of human colon, breast, cervical, and lung cancers (Webster et al., 1998) is yet to elicit commercial interest.

Nearly all *Photorhabdus* strains emit visible light. The expression of bioluminescence from *Photorhabdus luxAB* genes has utility as a marker for gene expression and regulation in the same manner as green florescent protein (GFP). Unlike conventional systems, expression of bioluminescence from the entire *luxCDABE* cassette does not require additional reagents, and allows for real-time monitoring superior to other *luxAB* systems at higher temperatures in both bacterial (Maoz et al., 2002) and eukaryotic (Gupta et al., 2007) systems.

7.7 Conclusion

In summary, the nematode and bacterial partners have a codependent relationship. The nematode is dependent upon the bacterium for quick killing of the insect host, suppressing secondary invaders, bioconversion of the insect biomass, provisioning, inhibition of the host immune response, and protection of the cadaver from scavengers. Conversely, the bacterium is dependent upon the nematode for protection from the soil environment, transportation, host recognition, penetration into the hemocoel, and inhibition of the host immune response.

The *Heterorhabditis/Photorhabdus* and *Steinernema/Xenorhabdus* complexes are emerging as model systems for the differentiation of both common and distinct factors involved in mutualism and pathogenesis. Like *Vibrio fischeri*, the bacterial symbiont of the Hawaiian bobtail squid, *Euprymna scolopes*, *Photorhabdus*, and *Xenorhabdus* are exposed to high stress and low nutrient colonization sites. However, unique to the nematode/bacteria complexes is the involvement of a third partner, the insect host, which provides a nutritionally rich environment. Because the bacteria undergo both symbiotic and pathogenic stages in their life cycle, new genes and regulation proteins are being elucidated in these systems.

Consider as an example virulence factors in *Photorhabdus* and *Xenorhabdus*. Tc's were first described in *Photorhabdus* (Bowen and Ensign, 1998; Bowen et al., 1998) and *Xenorhabdus* (Morgan et al., 2001), with similar toxins having subsequently been discovered in other bacteria including medically important species (Waterfield et al., 2002). Genetic studies in *Photorhabdus* have revealed that virulence in insects requires many of the same genes necessary for virulence in mammals (Derzelle et al., 2004; Bennett and Clarke, 2005). The recently completed genome sequences of *P. luminescens* (Duchaud et al., 2003) and *P. asymbiotica* (http://www.sanger.ac.uk/Projects/P_asymbiotica) and the soon to be completed sequences of *X. nematophila* and *X. bovienii* (http://www1.Xenorhabdus.org) have revealed a diverse array of genes that putatively encode potential virulence factors, and may contribute to understanding the process of virulence.

The *Photorhabdus* genome sequences coupled with the soon to be completed *H. bacteriophora* genome (http://genome.wustl.edu/genome) will open a new door to the analysis of tripartite mutualism, and are most likely to reveal parallels with other systems. This will help identify the basic principle necessary to maintain a mutualistic relationship while involving pathogenic factors. Our understanding of the nematode–bacterium mutualism is growing but requires further study to explain the provisions given by the symbiont and the resilience of the association over many generations.

ACKNOWLEDGMENT

We thank Albrecht Koppenhöfer for thoughtful discussion and comments on drafts of this chapter.

REFERENCES

Akhurst, R. J. 1980. Morphological and functional dimorphism in Xenorhabdus spp., bacteria symbiotically associated with the insect pathogenic nematodes Neoaplectana and Heterorhabditis. *Journal of General Microbiology* 121:303–309.

Akhurst, R. J. 1982. Antibiotic activity of *Xenorhabdus* spp., bacteria symbiotically associated with insect pathogenic nematodes of the families heterorhabditidae and steinernematidae. *Journal of General Microbiology* 128:3061–3065.

Akhurst, R. J. and N. E. Boemare. 1988. A numerical taxonomic study of the genus *Xenorhabdus* (*Enterobacteriaceae*) and proposed elevation of the subspecies of *X. nematophilus* to species. *Journal of General Microbiology* 134:1835–1845.

Akhurst, R. J., Boemare, N. E., Janssen, P. H., Peel, M. M., Alfredson, D. A., and C. E. Beard. 2004. Taxonomy of Australian clinical isolates of the genus *Photorhabdus* and proposal of *Photorhabdus asymbiotica* subsp. *asymbiotica* subsp. nov. and *P. asymbiotica* subsp. *australis* subsp. nov. *International Journal of Systematic and Evolutionary Microbiology* 54:1301–1310.

Baliadi, Y., Yoshiga, T., and E. Kondo. 2001. Development of *endotokia matricida* and emergence of originating infective juveniles of steinernematid and heterorhabditid nematodes. *Japanese Journal of Nematology* 31:26–36.

Banerjee, J., Singh, J., Joshi, M. C., Ghosh, S., and N. Banerjee. 2006. The cytotoxic fimbrial structural subunit of *Xenorhabdus nematophila* is a pore-forming toxin. *Journal of Bacteriology* 188:7957–7962.

Baur, M. E., Kaya, H. K., and D. R. Strong. 1998. Foraging ants as scavengers on entomopathogenic nematode-killed insects. *Biological Control: Theory and Applications in Pest Management* 12:231–236.

Bedding, R. A. and A. S. Molyneux. 1982. Penetration of insect cuticle by infective juveniles of *Heterorhabditis* spp. (Heterorhabditidae: Nematoda). *Nematologica* 28:354–359.

Bennett, H. P. J. and D. J. Clarke. 2005. The *PBGpe* operon in *Photorhabdus luminescens* is required for pathogenicity and symbiosis. *Journal of Bacteriology* 187:77–84.

Binnington, K. C. and L. Brooks. 1993. Fimbrial attachment of *Xenorhabdus nematophilus* to the intestine of *Steinernema carpocapsae*. In *Nematodes and the Biological Control of Insect Pests*, Bedding, R. A., Akhurst, R. J., and Kaya, H. K. (Eds.), pp. 147–155. East Melbourne: CSIRO.

Bintrim, S. B. and J. C. Ensign. 1998. Insertional inactivation of genes encoding the crystalline inclusion proteins of *Photorhabdus luminescens* results in mutants with pleiotropic phenotypes. *Journal of Bacteriology* 180:1261–1269.

Bird, A. F. and R. J. Akhurst. 1983. The nature of the intestinal vesicle in nematodes of the family Steinernematidae. *International Journal for Parasitology* 13:599–606.

Blackburn, M., Golubeva, E., Bowen, D., and R. H. Ffrench-Constant. 1998. A novel insecticidal toxin from *Photorhabdus luminescens*, toxin complex a (tca), and its histopathological effects on the midgut of *Manduca sexta. Applied and Environmental Microbiology* 64:3036–3041.

Boemare, N. 2002. Biology, taxonomy and systematics of *Photorhabdus* and *Xenorhabdus*. In: *Entomopathogenic Nematology*, Gaugler, R. (Ed.), pp. 35–36. Wallingford, U.K.: CABI Publishing.

Boemare, N. E., Boyer-Giglio, M. H., Thaler, J. O., Akhurst, R. J., and M. Brehélin. 1992. Lysogeny and bacteriocinogeny in *Xenorhabdus nematophilus* and other *Xenorhabdus* spp. *Applied and Environmental Microbiology* 58:3032–3037.

Boemare, N. E., Akhurst, R. J., and R. G. Mourant. 1993. DNA relatedness between *Xenorhabdus* spp. (*Enterobacteriaceae*), symbiotic bacteria of entomopathogenic nematodes, and a proposal to transfer *Xenorhabdus luminescens* to a new genus, *Photorhabdus gen. nov. International Journal of Systematic Bacteriology* 43:249–255.

Boemare, N., Givaudan, A., Brehelin, M., and C. Laumond. 1997. Symbiosis and pathogenicity of nematode-bacterium complexes. *Symbiosis* 22:21–45.

Bonifassi, E., Fischer-Le Saux, M., Boemare, N., Lanois, A., Laumond, C., and G. Smart. 1999. Gnotobiological study of infective juveniles and symbionts of *Steinernema scapterisci*: A model to clarify the concept of the natural occurrence of monoxenic associations in entomopathogenic nematodes. *Journal of Invertebrate Pathology* 74:164–172.

Bowen, D. J. and J. C. Ensign. 1998. Purification and characterization of a high-molecular-weight insecticidal protein complex produced by the entomopathogenic bacterium *Photorhabdus luminescens. Applied and Environmental Microbiology* 64:3029–3035.

Bowen, D., Rocheleau, T. A., Blackburn, M., Andreev, O., Golubeva, E., Bhartia, R., and R. H. Ffrench-Constant. 1998. Insecticidal toxins from the bacterium *Photorhabdus luminescens. Science* 280:2129–2132.

Bowen, D. J., Rocheleau, T. A., Grutzmacher, C. K., Meslet, L., Valens, M., Marble, D., Dowling, A., Ffrench-Constant, R., and Blight. 2003. Genetic and biochemical characterization of PrtA, an RTX-like metalloprotease from *Photorhabdus. Microbiology* 149:1581–1591.

Brachmann, A. O., Joyce, S. A., Jenke-Kodama, H., Schwär, G., Clarke, D. J., and H. B. Bode. 2007. A type II polyketide synthase is responsible for anthraquinone biosynthesis in *Photorhabdus luminescens. Chembiochem* 8:1721–1728.

Brillard, J., Ribeiro, C., Boemare, N., Brehélin, M., and A. Givaudan. 2001. Two distinct hemolytic activities in *Xenorhabdus nematophila* are active against immunocompetent insect cells. *Applied and Environmental Microbiology* 67:2515–2525.

Brillard, J., Duchaud, E., Boemare, N., Kunst, F., and A. Givaudan. 2002. The PhlA hemolysin from the entomopathogenic bacterium *Photorhabdus luminescens* belongs to the two-partner secretion family of hemolysins. *Journal of Bacteriology* 184:3871–3878.

Brown, S. E., Cao, A. T., Dobson, P., Hines, E. R., Akhurst, R. J., and P. D. East. 2006. Txp40, a ubiquitous insecticidal toxin protein from *Xenorhabdus* and *Photorhabdus* bacteria. *Applied and Environmental Microbiology* 72:1653–1662.

Brugirard-Ricaud, K., Givaudan, A., Parkhill, J., Boemare, N., Kunst, F., Zumbihl, R., and E. Duchaud. 2004. Variation in the effectors of the type III secretion system among *Photorhabdus* species as revealed by genomic analysis. *Journal of Bacteriology* 186:4376–4381.

Brugirard-Ricaud, K., Duchaud, E., Givaudan, A., Girard, P. A., Kunst, F., Boemare, N., Brehélin, M., and R. Zumbihl. 2005. Site-specific antiphagocytic function of the *Photorhabdus luminescens* type III secretion system during insect colonization. *Cellular Microbiology* 7:363–371.

Caldas, C., Cherqui, A., Pereira, A., and N. Simões. 2002. Purification and characterization of an extracellular protease from *Xenorhabdus nematophila* involved in insect immunosuppression. *Applied and Environmental Microbiology* 68:1297–1304.

Campbell, L. R. and R. Gaugler. 1991. Role of the sheath in desiccation tolerance of two entomopathogenic nematodes. *Nematologica* 37:324–332.

Campbell, J. F. and R. Gaugler. 1993. Nictation behaviour and its ecological implications in the host search strategies of entomopathogenic nematodes (Heterorhabditidae and Steinernematidae). *Behaviour* 126:3–4.

Ciche, T. A. and J. C. Ensign. 2003. For the insect pathogen *Photorhabdus luminescens*, which end of a nematode is out? *Applied and Environmental Microbiology* 69:1890–1897.

Ciche, T. A., Blackburn, M., Carney, J. R., and J. C. Ensign. 2003. Photobactin: A catechol siderophore produced by *Photorhabdus luminescens*, an entomopathogen mutually associated with *Heterorhabditis bacteriophora* NC1 nematodes. *Applied and Environmental Microbiology* 69:4706–4713.

Ciche, T. A., Darby, C., Ehlers, R.-U., Forst, S., and H. Goodrich-Blair. 2006. Dangerous liaisons: The symbiosis of entomopathogenic nematodes and bacteria. *Biological Control* 38:22–46.

Clarke, D. J. and B. C. A. Dowds. 1995. Virulence mechanisms of *Photorhabdus* sp. Strain K122 toward wax moth larvae. *Journal of Invertebrate Pathology* 66:149–155.

Couche, G. A., Lehrbach, P. R., Forage, R. G., Cooney, G. C., Smith, D. R., and R. P. Gregson. 1987. Occurrence of intracellular inclusions and plasmids in *Xenorhabdus* spp. *Journal of General Microbiology* 133:967–973.

Cowles, K. N. and H. Goodrich-Blair. 2005. Expression and activity of a *Xenorhabdus nematophila* haemolysin required for full virulence towards *Manduca sexta* insects. *Cellular Microbiology* 7:209–219.

Daborn, P. J., Waterfield, N., Silva, C. P., Au, C. P. Y., Sharma, S., and R. H. Ffrench-Constant. 2002. A single *Photorhabdus* gene, makes caterpillars floppy (mcf), allows *Escherichia coli* to persist within and kill insects. *Proceedings of the National Academy of Sciences USA* 99:10742–10747.

Derzelle, S., Duchaud, E., Kunst, F., Danchin, A., and P. Bertin. 2002. Identification, characterization, and regulation of a cluster of genes involved in carbapenem biosynthesis in *Photorhabdus luminescens. Applied and Environmental Microbiology* 68:3780–3789.

Derzelle, S., Ngo, S., Turlin, E., Duchaud, E., Namane, A., Kunst, F., Danchin, A., Bertin, P., and J. F. Charles. 2004. AstR-AstS a new two-component signal transduction system mediates swarming, adaptation to stationary phase and phenotypic variation in *Photorhabdus luminescens. Microbiology* 150:897–910.

Duchaud, E., Rusniok, C., Frangeul, L., Buchrieser, C., Givaudan, A., Taourit, S., Bocs, S., Boursaux-Eude, C., Chandler, M., Charles, J., Dassa, E., Derose, R., Derzelle, S., Freyssinet, G., Gaudriault, S., Médigue, C., Lanois, A., Powell, K., Siguier, P., Vincent, R., Wingate, V., Zouine, M., Glaser, P., Boemare, N., Danchin, A., and F. Kunst. 2003. The genome sequence of the entomopathogenic bacterium *Photorhabdus luminescens. Nature Biotechnology* 21:1307–1313.

Dunphy, G. B. 1995. Physicochemical properties and surface components of *Photorhabdus luminescens* influencing bacterial interaction with non-self response systems of nonimmune galleria mellonella larvae. *Journal of Invertebrate Pathology* 65:25–34.

Dunphy, G. B. and G. S. Thurston. 1990. Insect immunity. In *Entomopathogenic Nematodes in Biological Control*, Gaugler, R. and Kaya, H. K. (Eds.), pp. 301–323. Boca Raton: CRC Press.

Dunphy, G. B. and J. M. Webster. 1988. Lipopolysaccharides of *Xenorhabdus nematophilus* (*Enterobacteriaceae*) and their haemocyte toxicity in non-immune *Galleria mellonella* (Insecta: Lepidoptera) larvae. *Journal of General Microbiology* 134:1017–1028.

Dunphy, G. B. and J. M. Webster. 1991. Antihemocytic surface components of *Xenorhabdus nematophilus* var. Dutki and their modification by serum on nonimmune larvae of *Galleria mellonella. Journal of Invertebrate Pathology* 58:40–51.

Eleftherianos, I., Boundy, S., Joyce, S. A., Aslam, S., Marshall, J. W., Cox, R. J., Simpson, T. J., Clarke, D. J., Ffrench-Constant, R. H., and S. E. Reynolds. 2007. An antibiotic produced by an insect-pathogenic bacterium suppresses host defenses through phenoloxidase inhibition. *Proceedings of the National Academy of Sciences, USA* 104:2419–2424.

Endo, B. Y. and W. R. Nickle. 1991. Ultrastructure of the intestinal epithelium, lumen, and associated bacteria in *Heterorhabditis bacteriophora. Journal of the Helmithological Society of Washington* 58:202–212.

Ffrench-Constant, R. H., Waterfield, N., Daborn, P., Joyce, S., Bennett, H., Au, C., Dowling, A., Boundy, S., Reynolds, S., and D. Clarke. 2003. *Photorhabdus*: Towards a functional genomic analysis of a symbiont and pathogen. *FEMS Microbiology Reviews* 26:433–456.

Fischer-Le Saux, M., Viallard, V., Brunel, B., Normand, P., and N. E. Boemare. 1999. Polyphasic classification of the genus *Photorhabdus* and proposal of new taxa: *P. luminescens* subsp. *luminescens* subsp. nov., *P. luminescens* subsp. *akhurstii* subsp. nov., *P. luminescens* subsp. *laumondii* subsp. nov., *P. temperata* sp. nov., *P. temperata* subsp. *temperata* subsp. nov. and *P. asymbiotica* sp. nov. *International Journal of Systematic Bacteriology* 49:1645–1656.

Forst, S. and D. Clarke. 2002. Bacteria-nematode symbiosis. In *Entomopathogenic Nematology*, Gaugler, R. (Ed.), pp. 57–77. Wallingford, U.K.: CABI Publishing.

Forst, S. and K. Nealson. 1996. Molecular biology of the symbiotic-pathogenic bacteria *Xenorhabdus* spp. and *Photorhabdus* spp. *Microbiological Reviews* 60:21–43.

Forst, S., Dowds, B., Boemare, N., and E. Stackebrandt. 1997. *Xenorhabdus* and *Photorhabdus* spp.: Bugs that kill bugs. *Annual Review of Microbiology* 51:47–72.

Furgani, G., Böszörményi, E., Fodor, A., Máthé-Fodor, A., Forst, S., Hogan, J. S., Katona, Z., Klein, M. G., Stackebrandt, E., Szentirmai, A., Sztaricskai, F., and S. L. Wolf. 2008. *Xenorhabdus* antibiotics: A comparative analysis and potential utility for controlling mastitis caused by bacteria. *Journal of Applied Microbiology* 104:745–758.

Gaugler, R. 2002. *Entomopathogenic Nematology*, Gaugler, R. (Ed.), pp. 35–36. Wallingford, U.K.: CABI Publishing.

Georgis, R. 2002. The Biosys experiment: An insider's perspective. In *Entomopathogenic Nematology*, Gaugler, R. (Ed.), pp. 357–372. Wallingford, U.K.: CABI Publishing.

Gaugler, R., Campbell, J. F., Selvan, S., and E. E. Lewis. 1992. Large-scale inoculative releases of the entomopathogenic nematode *Steinernema glaseri*: Assessment 50 years later. *Biological Control* 2:181–187.

Gerrard, J. G., Joyce, S. A., Clarke, D. J., Ffrench-Constant, R. H., Nimmo, G. R., Looke, D. F. M., Feil, E. J., Pearce, L., and N. R. Waterfield. 2006. Nematode symbiont for *Photorhabdus asymbiotica*. *Emerging Infectious Diseases* 12:1562–1564.

Gerritsen, L. J. M. and P. H. Smits. 1993. Variation in pathogenicity of recombinations of *Heterorhabditis* and *Xenorhabdus luminescens* strains. *Fundamental and Applied Nematology* 16:367–373.

Gerritsen, L. J. M. and P. H. Smits. 1997. The influence of *Photorhabdus luminescens* strains and form variants on the reproduction and bacterial retention of *Heterorhabditis megidis*. *Fundamental and Applied Nematology* 20:317–322.

Givaudan, A. and A. Lanois. 2000. *flhdc*, the flagellar master operon of *Xenorhabdus nematophilus*: Requirement for motility, lipolysis, extracellular hemolysis, and full virulence in insects. *Journal of Bacteriology* 182:107–115.

Glaser, R. W. and H. Fox 1930. A nematode parasite of Japanese beetle (*Popillia japonica* Newm.). *Science* 70:16–17.

Goodrich-Blair, H. and D. J. Clarke. 2007. Mutualism and pathogenesis in *Xenorhabdus* and *Photorhabdus*: Two roads to the same destination. *Molecular Microbiology* 64:260–268.

Gupta, R. K., Patterson, S. S., Sayler, G. S., and S. A. Ripp. 2007. Lux expression in eukaryotic cells. United States Patent 7:300–792.

Hazir, S., Stackebrandt, E., Lang, E., Schumann, P., Ehlers, R. U., and N. Keskin. 2004. Two new subspecies of *Photorhabdus luminescens*, isolated from *Heterorhabditis bacteriophora* (Nematoda: Heterorhabditidae): *Photorhabdus luminescens* subsp. *kayaii* subsp. nov. and *Photorhabdus luminescens* subsp. *thracensis subsp. nov. Systematic and Applied Microbiology* 27:36–42.

Held, K. G., LaRock, C. N., D'Argenio, D. A., Berg, C. A., and C. M. Collins. 2007. A metalloprotease secreted by the insect pathogen *Photorhabdus luminescens* induces melanization. *Applied and Environmental Microbiology* 73:7622–7628.

Ji, D., Yi, Y., Kang, G.-H., Choi, Y.-H., Kim, P., Baek, N.-I., and Y. Kim. 2004. Identification of an antibacterial compound, benzylideneacetone, from *Xenorhabdus nematophila* against major plant-pathogenic bacteria. *FEMS Microbiology Letters* 239:241–248.

Johnigk, S. A. and R.-U. Ehlers. 1999. *Endotokia matricida* in hermaphrodites of *Heterorhabditis* spp. and the effect of the food supply. *Nematology: International Journal of Fundamental and Applied Nematological Research*. 1, pt. 7/8:717–726.

Kaya, H. K. and A. M. Koppenhöfer. 1999. Biology and ecology of insecticidal nematodes. *Proceedings Workshop: Optimal Use of Insecticidal Nematodes in Pest Management*, 28–30 August 1999, New Brunswick, NJ, pp. 1–8.

Kim, Y., Dongjin, J., Cho, S., and Y. Park. 2005. Two groups of entomopathogenic bacteria, *Photorhabdus* and *Xenorhabdus*, share an inhibitory action against phospholipase A2 to induce host immunodepression. *Journal of Invertebrate Pathology* 89:258–264.

Koppenhöfer, A. M., Baur, M. E., Stock, S. P., Choo, H. Y., Chinnasri, B., and H. K. Kaya. 1997. Survival of entomopathogenic nematodes within host cadavers in dry soil. *Applied Soil Ecology: A Section of Agriculture, Ecosystems and Environment* 6:231–240.

Lee, S. C., Stoilova-McPhie, S., Baxter, L., Fülöp, V., Henderson, J., Rodger, A., Roper, D. I., Scott, D. J., Smith, C. J., and J. A. W. Morgan. 2007. Structural characterization of the insecticidal toxin XptA1 reveals a 1.15 MDa tetramer with a cage-like structure. *Journal of Molecular Biology* 366:1558–1568.

Lengyel, K., Lang, E., Fodor, A., Szállás, E., Schumann, P., and E. Stackebrandt. 2005. Description of four novel species of *Xenorhabdus*, family *Enterobacteriaceae*: *Xenorhabdus budapestensis sp. nov.*, *Xenorhabdus ehlersii sp. nov., Xenorhabdus innexi sp. nov.*, and *Xenorhabdus szentirmaii sp. nov. Systematic and Applied Microbiology* 28:115–122.

Lewis, E. E., Gaugler, R., and R. Harrison. 1992. Entomopathogenic nematode host finding: Response to host contact cues by cruise and ambush foragers. *Parasitology* 105, pt. 2:309–315.

Lewis, E. E., Ricci, M., and R. Gaugler. 1996. Host recognition behavior predicts host suitability in the entomopathogenic nematode *Steinernema carpocapsae* (Rhabditida: Steinernematidae). *Parasitology* 113, pt. 6:573–579.

Li, J. X., Chen, G. H., Wu, H. M., and J. M. Webster. 1995. Identification of 2 pigments and a hydroxystilbene antibiotic from *Photorhabdus-luminescens. Applied and Environmental Microbiology* 61:4329–4333.

Li, J. X., Hu., K., and J. M. Webster. 1998. Antibiotics from *Xenorhabdus* spp. and *Photorhabdus* spp. (*Enterobacteriaceae*) (Review). *Chemistry of Heterocyclic Compounds* 34:1331–1339.

Liu, D., Burton, S., Glancy, T., Li, Z.-S., Hampton, R., Meade, T., and D. J. Merlo. 2003. Insect resistance conferred by 283-kDa *Photorhabdus luminescens* protein TcdA in *Arabidopsis thaliana. Nature Biotechnology* 21:1222–1228.

Maoz, A., Mayr, R., Bresolin, G., Neuhaus, K., Francis, K. P., and S. Scherer. 2002. Sensitive in situ monitoring of a recombinant bioluminescent *Yersinia enterocolitica* reporter mutant in real time on Camembert cheese. *Applied and Environmental Microbiology* 68:5737–5740.

Marokhazi, J., Waterfield, N., LeGoff, G., Feil, E., Stabler, R., Hinds, J., Fodor, A., and R. H. Ffrench-Constant. 2003. Using a DNA microarray to investigate the distribution of insect virulence factors in strains of *Photorhabdus* bacteria. *Journal of Bacteriology* 185:4648–4656.

Martens, E. C., Heungens, K., and H. Goodrich-Blair. 2003. Early colonization events in the mutualistic association between *Steinernema carpocapsae* nematodes and *Xenorhabdus nematophila* bacteria. *Journal of Bacteriology* 185:3147–3154.

McCoy, E. E. and R. W. Glaser. 1936. Nematode culture for Japanese beetle control. *New Jersey Department of Agriculture Circular* 265:3–9.

McInerney, B. V., Gregson, R. P., Lacey, M. J., Akhurst, R. J., Lyons, G. R., Rhodes, S. H., Smith, D. R. J., Engelhardt, L. M., and A. H. White. 1991a. Biologically active metabolites from *Xenorhabdus* spp., part 1. Dithiolopyrrolone derivatives with antibiotic activity. *Journal of Natural Products* 54:774–784.

McInerney, B. V., Taylor, W. C., Lacey, M. J., and R. J. Akhurst. 1991b. Biologically active metabolites from *Xenorhabdus* spp., part 2. Benzopyran-1-one derivatives with gastroprotective activity. *Journal of Natural Products* 54:785–795.

Mohd-Salleh, M. B. and L. C. Lewis. 1982. Feeding deterrent response of corn insects to beta-exotoxin of *Bacillus thuringiensis. Journal of Invertebrate Pathology* 39:323–328.

Morgan, J. A. W., Sergeant, M., Ellis, D., Ousley, M., and P. Jarrett. 2001. Sequence analysis of insecticidal genes from *Xenorhabdus nematophilus* PMFI296. *Applied and Environmental Microbiology* 67:2062–2069.

Nakayama, K., Takashima, K., Ishihara, H., Shinomiya, T., Kageyama, M., Kanaya, S., Ohnishi, M., Murata, T., Mori, H., and T. Hayashi. 2000. The r-type pyocin of *Pseudomonas aeruginosa* is related to p2 phage, and the f-type is related to lambda phage. *Molecular Microbiology* 38:213–231.

Nishimura, Y., Hagiwara, A., Suzuki, T., and S. Yamanaka. 1994. *Xenorhabdus japonicus sp. nov*. associated with the nematode *Steinernema kushidai. World Journal of Microbiology and Biotechnology* 10:207–210.

Paik, S., Park, M. K., Jhun, S. H., Park, H. K., Lee, C. S., Cho, B. R., Byun, H. S., Choe, S. B., and S. I. Suh. 2003. Isolation and synthesis of tryptamine derivatives from a symbiotic bacterium *Xenorhabdus nematophilus* PC. *Bulletin of the Korean Chemical Society* 24:623–626.

Park, D. and S. Forst. 2006. Co-regulation of motility, exoenzyme and antibiotic production by the EnvZ-OmpR-FlhDC-FliA pathway in *Xenorhabdus nematophila. Molecular Microbiology* 61:1397–1412.

Park, Y., Kim, Y., Putnam, S. M., and D. W. Stanley. 2003. The bacterium *Xenorhabdus nematophilus* depresses nodulation reactions to infection by inhibiting eicosanoid biosynthesis in tobacco hornworms, *Manduca sexta. Archives of Insect Biochemistry and Physiology* 52:71–80.

Park, Y., Herbert, E. E., Cowles, C. E., Cowles, K. N., Menard, M. L., Orchard, S. S., and H. Goodrich-Blair. 2007. Clonal variation in *Xenorhabdus nematophila* virulence and suppression of *Manduca sexta* immunity. *Cellular Microbiology* 9:645–656.

Paul, V. J., Frautschy, S., Fenical, W., and K. H. Nealson. 1981. Antibiotics in microbial ecology. Isolation and structure assignment of several new antibacterial compounds from the insect-symbiotic bacteria *Xenorhabdus* spp. *Journal of Chemical Ecology* 7:589–597.

Peel, M. M., Alfredson, D. A., Gerrard, J. G., Davis, J. M., Robson, J. M., McDougall, R. J., Scullie, B. L., and R. J. Akhurst. 1999. Isolation, identification, and molecular characterization of strains of *Photorhabdus luminescens* from infected humans in Australia. *Journal of Clinical Microbiology* 37:3647–3653.

Poinar, G. O., Jr. 1990. Biology and taxonomy of Steinernematidae and Heterorhabditidae. In *Entomopathogenic Nematodes in Biological Control*, Gaugler, R. and Kaya, H. K. (Eds), pp. 23–61. Boca Raton: CRC Press.

Poinar, G. O., Jr. 1993. Origins and phylogenetic relationships of the entomophilic rhabditids, *Heterorhabditis* and *Steinernema. Fundamental and Applied Nematology* 16:333–338.

Poinar, G. O., Jr. and G. M. Thomas. 1965. A new bacterium, *Achromobacter nematophilus sp. nov.* (*Achromobacteriaceae*: Eubacteriales), associated with a nematode. *International Bulletin of Bacteriological Nomenclature and Taxonomy* 15:249–252.

Poinar, G. O., Jr. and G. M. Thomas. 1966. Significance of *Achromobacter* Poinar and Thomas (*Achromobacteriaceae: Eubacteriales*) in the development of the nematode, DD-136 (*Neoplectana* sp. *Steinernematidae*). *Parasitology* 56:385–390.

Ribeiro, C., Vignes, M., and M. Brehélin. 2003. *Xenorhabdus nematophila* (*Enterobacteriaceae*) secretes a cation-selective calcium-independent porin which causes vacuolation of the rough endoplasmic reticulum and cell lysis. *Journal of Biological Chemistry* 278:3030–3039.

Richardson, W. H., Schmidt, T. M., and K. H. Nealson. 1988. Identification of an anthraquinone pigment and a hydroxystilbene antibiotic from *Xenorhabdus luminescens. Applied and Environmental Microbiology* 54:1602–1605.

Sergeant, M., Jarrett, P., Ousley, M., and J. A. W. Morgan. 2003. Interactions of insecticidal toxin gene products from *Xenorhabdus nematophilus* PMFI296. *Applied and Environmental Microbiology* 69:3344–3349.

Sharma, S., Waterfield, N., Bowen, D., Rocheleau, T., Holland, L., James, R., and R. Ffrench-Constant. 2002. The lumicins: Novel bacteriocins from *Photorhabdus luminescens* with similarity to the uropathogenic-specific protein (usp) from uropathogenic *Escherichia coli. FEMS Microbiology Letters* 214:241–249.

Sicard, M., Brugirard-Ricaud, K., Pagès, S., Lanois, A., Boemare, N. E., Brehélin, M., and A. Givaudan. 2004a. Stages of infection during the tripartite interaction between *Xenorhabdus nematophila*, its nematode vector, and insect hosts. *Applied and Environmental Microbiology* 70:6473–6480.

Sicard, M., Ferdy, J., Pagès, S., Le Brun, N., Godelle, B., Boemare, N., and C. Moulia. 2004b. When mutualists are pathogens: An experimental study of the symbioses between *Steinernema* (entomopathogenic nematodes) and *Xenorhabdus* (bacteria). *Journal of Evolutionary Biology* 17:985–993.

Sicard, M., Tabart, J., Boemare, N. E., Thaler, O., and C. Moulia. 2005. Effect of phenotypic variation in *Xenorhabdus nematophila* on its mutualistic relationship with the entomopathogenic nematode *Steinernema carpocapsae. Parasitology* 131:687–694.

Silva, C. P., Waterfield, N. R., Daborn, P. J., Dean, P., Chilver, T., Au, C. P. Y., Sharma, S., Potter, U., Reynolds, S. E., and R. H. Ffrench-Constant. 2002. Bacterial infection of a model insect: *Photorhabdus luminescens* and *Manduca sexta. Cellular Microbiology* 4:329–339.

Somvanshi, V. S., Lang, E., Ganguly, S., Swiderski, J., Saxena, A. K., and E. Stackebrandt. 2006. A novel species of *Xenorhabdus*, family *Enterobacteriaceae*: *Xenorhabdus indica sp. nov.*, symbiotically associated with entomopathogenic nematode *Steinernema thermophilum* Ganguly and Singh, 2000. *Systematic and Applied Microbiology* 29:519–525.

Stanley, D. W. and J. S. Miller. 2006. Eicosanoid actions in insect cellular immune functions. *Entomologia Experimentalis et Applicata* 119:1–13.

Sundar, L. and F. N. Chang. 1992. The role of guanosine-3′,5′-bis-pyrophosphate in mediating antimicrobial activity of the antibiotic 3,5-dihydroxy-4-ethyl-trans-stilbene. *Antimicrobial Agents and Chemotherapy* 36:2645–2651.

Sundar, L. and F. N. Chang. 1993. Antimicrobial activity and biosynthesis of indole antibiotics produced by *Xenorhabdus nematophilus. Journal of General Microbiology* 139:3139–3148.

Sztaricskai, F., Z. Dinya, Gy. Batta, E. Szallas, A. Szentirmai, and A. Fodor. 1992. Anthraquinones produced by enterobacters and nematodes. *Acta Chimica Hungarica Models in Chemistry* 129:697–707.

Tailliez, P., Pagès, S., Ginibre, N., and N. Boemare. 2006. New insight into diversity in the genus *Xenorhabdus*, including the description of ten novel species. *International Journal of Systematic and Evolutionary Microbiology* 56:2805–2818.

Thaler, J. O., Baghdiguian, S., and N. Boemare. 1995. Purification and characterization of xenorhabdicin, a phage tail-like bacteriocin, from the lysogenic strain f1 of *Xenorhabdus nematophilus*. *Applied and Environmental Microbiology* 61:2049–2052.

Thomas, G. M. and G. O. Poinar. 1979. *Xenorhabdus* gen-nov, a genus of entomopathogenic, nematophilic bacteria of the family *Enterobacteriaceae*. *International Journal of Systematic Bacteriology* 29:352–360.

Timper, P. and H. K. Kaya. 1989. Role of the second-stage cuticle of entomogenous nematodes in preventing infection by nematophagous fungi. *Journal of Invertebrate Pathology* 54:314–321.

Vigneux, F., Zumbihl, R., Jubelin, G., Ribeiro, C., Poncet, J., Baghdiguian, S., Givaudan, A., and M. Brehélin. 2007. The *xaxAB* genes encoding a new apoptotic toxin from the insect pathogen *Xenorhabdus nematophila* are present in plant and human pathogens. *Journal of Biological Chemistry* 282:9571–9580.

Visick, K. L. and E. G. Ruby. 2006. *Vibrio fischeri* and its host: It takes two to tango. *Current Opinion in Microbiology* 9:632–638.

Wang, J. and R. A. Bedding. 1996. Population development of *Heterorhabditis bacteriophora* and *Steinernema carpocapsae* in the larvae of *Galleria mellonella*. *Fundamental and Applied Nematology* 19:363–367.

Wang, Y. and R. Gaugler. 1999. *Steinernema glaseri* surface coat protein suppresses the immune response of *Popillia japonica* (Coleoptera: Scarabaeidae) larvae. *Biological control: Theory and Applications in Pest Management* 14:45–50.

Wang, Y., Campbell, J. F., and R. Gaugler. 1995. Infection of entomopathogenic nematodes *Steinernema glaseri* and *Heterorhabditis bacteriophora* against *Popillia japonica* (Coleoptera: Scarabaeidae) larvae. *Journal of Invertebrate Pathology* 66:178–184.

Waterfield, N. R., Daborn, P. J., and R. H. Ffrench-Constant. 2002. Genomic islands in *Photorhabdus*. *Trends in Microbiology* 10:541–545.

Waterfield, N. R., Daborn, P. J., Dowling, A. J., Yang, G., Hares, M., and R. H. Ffrench-Constant. 2003. The insecticidal toxin Makes caterpillars floppy 2 (Mcf2) shows similarity to HrmA, an avirulence protein from a plant pathogen. *FEMS Microbiology Letters* 229:265–270.

Watson, R. J., Joyce, S. A., Spencer, G. V., and D. J. Clarke. 2005. The *exbD* gene of *Photorhabdus temperata* is required for full virulence in insects and symbiosis with the nematode *Heterorhabditis*. *Molecular Microbiology* 56:763–773.

Webster, J. M., Chen, G., Hu, K., and J. Li. 2002. Bacterial metabolites. In *Entomopathogenic Nematology*, Gaugler, R. (Ed.), pp. 99–114. New York: CABI Publishing.

Webster, J. M., Chen, G., and J. Li. 1998. Parasitic worms: An ally in the war against the superbugs. *Parasitology Today* 14:161–163.

Yang, G., Dowling, A. J., Gerike, U., Ffrench-Constant, R. H., and N. R. Waterfield. 2006. *Photorhabdus* virulence cassettes confer injectable insecticidal activity against the wax moth. *Journal of Bacteriology* 188:2254–2261.

You, J., Liang, S., Cao, L., Liu, X., and R. Han. 2006. Nutritive significance of crystalline inclusion proteins of *Photorhabdus luminescens* in *Steinernema* nematodes. *FEMS Microbiology Ecology* 55:178–185.

Zhou, X., Kaya, H. K., Heungens, K., and H. Goodrich-Blair. 2002. Response of ants to a deterrent factor(s) produced by the symbiotic bacteria of entomopathogenic nematodes. *Applied and Environmental Microbiology* 68:6202–6209.

8

Interspecies Competition in a Bacteria–Nematode Mutualism

Nydia Morales-Soto, Holly Snyder, and Steven Forst

CONTENTS

8.1 Introduction

Host–microbe symbioses range from beneficial (mutualistic) interactions that enhance the fitness of both partners to detrimental (pathogenic) associations that compromise the fitness of the host. While pathogenic interactions have been studied extensively, mutualistic associations are less understood. Mutualistic interactions can vary considerably. Obligate associations, such as the intracellular bacteria *Buchnera* sp. and their aphid hosts, involve provisioning of essential nutrients to the host (Dale and Moran, 2006). In defensive mutualisms, as represented by fungal endophytes such as *Epichloë/Neotyphodium* that inhabit intercellular spaces of leaf structures of ryegrass, secondary metabolites are produced to repel herbivorous insects (Tanaka et al., 2005). The bioluminescent bacterium *Vibrio fischeri* is an example of a free-living microbe that colonizes its host, the bobtail squid, and produces light to counterilluminate shadows to help the squid avoid predation (Nyholm and McFall-Ngai, 2004). Bacterial species of the genus *Xenorhabdus* are unique in their ability to engage in both pathogenic and mutualistic associations with two different host organisms. *Xenorhabdus* sp. form mutualistic relationships with entomopathogenic nematodes (EPNs) and are pathogenic to a diverse array of insects. *Xenorhabdus* sp. are considered to be semi-obligate mutualists since they have not been isolated outside the host organisms but can be grown independently in the laboratory. During their mutualistic stage, *Xenorhabdus* sp. are carried in the intestine of the nematode. The transition to the pathogenic stage occurs when the bacteria are released into the hemocoel after the nematode enters the insect host (Forst and Clarke, 2002; Herbert and Goodrich-Blair, 2007). The nematode uses *Xenorhabdus* sp. for its pathogenic potential to kill the host as well as to provide nutrient sources for growth and development. The bacteria benefit by being vectored to the nutrient-rich source of the insect hemolymph. This chapter focuses on the competitive interactions that occur within the insect host and the mechanisms *Xenorhabdus* sp. employ to secure nutrient resources for themselves and their nematode partner. The contribution that the nematode makes to interspecies competition is less understood and therefore will not be covered in this chapter. For a detailed discussion of the molecular pathways involved in pathogenesis and mutualism in *Xenorhabdus* sp. and the sister taxa, *Photorhabdus* sp., the reader is referred to several excellent recent reviews on these topics (Joyce et al., 2006; Goodrich-Blair, 2007; Goodrich-Blair and Clarke, 2007; Herbert and Goodrich-Blair, 2007).

8.2 Life Cycle

Bacteria of the genus *Xenorhabdus* associate with soil EPNs of the family *Steinernematidae* (Figure 8.1). The bacteria–nematode complex has evolved as an intricate relationship in which both partners enjoy the benefits of mutualistic and parasitic associations (Forst and Clarke, 2002; Herbert and Goodrich-Blair, 2007). During the tripartite association, the bacteria–nematode pair acts as parasite of a variety of insects. The infective juvenile stage (IJ) of the nematode is the free-living, nonfeeding stage of the life cycle dedicated to finding and gaining access to insect larvae. *Steinernema* IJs carry their *Xenorhabdus* symbionts in a specialized region of the anterior intestine, commonly known as the receptacle or vesicle (Snyder et al., 2007). When the IJs encounter an insect larva, they gain access through natural openings (e.g., mouth, anus) or by penetrating the larval cuticle. Once inside the insect, the IJs make their way into the hemocoel releasing their bacterial symbionts into the nutrient-rich hemolymph. The outcome of the infection by the nematode–bacteria complex is insect death. *Xenorhabdus* sp. produce various toxins and cytotoxins that participate in the death of the insect (Forst and Clarke, 2002; Sergeant et al., 2006). As the bacteria reach high cell density they secrete exoenzymes that degrade insect tissues and macromolecules, and antibiotic compounds that suppress the growth of competitors (Webster

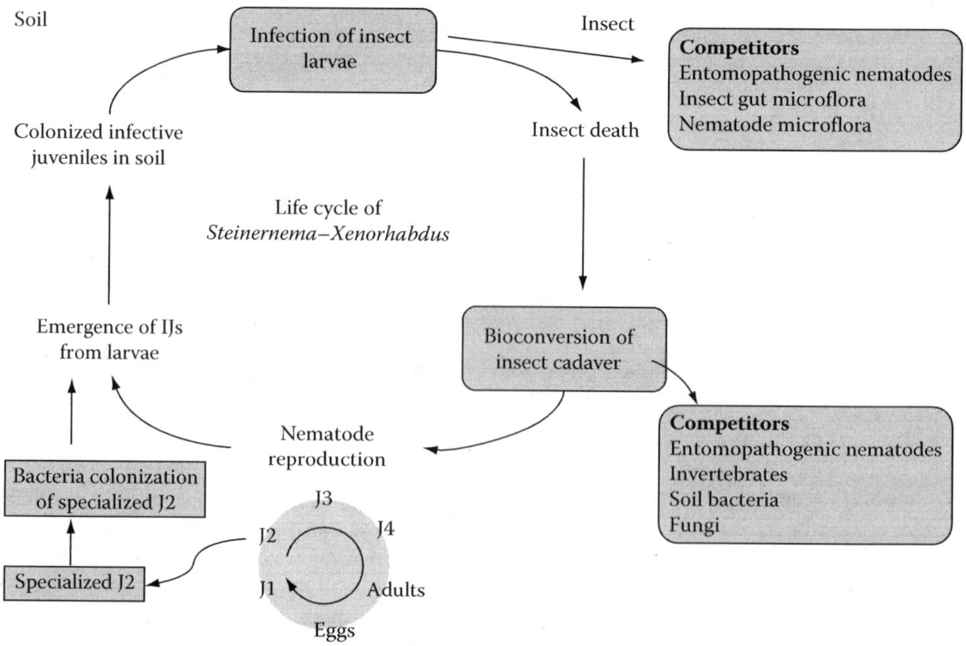

FIGURE 8.1 Schematic diagram of the *Steinernema–Xenorhabdus* life cycle.

et al., 2002; Park and Forst, 2006). The nematodes develop and reproduce in the hemocoel feeding on both bacteria and the nutrients derived from insect sources. With the depletion of nutrient supplies, the nematodes develop into the IJ stage, which are colonized by its *Xenorhabdus* partner. The IJs emerge from the cadaver into the soil where they search for a new insect host. Since the EPN–bacteria complex is pathogenic toward diverse insects, such as caterpillars, weevils, grubs, and borers, they are used as biological control agents against these agricultural pests (Grewal, 2002).

More than 50 *Steinernema* and 20 *Xenorhabdus* species have been characterized to date (Tailliez et al., 2006). Each nematode species is colonized by a specific *Xenorhabdus* sp. Some *Xenorhabdus*, such as *X. nematophila*, colonize a single *Steinernema* sp. while others, such as *X. bovienii*, form symbiotic associations with as many as seven different *Steinernema* sp. The *X. nematophila* and *S. carpocapsae* partnership is one of the most well-studied beneficial host–microbe associations. In the laboratory, it is possible to raise aposymbiotic *S. carpocapsae* nematodes and grow *X. nematophila* in pure culture making it an excellent system for genetic manipulation to study mutualistic relationships.

During the infectious process, the IJ penetrates the intestinal wall and invades the insect hemocoel allowing microbial competitors from the intestine as well as those carried on the nematode cuticle to infect the hemolymph. The presence of the nematode and bacteria presumably stimulate the host immune response thereby limiting growth of antagonistic competitors (Eleftherianos et al., 2006). Once in the hemocoel, exposure of the IJ to hemolymph triggers the process of recovery that results in the movement of *X. nematophila* out of the anterior intestine and release through the anus (Snyder et al., 2007). *X. nematophila* initially colonizes the muscles and connective tissue surrounding the anterior region of the midgut (Sicard et al., 2004a). Early in infection *X. nematophila* secretes compounds that inactivate the insect immune response by inhibiting hemocyte phospholipase A(2) activity and eicosanoid production that are required for immune system activation (Park and Kim, 2003; Park et al., 2003; Park et al., 2004). Inactivation of the immune system allows competitors that have entered the hemocoel to begin to grow and compete with *X. nematophila*. As *X. nematophila* reach higher cell density it secretes broadly active antibiotics that eliminate sensitive competitors. The types of microbial competitors *X. nematophila* encounter vary depending on the insect larva that they infect.

The soil environment presents numerous competitive challenges for the nematode–bacteria complex. In order to serve the metabolic needs of *S. carpocapsae* and *X. nematophila*, the insect cadaver must be protected from other opportunistic invaders such as soil bacteria and fungi, other nematodes, and invertebrate organisms (Kaya, 2002). Two or more EPN species can occur together in the same ecological niche and coinvade the same insect resulting in interspecies competition between the different nematode species and bacterial symbionts released into the hemocoel.

8.3 Interspecies Competition

8.3.1 Competitors from Insect Microflora

EPNs are pathogenic to several orders of insects including Lepidoptera, Diptera, Coleoptera, Orthoptera, and Hymenoptera (Laumond et al., 1979; Poinar, 1979). The nematodes encounter different consortia of commensal microorganisms as they migrate through the intestine of the diverse hosts. These microbes represent potential competitors for the nutrient resources of the insect if they gain access to the hemolymph as the nematode breaches the intestinal barrier. The intestinal microbiota of only a few insects has been studied. A survey of 37 wild insect taxa revealed that gram-positive *Enterococcus* sp. were highly represented in the insects analyzed (Martin and Mundt, 1972). In a recent survey of the microbiota of the Dipteran, *Drosophila melanogaster*, five different bacterial phyla were identified (Cox and Gilmore, 2007). The Firmicutes (low G+C gram-positive bacteria) accounted for 37% of the clones analyzed, with *Enterococcus faecalis* representing 90% of this phylum. Gram-negative proteobacteria accounted for 61% of the clones analyzed, the majority of which belonged to the alpha-proteobacteria (mainly *Acetobacter* sp.).

An extensive survey of cultured and uncultured midgut microbiota of the gypsy moth larvae (*Lymantria dispar*) grown on artificial media and different natural leaf sources demonstrated that the community structure varied substantially with changes in diet (Broderick et al., 2004). *E. faecalis* was the most prevalent species isolated from insects raised on artificial diet. In contrast, the gamma-proteobacterium, *Serratia marcescens,* was predominant in insects raised on aspen leaves while *E. faecalis* was a minor member of the community. Five or more bacterial genera were isolated under each of the dietary conditions examined.

Three different lepidopterans have been used as model insects to study the life cycle of EPNs: the wax worm (*Galleria mellonella*), tobacco hornworm (*Manduca sexta*), and common cutworm (*Spodoptera littoralis*). *Enterococcus* sp. were by far the most dominant bacteria isolated from *G. mellonella* (Lysenko and Weiser, 1974; Isaacson and Webster, 2002; Walsh and Webster, 2003). In an analysis identifying specifically gram-negative bacteria, three genera, *Salmonella*, *Pasteurella*, and *Xanthomonas*, were isolated from the intestine of *Galleria* (Gouge and Snyder, 2006). Examination of microbiota of *M. sexta* raised on standard laboratory diet containing antibiotics revealed predominantly the gram-positive bacteria, *Paenibacillus* and *Bacillus*, and the gram-negative bacterium, *Methylobacterium*. In contrast, the microbiota of *M. sexta* raised on standard diet without antibiotics contained predominantly gram-positive *Staphylococcus* and *Pediococcus* sp. and no gram-negative bacteria (van der Hoeven and Forst, unpublished). The microbiota of *Spodoptera* has not been analyzed yet. These findings indicate that the microbial diversity that the EPNs potentially encounter as they invade different host organisms is extensive. In addition, the effect of insect diet on the microbial community will further expand this diversity.

8.3.2 Nonsymbiotic Competitors from Entomopathogenic Nematodes

An additional source of potential competitors is nonsymbiotic bacteria carried between the juvenile cuticle and outer sheath of the nematode. Several gram-negative proteobacteria were shown to be associated with *Steinernema* sp. raised under laboratory conditions (Lysenko and Weiser, 1974; Boemare, 1983; Gouge and Snyder, 2006). The bacteria associated with *S. carpocapsae* included *Enterobacter gergoviae*, *S. marcescens*, *Pseudomonas* sp., and *Salmonella* sp. (Gouge and Snyder, 2006). In *G. mellonella* infected with *S. carpocapsae,* intestinal microflora were present early in infection, then declined to zero as *X. nematophila* grew to high levels. In contrast, *S. marcescens*, *Pseudomonas* sp., and *Salmonella* sp., were isolated from the hemolymph late in infection suggesting that they were resistant to the antimicrobial compounds produced by *X. nematophila.* Whether these or other bacterial species are associated with native EPNs in the soil environment, and compete for nutrient resources of the insect, remains to be determined.

Invasion of *Galleria* by other steinernematid nematodes yielded similar results. In insects infected with either *S. riobrave* (Isaacson and Webster, 2002) or *S. glaseri* or *S. feltiae* (Walsh and Webster, 2003) *Enterococcus* sp. were present early in infection and were subsequently eliminated as *Xenorhabdus* sp. grew to high levels. Later in infection, unidentified gram-negative species were detected and continued to increase to levels approaching that of *Xenorhabdus* (Walsh and Webster, 2003). In a recent study, it was shown that intestinal microflora appeared early in *Galleria* infected with either *S. riobrave* or *S. feltiae,* then disappeared as *Xenorhabdus* levels increased (Gouge and Snyder, 2006). Later in infection, gram-negative proteobacteria apparently derived from the invading EPN were isolated from the hemolymph. These competitors were presumably resistant to the antibiotics produced by *Xenorhabdus* sp. It is apparent from these findings that nonsymbiotic competitors are present during the period in which the IJ stage nematode is colonized by *Xenorhabdus* indicating that colonization does not take place under monoculture conditions. The cellular and molecular processes that account for species-specific colonization of the *Steinernema* nematode by its *Xenorhabdus* partner when other bacteria are present in the hemolymph are not presently understood.

8.3.3 Competition with Other *Xenorhabdus* Species

Aposymbiotic *S. carpocapsae* themselves are able to reproduce in *G. mellonella*, although at lower levels than symbiotic nematodes (Sicard et al., 2003). Co-injection of *Galleria* with aposymbiotic *S. carpocapsae* and nonnative *Xenorhabdus* sp. (e.g., *X. innexi* or *X. bovienii*) eliminated nematode reproduction indicating that the nonnative *Xenorhabdus* sp. were antagonistic to *S. carpocapsae* reproduction (Sicard et al., 2004b). Direct competition experiments between *X. nematophila* and different *Xenorhabdus* sp. co-injected with aposymbiotic *S. carpocapsae* were conducted to determine whether the presence of *X. nematophila* would override the antagonistic effect. Co-injection of *X. nematophila* with *X. innexi* did not restore the reproductive success of *S. carpocapsae* indicating that *X. nematophila* was unable to overcome the antagonistic effect of *X. innexi*. It was further shown that *X. innexi* was insensitive to R-type bacteriocins (see below) produced by *X. nematophila* (Sicard et al., 2005). In contrast, co-injection of *X. nematophila* with *X. bovienii* reduced the antagonistic effect such that *S. carpocapsae* reproduction approached normal levels. Interestingly, *X. bovienii* was shown to be sensitive to the R-type bacteriocins suggesting that the restoration of nematode reproduction was due to the elimination of the competitor, *X. bovienii*, by the R-type bacteriocins. Moreover, co-injection of both *X. nematophila* and *X. innexi* together with aposymbiotic *S. carpocapsae* and *S. scapterisci* (partner of *X. innexi*) resulted in an almost exclusive reproduction of the latter nematode (Sicard et al., 2006). These findings are consistent with the negative effect *X. innexi* had on *S. carpocapsae* reproduction (Sicard et al., 2004b) and the insensitivity of *X. innexi* to R-type bacteriocins of *X. nematophila* (Sicard et al., 2005). Whether *X. innexi* produces antimicrobial compounds or bacteriocins that suppress *X. nematophila* growth is not presently known.

8.3.4 Competition between *Steinernema* Species

Competition between *Xenorhabdus* sp. in nature depends on whether different steinernematid nematodes are able to coinvade an insect host. Since two or more EPN species can occupy the same ecological niche (Lewis, 2002) these competitive interactions are feasible. The outcome of the competition involves numerous factors, one of which is the degree and level of colonization of the respective nematodes. It has been shown that almost all *S. carpocapsae* IJs in a population are colonized while very few *S. scapterisci* IJs are colonized (Sicard et al., 2006). In competitions between symbiotic *S. carpocapsae* and *S. scapterisci*, >90% of the progeny IJs are *S. carpocapsae*. Thus, even though *X. innexi* is antagonistic to *S. carpocapsae* reproduction (see above), competition between the symbiotic nematodes strongly favors *S. carpocapsae* since it introduces large numbers of its bacterial symbiont relative to the low numbers of *X. innexi* introduced by *S. scapterisci*.

It has been shown that certain *Steinernema–Xenorhabdus* complexes are better suited for the infection of specific insect larvae. For instance, mortality of infected black cutworm (*Agrotis ipsilon*) was noticeably higher with *S. carpocapsae* (100%) compared with *S. glaseri* (83%) and *S. riobrave* (75%) (Koppenhöfer and Kaya, 1996). These differences in black cutworm mortality were further demonstrated in competition studies between *S. carpocapsae* and *S. glaseri*. *S. carpocapsae* was able to infect and control 67% of the larvae, whereas *S. glaseri* was obtained from 8% of the larvae while 25% of the infected larvae contained a mixture of *S. carpocapsae* and *S. glaseri*. On the other hand, *S. glaseri* was highly pathogenic (83% mortality) against the white grub larvae (*Cyclocephala hirta*) compared with *S. carpocapsae* (17% mortality) and *S. riobrave* (0% mortality). Finally, competition between *Steinernema* and *Heterorhabditis* can occur when a host is coinvaded by both nematodes (Alatorre-Rosas and Kaya, 1990, 1991). The multifaceted nature of the competitive interactions between EPNs in soil involve ecological factors (Lewis, 2002) and host foraging strategies (Campbell et al., 2003) that are just beginning to be understood.

8.4 Production of Antimicrobial Compounds and Suppression of Competition

Early studies on *Xenorhabdus* sp. proposed that they produced antimicrobial compounds to suppress growth of competitors to prevent putrefaction of the insect cadaver (Poinar et al., 1980). A subsequent survey of six isolates of *Xenorhabdus* and three isolates of *Photorhabdus* showed that antibiotic activity was high against gram-positive bacteria and more variable against gram-negative bacteria and fungi (Akhurst, 1982). A total of 20 *Xenorhabdus* sp. have been characterized, all of which produce antibiotics active against the indicator organism, *Micrococcus luteus* (Tailliez et al., 2006). This conservation of antibiotic production emphasizes the importance of these compounds in the life cycle of the *Xenorhabdus*–nematode complex.

8.4.1 Antibiotic Production in *Xenorhabdus nematophila*

Several unique antimicrobial compounds have been isolated from *X. nematophila* (Table 8.1). The most predominant were the water soluble benzopyran-1-one compounds called xenocoumacins (McInerney et al., 1991b) isolated from 48 h broth cultures of *X. nematophila* strain. All *X. nematophila* produce two forms of xenocoumacin, I and II, in approximately equal amounts. These compounds are structurally and pharmacologically similar to the amicoumacins produced by *Bacillus pumilus*. Both xenocoumacin I and II were active against low G+C gram-positive bacteria and *Escherichia coli* but were inactive against the other *Enterobacteriaceae* as well as *Pseudomonas aeruginosa*. Xenocoumacin I was active against several fungi species but was not active toward *Candida albicans* while xenocoumacin II did not display fungicidal activity. Structural analysis of xenocoumacin I predicted that its synthesis involves the incorporation of leucine and arginine residues and four acetate units (McInerney et al., 1991b). Nonribosomal peptide synthetases (nrps) and polyketide synthases (pks) are involved in incorporating amino acids and acetate units, respectively, into antibiotics and other secondary metabolites. The genome sequence of *X. nematophila* ATCC19061 has recently been completed (http://maizeapache.ddpsc.org/bact_db/). The nrps and pks family of genes was found to be expanded in *X. nematophila* relative to most other bacterial genomes. Current efforts are focused on identifying the nrps and pks genes involved in xenocoumacin synthesis.

Xenocoumacins were shown to be produced *in vivo*. Water extracts of macerated *G. mellonella* cadavers infected with *X. nematophila* strain DD136 displayed antibiotic activity against gram-positive species while the activity against gram-negative species was more variable (Maxwell et al., 1994). HPLC analysis identified the presence of xenocoumacin I and II in a ratio of 1:1 in these extracts. Xenocoumacin activity was initially detected 24 h after infection and reached maximal levels by 36 h. Antibiotic activity was not recovered by extraction of *X. nematophila*-infected *G. mellonella* with organic solvents such as ethanol, ethyl acetate, or chloroform.

Several nonpolar compounds possessing antibiotic activity have been isolated from broth culture supernatants of *X. nematophila* (Paul et al., 1981; Sundar and Chang, 1993; Ji et al., 2004). Two related indole-derived compounds were isolated by ethyl acetate extraction of cell-free supernatants from 48 h cultures (Sundar and Chang, 1993). These compounds were not detectable during log phase growth and reached high levels during stationary phase. Purified indole antibiotics were active against low G+C gram-positive bacteria, members of the *Enterobacteriaceae* and *Pseudomonas* sp. Whether these compounds have fungicidal activity is not presently known. Exposure to the indole antibiotics induced the

TABLE 8.1
Antibiotic Compounds of *Xenorhabdus* Species

Species	Strain	Compound	Common Name	Reference
X. nematophila	All	Benzopyran	Xenocoumacin	McInerney et al. (1991b)
	N/A[a]	Indole-type	None	Paul et al. (1981) Sundar and Chang (1993)
	BC1 D1 19061	Indole-type	Nematophin	Li et al. (1997)
	K1	Benzylidene acetone	None	Ji et al. (2004)
X. bovienii (*S. feltiae*)	T319	Dithiolpyrrolone	Xenorhabdin	McInerney et al. (1991a)
	N/A[a]	Oxidized dithiolpyrrolone	Xenorxide	Li et al. (1998)
X. miraniensis	Q1	Dithiolpyrrolone	Xenorhabdin	McInerney et al. (1991a)
	Q1	Benzopyran	Xenocoumacin	McInerney et al. (1991b)

[a] N/A, not available

accumulation of the regulatory nucleotide, ppGpp, in *E. coli* and *Pseudomonas putida* and dramatically inhibited RNA synthesis. A novel indole-type antibiotic called nematophin has been isolated from ethyl acetate extraction of culture supernatants of three different strains of *X. nematophila* (BC1, D1, ATCC19061; Li et al., 1997). Nematophin was active against *Bacillus* and *Staphylococcus* sp. but unlike other antibiotics of *X. nematophila,* it was not active against *M. luteus*. The concentration of nematophin was maximal after 48 h of incubation and varied greatly between the three different strains. Another nonpolar antibiotic, benzylideneacetone, was obtained from butanol extraction of 48 h cultures of *X. nematophila* K1 (Ji et al., 2004). It was active against some species of gram-negative plant pathogens while the activity against gram-positive bacteria and fungi was not assessed. Thus, four different types of antibiotic compounds have been isolated from different strains of *X. nematophila*. Whether a given strain produces all four compounds is not presently known. The expansion of the nrps and pks family of genes suggests that other as yet uncharacterized antimicrobial compounds may be produced by *X. nematophila*. The production of numerous antimicrobials presumably ensures that *X. nematophila* effectively competes against a broad spectrum of microorganisms it may encounter in the diverse insects it infects.

8.4.2 Antibiotic Production in *Xenorhabdus bovienii*

Unlike *X. nematophila* which has only been isolated from *S. carpocapsae*, *X. bovienii* has been isolated from seven different steinernematid species (Tailliez et al., 2006). A series of nonpolar dithiolopyrrolone compounds, referred to as xenorhabdins, have been characterized from *X. bovienii* strain T319 (McInerney et al., 1991a). This class of compounds is related to the dithiolopyrrolones found in *Streptomyces* sp. Antimicrobial activity of xenorhabdin 2 was shown to be active against a spectrum of gram-positive bacteria but was inactive against the gamma proteobacteria tested. Xenorhabdins were also shown to be produced by *X. miraniensis* (see below) but not by *X. nematophila*. *X. bovienii* also produce oxidized forms of xenorhabdins called xenorxides that possess both antimicrobial and antimycotic activity (Li et al., 1998). The genome sequence of *X. bovienii* isolated from *S. jollieti* has been determined (http://maizeapache.ddpsc.org/bact_db/) but the antimicrobial compounds from this bacterium have not been characterized yet. The genome contains numerous nrps and pks genes that are likely to be involved in antimicrobial syntheses.

8.4.3 Antibiotic Production in Other Species of *Xenorhabdus*

X. miraniensis strain Q1 produces both xenocoumacin and xenorhabdin antibiotics (McInerney et al., 1991 a,b). *X. nematophila* appears to have acquired the xenocoumacin gene cluster from *Bacillus* sp. while *X. bovienii* (*S. feltiae*) appears to have acquired the xenorhabdin cluster from *Streptomyces* sp. *X. miraniensis* may represent a species in which both clusters were acquired by lateral gene transfer. Whether these gene clusters have been acquired by other species of *Xenorhabdus* remains to be determined.

Antibiotic activity has been detected in macerated larval tissues of *G. mellonella* infected with different nematode species. Antibiotic activity derived from insects infected with *S. riobrave*/*X. cabanillasii* reached peak levels in approximately 5 days postinfection (Isaacson and Webster, 2002). In a separate study, antibiotic activity was determined in macerated tissues derived from insects infected with either *S. glaseri*/*X. poinarii* or *S. feltiae*/*X. bovienii* (Walsh and Webster, 2003). Maximal antibiotic activity was reached between 3–4 days after infection. The indicator strain used in both studies to measure antibiotic activity was *Bacillus subtilis*. The chemical nature of the antibiotic compounds produced and the spectrum of antibiotic activity were not determined in either of these studies.

8.4.4 Antibiotic Diversity in *Xenorhabdus* Species

The full spectrum of antimicrobial compounds produced by an individual species of *Xenorhabdus* remains to be determined. Based on studies in *X. nematophila* and *X. bovienii* (*S. feltiae*), individual species produce several unique compounds; *X. nematophila* produces xenocoumacins and nematophin while *X. bovienii* (*S. feltiae*) produces xenorhabdins and xenoxrides. All 20 species of *Xenorhabdus* identified so far produce antibiotic activity; however, the chemical structures of the antibiotic compounds have been determined only from three species. The gene clusters encoding the characterized antibiotics have apparently been acquired by lateral gene transfer from distantly related bacteria. These findings raise the possibility that other species of *Xenorhabdus* have acquired unique clusters of antibiotic genes from diverse genetic origins. Genomic analysis of *X. nematophila* and *X. bovienii* (*S. jollieti*) reveals a large number of nrps and pks genes suggesting that the repertoire of antibiotic production in these species is larger than currently recognized. Furthermore, growth of *Xenorhabdus* sp. under conditions other than the nutrient-rich broth cultures used so far may allow for the identification of new antibiotic compounds. Few studies have directly examined the role of antimicrobials in interspecies competition and mutualism. The availability of genome sequences for *Xenorhabdus* sp. will facilitate the identification of antibiotic genes and construction of mutant strains making the *Xenorhabdus*–nematode–insect model an excellent system to directly study the role of antimicrobials in interspecies competition and mutualistic host–microbe interactions.

8.5 R-type Bacteriocins

Bacteriocins are narrow spectrum antimicrobial compounds that target bacterial species closely related to the producer strain. Xenorhabdicins, the bacteriocins produced by *Xenorhabdus* sp., belong to the group of R-type bacteriocins (Ishii et al., 1965; Boemare et al., 1992). R-type bacteriocins, proteinaceous structures that resemble the tail portion of defective bacteriophages, have been identified from other gram-negative bacteria such as *P. aeruginosa* (pyocin R1), *Erwinia carotovora* (carotovoricin Er), and *Serratia plymithicum* J7 (serracin P) (Ishii et al., 1965; Itoh et al., 1978; Jabrane et al., 2002). Pyocin R1 from *P. aeruginosa* has been extensively studied. Absorption of pyocin R1 to the cell surface of sensitive related bacteria leads to contraction of the tail sheath and penetration of the tail tube through the outer membrane (Uratani and Hoshino, 1984). The outcome of this action is depolarization of the cytoplasmic membrane, inhibition of active transport, and increased permeability of the cell envelope to hydrophobic substances (Uratani, 1982; Uratani and Hoshino, 1984).

Bacteriocins produced by *X. nematophila* have been isolated from *in vitro* cultures induced with mitomycin C. *X. nematophila* bacteriocins are composed of two major subunits, the contractile tail sheath protein (43 kDa) and the tail tube protein (20 kDa), and five minor subunits that make up the baseplate

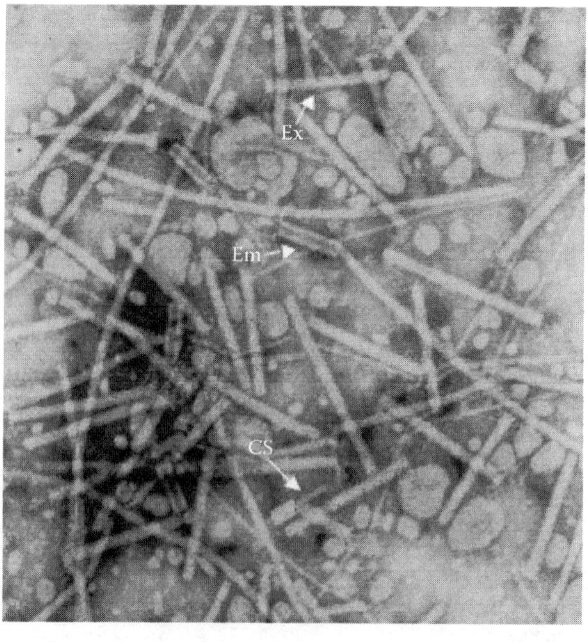

FIGURE 8.2 R-type bacteriocins from supernatant of noninduced *X. nematophila* ATCC19061. Ex, bacteriocin with extended tail sheath; Em, empty tail sheath; CS, contracted tail sheath with exposed tail tube.

and tail fiber structures (Thaler et al., 1995; Morales-Soto, unpublished data). The *in vitro* activity of the bacteriocins has only been tested against a small number of *Xenorhabdus* and *Photorhabdus* species. They were active against *X. beddingii*, *X. bovienii* (*S. feltiae*), and *Photorhabdus luminescens*, and were inactive against *X. cabanillasii* and *X. innexi* (Boemare et al., 1992; Sicard et al., 2005).

Since reproduction of *S. carpocapsae* was negatively affected when competing against either *X. innexi* or *X. cabanillasii* and was not affected when competing against *X. bovienii* or *P. luminescens* it was speculated that during interspecies competition, the fitness of *S. carpocapsae* was dependent on the ability of its bacterial symbiont (*X. nematophila*) to produce R-type bacteriocins that are active against the other *Xenorhabdus* or *Photorhabdus* species (Sicard et al., 2005; Sicard et al., 2006). This suggests that the production of R-type bacteriocins is essential for the survival of the *Steinernema–Xenorhabdus* complexes when insect larvae are co-infected by other *Steinernema* or *Heterorhabditis* nematodes. It is not presently understood why some related species are sensitive to *X. nematophila* bacteriocins while others are resistant.

Of the 20 *Xenorhabdus* sp. characterized to date only *X. nematophila*, *X. bovienii*, and *X. beddingii* have been shown to produce R-type bacteriocins upon induction with mitomycin C (Boemare et al., 1992; Baghdiguian et al., 1993). R-type bacteriocins are produced at lower levels under noninducing *in vitro* conditions (Figure 8.2). Unlike in *X. nematophila*, R-type bacteriocins of the sister taxa, *P. luminescens*, are not mitomycin C inducible (Gaudriault et al., 2004). The activity of the bacteriocins of *X. bovienii* and *X. beddingii* has not been tested against other *Xenorhabdus* species. It is not currently known whether other *Xenorhabdus* sp. produce active R-type bacteriocins. To understand better the role of these bacteriocins in the competitive abilities of the *Steinernema–Xenorhabdus* complexes it will be necessary to establish the full range of activity of bacteriocins produced by different *Xenorhabdus* sp.

ACKNOWLEDGMENTS

We thank members of the Forst laboratory for their helpful comments and critical reading of the manuscript. Work on the antibiotic genes clusters of *Xenorhabdus* has been supported by the Research Growth Initiative of the University of Wisconsin-Milwaukee.

REFERENCES

Akhurst, R. J. 1982. Antibiotic activity of *Xenorhabdus* spp., bacteria symbiotically associated with insect pathogenic nematodes of the families *Heterorhabditidae* and *Steinernematidae*. *J Gen Microbiol* 1982:3061–3065.

Alatorre-Rosas, R. and H. K. Kaya. 1990. Interspecific competition between entomopathogenic nematodes in the genera *Heterorhabditis* and *Steinernema* for an insect host in sand. *J Invertebr Pathol* 55:179–188.

Alatorre-Rosas, R. and H. K. Kaya. 1991. Interaction between two entomopathogenic nematode species in the same host. *J Invertebr Pathol* 57:1–6.

Baghdiguian, S., M. H. Boyer-Giglio, J. O. Thaler, G. Bonnot, and N. Boemare. 1993. Bacteriocinogenesis in cells of *Xenorhabdus nematophilus* and *Photorhabdus luminescens*. *Biol Cell* 79:177–185.

Boemare, N. E. 1983. Recherches sur les complexes nemato-bacteriens entomopathogenes: Etude bacteriologue, gnotoxenique et physiopathologique du mode d'action parasitaire de *Steinernema carpocapsae* Weiser (Rhabditida: Steinernematidae). PhD dissertation, University of Montpellier.

Boemare, N. E., M. H. Boyer-Giglio, J. O. Thaler, R. J. Akhurst, and M. Brehelin. 1992. Lysogeny and bacteriocinogeny in *Xenorhabdus nematophilus* and other *Xenorhabdus* spp. *Appl Environ Microbiol* 58:3032–3037.

Broderick, N. A., K. F. Raffa, R. M. Goodman, and J. Handelsman. 2004. Census of the bacterial community of the gypsy moth larval midgut by using culturing and culture-independent methods. *Appl Environ Microbiol* 70:293–300.

Campbell, J. F., E. E. Lewis, S. P. Stock, S. Nadler, and H. K. Kaya. 2003. Evolution of host search strategies in entomopathogenic nematodes. *J Nematol* 35:142–145.

Cox, C. R. and M. S. Gilmore. 2007. Native microbial colonization of *Drosophila melanogaster* and its use as a model of *Enterococcus faecalis* pathogenesis. *Infect Immun* 75:1565–1576.

Dale, C. and N. A. Moran. 2006. Molecular interactions between bacterial symbionts and their hosts. *Cell* 126:453–465.

Eleftherianos, I., J. Marokhazi, P. J. Millichap et al. 2006. Prior infection of *Manduca sexta* with non-pathogenic *Escherichia coli* elicits immunity to pathogenic *Photorhabdus luminescens*: Roles of immune-related proteins shown by RNA interference. *Insect Biochem Mol Biol* 36:517–525.

Forst, S. and D. Clarke. 2002. Bacteria-nematode symbiosis. In *Entomopathogenic Nematology*, Ed. R. Gaugler, pp. 57–77. Wallingford, U.K.: CABI Publishing.

Gaudriault, S., J. O. Thaler, E. Duchaud, F. Kunst, N. Boemare, and A. Givaudan. 2004. Identification of a P2-related prophage remnant locus of *Photorhabdus luminescens* encoding an R-type phage tail-like particle. *FEMS Microbiol Lett* 233:223–231.

Goodrich-Blair, H. 2007. They've got a ticket to ride: *Xenorhabdus nematophila–Steinernema carpocapsae* symbiosis. *Curr Opin Microbiol* 10:225–230.

Goodrich-Blair, H. and D. J. Clarke. 2007. Mutualism and pathogenesis in *Xenorhabdus* and *Photorhabdus*: Two roads to the same destination. *Mol Microbiol* 64:260–268.

Gouge, D. H. and J. L. Snyder. 2006. Temporal association of entomopathogenic nematodes (Rhabditida: *Steinernematidae* and *Heterorhabditidae*) and bacteria. *J Invertebr Pathol* 91:147–157.

Grewal, P. S. 2002. Formulation and application technology. In *Entomopathogenic Nematology*, Ed. R. Gaugler, pp. 265–287. Wallingford, U.K.: CABI Publishing.

Herbert, E. E. and H. Goodrich-Blair. 2007. Friend and foe: The two faces of *Xenorhabdus nematophila. Nat Rev Microbiol* 5:634–646.

Isaacson, P. J. and J. M. Webster. 2002. Antimicrobial activity of *Xenorhabdus* sp. RIO (*Enterobacteriaceae*), symbiont of the entomopathogenic nematode, *Steinernema riobrave* (Rhabditida: *Steinernematidae*). *J Invertebr Pathol* 79:146–153.

Ishii, S. I., Y. Nishi, and F. Egami. 1965. The fine structure of a pyocin. *J Mol Biol* 13:428–431.

Itoh, Y., K. Izaki, and H. Takahashi. 1978. Purification and characterization of a bacteriocin from *Erwinia caratovora. J Gen Appl Microbiol* 24:27–39.

Jabrane, A., A. Sabri, P. Compere, P. Jacques, I. Vandenberghe, J. Van Beeumen, and P. Thonart. 2002. Characterization of serracin P, a phage-tail-like bacteriocin, and its activity against *Erwinia amylovora*, the fire blight pathogen. *Appl Environ Microbiol* 68:5704–5710.

Ji, D., Y. Yi, G. H. Kang, Y. H. Choi, P. Kim, N. I. Baek, and Y. Kim. 2004. Identification of an antibacterial compound, benzylideneacetone, from *Xenorhabdus nematophila* against major plant-pathogenic bacteria. *FEMS Microbiol Lett* 239:241–248.

Joyce, S. A., R. J. Watson, and D. J. Clarke. 2006. The regulation of pathogenicity and mutualism in *Photorhabdus. Curr Opin Microbiol* 9:127–132.

Kaya, H. K. 2002. Natural enemies and other antagonists. In *Entomopathogenic Nematology*, Ed. R. Gaugler, pp. 189–203. Wallingford, U.K.: CABI Publishing.

Koppenhöfer, A. M. and H. K. Kaya. 1996. Coexistence of two steinernematid nematode species (Rhabditida:Steinernematidae) in the presence of two host species. *Appl Soil Ecol* 4:221–230.

Laumond, C., H. Mauléon, and A. Kermarrec. 1979. Données nouvelles sur le spectre d'hôtes et le parasitisme du nématode entomophage *Neoplectana carpocapsae. Entomophaga* 24:13–27.

Lewis, E. E. 2002. Behavioural ecology. In *Entomopathogenic Nematology*, Ed. R. Gaugler, pp. 205–223. Wallingford, U.K.: CABI Publishing.

Li, J., G. Chen, and J. M. Webster. 1997. Nematophin, a novel antimicrobial substance produced by *Xenorhabdus nematophilus* (*Enterobacteriaceae*). *Can J Microbiol* 43:770–773.

Li, J., K. Hu, and M. J. Webster. 1998. Antibiotics from *Xenorhabdus spp.* and *Photorhabdus spp.* (Enterobacteriaceae). *Chemistry of Heterocyclic Compounds* 34:1561–1570.

Lysenko, O. and J. Weiser. 1974. Bacteria associated with the nematode *Neoplectana carpocapsae* and the pathogenicity of this complex for *Galleria mellonella* larvae. *J Invertebr Pathol* 24:332–336.

Martin, J. D. and J. O. Mundt. 1972. Enterococci in insects. *Appl Microbiol* 24:575–580.

Maxwell, P. W., G. Chen, J. M. Webster, and G. B. Dunphy. 1994. Stability and activities of antibiotics produced during infection of the insect *Galleria mellonella* by two isolates of *Xenorhabdus nematophilus. Appl Environ Microbiol* 60:715–721.

McInerney, B. V., R. P. Gregson, M. J. Lacey et al. 1991a. Biologically active metabolites from *Xenorhabdus* spp., Part 1. Dithiolopyrrolone derivatives with antibiotic activity. *J Nat Prod* 54:774–784.

McInerney, B. V., W. C. Taylor, M. J. Lacey, R. J. Akhurst, and R. P. Gregson. 1991b. Biologically active metabolites from *Xenorhabdus* spp., Part 2. Benzopyran-1-one derivatives with gastroprotective activity. *J Nat Prod* 54:785–795.

Nyholm, S. V. and M. J. Mcfall-Ngai. 2004. The winnowing: Establishing the squid–*vibrio* symbiosis. *Nat Rev Microbiol* 2:632–642.

Park, D. and S. Forst. 2006. Co-regulation of motility, exoenzyme and antibiotic production by the EnvZ-OmpR-FlhDC-FliA pathway in *Xenorhabdus nematophila. Mol Microbiol* 61:1397–1412.

Park, Y. and Y. Kim. 2003. *Xenorhabdus nematophilus* inhibits p-bromophenacyl bromide (BPB)-sensitive PLA2 of *Spodoptera exigua*. *Arch Insect Biochem Physiol* 54:134–142.

Park, Y., Y. Kim, S. M. Putnam, and D. W. Stanley. 2003. The bacterium *Xenorhabdus nematophilus* depresses nodulation reactions to infection by inhibiting eicosanoid biosynthesis in tobacco hornworms, *Manduca sexta*. *Arch Insect Biochem Physiol* 52:71–80.

Park, Y., Y. Kim, H. Tunaz, and D. W. Stanley. 2004. An entomopathogenic bacterium, *Xenorhabdus nematophila*, inhibits hemocytic phospholipase A2 (PLA2) in tobacco hornworms *Manduca sexta*. *J Invertebr Pathol* 86:65–71.

Paul, V. J., S. Frautschy, W. Fenical, and K. H. Nealson. 1981. Antibiotics in microbial ecology, isolation and structure assignment of several new antibacterial compounds from the insect-symbiotic bacteria *Xenorhabdus* spp. *J Chem Ecol* 7:589–597.

Poinar, G. O. 1979. Biology and taxonomy of *Steinernematidae* and *Heterorhabditidae*. In *Entomopathogenic Nematodes in Biological Control*, Eds. R. Gaugler and H. K. Kaya, pp. 23–62. Boca Raton, FL: CRC Press.

Poinar, G. O., R. T. Hess, and G. Thomas. 1980. Isolation of defective bacteriophages from *Xenorhabdus* spp. (*Enterobacteriaceae*). *IRCS Med Sci* 8:141.

Sergeant, M., L. Baxter, P. Jarrett, E. Shaw, M. Ousley, C. Winstanley, and J. A. Morgan. 2006. Identification, typing, and insecticidal activity of *Xenorhabdus* isolates from entomopathogenic nematodes in United Kingdom soil and characterization of the *xpt* toxin loci. *Appl Environ Microbiol* 72:5895–5907.

Sicard, M., N. Le Brun, S. Pages, B. Godelle, N. Boemare, and C. Moulia. 2003. Effect of native *Xenorhabdus* on the fitness of their *Steinernema* hosts: Contrasting types of interaction. *Parasitol Res* 91:520–524.

Sicard, M., K. Brugirard-Ricaud, S. Pages, A. Lanois, N. E. Boemare, M. Brehelin, and A. Givaudan. 2004a. Stages of infection during the tripartite interaction between *Xenorhabdus nematophila*, its nematode vector, and insect hosts. *Appl Environ Microbiol* 70:6473–6480.

Sicard, M., J. B. Ferdy, S. Pages, N. Le Brun, B. Godelle, N. Boemare, and C. Moulia. 2004b. When mutualists are pathogens: An experimental study of the symbioses between *Steinernema* (entomopathogenic nematodes) and *Xenorhabdus* (bacteria). *J Evol Biol* 17:985–993.

Sicard, M., J. Tabart, N. E. Boemare, O. Thaler, and C. Moulia. 2005. Effect of phenotypic variation in *Xenorhabdus nematophila* on its mutualistic relationship with the entomopathogenic nematode *Steinernema carpocapsae*. *Parasitology* 131:687–694.

Sicard, M., J. Hinsinger, N. Le Brun, S. Pages, N. Boemare, and C. Moulia. 2006. Interspecific competition between entomopathogenic nematodes (*Steinernema*) is modified by their bacterial symbionts (*Xenorhabdus*). *BMC Evol Biol* 6:68.

Snyder, H., S. P. Stock, S. K. Kim, Y. Flores-Lara, and S. Forst. 2007. New insights into the colonization and release processes of *Xenorhabdus nematophila* and the morphology and ultrastructure of the bacterial receptacle of its nematode host, *Steinernema carpocapsae*. *Appl Environ Microbiol* 73:5338–5346.

Sundar, L. and F. N. Chang. 1993. Antimicrobial activity and biosynthesis of indole antibiotics produced by *Xenorhabdus nematophilus*. *J Gen Microbiol* 139:3139–3148.

Tailliez, P., S. Pages, N. Ginibre, and N. Boemare. 2006. New insight into diversity in the genus *Xenorhabdus*, including the description of ten novel species. *Int J Syst Evol Microbiol* 56:2805–2818.

Tanaka, A., B. A. Tapper, A. Popay, E. J. Parker, and B. Scott. 2005. A symbiosis expressed non-ribosomal peptide synthetase from a mutualistic fungal endophyte of perennial ryegrass confers protection to the symbiotum from insect herbivory. *Mol Microbiol* 57:1038–1050.

Thaler, J. O., S. Baghdiguian, and N. Boemare. 1995. Purification and characterization of xenorhabdicin, a phage tail-like bacteriocin, from the lysogenic strain F1 of *Xenorhabdus nematophilus*. *Appl Environ Microbiol* 61:2049–2052.

Uratani, Y. 1982. Dansyl chloride labeling of *Pseudomonas aeruginosa* treated with pyocin R1: Change in permeability of the cell envelope. *J Bacteriol* 149:523–528.

Uratani, Y. and T. Hoshino. 1984. Pyocin R1 inhibits active transport in *Pseudomonas aeruginosa* and depolarizes membrane potential. *J Bacteriol* 157:632–636.

Walsh, K. T. and J. M. Webster. 2003. Interaction of microbial populations in *Steinernema* (*Steinernematidae*, Nematoda) infected *Galleria mellonella* larvae. *J Invertebr Pathol* 83:118–126.

Webster, J. M., G. Chen, K. Hu, and J. Li. 2002. Bacterial metabolites. In *Entomopathogenic Nematology*, Ed. R. Gaugler, pp. 99–114. Wallingford, U.K.: CABI Publishing.

9

Defensive Symbionts in Aphids and Other Insects

Kerry M. Oliver and Nancy A. Moran

CONTENTS

9.1 Introduction

Symbiotic associations with microorganisms are ubiquitous in invertebrates, occurring in many phyla in both marine and terrestrial habitats. In terrestrial arthropods, the microorganisms are often heritable, being passed from mother to offspring with high fidelity (Buchner, 1965). Inherited symbiosis is particularly common in insects and most often involves an association with bacteria. Recent results indicate that some heritable symbionts in insects protect hosts from natural enemies. In this chapter, we briefly review the general phenomenon of heritable symbiosis in insects and summarize the results on cases in which these symbionts provide defenses. Although symbiont-based protection has been examined most extensively in aphids, the phenomenon is potentially common across many insect groups.

9.1.1 Background on Heritable Symbiosis in Insects

Heritable symbiosis can be documented using microscopy (e.g., Buchner, 1965; Fukatsu et al., 2007; Moran et al., 2005b) or, since 1990, molecular methods (e.g., Jeyaprakash and Hoy, 2000; O'Neill et al., 1992; Unterman et al., 1989; Wernegreen, 2002). However, further experimentation is required to establish the effects of symbionts on hosts. Often, these impacts remain unknown. Where effects on hosts have been studied, they can be divided into two broad categories: reproductive manipulation, in which the symbiont alters host reproduction, and mutualism, in which the symbiont confers an advantage to hosts in survival and/or reproduction. In each case, the effect enables the symbiont infection to spread within a host population by increasing the frequency of host matrilines infected with symbionts at the expense of uninfected matrilines.

Many insect groups that feed on nutrient-deficient diets possess obligate, "primary" symbionts that provide essential nutrients that are lacking in food. Thus, in sap-feeders such as aphids, primary symbionts provide amino acids that are lacking in plant sap; in blood-feeders such as tsetse flies, symbionts provide limiting vitamins that are rare or absent in vertebrate blood. Symbioses based in nutrition tend to be evolutionarily ancient, stable, and obligate with the symbionts occupying specialized organs known as bacteriomes (Baumann, 2005; Buchner, 1965; Wernegreen, 2002). Based on the number of insect groups with bacteriome structures and the expected number of species in these groups, about 10% of insects contain nutritional symbionts (Buchner, 1965; Douglas, 1989). Primary symbionts have not been reported to provide protection to hosts.

However, many heritable symbioses in insects are more labile within host lineages; although routinely inherited from mothers via infection of eggs within the mother's body, phylogenetic studies indicate that these infections can be lost and sometimes transferred among lineages within and between species. These associations are often referred to as "facultative" or "secondary" symbioses because they are generally not essential for the growth and reproduction of hosts. Unlike primary symbionts, facultative symbionts are not restricted to bacteriomes and instead inhabit a variety of cell types or live in the hemolymph. As a consequence, their frequency across hosts is not well known. For the reproductive manipulator, *Wolbachia pipientis*, estimates of frequency among insect lineages range from 17% to 75% (Jeyaprakash and Hoy, 2000; Werren and Windsor, 2000; Werren et al., 1995). Screens for other heritable symbionts are relatively limited but show that significant proportions of individuals can be infected within particular taxonomic groups (e.g., Kageyama et al., 2006; Mateos et al., 2006; Russell et al., 2003; Zchori-Fein and Brown, 2002). Besides *Wolbachia*, there are a large number of other facultative symbionts (e.g., overview in Baumann, 2005 for sap-feeding insects; Zchori-Fein and Perlman, 2004). For insect facultative symbionts, investigations of effects on hosts are few and mostly recent; for most cases, such effects are not known.

Among studied cases, defense against natural enemies is one of the major effects recently discovered for several insect–symbiont associations. Because specific natural enemies, including predators, parasites, parasitoid wasps as well as fungal and bacterial pathogens, are important selective forces in shaping the life history of many insects, we might expect that defensive symbionts are widespread in nature. Microorganisms, and bacteria in particular, harbor the potential to provide many types of defenses because collectively they have the capacity to make a wide variety of bioactive compounds, including diverse toxic molecules. Infection with symbionts can potentially provide an insect host lineage with novel defensive capabilities.

In this chapter, we summarize information on as many published examples of defensive mutualisms involving microbes in insects as we could find. Although relevant investigations are currently few and restricted to specific host groups, the recently described cases provide a guide informing studies of additional host groups which are likely to yield many more examples. We also summarize what is known about mechanisms of defense, including the role of bacteriophage and mobile genetic elements. These mechanisms will affect the stability of defenses within hosts, the ability for hosts to generate novel defenses in the face of novel challenges from enemies, and the propensity for symbionts and their defensive effects to be transferred among hosts of varying degrees of relatedness. Finally, we discuss the potential consequences of defensive symbionts in insect populations, including broader ecological impacts and practical implications for agriculture and medicine.

9.2 Examples of Defensive Symbionts in Insects

Despite long standing attention to the roles of inherited symbionts of insects as nutritional mutualists (e.g., Buchner, 1965; Douglas, 1989; Wernegreen, 2002) and reproductive parasites (e.g., Werren and O'Neill, 1997), the notion that symbionts may play a widespread role as defensive mutualists is a recent one. In this section, we will provide a review of known instances of symbiont-based protection in insects. While only a handful of cases have been reported, these examples demonstrate that infection with inherited microbes can serve as the basis for protection against a wide range of natural enemies.

9.2.1 Protection against a Parasitic Wasp in Alfalfa Weevils

The first report of an interaction between an insect symbiont and a host natural enemy was published as a section of a book chapter and has received little attention (Hsiao, 1996). The alfalfa weevil, *Hypera postica*, was accidentally introduced in the United States on three occasions, and these correspond to the three currently recognized strains: western, Egyptian, and eastern. Despite some intergraded areas, the three strains generally occupy separate geographical locations. The strains are morphologically very similar, but molecular information indicates that the eastern and Egyptian strains are more closely related (overview in Bundy et al., 2005). The western strain is fixed for infection with the common reproductive manipulator, *Wolbachia*, that causes cytoplasmic incompatibility (CI) when males of the western strain are mated with females of either the eastern or Egyptian strains (Hsiao and Hsiao, 1985a,b; Leu et al., 1989). This weevil is an important pest of alfalfa, and numerous parasitic wasps have been introduced to limit damage. The most effective parasitic wasp against the eastern and Egyptian strains, *Microctonus aethiopoides*, performs poorly when reared on the western strain and has failed to establish viable populations, despite numerous intentional releases, in areas inhabited by the western strain (review in Hsiao, 1996). Hsiao (1996) found that the western strain becomes highly susceptible to parasitism by *M. aethiopoides* after antibiotic treatment aimed at eliminating the CI-causing *Wolbachia*; 82% of parasitized, antibiotic-treated weevils produced an adult parasitic wasp, compared to only 11% of the symbiotic (nontreated) weevils. This rate of successful parasitism (82%) in antibiotic-treated western strain weevils was even higher than the rates observed in concurrent parasitism assays for eastern (56%) and Egyptian (69%) strain weevils. Hsiao (1996) interpreted these results to indicate that *Wolbachia* was the agent responsible for resistance to parasitism. These weevils, however, were not generally screened for symbionts (only diagnostic PCR was used to detect *Wolbachia*), and antibiotic-curing may have led to the removal of other bacterial symbionts. Therefore, the causative agent of resistance to parasitism has not been definitively determined in this interaction. Despite its prevalence among terrestrial invertebrates, there are no other reported instances in which *Wolbachia* is purported to influence interactions with natural enemies. This case is particularly interesting because it suggests that both host range and geographic range of natural enemies may be influenced by heritable symbionts.

A closely related parasitic wasp, *Bathyplectes curculionis*, is able to use western strain *H. postica* as a suitable host and has established viable populations in areas primarily inhabited by this strain after intentional introductions (e.g., Kingsley et al., 1993). This suggests that either the mechanism of resistance is specific to *M. aethiopoides* or *B. curculionis* has developed strategies to overcome symbiont-mediated protection (potential for this is discussed below). Interestingly, the eastern strain is also susceptible to *B. curculionis* but the closely related Egyptian strain exhibits very high levels of resistance (Maund and Hsiao, 1991). The Egyptian strain is not infected with *Wolbachia*; however, it has not been screened for other symbionts. It is possible that a microbial symbiont in the Egyptian strain contributes to the observed resistance to *B. curculionis*, a finding that would provide further evidence that host symbionts may influence the geographic range of particular natural enemies.

9.2.2 Protection against Spider Predation in *Paederus* Beetles

The next example of symbiont-based defense in insects, and the first to appear in the primary scientific literature, was discovered in the staphylinid beetle *Paederus*. Paederine beetles have long garnered attention because when their hemolymph contacts human skin, it causes painful lesions known as *Paederus* dermatitis (Gelmetti and Grimalt, 1993). On occasions, epidemics of *Paederus* dermatitis can occur, and there is speculation that several of the biblical plagues were caused by these beetles (Frank and Kanamitsu, 1987; Norton and Lyons, 2002). The chemical compound that produces the blistering, known as pederin, was first visualized in crystalline form in 1919 and later isolated in pure form in 1953, allowing its chemical structure to be determined in the 1960s as a polyketide amide (review in Frank and Kanamitsu, 1987). While the specific mechanism of action remains unknown, pederin is known to block mitosis through the inhibition of DNA and protein synthesis (e.g., Brega et al., 1968; Soldati et al., 1966).

While most of the interest in pederin is focused on its impact on humans, the role of this toxin as a potential defensive compound in *Paederus* beetles has also been addressed. Unlike true blister beetles (Meloidae), which secrete a vesicant toxin from their joints (called reflex bleeding), paederine beetles must be injured or crushed to expose their hemolymph toxin (Dettner, 1987). Unusual among other staphylinid beetles, *Paederus* beetles are active during daylight hours and exhibit orange and black markings. Given the known toxic effects of pederin, these colorful markings have been assumed to function as an advertisement of their unpalatability (aposematism) (Dettner, 1987; Frank and Kanamitsu, 1987). It was originally hypothesized that pederin functioned as a defense against vertebrate predators, but pederin does not appear to be distasteful to vertebrate predators or invoke negative associations resulting in reduced feeding (reviewed in Frank and Kanamitsu, 1987).

Pederin is synthesized and sequestered in adult females, accumulates in eggs, and is retained, with gradual dilution, throughout the larval development (Kellner, 1998; Kellner and Dettner, 1995). Interestingly, most but not all females possess pederin, and only pederin (+) females can transfer the toxin to their eggs (Kellner and Dettner, 1995). To determine if pederin functioned in defending against invertebrate predators, Kellner and Dettner (1996) conducted laboratory feeding assays to determine if the eggs and larvae of two *Paederus* species (*P. fuscipes* and *P. riparius*) were suitable prey for insect and arachnid predators. They reported that a range of insect predators, including beetles, ants, and heteropterans (true bugs), were not deterred by pederin (+) eggs or larvae. Diverse insect predators could feed on pederin (+) prey over long periods of time without exhibiting impairment (Kellner and Dettner, 1996). Several species of wolf spiders (Lycosidae) and one species of jumping spider (Saltididae), however, would never feed on pederin (+) larvae while often feeding on pederin (–) larvae (Kellner and Dettner, 1996). Furthermore, after rejecting pederin (+) larvae, spiders would often get engaged in stereotypical cleansing behavior that was also observed after encounters with pederin-fed *Drosophila*, suggesting that pederin is the basis for prey rejection (Kellner and Dettner, 1996). Wolf spiders are common in the natural environments of both species of *Paederus* (references in Kellner and Dettner, 1996) and potentially important natural enemies, although this has not been examined explicitly. The role of pederin as a potential defensive compound against parasitic wasps or entomopathogenic nematodes, both of which would be exposed to pederin during development within hosts, has not been examined.

Pederin remained a unique structure among natural products until 1988 when similar vesicant compounds were isolated from marine sponges (e.g., Perry et al., 1988; Sakemi et al., 1988). The discovery of similar compounds in distantly related taxa, combined with evidence of the pederin polymorphism in adult females and maternal inheritance patterns of this toxin, led Kellner to investigate whether pederin might be a product of symbiotic origin. He found that pederin (–) females could acquire the capacity to synthesize pederin by eating untreated pederin (+) eggs but not when fed on sterilized pederin (+) eggs (Kellner, 1999). He subsequently identified a bacterial symbiont present only in pederin (+) *P. sabaeus* females that is closely related to the human pathogen, *Pseudomonas aeruginosa* (Kellner, 2001a, 2002a). Furthermore, antibiotic treatment of pederin (+) eggs gave rise to adult females that could not synthesize pederin (Kellner, 2001b). The case for a symbiotic origin of pederin was cinched when Piel (2002) isolated a polyketide synthase cluster from a *Pseudomonas* symbiont present only in pederin (+) females. Interestingly, the pederin synthase genes are located on a horizontally transferred genomic region (symbiosis island), providing a likely mechanism for the movement of this complex toxin among distantly related symbiont taxa (Piel et al., 2004a).

Symbiont-mediated defense against spider predation has been experimentally demonstrated only for two *Paederus* species. However, it is likely a common phenotype among the more than 600 members of this species as nine additional *Paederus* and two *Paederidus* beetles are also known to contain this toxin (and another 22 are known to cause blistering characteristic of pederin) (Kellner, 2002b). This is the only demonstrated case of symbiont-based defense against predation in insects, but an example with striking similarities is found in marine invertebrates, as covered in another chapter in this volume (Lopanik et al., 2004).

9.2.3 Defense against Parasitic Wasps and Fungal Pathogens in Pea Aphids

Aphids have long been the model organisms for the study of heritable symbionts, and the pea aphid, *Acyrthosiphon pisum*, has become the most important insect system for studying symbionts that influence a range of ecological interactions, including those with natural enemies. In addition to their obligate primary symbiont, *Buchnera,* many aphids are also infected with inherited, facultative symbionts—known as secondary symbionts (SS) (e.g., Buchner, 1965; Fukatsu and Ishikawa, 1993). *Acyrthosiphon pisum*, for example, is known to be infected with at least five SS (Chen et al., 1996; Fukatsu et al., 2000, 2001; Sandström et al., 2001; Unterman et al., 1989), the three most common of which are members of the Enterobacteriaceae (Gammaproteobacteria) and have been named *Regiella insecticola, Serratia symbiotica*, and *Hamiltonella defensa* (Moran et al., 2005b). Aphids are ideal organisms for studying the effects of facultative symbionts; due to a largely parthenogenetic lifecycle, clonal lines can be maintained indefinitely in the laboratory and SS compositions can be manipulated by transfecting hosts using microinjection or selective curing with antibiotics (e.g., Chen and Purcell, 1997; Koga et al., 2003). In addition to their roles in defense, discussed below, *A. pisum* SS have also been found to provide thermal tolerance (e.g., Montllor et al., 2002) and influence performance on particular host plants (Tsuchida et al., 2004).

The first defensive role for SS identified in aphids involves interactions with parasitic wasps, which are often the most important natural enemies of aphids. Clonal lines of *A. pisum* had previously been shown to vary greatly with respect to susceptibility to their dominant natural enemy, the parasitic wasp, *Aphidius ervi* (Henter and Via, 1995). Solitary endoparasitic wasps, like *A. ervi,* locate and deposit an egg in a suitable aphid nymph, this egg hatches and the resulting larva develops within a living aphid, only killing the aphid upon pupation (Figures 9.1 and 9.2) (Godfray, 1994). To determine if *A. pisum* SS had a role in the observed variation in resistance to parasitism, Oliver et al. (2003) conducted

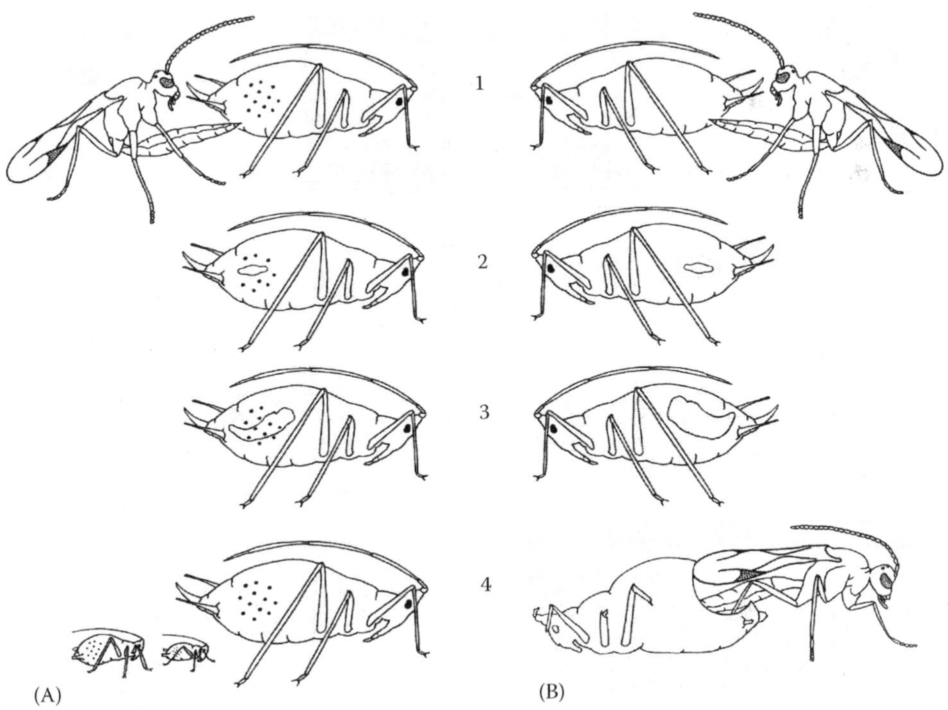

FIGURE 9.1 Effect of infection by *Hamiltonella* symbionts on the outcome of parasitization of *Acyrthosiphon pisum* aphids by *Aphidius ervi* wasps. Series (A) and (B) show progression for aphids with and without symbionts, respectively. 1: Adult female wasp deposits egg within aphid hemocoel. 2: Wasp egg hatches and larva begins development. 3: Wasp larva shows stunted development in presence of symbionts. 4: Aphids with symbionts overcome attack and proceed to reproduce while aphids without symbionts are killed, allowing wasp to pupate and emerge as a new adult. Details are discussed in text.

FIGURE 9.2 Photograph of *Aphidius ervi* ovipositing in *Acyrthosiphon pisum* host.

parasitism assays on experimental lines of *A. pisum* that were all of a single aphid genotype but differed with respect to SS infection status. They found that both *Hamiltonella* and *Serratia* conferred partial protection against attack by *A. ervi* by causing mortality to the developing wasp larvae. A third *A. pisum* symbiont, *Regiella*, did not reduce rates of successful parasitism (Oliver et al., 2003). A subsequent study found direct benefits of infection with *Hamiltonella* in the presence of parasitism; parasitized *Hamiltonella*-infected aphids produced far more offsprings than parasitized uninfected aphids (Oliver et al., 2006). In contrast, direct benefits of infection were not detected for aphids infected with *Serratia*. Other studies have found that *Serratia* confers thermal tolerance in *A. pisum* (Chen et al., 2000; Montllor et al., 2002; Russell and Moran, 2006), and these or other benefits, rather than protection from parasitoids, may explain the persistence of *Serratia* symbionts in aphid populations.

To determine if *Hamiltonella* generally confers resistance to parasitism, Oliver et al. (2005) investigated parasitism rates of multiple *Hamiltonella* strains in a single aphid clonal background. All isolates examined conferred resistance to parasitism, and levels of protection varied dramatically, from around 15% to nearly complete protection. Even a *Hamiltonella* strain from another aphid species, *Aphis craccivora*, conferred protection to *A. ervi* in *A. pisum* (this aphid is also attacked by *A. ervi*), suggesting that this symbiont likely plays a defensive role in other aphids (Oliver et al., 2005). Similar rates of successful parasitism were observed for a single *Hamiltonella* strain placed in multiple *A. pisum* clonal backgrounds, indicating little symbiont-by-aphid genotype interaction. Interestingly, all uninfected lines tested were highly and evenly susceptible to parasitism by *A. ervi* (Oliver et al., 2005; Kerry Oliver, unpublished). Insects normally defend themselves from endoparasitic wasps via a cellular encapsulation of wasp eggs or larvae (Salt, 1970); however, encapsulation responses are weak or nonexistent in aphids, including *A. pisum* (Bensadia et al., 2006; Carver and Sullivan, 1988). This observation combined with the finding that all SS-free *A. pisum* lines tested are highly susceptible to parasitism, while infection with SS confers resistance, suggests that *A. pisum* has largely outsourced defensive services to microbial symbionts. These results are corroborated by another set of studies that also reported large variation among numerous *A. pisum* clones in resistance to parasitism by *A. ervi* and its congener *A. eadyi* (Ferrari et al., 2001), and infection with *Hamiltonella* (called PABS in that study) was associated with lines resistant to attack by both *Aphidius* species (Ferrari et al., 2004).

Aphids are also subject to mortal infections from a range of fungal pathogens, the most common of which is *Pandora neoaphidis*. Aphids become infected when *P. neoaphidis* spores germinate on the aphid's cuticle. The resulting hyphae colonize the cuticle, eventually invading the aphid's body and killing the aphid. Ferrari et al. (2001, 2004) found variations among *A. pisum* clones in resistance to *P. neoaphidis*, and infection with *Regiella* (called PAUS) was strongly associated with resistant lines, though possible correlations with aphid genotypes were not ruled out. A subsequent investigation using five aphid clones, each with *Regiella*-infected and SS-free sublines, demonstrated that *Regiella* was the agent responsible for protection against *Pandora* (Scarborough et al., 2005).

Thus, in a single aphid species, it is the first one to have its SS considered for a defensive role, all three common SS (*Hamiltonella, Regiella*, and *Serratia*) have been demonstrated to influence interactions with natural enemies. Two of these, *Hamiltonella* and *Regiella*, have clear roles in defense against two major classes of enemies, parasitic wasps and fungal pathogens.

9.2.4 Protection against Fungal Pathogens in Wasps, Ants, and Beetles

Females of the digger wasp, *Philanthus triangulum* (European beewolf) hunt honeybees (*Apis mellifera*), paralyze them with venom, and embalm them with glandular secretions before depositing an egg on one or more bees and sealing each egg/prey combination within a brood cell (Strohm and Linsenmair, 1994). Brood cells, which are warm and humid, appear to offer a permissive environment for the development of pathogens. Kaltenpoth et al. (2005), however, observed that female digger wasps place copious amounts of a white antennal gland secretion into each brood cell and larval survival was much higher in cells containing this secretion. Using scanning electron microscopy, they observed cells and spores that were typical of actinomycetes bacteria in these antennal gland secretions. These bacteria were also present on cocoon walls and associated with reduced fungal growth (Kaltenpoth et al., 2005). These protective symbionts in the digger wasp have been named *Streptomyces philanthi* and are cultivated in antennal glands where they appear to be nutritionally provisioned by the wasp host (Goettler et al., 2007; Kaltenpoth et al., 2006).

A similar phenomenon has been recently described in the spruce beetle, *Dendroctonus rufipennis*, an important cause of mortality in mature spruce (Cardoza et al., 2006; Holsten et al., 1999). Female beetles bore through bark to create galleries in phloem tissue in which they deposit eggs. Larvae subsequently develop in a network of parental and larval galleries. Bark beetles, including the spruce beetle, are often associated with particular fungal symbionts, and effects range from mutualistic to parasitic (e.g., Hofstetter et al., 2006; Kopper et al., 2004). Four fungal species were found to invade *D. rufipennis* galleries and each conferred detrimental effects on beetle survival and reproduction (Cardoza et al., 2006). However, previously unknown oral secretions containing bacteria were found to limit fungal growth. A total of nine bacterial isolates were detected in oral secretions, each differing in capacity to inhibit growth of each of the four fungal associates on laboratory plates (Cardoza et al., 2006). Leafcutter ants are also known to use an actinomycete symbiont to protect their fungal gardens from fungal pathogens, a topic covered in a separate chapter in this book (Carafo and Currie, 2005; Currie et al., 1999). These cases illustrate that bacterial symbionts of diverse insect hosts can provide protection against diverse fungal pathogens, within a variety of ecological contexts.

9.3 Distribution and Genetic Features of Bacteria Providing Protection

9.3.1 Phylogenetic Distribution of Defensive Symbionts in Insects

Most known cases of defensive bacterial symbionts in insects fall in the Gammaproteobacteria and largely in the Enterobacteriaceae, which also includes many animal pathogens exemplified by *Yersinia pestis* (the agent of plague), *Salmonella enterica*, and pathogenic *Escherichia coli* and *Shigella*. *Pseudomonas*, the symbiont of *Paederus* beetles, is a more distant lineage within Gammaproteobacteria. The phylogenetic outlier is *Streptomyces,* the protective agent in brood cells of *Philanthus* wasps; this bacterium belongs to the Actinobacteria, a group of soil bacteria that has the capacity for the production of many antibiotic compounds including many used in medicine. Because defensive symbionts have been studied in very few cases, conclusions about their phylogenetic distribution are premature, though a clear cluster of insect defensive symbionts is found within the Enterobacteriaceae.

Even symbionts within Enterobacteriaceae correspond to several independent origins, based on phylogenies derived from ribosomal RNA sequences (Fukatsu, 2001; Russell et al., 2003). The lineages represented by *Hamiltonella* and *Regiella* are phylogenetically closest relatives, suggesting that their ancestor was also an insect symbiont and possibly a defensive symbiont. *Serratia symbiotica* represents a distinct origin of symbiosis, within the genus *Serratia*, that also includes pathogens of insects and other animals. In fact, an even larger number of symbiont lineages within the Enterobacteriaceae show similar patterns of erratic distribution among arthropod hosts, but most have not yet been studied with respect to their impact on host ecology. These include *Sodalis glossinidius* and related lineages (Novakova and Hypsa, 2007; Weiss et al., 2006), many secondary or facultative symbionts in psyllids, whiteflies, and scale insects (summary in Baumann, 2005), and others including *Arsenophonus* strains in a wide variety of insects (e.g., Dale et al., 2006). (*Arsenophonus* corresponds to a distinct lineage, related to *Proteus*). Potentially, a large proportion of these symbionts play a role in host protection; however, this possibility has rarely been tested.

9.3.2 Routes of Transmission among Hosts

As summarized above, all of these Enterobacteriaceae appear to be maternally transmitted with high fidelity, resulting in stable associations with host matrilines. However, they also appear to undergo some horizontal transmission both within species and between species, sometimes between very distantly related arthropods. The occurrence of horizontal transmission was first inferred by Chen and Purcell (1997), based on the observation of near identical 16S rRNA sequences for symbionts (PASS, now *S. symbiotica*) in *A. pisum* and the rose aphid, *Macrosiphum rosae*. Later studies on *Regiella*, *Hamiltonella*, *Serratia*, *Arsenophonus*, and other symbionts continued to show erratic distributions among host species and, often, variable presence within a species (Baumann, 2005; Clark et al., 1992; Darby et al., 2001; Hansen et al., 2007; Haynes et al., 2003; Russell et al., 2003; Sandström et al., 2001). Additionally, detailed phylogenetic study of strains of *Hamiltonella* in different hosts shows that different strains within *A. pisum* do not form a single cluster and related aphid species, such as members of the genus *Uroleucon*, harbor phylogenetically disparate *Hamiltonella* strains (Degnan and Moran, 2008). Thus, strict maternal transmission, which would produce phylogenetically clustered distributions rather than the observed pattern of closely related symbionts inhabiting distantly related hosts, can be ruled out. Although maternal transmission is observed to be nearly 100% in long-term laboratory strains under nonstress conditions, it may be lower in nature; for example, losses are associated with the annual sexual generation in aphids (Moran and Dunbar, 2006) and prolonged heat stress (Nancy Moran, unpublished data).

One mechanism by which symbionts associate with new matrilines, at least within species, is through paternal transmission (Moran and Dunbar, 2006). Both *Regiella* and *Hamiltonella* achieve high densities within accessory glands of male aphids and are transmitted to females during mating; subsequent sexually produced female progeny can be infected by the paternal symbiont, sometimes at high rates. This paternal transfer can result in infection of matrilines that were uninfected and produce coinfections in matrilines already possessing a different symbiont type. However, this mechanism does not provide an explanation for the distribution of closely related symbiont strains between distantly related host species, implying that other routes of transfer exist. These have proved elusive in laboratory trials. Although symbionts can be readily moved between aphid hosts through microinjection (Chen and Purcell, 1997; Oliver et al., 2003; Russell and Moran, 2005) or their addition to artificial diets (Darby and Douglas, 2003), they have not been observed to be taken up from food plants or transferred by interactions with natural enemies, for example, via the contaminated ovipositors of parasitic wasps (Darby and Douglas, 2003; Oliver unpublished). These routes of transfer may occur at rates too low to be readily detected in experiments. However, even low rates of horizontal transfer of these symbionts would suffice to explain their distributional patterns in host populations.

9.3.3 Gene Exchange, Bacteriophage, and Toxin-Encoding Genes in Defensive Symbionts of Insects

A uniting feature of facultative symbionts within Enterobacteriaceae, and a similarity to pathogenic Enterobacteriaceae, is that many appear to participate in exchange of genetic material through horizontal gene transfer. This transfer is typically associated with bacteriophage or transposases. The initial indication of active phage in insect symbionts was based on microscopy and genome sequencing for viral particles from an SS of *A. pisum* (van der Wilk et al., 1999); the bacteriophage was called APSE. This symbiont was later confirmed to be a strain of *Hamiltonella* (Degnan and Moran, 2008b). Variants of APSE were found in many strains of *Hamiltonella* and designated APSE-1, APSE-2, etc. based on near identical sequences for many genes (Degnan and Moran, 2008; Degnan and Moran, 2008b; Moran et al., 2005a; Sandström et al., 2001). As known for other lambdoid phages APSE genomes are mosaics of conserved regions associated with viral particle production and highly variable regions that differ in length often show no homology among different strains, and contain numerous coding gene regions (Moran et al., 2005a). Some of the genes within these variable regions show homology to well-characterized toxins that are associated with active or inactivated phage sequences in characterized pathogens of mammals. For example, APSE-1 contains a homolog of Shiga toxin, which is encoded by phage-derived regions of the *Shigella* genome, and other APSE contains homologs of genes encoding cytolethal distending toxin, also associated with phage in pathogenic *Salmonella* and *E. coli* (Moran et al., 2005a). In *Hamiltonella*, the toxin-encoding genes are highly expressed (Moran et al., 2005a). Because these genes are known to be restricted to eukaryotic targets, their presence in defensive symbionts raises the possibility that they are a part of the protective mechanism (Degnan and Moran, 2008; Moran et al., 2005a). This interpretation is further bolstered by the finding that, among a set of *Hamiltonella* strains identical for numerous chromosomal markers but differing in presence and type of APSE, the absence of APSE is associated with decreased protection (Degnan and Moran, 2008b). In a multilocus study of *Hamiltonella*, chromosomal genes showed largely clonal transmission, but phage-borne genes underwent recombination and could be found on different chromosome backgrounds (Degnan and Moran, 2008b). Because APSE and related phage undergo exchange among symbiont genomes, potentially extending to other lineages of Enterobacteriaceae, these observations suggest that key genes underlying defense are swapped between symbiont lineages. This scenario would enable a highly dynamic set of defensive mechanisms to be sampled by symbiont and host populations, permitting defensive toxins to rotate among ecologically remote situations.

Limited observations suggest that horizontal transfer is important in the movement of toxin-encoding genes among symbionts beyond the Enterobacteriaceae. As mentioned above, the *Pseudomonas* symbiont of *Paederus* beetles also participates in foreign gene uptake. The toxic polyketide pederin is itself produced by enzymes encoded by a horizontally transferred genomic region (Piel, 2002; Piel et al., 2004a). Furthermore, this genomic island within the *Paederus* symbiont is homologous to genomic regions of sponge endosymbionts underlying production of related toxins (Piel et al., 2004b). Because sampling is limited, one cannot conclude that gene transfer occurred directly between symbionts of marine sponges and terrestrial arthropods, but these results indicate that the same gene family has become incorporated in symbiont-based defense in these ecologically and phylogenetically remote animal hosts.

The most extensively studied genome for a facultative and mutualistic insect symbiont is that of *S. glossinidius*, the facultative horizontally transmitted symbiont with unknown roles in its tsetse hosts. *Sodalis* carries numerous phage-derived regions on its chromosome (Toh et al., 2006) as well as a number of plasmids, including at least one large plasmid originating from phage with substantial homology to genes of APSE (Clark et al., 2007) and plasmids bearing genes for toxin production (Darby et al., 2005). Numerous toxin-producing genes are found on the main chromosomes of *Hamiltonella* (Moran et al., 2005a).

In addition to toxin-encoding genes and bacteriophage, facultative symbionts of insects share numerous other genetic capabilities with pathogenic bacteria (overview in Pallen and Wren, 2007). Prominent among the common mechanisms employed by bacteria infecting eukaryotic hosts are type three secretion systems. These are complex molecular syringes that deliver proteins to host cell targets, that are required for pathogenesis in many bacteria infecting plants and animals, and have moved among bacterial lineages through horizontal transfer of DNA (Galán and Wolf-Watz, 2006; Pallen and Wren, 2007). Although they were first studied in mammalian pathogens, Type Three Secretion Systems have subsequently been identified as important agents of infection in symbionts as well as bacteria infecting a wide range of eukaryotic hosts. In insect facultative symbionts, these secretion systems are sometimes required for establishment of symbiosis, as shown for *Sodalis* (Dale et al., 2001). Intact Type Three Secretion Systems are found in the genomes of *Hamiltonella* and have been hypothesized to function in delivering toxins that act against parasitoid enemies of aphid hosts (Moran et al., 2005a; Degnan and Moran, 2008).

Thus, defensive insect symbionts are horizontally transmitted, can coinfect individual hosts, routinely exchange genetic material, and take up foreign genes, including many genes known to contribute to the infection of hosts by bacterial pathogens. Together, these observations suggest that facultative insect symbionts participate in an arena of gene exchange that allows swapping of diverse, ecologically potent genes. These genes are often borne on phage genomes, or flanked by phage genes on chromosomes, but may be permanently incorporated onto symbiont chromosomes. Toxin-encoding genes are prominent among those implicated in horizontal transfer in symbiont genomes. These toxins may be central to the generation of novel defenses, enabling hosts to respond rapidly as natural enemies change or become resistant to established defenses.

9.4 Impacts of Defensive Symbionts on Natural Populations and Communities

9.4.1 Maintenance of Defensive Symbionts in Aphid Populations

Worldwide surveys indicate that the defensive symbionts, *Hamiltonella* and *Regiella*, are found at intermediate frequencies in *A. pisum* populations (Oliver et al., 2006; Sandström et al., 2001; Simon et al., 2003; Tsuchida et al., 2002). The incidence of facultative symbiont infection in natural populations is determined by the relative fitness of uninfected vs. infected lineages, the rate of acquisition of SS by uninfected lineages, the rate of SS loss in infected lineages, and possible reproductive manipulation occurring during the aphid's sexual stage (Table 9.1). No evidence supports the role of aphid SS in reproductive manipulation, and opportunities for reproductive manipulation may be particularly limited in aphids in which populations are parthenogenetic for most generations. Loss of SS from infected lineages is uncommon, at least for infected lines reared in the laboratory, although heat or other stresses might result in some loss of SS in nature. While acquisition of SS via horizontal transmission within and among aphid species is indicated by phylogenetic analyses, the frequencies and mechanisms of such lateral transfers require further investigation, as discussed above.

TABLE 9.1

Factors Affecting the Frequency of Infection by Facultative Symbionts in *Acyrthosiphon pisum*

Factors increasing infection frequency of matrilines	*Benefits to host fitness conferred by symbiont infection (mutualism)*
	Protection against parasitic wasps (Oliver et al., 2003, 2005, 2008)
	Protection against fungal pathogens (Scarborough et al., 2005)
	Thermal tolerance (Chen et al., 2000; Montllor et al., 2002; Russell and Moran, 2006)
	Improved performance on particular food plants (Tsuchida et al., 2004)
	Symbiont acquisition in uninfected lines
	Paternal transfer into uninfected matrilines (Moran and Dunbar, 2006)
	Horizontal transfer via wasps, host plants, or other mechanisms, including interspecific transfer (Oliver et al., 2008)
Factors decreasing infection frequency of matrilines	Costs to host fitness resulting from infection (Oliver et al., 2008)
	Loss of symbionts from infected lines
	Losses in sexual generation (Moran and Dunbar, 2006)
	Losses due to heat or other stresses (Moran, unpublished)
	Competition with other symbionts infecting same host individuals
	Costs conferred to superinfected hosts (Oliver et al., 2006)
	Reduction of superinfection to single infection (Sandström et al., 2001)

Heritable symbionts can increase in frequency in the host populations by providing net benefits to hosts, i.e., through mutualism. Thus, the defensive benefits of infection with *Hamiltonella* and *Regiella* identified in laboratory assays provide a likely mechanism for the spread and persistence of these SS in natural populations. Using population cages, Oliver et al. (2008) confirmed that pressure from parasitic wasps can lead to a rapid increase in the proportion of aphids in a population infected with *Hamiltonella*. In control cages not exposed to parasitism, the frequency of *A. pisum* infected with *Hamiltonella* declined, indicating a cost of infection in the absence of parasitism (Oliver et al., 2008). This cost probably best explains why *Hamiltonella* remains at intermediate frequencies in natural populations.

Surveys also indicate that individual *A. pisum* can be infected with more than one SS (e.g., Tsuchida et al., 2002). One study found that an *A. pisum* line superinfected with both *Hamiltonella* and *Serratia* was even more resistant to parasitism by *A. ervi* than aphids singly infected with either strain (Oliver et al., 2006). A survey of *A. pisum* SS in natural populations, however, found that superinfection with these particular SS was less common than expected by chance, likely attributable to severe fitness costs detected in laboratory assays. In the same study, quantitative PCR estimation of SS indicated that titers of *Serratia* increased 20-fold when coinhabiting an aphid host also infected with *Hamiltonella* (Oliver et al., 2006). The increase in *Serratia* titers may be responsible for the poor performance of superinfected aphids and suggests that interactions among symbionts may also play a role in determining phenotypes and distributions of SS in nature.

9.4.2 Effects of Environmental Factors on Defensive Symbiosis

Environmental factors, such as temperature, may affect the performance of defensive symbionts and thereby influence the geographic distributions of symbionts. For instance, *A. pisum* lines that are resistant to parasitism by *A. ervi* at 20°C become susceptible to parasitism when reared at 25°C and 30°C (Bensadia et al., 2006). These resistant lines were infected with *Hamiltonella* that was likely the basis for the observed resistance (Bensadia et al., 2006). The symbionts themselves, or toxins involved in host protection, may suffer reduced performance at higher temperatures. Many toxins are known to be heat-labile (e.g., Johnson and Lior, 1988). Other environmental factors also likely influence the dynamics of symbiont-mediated defensive interactions, indicating the need to evaluate these interactions in natural communities.

9.4.3 Potential Responses by Natural Enemies to Overcome Symbiont-Based Defense

Herbivorous insects have developed strategies to overcome a range of plant defenses, including toxic secondary metabolites, so it should not be surprising to discover that natural enemies will develop strategies aimed at circumventing symbiont-based defenses. Microbial symbionts could influence parasitoid host location and acceptance by influencing cues associated with these behaviors. For example, bacteria in the noctuid moth *Acrolepiopsis assectella* produce the host-locating cues used by the parasitic wasp *Diadromus pulchellus* (Thibout et al., 1993). Also, the symbiotic bacteria of entomopathogenic nematodes are responsible for an "ant-deterrent factor" that serves to protect some nematodes from scavenging predators (Zhou et al., 2002). Microorganisms have also been found to affect the composition of aphids' honeydew (liquid excrement) (Davidson et al., 1994) that is often a host location cue used by natural enemies (e.g., Du et al., 1997). The defensive symbiont *Hamiltonella* has been detected in honeydew of *A. pisum* (Darby and Douglas, 2003). Wasps may be able to detect the presence of microbial symbionts, or changes in honeydew composition resulting from infection; if so, they may use this information in deciding whether to attack. Parasitic wasps may also have the option of overwhelming the symbiont-based defensive systems by depositing multiple offspring (superparasitism) in a single host. In solitary endoparasitic wasps, only one parasitoid will emerge from an aphid host regardless of the numbers of eggs deposited. In a superparasitized host, the surviving wasp may gain an advantage in the host-parasitoid conflict from extra doses of maternal factors such as venom.

Of course, natural enemies could also employ "offensive" symbionts to overcome symbiont-mediated defense. Parasitic wasps in the families Braconidae and Ichneumonidae are often associated with viral mutualists, called polydnaviruses, that assist the wasp in overcoming the immune functions of the parasitized host. Polydnaviruses are replicated and stably inherited in the wasp host and become pathogenic only in the parasitized host (overview in Kroemer and Webb, 2004). In another example, entomopathogenic nematodes (Steinernematidae and Heterorhabditidae) are infected with bacterial symbionts, *Xenorhabdus* and *Photorhabdus*, that produce a complex of toxins that render a wide range of insects suitable as hosts for nematode growth and reproduction (Bowen et al., 1998; Forst and Nealson, 1996; overview in Forst et al., 1997). Future research may reveal insect systems in which microbial warfare between offensive and defensive symbionts determines the outcome of host–parasite interactions.

9.4.4 Community Wide Effects of Defensive Symbionts

Defensive symbionts, although inconspicuous, are likely to have effects extending to natural communities. They can potentially reduce the amount or quality of resources available to the higher trophic levels as well as influence the host organism's impact on lower ones. In herbivorous insects, for example, defensive symbionts may increase the abundance and range of host insects with accompanying repercussions for food plants while decreasing the abundance of particular natural enemies limiting resources available for higher order natural enemies. A study investigating the effects of fungal endophytes, which protect grasses from herbivores, on natural communities found that endophytes influence food web interactions by limiting resource availability to higher levels (Omacini et al., 2001). Also, specific parasitoids and predators of herbivorous insects feeding on endophyte-infected grasses have been reported to suffer reduced performance, further evidence that effects of defensive symbionts may extend to higher trophic levels (e.g., Barker and Addison, 1996; Omacini et al., 2001; de Sassi et al., 2006).

9.5 Exploiting Defensive Symbionts for Applications in Medicine and Agriculture

The presence of defensive symbionts may affect the outcome of attempts to control populations of pest insects; alternatively, they might be exploited to manipulate populations in useful ways. Classical biological control programs aim to use natural enemies to limit populations of insect pests, including herbivores of crop plants and forests as well as vectors of diseases of plants, humans, and other animals. Biological control agents can include parasitoids, predators, or pathogens, and highly specialized agents are often preferred to limit the impact on nontarget species. The success of biocontrol efforts depends on susceptibility of the targets and the ability of targets to quickly evolve resistance. For example, variation in resistance to parasitism, as is observed among populations of *A. pisum* (Henter and Via, 1995; Hufbauer, 2001; Hufbauer and Via, 1999), *Glycaspis brimblecombei* (Hansen et al., 2007), and many others, may often reflect differences in symbiont's presence or symbiont strain. Symbiont-based resistance may be particularly labile during short-term evolutionary responses of populations since it can reflect the acquisition of novel genetic elements both by insects, which acquire new symbionts, and the symbionts, which can acquire novel genes as bacteriophage or other means. Symbiont-based resistance traits may provide explanation as to why biocontrol efforts range from highly successful (Stiling and Cornellissen, 2005) to unexpectedly ineffective (reviewed in Julien and Griffiths, 1998). Understanding this aspect of host defense, and the stability of defensive symbionts within target lineages, is critical to designing successful biological control programs. In programs using insect herbivores as agents to control invasive plants, defensive symbionts might be useful as protectors of the insect agents.

Numerous researchers have proposed facultative symbionts as a means for genetically modifying insect pests to reduce their negative impacts. For example, *Sodalis* can interact with trypanosomes within the tsetse host so as to affect the vector competency of the flies (Dale and Welburn, 2001). A requirement for modifying natural populations via symbiont introduction is some kind of selective force that can drive the symbiont into the population. Potentially, a defensive symbiont could be engineered to include desirable traits as well as fitness-enhancing resistance to natural enemies, allowing it to spread within target populations and thus neutralize them.

Defensive symbionts of insects present a source of diverse compounds potentially useful in medicine. As most of the enemies of the insects are animals, particularly parasitoids, many of these compounds will target animal cells, making them potential candidates for applications in chemotherapy of cancerous growth. Pederin, the antipredator toxin produced by the *Pseudomonas* symbiont of *Paederus* beetles, suppresses tumor growth in multiple cancer cell lines (e.g., Richter et al., 1997) and the cytolethal distending toxin that appears to contribute to the defensive role of *Hamiltonella* targets chromosomes of actively dividing eukaryotic cells, resulting in the arrest of cell division and cell death.

9.6 Extent and Importance of Defensive Symbiosis in Insect Ecology and Evolution

Facultative symbionts are potentially present in most species, and a defensive effect on hosts may be a very common mechanism enabling their spread. Bacteriome-associated symbionts, which are readily detected because they are abundant and occupy a special organ, occur in 10%–20% of insects (Buchner, 1965). Facultative symbionts, however, are almost certainly more prevalent though more difficult to detect. Current screening methods are likely to underestimate symbiont occurrence; usually, only one individual is sampled to represent a species, and facultative symbionts are ordinarily variable in presence. Furthermore, only the most distinctive symbiont clades can be reliably screened using diagnostic PCR, the easiest and most commonly used molecular method. Even for distinctive symbiont lineages such as *Wolbachia*, many false negatives result, depending on the choice of PCR primers and conditions (Jeyaprakash and Hoy, 2000). The few groups that are intensively studied, such as aphids and psyllids, have high frequencies of symbionts. Recall that *A. pisum*, which is probably the most intensively screened insect species, can have any of the five symbiont types. Intensive sampling of ladybird beetles revealed high frequencies of several symbiont types (Weinert et al., 2007). Particular symbiont groups are more common in some hosts than others. For example, a screen of 35 *Drosophila* species using universal bacterial primers on DNA from ovaries dissected from females detected only *Wolbachia* and *Spiroplasma* (Mateos et al., 2006).

What are these abundant facultative symbionts doing in hosts? Although some are known to spread within host populations through reproductive manipulation, most facultative symbionts have undetermined effects. Indeed, reproductive manipulation affecting sex ratio or mating compatibility can often be ruled out for insects that are maintained in lab colonies (e.g., Mateos et al., 2006; Weinert et al., 2007). In general, the high frequency of facultative symbionts in insects most likely reflects mutualistic effects on hosts. The ability of bacterial symbionts to act as portals for novel genetic material makes them an ideal source of defense against natural enemies that are often the most dynamic aspect of the host's environment. Antagonistic coevolution of insects with their natural enemies may be fueled by an input of defensive symbiont types, themselves sampling a variety of genes useful in defensive phenotypes.

9.7 Future Work

Investigations of defensive effects of insect symbionts are all recent and so far restricted to relatively few systems, with aphids and *Paederus* beetles being the only taxonomic groups receiving in-depth attention. In addition, while this chapter has focused on heritable symbionts that are passed through eggs or embryos, they are only one of the components of the microbiota associated with terrestrial invertebrates. Recent surveys of the gut microbiota of several insects suggest that many species harbor a specialized community of bacterial strains, characteristic of a host species and probably transferred through behavioral or ecological interactions (e.g., Cox-Foster et al., 2007; Schloss et al., 2006; Warnecke et al., 2007). The functional roles of these diagnostic microflorae are largely unknown, except in cases of specialized nutritional contributions. Because the gut is a primary entry point for opportunistic pathogens, these organisms may contribute to defense against harmful infections.

The integration of molecular studies and experimental work is the key to the next stage of research on defensive symbionts in insects. Potentially, we are seeing the first glimpse of an immense consortium of symbiotic microorganisms that jump among diverse arthropod hosts that sometimes settle into particular hosts and share genes encoding potent defenses. These may form a highly dynamic arsenal, tapped by many insect hosts. The extent of this phenomenon is not yet clear, and its elucidation will depend on tracking the symbionts within host populations, experimentally verifying the consequences of symbiont infections for hosts, and the comparative analysis of sequences from different symbiont types.

ACKNOWLEDGMENTS

We thank Patrick Degnan, Cara Gibson, and Martha Hunter for feedback on a draft of the manuscript. Cara Gibson provided the illustration for Figure 9.1 and Alex Wild provided the photograph of Figure 9.2. Becky Nankivell assisted with manuscript formatting. Financial support came from NSF grant 0313737 to NAM, and KMO is supported on a fellowship from the Postdoctoral Excellence in Research and Teaching program (National Institutes of Health) at the Center for Insect Science at the University of Arizona.

REFERENCES

Barker, G. M. and P. J. Addison. 1996. Influence of clavicipitaceous endophyte infection in ryegrass on development of the parasitoid *Microctonus hyperodae* loan (Hymenoptera: Braconidae) in *Listronotus bonariensis* (Kuschel) (Coleoptera: Curculionidae). *Biological Control* 7(3):281–287.

Baumann, P. 2005. Biology of bacteriocyte-associated endosymbionts of plant sap-sucking insects. *Annual Review of Microbiology* 59:155–189.

Bensadia, F., Boudreault, S., Guay, J. F., Michaud, D., and C. Cloutier. 2006. Aphid clonal resistance to a parasitoid fails under heat stress. *Journal of Insect Physiology* 52(2):146–157.

Bowen, D., Rocheleau, T. A., Blackburn, M., Andreev, O., Golubeva, E., Bhartia, R., et al. 1998. Insecticidal toxins from the bacterium *Photorhabdus luminescens*. *Science* 280(5372):2129–2132.

Brega, A., Falaschi, A., Decarli, L., and M. Pavan. 1968. Studies on mechanism of action of pederine. *Journal of Cell Biology* 36(3):485–496.

Buchner, P. 1965. *Endosymbiosis of animals with plant microorganisms* (Rev. Eng. [Translation by Bertha Mueller, with the collaboration of Frances H. Fockler] ed.). New York: Interscience Publishers.

Bundy, C. S., Smith, P. F., English, L. M., Sutton, D., and S. Hanson. 2005. Strain distribution of alfalfa weevil (Coleoptera: Curculionidae) in an intergrade zone. *Journal of Economic Entomology* 98(6):2028–2032.

Cafaro, M. J. and C. R. Currie. 2005. Phylogenetic analysis of mutualistic filamentous bacteria associated with fungus-growing ants. *Canadian Journal of Microbiology* 51(6):441–446.

Cardoza, Y. J., Klepzig, K. D., and K. F. Raffa. 2006. Bacteria in oral secretions of an endophytic insect inhibit antagonistic fungi. *Ecological Entomology* 31(6):636–645.

Carver, M. and D. J. Sullivan. 1988. Encapsulative defence reactions of aphids (Hemiptera: Aphididae) to insect parasitoids (Hymenoptera: Aphidiidae and Aphelinidae). In *Ecology and Effectiveness of Aphidophaga*, eds. A. F. G. Niemczyk, pp. 299–303. The Hague: SPB Academic Publishing.

Chen, D. Q., Campbell, B. C., and A. H. Purcell. 1996. A new *Rickettsia* from a herbivorous insect, the pea aphid *Acyrthosiphon pisum* (Harris). *Current Microbiology* 33(2):123–128.

Chen, D. Q., Montllor, C. B., and A. H. Purcell. 2000. Fitness effects of two facultative endosymbiotic bacteria on the pea aphid, *Acyrthosiphon pisum*, and the blue alfalfa aphid, *A. kondoi*. *Entomologia Experimentalis et Applicata* 95(3):315–323.

Chen, D. Q. and A. H. Purcell. 1997. Occurrence and transmission of facultative endosymbionts in aphids. *Current Microbiology* 34(4):220–225.

Clark, M. A., Baumann, L., Munson, M. A., Baumann, P., Campbell, B. C., Duffus, J. E., et al. 1992. The eubacterial endosymbionts of whiteflies (Homoptera, Aleyrodoidea) constitute a lineage distinct from the enodsymbionts of aphids and mealybugs. *Current Microbiology* 25(2):119–123.

Clark, A. J., Pontes, M., Jones, T., and C. Dale. 2007. A possible heterodimeric prophage-like element in the genome of the insect endosymbiont *Sodalis glossinidius*. *Journal of Bacteriology* 189:2949–2951.

Cox-Foster, D. L., Conlan, S., Holmes, E. C., Palacios, G., Evans, J. D., Moran, N. A., et al. 2007. A metagenomic survey of microbes in honey bee colony collapse disorder. *Science* 318(5848):283–287.

Currie, C. R., Scott, J. A., Summerbell, R. C., and D. Malloch. 1999. Fungus-growing ants use antibiotic-producing bacteria to control garden parasites. *Nature* 398(6729):701–704.

Dale, C., Beeton, M., Harbison, C., Jones, T., and M. Pontes. 2006. Isolation, pure culture, and characterization of "*Candidatus* Arsenophonus arthropodicus," an intracellular secondary endosymbiont from the hippoboscid louse fly *Pseudolynchia canariensis. Applied and Environmental Microbiology* 72(4):2997–3004.

Dale, C. and S. C. Welburn. 2001. The endosymbionts of tsetse flies: Manipulating host–parasite interactions. *International Journal of Parasitology* 31(5–6):628–631.

Dale, C., Young, S. A., Haydon, D. T., and S. C. Welburn. 2001. The insect endosymbiont *Sodalis glossinidius* utilizes a type III secretion system for cell invasion. *Proceedings of the National Academy of Sciences USA* 98(4):1883–1888.

Darby, A. C., Birkle, L. M., Turner, S. L., and A. E. Douglas. 2001. An aphid-borne bacterium allied to the secondary symbionts of whitefly. *FEMS Microbiology Ecology* 36(1):43–50.

Darby, A. C. and A. E. Douglas. 2003. Elucidation of the transmission patterns of an insect-borne bacterium. *Applied and Environmental Microbiology* 69(8):4403–4407.

Darby, A. C., Lagnel, J., Matthew, C. Z., Bourtzis, K., Maudlin, I., and S. C. Welburn. 2005. Extrachromosomal DNA of the symbiont *Sodalis glossinidius. Journal of Bacteriology* 187(14):5003–5007.

Davidson, E. W., Segura, B. J., Steele, T., and D. L. Hendrix. 1994. Microorganisms influence the composition of honeydew produced by the silverleaf whitefly, *Bemisia argentifoli. Journal of Insect Physiology* 40(12):1069–1076.

Degnan, P. H. and N. A. Moran. 2008. Evolutionary genetics of a defensive facultative symbiont of insects: Exchange of toxin-encoding bacteriophage. *Molecular Ecology* 17(3):916–929.

Degnan, P. H. and N. A. Moran. 2008b. Diverse phase-encoded toxins in a protective-insect endosymbiont. *Applied and Environmental Microbiology* 74(21):6781–6791.

Dettner, K. 1987. Chemosystematics and evolution of beetle chemical defenses. *Annual Review of Entomology* 32:17–48.

Douglas, A. E. 1989. Mycetocyte symbiosis in insects. *Biological Reviews of the Cambridge Philosophical Society* 64(4):409–434.

Du, Y. J., Poppy, G. M., Powell, W., and L. J. Wadhams. 1997. Chemically mediated associative learning in the host foraging behavior of the aphid parasitoid *Aphidius ervi* (Hymenoptera: Braconidae). *Journal of Insect Behavior* 10(4):509–522.

Ferrari, J., Darby, A. C., Daniell, T. J., Godfray, H. C. J., and A. E. Douglas. 2004. Linking the bacterial community in pea aphids with host-plant use and natural enemy resistance. *Ecological Entomology* 29(1):60–65.

Ferrari, J., Muller, C. B., Kraaijeveld, A. R., and H. C. J. Godfray. 2001. Clonal variation and covariation in aphid resistance to parasitoids and a pathogen. *Evolution* 55(9):1805–1814.

Forst, S., Dowds, B., Boemare, N., and E. Stackebrandt. 1997. *Xenorhabdus* and *Photorhabdus* spp.: Bugs that kill bugs. *Annual Review of Microbiology* 51:47–72.

Forst, S. and Nealson, K. 1996. Molecular biology of the symbiotic pathogenic bacteria *Xenorhabdus spp* and *Photorhabdus spp. Microbiological Reviews* 60(1):21–43.

Frank, J. H. and K. Kanamitsu. 1987. *Paederus*, sensu lato (Coleoptera, Staphylinidae)—natural history and medical importance. *Journal of Medical Entomology* 24(2):155–191.

Fukatsu, T. 2001. Secondary intracellular symbiotic bacteria in aphids in the genus *Yamotocallis* (Homoptera: Aphididae: Drepanosohinae). *Applied and Environmental Microbiology* 67(11):5315–5320.

Fukatsu, T. and H. Ishikawa. 1993. Occurrence of chaperonin-60 and chapereronin-10 in primary and secondary bacterial symbionts of aphids—implications for the evolution of an endosymbiotic system in aphids. *Journal of Molecular Evolution* 36(6):568–577.

Fukatsu, T., Koga, R., Smith, W. A., Tanaka, K., Nikoh, N., Sasaki-Fukatsu, K., et al. 2007. Bacterial endosymbiont of the slender pigeon louse, *Columbicola columbae*, allied to endosymbionts of grain weevils and tsetse flies. *Applied and Environmental Microbiology* 73(20):6660–6668.

Fukatsu, T., Nikoh, N., Kawai, R., and R. Koga. 2000. The secondary endosymbiotic bacterium of the pea aphid *Acyrthosiphon pisum* (Insecta: Homoptera). *Applied and Environmental Microbiology* 66(7):2748–2758.

Fukatsu, T., Tsuchida, T., Nikoh, N., and R. Koga. 2001. *Spiroplasma* symbiont of the pea aphid, *Acyrthosiphon pisum* (Insecta: Homoptera). *Applied and Environmental Microbiology* 67(3):1284–1291.

Galán, J. E. and H. Wolf-Watz. 2006. Protein delivery into eukaryotic cells by type III secretion machines. *Nature* 444(7119):567–573.

Gelmetti, C. and R. Grimalt. 1993. *Paederus* dermatitis—an easy diagnosable but misdiagnosed eruption. *European Journal of Pediatrics* 152(1):6–8.

Godfray, H. C. J. 1994. *Parasitoids: Behavioral and Evolutionary Ecology*. Princeton, NJ: Princeton University Press.

Goettler, W., Kaltenpoth, M., Herzner, G., and E. Strohm. 2007. Morphology and ultrastructure of a bacteria cultivation organ: The antennal glands of female European beewolves, *Philanthus triangulum* (Hymenoptera, Crabronidae). *Arthropod Structure and Development* 36(1):1–9.

Hansen, A. K., Jeong, G., Paine, T. D., and Stouthamer, R. 2007. Frequency of secondary symbiont infection in an invasive psyllid relates to parasitism pressure on a geographic scale in California. *Applied and Environmental Microbiology* 73(23):7531–7535.

Haynes, S., Darby, A. C., Daniell, T. J., Webster, G., van Veen, F. J. F., Godfray, H. C. J., et al. 2003. Diversity of bacteria associated with natural aphid populations. *Applied and Environmental Microbiology* 69(12):7216–7223.

Henter, H. J. and S. Via. 1995. The potential for coevolution in a host–parasitoid system. 1. Genetic-variation within an aphid population in susceptibility to a parasitic wasp. *Evolution* 49(3):427–438.

Hofstetter, R. W., Cronin, J. T., Klepzig, K. D., Moser, J. C., and M. P. Ayres. 2006. Antagonisms, mutualisms and commensalisms affect outbreak dynamics of the southern pine beetle. *Oecologia* 147(4):679–691.

Holsten, E. H., Their, R. W., Munson, A. S., and K. E. Gibson. 1999. *The Spruce Beetle* (vol. 127). Portland, OR: U. S. Dept. of Agriculture, Forest Service, Pacific Northwest Forest and Range Experiment Station.

Hsaio, T. H. 1996. Studies of interactions between alfalfa weevil strains, Wolbachia endosymbionts and parasitoids. In *The Ecology of Agricultural Pests: Biochemical Approaches*, eds. W. O. C. Symondson and J. E. Liddell, p. 517. London: Chapman and Hall.

Hsiao, C. and T. H. Hsiao. 1985a. *Ricketssia* as the cause of cytoplasmic incompatibility in the alfalfa weevil, *Hypera postica. Journal of Invertebrate Pathology* 45(2):244–246.

Hsiao, T. H. and C. Hsiao. 1985b. Hybridization and cytoplasmic incompatibility among alfalfa weevil strains. *Entomologia Experimentalis et Applicata* 37(2):155–159.

Hufbauer, R. A. 2001. Pea aphid-parasitoid interactions: Have parasitoids adapted to differential resistance? *Ecology* 82(3):717–725.

Hufbauer, R. A. and S. Via. 1999. Evolution of an aphid-parasitoid interaction: Variation in resistance to parasitism among aphid populations specialized on different plants. *Evolution* 53(5):1435–1445.

Jeyaprakash, A. and M. A. Hoy. 2000. Long PCR improves *Wolbachia* DNA amplification: *wsp* sequences found in 76% of sixty-three arthropod species. *Insect Molecular Biology* 9(4):393–405.

Johnson, W. M. and H. Lior. 1988. A new heat-labile cytolethal distending toxin (CLDT) produced by *Escherichia coli* isolates from clinical material. *Microbial Pathogenesis* 4:103–113.

Julien, M. H. and M. W. Griffiths. 1998. *Biological Control of Weeds: A World Catalogue of Agents and Their Target Weeds*. New York: CABI.

Kageyama, D., Anbutsu, H., Watada, M., Hosokawa, T., Shimada, M., and T. Fukatsu. 2006. Prevalence of a non-male-killing *Spiroplasma* in natural populations of *Drosophila hydei. Applied and Environmental Microbiology* 72(10):6667–6673.

Kaltenpoth, M., Goettler, W., Dale, C., Stubblefield, J. W., Herzner, G., Roeser-Mueller, K., et al. 2006. "*Candidatus* Streptomyces philanthi," an endosymbiotic streptomycete in the antennae of *Philanthus* digger wasps. *International Journal of Systematic and Evolutionary Microbiology* 56:1403–1411.

Kaltenpoth, M., Gottler, W., Herzner, G., and E. Strohm. 2005. Symbiotic bacteria protect wasp larvae from fungal infestation. *Current Biology* 15(5):475–479.

Kellner, R. L. L. 1998. When do *Paederus riparius* rove beetles (Coleoptera: Staphylinidae) biosynthesize their unique hemolymph toxin pederin? *Zeitschrift Fur Naturforschung C—a Journal of Biosciences* 53(11–12):1081–1086.

Kellner, R. L. L. 1999. What is the basis of pederin polymorphism in *Paederus riparius* rove beetles? The endosymbiotic hypothesis. *Entomologia Experimentalis et Applicata* 93(1):41–49.

Kellner, R. L. L. 2001a. Horizontal transmission of biosynthetic capabilities for pederin in *Paederus melanurus* (Coleoptera: Staphylinidae). *Chemoecology* 11(3):127–130.

Kellner, R. L. L. 2001b. Suppression of pederin biosynthesis through antibiotic elimination of endosymbionts in *Paederus sabaeus. Journal of Insect Physiology* 47(4–5):475–483.

Kellner, R. L. L. 2002a. Molecular identification of an endosymbiotic bacterium associated with pederin biosynthesis in *Paederus sabaeus* (Coleoptera: Staphylinidae). *Insect Biochemistry and Molecular Biology* 32(4):389–395.

Kellner, R. L. L. 2002b. Interspecific transmission of *Paederus* endosymbionts: Relationship to the genetic divergence among the bacteria associated with pederin biosynthesis. *Chemoecology* 12(3):133–138.

Kellner, R. L. L. and K. Dettner. 1995. Allocation of pederin during lifetime of *Paederus* rove beetles (Coleoptera: Staphylinidae): Evidence for polymorphism of hemolymph toxin. *Journal of Chemical Ecology* 21(11):1719–1733.

Kellner, R. L. L. and K. Dettner. 1996. Differential efficacy of toxic pederin in deterring potential arthropod predators of *Paederus* (Coleoptera: Staphylinidae) offspring. *Oecologia* 107(3):293–300.

Kingsley, P. C., Bryan, M. D., Day, W. H., Burger, T. L., Dysart, R. J., and C. P. Schwalbe. 1993. Alfalfa weevil (Coleoptera, Curculionidae) biological control—spreading the benefits. *Environmental Entomology* 22(6):1234–1250.

Koga, R., Tsuchida, T., and T. Fukatsu. 2003. Changing partners in an obligate symbiosis: A facultative endosymbiont can compensate for loss of the essential endosymbiont *Buchnera* in an aphid. *Proceedings of the Royal Society of London Series B—Biological Sciences* 270(1533):2543–2550.

Kopper, B. J., Klepzig, K. D., and K. F. Raffa. 2004. Components of antagonism and mutualism in *Ips pini*–fungal interactions: Relationship to a life history of colonizing highly stressed and dead trees. *Environmental Entomology* 33(1):28–34.

Kroemer, J. A. and B. A. Webb. 2004. Polydnavirus genes and genomes: Emerging gene families and new insights into polydnavirus replication. *Annual Review of Entomology* 49:431–456.

Leu, S. J. C., Li, J. K. K., and T. H. Hsiao. 1989. Characterization of *Wolbachia postica*, the cause of reproductive incompatibility among alfalfa weevil strains. *Journal of Invertebrate Pathology* 54(2):248–259.

Lopanik, N., Gustafson, K. R., and N. Lindquist. 2004. Structure of bryostatin 20: A symbiont-produced chemical defense for larvae of the host bryozoan, *Bugula neritina. Journal of Natural Products* 67(8):1412–1414.

Mateos, M., Catrezana, S. J., Nankivell, B. J., Estes, A., Markow, T. A., and N. A. Moran. 2006. Heritable endosymbionts of *Drosophila. Genetics* 174(1):363–376.

Maund, C. M. and T. H. Hsiao. 1991. Differential encapsulation of 2 *Bathyplectes* parasitoids among alfalfa weevil strains, *Hypera postica. Canadian Entomologist* 123(1):197–203.

Montllor, C. B., Maxmen, A., and A. H. Purcell. 2002. Facultative bacterial endosymbionts benefit pea aphids *Acyrthosiphon pisum* under heat stress. *Ecological Entomology* 27(2):189–195.

Moran, N. A., Degnan, P. H., Santos, S. R., Dunbar, H. E., and H. Ochman. 2005a. The players in a mutualistic symbiosis: Insects, bacteria, viruses, and virulence genes. *Proceedings of the National Academy of Sciences of the USA* 102(47): 16919–16926.

Moran, N. A. and H. E. Dunbar. 2006. Sexual acquisition of beneficial symbionts in aphids. *Proceedings of the National Academy of Sciences of the USA* 103(34):12803–12806.

Moran, N. A., Russell, J. A., Koga, R., and T. Fukatsu. 2005b. Evolutionary relationships of three new species of Enterobacteriaceae living as symbionts of aphids and other insects. *Applied and Environmental Microbiology* 71(6):3302–3310.

Norton, S. A. and C. Lyons. 2002. Blister beetles and the ten plagues. *Lancet* 359(9321):1950.

Nováková, E. and V. Hypsa. 2007. A new *Sodalis* lineage from bloodsucking fly *Craterina melbae* (Diptera, Hippoboscoidea) originated independently of the tsetse flies symbiont *Sodalis glossinidius. FEMS Microbiology Letters* 269(1):131–135.

Oliver, K. M., Campos, J., Moran, N. A., and M. S. Hunter. 2008. Population dynamics of defensive symbionts in aphids. *Proceedings of the Royal Society B—Biological Sciences* 275(1632):293–299.

Oliver, K. M., Moran, N. A., and M. S. Hunter. 2005. Variation in resistance to parasitism in aphids is due to symbionts not host genotype. *Proceedings of the National Academy of Sciences of the USA* 102(36):12795–12800.

Oliver, K. M., Moran, N. A., and M. S. Hunter. 2006. Costs and benefits of a superinfection of facultative symbionts in aphids. *Proceedings of the Royal Society B—Biological Sciences* 273(1591):1273–1280.

Oliver, K. M., Russell, J. A., Moran, N. A., and M. S. Hunter. 2003. Facultative bacterial symbionts in aphids confer resistance to parasitic wasps. *Proceedings of the National Academy of Sciences of the USA* 100(4):1803–1807.

Omacini, M., Chaneton, E. J., Ghersa, C. M., and C. B. Muller. 2001. Symbiotic fungal endophytes control insect host–parasite interaction webs. *Nature* 409(6816):78–81.

O'Neill, S. L., Giordano, R., Colbert, A. M. E., Karr, T. L., and H. M. Robertson. 1992. 16S ribosomal RNA phylogenetic analysis of the bacterial endosymbionts associated with cytoplasmic incompatibility in insects. *Proceedings of the National Academy of Sciences of the USA* 89(7):2699–2702.

Pallen, M. J. and B. W. Wren. 2007. Bacterial pathogenomics. *Nature* 449(7164):835–842.

Perry, N. B., Blunt, J. W., Munro, M. H. G., and Pannell, L. K. 1988. Mycalamide-A, an antiviral compound from a New Zealand sponge of the genus *Mycale*. *Journal of the American Chemical Society* 110(14):4850–4851.

Piel, J. 2002. A polyketide synthase-peptide synthetase gene cluster from an uncultured bacterial symbiont of *Paederus* beetles. *Proceedings of the National Academy of Sciences of the USA* 99(22):14002–14007.

Piel, J., Hofer, I., and D. Q. Hui. 2004a. Evidence for a symbiosis island involved in horizontal acquisition of pederin biosynthetic capabilities by the bacterial symbiont of *Paederus fuscipes* beetles. *Journal of Bacteriology* 186(5):1280–1286.

Piel, J., Hui, D. Q., Wen, G. P., Butzke, D., Platzer, M., Fusetani, N., et al. 2004b. Antitumor polyketide biosynthesis by an uncultivated bacterial symbiont of the marine sponge *Theonella swinhoei*. *Proceedings of the National Academy of Sciences of the USA* 101(46):16222–16227.

Richter, A., Kocienski, P., Raubo, P., and D. E. Davies. 1997. The in vitro biological activities of synthetic 18-O-methyl mycalamide B, 10-epi-18-O-methyl mycalamide B and pederin. *Anti-Cancer Drug Design* 12(3):217–227.

Russell, J. A., Latorre, A., Sabater-Muñoz, B., Moya, A., and N. A. Moran. 2003. Side-stepping secondary symbionts: Widespread horizontal transfer across and beyond the Aphidoidea. *Molecular Ecology* 12(4):1061–1075.

Russell, J. A. and N. A. Moran. 2005. Horizontal transfer of bacterial symbionts: Heritability and fitness effects in a novel aphid host. *Applied and Environmental Microbiology* 71(12):7987–7994.

Russell, J. A. and N. A. Moran. 2006. Costs and benefits of symbiont infection in aphids: Variation among symbionts and across temperatures. *Proceedings of the Royal Society B—Biological Sciences* 273(1586):603–610.

Sakemi, S., Ichiba, T., Kohmoto, S., Saucy, G., and T. Higa. 1988. Isolation and structure elucidation of onnamide-A, a new bioactive metabolite of a marine sponge, *Theonella* sp. *Journal of the American Chemical Society* 110(14):4851–4853.

Salt, G. 1970. *The Cellular Defense Reactions of Insects*. Cambridge: Cambridge University Press.

Sandström, J. P., Russell, J. A., White, J. P., and N. A. Moran. 2001. Independent origins and horizontal transfer of bacterial symbionts of aphids. *Molecular Ecology* 10(1):217–228.

de Sassi, C., Müller, C. B., and J. Krauss. 2006. Fungal plant endosymbionts alter life history and reproductive success of aphid predators. *Proceedings of the Royal Society B—Biological Sciences* 273(1591):1301–1306.

Scarborough, C. L., Ferrari, J., and H. C. J. Godfray. 2005. Aphid protected from pathogen by endosymbiont. *Science* 310(5755):1781.

Schloss, P. D., Delalibera, I., Handelsman, J., and K. F. Raffa. 2006. Bacteria associated with the guts of two wood-boring beetles: *Anoplophora glabripennis* and *Saperda vestita* (Cerambycidae). *Environmental Entomology* 35(3):625–629.

Simon, J. C., Carre, S., Boutin, M., Prunier-Leterme, N., Sabater-Muñoz, B., Latorre, A., et al. 2003. Host-based divergence in populations of the pea aphid: Insights from nuclear markers and the prevalence of facultative symbionts. *Proceedings of the Royal Society of London Series B—Biological Sciences* 270(1525):1703–1712.

Soldati, M., Fioretti, A., and M. Ghione. 1966. Cytotoxicity of pederin and some of its derivatives on cultured mammalian cells. *Experientia* 22(3): 176–178.

Stiling, P. and T. Cornelissen. 2005. What makes a successful biocontrol agent? A meta-analysis of biological control agent performance. *Biological Control* 34:236–246.

Strohm, E. and K. E. Linsenmair. 1994. Leaving the cradle—how beewolves (*Philanthus triangulum*) obtain the necessary spatial information for emergence. *Zoology—Analysis of Complex Systems* 98(3):137–146.

Thibout, E., Guillot, J. F., and J. Auger. 1993. Microorganisms are involved in the production of volatile kairomones affecting the host-seeking behavior of *Diandromus pulchellus*, a parasitoid of *Acrolepiopsis assectella*. *Physiological Entomology* 18(2):176–182.

Toh, H., Weiss, B. L., Perkin, S. A. H., Yamashita, A., Oshima, K., Hattori, M., et al. 2006. Massive genome erosion and functional adaptations provide insights into the symbiotic lifestyle of *Sodalis glossinidius* in the tsetse host. *Genome Research* 16(2):149–156.

Tsuchida, T., Koga, R., and T. Fukatsu. 2004. Host plant specialization governed by facultative symbiont. *Science* 303(5666):1989.

Tsuchida, T., Koga, R., Shibao, H., Matsumoto, T., and T. Fukatsu. 2002. Diversity and geographic distribution of secondary endosymbiotic bacteria in natural populations of the pea aphid, *Acyrthosiphon pisum*. *Molecular Ecology* 11(10):2123–2135.

Unterman, B. M., Baumann, P., and D. L. McLean. 1989. Pea aphid symbiont relationships established by analysis of 16S ribosomal RNAs. *Journal of Bacteriology* 171(6):2970–2974.

van der Wilk, F., Dullemans, A. M., Verbeek, M., and J. F. van den Heuvel. 1999. Isolation and characterization of APSE-1, a bacteriophage infecting the secondary endosymbiont of *Acyrthosiphon pisum. Virology* 262(1):104–113.

Warnecke, F., Luginbühl, P., Ivanova, N., Ghassemian, M., Richardson, T. H., Stege, J. T., et al. 2007. Metagenomic and functional analysis of hindgut microbiota of a wood-feeding higher termite. *Nature* 450(7169):560–565.

Weinert, L. A., Tinsley, M. C., Temperley, M., and F. M. Jiggins. 2007. Are we underestimating the diversity and incidence of insect bacterial symbionts? A case study in ladybird beetles. *Biology Letters* 3(6):678–681.

Weiss, B. L., Mouchotte, R., Rio, R. V., Wu, Y. N., Wu, Z., Heddi, A., et al. 2006. Interspecific transfer of bacterial endosymbionts between tsetse fly species: Infection establishment and effect on host fitness. *Applied and Environmental Microbiology* 72(11):7013–7021.

Wernegreen, J. J. 2002. Genome evolution in bacterial endosymbionts of insects. *Nature Reviews Genetics* 3(11):850–861.

Werren, J. H. and S. L. O'Neill. 1997 *Influential Passengers: Inherited Microorganisms and Arthropod Reproduction*, eds. S. L. O'Neill, A. A. Hoffmann, and J. H. Werren. Oxford: Oxford University Press.

Werren, J. H. and D. M. Windsor. 2000. Wolbachia infection frequencies in insects: Evidence of a global equilibrium? *Proceedings of the Royal Society of London Series B—Biological Sciences* 267(1450):1277–1285.

Werren, J. H., Windsor, D., and L. R. Guo. 1995. Distribution of *Wolbachia* among neotropical arthropods. *Proceedings of the Royal Society of London Series B—Biological Sciences* 262(1364):197–204.

Zchori-Fein, E. and Brown, J. K. 2002. Diversity of prokaryotes associated with *Bemisia tabaci* (Gennadius) (Hemiptera: Aleyrodidae). *Ecology and Population Biology* 95(6):711–718.

Zchori-Fein, E. and S. J. Perlman. 2004. Distribution of the bacterial symbiont *Cardinium* in arthropods. *Molecular Ecology* 13(7):2009–2016.

Zhou, X. S., Kaya, H. K., Heungens, K., and H. Goodrich-Blair. 2002. Response of ants to a deterrent factor(s) produced by the symbiotic bacteria of entomopathogenic nematodes. *Applied and Environmental Microbiology* 68(12):6202–6209.

10

Fungus-Growing Ant–Microbe Symbiosis: Using Microbes to Defend Beneficial Associations within Symbiotic Communities

Michael Poulsen, Ainslie E. F. Little, and Cameron R. Currie

CONTENTS

10.1 Introduction

Since the origin of life on Earth approximately 4.6 billion years ago, microscopic organisms have evolved into the most abundant and diverse life forms on the planet. They have played key roles in biogeochemical processes, including helping convert the Earth's atmosphere into an oxygen rich biosphere and recycling organic compounds in every ecosystem (Pace, 1997; Hoehler et al., 2001; Dawson and Pace, 2002; Baldauf, 2003). However, it is the capacity microbes have to form symbiotic associations with other organisms that has led to many major evolutionary transitions that shape our planet. Examples include the formation of the eukaryotic cell and the colonization of land by plants (Margulis, 1970; Selosse et al., 2004). In addition to these mutualistic host–microbe associations, parasitic microbes shape the ecology and evolution of every living organism, incur a strong selection pressure on their hosts, and are recognized as a major force driving host evolution and biological diversification (Jaenike, 1978; Hamilton, 1980; Price et al., 1986; Ewald, 1994). Thus, symbiotic microbes, both as parasites and mutualists, influence the biology of every living organism (Boucher et al., 1982; Ewald, 1994; Lutzoni et al., 2001; Jaenike and Perlman, 2002; Sanders, 2002).

A major theme of symbiosis is phylogenetically diverse macroorganisms engaging in symbiosis with specific lineages of microbes that have specialized metabolic capabilities. Examples include organisms forming mutualisms with photosynthetic microbes to obtain carbohydrates (Rudman, 1987; Douglas, 1994; Paracer and Ahmadjian, 1999; Knowlton and Rohwer, 2003), as well as plants and animals having symbiotic relationships with nitrogen-fixing bacteria to gain greater access to nitrogen (Goodman and Weisz, 2002; Zehr et al., 2003; Dixon and Kahn, 2004). As explored throughout this book, microbial symbionts can also mediate the interactions that occur between the hosts and their parasites and predators.

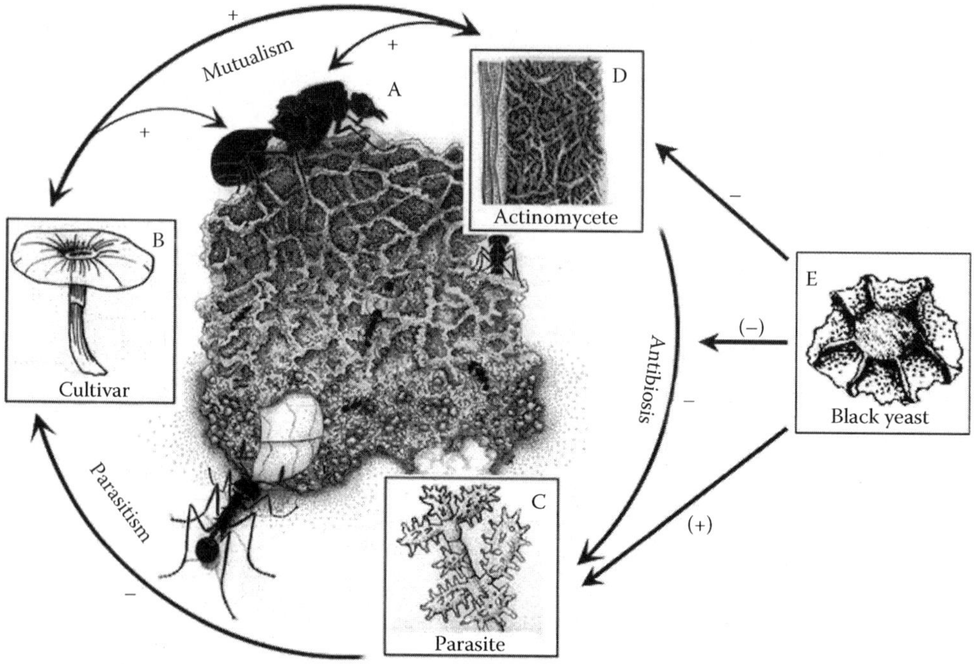

FIGURE 10.1 Fungus-growing ant–microbe symbiosis. A schematic diagram showing the five known symbionts of the fungus-growing ant–microbe symbiosis: attine ants, the mutualistic fungi they rear for food, the specialized parasites of the ant–fungus mutualism (*Escovopsis*), the actinomycete *Pseudonocardia* bacteria employed to defend against *Escovopsis*, and the black yeasts parasitizing the ant–actinomycete mutualism. Plusses and minuses signify the impact of individual symbionts on each other, and the signs in parentheses indicate indirect effects imposed. (Modified from Currie, C.R., *Ann. Rev. Microbiol.*, 55, 357, 2001a; black yeast drawing by Rebeccah Steffensen.)

Since species interactions are critical to the survival and reproduction of all organisms, it is perhaps not surprising that the hosts have evolved symbiotic associations with microbes for protection, and it is likely that these defensive mutualisms are more widespread than currently recognized.

Here we explore the theme of defensive mutualism within the fungus-growing ant–microbe symbiosis. Fungus-growing ants and their microbial partners form a complex symbiosis, involving at least five integrated, coevolved, and anciently associated lineages that span three taxonomic kingdoms. At the center of the symbiosis is the obligate mutualism between the ant farmers and the fungi they cultivate for food (Figure 10.1). This ant–fungus mutualism is exploited by microfungal parasites in the genus *Escovopsis*, and to defend against infection, the ants have a mutualistic association with antibiotic-producing filamentous bacteria in the genus *Pseudonocardia* (Figure 10.1). Similar to specialized parasites exploiting the ant–fungus mutualism, recent work has revealed that a black yeast symbiont exploits the ant–*Pseudonocardia* mutualism (Figure 10.1). In this chapter, we review the basic biology of the fungus-growing ant–microbe symbiosis, explore the ecology and evolution of the defensive mutualism, and suggest that defensive mutualisms involving actinomycete bacteria mediating beneficial and antagonistic symbiotic communities are likely more widespread than currently recognized.

10.2 Fungus-Growing Ants and Their Fungal Cultivar

Fungus-growing ants (Hymenoptera: Formicidae: Attini), as implied by their name, cultivate fungi for food. This obligate ant–fungus mutualism is believed to have originated ~50 million years ago in the Amazon Basin (Mueller et al., 2001; Schultz and Brady, 2008), occurs exclusively in the New World, and is most common in tropical ecosystems (Weber, 1972; Hölldobler and Wilson, 1990). Mature fungus-growing ant colonies produce alates (winged males and females), who leave the colony to mate during synchronized nuptial flights. A colony-founding queen collects a pellet of fungus from her natal nest before leaving for her mating flight, and this pellet is stored in her infrabuccal pocket (a pouch in the oral cavity). After mating, female alates select and excavate a suitable nest site, and the fungus from her infrabuccal pocket serves as the inoculum for a new garden (vertical transmission) (Hölldobler and Wilson, 1990, after von Ihering, 1898 and Autuori, 1956).

In addition to dispersing the fungus from parent to offspring colonies, fungus-growing ants engage in elaborate behaviors to promote fungal growth. Substrate, depending variably on the ant genus, is collected by foraging workers, pretreated, and subsequently used as manure to support the growth of the fungal cultivar (Quinlan and Cherrett, 1977; Mangone and Currie, 2007). As part of the process of substrate preparation, the ants place their fecal droplets on the collected organic (mostly plant) material before incorporating it in the fungus garden (e.g., Weber, 1966, 1972; Murakami and Higashi, 1997). Active enzymes, capable of degrading a range of plant polymers, have been found in the fecal droplets of various fungus-growing ant species, some of which have been shown to be identical to enzymes in cultured fungus mycelium (Boyd and Martin, 1975; Martin et al., 1975; D'Ettorre et al., 2002; Rønhede et al., 2004). Fecal droplet application allows for efficient distribution and recycling of fungal enzymes and likely facilitates faster breakdown of the substrate (e.g., Boyd and Martin, 1975). The ants also prune the fungus to maintain high productivity (Bass and Cherrett, 1996). In exchange for being cultivated, manured, and dispersed among the host ant generations, the fungus serves as the main food source for the ants and their brood (Quinlan and Cherrett, 1979; Cherrett et al., 1989). In the more derived fungus-growing ants (i.e., the four "higher" attine ant genera), the fungal cultivar produces specialized feeding structures: nutrient-rich swollen hyphal tips (gongylidia), which are fed to the ant larvae and consumed by adult ant workers and the queen (Quinlan and Cherrett, 1979; Cherrett et al., 1989; Bass and Cherrett, 1994; Schneider, 2000; Mueller et al., 2001). Hyphal swellings are also present in the fungi associated with the more phylogenetically basal groups of fungus-growing ants (Murakami and Higashi, 1997; Schultz et al., 2005), although their role as a specialized feeding source for the ants is unclear.

There are currently 234 described species of fungus-growing ants in 12 genera (Schultz and Meier, 1995; Wetterer et al., 1998; Brandão and Mayhé-Nunes, 2001; Schultz and Brady, 2008). A recent molecular phylogeny confirmed the monophyly of attine ants, and supports the presence of four major evolutionary transitions after the origin of ant agriculture (Schultz and Brady, 2008). "Lower" attine agriculture is the most basal form of ant fungiculture and is performed by the genera *Apterostigma*, *Cyphomyrmex*, *Mycetagroicus*, *Mycocepurus*, *Mycetosorites*, *Mycetophylax*, *Mycetarotes*, and *Myrmicocrypta*. The majority of species in this group of fungus-growing ants engages in the more ancestral form of the association, cultivating lepiotaceous fungi. Within "lower" attine agriculture, there are two evolutionary transitions: some *Apterostigma* species cultivate coral fungi in the family *Pterulaceae* and a group of *Cyphomyrmex* ants cultivate their fungi in a yeast form (Schultz and Brady, 2008). Most "lower" attine ants have inconspicuous colonies that are composed of a relatively small fungus garden with only dozens to a few thousand individuals, and mostly use plant detritus or insect feces as a substrate for fungiculture (Weber, 1966, 1972; Hölldobler and Wilson, 1990). The next recognized evolutionary transition is to "higher" attine agriculture, characterized by tighter associations with the cultivar, including more specialized substrate use for fungiculture. The "higher" attines consist of the genera *Trachymyrmex*, *Sericomyrmex*, *Acromyrmex*, and *Atta*. The latter two are known as leaf-cutting ants, due to their exclusive use of fresh plant material for fungiculture, and represent the latest evolutionary transition in ant fungiculture (Schultz and Brady, 2008). Colonies of these two genera have more elaborate social structure, including worker size polymorphism and task partitioning (Wilson, 1980a,b; Hart and Ratnieks, 2001), and also form larger colonies composed of tens of thousands of workers in *Acromyrmex* and millions in *Atta* (e.g., Weber, 1972; Hölldobler and Wilson, 1990).

Fungus-growing ants cultivate a phylogenetically diverse collection of fungi, which occur in two separate families in the Basidiomycota. Most of the cultivars belong to the Lepiotaceae, a family of litter-decomposing fungi (Agaricales: Lepiotaceae: *Leucocoprinus* and *Leucoagaricus*) (e.g., Weber, 1972; Chapela et al., 1994). Within this group, there are four major lineages of fungal mutualists: the slower-growing lepiotaceous fungi associated with some "lower" attines, the faster-growing "lower" attine fungi, the fungi associated with "higher" nonleaf-cutting ants (*Trachymyrmex* and *Sericomyrmex*), and the group of fungi associated with leaf-cutting ants (Chapela et al., 1994; Mueller et al., 1998) (Figure 10.3). As indicated above, some ant species in the genus *Apterostigma* cultivate coral fungi of the family Pterulaceae (Agaricales: *Pterula*) (Munkacsi et al., 2004; Villesen et al., 2004).

Although the cultivated fungi associated with fungus-growing ants are relatively diverse, most ant species associate with only a narrow range of cultivar strains. This fits the vertical mode of transmission of the fungus (i.e., from parent to offspring), which is expected to result in ancient clonally propagated symbiont lineages evolving in parallel with the ant host lineages (cf. Mueller et al., 2001). As expected and outlined above, broad groups of attine ants associate with specific corresponding groups of fungus cultivars, indicating some degree of ancient codiversification among the mutualist lineages (e.g., Chapela et al., 1994; Mueller et al., 2001; Currie et al., 2003b) (Figure 10.3). However, the broad-scale pattern of codiversification is disrupted at finer phylogenetic levels by switches of fungal cultivars between colonies (Mueller et al., 1998; Bot et al., 2001; Green et al., 2002; Mueller et al., 2004; Villesen et al., 2004; Mikheyev et al., 2006; Richard et al., 2007). For example, fungus-growing ants in the genera *Acromyrmex* and *Atta* are known to associate only with a specific cultivar clade (Figure 10.3), but sympatric ant species and genera exchange cultivar strains from within this clade (Bot et al., 2001; Mikheyev et al., 2006; Richard et al., 2007). Despite the presence of cultivar diversity within the population, and lateral transfers of cultivar strains between colonies, individual ant colonies appear to associate with a single fungus strain (Poulsen and Boomsma, 2005; Scott et al., in preparation). Genetic monoculture is maintained through both ant-mediated exclusion of nonnative fungal strains entering the colonies (Bot et al., 2001; Viana et al., 2001; Mueller et al., 2002) and by the resident fungus through the expression of incompatibility mechanisms (Poulsen and Boomsma, 2005).

10.3 Exploitation and Defense of the Ant–Fungus Mutualism

10.3.1 Specialized *Escovopsis* Parasites of the Ant–Fungus Mutualism

Escovopsis (Hypocreales: Ascomycota) is a morphologically diverse genus of microfungi that infects the gardens of fungus-growing ants (Currie et al., 1999a; Currie, 2001a). Infections by *Escovopsis* can be devastating, rapidly overgrowing whole fungus gardens (Currie et al., 1999a; C. R. Currie, personal observation). In some leaf-cutting ant populations more than 70% of colonies harbor infections with *Escovopsis*, indicating that the garden parasite is prevalent (Currie, 2001b). Furthermore, nest infections can be persistent, resulting in a dramatic reduction of the accumulation of garden substrate and the production of workers (Currie, 2001b). Consequently, *Escovopsis* exploitation of the ant–fungus mutualism can result in a significant reduction in ant–colony growth rate (Figure 10.2c), decreasing survival and reproduction, and even occasionally forcing the ants to abandon their nests (Currie et al., 1999a; Currie, 2001b). *Escovopsis* is a necrotrophic mycoparasite, targeting and consuming the ants' fungal cultivar (Reynolds and Currie, 2004). More specifically, it is attracted to the cultivar via chemotaxis and upon contact with the host fungus it secretes compounds that degrade the host cells, there after *Escovopsis* absorbs the released nutrients (Reynolds and Currie, 2004; Gerardo et al., 2006a) (Figure 10.2a and b).

Multiple lines of evidence suggest that the ant–fungus association has been exploited by *Escovopsis* from the earliest origin of the mutualism. This is supported by the presence of *Escovopsis* throughout the phylogenetic diversity of fungus-growing ants, including the most basal genera. The sister group to *Escovopsis* is the family Hypocreaceae (Currie et al., 2003b), which consists of known parasites of fungi, including species that infect fungi closely related to fungus-growing ant cultivars (Castle, 1998). This suggests that when the ants domesticated their fungal cultivar, they also, inadvertently, acquired *Escovopsis*. Furthermore, because *Escovopsis* has only been found associated with fungus-growing ant nests, it is unlikely that it recently invaded the mutualism and subsequently spread throughout the phylogenetic and geographic diversity of the association (Currie et al., 2003b). The strongest evidence for the early origin of *Escovopsis* is the parallel evolution of the parasite with both the ants and their fungal cultivars on a broad scale, with the *Escovopsis* phylogeny being composed of four major parasite clades, each of which is exclusively associated with a corresponding group of ants and mutualistic fungi (Currie et al., 2003b) (Figure 10.3).

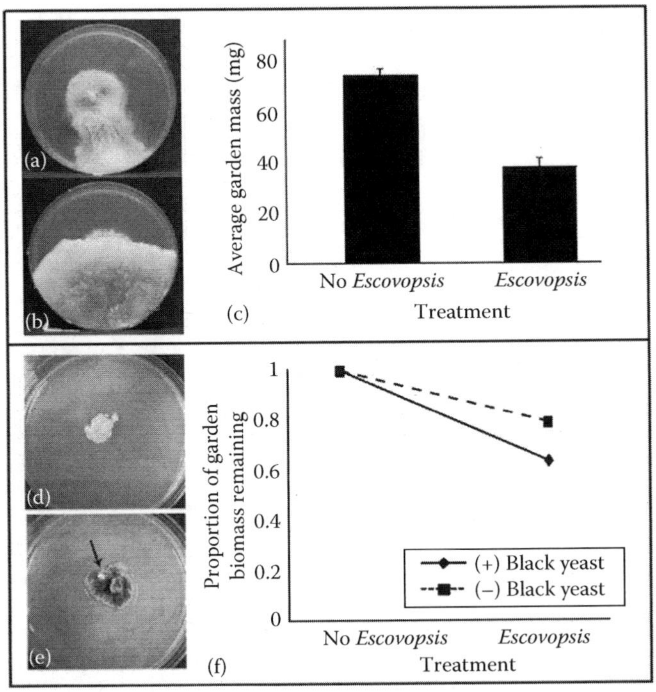

FIGURE 10.2 Parasites of fungus-growing ant mutualisms. (a–c) *Escovopsis* parasites of the ant–fungus mutualism. (a and b) Shows how *Escovopsis* displays directed growth toward the mutualistic fungus of the ants in vitro, ending a complete overgrowth and utilization of the cultivar fungus. (c) The effect of *Escovopsis* in vivo, showing the impact the parasite can have on garden mass of miniature ant nests (based on data presented in Figure 10.3b in Currie et al., 2003a). (d–f) Black yeast parasites of the ant–bacterium mutualism. (d and e) The growth of the actinomycetes in a Petri plate in the absence (d) and in the presence (e) of the black yeast, revealing that the bacterium is completely overgrown and that the black yeast utilizes the actinomycete for growth. (f) Shows how the presence of the black yeasts indirectly, negatively impact *Apterostigma pilosum* fungus-growing ant nests by increasing *Escovopsis*-induced garden morbidity. More specifically, in the absence of *Escovopsis*, the presence of the black yeast has no negative consequences for fungus garden size, but in the presence of both black yeast and *Escovopsis*, it exacerbates the effect of *Escovopsis* on garden biomass loss. (Modified from Little, A.E.F. and Currie, C.R., *Ecology*, 89, 1216, 2008.)

As mentioned above, *Escovopsis* infections of fungus gardens are limited to the combinations that lie within the broad groupings of corresponding cultivars and parasites (Currie et al., 2003b) (Figure 10.3). However, the fact that the garden parasite is horizontally transmitted between colonies leads to the prediction that individual ant colonies can be infected with diverse parasite strains, in addition to even hosting multiple *Escovopsis* strains simultaneously (Currie et al., 1999a; Currie, 2001a). Indeed, recent molecular phylogenetic analyses of leaf-cutting ant-associated *Escovopsis* by Taerum et al. (2007) revealed that *Escovopsis* strains isolated from the gardens of leaf-cutting ants form a monophyletic group of diverse *Escovopsis* parasites, they confirmed the lack of ant–parasite phylogenetic congruence at finer phylogenetic scales. Furthermore, simultaneous infection with diverse parasite strains in individual colonies has been confirmed (Gerardo et al., 2006b; Taerum et al., in preparation). Cultivar-*Escovopsis* specificity may occur at even finer scales within "lower" attine ants: Gerardo et al. (2004) demonstrated that *Escovopsis* strains from the nests of three sympatric species of *Cyphomyrmex* are specifically associated with particular cultivar strains, despite incongruence in the evolutionary histories of the ants and their fungal mutualists at more recent phylogenetic scales (Green et al., 2002). The reason for *Escovopsis*' inability to utilize hosts outside the broad groupings is unclear, but it is possible that the ability of the cultivar to inhibit other groups of *Escovopsis* drives the specificity observed across the broad codiversification pattern (Figure 10.3). In a recent cultivar-*Escovopsis* Petri plate bioassay, Gerardo et al. (2006a), pairing

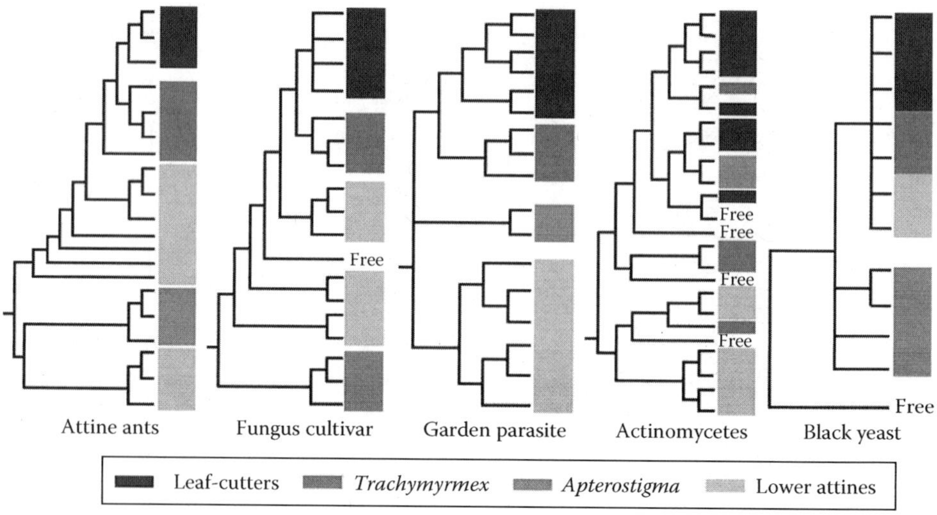

FIGURE 10.3 Coevolution of five symbionts associated with the fungus-growing ant symbiosis. The figure shows a schematic representation of the cophylogenetic patterns among the ants, the cultivar, *Escovopsis*, *Pseudonocardia*, and the black yeasts. Grey shading highlights the four recognized major groups of fungus-growing ants: the "lower" attine symbiosis (lightest grey), the *Apterostigma* symbiosis (light grey), the *Trachymyrmex* "higher" attine symbiosis (grey), and the leaf-cutting ant symbiosis (dark grey), showing a high degree of codiversification among all five symbionts. This indicates that the five lineages have undergone broad-scale coevolution since the origin of fungiculture in ants. Phylogenies are redrawn based on published evolutionary histories of the ants (Schultz and Brady, 2008), the cultivar (Mueller et al., 1998), the *Escovopsis* parasites (Currie et al., 2003b), the *Pseudonocardia* actinomycetes (Zhang et al., 2007; Poulsen et al., 2008; M. Cafaro et al., in preparation), and the black yeasts (Little and Currie, 2007). Free = free-living non-ant-associated microbes.

cultivar and parasite genotypes from both within and between the broad groupings of Figure 10.3, found that cultivars are rapidly overgrown by *Escovopsis* when challenged in natural symbiont pairings, but that cultivars can suppress *Escovopsis* only in host–parasite combinations that are not found in nature (i.e., between the broad groupings in Figure 10.3).

The ant–fungus mutualism is obligate and the fitness of both the partners is linked through vertical transmission of the fungus. Thus, defending the fungus garden from *Escovopsis* exploitation is crucial to both the mutualists. The fungal mutualist does not appear to contribute to defense against *Escovopsis*, which is perhaps surprising since it serves as the nutrient source for the parasite. The inability of the cultivar to defend itself against *Escovopsis* is evident from fungus gardens being rapidly overgrown by the parasite when ant workers are removed (Currie et al., 1999b; see Currie, 2001a for a review). Furthermore, as mentioned, in vitro bioassays have revealed that the cultivar is rapidly overgrown by *Escovopsis* when challenged in symbiont pairings (Gerardo et al., 2006a). In this study, and in another by Poulsen et al. (in preparation), the rapid overgrowth of the cultivar was observed virtually in every pairing involving different cultivar–parasite combinations. These findings are important because genes involved in defending the hosts from pathogens are expected to be under strong selection, which predicts the presence of variation in the outcome of different combinations of genetically distinct hosts and parasites. Since the cultivar does not defend itself from *Escovopsis,* and since this parasite can significantly affect the success of the ant–fungus mutualism, garden defense is a key component in this mutualism.

10.3.2 Specialized Ant Behaviors Help Protect the Mutualistic Fungus

Fungus-growing ants not only employ behaviors to promote growth of the fungal mutualism, they also engage in specialized behaviors to help protect the fungus garden from potential pathogens (see Poulsen and Currie, (2006) for a review). Like all ants, fungus growers conduct self- and allogrooming to help prevent infections of workers (Fernández-Marín et al., 2003; C. R. Currie and T. Murakami, personal observation), but they also groom their fungus garden. Workers detect the presence of alien fungi with their antennae and use their mouthparts to clean the mycelium of the mutualistic fungus (Currie and Stuart, 2001). The infrabuccal pocket serves as a reservoir for storing the groomed microbes, allowing workers to effectively quarantine potentially harmful microbes in a specialized, isolated, and enclosed location (Quinlan and Cherrett, 1978; Little et al., 2003, 2005; Poulsen and Currie, 2006). It has been shown that *Escovopsis* spores present in the infrabuccal pockets are not viable (Little et al., 2005). During grooming, workers actively use secretions from their metapleural glands, which are near the hind legs of ants and are known to produce antibiotics (Fernández-Marín et al., 2006), and may aid in sterilizing the spores that are groomed off the fungus garden and the ants themselves. Grooming and metapleural gland secretions contribute to reducing the risk of infections establishing within fungus gardens; however, if infections overcome these defenses, fungus-growing ants can excise and discard larger pieces of infected garden material. This behavior, referred to as weeding, has been empirically shown to help reduce *Escovopsis* infection spreading within fungus gardens in *Atta* leaf-cutting ants (Currie and Stuart, 2001).

10.3.3 Ant–Bacterium Defensive Mutualism of the Ant–Fungus Mutualism

Although behavioral defenses are important (Currie and Stuart, 2001), one of the primary mechanisms used by fungus-growing ants to deal with *Escovopsis* is through a mutualistic association with bacteria (Currie et al., 1999b, 2003a). The bacteria are filamentous gram-positive actinomycetes in the genus *Pseudonocardia* (Pseudonocardiaceae: Actinomycetales) (Currie et al., 1999b; Cafaro and Currie, 2005), which secrete antibiotic compounds that inhibit the garden parasite *Escovopsis* (Currie et al., 2003a; Cafaro and Currie, 2005; Poulsen et al., in preparation) (Figure 10.4). The *Pseudonocardia* bacterium grows on genus-specific locations on the ant cuticle, and is vertically transmitted between colonies on the cuticle of colony founding queens (Currie et al., 1999b). Vertical transmission ensures an efficient defense for newly founded fungus gardens and at the same time, aligns the reproductive interests of the ants and the bacterium (Currie et al., 1999b; Herre et al., 1999; cf. Frank, 2003; cf. Poulsen et al., 2005) (Figure 10.4). Within colonies, the abundance of the bacterium is dependent on both caste and age of the ants (Currie et al., 1999b, 2003a; Poulsen et al., 2002, 2003b), and bacterial growth is inducible under stressful conditions such as garden decline or *Escovopsis* infection (Currie et al., 2003a). The

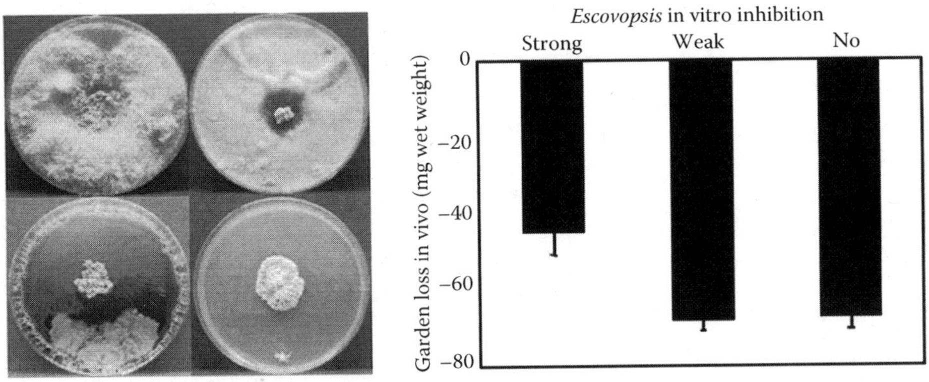

FIGURE 10.4 (See color insert following page 206.) Ant–bacterium mutualism for defense against *Escovopsis*. The figure illustrates bioassay pairings that display varying degree of inhibition of *Escovopsis* (inoculated at the edge of the Petri plate) by ant-associated *Pseudonocardia* bacteria (inoculated two weeks prior to *Escovopsis* at the centre of the Petri plate). The varying degrees of inhibition in vitro (left) result in varying degrees of in vivo fungus garden morbidity caused by *Escovopsis*: infections with resistant parasites (weakly or not inhibited by the actinomycetes: zone of inhibition less than 1 cm) result in significantly higher morbidity than when the resident actinomycetes bacterium has strong inhibitory capabilities against the infection parasite strain (in vitro inhibition resulting in zone of inhibition being more than 1 cm). (Modified from Poulsen, M. Cafaro, M., Erhardt, D.P., Little, A.E.F., Gerardo, N.M., and Currie, C.R., 2008, in preparation.)

presence of elaborate cuticular crypts supported by unique exocrine glands (Currie et al., 2006), in addition to metabolic costs associated with carrying a full actinomycete cover major workers in *Acromyrmex* (Poulsen et al., 2003a), supports the hypothesis that the ants actively allocate nutrients for bacterial growth. Although the nutrients remain unknown, the presence of specialized structures across all attine genera studied suggests an ancient and highly evolved association (Currie et al., 2006).

In addition to demonstrating in vitro inhibition of *Escovopsis* by *Pseudonocardia,* studies have verified the role of the bacterial mutualist in helping suppress infections by the garden parasite in vivo (Figure 10.4). Currie et al. (2003b) conducted a two-by-two factorial experiment crossing the presence/absence of the garden parasite with the presence/absence of the bacteria on workers in *Acromyrmex octospinosus* subcolonies. Subcolonies with fungus gardens tended by bacteria-free workers exhibit significantly more severe negative impact from *Escovopsis* infection than subcolonies with the bacteria still present on the ants (Currie et al., 2003b). In a recent study, Poulsen et al. (in preparation) evaluated the role of inhibitory capabilities of the resident bacterial strain in helping defend the ant–fungus mutualism during *Escovopsis* infection, linking observations of in vitro inhibition of the garden parasite to in vivo suppression of *Escovopsis* infection within the subcolonies (Figure 10.4). This study found that *Escovopsis* strains susceptible to the *Pseudonocardia*-derived compounds in vitro had a significantly lower impact on fungus garden biomass in vivo as compared to strains resistant to the antibiotics (Poulsen et al., in preparation) (see Figure 10.4).

As predicted by the vertical mode of transmission of *Pseudonocardia* by incipient queens (Currie et al., 1999a), phylogenetic comparisons between *Pseudonocardia* and fungus-growing ants reveal some degree of broad-scale codiversification among the mutualists (Cafaro and Currie, 2005; Poulsen et al., 2007; Cafaro et al., in preparation) (Figure 10.3). Strict cocladogenesis between the ants and the bacteria is, however, disrupted by *Pseudonocardia* strains switching between the attine ant species (Poulsen et al. 2005; Cafaro et al., in preparation) (Figure 10.3). Evidence for *Pseudonocardia* switches between sympatric (Poulsen et al., 2005) and allopatric (Poulsen et al., in preparation) ant species within the genus *Acromyrmex* suggests that horizontal symbiont transmission may be common. While such switches disrupt phylogenetic congruence, ant preference for native over nonnative strains (Zhang et al., 2007), in addition to active exclusion of foreign strains by means of ant grooming (M. Poulsen and C. R. Currie, unpublished data), likely facilitates the rearing of a single strain of bacterium within the nests (cf. Poulsen et al., 2005). At the same time, the ability to recognize closely related bacterial strains might allow the ants to acquire novel bacterial strains when confronted with infections by strains of *Escovopsis* parasites resistant to the antibiotics produced by their resident strain of *Pseudonocardia*.

10.3.4 Sustainable Use of *Pseudonocardia* for Defense

Recent work has revealed that over the history of this ancient ant–microbe association the garden parasite *Escovopsis* has repeatedly evolved resistance toward the *Pseudonocardia*-derived antibiotics (Poulsen et al., in preparation). This was detected in a Petri plate bioassay experiment pairing *Escovopsis* and *Pseudonocardia* symbiont strains isolated from phylogenetically diverse fungus-growing ant nests. As noted above, when subcolonies of *Acromyrmex* leaf-cutting ants are infected with *Escovopsis* strains resistant to the antibiotics produced by their resident actinomycete, the garden parasite causes significantly higher garden morbidity compared to infections with parasite strains susceptible to the antibiotics (Poulsen et al., in preparation) (Figure 10.4). These findings indicate that *Escovopsis* imposes a constant selective pressure on the ant–fungus–bacterium mutualism to counteract resistance. This appears to occur through the acquisition of novel antibiotics, either by the ant-associated *Pseudonocardia* evolving the ability to produce new compounds or by the ants acquiring novel actinomycete strains. These bioassay pairings involving diverse strains of *Escovopsis* and *Pseudonocardia*, revealed a high degree of variation in the inhibition profiles of individual strains of ant-associated bacteria (Poulsen et al., in preparation). The presence of abundant diversity in the chemistry of the antibiotics within this symbiosis, paired with the presence of occasional *Escovopsis* resistance to the antibiotics, suggests that *Pseudonocardia–Escovopsis* interactions are shaped by antagonistic coevolution. Indeed, this association may represent an evolution arms-race, with the continuous evolution of resistance in the parasite on one side and the emergence of novel antibiotics in the actinomycetes on the other (Poulsen et al., in preparation). This observed wide variation in antibiotic secretions between even very closely related *Pseudonocardia* strains supports this idea (Poulsen et al., in preparation). The benefit of this defensive mutualism may thus ultimately be the ability of the actinomycete bacterium to counteract parasite evolution (i.e., the emergence of antibiotic resistance), which the ant–fungus mutualism alone would be unlikely to achieve because of the relatively long generation time of the ants and the low genetic diversity of the cultivar.

10.3.5 Black Yeast Exploits the Ant–*Pseudonocardia* Mutualism

The evolution of resistance in *Escovopsis* is not the only challenge imposed on the ant–actinomycete defensive mutualism and, consequently, the success of the ant–fungus mutualism. A recent discovery in the fungus-growing ant–microbe symbiosis is an antagonistic symbiont of the ant–bacterial mutualism: black yeasts (Ascomycota, Herpotrichiellaceae, *Phialophora*) (Little and Currie, 2007) (Figure 10.2d through f). Black yeasts are typically able to degrade chitin and cellulose and are found associated with soil or decaying plant material (deHoog, 1977). Also, species in the genera *Exophiala* and *Phialophora* (the closest relatives of the ant-associated black yeasts) have been isolated from wood beetle galleries (Kerrigan and Rogers, 2003), *Capronia* sp. has been found in the guts of beetles (Zhang et al., 2003), and there is speculation that insects might be involved in the transmission of black yeasts found in cacti (Rosa et al., 1994). Finally, black yeasts are also the common opportunistic pathogens of humans who have undergone organ transplants (deHoog, 1977). In fungus-growing ants, black yeasts are readily cultured from workers in one of the more basal lineages of fungus-growing ants (*Apterostigma*) and are localized on the cuticle of these ants in the same region where they house their bacterial mutualist (under the forelegs on the propleura) (Little and Currie, 2007). Although attempts to culture black yeasts from other genera of fungus-growing ants have been unsuccessful, they are readily detectable by culture-independent methods. Black yeasts associate with attine ants across a wide geographic distribution, including locations to the east and west of the Panama Canal, Amazonian Ecuador, and Peru, and can infect 50%–100% of colonies within ant populations (Little and Currie, 2007).

Isolations from different locations on the surface of *Apterostigma* ants revealed that black yeasts are most common on the ventral surface of the thorax (e.g., under the forelegs) (Little and Currie, 2007), which is the location of actinomycete growth in *Apterostigma*. The shared location of black yeasts and mutualistic actinomycetes led to the hypothesis that the two microbes interact. Bioassay challenges between the black yeast and *Pseudonocardia* on artificial medium suggest that black yeasts exploit the bacterium: black yeast growth significantly increases in the presence of *Pseudonocardia*, at the cost of bacterial growth (Little and Currie, 2008) (Figure 10.2d and e). Since the black yeast derives nutrients from *Pseudonocardia*, it has the potential to impact the health of the ants by disrupting their ability to inhibit the parasite *Escovopsis* via their antibiotic-producing bacteria. Through experimental manipulation of *Apterostigma pilosum* subcolonies, Little and Currie (2008) found that significantly more fungal garden biomass was lost in colonies treated with both black yeast and *Escovopsis* than in other treatments (Figure 10.2f). Thus, when fungus-growing ant colonies are infected with the garden parasite *Escovopsis*, the presence of black yeasts reduces the ants' ability to suppress fungus garden infection. This decreases the fitness of the ant–fungus mutualism, while synergistically benefiting the garden parasite by allowing more severe garden infections (Figure 10.2f). The causal reason for the indirect benefit for *Escovopsis* by the black yeasts is unclear. Black yeasts do not significantly promote the growth of *Escovopsis*, they do not suppress or kill the fungal cultivar, and a series of in vitro experiments were unable to detect any significant difference in *Pseudonocardia–Escovopsis* interactions in the presence/absence of *Phialophora*, indicating that *Phialophora* does not inhibit antibiotic production by the bacteria (Little and Currie, 2008). However, the reduction of *Pseudonocardia* bacterial biomass on the cuticle of the ants caused by black yeast symbionts likely reduces the quantity of antibiotics available to inhibit the garden pathogen.

The high prevalence, broad geographic distribution, specialized localization on the ant body, and presence in both ancient and more derived lineages of fungus-growers indicate that the black yeasts are symbionts of fungus-growing ants rather than being transiently associated with the ants. A phylogeny based on 28S rDNA sequences showed that fungus-growing ant-associated black yeasts form a well-supported monophyletic group, indicating an ancient association that originated only once (Figure 10.3).

Apterostigma-associated black yeasts appear to form a basal group in the phylogeny, while black yeasts from *Cyphomyrmex*, *Trachymyrmex*, and *Acromyrmex* form a derived clade (Figure 10.3). The limited variation in 28S does not allow for a separation of strains from the latter three ant genera, so whether black yeasts have coevolved at the ant genus level is unclear. Future studies, including on black yeasts from additional ant genera and species, will help resolve the degree to which this fifth symbiont has coevolved with the other symbionts of the system.

10.4 Exploitation and Defense of Mutualism and Beyond

After having reviewed the fungus-growing ant–microbe symbiosis, we will now argue that this type of defensive mutualism is likely more widespread than currently recognized. The foundation of our argument is that microbial exploiters are likely important and mostly overlooked components in many mutualisms. In addition, the use of antibiotic-producing actinomycetous bacteria to mediate fungal symbiont communities has recently been discovered in another insect–fungal mutualism. Finally, we argue that the physiological capacity of actinomycetous bacteria to produce antibiotics, their short generation time, and their omnipresence and diversity in ecosystems make actinomycetes excellent candidates as defensive symbionts with diverse macroorganisms.

The exploitation of mutualisms is theoretically expected because a fundamental feature of mutualisms is that they involve the exchange of "goods or services," which are potentially valuable to other organisms and, if unprotected, could be exploited by "third" parties. Mutualists, or species derived from them, obtaining benefits from their partners without providing rewards in return (cheaters) have been identified in diverse systems (for a review, see Bronstein, 2001). However, the presence of parasites unrelated to either mutualist partner exploiting mutualisms has received little attention. The discovery of two specialized and potentially virulent microbial parasites of the ant–fungus and ant–bacterium mutualisms, respectively, highlights the potential for "third" parties to negatively impact mutualistic associations. In addition, because the fungus-growing ant–microbe symbiosis is a relatively well-studied mutualism, the fact that these antagonists have only recently been identified suggests that such associations are likely overlooked in many systems. The reason for this may be lack of necessary molecular and microbiological techniques, without which identification of these additional symbionts has not been possible. In addition, determining the impact of exploitation is often difficult, especially if symbiont effects are viewed outside a proper community context. The discovery of the black yeast symbiont in the fungus-growing symbiosis illustrates this well, as the black yeast appears to be unculturable from most of the diversity of the mutualism, and the negative impact it has on the mutualism is only obvious in the presence of *Escovopsis*.

Until recently, the use of a microbe to defend a mutualism from exploitation had only been described in the fungus-growing ant–microbe symbiosis. However, a parallel defensive mutualism involving an actinomycetous bacterium is now known to occur in another insect–fungus mutualism. More specifically, Scott et al. (2008) discovered that southern pine beetles (*Dendroctonus frontalis*) have a mutualistic association with a species of *Streptomyces* (actinomycete) that helps protect their beneficial fungus from exploitation. Southern pine beetles attack and kill live pine trees, and are aided by a mutualistic fungus (*Entomocorticium* sp. A) that helps the beetles overcome host tree defenses. The fungus also serves as an important food source for the beetle larvae. An antagonistic fungus, *Ophiostoma minus*, can outcompete the beetle's beneficial fungus, thereby disrupting beetle larval development and directly impact the fitness of the beetles (Klepzig and Wilkins, 1997; Klepzig et al., 2001; Six and Klepzig, 2004; Hofstetter et al., 2006). The *Streptomyces* sp. grows abundantly within the beetle galleries and also occurs in the mycangium. The mycangium is involved in housing the *Entomocorticium* sp. A, indicating that the actinomycete may also be vertically transmitted between host trees by the beetles. The bacterium secretes a polyene peroxide compound that selectively inhibits the antagonistic fungus, but not the beneficial fungus (Scott et al., 2008). Thus, this beetle–fungus–bacterium tripartite mutualism is a second example of an insect engaging in a mutualism with actinomycetes to mediate the interactions between their beneficial and antagonistic fungal symbionts. Actinomycetous bacteria have also been isolated from fungus-growing termite colonies (Batra and Batra, 1979; Ketch et al., 2001), suggesting that the termite–fungus mutualism may also be a tripartite mutualistic association between an insect, a fungus, and a bacterium.

Maintaining the actinomycete bacteria as defensive mutualists may be particularly beneficial because of their ability to produce diverse compounds with antimicrobial properties. Actinomycetes are ubiquitous in soil and produce the majority of known antibiotics (Tanaka and Omura, 1988), which play a key role in competition and defense of the bacteria themselves in their environment (e.g., Clardy et al., 2005). In particular, the genus *Streptomyces* is a significant source of novel antibiotics (Watve et al., 2001). The perhaps best-known antibiotic from this genus, streptomycin, was discovered in 1944, and its finding initiated an extensive search for secondary metabolites with antibiotic properties produced by *Streptomyces* (Watve et al., 2001). Human societies have benefited profoundly from the secondary metabolites produced by actinomycetes, and fungus-growing ants and beetles benefit from their mutualistic associations with actinomycetes by gaining access to their ability to produce potent antimicrobial compounds (Poulsen et al., 2008; Scott et al., 2008). A key feature of the use of antibiotics from actinomycetes by both human and ant is the acquisition of new bacterial strains for novel chemistry: humans continue to explore actinomycetes for novel antibiotics to replace compounds that have become ineffective due to the evolution of parasite resistance (Watve et al., 2001; Clardy et al., 2005). Similarly, the ant system is characterized by events of actinomycete acquisition through between-colony switching as a possible means of rapid acquisition of new compounds (Poulsen et al., in preparation; Cafaro et al., in preparation). Whether similar dynamics are present in the beetle–bacterium mutualism remains to be investigated.

Another predicted benefit of employing actinomycete bacteria for antibiotic production is their relatively short generation time, facilitating faster accumulation of mutations in their genomes, and their ability to frequently exchange the genetic material (e.g., Waksman 1959; Kinoshita-Iramina et al., 1997; Ravel et al., 2000). This may be particularly important because the theory predicts that microbial parasites are able to rapidly adapt to a genetically homogeneous hosts due to their smaller size and shorter generation time (e.g., Hamilton, 1980; Hamilton et al., 1990), which may compromise host defense. In some groups, the slower generation time in the host as compared to pathogens is counteracted by adaptive immune systems. However, mediation of microbial symbionts by host immune systems may not be possible in many mutualisms. Engaging in defensive mutualisms with organisms, like actinomycetes, that have the ability to rapidly change their secondary metabolism in response to parasite evolution, may therefore provide a particularly efficient strategy for defense against microbial parasites of mutualisms.

As discussed above, the discoveries that both fungus-growing ants and the southern pine beetles engage in defensive mutualisms with actinomycetous bacteria suggest that actinomycetes are valuable defensive mutualists mediating symbioses between insects and fungi (see Sections 10.3.3 and 10.4). However, given the metabolic capacity of actinomycetes to produce antibiotics and the importance of parasites to all hosts, the benefits received from engaging in this type of defensive mutualism with actinomycetes likely extends beyond the protection of mutualisms. Indeed, recent findings have documented the presence of actinomycetes with antibiotic properties in other organisms (e.g., Araujo et al., 2001; Cao et al., 2004; Kaltenpoth et al., 2005). The documented cases include an association between actinomycetes and solitary wasps: Kaltenpoth et al. (2005) found that the European beewolf (*Philanthus triangulum*) has a specialized symbiotic association with *Streptomyces* bacteria that occur in specific antennal glands. The presence of these *Streptomyces* bacteria appears to help protect developing larvae during their 9 month over-wintering stage (Kaltenpoth et al., 2005). Similarly, actinomycetes in the genera *Microbispora*, *Streptomyces*, and *Streptosporangium* have also been identified as endophytes in plants, where they may play a role in defending the host from plant pathogens (e.g., Araujo et al., 2001, 2002; Cao et al., 2004). Future work will likely reveal that actinomycete bacteria contributing antibiotics for defense against pathogens commonly occur as symbionts of diverse hosts.

10.5 Conclusions

The fungus-growing ant–microbe association represents a paradigmatic example of the complexity of symbiosis and the benefit of engaging in a defensive mutualism. As we have reviewed, the system is composed of two ancient coevolved (ant–fungus and ant–bacterium) mutualisms, both of which are exploited by their respective specialized parasites, *Escovopsis* and black yeasts. The presence of these two antagonists illustrates that the resources exchanged within mutualistic associations are prone to exploitation by microbes, and thus require defense against exploitation. The role microbes play in defending hosts, as highlighted throughout this book, the likely feature of mutualisms facing microbial exploitation, and the role bacteria play in mediating antagonistic and beneficial symbionts of fungus-growing ants and southern pine beetles, suggest that tripartite mutualisms involving microbial symbionts for defense is likely widespread. In addition, the findings that actinomycetes are employed in defensive mutualisms and in defense of diverse macroorganisms themselves, suggest that actinomycete bacteria play an important role as defensive mutualists more extensively.

REFERENCES

Araujo, W. L., Maccheroni, W., Jr., Aguilar-Vildoso, C. I., Barroso, P. A. V., Saridakis, H. O., and J. L. Azevedo. 2001. Variability and interactions between endophytic bacteria and fungi isolated from leaf tissues of citrus rootstocks. *Canadian Journal of Microbiology* 47:229–236.

Araujo, W. L., Marcon, J., Maccheroni, W., Jr., Van Elsas, J. D., Van Vuurde, J. W. L., and J. L. Azevedo. 2002. Diversity of endophytic bacterial populations and their interaction with *Xylella fastidiosa* in citrus plants. *Applied and Environmental Microbiology* 68:4906–4914.

Autuori, M. 1956. La fondation de sociétés chez les fourmis champignonnistes du genre "*Atta.*" In *L'Instinct dans le comportement des animaux et de l'homme*, eds. M. Autuori, M. P. Bénassy, and J. Benoit, Masson et Cie, Paris.

Baldauf, S. L. 2003. The deep roots of eukaryotes. *Science* 300:1703–1706.

Bass, M. and J. M. Cherrett. 1994. The role of leaf-cutting ant workers (Hymenoptera: Formicidae) in the fungus garden maintenance. *Ecological Entomology* 19:215–220.

Bass, M. and J. M. Cherrett. 1996. Leaf-cutting ants (Formicidae, Attini) prune their fungus to increase and direct its productivity. *Functional Ecology* 10:55–61.

Batra, L. R. and S. W. T. Batra. 1979. Termite–fungus mutalism. In *Insect-Fungus Symbiosis*, ed. L. R. Batra, Allanheld, Osmun and Co, Monclair.

Bot, A. N. M., Rehner, S. A., and J. J. Boomsma. 2001. Partial incompatibility between ants and symbiotic fungi in two sympatric species of *Acromyrmex* leaf-cutting ants. *Evolution* 55:1980–1991.

Boucher, D. H., James, S., and K. H. Keeler. 1982. The ecology of mutualism. *Annual Review of Ecology and Systematics* 13:315–347.

Boyd, N. D. and M. M. Martin. 1975. Fecal proteinases of the fungus-growing ant *Atta texana*: Their fungal origin and ecological significance. *Journal of Insect Physiology* 21:1815–1820.

Brandão, C. R. F. and A. J. Mayhé-Nunes. 2001. A new fungus-growing ant genus, *Mycetagroicus*, with the description of three new species and comments on the monophyly of the Attini (Hymenoptera: Formicidae). *Sociobiology* 38:639–665.

Bronstein, J. L. 2001. The exploitation of mutualisms. *Ecology Letters* 4:277–287.

Cafaro, M. and C. R. Currie. 2005. Phylogenetic analysis of mutualistic filamentous bacteria associated with fungus-growing ants. *Canadian Journal of Microbiology* 51:441–446.

Cao, L., Qiu, Z., You, J., Tan, H., and S. Zhou. 2004. Isolation and characterization of endophytic *Streptomyces* strains from surface-sterilized tomato (*Lycopersicon esculentum*) roots. *Letters in Applied Microbiology* 39:425–430.

Castle, A. 1998. Morphological and molecular identification of *Trichoderma* isolates on North American mushroom farms. *Applied and Environmental Microbiology* 64:133–137.

Chapela, I. H., Rehner, S. A., Schultz, T. R., and U. G. Mueller. 1994. Evolutionary history of the symbiosis between fungus-growing ants and their fungi. *Science* 266:1691–1694.

Cherrett, J. M., Powell, R. J., and D. J. Stradling. 1989. The mutualism between leaf-cutting ants and their fungi. In *Insect–Fungus Interactions*, eds. N. Wilding, N. M. Collins, P. M. Hammond, and J. F. Webber, Academic Press, London.

Clandy, J., Fishback, M. A., and C. T. Walsh. 2005. New antibiotics from bacterial natural products. *Nature Biotechnology* 24:1541–1550.

Currie, C. R. 2001a. A community of ants, fungi, and bacteria: A multilateral approach to studying symbiosis. *Annual Review of Microbiology* 55:357–380.

Currie, C. R. 2001b. Prevalence and impact of a virulent parasite on a tripartite mutualism. *Oecologia* 128:99–106.

Currie, C. R., Bot, A. N. M., and J. J. Boomsma. 2003a. Experimental evidence of a tripartite mutualism: Bacteria protect ant fungal gardens from specialized parasites. *Oikos* 101:91–102.

Currie, C. R., Mueller, U. G., and D. Malloch. 1999a. The agricultural pathology of ant fungus gardens. *Proceedings of the National Academy of Sciences USA* 96:7998–8002.

Currie, C. R., Poulsen, M., Mendenhall, J., Boomsma, J. J., and J. Billen. 2006. Coevolved crypts and exocrine glands support mutualistic bacteria in fungus-growing ants. *Science* 311:81–83.

Currie, C. R. Scott, J. A., Summerbell, R. C., and D. Malloch. 1999b. Fungus-growing ants use antibiotic-producing bacteria to control garden parasites. *Nature* 398:701–704.

Currie, C. R. and A. E. Stuart. 2001. Weeding and grooming of pathogens in agriculture by ants. *Proceedings of the Royal Society of London B* 268:1033–1039.

Currie, C. R., Wong, B., Stuart, A. E., Schultz, T. R., Rehner, S. A., Mueller, U. G., Sung, G.-H., Spatafora, J. W., and N. A. Straus. 2003b. Ancient tripartite coevolution in the attine ant–microbe symbiosis. *Science* 299:386–388.

Dawson, S. C. and N. R. Pace. 2002. Novel kingdom-level eukaryotic diversity in anoxic environments. *Proceedings of the National Academy of Sciences USA* 99:8324–8329.

deHoog, G. S. 1977. The black yeasts and allied hyphomycetes. *Studies in Mycology* 15:1–233.

D'Ettorre, P., Mora, P., Dibangou, V., Rouland, C., and C. Errard. 2002. The role of the symbiotic fungus in the digestive metabolism of two species of fungus-growing ants. *Journal of Comparative Physiology B* 172:169–176.

Dixon, R. and D. Kahn. 2004. Genetic regulation of biological nitrogen fixation. *Nature Reviews Microbiology* 2:621–631.

Douglas, A. E. 1994. *Symbiotic Interactions,* Oxford University Press, Oxford.

Ewald, P. W. 1994. *Evolution of Infectious Diseases*, Oxford University Press, Oxford.

Fernández-Marín, H., Zimmerman, J. K., Rehner, S. A., and W. T. Wcislo. 2006. Active use of the metapleural glands by ants in controlling fungal infection. *Proceedings of the Royal Society of London B* 273:1689–1695.

Fernández-Marín, H., Zimmerman, J. K., and W. T. Wcislo. 2003. Nest-founding in *Acromyrmex octospinosus* (Hymenoptera, Formicidae, Atini): Demography and putative profylactic behaviors. *Insectes Sociaux* 50:304–308.

Frank, S. A. 2003. Perspective: Repression of competition and the evolution of cooperation. *Evolution* 57:693–705.

Gerardo, N. M., Jacobs, S. R., Currie, C. R., and U. G. Mueller. 2006a. Ancient host–pathogen associations maintained by specificity of chemotaxis and antibiosis. *Public Library of Science Biology* 4:1358–1363.

Gerardo, N. M., Mueller, U. G., and C. R. Currie. 2006b. Complex host-pathogen coevolution in the *Apterostigma* fungus-growing ant–microbe symbiosis. *BMC Evolutionary Biology* 6:88.

Gerardo, N. M., Mueller, U. G., Price, S. L., and C. R. Currie. 2004. Exploiting a mutualism: Parasite specialization on cultivars within the fungus-growing ant symbiosis. *Proceedings Of the Royal Society of London B* 271:1791–1798.

Green, A. M., Mueller, U. G., and R. M. M. Adams. 2002. Extensive exchange of fungal cultivars between sympatric species of fungus-growing ants. *Molecular Ecology* 11:191–195.

Goodman, R. M. and J. B. Weisz. 2002. Plant–microbe symbioses: An evolutionary survey. In *Biodiversity of Microbial Life*, eds. J. T. Staley and A. L. Reysenbach, Wiley-Liss Inc., New York.

Hamilton, W. D. 1980. Sex versus non-sex versus parasite. *Oikos* 35:282–290.

Hamilton, W. D., Axelrod, R., and R. Tanese. 1990. Sexual reproduction as an adaptation to resist parasites (A review). *Proceedings of the National Academy of Science USA* 87:3566–3573.

Hart, A. G. and F. L. W. Ratnieks. 2001. Task partitioning, division of labour, and nest compartmentalisation collectively isolate hazardous waste in the leaf-cutting ant *Atta cephalotes. Behavioral Ecology and Sociobiology* 49:387–392.

Herre, E. A., Knowlton, N., Mueller, U. G., and S. A. Rehner. 1999. The evolution of mutualisms: Exploring the paths between conflicts and cooperation. *Trends in Ecology and Evolution* 14:49–53.

Hoehler, T. M., Bebout, B. M., and D. J. des Marais. 2001. The role of microbial mats in the production of reduced gases on the early Earth. *Nature* 412:324–327.

Hofstetter, R. W., Cronin, J. T., Klepzig, K. D., Moser, J. C., and M. P. Ayres. 2006. Antagonisms, mutualisms and commensalisms affect outbreak dynamics of the southern pine beetle. *Oecologia* 147:679–691.

Hölldobler, B. and E. O. Wilson. 1990. *The Ants*, Springer Verlag, Berlin.

Jaenike, J. 1978. A hypothesis to account for the maintenance of sex within populations. *Evolutionary Theory* 3:191–194.

Jaenike, J. and S. J. Perlman. 2002. Ecology and evolution of host–parasite associations: Mycophagous *Drosophila* and their parasitic nematodes. *American Naturalist* 160(suppl.):S23–S39.

Kaltenpoth, M., Gottler, W., Herzner, G., and E. Strohm. 2005. Symbiotic bacteria protect wasp larvae from fungul infestation. *Current Biology* 15:475–479.

Kerrigan, J. and J. D. Rogers. 2003. Microfungi associated with the wood-boring beetles *Saperda calcarata* (poplar borer) and *Cryptorhynchus lapathi* (poplar and willow borer). *Mycotaxon* 86:1–18.

Ketch, L. A., Malloch, D., Mahaney, W. C., and M. A. Huffman. 2001. Comparative microbial analysis and clay mineralogy of soils eaten by chimpanzees (*Pan troglodytes schweinfurthii*) in Tanzania. *Soil Biology and Biochemistry* 33:199–203.

Kinoshita-Iramina, C., Kitahara, M., Doi, K., and S. Ogata. 1997. A conjugative linear plasmid in *Streptomyces laurentii* ATCC31255. *Bioscience, Biotechnology and Biochemistry* 61:1469–1473.

Klepzig, K. D., Moser, J., Lombardero, L., Hofstetter, R., and M. Ayres. 2001. Symbiosis and competition: Complex interactions among beetles fungi and mites. *Symbiosis* 30:83–96.

Klepzig, K. D. and R. T. Wilkins. 1997. Competitive interactions among symbiotic fungi of the southern pine beetle. *Applied and Environmental Microbiology* 63:621–627.

Knowlton, N. and F. Rohwer. 2003. Microbial mutualisms on coral reefs: The host as a habitat. *American Naturalist* 162:S51-S62.

Little, A. E. F. and C. R. Currie. 2007. Symbiotic complexity: Discovery of a fifth symbiont in the attine ant–microbe symbiosis. *Biology Letters* 3:501–504.

Little, A. E. F. and C. R. Currie. 2008. Black yeast symbionts compromise the efficiency of antibiotic defense in fungus-growing ants. *Ecology* 89(5):1216–1222.

Little, A. E. F., Murakami, T., Mueller, U. G., and C. R. Currie. 2003. The infrabuccal pellet piles of fungus-growing ants. *Naturwissenschaften* 90:558–562.

Little, A. E. F., Murakami, T., Mueller, U. G., and C. R. Currie. 2005. Defending against parasites: Fungus-growing ants combine specialized behaviours and microbial symbionts to protect their fungus gardens. *Biology Letters* 2:12–16.

Lutzoni, F., Pagel, M., and V. Reeb. 2001. Major fungal lineages are derived from lichen symbiotic ancestors. *Nature* 411:937–940.

Mangone, D. M. and C. R. Currie. 2007. Garden substrate preparation behaviours in fungus-growing ants. *Canadian Entomologist* 139:841–849.

Margulis, L. 1970. *Origin of Eukaryotic Cells*. Yale University Press, New Haven, CT.

Martin, M. M., Boyd, N. D., Gieselmann, M. J., and R. G. Silver. 1975. Activity of faecal fluids of a leaf-cutting ant toward plant cell wall polysaccharides. *Journal of Insect Physiology* 21:1887–1892.

Mikheyev, S., Mueller, U. G., and J. J. Boomsma. 2006. Population-genetic signatures of diffuse coevolution between Panamanian leaf-cutter ants and their cultivar fungi. *Molecular Ecology* 16:209–216.

Mueller, U. G., Poulin, J., and R. M. M. Adams. 2004. Symbiont choice in a fungus-growing ant (Attini, Formicidae). *Behavioral Ecology* 15:357–364.

Mueller, U. G., Rehner, S. A., and T. R. Schultz. 1998. The evolution of agriculture in ants. *Science* 281:2034–2038.

Mueller, U. G., Schultz, T. R., Currie, C. R., Adams, R. M. M., and D. Malloch. 2001. The origin of the attine ant–fungus mutualism. *Quarterly Review of Biology* 76:169–197.

Munkacsi, A. B., Pan, J. J., Villesen, P., Mueller, U. G., Blackwell, M., and D. J. McLaughlin. 2004. Convergent coevolution in the domestication of coral mushrooms by fungus-growing ants. *Proceedings of the Royal Society of London B* 271:1777–1782.

Murakami, T. and S. Higashi. 1997. Social organization in two primitive attine ants, *Cyphomyrmex rimosus* and *Myrmicocrypta ednaella*, with reference to their fungus substrates and food sources. *Journal of Ethology* 15:17–25.

Pace, N. R. 1997. A molecular view of microbial diversity and the biosphere. *Science* 276:734–740.

Paracer, S. and V. Ahmadjian. 1999. *Symbiosis: An Introduction to Biological Associations*, Oxford University Press, Oxford.

Poulsen, M. and J. J. Boomsma. 2005. Mutualistic fungi control crop diversity in fungus-growing ants. *Science* 307:741–744.

Poulsen, M., Bot, A. N. M, Currie, C. R., and J. J. Boomsma. 2002. Mutualistic bacteria and a possible trade-off between alternative defence mechanisms in *Acromyrmex* leaf-cutting ants. *Insectes Sociaux* 49:15–19.

Poulsen, M., Bot, A. N. M, Currie, C. R., Nielsen, M. G., and J. J. Boomsma. 2003a. Within colony transmission and the cost of a mutualistic bacterium in the leaf-cutting ant *Acromyrmex octospinosus. Functional Ecology* 17:260–269.

Poulsen, M., Bot, A. N. M., and J. J. Boomsma. 2003b. The effect of metapleural gland secretion on the growth of a mutualistic bacterium on the cuticle of leaf-cutting ants. *Naturwissenschaften* 90:406–409.

Poulsen, M., Cafaro, M., Boomsma, J. J., and C. R. Currie. 2005. Specificity of the mutualistic association between actinomycete bacteria and two sympatric species of *Acromyrmex* leaf-cutting ants. *Molecular Ecology* 14:597–3604.

Poulsen, M., Cafaro, M., Erhardt, D. P., Little, A. E. F., Gerardo, N. M., and C. R. Currie. 2008. Ancient antibiotic use in ants, in preparation.

Poulsen, M. and C. R. Currie. 2006. Complexity of insect–fungal associations: Exploring the influence of microorganisms on the attine ant–fungus symbiosis. In *Insect Symbiosis*, vol. II, eds. K. Bourtzis and T. Miller, CRC Press, Boca Raton, FL.

Poulsen, M., Erhardt, D. P., Molinaro, D. T., Lin, T. L., and C. R. Currie. 2007. Antagonistic bacterial interactions help shape host-symbiont dynamics within the fungus-growing anti-microbe mutualism. *PLoS ONE* 2:e960.

Quinlan, R. J. and J. M. Cherrett. 1977. The role of substrate preparation in the symbiosis between the leaf-cutting ant *Acromyrmex octospinosus* (Reich), and its food fungus. *Ecological Entomology* 2:161–170.

Quinlan, R. J. and J. M. Cherrett. 1978. Aspects of the symbiosis of the leaf-cutting ant *Acromyrmex octospinosus* (Reich) and its food fungus. *Ecological Entomology* 3:221–230.

Quinlan, R. J. and J. M. Cherrett. 1979. The role of fungus in the diet of the leaf-cutting ant *Atta cephalotes* (L.). *Ecological Entomology* 4:151–160.

Ravel, J., Wellington, E. M., and R. T. Hill. 2000. Interspecific transfer of *Streptomyces* giant linear plasmids in sterile amended soil microcosms. *Applied and Environmental Microbiology* 66:529–534.

Reynolds, H. T. and C. R. Currie. 2004. Pathogenecity of *Escovopsis weberi*: The parasite of the attine ant–microbe symbiosis directly consumes the ant-cultivated fungus. *Mycologia* 96:955–959.

Richard, F.-J., Poulsen, M., Hefetz, A., Errard, C., Nash, D. R., and J. J. Boomsma. 2007. The origin of chemical profiles of fungal symbionts and their significance for nest-mate recognition in *Acromyrmex* leaf-cutting ants. *Behavioral Ecology and Sociobiology* 61:1637–1649.

Rønhede, S., Boomsma, J. J., and S. Rosendahl. 2004. Fungal enzymes transferred by leaf-cutting ants initiate the degradation of plant material in the fungus garden. *Mycological Research* 108:101–106.

Rosa, C. A., Morais, P. B., Hagler, A. N., Mendoca-Hagler, L., and R. F. Monteiro. 1994. Yeast communities of the cactus *Pilosocerus arrabidae* and associated insects in the Sandy Costal Plains of Southeast Brazil. *Antonie van Leeuwenhoek* 65:55–62.

Rudman, W. B. 1987. Solar-powered animals. *Natural History* 96:50–53.

Sanders I. R. 2002. Ecology and evolution of multigenomic arbuscular mycorrhizal fungi. *The American Naturalist* 160:128–141.

Schneider, M. 2000. Observations on brood care behaviour of the leafcutting ant *Atta sexdens* L. (Hymenoptera: Formicidae). *Abstracts of the XXI International Congress of Entomology*, vol. 2 (August 20–26, 2000, Foz do Iguassu), p. 895, Embraja Sojo, London.

Schultz, T. R. and S. G. Brady. 2008. Major evolutionary transitions in ant agriculture. *Proceedings of the National Academy of Science USA*. PNAS Early Edition 1–6, www.pnas.org/cgi/doi/10.1073/pnas.0711024105.

Schultz, T. R. and R. Meier. 1995. A phylogenetic analysis of the fungus-growing ants (Hymenoptera: Formicidae: Attini) based on morphological characters of the larvae. *Systematic Entomology* 20:337–370.

Schultz, T. R., Mueller, U. G., Currie, C. R., and S. A. Rehner. 2005. Reciprocal illumination: A comparison of agriculture in humans and fungus-growing ants, In *Insect–Fungal Associations: Ecology and Evolution*, eds. F. Vega and M. Blackwell, Oxford University Press, Oxford.

Scott, J. J., Oh, D. C., Yuceer, M. C., Klepzig, K. D., Clardy, J., and C. R. Currie. 2008. Bacterial protection of beetle, fungus mutualism. *Science* 322:63.

Selosse, M.-A., Baudoin, E., and Vandenkoornhuyse, P. 2004. Symbiotic microorganisms, a key for ecological success and protection of plants. *Comftes Rendus Biologies* 327:639–648.

Six, D. L. and K. D. Klepzig. 2004. Dendroctonus bark beetles as model systems for studies on symbiosis. *Symbiosis* 37:207–232.

Taerum, S. J., Cafaro, M. J., Little, A. E. F., Schultz, T. R., and C. R. Currie. 2007. Low host–pathogen specificity in the leaf-cutting ant–microbe symbiosis. *Proceedings of the Royal Society of London B* 274:1971–1978.

Tanaka, Y. and S. Omura. 1988. Regulation of biosynthesis of polyketide antibiotics (Review article). In *Biology of Actinomycetes '88*, eds. Y. Okami, T. Beppu, and H. Ogawara, pp. 418–423. Japanese Scientific Society Press, Japan.

Viana, A. M. M., Frézard, A., Malosse, C., Della Lucia, T. M. C., Errard, C., and A. Lenoir. 2001. Colonial recognition of fungus as the fungus-growing ant *Acromyrmex subterraneus* (Hymenoptera: Formicidae). *Chemoecology* 11:29–36.

Villesen, P., Mueller, U. G., Adams, R. M. M., and A. C. Bouck. 2004. Evolution of ant–cultivar specialization and cultivar switching in *Apterostigma* fungus-growing ants. *Evolution* 58:2252–2265.

von Ihering, H. 1898. Die Anlagen neue Colonien und Pilzgärten bei *Atta sexdens. Zoologischer Anzeiger* 21:238–245.

Waksman, S. A. 1959. *The Actinomycetes*, The Williams and Wilkins Company, Baltimore, MD.

Watve, M. G., Tickoo, R., Jog, M. M., and B. D. Bhole. 2001. How many antibiotics are produced by the genus *Streptomyces*? *Archives of Microbiology* 176:386–390.

Weber, N. A. 1966. Fungus-growing ants. *Science* 153:587–604.

Weber, N. A. 1972. *Gardening Ants: The Attines*, American Philosophical Society, Philadelphia.

Wetterer, J. K., Schultz, T. R., and R. Meier. 1998. Phylogeny of the fungus-growing ants (tribe Attini) based on mtDNA sequence and morphology. *Molecular Phylogenetics and Evolution* 9:42–47.

Wilson, E. O. 1980a. Caste and division of labor in leaf-cutter ants (Hymenoptera: Formicidae: *Atta*). I. The overall pattern of *Atta sexdens. Behavioral Ecology and Sociobiology* 7:143–156.

Wilson, E. O. 1980b. Caste and division of labor in leaf-cutter ants (Hymenoptera: Formicidae: *Atta*). II. The ergonomic optimization of leaf cutting. *Behavioral Ecology and Sociobiology* 7:157–165.

Zehr, J. P., Jenkins, B. D., Short, S. M., and G. F. Steward. 2003. Nitrogenase gene diversity and microbial community structure: A cross-system comparison. *Environmental Microbiology* 5:539–554.

Zhang, M., Poulsen, M., and C, R. Currie. 2007. Symbiont recognition of mutualistic bacteria by *Acromyrmex* leaf-cutting ants. *The International Society of Microbial Ecology Journal* 1:313–320.

Zhang, N., Shu, S., and M. Blackwell. 2003. Microorganisms in the guts of beetles: Evidence from molecular cloning. *Journal of Invertebrate Pathology* 84:226–233.

Part III

Eukaryotic Defensive Symbionts

11

Chemical Defense in Lichen Symbioses

James D. Lawrey

CONTENTS

11.1 Introduction

Lichens are slow-growing, long-lived, plant-like organisms that face environmental challenges unusual for most fungi. As a consequence, unique ecophysiological adaptations can be observed in lichens, among them are the mechanisms that offer protection against physical and biotic environmental hazards. This chapter begins with an overview of the lichen symbiosis and the various environmental threats commonly faced by lichens. It then discusses the evidence for hypothesized defensive mechanisms to counter these threats.

11.2 Lichen Symbiosis

Lichenization is the most successful form of obligate mutualism involving fungi and photosynthetic microorganisms. It is also the only mutualism that develops a vegetative body unique to the symbiosis and not observed in the isolated symbiotic partners. Approximately 20% of the described fungi form lichens with either chlorophyte or cyanobacterial photobionts. Most (>99%) of the 14,000 species of lichens are members of the Ascomycota, and nearly half of the described ascomycetes form lichens. Although they are far less common in the Basidiomycota, lichen-formers have evolved in at least three major clades of the mushroom-forming Agaricomycetes (Lawrey et al., 2007).

As a nutritional category of fungi, lichen-formers are common, widespread, and successful. Lichens are floristic components of most terrestrial ecosystems on every continent, and the dominant vegetation on about 8% of the earth's surface (Nash, 1996). It is impossible to generalize about lichen ecology since lichens inhabit terrestrial habitats ranging from arctic tundra to tropical rainforests. As "plant-like" fungi, however, all lichens require light, so typical substrates include rocks, barks, leaves, and soil—nearly any surface that is exposed to adequate amount of light. This requirement for light means that lichen fungi are frequently exposed to conditions uncommon for normal fungi. Lichens are especially noteworthy for their tolerance of extreme temperatures and irradiation levels. For example, lichens are one of only a few groups of organisms that colonize substrates in hot and cold deserts, where they may live both on the surface and even inside of rocks as endolithic organisms (Friedmann, 1981, 1982; Green et al., 1999; Wynn-Williams et al., 2000; Galun, 2001; de los Ríos et al., 2005). A recent European Space Agency experiment even demonstrated that certain lichens (*Rhizocarpon geographicum* and *Xanthoria elegans*) exhibit no ill effects after a 15-day exposure to outer space conditions (Sancho et al., 2007).

Lichens typically consist of layers of tissues, usually including (1) upper and lower cortical layers made up of thick, protective conglutinated masses of fungal hyphae, (2) a thin layer of photobiont cells just under the upper cortex, and (3) a loosely woven, central mass of medullary hyphae that support the photobiont layer and facilitate internal gas exchange. Photobiont cells typically make up less than 10% of the total thallus mass, but as the only source of fixed carbon for both partners, they are positioned in ways that optimize photosynthesis. Although great morphological variation exists among lichens, all of them exhibit modifications that derive from the unusual (for fungi) requirements for photosynthesis.

The nature of lichen symbiosis has been the subject of several excellent reviews (Honegger, 1991, 1992, 1998; Ahmadjian, 1993). Much of the debate in the literature over lichen symbiosis centers on the degree to which captured photobiont cells benefit from the association. Lichen associations are considered to be mutualistic since free-living fungal partners (mycobionts) and photosynthetic partners (photobionts) are rare in the most advanced groups. The associations of many of the more primitive crustose lichens are probably best considered controlled parasitisms since they frequently result in damage to the photobionts. Technically, since all lichen photobionts are nutritional hosts for the more dominant fungal partner, the association is always parasitic to some extent. Although most lichen fungi are found in nature almost exclusively in the lichenized state, each of the symbionts can be isolated and maintained in culture. In both states, symbiotic and aposymbiotic, lichen fungi are unusually slow-growing.

Lichens are not unlike other common fungal mutualisms. As Honegger (1991) has pointed out, lichens are functionally similar to ectomycorrhizae, in which the photosynthetic partner obtains water and mineral nutrients from the fungus, which in return obtains fixed carbon. However, unlike mycorrhizal fungi that enclose a nonphotosynthetic underground organ of a plant that produces photosynthetic tissue aboveground, lichen fungi enclose minute photosynthetic cells and must provide a suitable environment for photosynthesis and gas exchange.

The process of lichenization has significant consequences to each of the associated partners, but especially to the mycobiont, which develops few of its symbiotic characteristics in the aposymbiotic state. In addition to developing complex morphological structures arranged to optimize photosynthetic activities of the captured photobiont, lichens produce unusual secondary metabolites. Of the many functions attributed to these compounds, many of which are produced only by lichenized fungi, a general defensive role is supported by a growing body of evidence. This chapter discusses the defensive nature of lichen metabolites and some of the recent research on lichen chemical defense.

11.3 Secondary Metabolites and Lichen Defense

One of the unique characteristics of many lichens is their production of secondary metabolites. Over 800 of these compounds have now been described, and although some are known from plants and nonlichen fungi, many are unique to lichens (Huneck and Yoshimura, 1996; Huneck, 2001). Most of these compounds are weak phenolic acid derivatives of the acetate–polymalonate pathway, including depsides, depsidones, simple phenolic acids, aliphatic acids, dibenzofurans, esters, chromones, xanthones, anthraquinones, and naphthoquinones. A smaller number are derivatives of the mevalonic acid pathway, including carotenoids and triterpenoids, and the shikimic acid pathway, including yellow-pigmented pulvinic acid derivatives. Details of the diversity, structure, and synthesis of lichen substances can be found in several excellent reviews (Culberson and Elix, 1989; Crittenden and Porter, 1991; Fahselt, 1994; Elix, 1996; Huneck and Yoshimura, 1996; Huneck, 1999).

Most typical lichen compounds are fungal in origin and their biosynthesis is analogous to processes that go on in nonlichen fungi. Photobionts are apparently not required for synthesis since isolated lichen mycobionts can produce typical compounds under certain conditions. However, most compounds are produced only by lichenized mycobionts, and some isolated mycobionts produce compounds different from the symbiotic fungi (Ahmadjian, 1993), all of which suggests that the presence of photobionts creates conditions necessary for proper synthesis of many compounds.

Lichen products are typically deposited as water-insoluble crystals on the outer surfaces of fungal hyphae, which can be seen clearly in scanning electron micrographs (Fahselt, 1994; Honegger, 1998). They can be extracted in organic solvents, in some cases without harming the metabolic function of the symbionts (Solhaug and Gauslaa, 2001). Their distribution is not uniform throughout the thallus. For example, some compounds are found only in the cortical tissues above the photobiont layer, while others are restricted to the internal medullar tissues, or to specific vegetative or reproductive structures. Concentrations vary within thalli, especially in tissues of different ages, but usually range from 1% to 5% thallus dry wt. Unusually, high concentrations (up to 30% dry wt.) have been measured in some cases (Huneck, 1973).

Lichen compounds have long been used in the lichen taxonomy as an aid to the identification of groups, and abundant chemical data on lichens have been collected. Indeed, more than one-third of all described lichens have been characterized chemically, making them one of the best-studied groups of fungi in this regard.

11.3.1 Adaptive Roles of Secondary Metabolites

What do lichen compounds do? Most lichenologists dismiss the idea that they are waste products with little adaptive value, especially given the energetic investment necessary to manufacture them. The fact that many are unique to lichenized fungi suggests they are somehow essential for the maintenance of a symbiotic state. This hypothesis is attractive but not supported by the available evidence. First, lichen compounds are made by lichen fungi in both the symbiotic and isolated aposymbiotic states, although production of most lichen compounds is much greater in lichenized fungi. Also, certain lichen compounds, estimated to be around 10% of the described lichen substances, are not unique to lichens but are produced by many nonlichen fungi and vascular plants as well (Elix, 1996), and many lichens produce no compounds at all. Lichen compounds probably play important physiological roles in regulating the lichen symbiosis (Honegger, 1991, 1992, 1998; Armaleo, 1993), but if they were essential for the origin and maintenance of the symbiotic state, there would be stronger evidence for this.

Various adaptive functions have been proposed for these compounds, the most commonly discussed of which are

1. Regulators of internal water relations. Since most lichen compounds are practically insoluble in water, their presence on the surface of medullary hyphae may serve to maintain air pockets for gas exchange around photobiont cells, and it may also seal water inside of fungal hyphae (Armaleo, 1993; Lange et al., 1997; Honegger, 1998).
2. Regulators of photobiont metabolism. Though not directly involved in primary metabolism, lichen compounds are known to suppress algal growth and metabolic activity (Follmann and Villagrán, 1965; Kinraide and Ahmadjian, 1970; Honegger, 1987), and may regulate enzymes important in primary metabolism (Blanco et al., 1984; Vicente, 1985; Legaz and Vicente, 1989; Perez-Urria and Vicente, 1989).
3. Mineralization of essential elements. Some lichen compounds form complexes with mineral elements, which may result in mineralization of essential elements for uptake by lichens (Schatz, 1962; Syers and Iskandar, 1973; Purvis et al., 1987, 1990; Jones, 1988).
4. Light screens. Compounds produced in the upper cortical tissues are able to provide light-screening protection for the underlying photobiont cells, especially from harmful UV radiation (Fahselt, 1994).
5. Allelopathic agents. The general antibiotic nature of lichen compounds suggests they may function as aggressive inhibitors of potential competitors in the habitat. Since lichens are generally far more slow-growing than their competitors, allelopathic agents provide a

competitive advantage in habitats where the competition is most intense. Chemical inhibition of bryophytes, vascular plants, and other lichens has been documented experimentally (review by Lawrey, 1995a), and the lichen compounds are also known to inhibit mycorrhizal fungi that indirectly affects plant competitors. Epiphytic lichens that live on the surface of other lichens may also be distributed in nonrandom ways as a result of lichen chemistry. For example, an epiphytic *Lepraria* sp. was observed to colonize the foliose lichen *Neofuscelia verruculifera* more frequently than the morphologically similar but chemically different *N. loxodes*, a nonrandom distribution that was attributed to allelopathic inhibition (Culberson et al., 1977).

6. Defense from pathogens and predators. The antibiotic nature of lichen compounds suggests they may defend lichen thalli from a variety of biotic threats, including microorganisms and invertebrate consumers (reviews by Lawrey, 1984, 1986). A large and growing literature on the subject provides strong support for such an adaptive role.
7. Stress-induced response. As long-lived, slow-growing organisms, lichens are subjected to numerous biotic and abiotic stresses, and the production of secondary metabolites may be one of the many ways they cope with these stresses (Lange, 1992). Evidence suggests the induction of compounds may be a general stress response in lichens inasmuch as various sorts of stresses, including air pollution, can stimulate the production of defense compounds, which subsequently function in a variety of ways to lessen the damage caused by stress.

Although not all hypothesized roles are defensive, the evidence in favor of a defensive role for these compounds is growing, and several interesting new studies have appeared in the literature. In the sections that follow, the most widely discussed defensive roles of lichen compounds, photoprotection, defense against pathogens and predators, and stress responses will be discussed, with emphasis given to recent research.

11.3.2 Lichen Chemical Defense: UV Protection of Photobiont Cells

The fact that lichen thalli are commonly yellow, orange, gray, or brown and not bright green like most plants is due to the presence of lichen compounds that are restricted to the upper cortex just above the photobiont cells (Rikkinen, 1995). These compounds are assumed to play an important adaptive role as light-screening agents. The most common cortical compounds are the depsides atranorin and chloroatranorin, the usnic acids, anthraquinones, xanthones, and pulvinic acid derivatives (Elix, 1996). All of these compounds are pigments capable of significantly filtering sunlight reaching sensitive photobiont cells underneath. Variations in cortical chemistry are known to correlate directly with light intensity. The ground-dwelling *Cladonia subtenuis*, for example, produces higher concentrations of usnic acid in high light environments (Rundel, 1969, 1978). Similarly, thalli of the rock-dwelling *Xanthoria parietina* in exposed habitats have higher concentrations of the anthraquinone parietin (Hill and Woolhouse, 1966; Solhaug and Gauslaa, 1996; Gauslaa and Ustvedt, 2003).

Some lichens also produce light-screening melanins, the concentrations of which have been shown to be correlated with light intensity and capable of screening both photosynthetically active radiation and UV-B (Gauslaa and Solhaug, 2001; Rancan et al., 2002; Nybakken et al., 2004). For example, when transplanted from shaded to more exposed habitats, thalli of *Lobaria pulmonaria* develop darker upper cortical tissues caused by the presence of melanins. By contrast, *Cetraria islandica* exhibited no measurable differences in melanin content or cortical screening ability along a latitudinal UV-B gradient (Nybakken et al., 2004).

The fact that many lichen compounds strongly absorb radiation in the UV range, especially UV-B and UV-C, suggests they may play a role as protective screens even in the medulla (Fahselt, 1994). For example, lichens exposed to UV-filtered light exhibit reduced phenolic acid concentration over time (Bachereau and Asta, 1997, 1998). In addition, certain cortical compounds may be induced by UV radiation under both natural conditions (Solhaug et al., 2003; Nybakken et al., 2004; McEvoy et al., 2006) and in the laboratory (Swanson and Fahselt, 1997; Solhaug and Gauslaa, 2004). In the field study of Nybakken et al. (2004), thalli of *Xanthoria parietina* were extracted in acetone to remove the photoprotectants and then transplanted with and without filters to habitats exposed to different amounts of UV radiation. The results suggest that UV-B is necessary to stimulate the resynthesis of parietin, and that resynthesis is faster in higher light intensities. The laboratory study of Solhaug and Gaulsaa (2004) showed that this response is rather complex since it is enhanced by the additions of ribitol, the polysaccharide most commonly produced by lichen photobionts. This combination of UV light and ribitol also significantly stimulates the resynthesis of the cortical compound usnic acid and an additional unidentified blue pigment, possibly a melanin of some sort, in the thalli of *Xanthoparmelia stenophylla* exposed to UV light (McEvoy et al., 2006).

Production of photoprotective cortical compounds by lichenized thalli therefore appears to be somewhat responsive to changes in light conditions. Isolated asymbiotic mycobionts also produce these compounds, but the production appears to be less responsive to light. For example, isolated lichen mycobionts of *Usnea strigosa* produce usnic acid in low light (Ahmadjian and Jacobs, 1985), and isolated mycobionts of *Ramalina siliquosa* maintained in different light conditions exhibit no change in usnic acid concentration (Hamada, 1991). Resynthesized lichens (*Cladonia cristatella*) maintained in various controlled environmental conditions produced no usnic acid at all, and the concentrations of other compounds varied little under varying light conditions (Culberson et al., 1983). So many cortical compounds appear to be constitutive photoprotectants capable of responding slightly to changes in light intensity. However, this responsiveness may be more characteristic of the intact symbiotic lichen in its natural habitat than of the isolated aposymbiotic mycobiont in the laboratory.

11.3.3 Lichen Chemical Defense: Avoidance of Predators and Pathogens

Controversy over the ecological role of lichen compounds began over 100 years ago with Zukal's (1895) proposal that compounds defended lichens from invertebrates, an argument countered almost immediately by Zopf (1896), who, after testing various lichens and lichen compounds as food for snails, concluded they had no such function. Early support for Zukal's hypothesis was provided by Stahl (1904), who observed that lichens avoided by snails were readily consumed if they were first treated in a weak soda solution to remove phenolic compounds.

Lichen compounds have also been studied as a source of antimicrobial compounds. Beginning in the 1940s (Burkholder et al., 1944; Stoll et al., 1947), lichen compounds have been shown to exhibit antibiotic activities (reviews by Vartia, 1973; Crittenden and Porter, 1991; Huneck, 1999), and considerable evidence for clinically significant bioactivities for these compounds has accumulated since (e.g., Aslan et al., 2001). Some compounds, notably usnic acid, have long been recognized to have antibiotic properties (review by Cocchietto et al., 2002), and this has led to the use of these compounds in commercial preparations such as the antiseptic cream "Usno," and to recommendations that they be used in new clinical applications, for example, in the treatment of methicillin-resistant strains of *Staphylococcus aureus* (Elo et al., 2007). Usnic acids are known to have tumor-inhibiting, antihistamine, spasmolytic, and antiviral properties (Hawksworth and Hill, 1984). If lichen compounds have such bioactivities in the laboratory, they may be capable of protecting lichens against microbial attacks in nature, and there is some evidence to support this hypothesis.

The following sections review recent studies on the defensive role of lichen compounds against microorganisms and against invertebrate consumers.

11.3.3.1 Defense against Microorganisms

Lichens appear to be generally well-defended against microorganisms in nature, a possible reason for their remarkable longevities. Studies, during the 1940s and 1950s, of lichens as a potential source of antibiotics, combined with recent research, reveal several general characteristics about lichen substances as antibiotics: (1) more than half of all the lichens tested exhibit antibiotic properties, (2) the most effective substances are usnic acids, pulvinic acid derivatives, aliphatic acids, anthraquinones, and orcinol-type depsides and depsidones, and (3) microbial groups sensitive to lichen compounds include gram-positive bacteria, fungi, and viruses (Lawrey, 1986; Crittenden and Porter, 1991; Huneck, 1999).

Lichens are exposed to countless microorganisms in nature. Indeed, it is becoming more apparent that lichen thalli harbor unusually diverse communities of bacteria (Cardinale et al., 2006) and fungi (Petrini et al., 1990; Girlanda et al., 1997; Li et al., 2007), many of which are undescribed and all of which apparently reside inside lichen thalli without causing any obvious symptoms. Some are generalized and opportunistic saprobes not unique to lichens (Petrini et al., 1990), but will grow rapidly on damaged thalli. Others are apparently obligate endolichenic organisms (Miadlikowska et al., 2004) that live inside lichen thalli without any outward indication of their presence. For lichen thalli to harbor such diverse assemblages of microorganisms without ill effects suggests that they are able to defend themselves against the vast majority of them and the presence of antibiotic compounds may be involved in this.

In addition to the nearly invisible endolichenic microflora, there are lichenicolous fungi that live obligately in or on lichens and form obvious structures (reviews by Clauzade et al., 1989; Richardson, 1999; Lawrey and Diederich, 2003). There are an estimated 3000 species, about half of which have been described, mostly ascomycetes with unusually high host specificities. The chemical ecology of these interactions has not been investigated much at all, but there is some evidence that host preferences are due in part to the presence of lichen compounds (Lawrey, 1993, 1995b, 2000; Torzilli and Lawrey, 1995; Lawrey et al., 1999; Torzilli et al., 1999), and that many lichenicolous fungi exhibit marked tolerances to lichen defense compounds (Lawrey, 1995b). For example, an undescribed lichenicolous species of *Fusarium* grows on a variety of chemically different lichens in nature, and also on lichen tissues in the laboratory. It is not known to be inhibited by any lichen extract or compound, and is even capable of degrading certain lichen compounds known to be inhibitory to other fungi (Torzilli et al., 2002). Normally, well-defended lichens exposed to this *Fusarium* in nature rapidly lose their chemical defenses, resulting in subsequent colonization by a variety of intolerant fungi (Lawrey et al., 1999; Lawrey, 2000). Given the diversity of microorganisms known to associate with lichens, the potential exists for a wide variety of interesting chemical interactions like this to take place in lichens.

11.3.3.2 Defense against Vertebrate and Invertebrate Consumers

In boreal zones, lichens are an important food for reindeer and caribou, whose winter diet is 90% ground-dwelling "reindeer" lichens of the genus *Cladonia* (Brodo et al., 2001). Other ungulates, such as deer, mountain goats, and moose, consume lichens to varying degrees, apparently without much regard to chemistry. Lichen fodder is low in proteins and minerals, which must be supplied by other foods, but certain animals, especially caribou, are lichen specialists which cannot survive long without access to lichens (Brodo et al., 2001). For these animals, lichen chemistry appears not to be an important deterrent.

Lichens are also consumed by a wide variety of invertebrates, including mites, nematodes, rotifers, gastropods, and insects (Thysanura, Collembola, Psocoptera, Protura, Embioptera, Lepidoptera), and several reviews are available (Peake and James, 1967; Sowter, 1971; Gerson, 1973; Richardson, 1975; Gerson and Seaward, 1977; Lawrey, 1987; Sharnoff Web site http://www.lichen.com/invertebrates.html). The extent to which invertebrates significantly damage lichens is difficult to gauge from the limited observational evidence. There are reports of extensive damage to lichens by springtails (Hale, 1972), psocids (Broadhead and Thornton, 1955; Broadhead, 1958; Seaward, 1965; Laundon, 1971; Richardson, 1975), moth larvae (Richardson, 1975; Sigal, 1984), mites (Travé, 1963, 1969; Gjelstrup and Søchting, 1979; Colloff, 1983; Seyd and Seaward, 1984; Reutimann and Scheidegger, 1987), and gastropods (Peake and James, 1967; Yom-Tov and Galun, 1971; Fröberg et al., 1993; Baur et al., 1995; Benesperi and Tretiach, 2004). In general, however, lichens are clearly not consumed by invertebrates to the same extent as vascular plants are, suggesting that lichen chemistry may play an antiherbivore role in many of these interactions.

Evidence for chemical defense against invertebrates comes largely from field observations and laboratory experiments involving terrestrial gastropods (Yom-Tov and Galun, 1971; Lawrey, 1980, 1983; Fröberg et al., 1993; Baur et al., 1994; Gauslaa, 2005), polyphagous insects (Slansky, 1979; Stephenson and Rundel, 1979; Rambold, 1985; Blewitt and Cooper-Driver, 1990; Emmerich et al., 1993; Giez et al., 1994; Hesbacher et al., 1994, 1995; Nimis and Skert, 2006), or oribatid mites (Reutimann and Scheidegger, 1987), whose varied diets make them good subjects for preference tests. In laboratory feeding experiments, animals avoid lichens or foods treated with their secondary compounds, and when forced to eat treated foods they frequently exhibit significant reductions in growth and longevity (Slansky, 1979; Hätscher et al., 1991; Emmerich et al., 1993; Giez et al., 1994).

Unlike the photoprotective cortical compounds that vary in concentration in different light intensities, medullary lichen compounds appear to be constitutive defense compounds that vary little in response to herbivory. One interesting exception is a study by Timdal (1989), who noted that *Cladonia bacilliformis* and *C. norvegica* produced rhodocladonic acid when attacked by a lichenicolous mite, which may thereby defend against further attack. Differences in the concentrations of various compounds appear to be adaptive, however, insofar as they protect life stages and tissues in a way that maximizes individual fitness. For example, juvenile thalli (Golojuch and Lawrey, 1988) and reproductive tissues (Hyvärinen et al., 2000) are frequently endowed with higher concentrations of defensive compounds. But few studies of intrathalline variations in lichen compound concentrations have been done.

What compounds are most effective against invertebrates? Since polyphagous invertebrates make different choices in different feeding experiments, and most studies offer only limited number of choices to start with, no single lichen compound or category of compounds can be identified as the most effective deterrent to grazing. Two recent studies offer the best overall assessment of effectiveness, one (Gauslaa, 2005) a study of a generalist herbivore snail (*Cepaea hortensis*) and the other (Nimis and Skert, 2006) of a polyphagous coleopteran (*Lasioderma serricorne*). Gauslaa (2005) tested 17 species of lichens for levels of chemical defense, offering snails a choice of two halves of a lichen thallus, one-half left untreated and the other extracted in acetone to remove phenolic compounds. In each case, the extracted thallus pieces were much preferred over natural lichens, indicating that lichens are generally well-defended chemically. However, the phenolic compounds produced by lichens in the chemically rich family Parmeliaceae were noticeably more inhibitory than the compounds produced by members of the Physciaceae and Teloschistales. Since members of the Parmeliaceae are most common in heavily leached, nutrient-poor habitats, and the Physciaceae and Teloschistales are generally more nitrophilous, Gauslaa suggests that the effectiveness and chemical basis of deterrent compounds in each group may be related to nutrient availability.

The unfortunate discovery of a beetle (*Lasioderma serricorne*) infestation in the lichen herbarium at the University of Trieste, presented Nimis and Skert (2006) with the opportunity to assess the levels of damage to different species of lichens and especially to see if damage was related in any way to lichen chemistry. Nearly 1500 specimens representing 50 lichen species were analyzed to see if beetle damage was species-specific or compound-specific, and a number of interesting patterns emerged. First, just as Gauslaa (2005) discovered, the simple presence of lichen compounds significantly deterred feeding by beetles, even in long-dead herbarium specimens. The effectiveness of various compounds differed significantly, and compounds known to be effective against other polyphagous invertebrates were found to be more or less so in this case. Also, lichen compounds such as zeorin, which were assumed to have antiherbivore properties (Mattsson, 1987), were confirmed as antifeedants in this study, and pruinose lichens that produce a distinctive layer of calcium oxalate crystals on the surface of lichen thalli were also found to be well-defended.

Studies of polyphagous herbivores not known to consume lichens in nature provide opportunities to assess the defensive attributes of a broad range of lichens and lichen compounds, but it has been argued that any observed chemical deterrence in these animals is of doubtful significance in nature. Although few detailed studies of lichenivorous invertebrates have been done, there is sufficient evidence to indicate that specialized lichen feeders make food choices based on lichen chemistry just as generalists do, which supports the hypothesis that chemical defense against herbivores is real, although the extent to which lichen chemical defense affects lichen fitness depends on the herbivore studied.

In western Virginia, the fungivorous slug species *Pallifera varia* is a common lichen feeder, although it also consumes a variety of nonlichen fungi. A series of field and laboratory studies (Lawrey, 1980, 1983) demonstrated that food choices made by *P. varia* are nonrandom, predictable, and based in large part on lichen chemistry. In the field, slugs were consistently observed grazing on certain dominant lichen species rather than others, and preference for these lichens continued in the laboratory. To see if lichen chemistry was responsible for these feeding choices, lichen compounds were extracted and added to baited filter paper disks, which were then offered to slugs as food under controlled conditions. Compounds from lichens avoided in the field consistently elicited avoidance responses by slugs in the laboratory.

Lichen compounds are only one of the many factors that influence gastropod lichen grazers, however. An illustration of this comes from a series of studies on the feeding ecology of land snails from the Baltic Island of Öland, Sweden (Baur et al., 1992; Fröberg et al., 1993; Baur et al., 1995). The habitat is a limestone grassland with 108 species of calcicolous lichens and 17 species of snails. Four of these species are common consumers of rock-inhabiting lichens and cyanobacteria in the habitat, and snails maintained under laboratory conditions grow and reproduce when fed a diet consisting of only lichens and cyanobacteria. For all the lichen-feeding snails, certain species were preferred over others, although preferences were not identical for all snails and the factors that determined preferences included physical as well as chemical deterrents. Certain species exhibited few preferences that could be attributed solely to chemistry. Some preferred reproductive to vegetative tissues, and some the opposite. Endolithic lichens that live inside rocks were more protected from grazing than epilithic ones, as would be expected. Finally, a lichen that does not produce secondary metabolites at all, *Verrucaria nigrescens*, is least preferred by snails, possibly because the carbonized tissues produced by these lichens are less attractive or palatable than the tissues of other lichens. Many snails also appear to readily consume free-living epilithic cyanobacteria, but not lichens that have cyanobacterial photobionts (such as *Collema fuscovirens*). Avoidance of cyanolichens by snails has been observed in other studies (Gerson, 1973; James and Henssen, 1976), suggesting a possible defensive characteristic of these lichens, some of which may be chemical. Cyanobacteria are known to produce toxic compounds (Briand et al., 2003), some of which may have clinical applications (Burja et al., 2001), and lichens are known to harbor toxic cyanobacteria (Oksanen et al., 2004), which indicates a potential antiherbivore role of cyanolichen secondary compounds.

The genus *Peltigera* includes many species that associate with cyanobacteria, either as the primary photobiont, or as N-fixing cephalodia on the thallus surface. An interesting recent study of land snail damage to four species of *Peltigera* (Benesperi and Tretiach, 2004) indicates a strong preference for lichens lacking secondary compounds. In both field observations and laboratory experiments, two gastropod grazers (*Cantareus aspersa* and *Limax* sp.) readily consumed the thalli of unprotected species (*P. degenii*, *P. praetextata*), but avoided the thalli of chemically well-endowed species (*P. horizontalis*, *P. neckeri*). The well-defended species were similar chemically, indicating a potential deterrent role of gyrophoric acid and its derivatives, and various terpenoids, especially zeorin, a compound found to be an effective deterrent to the polyphagous beetle *Lasioderma serricorne* (Nimis and Skert, 2006).

A number of interesting studies have been done recently on arctiid moth larvae in the subfamily Lithosiinae (Pöykko and Hyvärinen 2003; Pöykkoet al., 2005), several species of which feed exclusively on lichens. In controlled laboratory experiments, larvae of the species *Eilema depressum* and *E. complanum* exhibited clear preferences for lichens that do not contain secondary metabolites (*Melanelia exasperata*) over lichens that produce them (*Vulpicida pinastri*, *Hypogymnia physodes*, *Xanthoria parietina*). Of the avoided lichens, some (*V. pinastri and H. physodes*) were subsequently found to be quite toxic whereas others (*X. parientina*) had little effect on survival of larvae. When *V. pinastri* and *H. physodes* thalli were extracted in acetone to remove phenolic compounds and offered as food, the larvae readily consumed all of them and exhibited near-normal rates of growth and survival. Even for specialist lichen feeders, then, lichen compounds vary in effectiveness against grazing.

11.3.3.3 Different Compounds Have Different Functions: The Same Compounds Have Different Functions

The debate over the adaptiveness of lichen compounds sometimes makes it appear that all should have a single function, but most authors recognize that lichen compounds are probably adaptive for a wide variety of reasons and each compound may play a variety of roles. Some research projects consider the evidence about multiple functions. For example, Lawrey (1989) found that lichen extracts that were most effective against generalist herbivores were also most effective against gram-positive bacteria. So the suite of lichen compounds produced by well-defended lichens is not specifically targeting any particular biotic threat. Another recent study (McEvoy et al., 2007) investigated the changes in pools of medullary depsidones and cortical melanins produced by thalli of *Lobaria pulmonaria* transplanted to habitats that differed significantly in exposure to light. Concentrations of seven medullary depsidones remained unchanged along light gradients, but cortical melanins responded significantly to altered light conditions. The authors concluded that such responses suggest a photoprotective role for the cortical substances and an herbivore defense role for the medullary substances. Future studies of lichen chemical ecology would best be designed to consider the potential significance of the widest array of functional roles.

11.3.4 Stress-Induced Changes to Lichen Chemistry

There is overwhelming evidence that lichen compounds serve as one of the many defensive measures to protect lichens from physical and biotic threats. Taken together, the evidence indicates that the chemistry is constitutive and of a generally defensive nature, not targeting any particular threat. However, there is potential for lichens to respond chemically to changes in their environment. For example, there is some evidence that lichen compounds are induced in lichens transplanted to different environments, and lichens are known to be able to replace compounds that are removed experimentally. This flexibility in compound production is highly adaptive in long-lived, stress-tolerant organisms that face a variety of unpredictable threats. Lange (1992) has hypothesized that the production of secondary metabolites by lichens may be a generalized stress adaptation to life in extreme environments, and there is some evidence that production of compounds can be stimulated by stress, including stresses not faced by lichens in their evolutionary past. This provides support for the hypothesis that chemical defense is stress-induced and not "for" any particular sort of threat.

All of the studies cited in this chapter provide general support for this idea, but two recent studies provide evidence that stress by itself is an important factor in lichen chemistry. Follmann and Schulz (1993) discovered a remarkable range of stress-induced chemical changes in a variety of phylogenetically different lichens. Stresses of various sorts, both physical and biotic, induced qualitative, quantitative, and positional changes in lichen chemistry, suggesting for the first time that stresses of various sorts can elicit similar responses in lichens. Even anthropogenic stresses such as air pollution can induce alterations in lichen chemistry. For example, Białońska and Dayan (2005) found that concentrations of physodalic acid increased significantly in thalli of *Hypogymnia physodes* transplanted from unpolluted to polluted environments, suggesting a possible pollution ameliorating role for this compound. A later study (Hauck and Huneck, 2007a) confirmed that physodalic acid can bind harmful metals to fungal cell walls, thereby preventing entry into the cells. Fumarprotocetraric acid appears to also provide protection from metals in the pollution-tolerant lichen *Lecanora conizaeoides* (Hauck and Huneck, 2007b). Enhanced production of lichen compounds in these situations is probably not triggered specifically by pollution, but rather represents a general response to stress that is clearly adaptive. The actual mechanisms by which such responses are induced remain to be discovered, but there appears to be evidence from several sources that compound production is one of many stress responses in lichens.

11.4 Lichens as Defensive Mutualisms

As textbook examples of mutualism, lichens have long been studied to understand the consequences of transition to a mutualistic mode of existence. Symbiotic and aposymbiotic states of lichen-forming fungi are very different in appearance, chemistry, and behavior, and the transition to a fully lichenized thallus involves a complex series of interactions that utterly transforms the fungal partner. Nevertheless, even phylogenetically different lichen-forming fungi behave in similar ways during the process of lichenization, creating fully symbiotic lichens that are remarkably similar physiologically, ecologically, and chemically. It is impossible to dismiss the nearly universal and defining characteristics of symbiotic lichens as random accidents. Assuming they are adaptive, what is the adaptive significance of these changes?

It has been argued that lichenization transforms a "normal" absorptive fungal mycelium living inside or on its food into a fungal cultivator of algae (Sanders, 2001). This creates for the individual fungus an entirely different set of problems to solve. First, the fungus must recognize and incorporate into its body appropriate photosynthetic cells to serve as a food supply. Then, it must develop thallus structures that permit these cells to acquire light for photosynthesis, and finally it must colonize and maintain itself on an appropriately illuminated substrate. This transition to a lichenized state creates physiological problems not faced by the aposymbiotic fungus, including extreme temperatures, drought, nutrient deficiencies, and exposure to pathogens and predators. Stress-adapted organisms are usually capable of withstanding these extreme conditions, but the cost is slow growth in relatively unpredictable environments. Adaptive responses typically include physiological plasticity, tolerance of extremes, and investment in defense.

Lichens exhibit many of these traits and can therefore be considered defensive mutualisms insofar as they produce stress-adapted characters when lichenized.

The question of which came first, defenses or symbiosis, is purely academic for most lichens. They are members of ancient lineages that are entirely lichenized, so this condition has probably persisted for millions of years. Indeed, it may be the basal condition for most major clades of ascomycetes (Lutzoni et al., 2001). Production of defenses by the earliest lichens may have been an important precondition for the transition to mutualism, but this idea cannot be rigorously tested. Still, the prevalence of various defensive adaptations in lichens makes clear the significance of defense in the unparalleled success of lichen mutualisms.

REFERENCES

Ahmadjian, V. 1993. *The Lichen Symbiosis*. New York: Wiley.

Ahmadjian, V. and J. B. Jacobs. 1985. Artificial reestablishment of lichens. IV. Comparison between natural and synthetic thalli of *Usnea strigosa. Lichenologist* 17:149–165.

Armaleo, D. 1993. Why do lichens make secondary products? *XV International Botanical Congress Abstracts*, Congress Center of Pacifico Yokohama, Japan, p. 11.

Aslan, A, Güllüce, M., and E. Atalan. 2001. A study of antimicrobial activity of some lichens. *Bulletin of Pure and Applied Sciences* B 20:23–26.

Bachereau, F. and J. Asta. 1997. Effects of solar ultraviolet radiation at high altitude on the physiology and the biochemistry of a terricolous lichen (*Cetraria islandica* (L.) Ach.). *Symbiosis* 23:197–217.

Bachereau, F. and J. Asta. 1998. Effects of solar ultraviolet radiation at high altitude on the phenolic compounds contents of *Cetraria islandica* (L.) Ach. (terricolous lichen). *Ecologie* 29:267–270.

Baur, A., Baur, B., and L. Fröberg. 1992. The effect of lichen diet on growth rate in the rock-dwelling land snails *Chondrina clienta* (Westerlund) and *Balea perversa* (Linnaeus). *Journal of Molluscan Studies* 58:245–247.

Baur, A., Baur, B., and L. Fröberg. 1994. Herbivory on calcicolous lichens: different food preferences and growth rates in two co-existing land snails. *Oecologia* 98:313–319.

Baur, B., Fröberg, L., and A. Baur. 1995. Species diversity and grazing damage in calcicolous lichen communities on stone walls in land Sweden. *Annales Botanici Fennici* 32:239–250.

Benesperi, R. and M. Tretiach. 2004. Differential land snail damage to selected species of the lichen genus *Peltigera. Biochemical Systematics and Ecology* 32:127–138.

Bialonska, D. and F. E. Dayan. 2005. Chemistry of the lichen *Hypogymnia physodes* transplanted to an industrial region. *Journal of Chemical Ecology* 31:2975–2991.

Blanco, M. J., Suarez, C., and C. Vicente. 1984. The use of urea by *Evernia prunastri* thalli. *Planta* 162:305–310.

Blewitt, M. R. and G. A. Cooper-Driver. 1990. The effects of lichen extracts on feeding by gypsy moths (*Lymantria dispar*). *Bryologist* 93:220–221.

Briand, J. F., Jacquet, S., Bernard, C., and J. F. Humbert. 2003. Health hazards for terrestrial vertebrates from toxic cyanobacteria in surface water ecosystems. *Veterinary Research* 34:361–377.

Broadhead, E. 1958. The psocid fauna of larch trees in northern England. An ecological study of mixed species populations exploiting a common resource. *Journal of Animal Ecology* 27:217–263.

Broadhead, E. and I. W. B. Thornton. 1955. *Elipsocus mclachlani* feeding on lichens. *Oikos* 6:1–50.

Brodo, I. M., Sharnoff, S. D., and S. Sharnoff. 2001. *Lichens of North America*. New Haven, CT: Yale University Press.

Burja, A. M., Banaigs, B., Abou-Mansour, E., Burgess, J. G., and P. C. Wright. 2001. Marine cyanobacteria: A prolific source of natural products. *Tetrahedron* 57:9347–9377.

Burkholder, P. R., Evans, A. W., McVeigh, I., and H. K. Thornton. 1944. Antibiotic activity of lichens. *Proceedings of the National Academy of Sciences USA* 30:250–255.

Cardinale, M., Puglia, A. M., and M. Grube. 2006. Molecular analysis of lichen-associated bacterial communities. *FEMS Microbiology Ecology* 57:484–495.

Clauzade, G., Diederich, P., and C. Roux. 1989. Nelikenigintaj fungoj likenlogaj–Ilustrita determinlibro. *Bulletin de la Société Linnéenne de Provence, Numéro spécial* 1:1–142.

Cocchietto, M., Skert, N., Nimis, P. L., and G. Sava. 2002. A review on usnic acid, an interesting natural compound. *Naturwissenschaften* 89:137–146.

Colloff, M. J. 1983. Oribatid mites associated with marine and maritime lichens on the island of Great Cumbrae. *Glasgow Naturalist* 20:347–359.

Crittenden, P. D. and N. Porter. 1991. Lichen forming fungi: Potential sources of novel metabolites. *Trends in Biotechnology* 9:409–412.

Culberson, C. F., Culberson, W. L., and A. Johnson. 1977. Nonrandom distribution of an epiphytic *Lepraria* on two species of *Parmelia. Bryologist* 80:201–203.

Culberson, C. F., Culberson, W. L., and A. Johnson. 1983. Genetic and environmental effects on growth and production of secondary compounds in *Cladonia cristatella. Biochemical Systematics and Ecology* 11:77–84.

Culberson, C. F. and J. A. Elix. 1989. Lichen substances. In *Methods in Plant Biochemistry*, Eds. P. M. Dey and J. B. Harborne, Vol I, pp. 509–535. London: Academic Press.

de los Ríos, A., Wierzchos, J., Sancho, L. G., Green, T. G. A., and C. Ascaso. 2005. Ecology of endolithic lichens colonizing granite in continental Antarctica. *Lichenologist* 37:383–395.

Elix, J. A. 1996. Biochemistry and secondary metabolites. In *Lichen Biology*, Ed. T. H. Nash III, pp. 154–180. New York: Cambridge University Press.

Elo, H., Matikainen, J., and E. Pelttari. 2007. Potent activity of the lichen antibiotic (+)-usnic acid against clinical isolates of vancomycin-resistant enterococci and methicillin-resistant *Staphylococcus aureus. Naturwissenschaften* 94:465–468.

Emmerich, R., Giez, I., Lange, O. L., and P. Proksch. 1993. Toxicity and antifeedant activity of lichen compounds against the polyphagous herbivorous insect *Spodoptera littoralis. Phytochemistry* 33:1389–1394.

Fahselt, D. 1994. Secondary biochemistry of lichens. *Symbiosis* 16:117–165.

Follmann, G. and M. Schulz. 1993. Stress-induced changes in the secondary products of lichen thalli. In *Phytochemistry and Chemotaxonomy of Lichenized Ascomycetes—A Festschrift in Honour of Siegfried Huneck*, Eds. G. B. Feige and H. T. Lumbsch, pp. 75–86. J. Cramer, Berlin, Stuttgart: Bibliotheca Lichenologica.

Follmann, G. and V. Villagrán. 1965. Flechtenstoffe und Zellpermeabilität. *Zeitschrift für Naturforschung* 20B:723.

Friedmann, E. I. 1981. Endolithic microorganisms in the dry valleys of Antarctica. *Antarctic Journal of the United States* 16:174–175.

Friedmann, E. I. 1982. Endolithic microorganisms in the Antarctic cold desert. *Science* 215:1045–1053.

Fröberg, L., Baur, A., and B. Baur. 1993. Differential herbivore damage to calicicolous lichens by snails. *Lichenologist* 25:83–95.

Galun, M. 2001. Endolithic microorganisms. In *Trichomycetes and other Fungal Groups. Robert W. Lichtwardt Commemoration Volume*, Eds. J. K. Misra and B. W. Horn, Enfield, pp. 209–223. New Hampshire & Plymouth, U.K.: Science Publishers, Inc.

Gauslaa, Y. 2005. Lichen palatability depends on investments in herbivore defence. *Oecologia* 143:94–105.

Gauslaa, Y. and K. A. Solhaug. 2001. Fungal melanins as a sun screen for symbiotic green algae in the lichen *Lobaria pulmonaria. Oecologia* 126:462–471.

Gauslaa, Y. and E. M. Ustvedt. 2003. Is parietin a UV-B or a blue-light screening pigment in the lichen *Xanthoria parietina*? *Photochemical and Photobiological Sciences* 2:424–432.

Gerson, U. 1973. Lichen-arthropod associations. *Lichenologist* 5:434–443.

Gerson, U. and M. R. D. Seaward. 1977. Lichen-invertebrate associations. In *Lichen Ecology*, Ed. M. R. D. Seaward, pp. 69–119. London: Academic Press.

Giez, I., Lange, O. L., and P. Proksch. 1994. Growth retarding activity of lichen substances against the polyphagous herbivorous insect *Spodoptera littoralis. Biochemical Systematics and Ecology* 22:113–120.

Girlanda, M., Isocrono, D., and A. M. Luppi-Mosca. 1997. Two foliose lichens as microfungal ecological niches. *Mycologia* 89:531–536.

Gjelstrup, P. and U. Søchting. 1979. Cryptostigmatid mites (Acarina) associated with *Ramalina siliquosa* (Lichenes) on Bornholm in the Baltic. *Pedobiologia* 19:237–245.

Golojuch, S. T. and J. D. Lawrey. 1988. Quantitative variation in vulpinic and pinastric acids produced by *Tuckermannopsis pinastri* (lichen-forming Ascomycotina, Parmeliaceae). *American Journal of Botany* 75:1871–1875.

Green, T. G. A., Schroeter, B., and L. G. Sancho. 1999. Plant life in Antarctica. In *Handbook of Functional Plant Ecology*, Eds. F. I. Pugnaire and F. Valladares, pp. 495–543. New York: Marcel Dekker.

Hale, Jr. M. E. 1972. Natural history of Plummers Island, Maryland. XXI. Infestation of the lichen *Parmelia baltimorensis* Gyel. & For. by *Hypogastrura packardi* Folsom (Collembola). *Proceedings of the Biological Society of Washington* 85:287–296.

Hamada, N. 1991. Environmental factors affecting the content of usnic acid in the lichen mycobiont of *Ramalina siliquosa. Bryologist* 94:57–59.

Hätscher, I., Veit, M., Proksch, P., Lange, O. L., and H. Zellner. 1991. Feeding deterrency and growth retarding activity of lichen substances against *Spodoptera littoralis. Planta Medica* 57 (Suppl. 2):A26.

Hauck, M. and S. Huneck. 2007a. Lichen substances affect metal absorption in *Hypogymnia physodes. Journal of Chemical Ecology* 33:219–223.

Hauck, M. and S. Huneck. 2007b. The putative role of fumarprotocetraric acid in the manganese tolerance of the lichen *Lecanora conizaeoides. Lichenologist* 39:301–304.

Hawksworth, D. L. and D. J. Hill. 1984. *The Lichen-Forming Fungi.* New York: Chapman and Hall.

Hesbacher, S., Baur, B., Baur, A., and P. Proksch. 1994. Sequestration of lichen compounds by terrestrial snails. *Journal of Chemical Ecology* 21:233–246.

Hesbacher, S., Giez, I., Embacher, G., Fiedler, K., Max, W., Trawöger, A., Türk, R., Lange, O. L., and P. Proksch. 1995. Sequestration of lichen compounds by lichen-feeding members of the Arctiidae (Lepidoptera). *Journal of Chemical Ecology* 21:2079–2089.

Hill, D. J. and H. W. Woolhouse. 1966. Aspects of the autecology of *Xanthoria parietina agg. Lichenologist* 3:207–214.

Honegger, R. 1987. Questions about pattern formation in the algal layer of lichens with stratified (heteromerous) thalli. In *Progress and Problems in Lichenology in the Eighties*, Ed. E. Peveling, pp. 59–71. Bibliotheca Lichenologica No. 25. J. Cramer: Berlin-Stuttgart.

Honegger, R. 1991. Functional aspects of the lichen symbiosis. *Annual Review of Plant Physiology and Plant Molecular Biology* 42:553–578.

Honegger, R. 1992. Lichens: mycobiont-photobiont relationships. In *Algae and Symbioses: Plants, Animals, Fungi, Viruses, Interactions Explored*, Ed. W. Reisser, pp. 255–270. England: Biopress Limited.

Honegger, R. 1998. The lichen symbiosis—what is so spectacular about it? *Lichenologist* 30:193–212.

Huneck, S. 1973. Nature of lichen substances. In *The Lichens*, Eds. V. Ahmadjian and Jr. M. E. Hale, pp. 495–522. New York: Academic Press.

Huneck, S. 1999. The significance of lichens and their metabolites. *Naturwissenschaften* 86:559–570.

Huneck, S. 2001. New results on the chemistry of lichen substances. In *Fortschritte der Chemie Organischer Naturstoffe, 81*, Eds. W. Herz, H. Falk, G. W. Kirby, and R. E. Moore, pp. 1–276. Wein: Springer-Verlag.

Huneck, S. and Y. Yoshimura. 1996. *Identification of Lichen Substances.* Berlin: Springer.

Hyvärinen, M., Koopmann, R., Hormi, O., and J. Tuomi. 2000. Phenols in reproductive and somatic structures of lichens: a case of optimal defence? *Oikos* 91:371–375.

James, P. W. and A. Henssen. 1976. The morphological and taxonomic significance of cephalodia. In *Lichenology: Progress and Problems*, Eds. D. H. Brown, D. L. Hawksworth, and R. H. Bailey, pp. 27–77. London: Academic Press.

Jones, D. 1988. Lichens and pedogenesis. In *Handbook of Lichenology*, Vol III, Ed. M. Galun, pp. 109–124. Boca Raton, FL: CRC Press.

Kinraide, W. T. B. and V. Ahmadjian. 1970. The effects of usnic acid on the physiology of two cultured species of the lichen alga *Trebouxia* Puym. *Lichenologist* 4:234–247.

Lange, O. L. 1992. Pflanzenleben unter Streß. Flechten als Pioniere der Vegetation an Extremstandorten der Erde. Würzburg: University of Würzburg.

Lange, O. L., Green, T. G. A., Reichenberger, H., Hesbacher, S., and P. Proksch. 1997. Do secondary substances in the thallus of a lichen promote CO_2 diffusion and prevent depression of net photosynthesis at high water content? *Oecologia* 112:1–3.

Laundon, J. R. 1971. Lichen communities destroyed by psocids. *Lichenologist* 5:177.

Lawrey, J. D. 1980. Correlations between lichen secondary chemistry and grazing activity by *Pallifera varia. Bryologist* 83:328–334.

Lawrey, J. D. 1983. Lichen herbivore preference: a test of two hypotheses. *American Journal of Botany* 70:1188–1194.

Lawrey, J. D. 1984. *Biology of Lichenized Fungi*. New York: Praeger Publishers.

Lawrey, J. D. 1986. Biological role of lichen substances. *Bryologist* 89:111–122.

Lawrey, J. D. 1987. Nutritional ecology of lichen/moss arthropods. In *Nutritional Ecology of Insects, Mites, Spiders, and Related Invertebrates*, Eds. F. Slansky and J. G. Rodriguez, pp. 209–233. New York: Wiley.

Lawrey, J. D. 1989. Lichen secondary compounds: evidence for a correspondence between antiherbivore and antimicrobial function. *Bryologist* 92:326–328.

Lawrey, J. D. 1993. Chemical ecology of *Hobsonia christiansenii*, a lichenicolous hyphomycete. *American Journal of Botany* 80:1109–1113.

Lawrey, J. D. 1995a. Lichen allelopathy: A review. In *Allelopathy*, Eds. Inderjit, Dakshinin KMM, pp. 26–38. Washington, DC: Einhellig FA. American Chemical Society.

Lawrey, J. D. 1995b. The chemical ecology of lichen mycoparasites. *Canadian Journal of Botany* 73(Suppl. 1): 603–608.

Lawrey, J. D. 2000. Chemical interactions between two lichen-degrading fungi. *Journal of Chemical Ecology* 26:1821–1831.

Lawrey, J. D., Binder, M., Diederich, P., Molina, M. C., Sikaroodi, M., and D. Ertz. 2007. Phylogenetic diversity of lichen-associated homobasidiomycetes. *Molecular Phylogenetics and Evolution* 44:778–789.

Lawrey, J. D. and P. Diederich. 2003. Lichenicolous fungi: Interactions, evolution, and biodiversity. *Bryologist* 106:80–120.

Lawrey, J. D., Torzilli, A. P., and V. Chandhoke. 1999. Destruction of lichen chemical defenses by a fungal pathogen. *American Journal of Botany* 86:184–189.

Legaz, M. E. and C. Vicente. 1989. Regulation of urease activity of *Cladina dendroides* and its photobiont by lichen phenols. *Plant Science* 63:15–24.

Li, W. C., Zhou, J., Guo, S. Y., and L. D. Guo. 2007. Endophytic fungi associated with lichens in Baihua mountain of Beijing, China. *Fungal Diversity* 25:69–80.

Lutzoni, F., Pagel, M., and V. Reeb. 2001. Major fungal lineages are derived from lichen symbiotic ancestors. *Nature* 411:937–940.

Mattsson, J. E. 1987. Zeorin in *Cetraria pinastri* from an unusual habitat. *Bibliotheca Lichenologica* 25:207–208.

McEvoy, M., Nybakken, L., Solhaug, K. A., and Y. Gauslaa. 2006. UV triggers the synthesis of the widely distributed secondary compound usnic acid. *Mycological Progress* 5:221–229.

McEvoy, M., Gauslaa, Y., and K. A. Solhaug. 2007. Changes in pools of depsidones and melanins, and their function, during growth and acclimation under contrasting natural light in the lichen *Lobaria pulmonaria*. *New Phytologist* 175:271–282.

Miadlikowska, J., Arnold, A. E., Hofstetter, V., and F. Lutzoni. 2004. High diversity of cryptic fungi inhabiting healthy lichen thalli in a temperate and tropical forest. In *Book of Abstracts of the 5th IAL Symposium. Lichens in Focus*, Eds. T. Randlane and A. Saag, p. 43. Estonia: Tartu University Press.

Nash III, T. H. 1996. Photosynthesis, respiration, productivity and growth. In *Lichen Biology*, Ed. T. H. Nash III, pp. 88–120. New York: Cambridge University Press.

Nimis, P. L. and N. Skert. 2006. Lichen chemistry and selective grazing by the coleopteran *Lasioderma serricorne*. *Environmental and Experimental Botany* 55:175–182.

Nybakken, L., Solhaug, K. A., Bilger, W., and Y. Gauslaa. 2004. The lichens *Xanthoria elegans* and *Cetraria islandica* maintain a high protection against UV-B radiation in Arctic habitats. *Oecologia* 140:211–216.

Oksanen, I., Jokela, J., Fewer, D. P., Wahlsten, M., Rikkinen, J., and K. Sivonen. 2004. Discovery of rare and highly toxic microcystins from lichen-associated cyanobacterium *Nostoc* sp. strain IO-102-I. *Applied and Environmental Microbiology* 70:5756–5763.

Peake, J. F. and P. W. James. 1967. Lichens and mollusca. *Lichenologist* 3:425–427.

Perez-Urria, E. and C. Vicente. 1989. Purification and some properties of a secreted urease from *Evernia prunastri* thallus. *Journal of Plant Physiology* 133:692–695.

Petrini, O., Hake, U., and M. M. Dreyfuss. 1990. An analysis of fungal communities isolated from fruticose lichens. *Mycologia* 82:444–451.

Pöykkö, H. and M. Hyvärinen. 2003. Host preference and performance of lichenivorous *Eilema* spp. larvae in relation to lichen secondary metabolites. *Journal of Animal Ecology* 72:383–390.

Pöykkö, H., Hyvärinen, M., and M. Bačkor. 2005. Removal of lichen secondary metabolites affects food choice and survival of lichenivorous moth larvae. *Ecology* 86:2623–2632.

Purvis, O. W., Elix, J. A., Broomhead, J. A., and G. C. Jones. 1987. The occurrence of copper-norstictic acid in lichens from cupriferous substrata. *Lichenologist* 19:193–203.

Purvis, O. W., Elix, J. A., and K. L. Gaul. 1990. The occurrence of copper-psoromic acid in lichens from cupriferous substrata. *Lichenologist* 22:345–354.

Rambold, G. 1985. Fütterungsexperimente mit den an Flechten fressenden Raupen von *Setina aurita* Esp. (Lepidoptera, Arctiidae). *Nachrichtenblatt der Bayerischen Entomologen* 34:82–90.

Rancan, F., Rosan, S., Boehm, K., Fernández, E., Hidalgo, M. E., Quilhot, W., Rubio, C., Boehm, F., Piazena, H., and U. Oltmanns. 2002. Protection against UVB irradiation by natural filters extracted from lichens. *Journal of Photochemistry and Photobiology B: Biology* 68:133–139.

Reutimann, P. and C. Scheidegger. 1987. Importance of lichen secondary products in food choice of two oribatid mites (Acari) in an alpine meadow ecosystem. *Journal of Chemical Ecology* 13:363–369.

Richardson, D. H. S. 1975. *The Vanishing Lichens*. Vancouver: David and Charles.

Richardson, D. H. S. 1999. War in the world of lichens: parasitism and symbiosis as exemplified by lichens and lichenicolous fungi. *Mycological Research* 103:641–650.

Rikkinen, J. 1995. What's behind the pretty colours? A study on the photobiology of lichens. *Bryobrothera* 4:1–239.

Rundel, P. W. 1969. Clinal variation in the production of usnic acid in *Cladonia subtenuis* along light gradients. *Bryologist* 72:40–44.

Rundel, P. W. 1978. The ecological role of secondary lichen substances. *Biochemical and Systematic Ecology* 6:157–170.

Sancho, L. G., de la Torre, R., Horneck, G., Ascaso, C., de los Ríos, A., Pintado, A., Wierzchos J., and M. Schuster. 2007. Lichens survive in space: Results from the 2005. LICHENS experiment. *Astrobiology* 7:443–454.

Sanders, W. B. 2001. Lichens: The interface between mycology and plant morphology. *BioScience* 51:1025–1035.

Schatz, A. 1962. Pedogenic (soil-forming) activity of lichen acids. *Naturwissenschaften* 49:518.

Seaward, M. R. D. 1965. Lincolnshire psocids. *Transactions of the Lincolnshire Naturalists' Union* 16:99–100.

Seyd, E. L. and M. R. D. Seaward. 1984. The association of oribatid mites with lichens. *Zoological Journal of the Linnean Society* 80:369–420.

Sharnoff, S. http://www.lichens.com/invertebrates.html

Sigal, L. L. 1984. Of lichens and lepidopterans. *Bryologist* 87:66–68.

Slansky, Jr. F. 1979. Effects of the lichen chemicals atranorin and vulpinic acid upon feeding and growth of larvae of the yellow-striped armyworm, *Spodoptera ornithogalli. Environmental Entomology* 8:865–868.

Solhaug, K. A. and Y. Gauslaa. 1996. Parietin, a photoprotective secondary product of the lichen *Xanthoria parietina. Oecologia* 108:412–418.

Solhaug, K. A. and Y. Gauslaa. 2001. Acetone rinsing—a method for testing ecological and physiological roles of secondary compounds in living lichens. *Symbiosis* 30:301–315.

Solhaug, K. A. and Y. Gauslaa. 2004. Photosynthates stimulate the UV-B induced fungal anthraquinone synthesis in the foliose lichen *Xanthoria parietina. Plant, Cell, and Environment* 27:167–176.

Solhaug, K. A., Gauslaa, Y., Nybakken L., and W. Bilger. 2003. UV-induction of sun-screening pigments in lichens. *New Phytologist* 158:91–100.

Sowter, F. A. 1971. Mites (Acari) and lichens. *Lichenologist* 5:176.

Stahl, G. E. 1904. Die Schutzmittel der Flechten gegen Tierfrass. In *Festschrift zum Siebzigsten Geburtstage von Ernst Haekel*, pp. 357–375, Gustav Fischer, Jena.

Stephenson, N. L. and P. W. Rundel. 1979. Quantitative variation and the ecological role of vulpinic acid and atranorin in the thallus of *Letharia vulpina. Biochemical Systematics and Ecology* 7:263–267.

Stoll, A., Renz, J., and A. Brack. 1947. Antibiotika aus Flechten (Vierte Mitteilung über antibakterielle Stoffe). *Experientia* 3:111–113.

Swanson, A. and D. Fahselt. 1997. Effects of ultraviolet on polyphenolics of *Umbilicaria Americana. Canadian Journal of Botany* 75:284–289.

Syers, J. K. and I. K. Iskandar. 1973. Pedogenic significance of lichens. In *The Lichens*, Eds. V. Ahmadjian and M. E. Hale Jr., pp. 224–248. New York: Academic Press.

Timdal, E. 1989. The production of rhodocladonic acid in *Cladonia bacilliformis* and *C. norvegica* triggered by the presence of a lichenicolous mite. *Graphis Scripta* 2:125–127.

Torzilli, A. P., Balakrishna, S., O'Donnell, K., and J. D. Lawrey. 2002. The degradative activity of a lichenicolous *Fusarium* sp. compared to related entomogenous species. *Mycological Research* 106:1204–1210.

Torzilli, A. P. and J. D. Lawrey. 1995. Lichen metabolites inhibit cell wall-degrading enzymes produced by the lichen parasite *Nectria parmeliae. Mycologia* 87:841–845.

Torzilli, A. P., Mikelson, P. A., and J. D. Lawrey 1999. Physiological effect of lichen secondary metabolites on the lichen parasite *Marchandiomyces corallinus. Lichenologist* 31:307–314.

Travé, J. 1963. Ecologie et biologie des Oribates (Acariens) saxicoles et arboricoles. *Vie Milieu* (Suppl.) 14:1–267.

Travé, J. 1963. Sur le peuplement des lichens-crustacés des Iles Salvges par les Oribates (Acariens). *Revue d'écologie et de Biologie du sol* 6:239–248.

Vartia, K. O. 1973. Antibiotics in lichens. In *The Lichens*, Eds. V. Ahmadjian, and M. E. Hale Jr., pp. 547–561. New York: Academic Press.

Vicente, C. 1985. Surface physiology of lichens: Facts and Concepts. In *Surface Physiology of Lichens*, Eds. C. Vicente, D. H. Brown, and M. E. Legaz, pp. 11–24. Universidad Complutense de Madrid, Madrid.

Wynn-Williams, D. D., Holder, J. M., and H. G. M. Edwards. 2000. Lichens at the limits of life: Past perspectives and modern technology. In *New Aspects in Cryptogamic Research*, Eds. B. Schroeter, M. Schlensog, and T. G. A. Green, pp. 275–288. Contributions in Honour of Ludger Kappen. Bibliotheca Lichenologica, J. Cramer, Berlin, Stuttgart.

Yom-Tov, Y. and M. Galun. 1971. Note on feeding habits of the desert snails *Sphincterochila boissieri* charpentier and *Trochidea (Xenoorassa) seetzeni* Charpentier. Veliger 14:86–88.

Zopf, W. 1896. Zur biologischen Bedeutung der Flechtensäuren. *Biologisches Zentralblatt* 16:593–610.

Zukal, H. 1895. Morphologische und biologische Untersu chungen uber die Flechteu Sitzungsberichte. *Königliche Böhmische Gesellschaft der Wissenschaften, Mathematische Naturalische Klasse* 104:1303–1395.

12

Arbuscular Mycorrhizae as Defense against Pathogens

José Manuel García Garrido

CONTENTS

12.1 Introduction

Probably the most relevant and widespread mutualistic symbiosis in the plant kingdom is the arbuscular mycorrhiza (AM). This association is formed between arbuscular mycorrhizal fungi (AMF) and the vast majority of plants present in natural and cultivated systems (Jeffries and Barea, 2001). The establishment of this mutualistic association is a successful strategy to improve the nutritional status of both partners. The fungi, considered as biotrophic microorganisms, receive fixed carbon compounds from the host plant, while the plant benefits from the association by the increase in nutrient uptake via the fungal mycelium (Smith and Read, 1997).

Fungal penetration and establishment in the host roots involve a complex sequence of events and intracellular modifications, including signal release and perception, fungal penetration and inter and intracellular colonization of the root cortex, and formation of the arbuscules, which are highly branched fungal structures specialized in nutrient exchange (Bonfante-Fasolo and Perotto, 1992). Briefly, the symbiosis can be divided into two stages (Figure 12.1). The precolonization stage, where the AM spore in the soil germinates and hyphal growth and differentiation occur, resulting in the formation of appresorium,

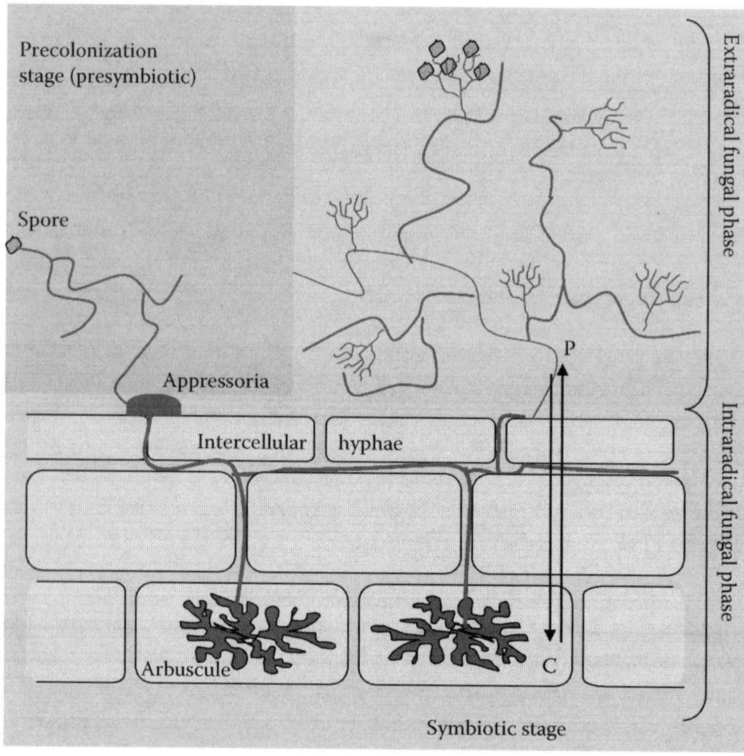

FIGURE 12.1 Drawing of AM showing the most relevant fungal structures. In the precolonization stage, after spore germination, AMF grow toward the root and form the appresorium at the root surface. In the symbiotic stage, the fungi form inter and intracellular hyphae and arbuscules in their intraradical phase and develop an external mycelium in the soil in their symbiotic extraradical phase.

the AMF structures for root penetration, and the symbiotic stage, where the AMF has penetrated the root, formed its intraradical structures, e.g., inter- and intracellular hyphae and arbuscules, and an exchange of nutrients between the host and the symbiont occurs. In the symbiotic stage and when the AM is well established, two different phases in the AMF can be distinguished: the intraradical fungal phase, when the fungal mycelium biotrophically colonizes the root, and the extraradical phase formed by the external mycelium developed in the soil.

Both phases of the mycelia are important to complete functionality of AM. The external mycelium takes up mineral nutrients that are transferred to the plant through the fungal intraradical structures, and it has a key ecological role as a living component of the rhizosphere (Harley and Smith, 1983). On the other hand, the AMF penetration and growth into the roots lead to the activation of the symbiotic programs for functional mycorrhization in both symbionts, with great alterations in the morphology and physiology of roots and fungal hyphae (Bonfante-Fasolo and Perotto, 1992; Parniske, 2004). In addition, changes in the hormonal, nutritional, and defensive molecular pathways are necessary in the plant for the regulation and functioning of AM (García-Garrido and Ocampo, 2002). These particular morphological and molecular features of AM symbiosis inside and outside the root support other important beneficial effects on plant fitness, apart from the nutritional improvement, such as the higher stress-adaptation capacity, and the effective bioprotection against plant pathogens (Whipps, 2004; Pozo and Azcón-Aguilar, 2007).

Many studies describe particular cases of the bioprotective effect of mycorrhization against biotic stress, and hypothetical modes of action have been postulated. However, there is little information about the causal relationships between the morphological and molecular features of AM symbiosis and the mechanism implicated in the bioprotectional effect. Supported by recent advances in the signaling processes of plant defense response and plant–microbe interactions, the current research on AM plant–pathogen interactions focuses on the identification at the molecular and genetic levels of key regulator stages associated with the bioprotective effect.

12.2 Bioprotective Effect of Arbuscular Mycorrhizae

The analysis of the published information on AM bioprotection shows that a great number of pathogen and plant species combinations and a smaller number of AM fungal species are involved in the biocontrol (more than 80% of the studies on the bioprotectional effect of mycorrhization have been performed with the genus *Glomus*). This probably reflects the relatively natural occurrence of these AM fungal species or, more probably, their generalized use as model species in mycorrhizal studies. Likewise, most of the studies on protection have been carried out in controlled environmental conditions, and field experiments have been relatively few. Most data about bioprotection due to mycorrhization are available from soil-borne fungal pathogens, but examples of control of soil bacterial pathogens, foliar pathogens, root nematodes, and phytophagous insects are known. Most of those data have been compiled in excellent reviews on this subject (Linderman, 1994; Singh et al., 2000; Azcón et al., 2002; Whipps, 2004; Xavier and Boyetchko, 2004). Table 12.1 shows most representative effects of arbuscular mycorrhizae exhibiting control of plant diseases.

The effects of AM on bacterial diseases have not been intensively investigated. Some studies have shown that *Glomus mosseae* protects tomato plants against the damage caused by *Pseudomonas syringae* and *Erwinia carotovora* inoculated in soil (García-Garrido and Ocampo, 1988, 1989). The protection was more effective when the pathogen is added 3 weeks after the AMF inoculation, and the pathogen survival in the rhizosphere is lower in mycorrhizal plants than in nonmycorrhizal plants. Interestingly, a recent study showed the first experimental evidence that AM might display enhanced disease resistance to shoot pathogenic bacteria. In this report, mycorrhizal *Medicago truncatula* plants showed less incidence of disease from a virulent bacterial pathogen, *Xhantomonas campestris,* than was the case with nonmycorrhizal plants (Liu et al., 2007). The increased resistance in shoots of mycorrhizal plants was parallel to the systemic induction of many genes thought to be involved in stress or defense responses (Liu et al., 2007).

In the case of pathogenic root nematodes, the inoculation of AMF generally results in a reduction of disease severity, but the responses may vary and the mechanisms implicated are controversial (Hussey and

TABLE 12.1

Effects of AM Exhibiting Control of Plant Diseases

Pathogen/Pest Type	Disease	Bioprotective Effect
Bacterial	Root necrosis	No effect
	Bacterial leaf blight	Reduction of pathogen survival
		Reduction in disease incidence and severity
Root nematodes	Root knot	No effect
		General increase of disease tolerance
		Reduction of pathogen survival in soil
Foliar pathogens	Viral infections, powdery mildew, and rust fungi	No effect
		General increase of disease tolerance and in some cases increased susceptibility
Root fungal pathogens	Root rot or wilting	General reduction of disease incidence and severity
		Increase of resistance
Phytophagous insects	Wounding	Effect generally depends on the insect lifestyle
		Reduction in the incidence
		Induction of indirect defense

Roncadori, 1982; Ingham, 1988). It seems that AM plants develop an increased tolerance to certain pathogen nematodes and in some cases the nematode populations, measured as numbers of galls or eggs per unit of root length, are reduced (Hussey and Roncadori, 1978; Calvet et al., 1995; Jaizme-Vega et al., 1997).

The effects of arbuscular mycorrhizal symbiosis on aboveground herbivores have been investigated with contrasting results. Although information is scare and few studies have been done, it seems that AM also have significant consequences for the growth and survival of phytophagous insects (Gange, 2006). Again, as with pathogenic root nematodes, AM-symbioses consistently reduce attacks by root-feeding insects. Nevertheless, the effects on foliar-feeding ones are more variable and depend on the aspects related to changes produced to the leaves by AM colonization which affect their suitability as food for the insects such as the alteration of the carbon: nutrient balance, the increase in the C allocation to roots, and changes in leaf morphology. The effect generally depends on the insect lifestyle and degree of specialization: AM reduces the incidence of generalist chewing insects, while sap sucking thrives better on mycorrhizal plants (Gange, 2006). Interestingly, an induced indirect defense has been evidenced in AM plant attacks by herbivorous insects (Guerrieri et al., 2004). Indirect defenses typically involve the production of volatile semiochemicals that are attractive to the natural enemies of herbivorous insects. The study demonstrated that tomato plants colonized by the arbuscular mycorrhizal fungus *G. mosseae* have enhanced attractiveness for the parasitic wasp *Aphidius ervi*, the natural enemy of the phytophagous aphid *Macrosiphym euphorbiae*. Additionally, the arbuscular colonization has a negative effect on *M. euphorbiae* insect populations (Guerrieri et al., 2004).

Most studies on protection by AM deal with the reduction of incidence and severity of soil-borne diseases, including root rot or wilting caused by fungi such as *Rhizoctonia*, *Fusarium*, or *Verticillium*, and root rot caused by oomycetes such as *Phytophthora*, *Pythium*, and *Aphanomyces*. Numerous reports show a clear localized protective effect (reviewed by Singh et al., 2000; Azcón et al., 2002; Xavier and Boyetchko, 2004; St-Arnaud and Vujanovic, 2007) and recently, several studies also found a systemic protective effect against these soil-borne pathogens (Cordier et al., 1998; Pozo et al., 2002; Khaosaad et al., 2007). In general, it is assumed that a mycorrhizal protection against root fungi pathogens required a high degree of AM root colonization, whereas intermediate and low levels of AM root colonization showed no bioprotectional effect. Apparently, a critical level of AM root colonization is needed to provide bioprotection for mycorrhizal plants. In a recent report, it has been demonstrated that not only the local, but also the systemic bioprotectional effect of mycorrhization depends on the degree of AM root colonization (Khaosaad et al., 2007).

While soil-borne diseases caused by fungal pathogens and nematodes are most often reduced by AM, the mycorrhizal effects on foliage diseases are contradictory and less conclusive. AM symbioses have been associated with enhanced susceptibility to biotrophic pathogens including viruses, powdery mildew, and rust fungi (*Blumeria, Oidium, Uromyces*) (Shaul et al., 1999; Gernns et al., 2001; Whipps, 2004). In some cases, the enhanced development of the foliar pathogen was due to increased tolerance rather than increased incidence or frequency of infections, and was often correlated with improvement of plant mass and yield in mycorrhizal plants. Nevertheless, mycorrhization reduces disease symptoms caused by phytoplasma (Lingua et al., 2002), and provides protection against the necrotroph fungus *Alternaria solani* (Fritz et al., 2006) and the shoot pathogenic bacteria *X. campestris* (Liu et al., 2007).

12.3 Factors Affecting the Bioprotective Effect of Arbuscular Mycorrhizae

Biocontrol involves the use, management, and harnessing of interactions among the plant, pathogen, biocontrol agent, and environment to improve plant health (Baker, 1987). Even in the simplest model, the biocontrol studies imply at least interactions among the three organisms and the physical, chemical, and microbiological environment. In the case of arbuscular mycorrhizae, these interactions can be complex because they occur in the rhizosphere, a dynamic environment, which is characterized by rapid change and intense microbial activity affected by the roots (Whipps, 2001). Bearing this in mind, it is necessary to focus on research into the key factors that affect the efficiency of arbuscular mycorrhizae biocontrol in order to understand and make possible the practical use of AM in the management of plant disease in agroecosystems. These are (1) pathogen cell life cycle and epidemiology,

(2) plant genotype, and (3) the extent of mycorrhizal infection and AM fungus isolation efficiency. Apart from these biotic factors, the bioprotective effect of AM seems to depend on several other abiotic factors affecting the biocontrol efficacy of the AMF such as temperature, soil moisture, and soil P-content (for details see Singh et al., 2000).

12.3.1 Pathogen Etiology and Epidemiology

As for any biocontrol agent, it should be noted that the potential for biological control is directly related to the inoculum potential, virulence, and lifestyle of the pathogen. The protection due to AM is not effective for all pathogens, and most data about its bioprotective effects indicate that AM colonization affects mainly the incidence and severity of root diseases (Singh et al., 2000; Azcón et al., 2002; Xavier and Boyetchko, 2004; Whipps, 2004; St-Arnaud and Vujanovic, 2007). In the case of fungal diseases, AM reduces soil-borne fungal pathogens and has no effect or susceptibility to foliar fungal pathogens. Root fungal pathogens can be broadly classified by the mode of feeding as either root necrotrophs, which cause extensive tissue destruction, or wilt fungi, which invade the vessels and cause plants to wilt (Parbery, 1996). In the case of fungal foliar diseases, such as powdery mildew and rust fungi, they are associated with biotrophic fungi (*Blumeria*, *Oidium*, *Uromyces*). However, biotrophy and necrotropphy are the extremes of a continuum, and close examination of symptoms allows their reclassification as predominantly biotrophic hemibiotrophs, and predominantly necrotrophic hemibiotrophs (Parbery, 1996).

As with fungal diseases, the outcome for nematodes depends upon their mode of feeding. The data obtained show that AMF generally harms sedentary endoparasitic nematodes and improves the growth of migratory nematodes (Borowicz, 2001), but these results may vary and the different responses may be due to differences in AMF behavior and mechanisms involved in protection. As previously mentioned, reports on mycorrhizal effects on herbivorous insect diseases are scarcer and less conclusive, and the bioprotective effect of AM generally depends on the insect's lifestyle (Gange, 2006).

It is interesting to note here that plant defenses against pathogens and herbivorous insects form a comprehensive network of interacting signal transduction pathways, and that these different pathways are generally related to the attacker's lifestyle. A salicylic acid (SA) pathway provides broad-spectrum resistance to biotrophic pathogens that feed on a living host cell. In a distinct signaling process, jasmonic acid (JA) protects the plant from insect infestation and necrotrophic pathogens that kill the host cell before feeding. As we shall see later, this last model mediated by JA could explain the spectrum of effectiveness described for AM bioprotection, mainly in the case of induced resistance.

12.3.2 Plant Genotype

In a number of studies it has been demonstrated that, depending on the host genotype, the degree of AM root colonization and the effect of mycorrhization on plant growth can vary. It has been suggested that AM development and its effect on the host plant are at least partially under the genetic control of the host (Lackie et al., 1987; Vierheilig and Ocampo, 1991; Hetrick et al., 1993). The host genome also seems to affect the degree of protection provided by the AMF, as variation in the host genotype seems to result in a differing bioprotective response by the mycorrhizal association. Different levels of control of the same pathogen can be found with the same AM fungus on differing cultivars of plants. For example, depending on the genotype, mycorrhizal strawberry shows a variation in susceptibility to *Phytophthora fragariae* (Mark and Cassells, 1996). This example illustrates the importance of the choice of test-plant material in any screening procedure of AMF for biocontrol activity.

12.3.3 AM Fungi Isolate Efficiency and the Extent of Mycorrhizal Infection

The ability to enhance resistance/tolerance differs among AMF isolates. However, only a few studies have been carried out in order to compare the efficiency of different isolates of AMF in biocontrol processes. For example, *Glomus intraradices* has been found not to protect clover against nematode infection (Habte et al., 1999) and was not effective in reducing the disease symptoms produced by *Phytophthora parasitica* infection in tomato (Pozo et al., 2002). In contrast, root colonization by *G. mosseae* resulted in a clear protective effect against nematode and *P. parasitica* infection (Habte et al., 1999; Pozo et al., 2002). The data on the bioprotective effects of mycorrhization mostly come from the genus *Glomus* (80% of the studies) and particularly from *G. mosseae* (Singh et al., 2000; Whipps, 2004). This reflects their use as model species in mycorrhizal studies rather than any particular enhanced biocontrol properties, and reveals the need for screening and selection of AM fungal isolates for biocontrol studies. As some AMF exhibit a clear bioprotective effect, whereas other AMF do not affect the plant–pathogen interaction in terms of bioprotection, it is clear that more comparative studies are needed with a greater variety of AMF.

After plant inoculation with an AMF, the first signs of root colonization are visible within a few days. Thereafter, AM root colonization can increase drastically until it reaches the final plateau (e.g., for tomato and soybean around 60%; Wyss et al., 1991; Vierheilig et al., 1994). During the different stages of AM fungal penetration and root colonization, a number of physiological and morphological changes occur in the host plant and most of these changes can influence the induced increase in resistance or tolerance. In general, a local bioprotective effect has been linked with a high degree of AM root colonization, whereas intermediate and low levels of AM root colonization showed no bioprotection effect. Apparently, a critical level of AM root colonization is needed to provide bioprotection for mycorrhizal plants, especially against fungal and Oomycetes root pathogens. In mycorrhizal tomato plants, a bioprotectional effect against *P. parasitica* (Cordier et al., 1998) and *Fusarium oxysporum* (Caron et al., 1986), and in wheat plants, against *Gaeumannomyces graminis* (Graham and Menge, 1982) could only be observed when the roots were heavily colonized by the AMF, but low mycorrhization levels resulted in no bioprotection. It has been demonstrated that not only the local, but also the systemic bioprotective effect of mycorrhization depends on the degree of AM root colonization (Khaosaad et al., 2007).

If early AM colonization is required for disease suppression and the suppression is not only local but also systemic, indirect effects on host physiology must be involved in the bioprotective effect. In this sense, the importance of most of these physiological changes is still unclear. However, the benefit of mycorrhization is among others, a bioprotective effect against soil-borne fungal pathogens (Singh et al., 2000; Azcón et al., 2002; Xavier and Boyetchko, 2004; St-Arnaud and Vujanovic, 2007).

12.4 Mechanisms of Biocontrol Activated in Arbuscular Mycorrhizae

Several modes of action of AMF as biocontrol agents have been postulated, and as with other microbial biocontrol agents the effective protection is probably a consequence of the action and interaction of several mechanisms, none of which are mutually exclusive. A number of mechanisms have been suggested to be involved in the bioprotective effect of mycorrhization against pathogens. However, hard experimental data are not available for all of them yet. Two major groups of mycorrhizal mode-of-action mechanisms that mediate bioprotection are considered here (1) direct modes of action, including direct competition and inhibition and (2) indirect modes of action, where the mechanism is mediated by alteration in the plant or by rhizosphere modifications (Figure 12.2). This classification excludes the effect of enhanced nutrition, which obviously increases the capacity of the AM plant to compensate for the damage caused by the pathogen's attack and so there is no specific mechanism linked to disease resistance.

12.4.1 Direct Modes of Action

12.4.1.1 Competition

Two levels of competition between AMF and root pathogens have been suggested: a competition for nutrients and for colonization sites. While the physical competition for space has been demonstrated in some cases, there is no direct evidence to support the competition for nutrients with root pathogens, although both competition for photosynthate or carbon in roots (Morandi et al., 2002) and competition for exudates external to the root (Norman and Hooker, 2000) have been postulated as potential mechanisms of disease biocontrol caused by arbuscular mycorrhizal.

Histochemical and structural studies on the interactions between *Phytophthora* and AMF reveal that the cells in the root already occupied by the AMF cannot be further colonized by the pathogens

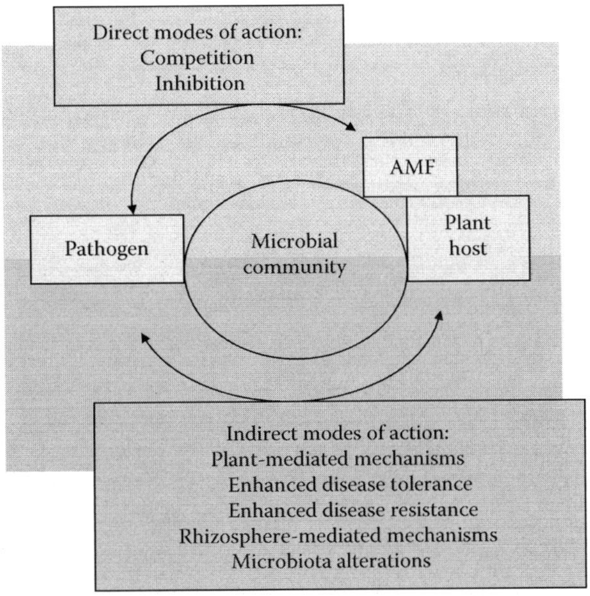

FIGURE 12.2 Modes of action associated with the biological control of plant disease by AM.

(Cordier et al., 1996, 1998), suggesting a physical exclusion exerted by the AMF. However, this is not a general rule and some fungal root pathogens and certain nematodes can occupy the same sites as AMF, within the root cortex (Dehne, 1982; Calvet et al., 1995).

Less conclusive and more contradictory is the suggestion that the carbon availability in mycorrhizal plants could explain the biocontrol effect of mycorrhization. Theoretically, once a plant is colonized by AMF, due to the carbohydrates used by the symbiotic AM fungus, less carbon could be available for a root colonizing fungal pathogen (Singh et al., 2000; Azcón et al., 2002; Xavier and Boyetchko, 2004). Different AMF have been reported to exhibit a different carbon sink strength in mycorrhizal roots (Lerat et al., 2003a,b), and thus, should exhibit a different biocontrol effect. However, no correlation was found between the carbon sink strength showed by different AMF and their biocontrol capacity. *G. mosseae* (BEG 12) has been shown in different plant systems not exhibiting any carbon sink strength and it has been implicated in different biocontrol processes. Moreover, despite its high carbon sink strength the AMF *G. intraradices* did not provide bioprotection against *P. parasitica* (Pozo et al., 2002). From these data it seems that carbon competition between AMF and a pathogen can be disregarded as a factor involved in mycorrhizal bioprotection.

12.4.1.2 Inhibition

It has been shown that AMF have a direct inhibiting effect on the growth of fungal and bacterial pathogens in soil (García-Garrido and Ocampo, 1989; St-Arnaud et al., 1995; Filion et al., 1999). Nevertheless, the possible mechanisms of direct action such as the production of specific antibiotics or inhibitory compounds by AMF have not been found, though the production by AM roots of exudates with the capacity to inhibit pathogens has been shown (Norman and Hooker, 2000; Filion et al., 2003).

12.4.2 Indirect Modes of Action

12.4.2.1 Rhizosphere-Mediated Mechanisms

There is strong evidence that mycorrhization induces qualitative and quantitative changes in the rhizosphere as a consequence of both the alterations of the root physiology and exudation pattern and the formation of the mycelium extraradical phase of the AMF. The alterations in the plant root could act on the pathogen directly by altering the composition of the root exudates, thus reducing the levels of stimulatory compounds and the production of inhibitory compounds (see above), or indirectly, through the alteration of the physio-chemical and biological conditions of the rhizosphere. Consequently, some microorganisms can be specifically stimulated or inhibited to grow in the rhizosphere of AM plants. Changes in rhizosphere soil pH is influenced by the mycorrhizal establishment and mycelium development (Bago et al., 1996; Villegas et al., 1996). However, no data are yet available on how these pH changes in the rhizosphere affect root pathogens. Nevertheless, there are a number of reports showing clearly that AM roots and AM fungal exudates affect bacteria (Sood, 2003), fungi (St-Arnaud et al., 1995; Filion et al., 1999; Norman and Hooker, 2000; Lioussanne et al., 2003; Scheffknecht et al., 2006, 2007), and nematodes (Ryan and Jones, 2004).

The studies concerning microbial changes in the AM rhizosphere frequently reported that a potential mode of action of arbuscular mycorrhizal biocontrol is the stimulation of the development of antagonistic soil microbiota. This selective alteration of the populations of antagonistic soil microorganisms is variable among different AM isolates and is sometimes dependent on the nature of the growth substrate (Caron et al., 1985). Selective increases of antagonists to pathogens are only possible if the antagonists are present in the soil and consequently, the AM fungus increases their density. In this context, it is interesting to note the hypothesis that the disease suppression could specifically depend on the ability of the AM fungal strain or species to increase, decrease, or have no effect on the populations of antagonists and deleterious microbes preexisting in the soil.

The evidence that AMF modify microbial population in the rhizosphere further suggests that AMF are highly competitive in this environment. Some results confirm this suggestion since AMF are relatively tolerant to antagonists that inhibit fungal pathogens and some of these antagonists even stimulate mycorrhizal infection (Barea et al., 1998; Budi et al., 1999).

12.4.2.2 Plant-Mediated Mechanisms

12.4.2.2.1 Improve Nutrition and Morphological Alterations of the Root

It has been suggested that the improved nutrient status of mycorrhizal plants and the increase in biomass make them more tolerant to damages caused by pathogens and carbon drain from the plant to the pathogen. However, there is strong evidence that the nutritional effect of the AM symbiosis is only one of several aspects of the mycorrhizal effect on pathogens (Trotta et al., 1996) and does not explain the biocontrol effect. Moreover, it has been suggested that the increase in nutrient uptake together with the hormonal and nutritional changes in AM roots could compensate for reduced disease symptoms.

Due to mycorrhization the morphology of the root changes (Berta et al., 1995; Copetta et al., 2006), and consequently, the root-disease development pattern could be modified. However, the real significance in terms of bioprotection of these changes in morphology and architecture in the root system has not been investigated, and no clear correlation to a bioprotectional effect of mycorrhization has been found yet. On the other hand, some evidence obtained from the studies using *Phytophthora* reveals that root architecture changes in mycorrhizal plants were not directly implicated in the bioprotection effect (Norman et al., 1996; Vigo et al., 2000). More research linking AM alterations in root parameters such as root fresh weight, total root length, number of root tips, and the degree of branching with data on bioprotective effects is necessary to demonstrate whether morphological alterations of the root due to mycorrhization are really involved in the bioprotectional effect of mycorrhization.

12.4.2.2.2 Induced Resistance

Plants possess a variety of constitutive and inducible defense mechanisms conferring protection against a broad spectrum of pathogenic microorganisms. Some of these mechanisms include the production of toxic metabolites, the structural reinforcement of cell structures, and the induction of specific protein products (known as pathogenesis-related proteins or PR) that specifically inhibit the pathogens. Moreover, the action of these mechanisms can be local (in the pathogen infection area) or systemic, and their correct functionality depends on a comprehensive network of interacting signal transduction pathways. These different pathways are generally related to the attacker's lifestyle and are orchestrated by signals that regulate differential sets of defense-related genes (De Vos et al., 2005). In this respect, SA coordinates defense mechanisms that are generally effective against biotrophic pathogens, whereas jasmonates (JA) regulate wounding responses and resistance against necrotrophs (Glazebrook, 2005).

Interestingly, both SA and JA are also the signals responsible for the plant's inducible resistance and defense mechanisms. SA is involved in signaling processes providing systemic acquired resistance (SAR), protecting the plant from further infection after an initial pathogen attack (Ryals et al., 1996). SAR is long lasting and provides broad spectrum resistance mainly to biotrophic pathogens. JA is implicated in the signaling process of the so-called induced systemic resistance (ISR), activated in the plant after colonization by nonpathogenic rhizobacteria (root-surface-attached bacteria, plant growth-promoting rhizobacteria or PGPR) thereby providing local and systemic protection against necrotrophic pathogens (Van Loon et al., 1998).

As with SAR or ISR, a phenomenon called mycorrhiza-induced resistance (MIR) has been proposed. This mechanism of bioprotection involves physiological and biochemical alterations of the host defense responses after mycorrhization. In this respect, during the mycorrhizal symbiosis, defense responses appear to be activated and strictly modulated (García-Garrido and Ocampo, 2002). In the early stages of root colonization, only a weak and transient activation of defense responses in the host plant takes place, and later, in a full established mycorrhizal symbiosis, defense responses are restricted to the arbuscule-containing cells. Some of these responses are possibly linked with a protective effect of the mycorrhizal plant against pathogens, e.g., the induction of hydrolytic enzymes (Pozo et al., 1999), the accumulation of phytoalexins (Morandi, 1989; Harrison and Dixon, 1993; Larose et al., 2002) and callose (Cordier et al., 1998), the accumulation of SA (Blilou et al., 1999, 2000) and JA (Hause et al., 2007) and reactive oxygen species (Salzer et al., 1999). During AM development, there is evidence that these defensive responses occur (García-Garrido and Ocampo, 2002) and they are strongly stimulated when a subsequent attack by a pathogen takes place. It is evident that the mechanisms of plant defense are activated more quickly and to a greater degree in AM plants than in non-AM plants, and it has been suggested that AM colonization acts as a priming system for the local and systemic process of pathogen resistance (Pozo and Azcón-Aguilar, 2007).

Some results showing stronger defensive reactions observed in AM plants upon pathogen attacks support this hypothesis. For example, AM protection against *P. parasitica* infection was systemic and mediated by the enhanced capacity in AM plants to form papilla-like structures around the sites of pathogen infection and to accumulate PR-1a and basic ß-1,3 glucanases upon *Phytophthora* infection (Pozo et al., 1999, 2002). Similarly, mycorrhizal Ri T-DNA transformed carrot roots showed a stronger defense reaction against *Fusarium* than nonmycorrhizal roots (Benhamou et al., 1994). Other examples such as the accumulation of phytoalexins in AM *Rhizoctonia*-infected potato plantlets (Yao et al., 2003), or the expression of a chitinase gene in response to the nematode *Meloidogyne incognita* (Li et al., 2006) has also been associated with the primed effect in AM plants. Some recently obtained results illustrate that primed responses in AM are not only localized in the colonized areas, but also they are systemic. In this respect, small but significant systemic increase in defense response transcripts in the shoots of tomato and *M. truncatula* plants is correlated with primed defenses in shoots upon challenge by pathogenic bacteria (Liu et al., 2007; Pozo et al., 2007).

On the other hand, in the most recent years it has been reported that once plants are colonized by AMF, further root colonization by AMF is regulated (Vierheilig, 2004). By analogy with the rhizobial autoregulatory mechanism in legume plants, this AMF phenomenon has been named "autoregulation of mycorrhization." Recently, it has been suggested that the bioprotective effect of mycorrhization and the autoregulation of mycorrhization are possibly two sides of the same coin (Vierheilig et al., 2008). In this sense, the systemic effect of both mycorrhizal biocontrol and mycorrhizal autoregulation, and their regulation by the degree of AM root colonization and their alterations by the roots exudate a point toward "one mechanism, two effects." It seems plausible that a mycorrhizal plant develops only one mechanism to repulse further colonization by fungi and does not discriminate between AMF and soil-borne pathogenic fungi (Vierheilig and Piché 2002; Vierheilig, 2004).

The phenomena of SAR and ISR are mediated by the SA and JA pathway, respectively, which regulate differential sets of defense-related genes (De Vos et al., 2005). Similarly, it is tempting to speculate that

MIR is mediated by similar endogenous signals and some evidence supports this hypothesis. Both SA and JA have been shown as signal molecules associated with a correct mycorrhization, and the modulation of SA and JA levels in mycorrhizal roots has been observed (Hause et al., 2007). In mycorrhiza-defective *Myc*-mutants (unable to form AM), SA levels are enhanced in response to AMF, while the accumulation is low and transient in mycotrophic wild-type plants (Blilou et al., 1999). In concordance with the SA levels, only local, weak, and transient defense responses are activated during the early steps of compatible AM interactions (García-Garrido and Ocampo, 2002; Liu et al., 2003), while stronger defense responses occur in *Myc*-mutants (Gollotte et al., 1993), suggesting that the mycorrhization leads to a partial suppression of SA dependent responses in the plant. Equally, the importance of jasmonate in AM development has been proven (Hause et al., 2007). JA-responsive genes and genes involved in JA biosynthesis are expressed in arbuscule-containing cells, and mycorrhizal roots are associated with increased JA levels (Hause et al., 2002). This increase occurs after the onset of mycorrhization and is probably associated with fully established mycorrhiza (Hause et al., 2002). Several mechanisms of action for jasmonates, including induction of flavonoid biosynthesis, reorganization and alterations in the cytoskeleton, regulation of the sink status of roots, and an increase in the plant fitness could be implicated in the development of AM.

Therefore, it is logical to speculate on the possibility that functional mycorrhiza implies partial suppression of SA dependent responses in the plant and activation of those dependent on JA (Pozo and Azcón-Aguilar, 2007), and as with rhizobacteria-mediated ISR, MIR is based on the priming of JA-regulated responses and is mainly effective against necrotrophic pathogens.

12.5 Potentialities and Considerations in the Use of AM Fungi as Biocontrol Agents

Once laboratory research has demonstrated that AMF are capable of controlling diseases caused by specific soil-borne plant pathogens, the question arises of the feasibility of using them commercially as biocontrol agents, suitable for plant production systems in agricultural practices. However, due to the peculiarities of AMF biology and AMF culture management, some considerations and requirements must be borne in mind for a successful application.

12.5.1 Suitable Screening Procedure

Plant disease biocontrol by AMF involves interactions among the plant, the pathogen, the AMF, and the physical, chemical, and microbiological environment. Therefore, there is a great number of variables that may be involved in the process of biocontrol, and in each particular situation (pathogen–crop–environment) the ability to enhance resistance/tolerance can differ among AMF isolates. This then underlines the need and importance for screening and selection of AMF isolates for biocontrol studies, in each particular situation. The ideal situation would be to carry out the screening and selection of the more efficient AM genotype for a particular disease in a given agrotechnological situation or in the suppressive soils where the pathogen is present but the disease is not expressed. In this sense, the need for extensive screening programs to detect effective mycorrhizal isolates for a range of activities in a range of environments has been recognized (see Whipps, 2004). Nevertheless, the majority of the studies carried out in this respect have been centered on the screening of the available culture collections of AMF and no extensive screenings have been done.

12.5.2 Effective Bioassay and Large-Scale Validation

Suitable screening procedures require the use of appropriate bioassays that reflect the ecological situation where the biocontrol is needed, and allow the correct and reproducible assessment of parameters of disease control. In the case of AMF, most of the studies of bioprotection have been carried out under environmentally controlled conditions, using sterilized media. In this situation, it is possible to assure a correct mycorrhization and effective biocontrol, but in some cases no correlation was found between the results obtained in field and in laboratory conditions (Bødker et al., 2002), which highlights the importance of field experiments to validate the AMF capacity as a biocontrol agent.

On the other hand, successful exploitation, manipulation, and management of AMF to protect plants depend on the increased understanding of the mechanisms and mode of action governing each particular plant/pathogen/environment/AM isolate interaction. Therefore, it is important to know whether the effects are due to direct interactions with the pathogen or are mediated by the plant. In the latter case, the importance of the plant genotype is crucial.

Other criteria of selection and screening of AMF are also important to evaluate their efficiency and must be considered where possible. These are: the capacity of saprophytic competition, the interspecific compatibility, the genetic stability, the industrial viability, and the survival and tolerance to biotic and abiotic factors.

The sequence of inoculation has also been suggested as an important factor in mycorrhizal bioprotection. In general, it has been postulated that the inoculation with AMF has to be prior to inoculation with the soil-borne pathogen (Singh et al., 2000; Azcón et al., 2002; Xavier and Boyetchko, 2004). However, this aspect is probably closely linked with the degree of mycorrhization. The earlier the AMF colonizes the root, the higher the mycorrhization level will be before a pathogen infection.

12.5.3 Inoculum Production and Formulation

Due to the difficulty of cultivating AMF in vitro, the use of AMF inoculum is not very widespread, and although many companies throughout the world produce this inoculum, none of them sells it on a large scale and only specific applications of AMF such as a plant growth promoter for nursery plants in horticulture, in potting cultures or in plant-propagation processes have been successful.

The advances in the technologies of AMF culture and inoculum production have been parallel to the increase in the scientific knowledge on the biology of AMF and the AM symbiosis. In this way, we have progressed from technologies based on the use of sterile or inert substrates as a base for the product, to the utilization of technologies for in vitro culture and bioreactor systems which produce large quantities of inoculum (Habte and Osorio, 2004; Declerck et al., 2005). Therefore, the above-mentioned difficulties in inoculum production and formulation, together with the specific legal requirements for the registration of mycorrhizal fungi as biocontrol agents, such as information on efficiency, purity, concentration, and the environmental impact and risk of the inoculants mean that, at present, there is no commercial AMF inoculum with a specific application in plant disease biocontrol (Whipps, 2004).

12.5.4 Combined Strategies of Biocontrol

The positive contributions that the plants gain from their association with AMF mean that the AM symbiosis constitutes one of the major potential tools for sustainable agriculture. The use of AM is compatible with the environment and avoids the negative impact of the chemicals currently used in intensive agriculture. AMF have great potential for use in integrated disease-control strategies that combine mycorrhizal inoculation with other sustainable practices.

One approach that improves the biocontrol activity of AMF is to increase the natural mycorrhizal populations through processes that enhance the establishment of the natural mycorrhization and consequently, the possibility of disease control. In this sense, the use of plant cultivars that are more responsive to AMF, the application of organic amendments, the reduction of inorganic fertilizers, and the application of natural plant-derivate AM stimulants have been shown to be the factors that positively influence the AM populations in soil (Aikama et al., 2000; Ryan and Graham, 2002).

Other proposed alternatives combine the application of a mycorrhizal inoculum with either biological or physiochemical treatments. The biological treatments that have been tested include the coinoculation with formulated microbial antagonists (Waschkies et al., 1994; Datnoff et al., 1995) or with plant-resistance inducers (Tosi and Zazzerini, 2000). The physiochemical practices such as soil solarization (Afek et al., 1991) and coinoculation of AM with low doses of specific fungicides have also been done successfully (Hwang et al., 1993).

12.6 Conclusions

There is ample scientific evidence that AMF play a part in the protection of host plants against pathogens. The association generally leads to a reduction in the damage caused by soil-borne pathogens, but the effect on shoot-targeting pathogens depends greatly on the attacker's lifestyle. Direct and indirect modes of action have been suggested to be involved in the bioprotective effect of mycorrhization against pathogens, and recent advances regarding signalling processes in mutualistic and pathogenic associations have allowed us to define a specific mechanism of induction of resistance by arbuscular mycorrhizae (MIR). Future advances should allow the identification at the molecular and genetic levels of key regulator stages associated with the bioprotective effect and facilitate practical exploitation in plant protection.

On the other hand, the practical application and commercialization of AMF as biocontrol agents seem to be more distant. Large-scale screening procedures are required to find more efficient isolates, and technical difficulties in inoculum production and formulation still exist. Additionally, the specific legal requirements for the registration of mycorrhizal fungi as biocontrol agents mean that its use is not yet a viable option for industrial producers of inoculants.

ACKNOWLEDGMENT

I would like to thank the Spanish Ministry of Education and Science for financial support (AGL2005–0639).

REFERENCES

Afek, U., Menge, J. A., and E. L. V. Johnson. 1991. Interaction among mycorrhizae, soil solarization, metalaxyl, and plants in the field. *Plant Disease* 75:665–671.

Aikama, J., Ishii, T., Kuramoto, M., and K. Kadoya. 2000. Growth stimulants for vesicular-arbuscular mycorrhizal fungi in satsuma mandarin pomace. *Journal of Japanese Society of Horticultural Sciences* 69:385–389.

Azcón, C., Jaizme-Vega, M. C., and C. Calvet. 2002. The contribution of arbuscular mycorrhizal fungi to the control of soil-borne plant pathogens. In: Gianinazzi, S., Schüepp, H., Barea, J. M., and Haselwandter, K. (Eds.), *Mycorrhizal Technology in Agriculture*. Birkhäuser Verlag, Switzerland, pp. 187–197.

Bago, B., Vierheilig, H., Piché, Y., and C. Azcón. 1996. Nitrate depletion and pH changes induced by the extraradical mycelium of the arbuscular mycorrhizal fungus *Glomus intraradices* grown in monoxenic culture. *New Phytologist* 133:273–280.

Baker, K. F. 1987. Evolving concepts of biological control of plant pathogens. *Annual Review of Phytopathology* 25:67–85.

Barea, J. M., Andrade, G., Bianciotto, V., Dowling, D., Lohrke, S., Bonfante, P., O'Gara, F., and C. Azcon-Aguilar. 1998. Impact on arbuscular mycorrhiza formation of *Pseudomonas* strains used as inoculants for biocontrol of soil-borne fungal plant pathogens. *Applied and Environmental Microbiology* 64:2304–2307.

Benhamou, N., Fortin, J. A., Hamel, C., Starnaud, M., and A. Shatilla. 1994. Resistance responses of mycorrhizal Ri T-DNA-transformed carrot roots to infection by *Fusarium oxysporum* f sp *chrysanthemi*. *Phytopathology* 84:958–968.

Berta, G., Trotta, A., Fusconi, A., Hooker, J. E., Munro, M., Atkinson, D., Giovannetti, M., Morini, S., Fortuna, P., Tisserant, B., Gianinazzi-Pearson, V., and S. Gianinazzi. 1995. Arbuscular mycorrhizal induced changes to plant growth and root system morphology in *Prunus cerasifera*. *Tree Physiology* 15:281–293.

Blilou, I., Ocampo, J. A., and J. M. García-Garrido. 1999. Resistance of pea roots to endomycorrhizal fungus or *Rhizobium* correlates with enhanced levels of endogenous salicylic acid. *Journal of Experimental Botany* 50:1663–1668.

Blilou, I., Ocampo, J. A., and J. M. García-Garrido. 2000. Induction of *Ltp* (lipid transfer protein) and *Pal* (phenylalanine ammonia-lyase) gene expression in rice roots colonized by the arbuscular mycorrhizal fungus *Glomus mosseae*. *Journal of Experimental Botany* 51:1969–1977.

Bødker, L., Kjøller, R., Kristensen, K., and S. Rosendahl. 2002. Interactions between indigenous arbuscular mycorrhizal fungi and *Aphanomyces euteiches* in field-grown pea. *Mycorrhiza* 12:7–12.

Bonfante-Fasolo, P. and S. Perotto. 1992. Plant and endomycorrhizal fungi: The cellular and molecular basis of their interaction. In: Verma, D. (Ed.), *Molecular Signals in Plant-Microbe Communications*. CRC Press, Boca Raton, FL, pp. 445–470.

Borowicz, V. A. 2001. Do arbuscular mycorrhizal fungi alter plant-pathogen relations? *Ecology* 82:3057–3068.

Budi, S. W., van Tuinen, D., Martinotti, G., and S. Gianinazzi. 1999. Isolation from the *Sorghum bicolor* mycorrhizosphere of a bacterium compatible with arbuscular mycorrhiza development and antagonistic towards soilborne fungal Pathogens. *Applied and Environmental Microbiology* 65:5148–5150.

Calvet, C., Pinochet, J., Camprubí, A., and C. Fernández. 1995. Increased tolerance to the root lesion nematode *Pratylenchus vulnus* in mycorrhizal micropropagated BA-29 quince rootstock. *Mycorrhiza* 5:253–258.

Caron, M., Fortin, J. A., and C. Richard. 1985. Influence of substrate on the interaction of *Glomus intraradices* and *Fusarium oxysporum* f.sp.*radicis-lycopersici* on tomatoes. *Plant and Soil* 87:233–239.

Caron, M., Fortin, J. A., and C. Richard. 1986. Effect of inoculation sequence on the interaction between *Glomus intraradices* and *Fusarium oxysporum* f. sp. *radicis-lycopersici* in tomatoes. *Canadian Journal of Plant Pathology* 8:12–16.

Copetta, A., Lingua, G., and G. Berta. 2006. Effects of three AM fungi on growth, distribution of glandular hair, and of essential oil production in *Ocimum basilicum* L. var. Genovese. *Mycorrhiza* 16:485–494.

Cordier, C., Gianinazzi, S., and V. Gianinazzi-Pearson. 1996. Colonization pattern of root tissue by *Phytophthora nicotianae* var. *parasitica* related to reduced disease in mycorrhizal tomato. *Plant and Soil* 185:223–232.

Cordier, C., Pozo, M. J., Barea, J. M., Gianinazzi, S., and V. Gianinazzi-Pearson. 1998. Cell defense responses associated with localized and systemic resistance to *Phytophthora parasitica* induced in tomato by an arbuscular mycorrhizal fungus. *Molecular Plant-Microbe Interactions* 11:1017–1028.

Datnoff, L. E., Nemec, S., and K. Pernezny. 1995. Biological control of *Fusarium* crown and root rot of tomato in Florida using *Trichoderma harzianum* and *Glomus intraradices. Biological Control* 5:427–431.

Declerck, S., Strullu, G., and J. A. Fortin. 2005. *In Vitro Culture of Mycorrhizas*. Springer–Verlag, Berlin, Heidelberg.

Dehne, H. W. 1982. Interactions between vesicular-arbuscular mycorrhizal fungi and plant pathogens. *Phytopathology* 72:1115–1119.

De Vos, M., Van Oosten, V. R., Van Poecke, R. M. P., Van Pelt, J. A., Pozo, M. J., Mueller, M. J., Buchala, A. J., Metraux, J. P., Van Loon, L. C., Dicke, M., et al. 2005. Signal signature and transcriptome changes of *Arabidopsis* during pathogen and insect attack. *Molecular Plant-Microbe Interactions* 18:923–937.

Filion, M., St-Arnaud, M., and J. A. Fortin. 1999. Direct interaction between the arbuscular mycorrhizal fungus *Glomus intraradices* and different rhizosphere microorganisms. *New Phytologist* 141:525–533.

Filion, M., St-Arnaud, M., and S. H. Jbaji-Hare. 2003. Quantification of *Fusarium solani* f. sp. *phaseoli* in mycorrhizal bean plants and surrounding mycorrhizosphere soil using real-time polymerase chain reaction and direct isolations on selective media. *Phytopathology* 93:229–235.

Fritz, M., Jakobsen, I., Lyngkjaer, M. F., Thordal-Christensen, H., and J. Pons-Kuehnemann. 2006. Arbuscular mycorrhiza reduces susceptibility of tomato to *Alternaria solani. Mycorrhiza* 16:413–419.

Gange, A. C. 2006. Insect-mycorrhizal interactions: Patterns, processes, and consequences. In: Ohgushi, T., Craig, T. P., and Price, P. W. (Eds.), *Indirect Interaction Webs: Nontrophic linkages through Induced Plant traits*. Cambridge University Press, Cambridge, pp. 124–144.

García-Garrido, J. M. and J. A. Ocampo. 1988. Interaction between *Glomus mosseae* and *Erwinia carotovora* and its effects on the growth of tomato plants. *New Phytologist* 110:551–555.

García-Garrido, J. M. and J. A. Ocampo. 1989. Effect of VA mycorrhizal infection of tomato on damage caused by *Pseudomonas syringae. Soil Biology and Biochemistry* 21:65–167.

García-Garrido, J. M. and J. A. Ocampo. 2002. Regulation of the plant defence response in arbuscular mycorrhizal symbiosis. *Journal of Experimental Botany* 53:1377–1386.

Gernns, H., von Alten, H., and H. M. Poehling. 2001. Arbuscular mycorrhiza increased the activity of a biotrophic leaf pathogen—is a compensation possible? *Mycorrhiza* 11:237–243.

Glazebrook, J. 2005. Contrasting mechanisms of defense against biotrophic and necrotrophic pathogens. *Annual Review of Phytopathology* 43:205–227.

Gollotte, A., Gianinazzi-Pearson, V., Giovannetti, M., Sbrana, C., Avio, L., and S. Gianinazzi. 1993. Cellular localization and cytochemical probing of resistance reactions to arbuscular mycorrhizal fungi in a "locus a" *myc*-mutant of *Pisum sativum* L. *Planta* 191:112–122.

Graham, J. H. and J. A. Menge. 1982. Influence of vesicular-arbuscular mycorrhizae and soil phosphorous on take-all disease of wheat. *Phytopathology* 72:95–98.

Guerrieri, A., Lingua, G., Digilio, M. C., Massa, N., and G. Berta. 2004. Do interactions between plant roots and the rhizosphere affect parasitoid behaviour? *Ecological Entomology* 29:753–756.

Habtc, M. and N. W. Osorio. 2001. Producing and applying arbuscular mycorrhizal inoculum. In: Elevitch, C.R. (Ed), *The Overstory Book, Cultivating Connections with Trees*, 2nd edn., Permanent Agriculture Resources (agroforestry.net), Holualoa, HI, pp. 68–73.

Habte, M., Zhang, Y. C., and D. P. Schmitt. 1999. Effectiveness of *Glomus* species in protecting white clover against nematode damage. *Canadian Journal of Botany* 77:135–139.

Harley, J. L. and S. E. Smith. 1983. *Mycorrhizal Symbiosis*. Academic Press, New York.

Harrison, M. and R. Dixon. 1993. Isoflavonoid accumulation and expression of defense gene transcripts during the establishment of vesicular arbuscular mycorrhizal associations in roots of *Medicago truncatula*. *Molecular Plant-Microbe Interactions* 6:643–659.

Hause, B., Maier, W., Miersch, O., Kramell, R., and D. Strack. 2002. Induction of jasmonate biosynthesis in arbuscular mycorrhizal barley roots. *Plant Physiology* 130:1213–1220.

Hause, B., Mrosk, C., Isayenkov, S., and D. Strack. 2007. Jasmonates in arbuscular mycorrhizal interactions. *Phytochemistry* 8:101–110.

Hetrick, B. A. D., Wilson, G. W. T., and T. S. Cox. 1993. Mycorrhizal dependence of modern wheat cultivars and ancestors: A synthesis. *Canadian Journal of Botany* 71:512–518.

Hussey, R. S. and R. W. Roncadori. 1978. Interaction of *Pratylenchus branchyrus* and *Gigaspora margarita* on cotton. *Journal of Nematology* 10:18–20.

Hussey, R. S. and R. W. Roncadori. 1982. Vesicular-arbuscular mycorrhizae may limit nematode activity and improve plant growth. *Plant Disease* 66:9–14.

Hwang, S. F., Chakravarty, P., and D. Prevost. 1993. Effects of rhizobia, metalaxyl, and VA mycorrhizal fungi on growth, nitrogen fixation, and development of Phytium root rot of sainfoin. *Plant Disease* 77:1093–1098.

Ingham, R. E. 1988. Interactions between nematodes and VA mycorrhizae. *Agricultural Ecosystems and Environment* 24:169–182.

Jaizme-Vega, M. C., Tenoury, P., Pinochet, J., and M. Jaumot. 1997. Interactions between the root-knot nematode *Meloidogyne incognita* and *Glomus mosseae* in banana. *Plant and Soil* 196:27–35.

Jeffries, P. and J. M. Barea. 2001. Arbuscular mycorrhiza a key component of sustainable plant-soil ecosystems. In: Hock. B. (Ed.), *The Mycota. Vol IX Fungal Association*. Spinger-Verlag, Berlin, Heidelberg, pp. 95–113.

Khaosaad, T., García Garrido, J. M., Steinkellner, S., and H. Vierheilig. 2007. Take-all disease is systemically reduced in roots of mycorrhizal barley plants. *Soil Biology and Biochemistry* 39:727–734.

Lackie, S. M., Garriock, M. L., Peterson, R. L., and S. R. Bowley. 1987. Influence of host plant on the morphology of the vesicular-arbuscular mycorrhizal fungus *Glomus versiforme* (Daniels and Trappe) Berch. *Symbiosis* 3:147–158.

Larose, G., Chenevert, R., Moutoglis, P., Gagne, S., Piché, Y., and H. Vierheilig. 2002. Flavonoid levels in roots of *Medicago sativa* are modulated by the developmental stage of the symbiosis and the root colonizing arbuscular mycorrhizal fungus. *Journal of Plant Physiology* 159:1329–1339.

Lerat, S., Lapointe, L., Gutjahr, S., Piché, Y., and H. Vierheilig. 2003a. Carbon partitioning in a split-root system of arbuscular mycorrhizal plants is fungal and plant species dependent. *New Phytologist* 157:589–595.

Lerat, S., Lapointe, L., Piché, Y., and H. Vierheilig. 2003b. Variable carbon sink strength of different *Glomus mosseae* strains colonizing barley roots. *Canadian Journal of Botany* 81:886–889.

Li, H. Y., Yang, G. D., Shu, H. R., Yang, Y. T., Ye, B. X., Nishid, I., and C. C. Zheng. 2006. Colonization by the arbuscular mycorrhizal fungus *Glomus versiforme* induces a defense response against the root-knot nematode *Meloidogyne incognita* in the grapevine (*Vitis amurensis* Rupr.), which includes transcriptional activation of the class III chitinase gene VCH3. *Plant Cell Physiology* 47:154–163.

Linderman, R. G. 1994. Role of VA fungi in biocontrol. In: Pfleger, F. L. and Linderman, R. G. (Eds.), *Mycorrhizae and Plant Health*. APS Press, St Paul, pp. 1–26.

Lingua, G., D'Agostino, G., Massa, N., Antosiano, M., and G. Berta. 2002. Mycorrhiza-induced differential response to a yellow disease in tomato. *Mycorrhiza* 12:191–198.

Lioussanne, L., Jolicoeur, M., and M. St. Arnaud. 2003. Effects of the alteration of tomato root exudation by *Glomus intraradices* colonization on *Phytophthora parasitica* var. Nicotianae zoospores. Abstract No. 253, Abstract Book ICOM 4; Montreal/Canada Page 291.

Liu, J., Blaylock, L. A., Endre, G., Cho, J., Town, C. D., Van den Bosch, K. A., and M. J. Harrison. 2003. Transcript profiling coupled with spatial expression analyses reveals genes involved in distinct developmental stages of an arbuscular mycorrhizal symbiosis. The *Plant Cell* 215:2106–2123.

Liu, J., Maldonado-Mendoza, I., López-Meyer, M., Cheung, F., Town, C. D., and M. J. Harrison. 2007. Arbuscular mycorrhizal symbiosis is accompanied by local and systemic alterations in gene expression and an increase in disease resistance in the shoots. *The Plant Journal* 50:529–544.

Mark, G. L. and A. C. Cassells. 1996. Genotype-dependence in the interaction between *Glomus fistulosum*, *Phytophthora fragariae* and the wild strawberry (*Fragaria vesca*). *Plant and Soil* 185:233–239.

Morandi, D. 1989. Effect of xenobiotics on endomycorrhizal infection and isoflavonoid accumulation in soybean roots. *Plant Physiology and Biochemistry* 27:697–770.

Morandi, D., Gollotte, A., and P. Camporota. 2002. Influence of an arbuscular mucorrhizal fungus on the interactions of a bionucleate *Rhizoctonia* species with Myc+ and Myc- pea roots. *Mycorrhiza* 12:97–102.

Norman, J. R. and J. E. Hooker. 2000. Sporulation of *Phytophthora fragariae* shows greater stimulation by exudates of non-mycorrhizal than by mycorrhizal strawberry roots. *Mycological Research* 104:1069–1073.

Norman, J. R., Atkinson, D., and. J. E. Hooker. 1996. Arbuscular mycorrhizal fungal-induced alteration to root architecture in strawberry and induced resistance to the root pathogen *Phytophthora fragariae. Plant and Soil* 185:191–198.

Parbery, D. G. 1996. Tropism and the ecology of fungi associated with plants. *Biological Reviews* 71:473–527.

Parniske, M. 2004. Molecular genetics of the arbuscular mycorrhizal symbiosis. *Current Opinion in Plant Biology* 7:414–421.

Pozo, M. J. and C. Azcón-Aguilar. 2007. Unraveling mycorrhiza-induced resistance. *Current Opinion in Plant Biology* 4:393–398.

Pozo, M. J., Azcón-Aguilar, C., Dumas-Gaudot, E., and J. M. Barea. 1999. ß-1,3-glucanase activities in tomato roots inoculated with arbuscular mycorrhizal fungi and/or *Phytophthora parasitica* and their possible involvement in bioprotection. *Plant Science* 141:149–157.

Pozo, M. J., Cordier, C., Dumas-Gaudot, E., Gianinazzi, S., Barea, J. M., and C. Azcon-Aguilar. 2002. Localized versus systemic effect of arbuscular mycorrhizal fungi on defence responses to *Phytophthora* infection in tomato plants. *Journal of Experimental Botany* 53:525–534.

Pozo, M. J., Verhage, A., García, J., García-Andrade, J., and C. Azcón-Aguilar. 2007. Signalling pathways involved in mycorrhiza-induced systemic resistance. Abstract Book, XIII International congress on Molecular Plant-Microbe Interactions. Sorrento Italia, July.

Ryals, J., Neuenschwander, U., Willits, M. G., Molina, A., Steiner, H-Y., and M. D. Hunt. 1996. Systemic acquired resistance. *The Plant Cell* 8:1809–1819.

Ryan, M. H. and J. H. Graham. 2002. Is there a role for arbuscular mycorrhizal fungi in production agriculture? *Plant and Soil* 244:263–271.

Ryan, A. and P. Jones. 2004. The effect of mycorrhization of potato roots on the hatching chemicals active towards the potato cyst nematodes, *Globodera pallida* and *G. rostochiensis. Nematology* 6:335–342.

Salzer, P., Corbière, H., and T. Boller. 1999. Hydrogen peroxide accumulation in *Medicago truncatula* roots colonized by the arbuscular mycorrhiza-forming fungus *Glomus mosseae. Planta* 208:319–325.

Scheffknecht, S., St-Arnaud, M., Khaosaad, T., Steinkellner, S., and H. Vierheilig. 2007. An altered root exudation patler through my corrhization affecting microconidia germination of the highly specialized tomato pathogen *Fusarium Oxysporum* F. sp. *lycopersici* (Fol) is not tomato specific but also occurs in *Fol* nonhost plants. *Canadian Journal of Botany* 85:347–351.

Scheffknecht, S., Mammerler, R., Steinkellner, S., and H. Vierheilig. 2006. Root exudates of mycorrhizal tomato plants exhibit a different effect on microconidia germination of *Fusarium oxysporum* f. sp. *lycopersici* than root exudates from non-mycorrhizal tomato plants. *Mycorrhiza* 16:365–370.

Shaul, O., Galili, S., Volpin, H., Ginzberg, I., Elad, Y., Chet, I., and Y. Kapulnik. 1999. Mycorrhiza induced changes in disease severity and PR protein expression in tobacco leaves. *Molecular Plant-Microbe Interactions* 12:1000–1007.

Singh, R., Adholeya, A., and K. G. Mukerji. 2000. Mycorrhiza in control of soil-borne pathogens. In: Mukerji, K. G., Chamola. B. P., Singh, J. (Eds.), *Mycorrhizal Biology*. Kluwer Academic/Plenum Publishers, New York, USA, pp. 173–196.

Smith, S. E. and D. J. Read. 1997. *Mycorrhizal Symbiosis*. Academic Press, London.

Sood, S. G. 2003. Chemotactic response of plant-growth-promoting bacteria towards roots of vesicular-arbuscular mycorrhizal tomato plants. *FEMS Microbiology and Ecology* 45:219–227.

St-Arnaud, M., Hamel, C., Vimard, B., Caron, M., and J. A. Fortin. 1995. Altered growth of *Fusarium oxysporum* f. sp. *chrysanthemi* in an in vitro dual culture system with the vesicular arbuscular mycorrhizal fungus *Glomus intraradices* growing on *Daucus carota* transformed roots. *Mycorrhiza* 5:431–438.

St-Arnaud, M. and V. Vujanovic. 2007. Effect of the arbuscular mycorrhizal symbiosis on plant diseases and pests. In: Hamel, C. and Plenchette, C. (Eds.), *Mycorrhizae in Crop Production: Applying knowledge.* Haworth Press, Binghampton, NY pp. 67–122.

Tosi, L. and A. Zazzerini. 2000. Intaractions between *Plasmopara helianthi, Glomus mosseae* and two plant activators in sunflower plants. *European Journal of Plant Pathology* 106:735–744.

Trotta, A., Varese, G. C., Gnavi, E., Fusconi, A., Sampo, S., and G. Berta. 1996. Interactions between the soil-borne root pathogen *Phytophthora nicotianae* var. *parasitica* and the arbuscular mycorrhizal fungus *Glomus mosseae* in tomato plants. *Plant and Soil* 185:199–209.

Van Loon, L. C., Bakker, P. A. H. M., and C. M. J. Pieterse. 1998. Systemic resistance induced by rhizosphere bacteria. *Annual Review of Phytopathology* 26:453–483.

Vierheilig. H. 2004. Further root colonization by arbuscular mycorrhizal fungi in already mycorrhizal plants is suppressed after a critical level of root colonization. *Journal of Plant Physiology* 161:339–341.

Vierheilig, H., Alt, M., Mohr, U., Boller, T., and A. Wiemken. 1994. Ethylene biosynthesis and activities of chitinase and ß-1,3-glucanase in the roots of host and non-host plants of vesicular-arbuscular mycorrhizal fungi after inoculation with *Glomus mosseae. Journal of Plant Physiology* 143:337–343.

Vierheilig, H. and J. A. Ocampo. 1991. Receptivity of various wheat cultivars to infection by VA-mycorrhizal fungi as influenced by inoculum potential and the relation of VAM effectiveness to succinic dehydrogenase activity of the mycelium in the roots. *Plant and Soil* 133:291–296.

Vierheilig, H. and Y. Piché. 2002. Signalling in arbuscular mycorrhiza: Facts and hypotheses. In: Buslig, B. and Manthey, J. (Eds.), *Flavonoids in Cell Function*. Kluwer Academic/Plenum Publishers, New York, pp. 23–39.

Vierheilig, H., Steinkellner, S., Khaosaad, T., and J. M. García-Garrido. 2008. The biocontrol effect of mycorrhization on soil-borne fungal pathogens and the autoregulation of the AM symbiosis: One mechanism, two effects? In: Varma. V. (Ed.), *Mycorrhiza: Biology, Genetics, Novel Endophytes and Biotechnology*. Springer-Verlag, Heidelberg, Germany pp. 307–320.

Vigo, C., Norman, J. R., and J. E. Hooker. 2000. Biocontrol of the pathogen *Phytophthora parasitica* by arbuscular mycorrhizal fungi is a consequence of effects on infection loci. *Plant Pathology* 49:509–514.

Villegas, J., Williams, R. D., Archambault, J., and J. A. Fortin. 1996. Effects of N source on pH and nutrient exchange of extramatrical mycelium in a mycorrhizal Ri T-DNA transformed root system. *Mycorrhiza* 6:247–251.

Waschkies, C., Schropp, A., and H. Marschner. 1994. Relations between grapevine replant disease and root colonization of grapevine (*Vitis* sp.) by fluorescents pseudomonas and endomycorrhizal fungi. *Plant and Soil* 162:219–227.

Whipps, J. M. 2001. Microbial interactions and biocontrol in the rhizospherre. *Journal of Experimental Botany* 52:487–511.

Whipps, J. M. 2004. Prospects and limitations for mycorrhizas in biocontrol of root pathogens. *Canadian Journal of Botany* 82:1198–1227.

Wyss, P., Boller, T., and A. Wiemken. 1991. Phytoalexin response is elicited by a pathogen (*Rhizoctonia solani*) but not by a mycorrhizal fungus (*Glomus mosseae*) in bean roots. *Experientia* 47:395–399.

Xavier, L. J. C. and S. M. Boyetchko. 2004. Arbuscular mycorrhizal fungi in plant disease control. In: Arora, D. K. (Ed.), *Fungal Biotechnology in Agricultural, Food, and Environmental Applications*. Mycology series, Vol. 21. CRS press, Boca Raton, FL, pp. 183–194.

Yao, M. K., Desilets, H., Charles, M. T., Boulanger, R., and R. J. Tweddell. 2003. Effect of mycorrhization on the accumulation of rishitin and solavetivone in potato plantlets challenged with *Rhizoctonia solani. Mycorrhiza* 13:333–336.

13

Evaluation of Mycorrhizal Symbioses as Defense in Extreme Environments

John Dighton

CONTENTS

13.1 Introduction to Mycorrhizal Symbiosis

Evolutionally, terrestrial fungi appeared at about the same time as land plants emerged. In addition to their role as saprotrophs, some fungi became intimately associated with roots of plants, enhancing their abilities to sequester nutrie nt elements; the mycorrhizal symbiotic association. Fossil records show these associations as possibly primitive endomycorrhizae in the Rhynie cherts (410–360 mya) and as ectomycorrhizae of pines in the Princeton cherts (50 mya) (Pirozynski and Malloch, 1975; Halling, 2001; Helgason and Fitter, 2005). Mycorrhizal associations now occur in some 85% of all the plant species on this planet. Recent estimates suggest that some 3617 plant species of 263 families have a mycorrhizal association (Wang and Qiu, 2006). Thus, it is regarded that the mycorrhizal condition is the most prevalent symbiotic condition on earth. Excellent discussions of mycorrhizae and mycorrhizal ecology can be found in Smith and Read (1997), Van der Heijden and Sanders (2002), and Peterson et al. (2004) and will not be elaborated upon here.

There are varying degrees in specificity of plant/fungal association in mycorrhizae and dependency of the plant on mycorrhizal associations. A number of plant species and families will associate only with a limited number of fungal species, leading to specific mycorrhizal types, such as ericoid, arbutoid, orchid, and monotropoid mycorrhizal associations being highly specific to limited plant families. A large number of grasses, herbs, and trees form associations with a relatively restricted fungal flora (Glomeromycota) to form arbuscular mycorrhizal associations. In contrast, a more limited set of plant species (mainly trees) associate with a vast diversity of fungal species (Basidiomycota and Ascomycota) in the ectomycorrhizal state. Even within these broad categories of specificity there is often greater specificity within plant and fungal families. Some plants are heavily or entirely dependent upon mycorrhizal associations for their survival (obligate mycosymbionts), whereas others may associate only with mycorrhizal fungi when needed (facultative mycosymbionts). Certain fungal species may associate only with one plant species (e.g., the European larch will associate only with *Suillus grevillei*), whereas others may have broad host specificity. The factors determining these degrees of specificity are not clearly understood, but it is likely that a combination of genetics, evolution, and environmental factor is involved. It is in extreme environments, where it is expected that obligate mycorrhizal associations are most likely to exist, the fungal symbionts are of greatest benefit to their host plant.

13.2 Stress and Extreme Environment

In their discussion of microbial life in extreme soils, Torsvig and Øvreås (2008) identified water, salinity, temperature, pH, radiation, low nutrients, and pollution as physicochemical factors limiting to life. It is in these extreme environments where a sustained stress or multiple stressors exist which stress (S-) tolerant strategists (sensu Grime, 1977) that arise in the community. Upon removal of a sustained stress, these species will be replaced by either ruderals (R-) or combatitive (C-) strategists, depending on the trajectory of change in the environmental conditions. In the context of mycorrhizae it is likely that stress in arbuscular mycorrhizae will lead to a change in root colonization as a measure of plant dependence on the association. In the ectomycorrhizal context, there is likely to be a mixture of change in root colonization, combined with a change in fungal community composition. This results from the fact that there are a limited number of arbuscular mycorrhizal fungal species and that their functional diversity is likely smaller than that of the more specious ectomycorrhizal fungal flora. The functional shift attributed to the change in root colonization and/or fungal community composition will provide some degree of defense to the host plant against the stressor. Colpaert and Van Tichelen (1996) took the model of Andersen and Rygiewicz (1991) to suggest that the environmental stress will differentially impact the fungal and plant component of the symbiosis and, through a series of metabolic feedbacks, the symbiosis will establish a new steady state, which will be best adapted to the current conditions. Some examples of these adaptations will be explored below.

13.3 Lessons from the Global Distribution of Mycorrhizal Types

Naturally, the distribution of mycorrhizal types closely follows the global distribution of appropriate plant host species. However, Read (Smith and Read, 1997) suggested an underlying factor that linked both plant and mycorrhizal symbionts with soil factors. Using the analogy of an altitudinal cline up a mountain to the variability in soil characteristics from poles to the equator, Read suggested that the dominant mycorrhizal types (ericoid-, ecto-, and arbuscular-mycorrhizae) were closely related to the nutrient sources available in the soil and the enzymatic competence of the mycorrhizal community (Read, 1991a, 1993; Smith and Read, 1997). In the extreme cold and wet environments of high latitude or altitude, litter quality is reduced by the presence of complex chemistry and toxic secondary plant metabolites, and the decomposition is climatically limited. Organic matter accumulates to provide high soil nutrient capital, but low availability. The dominant plant species are ericaceous with ericoid mycorrhizae which possess protease enzymes that allow them to directly access organic forms of nitrogen (Went and Stark, 1968). More mesic environments support ectomycorrhizal coniferous, mixed coniferous, and deciduous forest which access organic forms of both nitrogen and phosphorus by the production of protease and acid phosphatase enzymes. In contrast, nutrient mineralization is more rapid in deciduous forests and grasslands of lower altitudes and in the tropics, where the dominant arbuscular mycorrhizal symbionts are adapted for efficient acquisition of mineral nutrients and have limited enzymatic capabilities. Thus, the link between host plant and mycorrhizal type appears to be a mixture of fungal/host compatibility and environmental limitations. As the stressors in the environment become more extreme, it is anticipated that highly specific mycorrhizal associations will evolve. This is, however, an area of mycorrhizal research that has received little attention, and the following will describe a number of specific interactions that have been reported in the literature and from which we will make some general models.

13.4 Cold Environments

The importance of, particularly, arbuscular mycorrhizae in extremely cold ecosystems was highlighted by Haselwandter and Read (1980), Väre et al. (1992), and Gardes and Dahlberg (1996), who reported that many plant species in these arctic and alpine ecosystems were either nonmycorrhizal or associated with arbuscular mycorrhizae. In particular, they highlighted the possible importance of dark-septate fungal forms in these ecosystems. Root colonization by fungi is, however, low for both arbuscular and dark-septate fungi. Despite the low host tree species diversity, Gardes and Dahlberg (1996) reported a relatively high ectomycorrhizal fungal diversity, possibly related to microsite heterogeneity. Indeed, Olssen et al. (2004) showed that high arctic sites had a high arbuscular mycorrhizal propagule density in which the fine endophytic forms dominated. The colonization of roots of arbuscular mycorrhizal host plants declined with higher latitude until in the most extreme sites almost all plant species were nonmycorrhizal. This indicates that there is a northernmost limitation to the survival of mycorrhizal fungi (76°N–78°N) and, possibly, an altitudinal limit, where arbuscular mycorrhizal colonization of *Polygonum*, *Cirsium*, and *Clematis* declined with increasing altitude on Mount Fuji (Wu et al., 2004). Arbuscular mycorrhizal fungi in arctic conditions are speculated to function as saprotrophs more than obligate mycorrhizae, and the higher incidence of occurrence of dark-septate endophytes, such as *Phialocephala fortinii* in arctic species, may result from their greater influence on enhancing plant nutrition (N) than arbuscular mycorrhizae (Ruotsalainen and Kytöviita, 2004).

Glacial forefronts provide us with a good example of a successional environment. The invasion of plants and their mycorrhizal associations in such ecosystems has been explored by Jumpponen and Egerton-Warburton (2005). In these extreme environments, successful colonization in the new substrate (glacial outwash material) is dependent upon the ability of both plants and fungal propagules to disperse into the area and for the two to meet and form successful associations. In the absence of abundant mycorrhizal propagules, colonization is low (Trappe and Strand, 1969; Mikola, 1970), but mycorrhizal colonization increases over time as ectomycorrhizal propagule density increases (Jumpponen et al., 2002) or as arbuscular mycorrhizal propagule density increases on sand dunes (Nicolson and Johnston, 1979) or mine tailings (Zak and Parkinson, 1983). Jumpponen and Egerton-Warburton (2005) place this in the context of the interaction between environmental and biotic filters, which regulate the degree of mycorrhizal association between host plants and fungi. The host filter is the nature of the colonizing plant community and its mycorrhizal dependency (Vandenkoornhuyse et al., 2003) and the environmental filters being the multidimensional niche characterized by physical, chemical, and biological variables affecting the target species (Morin, 1999). Changes in soil nutrients over time (Jumpponen et al., 1998; Ohtonen et al., 1999), temperature and moisture (Pringle and Bever, 2002), organic matter, pH, etc. (Johnson and Wedin, 1997; Erland and Taylor, 2002; Neville et al., 2002) all act to influence the mycorrhizal community established on specific hosts at specific times during the succession.

Physiological adaptations of mycorrhizal fungi in cold ecosystems include the production of mannitol and trehalose by ectomycorrhizal fungal hyphae which aid in cryoprotection, allowing their survival during the coldest conditions and maintaining their presence to help facilitate nutrient and water uptake in their hosts (Tibbett et al., 2002).

In mixed species communities in tundra ecosystems there appears to be a tight, synergistic interaction between plant species of contrasting mycorrhizal association. In a plant removal experiment, Urcelay et al. (2003) showed that removal of dominant ericoid mycorrhizal plant species (*Ledum palustre*) significantly decreased the ectomycorrhizal colonization of *Betula nana*, but did not decrease ectomycorrhizal richness or diversity. However, reciprocal removal of birch did not affect the mycorrhizal status of *Ledum*. There is a suggestion that the close interaction between ericoid and ectomycorrhizae in these extreme environments is a self-sustaining system to optimize nutrient uptake by the plant community, with the ericaceous community regulating nutrient availability to the trees.

13.5 Drought

The ability of mycorrhizal extraradical hyphae and, in particular, rhizomorphs of ectomycorrhizae to translocate nutrients from soil to plant also enables them to translocate water. This has also been found to be of great importance for plant growth in dry environments. Augé (2001) comprehensively reviewed the influence of arbuscular mycorrhizae on drought tolerance in plants. He provides tables showing species of some 34 plant genera having the mycorrhizal benefit mediated by improved phosphorus nutrition and species of 27 (overlapping) genera showing mycorrhizal-mediated drought tolerance without improved phosphorus nutrition. Obviously the effect of mycorrhizae on drought tolerance is both plant and fungal symbiont dependent and is also modified by host plant nutrition and other factors. For example, presence of arbuscular mycorrhizae on *Acacia* did not improve the plant growth under drought stress in both presence and absence of additional phosphorus fertilizer. However, in another tropical tree, *Leucaena*, the presence of mycorrhizae significantly improved plant growth under drought conditions at both levels of phosphorus supply (Michelsen and Rosendahl, 1990). The finding of mycorrhizal associations of *Acacia* roots at depths of 30 m in Senegal is likely to be more associated with water acquisition than that of nutrients (J. Wilson, personal communication).

The benefits of increased plant nutrient content and host plant performance of pea as a result of mycorrhizal colonization was also shown to be more beneficial under plant water stress than under adequate water supply (Kristek et al., 2007). Mena-Violante et al. (2006) showed that fruit production of pepper plants was similar in droughted mycorrhizal plants and nondroughted nonmycorrhizal plants. Cruz et al. (2004) showed that the leaf water potential in drought-stressed papaya plants was more severe in nonarbuscular mycorrhizal plants due to an increased production of ethylene and 1-animocyclopropane-1-carboxylic acid in mycorrhizal plants.

In desert ecosystems, the need for effective drought tolerance is greater than in seasonally dry environments. Collier et al. (2003) surveyed plants in the Chihuahuan desert and found that they could be divided into two groups (1) annuals with thin roots (118–375 μm diameter) and low degrees of root colonization by arbuscular mycorrhizae (6.25%) and (2) perennials with thicker roots (202–818 μm diameter) and high levels of root colonization (72%). It is suggested that during their short growth period, ephemerals may not have time to benefit from mycorrhizal associations, whereas the longer-lived perennials benefit from the water acquisition properties of their arbuscular mycorrhizal associations and, therefore, establish more obligate mycorrhizal associations.

Seasonal degree of root colonization by arbuscular mycorrhizae of *Zygophyllum dumosum* in the Negev desert was positively related to soil moisture showing that drought condition reduces mycorrhization (He et al., 2002). Similar correlations between soil moisture and mycorrhizal colonization were seen in grasses of the Namibian desert, where moisture had a strong effect of root colonization (multiple regression colonization = 15.2 (moisture) − 16.57P = 0.002 and = 15.73 (moisture) + 28.6 (stability) − 42.4P = 0.001) whereas sand stability had a greater influence on mycorrhizal spore abundance than did moisture (spore abundance = 408 (stability) − 213.8P = 0.001 and = 26.8 (moisture) + 413.6 (stability) − 287.8P = 0.001) (Jacobson, 1997). Under drought stress of a matric potential of −0.5 MPa, *Helianthemum almeriense* and associated mycorrhizal fungus, the desert truffle *Terfezia claveryi*, showed increased survival, 26% greater water potential, 92% higher transpiration rate, 45% higher stomatal conductance, and 88% higher photosynthesis in mycorrhizal plants compared to nonmycorrhizal plants (Morte et al., 2000).

Ectomycorrhizal trees also grow in seasonally droughty conditions. Shi et al. (2002) showed that the ectomycorrhizal community composition of drought-stressed beech trees were significantly different from unstressed trees. Drought caused the accumulation of sucrose, glucose and fructose, and mannitol and arabitol in mycorrhizal roots at the expense of trehalose. Accumulation of sugar alcohols, known to counteract drought stress, were mycorrhizal–alcohol species-specific (*Xerocomus chrysenteron* formed large amounts of arabitol, while *Lactarius subdulcis* accumulated mannitol). Swaty et al. (2004)

showed that when forests of pinyon pine had experienced a severe drought year, various patterns of ectomycorrhizal colonization of surviving trees occurred. Surviving trees in severe drought regions had 50% less mycorrhizal colonization and altered community composition, relative to low mortality sites. Where the drought stress was less, trees showed a twofold increase in mycorrhizal colonization compared to low mortality sites. This suggests a quantitative and qualitative response of the mycorrhizal community to drought with an increase in abundance of mycorrhizae to support recovery under moderate drought conditions, but a decline in extreme conditions.

Defensive strategies to drought appear to range from none, in ephemeral annuals, to an increase in dependency of arbuscular mycorrhizae in drought-stressed perennials and altered ectomycorrhizal communities in tree species. There is a favorable selection of fungal symbionts able to produce sugar alcohols, which protect the fungus/root symbioses. There is an obvious benefit of mycorrhizal associations in droughted systems, and their presence is key to the rehabilitation of desertified ecosystems (Jeffries et al., 2002).

13.6 Salinity

In some agricultural contexts, particularly in tropical areas, frequent crop irrigation leads to an accumulation of salt and subsequently the development of highly saline soils. In these conditions arbuscular mycorrhizae have been shown to have significant benefit for crop production enabling the plant growth more effectively under these stressed conditions. This allows crop production to continue in areas that would otherwise be abandoned.

Johnson-Green et al. (2001) showed that the arbuscular colonization of plants in an extremely saline part of an inland salt pan was significantly reduced compared to other saline areas, suggesting that the fungi had a lower salinity tolerance than halophytic host plants such as *Puccinellia nuttalliana*. Gupta and Krishnamurthy (1996) showed that the arbuscular mycorrhizal colonization of peanut roots was significantly reduced at 5% NaCl compared to 1%, but plant performance and root nodulation with nitrogen-fixing bacteria were enhanced at both the salinity levels in the presence of mycorrhizae.

In an attempt to mimic salt marsh conditions in a multifactorial greenhouse study, McHugh and Dighton (2004) showed that increased salinity (0 to 7 ppt NaCl) significantly reduced the arbuscular mycorrhizal colonization of *Spartina alterniflora* but not that of *S. cynosuroides* plants on a greenhouse study.

Using a split root technique, Füzy et al. (2007) examined the effects of soil salinity on mycorrhizal colonization of clover. The nonsalinized half root system of a plant had the same degree of arbuscular mycorrhizal colonization as control plants (65%), whereas the salinized half has a significantly reduced mycorrhizal colonization (45%). In contrast, if both halves of the root system were in salty soil, the degree of colonization was significantly increased (76%). These results suggest that in a heterogenous environment (differences between the two chamber halves), the plant can compensate for adverse condition by reducing investment in plant/mycorrhizal resources in the adverse location. When the whole plant is stressed, there is a greater investment in the mycorrhizal component in an attempt to overcome the stress.

13.7 Herbivory and Defoliation

A major factor in the maintenance of a mycorrhizal symbiosis is the supply of carbohydrates from the host plant to the fungus. Any disruption of this supply is likely to create stress for the fungal symbionts. Aboveground herbivory of plant parts by either vertebrates or invertebrates may reduce net carbon assimilation by plants and may affect mycorrhizal colonization of roots. In arbuscular mycorrhizal plant species, Allsop (1998) showed differences between grazing-adapted and nonadapted plant species in their mycorrhizal response to herbivory. Artificially an intense-imposed grazing pressure on *Themeda trianda* reduced the leaf growth and phosphorus accumulation and significantly reduced the arbuscular mycorrhizal colonization of roots and the extension of extraradical hyphae. In contrast, although mycorrhizal colonization rate declined, there was no reduction in the extraradical hyphal density of the grazing-adapted plant species of *Lolium perenne* and *Digitaria eriantha*. Similar responses were seen in a comparison between the grazing-tolerant grass *Zoysia japonica* and the intolerant grass *Miscanthus sinensis* (Saito et al., 2004), where change in the mycorrhizal community of *Miscanthus* under simulated grazing was attributed to a reduction in the supply of soluble, nonstructural carbohydrates to sustain mycorrhizae (Gange et al., 2002). Grazing-induced changes in the mycorrhizal species composition or efficiency of mycorrhizal symbionts may affect plant performance in terms of growth and/or nutrient content (Hokka et al., 2004), thus influencing competitive ability and fecundity (Bever et al., 2002; Van der Heijden, 2002; Bever and Schultz, 2005), thus influencing plant community composition. For example, Mikola et al. (2005) showed that in a three plant species mixture, defoliation of *Trifolium repens* increased the arbuscular mycorrhizal colonization, but there was no increase in *Phleum preatense*. Differential effects of mycorrhizal status of plants also influenced the proportion of trophic groups of soil nematodes. The increased nutrient content of mycorrhizal plants could be part of the explanation of the selective grazing of mycorrhizal plant species by grasshoppers in tallgrass prairies (Kula et al., 2005), where enhanced compensatory growth of strong mycorrhizal plants maintained their presence in the community despite strong selection of these species by the grazers. In contrast, Palermo et al. (2003) could find no influence of mycorrhizae on palatability of Douglas fir to western spruce budworm in comparison with nonmycorrhizal or fertilized tree seedlings. However, there were differences in the nutritional status (increased foliar P and Mg concentrations) in *Laccaria bicolor*-mycorrhizal herbivore resistant seedlings compared to nonresistant seedlings. Hence the role of mycorrhizae in aboveground herbivory is far from clear.

In ectomycorrhizal associations, herbivory also influences the mycorrhizal status of host plants. Markkola et al. (2004) showed that the root:shoot ratio and root ergosterol content (measure of mycorrhizal biomass) of severely (50%–100%) defoliated birch seedlings decreased significantly and led to impaired growth in the subsequent years. An artificially imposed 50% defoliation of lodgepole pines in Yellowstone National park was found to affect the ectomycorrhizal community of both the defoliated tree species and the nondefoliated cohabiting Engleman spruce (Cullings et al., 2001). The relative abundance of mycorrhiza on pine:spruce was reduced from 6:1 in control plots to 1:1 in defoliated plots. Dominant *Inocybe* species in undefoliated plots was replaced by Agaricoid and Suilloid species in defoliated plots, possibly indicating a response in fungal species carbohydrate requirements. Despite repeated gypsy moth infestations over 3 years and defoliation levels between 10% and 100%, Kosala et al. (2004) reported no change in either the arbuscular mycorrhizal colonization of ectomycorrhizal community of polar roots. It is suggested that these early successional tree species are more resilient to defoliation, irrespective of mycorrhizal status.

Kytöviita and Sarjala (1997) related the defensive role of mycorrhiza to tree defoliation to the production of polyamines in the roots. In both pine seedlings in association with the ectomycorrhizal fungus *Suillus variegatus* and birch seedlings with *Paxillus involutus*, the concentrations of putrescine significantly declined in nonmycorrhizal roots as a result of defoliation, but remained the same in mycorrhizal plants. They suggest that putrescine has a buffering effect on the root system by binding excess H^+ ions, stabilizing pH, and positively influencing long-distance nutrient transport within the plant, thus minimizing the effect of defoliation. Subsequent studies on birch mycorrhizae (Kytöviita, 2005) showed that there was reciprocal exchange of N and C through the mycorrhizal association through which the effect of defoliation manifest itself in a reduction of carbohydrate to support ectomycorrhizal fungal mycelium, but not root demand. This data supports that of Markkola et al. (2004) who showed that defoliation of birch influenced the ectomycorrhizal fungal biomass associated with roots more in the year of defoliation than defoliation a year earlier, whereas the effects on reducing plant growth were seen by defoliation in both the same year and the pervious year. Thus, there are largely negative effects of defoliation on the mycorrhizal status of plants. However, the shift in mycorrhizal community may adapt the plant for enhanced nutrient acquisition and alter the palatability of the aboveground parts to their grazers.

13.8 Acidifying Pollutants

In the late 1970s, soil ecologists became involved in the research on acid rain (sulfuric acid from dissolved SO_2 in rain) in relation to the "Waldsturben" effect of the forest dieback in Bavarian forests, where the observations of Ulrich (1979), Hütermann (1982, 1985), Blaschke (1988), and Jansen et al. (1988)

alerted researchers to the fact that acid rain was affecting root growth and the ectomycorrhizal status of trees. The soil-mediated effect of acid rain caused reduction in the Ca:Al ratio in soils, leading to aluminum toxicity. This resulted in the death of roots and their associated ectomycorrhizae. At the same time, acid etching of leaves reduced their photosynthesis and increased the leaching of essential nutrients, leading to a reduction in photoassimilates to roots in a similar way to herbivory. A dual impact model of acidifying pollutants on forest ectomycorrhizae was developed. In this model, reduction of mycorrhizal associations of the plant roots occurred: (1) via a reduction in photosynthesis in the tree canopy, reducing the energy supply to roots and (2) via acid induced increase in the availability of toxic metal ions (aluminum, manganese, and magnesium) in the soil, resulting in root damage and loss or changes in the community composition of ectomycorrhizal fungal symbionts (Dighton and Jansen, 1991). This model has a confirmation from a number of studies showing that ectomycorrhizal community composition changes with increasing acid rain and evidences that the photosynthesis is significantly reduced by acid rain. A summary of research on acid rain effects on mycorrhizae can be found in Jansen and Dighton (1990).

The overall effect of acid rain on ectomycorrhizae is to reduce fungal species diversity. This appears particularly to be a reduction in those fungal species which produce large amounts of extraradical hyphal structures and a more complex sheath (Dighton and Skeffington, 1987). In conjunction with observations of decline in specific species of basidiomycete fruitbodies (Arnolds, 1988), it is suggested that acid rain largely impacts K-selected fungal species. Thus, the overall effect of acid rain is negative on the mycorrhizal association, with potential loss of diversity of ecosystem function of the K-strategist mycorrhizae (Dighton and Mason, 1985; Last et al., 1987).

13.9 Fertigation: N Deposition

Using historical mushroom foray records, Arnolds (1991) detected changes in the ratio of ectomycorrhizal to saprotrophic basidiomycete fruitbody abundance in the Netherlands over recent years. He attributed this to enhanced nitrogen deposition that acted both as a fertilizer (reducing the effectiveness of mycorrhizae for nitrogen uptake into host plants) and as an acidifying pollutant.

European studies have shown that N deposition generally reduces the ectomycorrhizal fungal dependency of host plants (Arnolds, 1991; Brandrud, 1995; Kårén and Nylund, 1997; Brandrud and Timmermann, 1998; Jonsson et al., 2000; Taylor et al., 2000; Peter et al., 2001). As a result, fungal species are emerging as likely indicators of nitrogen deposition (Dighton et al., 2004). Ectomycorrhizal fungal species are more sensitive to N deposition than saprotrophic species (Arnolds, 1991; Peter et al., 2001); and ectomycorrhizal fungi forming resupinate fruitbodies become more abundant than large mushrooms under high N deposition (Peter et al., 2001). Genera such as *Russula* and *Cortinarius* have been identified as nitrophobic species (Arnolds, 1991; Brandrud, 1995; Taylor et al., 2000; Peter et al., 2001), along with the genera *Tricholoma*, *Lactarius*, and *Hebeloma*, as mushrooms (Lilleskov et al., 2001), and ectomycorrhizae of *Piloderma* spp., *Amphinema byssoides*, *Cortinarius* spp., and *Tomentella* spp. (Lilleskov et al., 2002). Mycorrhizal taxa considered nitrophilic include *Cantherellus*, *Lactarius theiogalus*, *Lactarius rufus*, *Laccaria*, *Paxillus involutus*, *Hygrophorus olivaceoalbus*, *and Tylospora fibrillosa* (Brandrud, 1995; Taylor et al., 2000; Lilleskov et al., 2001; Erland and Taylor, 2002) along with *Tomentella sublilacina* and *Thelephora terrestris*, as identified from root tips (Lilleskov et al., 2002). In contrast, Avis et al. (2003) reported increases in the fruiting of *Russula* species in the presence of 17 g N m^{-2} y^{-1} and decline in that of *Cortinarius* species in particular and a general 50% reduction in ectomycorrhizal fruitbody richness in an oak savanna. Fungi associated with hardwood species may be more tolerant than those of coniferous species (Arnolds, 1991; Wallander and Kottke, 1998; Taylor et al., 2000), because hardwood stands typically have higher N availability and association with nitrophillic symbionts.

With abundant inorganic nitrogen supply and a high carbon cost of N assimilation amino acids (Wallander, 1995), changes in the ectomycorrhizal fungal community may be toward species that either are more carbon use efficient or to species that are C parasites (Lilleskov and Bruns, 2001). Also, increased inorganic N availability causes greater fungal allocation of C to amino acid production instead of carbohydrate production leading to reduced root growth (Peter et al., 2001) and reduced growth of extraradical hyphae (Arnebrandt, 1994). The physical structure of ectomycorrhizae is also changed under high N loading (Arnebrandt and Soderstrom, 1992; Brunner and Scheidegger, 1994; Kårén and Nylund, 1997).

These shifts in mycorrhizal community composition may also be related to the response of mycorrhizae to altered stoichiometry of available nutrients; following Leibig's Law of the Minimum (Read, 1991b) as it has been shown that N saturated stagnopodsols, supporting Sitka spruce forest in Wales, showed enhanced potassium and phosphorus deficiency (Harrison et al., 1995). This change in relative availability of nutrients may be a driving force for ectomycorrhizal fungal community changes to those optimized for P and K acquisition.

13.10 Radionuclide Pollutants

Accumulation of radionuclides into ectomycorrhizal basidiomycete fruitbodies is often higher than that of surrounding soils and has been recently reviewed by Dighton et al. (2008a,b). These fungi often show 10 times the concentration of radiocesium than leaf litter (Malinowska, 2006), whereas saprotrophic fungal fruitbodies had half that level of accumulation (Witkamp, 1968; Witkamp and Barzansky, 1968). High levels of accumulation of radionuclides in mushrooms have been recorded many times, especially following the Chernobyl accident (Haselwandter, 1978; Ijpelaar, 1980; Eckl et al., 1986; Guillitte et al., 1987; Dighton and Horrill, 1988; Haselwandter et al., 1988; Oolberkkink and Kuyper, 1989; Muramatsu et al., 1991; Guillette et al., 1994; Haselwandter and Berreck, 1994; Mietelski, 1994; Yoshida and Muramatsu, 1994; Grodzinskaya et al., 1995; Malinowska, 2006). Many of these hyperaccumulator ectomycorrhizal species are members of the Cortinariacea.

Analysis of the isotope ratio of radiocesium from Chernobyl (^{137}Cs:^{134}Cs) in fruitbodies of ectomycorrhizal basidiomycete fungi indicates that a large proportion (25% to 92%) of ^{137}Cs was accumulated that originated from sources occurring prior to the accident at Chernobyl (Byrne, 1988; Dighton and Horril, 1988; Giovani et al., 1990). This suggests long-term accumulation of radionuclides by these fungi. Research suggests that this is achieved by internal translocation within the mycelium and directional transport to fruiting bodies, along with other nutrients (Gray et al., 1995, 1996; Connolly et al., 1998; Baeza et al., 2002).

Due to the fact that mycorrhizal fungi, in association with their host plants, have the capacity to absorb and retain high levels of heavy metals and radionuclides, there are suggestions that they may be beneficial in the restoration and remediation of polluted environments. In particular, the high accumulation into basidiomycete fruiting structures in ectomycorrhizal fungi could be used as a harvestable product that could be exported from the site, resulting in a net export of the pollutant chemical.

13.11 Fungal Regulation of Radionuclide Uptake by Plants

In both ericoid and arbuscular mycorrhizal host plants, there appears to be a mycorrhizal effect of enhancing radionuclide uptake by the plant, but a significant change in the internal allocation of those radionuclides. Radionuclides appear to be prevented from being translocated into the shoots of plants by these mycorrhizae, a possible defense mechanism.

However, evidence for enhanced uptake of radioniclides by arbuscular mycorrhizae is somewhat conflicting (Haselwandter and Berreck, 1994). Sweet clover and Sudan grass showed a slight, but statistically insignificant, increase in the uptake of ^{137}Cs and ^{60}Co by mycorrhizal plants, but mycorrhizal *Agrostis tenuis* by *Glomus mosseae* significantly decreased the Cs content of shoots (Berreck and Haselwandter, 2001). The protective effect of mycorrhizae was also shown in transformed carrot root in association with *Glomus* mycorrhizae, where uptake of radiocesium was strongly correlated to fungal hyphal length. These mycorrhizal fungal hyphae have a greater uptake capacity than the root, resulting in preferential hyphal accumulation in the fungus than in the host plant (de Boulois et al., 2005). But there was no mycorrhizal enhancement of ^{134}Cs uptake in clover, *Eucalyptus*, or maize but mycorrhizal enhanced uptake of ^{65}Zn in maize (Joner et al., 2004). Similarly, mycorrhizae increased ^{137}Cs uptake by leek, but not by ryegrass (Rosén et al., 2005). In grasses, similar enhanced

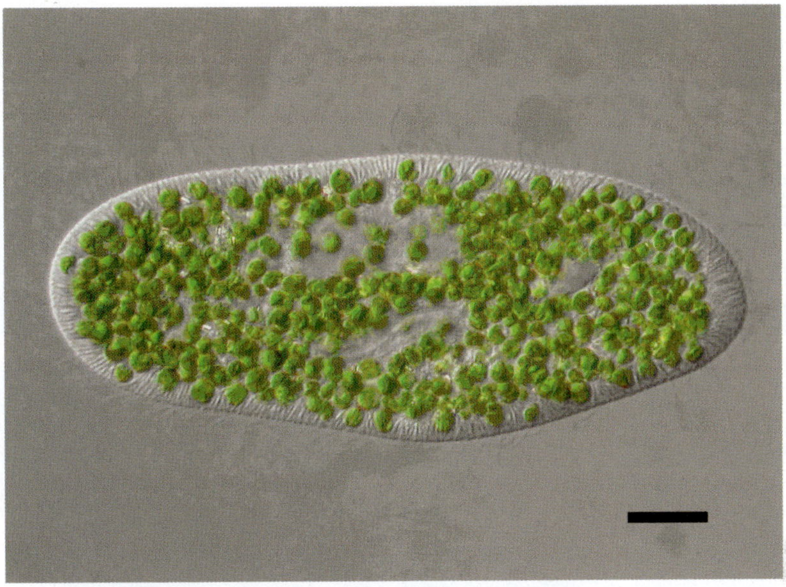

FIGURE 4.3 *P. bursaria* with *Chlorella* symbionts. Phase contrast. Bar = 10 μm.

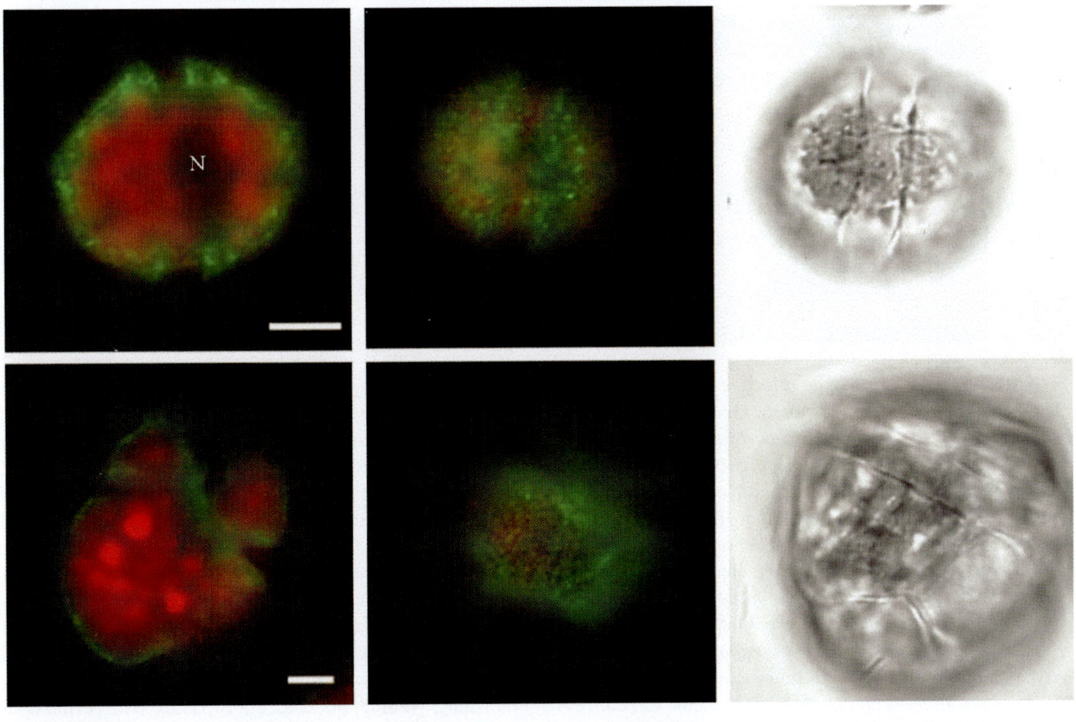

FIGURE 4.7 Immunolocalization of saxitoxin (STX/neoSTX) in *A. lusitanicum* (upper row) and *Alexandrium fundyense* (bottom row). Right micrographs are differential interference contrast images. Bars = 10 μm.

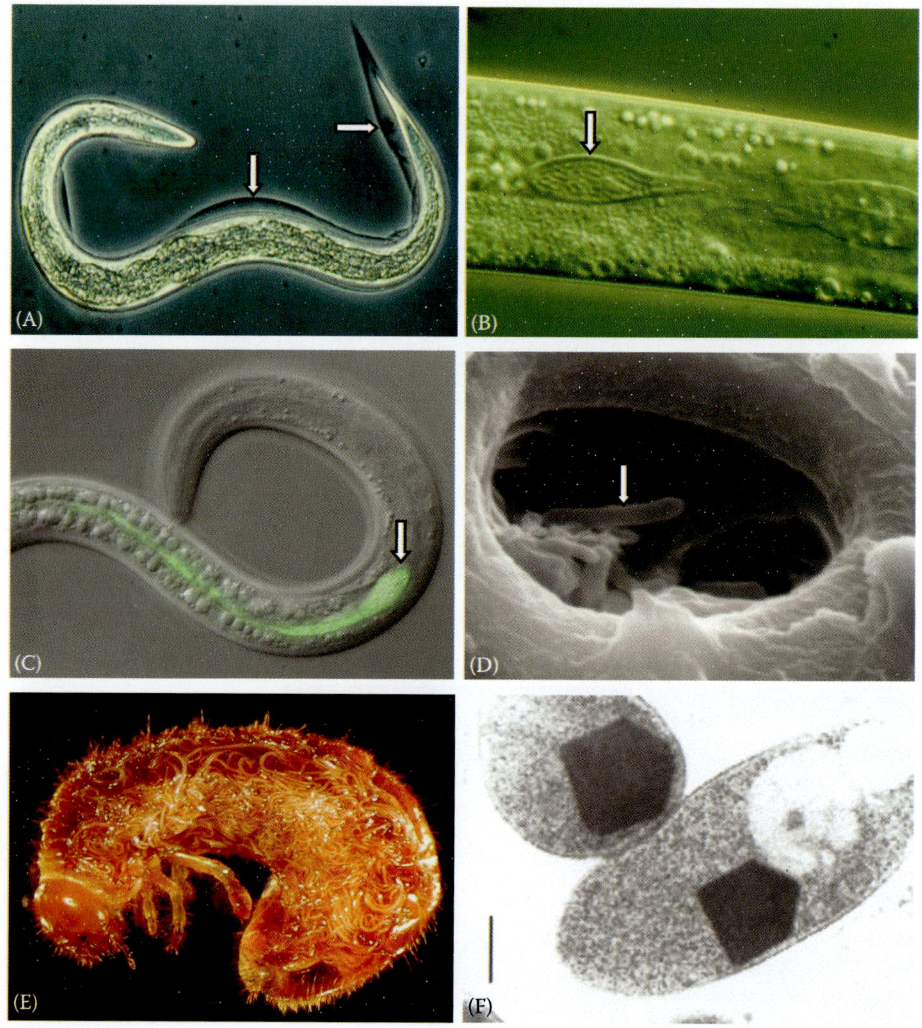

FIGURE 7.1 (A) Third-stage infective juvenile of *Steinernema carpocapsae* nematode. Note protective sheath (arrows) encasing the nematode. (Photo by Gaugler, R. *Entomopathogenic Nematology*, Gaugler, R. (Ed.), CABI Publishing, Wallingford, U.K., 35–36. With permission.) (B) Anterior region of the intestine of infective juvenile of *Steinernema carpocapsae* showing the vesicle (arrow) which is colonized by cells of the nematode's bacterial symbiont, *X. nematophila*. (Courtesy of Christian Laumond and Noël Boemare INRA, France.) (C) Bolus of GFP-labeled *Photorhabdus luminescens* in the intestine of infective juvenile of *Heterorhabditis bacteriophora* nematode immediately prior to regurgitation. (From Ciche, T.A., Darby, C., Ehlers, R.-U., Forst, S., and Goodrich-Blair, H., *Biological Control*, 38, 22, 2006. With permission; Courtesy of T. Ciche.) (D) Scanning electron micrograph of the oral cavity of *Steinernema carpocapsae* showing cells of its associated symbiont, *X. nematophila* (arrow). (Courtesy of L. LeBeck.) (E) Larvae of the Japanese beetle (*Popillia japonica*) with the cuticle excised to reveal the developing adult nematodes of *Heterorhabditis bacteriophora* within. (Photos by Wang, Y. and Gaugler, R., *Entomopathogenic Nematology*, Gaugler, R. (Ed.), CABI Publishing, Wallingford, U.K., 35–36. With permission.) (F) Crystalline inclusion bodies (black areas) in cells of *Photorhabdus temperata*. (From Boemare, N., in *Entomopathogenic Nematology*, Gaugler, R. (Ed.), CABI Publishing, Wallingford, U.K., 2002, 35–36. With permission; Courtesy of EMIP, INRA, Montpellier, France.)

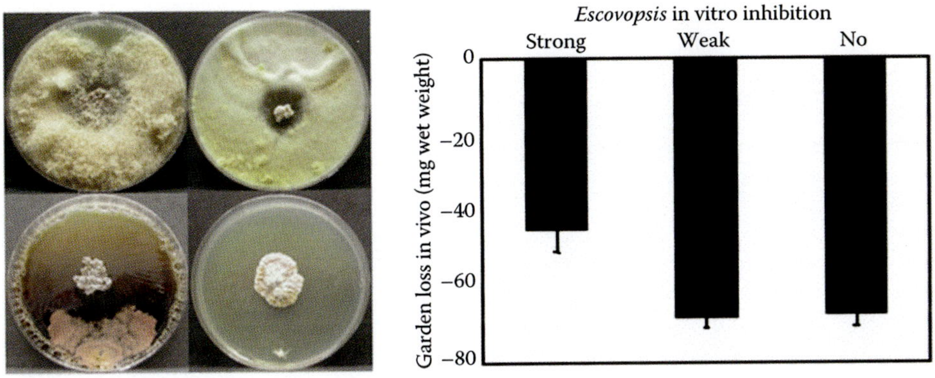

FIGURE 10.4 Ant–bacterium mutualism for defense against *Escovopsis*. The figure illustrates bioassay pairings that display varying degree of inhibition of *Escovopsis* (inoculated at the edge of the Petri plate) by ant-associated *Pseudonocardia* bacteria (inoculated two weeks prior to *Escovopsis* at the centre of the Petri plate). The varying degrees of inhibition in vitro (left) result in varying degrees of in vivo fungus garden morbidity caused by *Escovopsis*: infections with resistant parasites (weakly or not inhibited by the actinomycetes: zone of inhibition less than 1 cm) result in significantly higher morbidity than when the resident actinomycetes bacterium has strong inhibitory capabilities against the infection parasite strain (in vitro inhibition resulting in zone of inhibition being more than 1 cm). (Modified from Poulsen, M. Cafaro, M., Erhardt, D.P., Little, A.E.F., Gerardo, N.M., and Currie, C.R., 2008, in preparation.)

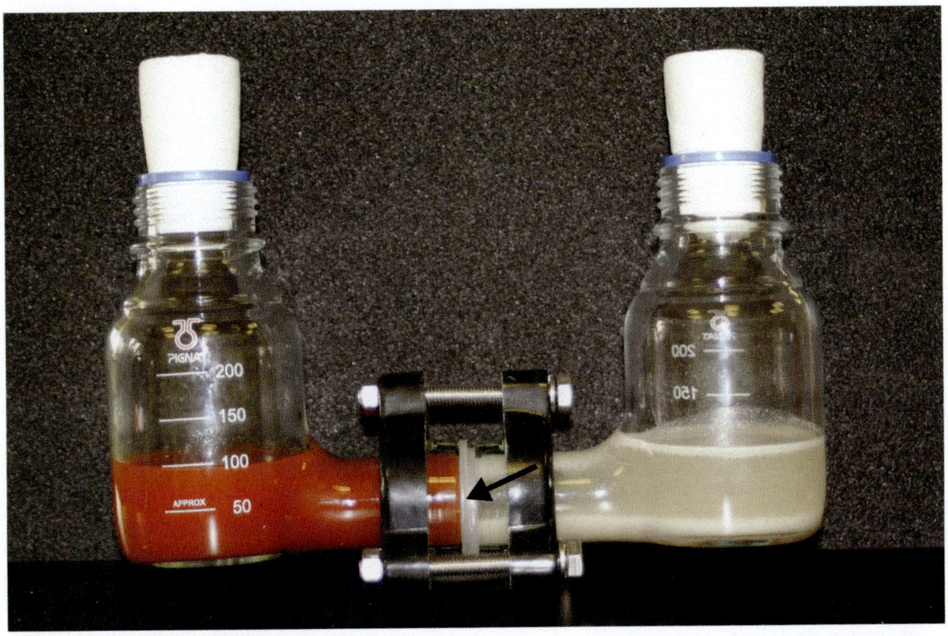

FIGURE 17.1 Prototype for a shaker-mounted dual fermentation vessel. The design was inspired by dual culture systems for *Symbiobacterium thermophilum* (Ohno et al., 1999). Each fermentation bottle is separated by a dialysis membrane mounted at the junction between the bottles arrow.

translocation of radiocesium into shoots of arbuscular mycorrhizal *Festuca ovina* were shown but not into clover (*Trifolium repens*) (Dighton and Terry, 1996), and a decrease in the Cs translocated to the shoots of *Agrostis tenuis* in the presence of arbuscular mycorrhizae (Berreck and Hasselwandter, 2001).

Ericoid mycorrhizal heather plants (*Calluna vulgaris*) accumulated less radiocesium than nonmycorrhizal plants (Clint and Dighton, 1992), indicative of accumulation in fungal components in the root. However, mycorrhizal plants allowed a greater proportion of the Cs to be translocated to shoots compared to nonmycorrhizal plants. A similar 18% higher shoot accumulation of ^{137}Cs in mycorrhizal heather, compared to nonmycorrhizal plants, was observed by Strandberg and Johansson (1998). Elevated levels of radionuclide in roots compared to shoots in mycorrhizal plants suggests that the mycorrhizal fungi accumulate radiocesium in the fungal tissue in a similar manner to that shown for heavy metals and ectomycorrhizae (Denny and Wilkins, 1987a,b). This concept has been reviewed in terms of competition between radiocesium and potassium in mycorrhizal *Agrostis tenuis* symbioses (Berreck and Haselwandter, 2001). They showed that mycorrhizal development in plant roots reduced Cs uptake by the plant at moderate nutrient levels in the soil, which suggests that the protective mechanism is due to sequestration of Cs in the extraradical hyphae of the mycorrhizal fungus and a reduced translocation into the host plant. They also demonstrated that, for this fungal plant interaction, there was no benefit of adding potassium to reduce Cs uptake.

13.12 Climate Change

Climate change can encompass a number of changes in our environment including increase in atmospheric carbon dioxide (CO_2) levels, increased temperature, changes in rainfall amount, and distribution pattern (both spatial and temporal) and has been reviewed by Treseder (2005). The impact of these factors on mycorrhizae is in its infancy, but it is clear that changes in CO_2 concentrations in the atmosphere influence the C:nutrient ratio of plant material that enters the decomposition pathways. As a result, the change in nutrient availability causes changes in both the composition of mycorrhizal communities as nutrient content is reduced. The response of mycorrhizae to rainfall, ozone, UV light, and CO_2 appears to be unclear from the few studies carried out to date.

In a meta-analysis of information regarding the interaction of mycorrhizae with nitrogen, phosphorus, and elevated CO_2, Treseder (2004) evaluated that there was on average a 15% reduction of mycorrhizal abundance under elevated N, 32% reduction under elevated phosphorus levels, and 47% increase due to increases in CO_2 level. A mechanistic model of elevated CO_2 effects on arbuscular mycorrhizal symbioses proposes a positive feedback of enhanced photosynthesis providing greater carbohydrate to roots (Lewis et al., 1994; Gorisen and Kuyper, 2000) to promote increased ectomycorrhizal density (O'Neill et al., 1987) and extraradical fungal hyphal development (Rillig and Allen, 1998; Sanders et al., 1998; Rillig et al., 2000), greater phosphate uptake (Rouhier and Read, 1998), and greater plant growth (Fitter et al., 2000). Additionally, the ephemeral nature of arbuscular mycorrhizal extraradical hyphae is thought to have a high turnover rate, thus contributing more to soil respiration. Thus it is suggested that arbuscular mycorrhizal fungi may play an important role in the fast cycling of carbon in the ecosystems in which they dominate.

In a similar way, ectomycorrhizal fungi receive a greater carbohydrate supply under elevated CO_2 levels (Gorisen and Kuyper, 2000), but the way in which this carbon is utilized is much more species-specific than in arbuscular mycorrhizae. In Scots pine seedling experiments, Gorisen and Kuyper (2000) showed that additional carbon was incorporated into the fungal component of *Suillus bovinus* mycorrhizae, hence lowering shoot:root ratios, whereas no such incorporation occurred with *Laccaria laccata*, where excess carbon was respired rather than incorporated. This highlights one of the possibly more important differences between arbuscular and ecto-mycorrhizae, where responses in ectomycorrhizal systems may lead to a greater shift in fungal community composition than in arbuscular mycorrhizae, where degree of root colonization varies. Our understanding of the functional changes induced by changes in fungal community composition on a root system is far from complete and, although we can speculate on distinct differences in physiological aptitude between fungal species, the combined effect of a mixed species community may depend on the actual species richness as much as the species identity within the community (Baxter and Dighton, 2001, 2005a,b).

13.13 Restoration and Urbanization

13.13.1 Restoration

Quilambo (2003) briefly discusses the role of mycorrhizae in site restoration, suggesting that the influence of mycorrhizae is mainly to aid in soil stability and drought tolerance of colonizing plants.

The ability of mycorrhizae to promote enhanced growth and establishment of plants in disturbed habitats suggests that mycorrhiza can be important for restoration projects. As we see below, many mycorrhizae have the ability to protect the host plant from pollutants, but in the context of restoration, the ability of mycorrhizal association to improve soil structure may be of equal importance. Firstly, mycorrhizal fungi assist the establishment of vegetation on, usually, poor soils in a restoration site. This allows the return of plant parts to soil, establishing the organic component of a more healthy soil. Secondly, the ability of fungi to bind soil mineral particles together either by physical means or by the production of the sticky glycoprotein, glomalin, by some arbuscular mycorrhizal fungal species increases soil stability (Miller and Jastrow, 1992; Wright and Upadhyaya, 1998) and prevents soil erosion. The increased sand binding by arbuscular mycorrhizal fungal hyphae and subsequent stability of soil have been shown to enhance the establishment of other plant species in desert restoration (Carillo-Garcia et al., 1999), with the initial mycorrhizal plants acting as nurse species for later successional trees. The factors of enhanced plant establishment, maintenance of diversity, increased efficiency of nutrient cycling, and soil stability have been attributed to the presence of mycorrhizae in the rehabilitation of desert areas (Jeffries et al., 2002).

Although growth and development of late successional plant communities are more likely to be responsive to arbuscular mycorrhizal colonization than early successional species (Pezzani et al., 2006), it would seem that the presence of mycorrhizae in early successional communities provides assistance with soil development and stability, which allows sequential plant communities to establish.

The combination of both factors increases soil aggregation and the storage of carbon as protected organic matter within soil aggregates. Mycorrhizal fungal-assisted long-term carbon sequestration in soil aggregates is a potentially valuable resource in a world of increasing atmospheric CO_2.

Given the ability of some mycorrhizal species to provide protection against heavy metals (Marx, 1975, 1980; Denny and Wilkins, 1987; Fomina et al., 2005; Gadd, 2007), there has been interest in using mycorrhizal inoculum to aid plant establishment on degraded and often polluted sites. This has led to the establishment of a number of companies producing commercial mycorrhizal inocula for both horticulturalists and restoration ecologists. Indeed the value of arbuscular mycorrhizal inoculation of plants for restoration in polluted sites (bioremediation) as defense against the heavy metal, polycyclic aromatic hydrocarbons, and radionuclide pollutants has been reviewed by Leyval et al. (2002).

13.13.2 Urbanization

In some respects, there is little separation between urbanization and restoration as most restoration sites are an artifact of human disturbance. Much vegetation in an urban setting is in the form of widely dispersed ruderal species occupying rudimentary soils in sidewalk cracks or degraded concrete or asphalt. Similar to the plants in most extreme cold conditions, many of these species are nonmycorrhizal (Wang and Qui, 2006). Cousins et al. (2003) compared the arbuscular mycorrhizal association of plants in the Sonoran desert, where the desert was converted to agriculture or urban areas. Fungal spore density was significantly higher in the desert than in agricultural or urban soils; however, fungal species richness was greatest in urban nonresidential soils and lowest in agricultural soils. There was a significant separation of fungal community composition between one group comprising only urban soils and another community that was predominantly native of agricultural soils. Stressors in these ecosystems include low nutrient availability, drought, extreme temperature fluctuation, physical disturbance, and pollutants. There is little information about the ecology of mycorrhizae in urban settings.

13.14 Conclusions

There seems to be two contrasting strategies between endomycorrhizae (arbuscular and ericaceous) and ectomycorrhizae in their response to extreme environments. Given the restricted diversity of arbuscular and ericoid mycorrhizae, their response tends to be either an increase or decrease in root colonization. Within the greater diversity of ectomycorrhizal fungal diversity, the response here is a combination of increase or reduction in root colonization combined with a change in species composition of the mycorrhizal community. How does this translate into defensive functions and can these changes really be regarded as defensive mechanisms?

The initiation of a mycorrhizal root system relies on the combination of recognition systems or biosignaling systems that ensure compatibility between the fungal and host plant. This is then modified by edaphic factors that appear to select for the most optimum fungal species or exclude fungal species that are sensitive to adverse conditions in the extreme environment. For example, there are many studies showing a change in ectomycorrhizal community composition with acid rain or nitrogen deposition (Bradrud and Timmerman, 1998; Lilleskov et al., 2002; Avis et al., 2003). These changes have been related to changes in the Ca:Al ratio and Al toxicity in the case of acid rain (Thompson and Medve, 1984; Jongbloed and Borst-Pauwels, 1988) and known tolerances of mycorrhizae to nitrogen (nitrophiles and nitrofuges; Dighton et al., 2004). In climatically limited environments, there is a reduction in the overall species composition, favoring those species with optimal strategies for obtaining nutrients for oligotrophic soils and resisting freeze/thaw cycles. Although we know something about the physiology and efficacy of some ectomycorrhizal fungal species, we have only limited information about their functional interactions at the community scale. Do they all do the same job with the same efficiency? Do they influence each other in their function? It is clear that different fungal species have different properties, such as exoenzyme capacity (Cairney, 1999; Tarvainen et al., 2004; Dong et al., 2007) and their relative efficiency in nutrient acquisition is variable (Dighton et al., 1990, 1993). Thus the species composition of a mycorrhizal community on a host plant will determine the actual range and process rate of physiological functions. However, preliminary evidence also suggests that diversity per se and not the species composition of a mycorrhizal community can also affect function (Baxter and Dighton, 2001, 2005), where nutrient influx into host plants increases with increased fungal species richness, irrespective of the actual species composition.

In many of the examples shown above, the responses of ericaceous or arbuscular mycorrhizae to environmental changes have been changed in degree of root colonization by fungal structures. However, there are examples where the species composition of those fungal communities usually changes with a reduction in species diversity (Lilleskov, 2005; Treseder, 2005). This change in either species diversity or degree of mycorrhizal colonization at the larger scale of landscapes or ecosystems may be a reflection of the responses of these communities seen at smaller scales of resolution within ecosystems in terms of resource heterogeneity (Olsson et al., 2002), where the dynamic nature of local scale changes in mycorrhizal diversity and function is influenced by the microscale changes in resources available to mycorrhizal roots.

So, are the changes in mycorrhizal responses to either local or landscape level changes in resources really defensive interactions or do they merely reflect the optimization of utilization of the resources available at the moment? There certainly seem to be instances where defensive mechanisms can be evoked, particularly in the interactions between mycorrhizae and heavy metals and radionuclides (Clint and Dighton, 1992; Dighton and Terry, 1996; Berreck and Haselwandter, 2001), where immobilization of toxic material in fungal tissue associated with the roots restricts the uptake into aboveground plant parts. However, other than these situations, it is my belief that we need further studies to elucidate actual defensive mechanisms in the functions of mycorrhizal associations in other situations where we now consider the changes in community composition or root colonization as being an optimization of the symbiosis for that place at that particular time.

REFERENCES

Allsop, N. 1998. Effect of defoliation on the arbuscular mycorrhizas of three perennial pasture and rangeland grasses. *Plant and Soil* 202:117–124.

Andersen, C. P. and P. T. Rygiewicz. 1991. Stress interactions and mycorrhizal plant response: Understanding carbon allocation priorities. *Environmental Pollution* 73:217–244.

Arnebrant, K. 1994. Nitrogen amendments reduce the growth of extrametrical ectomycorrhizal mycelium. *Mycorrhiza* 5:7–15.

Arnebrant, K. and B. Soderstrom. 1992. Effects of fertilizer treatments on ectomycorrhizal colonization potential in two Scots pine forests in Sweden. *Forest Ecology and Management* 53:77–89.

Arnolds, E. 1991. Decline of ectomycorrhizal fungi in Europe. *Agriculture Ecosystems and Environment* 35:209–244.

Arnolds, E. 1988. The changing macromycete flora in the Netherlands. *Transactions of the British Mycological Society* 90:391–406.

Augé, R. M. 2001. Water relations, drought and vesicular-arbuscular mycorrhizal symbiosis. *Mycorrhiza* 11:3–42.

Avis, P. G., McLaughlin, D. J., Dentinger, B. C., and P. B. Reich. 2003. Long-term increase in nitrogen supply alters above- and below-ground ectomycorrhizal communities and increases the dominance of *Russula* spp. in a temperate oak savanna. *New Phytologist* 160:239–253.

Baeza, A., Guillén, F. J., and S. Hernández. 2002. Transfer of ^{134}Cs and ^{85}Sr to *Pleurotus eryngii* fruiting bodies under laboratory conditions: A compartmental model approach. *Bulletin of Environmental Contamination Toxicology* 69:817–828.

Baxter, J. W. and J. Dighton. 2001. Ectomycorrhizal diversity alters growth and nutrient acquisition of gray birch (*Betula populifolia* Marshall) seedlings in host-symbiont culture conditions. *New Phytologist* 152:139–149.

Baxter, J. W. and J. Dighton. 2005. Phosphorus source alters host plant response to ectomycorrhizal diversity. *Mycorrhiza* 15:513–523.

Berreck, M. and K. Haselwandter. 2001. Effect of the arbuscular mycorrhizal symbiosis upon uptake of cesium and other cations by plants. *Mycorrhiza* 10:275–280.

Bever, J. D., Pringle, A., and P. A. Schultz. 2002. Dynamics within the plant-mycorrhizal fungal mutualism: Testing the nature of community feedback. In *Mycorrhizal Ecology*, Eds. M. G. A. van der Heijden and I. R. Sanders, pp. 267–294. Berlin: Springer.

Bever, J. D. and P. A. Schultz. 2005. Mechanisms of arbuscular mycorrhizal mediation of plant–plant interactions. In *The Fungal Community: Its Organization and Role in the Ecosystem*, 3rd edn., Eds. J. Dighton, J. F. Jr. White, and P. Oudemans, pp. 443–459. Baton Rouge: Taylor and Francis.

Blaschke, H. 1988. Mycorrhizal infection and changes in fine root development of Norway spruce influenced by acid rain in the field. In *Ectomycorhiza and Acid Rain*, Eds. A. E. Jansen, J. Dighton, and A. H. M. Bresser, pp. 112–115. Bilthoven, the Netherlands: CEC.

Brandrud, T. E. 1995. The effects of experimental nitrogen addition on the ectomycorrhizal fungus flora in an oligotrophic spruce forest at Gardsjon, Sweden. *Forest Ecology and Management* 71:111–122.

Brandrud, T. E. and V. Timmermann. 1998. Ectomycorrhizal fungi in the NITREX site at Garsjon, Sweden; below and above-ground responses to experimentally-changed nitrogen inputs 1990–1995. *Forest Ecology and Management* 101:207–214.

Byrne, A. R. 1988. Radioactivity in fungi in Slovenia, Yugoslavia, following the Chernobyl accident. *Journal of Environmental Radioactivity* 6:177–183.

Cairney, J. W. G. 1999. Intraspecific physiological variation: Implications for understanding functional diversity in ectomycorrhizal fungi. *Mycorrhiza* 9:125–135.

Carrillo-Garcia, A., León de la Luz, J.-L., Bshan, Y., and G. J. Bethlenfalvay. 1999. Nurse plants, mycorrhizae, and plant establishment in a disturbed area of the Sonoran desert. *Restoration Ecology* 7:321–335.

Clint, G. M. and J. Dighton. 1992. Uptake and accumulation of radiocaesium by mycorrhizal and non-mycorrhizal heather plants. *New Phytologist* 122:555–561.

Collier, S. C., Yarnes, C. T., and R. P. Herman. 2003. Mycorrhizal dependency of Chihuahuan desert plants is influenced by life history strategy and root morphology. *Journal of Arid Environments* 55:223–229.

Colpaert, J. V. and K. K. Van Tichelen. 1996. Mycorrhizas and environmental stress. In *Fungi and Environmental Change*, Eds. J. C. Frankland, N. Magan, and G. M. Gadd, pp. 109–128. Cambridge: Cambridge University Press.

Connolly, J. H., Shortle, W. C., and J. Jellison. 1998. Translocation and incorporation of strontium carbonate derived strontium into calcium oxalate crystals by the wood decay fungus *Resinicium bicolor. Canadian Journal of Botany* 77:179–187.

Cousins, J. R., Hope, D., Gries, C., and J. C. Stutz. 2003. Preliminary assessment of arbuscular mycorrhizal fungal diversity and community structure in an urban ecosystem. *Mycorrhiza* 13:319–326.

Cruz, A. F., Ishii, T., and K. Kadoya., 2000. Effects of arbuscular mycorrhizal fungi on tree growth, leaf water potential, and levels of 1-aminocyclopropane-1-carboxylic acid and ethylene in the roots of papaya under water-stress conditions. *Mycorrhiza* 10:121–123.

Cullings, K. W., Vogler, D. R., Parker, V. T., and S. Makhija. 2001. Defoliation effects on the ectomycorrhizal community of a mixed *Pinus contorta/Picea engelmannii* stand in Yellowstone Park. *Oecologia* 127:533–539.

de Boulois, H. D., Delvaux, B., and S. Declerck. 2005. Effects of arbuscular mycorrhizal fungi on the root uptake and translocation of radiocaesium. *Environmental Pollution* 134:515–524.

Denny, H. J. and D. A. Wilkins. 1987a. Zinc tolerance in *Betula* spp. I. Effects of external concentration of zinc on growth and uptake. *New Phytologist* 106:517–524.

Denny, H. J. and D. A. Wilkins. 1987b. Zinc tolerance in *Betula* spp. IV. The mechanism of ectomycorrhizal amelioration of zinc toxicity. *New Phytologist* 106:545–553.

Dighton, J. and A. D. Horrill. 1988. Radiocaesium accumulation in the mycorrhizal fungi *Lactarius rufus* and *Inocybe longicystis*, in upland Britain. *Transactions of the British Mycological Society* 91:335–337.

Dighton, J. and A. E. Jansen. 1991. Atmospheric pollutants and ectomycorhizas: More questions than answers? *Environmental Pollution* 73:179–204.

Dighton, J. and R. A. Skeffington. 1987. Efects of artificial acid precipitation on the mycorrhizas of Scots pine seedlings. *New Phytologist* 107:191–202.

Dighton, J. and P. A. Mason. 1985. Mycorrhizal dynamics during forest tree development. In *Developmental Biology of Higher Fungi*, Eds. D. Moore, L. A. Casselton, D. A. Wood, and J. C. Frankland, pp. 163–171. Cambridge: Cambridge University Press.

Dighton, J., Mason, P. A., and J. M. Poskitt. 1990. Field use of 32P tracer to measure phosphate uptake by birch mycorrhizas. *New Phytologist* 116:655–661.

Dighton, J., Poskitt, J. M., and T. K. Brown. 1993. Phosphate influx into ectomycorrhizal and saprotrophic fungal hyphae in relation to phosphorus supply: A potential method for selection of efficient mycorrhizal species. *Mycological Research* 97:355–358.

Dighton, J. and G. M. Terry. 1996. Uptake and immobilization of caesium in UK grassland and forest soils by fungi following the Chernobyl accident. In *Fungi and Environmental Change*, Eds. J. C. Frankland, N. Magan, and G. M. Gadd, pp. 184–200. Cambridge: Cambridge University Press.

Dighton, J., Tuininga, A. R., Gray, D. M., Huskins, R. E., and T. Belton. 2004. Impacts of atmospheric deposition on New Jersey pine barrens forest soils and communities of ectomycorrhizae. *Forest Ecology and Management* 201:131–144.

Dighton, J., Tugay, T., and N. N. Zhdanova. 2008a. Interactions of fungi and radionuclides in soil. In *Microbiology of Extreme Soils, Soil Biology*, vol. 13, Eds. P. Dion and C. S. Nautiyal, pp. 333–355. Heidelberg: Springer-Verlag.

Dighton, J., Tugay, T., and N. N. Zhdanova. 2008b. Fungi and ionizing radiation from radionuclides. *FEMS Microbial Letters* 281:109–120.

Dong, S., Brooks, D., Jones, M. D., and S. J. Grayston. 2007. A method for linking *in situ* activities of hydrolytic enzymes to associated organisms in forest soils. *Soil Biology and Biochemistry* 39:2414–2419.

Eckl, P., Hoffman, W., and R. Turk. 1986. Uptake of natural and man-made radionuclides by lichens and mushrooms. *Radiation and Environmental Biophysics* 25:43–54.

Erland, S. and A. F. S. Taylor. 2002. Diversity of ecto-mycorrhizal fungal communities in relation to the abiotic environment. In *Ecological Studies vol. 157: Mycorrhizal Ecology*, Eds. M. G. A. van der Heijden and I. R. Sanders, pp. 164–200. New York: Springer.

Fitter, A. H., Heinemeyer, A., and P. L. Staddon. 2000. The impact of elevated CO_2 and global climate change on arbuscular mycorrhiza: A mycocentric approach. *New Phytologist* 147:179–187.

Fomina, M., Burford, E. P., and G. M. Gadd. 2005. Toxic metals and fungal communities. In *The Fungal Community: Its Organization and Role in the Ecosystem*, 3rd edn., Eds. J. Dighton, J. F. White, and P. Oudemans, pp. 733–758. Baton Rouge: Taylor and Francis.

Füzy, A., Tóth, T., and B. Biró. 2007. Mycorrhizal colonization can be altered by the direct and indirect effect of drought and salt in a split root experiment. *Cereal Research Communication* 35:401–404.

Gadd, G. M. 2007. Geomycology: Biogeochemical transformations of rocks, minerals, metals and radionuclides by fungi, bioweathering and bioremediation. *Mycological Research* 111:3–50.

Gange, A. C., Bower, E., and V. K. Brown. 2002. Differential effects of insect herbivory on arbuscular mycorrhizal colonization. *Oecologia* 131:103–112.

Gardes, M. and A. Dahlberg. 1996. Mycorrhizal diversity in arctic and alpine tundra: An open question. *New Phytologist* 133:147–157.

Giovani, C., Nimis, P. L., Land, P., and R. Padovani. 1990. Investigation of the performance of macromycetes as bioindicators of radioactive contamination. In *Transfer of Radionuclides in Natural and Semi-Natural Environments*, Eds. G. Desmet, P. Nassimbeni, and M. Belli, pp. 485–491. London: Elsevier Applied Science.

Gorisen, A. and T. W. Kuyper. 2000. Fungal species-specific response of ectomycorrhizal Scots pine (*Pinus sylvestris*) to elevated CO_2. *New Phytologist* 146:163–168.

Gray, S. N., Dighton, J., Olsson, S., and D. H. Jennings. 1995. Real-time measurement of uptake and translocation of ^{137}Cs within mycelium of *Schizophyllum commune* Fr. by autoradiography followed by quantitative image analysis. *New Phytologist* 129:449–465.

Gray, S. N., Dighton, J., and D. H. Jennings. 1996. The physiology of basidiomycete linear organs III. Uptake and translocation of radiocaesium within differentiated mycelia of *Armillaria* spp. growing in microcosms and in the field. *New Phytologist* 132:471–482.

Grime, J. P. 1997. Evidence for the existence of three primary strategies in plants and its relevance to ecological and evolutionary theory. *American Naturalist* 111:1169–1194.

Grodzinskaya, A. A., Berreck, M., Wasser, S. P., and K. Haselwandter. 1995. Radiocaesium in fungi: Accumulation pattern in the Kiev district of Ukraine including the Chernobyl zone. *Sydowia* 10:88–96.

Guillitte, O., Gasia, M. C., Lambinon, J., Fraiture, A., Colard, J., and R. Kirchmann. 1987. La radiocontamination des champignons sauvages en Belgique et au Grand-Duche de Luxembourg apres l' accident nucleaire de tchernobyl. *Memoirs Societie Royale Botanique Belgique* 9:79–93.

Guillette, O., Melin, J., and L. Wallberg. 1994. Biological pathways of radionuclides originating from the Chernoyl fallout in a boreal forest ecosystem. *Science of the Total Environment* 157:207–215.

Gupta, R. and K. V. Krishnamurthy. 1996. Response of mycorrhizal and nonmycorrhizal *Arachis hypogea* to NaCl and acid stress. *Mycorrhiza* 6:145–149.

Halling, R. E. 2001. Ectomycorrhizae: Co-evolution, significance, and biogeography. *Annals of the Misouri Botanical Gardens* 88:5–13.

Harrison, A. F., Stevens, P. A., Dighton, J., Quarmby, C., Dickinson, A. L., Jones, H. E., and D. M. Howard. 1995. The critical load of nitrogen for Sitka spruce forests on stagnopodsols in Wales: Role of nutrient limitations. *Forest Ecology and Management* 76:139–148.

Haselwandter, K. 1978. Accumulation of the radioactive nuclide ^{137}Cs in fruitbodies of basidiomycetes. *Health Physics* 34:713–715.

Haselwandter, K. and M. Berreck. 1994. Accumulation of radionuclides in fungi. In *Metal Ions in Fungi*, Eds. G. Winkelmann and D. R. Winge, pp. 259–277. New York: Marcel Dekker.

Haselwandter, K., Berreck, M., and P. Brunner. 1988. Fungi as bioindicators of radiocaesium contamination. Pre- and post Chernobyl activities. *Transactions of the British Mycological Society* 90:171-176.

Haselwandter, K. and D. J. Read. 1980. Fungal associations of roots of dominant and sub-dominant plants in high-alpine vegetation systems with special reference to mycorrhiza. *Oecologia* 45:57–62.

He, X., Mouratov, S., and Y. Steinberger. 2002. Temporal and spatial dynamics of vesicular-arbuscular mycorrhizal fungi under the canopy of *Zygophyllum dumosum* Boiss. In the Negev Desert. *Journal of Arid Environments* 52:379–287.

Helgason, T. and A. Fitter. 2005. The ecology and evolution of the arbuscular mycorrhizal fungi. *The Mycologist* 19:96–101.

Hokka, V., Mikola, J., Vestberg, M., and H. Setälä. 2004. Interactive effects of defoliation and an AM fungus on plants and soil organisms in experimental legume-grass communities. *Oikos* 106:73–84.

Hütterman, A. 1982. Fruhdiagnose von Immissionsschaden im Wurzelbereich von Waldbaumen. Immissionbelastungen von Waldökosystemen. *Landesanst. f. Okologie, Landschaftsentw. u. Forstpl. Nordhein-Westfalen*, pp. 26–31.

Hüttermann, A. 1985. The effects of acid deposition on the physiology of the forest ecosystem. *Experientia* 41:585–590.

Ijpelaar, P. 1980. Het Caesium-137 gehalte van verschillende paddestoelsoorten. *Coolia* 23:86–91.

Jacobson, K. M. 1997. Moisture and substrate determine VA-mycorrhizal fungal community distribution and structure in an arid grassland. *Journal of Arid Environments* 35:59–75.

Jansen, A. E. and J. Dighton. 1990. *Effects of Air Pollutants on Ectomycorrhizas*, Air Pollution Report 30, CEC, Brussels, p. 58.

Jansen, A. E., Dighton, J., and A. H. M. Bresser. 1988. *Ectomycorrhiza and Acid Rain*. Brussels: *CEC Air Pollution Research Report 12*.

Jeffries, P., Craven-Griffiths, A., Barea, J. M., Levy, Y., and J. C. Dodd. 2002. Application of arbuscular mycorrhizal fungi in the revegetation of desertified Mediterranean ecosystems. In *Mycorrhizal Technology in Agriculture*, Eds. S. Gianinazi, H. Schüepp, J. M. Barea, and K. Haselwandter, pp. 151–174. Switzerland: Birkhäuser Verlag.

Johnson, N. C. and D. A. Wedin. 1979. Soil carbon, nutrients, and mycorrhizae during conversion of a dry tropical forest to grassland. *Ecological Applications* 7:171–182.

Johnson, N. C. and D. A. Wedin. 1997. Soil carbon, nutrients, and mycorrhizae during conversion of a dry tropical forest to grassland. *Ecological Applications* 7:171–182.

Johnson-Green, P., Kenkel, N. C., and T. Booth. 2001. Soil salinity and arbuscular mycorrhizal colonization of *Puccinella nuttallana*. *Mycological Research* 105:1094–1110.

Joner, E. J., Roos, P., Jansa, J., Frossard, E., Leyval, C., and I. Jakobsen. 2004. No significant contribution of arbuscular mycorrhizal fungi to transfer of radiocesium from soil to plants. *Applied and Environmental Microbiology* 70:6512–6517.

Jongbloed, R. H. and G. W. F. H. Borst Pauwels. 1988. Efects of Al^{3+} and NH_4^+ on growth and uptake of K^+ and $H_2PO_4^-$ by three ectomycorrhizxal fungi in pure culture. In *Ectomycorhiza and Acid Rain*, Eds. A. E. Jansen, J. Dighton, and A. H. M. Bresser, pp. 47–52. Bilthoven, the Netherlands: CEC.

Jonsson, L., Dahlberg, A., and T. E. Brandrud. 2000. Spatiotemporal distribution of an ectomycorrhizal community in an oligotrophic Swedish *Picea abies* forest subjected to experimental nitrogen addition: Above- and below-ground views. *Forest Ecology and Management* 132:143–156.

Jumpponen, A., Mattson, K. G., and J. M. Trappe. 1998. Mycorrhizal functioning of *Phialocephala fortinii* with *Pinus contorta* on glacier forefront soil: Interactions with soil nitrogen and organic matter. *Mycorrhiza* 7:261–265.

Jumpponen, A., Trappe, J. M., and E. Cázares. 2002. Occurrence of ectomycorrhizal fungi on the forefront of retreating Lyman Glacier (Washington, DC) in relation to time since deglaciation. *Mycorrhiza* 12:43–49.

Jumpponen, A. and L. M. Egerton-Warburton. 2005. Mycorrhizal fungi in successional environments: A community assembly model incorporating host plant, environmental, and biotic factors. In *The Fungal Community: Its Organization and Role in the Ecosystem*, 3rd edn., Eds. J. Dighton, J. F. Jr. White, and P. Oudemans, pp. 139–168. Boca Raton: Taylor & Francis.

Kårén, O. and J. E. Nylund. 1997. Effects of ammonium sulphate on the community structure and biomass of ectomycorrhizal fungi in a Norway spruce stand in southwestern Sweden. *Canadian Journal of Botany* 75:1628–1642.

Kosala, K. R., Dickmann, D. I., Paul, E. A., and D. Parry. 2004. Repeated insect defoliaton effects on growth, nitrogen aquisition, carbohydrates, and root demography of poplars. *Oecologia* 129:65–74.

Kristek, S., Kristek, A., Guberac, V., Stanisavjevic, A., and S. Rasic. 2007. The influence of mycorrhizae on pea yield and quality in drought caused stress conditions. *Cereal Research Communication* 35:681–684.

Kula, A. A. R., Hartnett, D. C., and G. W. T. Wilson. 2005. Effects of mycorrhizal symbiosis on tallgrass prairie plant–herbivore interactions. *Ecological Letters* 8:61–69.

Kytöviita, M.-M. 2005. Role of nutrient level and defoliation on symbiotic function: Experimental evidence by tracing $^{14}C/^{15}N$ exchange in mycorrhizal birch seedlings. *Mycorrhiza* 15:65–70.

Kytöviita, M.-M. and T. Sarjala. 1997. Effects of defoliation and symbiosis on polyamine levels in pine and birch. *Mycorrhiza* 7:107–111.

Last, F. T., Dighton, J., and P. A. Mason. 1987. Successions of sheathing mycorrhizal fungi. *Trends in Ecology and Evolution* 2:157–161.

Lewis, D. J., Thomas, R. B., and B. R. Strain. 1994. Effect of elevated CO_2 on mycorrhizal colonization of loblolly pine (*Pinus taeda* L.) seedlings. *Plant and Soil* 165:81–88.

Leyval, C., Joner, E. J., del Val, C., and K. Haselwandter. 2002. Potential of arbuscular mycorrhizal fungi for bioremediation. In *Mycorrhizal Technology in Agriculture*, Eds. S. Gianinazi, H. Schüepp, J. M. Barea, and K. Haselwandter, pp. 151–174. Switzerland: Birkhäuser Verlag.

Lilleskov, E. A. and T. D. Bruns. 2001. Nitrogen and ectomycorrhizal fungal communities: What we know, what we need to know. *New Phytologist* 149:154–158.

Lilleskov, E. A., Fahey, T. J., Horton, T. R., and G. M. Lovett. 2001. Ectomycorrhizal fungal aboveground community change over an atmospheric nitrogen deposition gradient. *Ecological Applications* 11:397–410.

Lilleskov, E. A., Fahey, T. J., Horton, T. R., and G. M. Lovett. 2002. Belowground ectomycorrhizal fungal community change over a nitrogen deposition gradient in Alaska. *Ecology* 83:104–115.

Lilleskov, E. A. 2005. How do composition, structure, and function of mycorrhizal fungal communities respond to nitrogen deposition and ozone exposure? In *The Fungal Community: Its Organization and Role in the Ecosystem*, 3rd edn., Eds. J. Dighton, J. W. Jr. White, and P. Oudemans, pp. 769–801. Baton Rouge: Taylor and Francis.

Malinowska, E., Szefer, P., and Bojanowski, R. (2006). Radionuclide content in *Xercomus badius* and other commercial mushrooms from several regions in Poland. *Food Chemistry* 97:19–24.

Markkola, A., Kuikka, K., Rauito, P., Härmä, E., Roitto, M., and J. Tuomi. 2004. Defoliation increases limitation in ectomycorrhizal symbiosis of *Betula pubescens. Oecologia* 140:234–240.

Marx, D. H. 1975. Mycorrhiza and establishment of trees on strip-mined land. *Ohio Journal of Science* 75:288–297.

Marx, D. H. 1980. Role of mycorrhizae in forestation of surface mines. Compact Commission and USDA Forest Service. Trees for Reclamation; Lexington, Kentucky, USA. Interstate Mining Compact Commission and U.S. Department of Agriculture, Forest Service; pp. 109–116.

McHugh, J. M. and J. Dighton. 2004. Influence of mycorrhizal inoculation, inundation period, salinity and phosphorus availability on the growth of two salt marsh grasses, *Spartina alterniflora* Lois. and *Spartina cynosuroides* (L.) Roth. in nursery system. *Restoration Ecology* 12:533-545.

Mena-Violante, H., Ocampo-Jiménez, O., Dendooven, L., Martinéz-Soto, G., González-Castañeda, J., Davies, F. T. Jr., and V. Olalde-Portugal. 2006. Arbuscular mycorrhizal fungi enhance fruit growth and quality of chile ancho (*Capsicum annum* L cv San Luis) plants exposed to drought. *Mycorrhiza* 16:261–267.

Michelsen, A. and Rosendahl, S. 1990. The effect of VA mycorrhizal fungi, phosphorus and drought stress on the growth of *Acacia nilotica* and *Leucana leucocephala* seedlings. *Plant and Soil* 124:7–13.

Mietelski, J. W., Jasinska, M., Kubica, B., Kozak, K., and P. Macharski. 1994. Radioactive contamination of Polish mushrooms. *Science of the Total Environment* 157:217–226.

Mikola, P. 1970. Mycorrhizal inoculation in afforestation. *International Review of Forest Research* 3:123–196.

Mikola, J., Nieminen, M., Ilmarinen, K., and W. Silvester. 2005. Belowground responses by AM fungi and animal trophic groups to repeated defoliation in an experimental grassland community. *Soil Biology and Biochemistry* 37:1630–1639.

Miller, R. M. and J. D. Jastrow. 1992. The role of mycorrhizal fungi in soil conservation. In *Mycorrhizae in Sustainable Agriculture*, Eds. G. J. Bethlanfalvay and R. G. Linderman, pp. 29–44. Madison, Wisconsin: American Society of Agronomy, Special Publication No. 54.

Morin, P. J. 1999. *Community Ecology*. Oxford: Blackwell Science Inc.

Morte, A., Lovisolo, C., and A. Schubert. 2000. Effect of drought stress on growth and water relations of the mycorrhizal association *Helianthemum almeriense—Terfezia claveryi. Mycorrhiza* 10:115–119.

Muramatsu, Y., Yoshida, S., and M. Sumia. 1991 Concentrations of radiocesium and potassium in basidiomycetes collected in Japan. *Science of the Total Environment* 105:29–39.

Neville, J., Tessier, J. L., Morrison, I., Scaratt, J., Canning, B., and J. N. Klironomos. 2002. Soil depth distribution of ecto- and arbuscular mycorrhizal fungi associated with *Populus tremuloides* within a 3-year-old boreal forest clear-cut. *Applied Soil Ecology* 19:209–216.

Nicholson, T. H. and C. Johnston. 1979. Mycorrhiza in the Graninae. III. *Glomus fasciculatus* and the endophyte of pioneering grasses in maritime sand dunes. *Transactions of the Britis Mycological Society* 72:261–268.

Ohtonen, R., Fritze, H., Pennanen, T., Jumpponen, A., and J. M. Trappe. 1999. Ecosystem properties and microbial community changes in primary succession on a glacier forefront. *Oecologia* 119:239–246.

Olssen, P.A., Eriksen, B., and A. Dahlberg. 2004. Colonization by arbuscular mycorrhizal and fine endophytic fungi in herbaceous vegetation in the Canadian high arctic. *Canadian Journal of Botany* 82:1547–1556.

Olsson, P. A., Jakobsen, I., and H. Wallander. 2002. Foraging and resource allocation strategies of mycorrhizal fungi in a patchy environment. In *Mycorrhzial Ecology*, Eds. M. G. A. Van der Heijden and I. R. Sanders, pp. 93–115. Berlin: Springer-Verlag.

O'Neill, E. G., Luxmoore, R. J., and R. J. Norby. 1987. Increases in mycorrhizal colonization and seedling growth in *Pinus echinata* and *Quercus alba* in an enriched CO_2 atmosphere. *Canadian Journal of Forest Research* 17:878–883.

Oolbekkink, G. T. and T. W. Kuyper. 1989. Radioactive caesium from Chernobyl in fungi. *The Mycologist* 3:3–6.

Palermo, B. L., Clancy, K. M., and G. W. Koch. 2003. The potential of ectomycorrhizal fungi in determining Douglas-fir resistance to defoliation by the Western spruce budworm (Lepidoptera: Totricidae). *Journal of Economic Entomology* 96:783–791.

Peter, M., Ager, F., and S. Egli. 2001. Nitrogen addition in a Norway spruce stand altered macromycete sporocarp production and below-ground ectomycorrhizal species composition. *New Phytologist* 149:311–325.

Peterson, R. L., Massicotte, H. B., and L. H. Melville. 2004. *Mycorrhizas: Anatomy and Cell Biology*, pp. 173. Wallingford: NRC Research Press, CABI Publishing.

Pezzani, F., Montaña, G., and R. Guevara. 2006. Associations between arbuscular mycorrhizal fungi and grasses in the successional context of a two-phase mosaic in the Chihuahuan Desert, *Mycorrhiza* 16:285–295.

Pirozynski, K. A. and D. W. Malloch. 1975. The origin of land plants: A matter of mycotropism. *Biosystems* 6:153–164.

Pringle, A. and J. D. Bever. 2002. Divergent phenologies may facilitate the coexistence of arbuscular mycorrhizal fungi in a North Carolina grassland. *American Journal of Botany* 89:1439–1446.

Quilambo, O. A. 2003. The vesicular-arbuscular mycorrhizal symbiosis. *African Journal of Biotechnology* 2:539–546.

Read, D. J. 1991a. Mycorrhizas in ecosystems. *Experientia* 47:376–391.

Read, D. J. 1991b. Mycorrhizas in ecosystems—nature's response to the "law of the minimum." In *Frontiers in Mycology*, Ed. D. L. Hawksworth. Walingford, U.K.: CAB International.

Read, D. J. 1993. Mycorrhiza in plant communities. *Advances in Plant Pathology* 9:1–31.

Rillig, M. and M. F. Allen. 1998. Arbuscular mycorrhizae of *Gutierrezia sarothrae* and elevated carbon dioxide: Evidence for shifts in C allocation to and within the mycobiont. *Soil Biology and Biochemistry* 30:2001–2008.

Rillig, M. C., Hernandez, G. Y., and P. C. D. Newton. 2000. Arbuscular mycorrhizae respond to elevated atmospheric CO_2 after long-term exposure: Evidence from a CO_2 spring in New Zealand supports the resource balance model. *Ecology Letters* 3:475–478.

Rosén K. Zhong, Weiliang, Z., and A. Mårtensson. 2005. Arbuscular mycorrhizal fungi mediated uptake of ^{137}Cs in leek and ryegrass. *Science of the Total Environment* 338:283–290.

Rouhier, H. and D. J. Read. 1998. The role of mycorrhiza in determining the response of *Plantago lanceolata* to CO_2 enrichment. *New Phytologist* 139:367–373.

Ruotsalainen, A. L. and M. M. Kytöviita. 2004. Mycorrhiza does not alter low temperature impact on *Gnaphalium norvegicum. Oecologia* 140:226–233.

Saito, K., Suyama, Y., Sato, S., and K. Sugawara. 2004. Defoliation effects on the community structure or arbuscular mycorrhizal fungi based on 18S rDNA sequences. *Mycorrhiza* 14:363–373.

Sanders, I. R., Strietwolf-Engel, R., van der Heijden, M. G. A., Boller, T., and A. Wiemken. 1998. Increased allocation to external hyphae of arbuscular mycorrhizal fungi under CO_2 enrichment. *New Phytologist* 117:496–503.

Shi, L., Guttenberger, M., Kottke, I., and R. Hampp. 2002. The effect of drought on mycorrhizas of beech (*Fagus sylvatica* L.): Changes in community structure, and the content of carbohydrates and nitrogen storage bodies of the fungus. *Mycorrhiza* 12:303–311.

Smith, S. E. and D. J. Read. 1997. *Mycorrhizal Symbiosis* (2nd Edition). London: Academic Press. pp. 605.

Strandberg, M. and M. Johansson. 1998.^{134}Cs in heather seed plants grown with and without mycorrhiza. *Journal of Environmental Radioactivity* 40:175–184.

Swaty, R. L., Deckert, R. J., Whitham, T. G., and C. A. Ghering. 2004. Ectomycorrhizal abundance and community composition shifts with drought: Predictions from tree rings. *Ecology* 85:1072–1084.

Tarvainen, O., Markkola, A. M., Ahonen-Jonnarth, U., Jumpponen, A., and R. Strommer. 2004. Changes in ectomycorrhizal colonization and root peroxidase activity in *Pinus sylvestris* nursery seedlings planted in forest humus. *Scandinavian Journal of Forest Research* 19:400–408.

Taylor, A. F. S., Martin, F., and D. J. Read. 2000. Fungal diversity in ectomycorrhizal communities of Norway spruce [*Picea abies* (L.) Karst.] and Beech (*Fagus sylvatica* L.) along north–south transects in Europe. In *Ecological Studies vol. 142: Carbon and Nitrogen Cycling in European Forest Ecosystems*, Ed. Schulze, E.-D. New York: Springer.

Thompson, G. W. and R. J. Medve. 1984. Effects of aluminum and manganese on the growth of ectomycorrhizal fungi. *Applied and Environmental Microbiology* 48:556–560.

Tibbett, M., Sanders, F. E., and J. W. G. Cairney. 2002. Low-temperature-induced changes in trehalose, mannitol and arabitol associated with enhanced tolerance to freezing in ectomycorrhizal basidiomycetes (*Hebeloma* spp.). *Mycorrhiza* 12:249–255.

Torsvig, V. and L. Øvreås. 2008. Microbial diversity, life strategies, and adaptation to life in extreme soils. In *Microbiology of Extreme Soils. Soil Biology,* vol. 13, Eds. P. Dion and C. S. Nautiyal, pp. 15–43. Berlin: Springer-Verlag.

Trappe, J. M. and R. F. Strand. 1969. Mycorrhizal deficiency in a Douglas-fir region nursery. *Forest Science* 15:381–389.

Treseder, K. 2004. A meta-analysis of mycorrhizal responses to nitrogen, phosphorus, and atmospheric CO_2 in field studies. *New Phytologist* 164:347–355.

Treseder, K. A. 2005. Nutrient acquisition strategies of fungi and their relation to elevated atmospheric CO_2. In *The Fungal Community: Its Organization and Role in the Ecosystem*, 3rd edn., Eds. J. Dighton, J. F. Jr. White, and P. Oudemans, pp. 713–731. Boca Raton: Taylor & Francis.

Ulrich, B., Mayer, R., and P. K. Khanna. 1979. Deposition von Luftverunreinigungen und ihre Auswirkungen in Waldökosystemen im Solling. J. D. Sauerlanders Verlag, Frankfurt: Schriften aus der Forstlichen Fakultät der Üniversität Göttingen 58.

Urcelay, C., Bret-Harte, M. S., Díaz, S., and F. S. Chapin. 2003. Mycorrhizal colonization mediated by species interactions in arctic tundra. *Oecologia* 137:339–404.

Van der Heijden, M. G. A. 2002. Arbuscular mycorrhizal fungi as a determinant of plant diversity: In search of underlying mechanisms and general principles. In *Mycorrhzial Ecology*, Eds. M. G. A. Van der Heijden and I. R. Sanders, pp. 243–266. Berlin: Springer-Verlag.

Van der Heijden, M. G. A. and I. R. Sanders. 2002. *Mycorrhzial Ecology*, p. 469. Berlin: Springer-Verlag.

Vandenkoornhuyse, P., Ridgway, K. P., Watson, I. J., Fitter, A. H., and J. P. W. Young. 2003. Co-existing grass species have distinctive mycorrhizal communities. *Molecular Ecology* 12:3085–3096.

Väre, H., Vestberg, M., and S. Eurola. 1992. Mycorrhiza and root associated fungi in Spitsbergen. *Mycorrhiza* 1:93–104.

Wallander, H. 1995. A new hypothesis to explain allocation of dry-matter between mycorrhizal fungi and pine-seedlings in relation to nutrient supply. *Plant and Soil* 169:243–248.

Wallander, T. and I. Kottke. 1998. Nitrogen deposition and ectomycorrhizas. *New Phytologist* 139:169–187.

Wang, B. and Y. L. Qui. 2006. Phylogenetic distribution and evolution of mycorrhizas in land plants. *Mycorrhiza* 16:299–363.

Went, F. W. and N. Stark. 1986. The biological and mechanical role of soil fungi. *Proceeding of the National Academy of Sciences USA* 60:497–504.

Witkamp, M. 1968. Accumulation of ^{137}Cs by *Trichoderma viride* relative to ^{137}Cs in soil organic matter and soil solution. *Soil Science* 106:309–311.

Witkamp, M. and B. Barzansky. 1968. Microbial immobilization of ^{137}Cs in forest litter. *Oikos* 19:392–395.

Wright, S. F. and A. Upadhyaya. 1998. A survey of soils for aggregate stability and glomalin, a glycoprotein produced by hyphae of arbuscular mycorrhizal fungi. *Plant and Soil* 198:97–107.

Wu, B., Isobe, K., and R. Ishii. 2004. Arbuscular mycorrhizal colonization of the dominant plant species in primary suvccessional volcanic deserts on the Southeastern slope of Mount Fuji. *Mycorrhiza* 14:391–395.

Yoshida, S. and Y. Muramatsu. 1994. Accumulation of radiocesium in basidiomycetes collected from Japanese forests. *Science of the Total Environment* 157:197–205.

Zak, J. C. and D. Parkinson. 1983. Effects of surface amendation of two mine spoils in Alberta, Canada: Vesicular-arbuscular mycorrhizal development of slender wheatgrass: A 4-year study. *Canadian Journal of Botany* 61:798–803.

14

Effect of Arbuscular Mycorrhizal Symbiosis on Enhancement of Tolerance to Abiotic Stresses

Hinanit Koltai and Yoram Kapulnik

CONTENTS

14.1 Introduction

Most plant species harbor symbioses with soil microbes in their roots. Mycorrhizal symbiosis plays a major role in ecosystem nutrient cycling, while providing plants with essential nutrients. The most common mycorrhizal symbioses are those with arbuscular mycorrhizal fungi (AMF). The potential of AMF as biofertilizers for the enhancement of crop productivity is well recognized and continues to be exploited (Azcon-Aguilar and Barea, 1997). The AMF are members of the fungal phylum *Glomeromycota* (Schüssler et al., 2001) and form symbiotic associations with most terrestrial vascular flowering plants (Smith and Read, 1997). The abundance of mycorrhizal symbioses suggests their evolutionary success, as evidenced by the evolution of efficient survival strategies to increase the probability of meeting the host's roots and by the development of specialized symbiotic structures (reviewed by Harrison, 2005).

For clarity, the AMF–host symbiotic association can be functionally divided into two stages:

1. Presymbiotic stage: during this stage the fungal spore germinates, and the emerging hyphae develop, with limited growth. Once the hyphae come into contact with the host roots, an appressorial contact structure is formed by the fungus, and a prepenetration apparatus is formed by the host plant in a cell-layer-controlled fashion (Genre et al., 2005; Siciliano et al., 2007). During this presymbiotic stage, a reciprocal exchange of signals between the partners is required to effect significant morphological changes (reviewed by Paszkowski, 2006; Giovannetti et al., 1994; Bonfante et al., 2000).
2. Symbiotic stage: following a successful presymbiotic stage, the fungus penetrates the root epidermis and grows within the root cortex, developing characteristic functional structures called arbuscules. The fungal arbuscules remain in the apoplastic compartment surrounded by an interfacial matrix and a periarbuscular membrane that protrudes into the plant cell. These constructions are the site of bidirectional exchange between the fungus and the host; the fungus translocates mineral elements and in turn absorbs carbon from the host plant (reviewed by Harrison, 2005). In parallel, the fungus produces an extraradical mycelium from which spores are eventually formed (Smith and Read, 1997). Growth and differentiation of the plant root and fungal hyphae are tightly coordinated during mycorrhizal symbiosis, suggesting that reciprocal recognition and regulation are required, either via direct cell-to-cell interaction or via exchange of diffusible signals between the two symbionts.

AM symbiosis is recognized as beneficial to the plant host, in that it serves to increase its supply of mineral nutrients, especially under nutrient-limiting conditions. However, other key contributions have been recorded, including improved rooting and plant establishment, improved uptake of ions with low mobility, and accelerated budding and flowering (Smith and Read, 1997). Moreover, improved plant tolerance to biotic and abiotic stresses has also been recognized. This latter phenomenon is the subject of present review.

According to the ecophysiology literature, stress is composed of two processes. The first is the shortage of available resources. The second is damage to the organism's biological structures, such as cellular organelles, due to membrane, protein, or nucleic acid injury. This physiological state may lead to a significant departure from optimal metabolism, and this provokes a response in the stressed organism that is designed to limit or reduce the stress's effect on biological structures (reviewed by Pierce et al., 2005).

Mycorrhiza has been shown to promote plant resilience to a number of abiotic stress conditions. Here we describe some of the tightly coordinated events occurring between plants and AMF during the mycorrhitic association. We review studies that have demonstrated the effects of AMF on the host plant's responses under conditions of mineral depletion, i.e., when soil nutrients are not readily mobile or are present in extremely low concentrations, under conditions of drought and high soil salinity, in the presence of heavy metals in the soil, extreme soil pH, extreme temperatures, and low light intensity. We discuss the mechanisms that govern each of the AM-enhanced plant responses to these abiotic stresses, the cost vs. benefit under stress conditions, and future implications for ecosystems and sustainable agriculture.

14.2 Arbuscular Mycorrhizal Symbiosis

The biological and chemical characteristics of AMF have been in the subject of a substantial number of studies, and these have been discussed in a number of recent reviews (e.g., Gianinazzi-Pearson and Brechenmacher, 2004; Balestrini and Bonfante, 2005; Genre and Bonfante, 2005; Harrison, 2005; Hause and Fester, 2005; Karandashov and Bucher, 2005; Paszkowski, 2006; Bucher, 2007; Reinhardt, 2007). Moreover, numerous cellular and molecular events have been identified during the symbiotic AMF–plant association. For example, many mycorrhiza-induced genes have been identified as being induced both locally and systemically in the host and the fungus, including stress-associated genes and those involved in enzymatic activities, signaling, and transport (e.g., Liu et al., 2003, 2004, 2007; Wulf et al., 2003; Brechenmacher et al., 2004;

Weidmann et al., 2004; Balestrini et al., 2005; Güimil et al., 2005; Hohnjec et al., 2005; Kistner et al., 2005; Balestrini and Lanfranco, 2006 and references therein; Maeda et al., 2006; Franken et al., 2007; Javot et al., 2007; Krajinski and Frenzel, 2007; Massoumou et al., 2007).

The coordinated biological processes occur in both the fungus and the plant, suggesting that the symbiosis is a highly controlled process. This notion is strengthened by the identification of mutants that "resist" colonization by the fungi (David-Schwartz et al., 2001, 2003; Gadkar et al., 2001; Paszkowski et al., 2006; Reddy et al., 2007); the existence of such mutants suggests genetic control of mycorrhizal establishment by the host plant. Here, we briefly describe some of the symbiotic events that are tightly coordinated between the two partners.

During the presymbiotic stage, AMF colonization initiates with hyphae that arise from soil-borne propagules—resting spores (most AMF species form spores in the soil) or mycorrhizal root fragments—or from AM plants growing in the vicinity. Spores are capable of germinating in the absence of host-derived signals, but they grow only to a limited extent and are unable to produce extensive mycelia and to complete their life cycle without establishing a functional symbiosis with a host plant. Nuclear division, nuclear DNA replication (Burggraaf and Beringer, 1989), and synthesis of mitochondrial DNA, RNA, and proteins can all occur at this precontact stage in the fungal hyphae (Tamasloukht et al., 2003).

The germinating hyphae respond to the presence of nearby roots (Mosse and Hepper, 1975; Becard and Pichè, 1989; Gianinazzi-Pearson et al., 1989; Giovannetti et al., 1993a,b, 1994; Buée et al., 2000; Nagahashi and Douds, 2000; Vierheilig and Piché, 2002; Becard et al., 2004; recently reviewed by Requena et al., 2007). Plant roots release a wide range of compounds (Walker et al., 2003), many of which are involved in complex communication processes between the root and the rhizosphere microbiota. Volatile root product(s) are considered to be AMF growth stimulators (Bécard and Piché, 1989), as well as several hydrophilic and hydrophobic compounds of root exudates (Nagahashi and Douds, 2000; Buée et al., 2000). Among the latter are flavonoids, key signaling compounds in rhizobium–legume symbiosis, and stimulators of AMF symbiosis (Bécard and Piché, 1989; Gianinazzi-Pearson et al., 1989; Ruan et al., 1995; Poulin et al., 1997; Requena et al., 2007).

Strigolactones secreted by plant roots have recently recognized as important "rhizospheric plant signals" involved in stimulating the presymbiotic growth of AMF at different stages, i.e., during spore germination and during hyphal growth and branching (Matusova et al., 2005; Besserer et al., 2006; Gomez-Roldan et al., 2007). Strigolactones are present in root exudates in extremely low concentrations (Akiyama and Hayashi, 2006), and their concentration tends to rise under the suboptimal growth conditions (limited nutrition, etc.) that favor AMF colonization (Yoneyama et al., 2007).

In addition to root exudates, other factors within the rhizosphere, such as soil pH, temperature, moisture, mineral and organic nutrients, host plants, and microorganisms, influence the efficiency of AMF spore germination and establishment in the root system (Giovannetti, 2000).

Following the establishment of mycorrhizal symbiosis, nutrients are exchanged between the AMF and the host plant. As obligate symbionts, the fungi require fixed carbon from their host. In return, the fungi provide the plants with a variety of minerals, primarily phosphate (discussed in Section 14.3.1).

It was discovered that AMF mediate carbon partitioning in the plant host. They affect the carbon source, for example, by increasing leaf area and phosphate content; they affect whole-plant carbon partitioning by reducing the root-to-shoot ratio; and they affect carbon partitioning within the root, by affecting the level of carbon there and its distribution between soluble and insoluble forms. The fungi may take up glucose in either the intracellular hyphae or the arbuscules, and this uptake is estimated at between 4% and 20% of the plant's total photosynthetic products (reviewed by Douds et al., 2000). Hence, the costs and benefits of the symbiosis must be weighed by the host, and their relative weights may change under stress conditions; these and other considerations are discussed further on.

14.3 Effect of Arbuscular Mycorrhizal Symbiosis on Plant Stress Responses

One of the more intriguing effects of mycorrhizal symbiosis is a recorded elevation in the resilience of the host plants to environmental conditions. In the following, we describe several of these reports and give some insight into possible associated mechanisms.

14.3.1 Mineral Depletion

One of the most common abiotic stresses experienced by plants during their growth in soil is mineral deficiency. This is due to the fact that in most soils, the accessibility of mineral elements is lower than that required by the plant for optimal growth.

In terms of nutrient supply to the roots, two factors may be considered. One is the concentration of nutrients in the soil solution. Soil-solution mineral contents vary widely, depending on many factors, such as soil moisture, depth, pH, cation-exchange capacity, redox potential, organic matter content, and microbial activity (Marschner, 1995).

The other, in many cases, is the availability of minerals for direct uptake in the plant rhizosphere influencing the plant's ability to grow optimally or even survive. For example, an increase in the rhizospheric soil pH might alter the availability of major soil minerals such as phosphorus (P), calcium (Ca), aluminum (Al), and iron (Fe) phosphates, all of which are important for plant growth in certain quantities.

Supplementing the soil with chemical fertilizers may rectify the nutrient depletion. However, to genuinely meet the nutrient demand of soil-grown plants, nutrients must reach the root surface (rhizosphere). When the roots are active in mineral and water uptake, the mineral concentrations in close proximity to the root surfaces decrease rapidly, leading to the development of a "depletion zone." If not replenished, conditions of nutritional stress are created. AMF aid the plants in bridging the depleted zone, by promoting the uptake of mineral nutrients.

One of the possible mechanisms by which AMF bridge this depletion zone is by providing a greater root surface area via its association with the AMF hyphae, which can exploit larger volumes of soil, and hence increase the amount of nutrients, being presented to the mineral uptake apparatus. This greater root surface area may enable the plant to overcome the deficits in mineral nutrients and water typically found in the depletion zone, near the root surface. The small diameter of the fungal hyphae (averaging 3 or 4 μm) enables them to penetrate soil pores and contact soil particles/cavities that are inaccessible to roots and/or root hairs (the latter having an average diameter of >10 μm). Fungal hyphae normally transport mineral nutrients further away than nonmycorrhizal roots, thus extending the absorption zone by several millimeters to upto 10 cm (Rausch and Bucher, 2002).

The most pronounced growth enhancement elicited by AMF is due to the improved supply of several low-mobility minerals from the soil solution, predominantly P (Smith and Read, 1997), a major structural element in nucleic acids, phospholipids, and several enzymes and coenzymes. Phosphorus is also involved in energy metabolism and signal-transduction cascades. The P-absorption rate by roots is higher than its diffusion rate in the soil, leading to the creation of a depletion zone at the root surface and, as a result, to P constraints on the plant (Marschner, 1995).

External hyphae can absorb and translocate P to the host from beyond the depletion zone of nonmycorrhizal roots, and the uptake rate of P per unit root length is two or three times higher in mycorrhizal vs. nonmycorrhizal plants (Smith and Read, 1997). AMF hyphae owe their high P-uptake efficacy not only to their small diameter and large surface area but also to the accumulation/storage of polyphosphates in their vacuoles (Poirier and Bucher, 2002) and to the presence of specific P transporters. The efficacy of AMF in providing P to the host plants strongly depends on the AMF species, and even varies among different ecotypes of the same species, as well as among the plant species tested (Smith and Read, 1997).

It has been widely demonstrated that under suboptimal nutritional P levels, mycorrhizal plants are larger and exhibit greater total P uptake than nonmycorrhizal plants. The ability of AMF to contribute to plant P absorption, especially under low nutritional P levels, was termed by Smith et al. (2003) as a "direct" uptake mechanism by specific receptors of the root cell, through the root epidermis. These plant P transporters, upon which the direct P-absorption mechanism depends, may be either constitutively expressed or induced only upon AMF colonization (Rausch et al., 2001; Harrison et al., 2002; Paszkowski et al., 2002).

In addition, a P-uptake mechanism was defined via the fungal mycelium. For the uptake facilitated by the mycorrhizal fungi, two AMF P transporters, exhibiting high affinity toward P, were identified: GvPT and GiPT from *Glomus versiforme* and *G. intraradices*, respectively (Harrison and van Buuren, 1995; Maldonado-Mendoza et al., 2001; Benedetto et al., 2005). These fungal P transporters were shown to be located on the external hyphae, which are responsible for the uptake of rhizospheric P and its translocation to the host plant (Brundrett, 2002). It was suggested that most P can be taken up via the mycorrhizal uptake pathway (Smith et al., 2003, 2004).

Nevertheless, the mechanism governing P translocation from the fungus to the plant is currently unknown. It might be that translocation of phosphate ions through the fungal plasma membrane into the host cells is facilitated by the plant P transporters—perhaps those that are induced by the AMF (Harrison and van Buuren, 1995; Maldonado-Mendoza et al., 2001). This notion is supported by the finding that induced plant (*Medicago*) P transporters exhibited higher affinity toward phosphate than the fungal P transporter (Harrison et al., 2002), which may lead to P translocation from fungi to plant cells.

In conclusion, due to the abundance of mycorrhizal associations with a variety of plant species, in most natural and agricultural ecosystems, the AMF-induced P transporters—of either plant or fungal origin—may be pivotal in determining P acquisition for plant productivity and fitness.

Apart from P, which has been the most intensively studied, AMF have been shown to take up other elements, including nitrogen (N), zinc (Zn), copper (Cu), calcium (Ca), magnesium (Mg), potassium (K), sulfur (S), and manganese (Mn), all of which are found at distinctly higher levels in mycorrhizal vs. nonmycorrhizal plants (Raghothama, 2000).

Apart from the activity of P transporters, the mechanism that governs mineral absorption is not well understood. Recently, specialized structures, termed branch-absorbing structures, have been implicated in mineral absorption (Cavagnaro et al., 2001). Other mechanisms have been proposed, such as the release of specific phosphatases, metabolites, and siderophores capable of solubilizing and promoting the uptake of mineral elements from the rhizosphere, but with no supporting empirical data. Another possible mechanism for the enhancement of mineral acquisition by mycorrhitic plants may relate to the architecture of their roots, which are shorter and more branched than nonmycorrhitic roots; this architecture increases the total root-elongation zone, which is the site of mineral acquisition (Marschner, 1995).

The above described studies show that AMF can help relieve mineral stress in plants. However, high mineral nutrient levels, as found in highly fertile soils, for example, can inhibit fungal spore germination and growth. Moreover, it has been conclusively shown that high P levels strongly inhibit various stages of the spore's ontogenic cycle and do not result in the establishment of symbiosis, whereas high levels of organic amendment lead to the production of a toxic compound, such as ammonia, which suppresses the soil biota, including AMF (Pfleger and Linderman, 1994).

14.3.2 Drought

Drought and salt are considered the most important abiotic stress conditions in terms of their significant inhibition of plant growth, development, and reproducibility (Kramer and Boyer, 1997). Both drought and salinity conditions impose osmotic stress on the plant; this stress is perceived by the plant and provokes a defensive response (reviewed by Pierce et al., 2005). In addition, salinity imposes solute toxicity (discussed below).

AM symbiosis has been shown in a number of studies to improve plant resistance to drought (e.g., Subramanian and Charest, 1998; Ruiz-Lozano and Azcón, 2000; Porcel et al., 2003; Bolandnazar et al., 2007), with both increased dehydration-avoidance and tolerance being reported (Allen and Boosalis, 1983; Davies et al., 1993; Augé, 2001). In contrast, other studies have shown little or no AM enhancement of drought resistance (e.g., Hetrick et al., 1987; Simpson and Daft, 1991). In the following, we present several studies that suggest possible mechanisms underlying this effect.

14.3.2.1 Effect on Water Movement

AMF have been suggested to affect the rate of water movement into, within, and out of the plant (reviewed by Ruiz-Lozano, 2003); this ability has been shown to be unrelated to improvements in P acquisition (Augé, 2001). The effect of AMF on plant water movement was suggested to be mediated via their effect on osmotic adjustments in the plant. Osmotic adjustment, or osmoregulation, is the ability of cells to decrease their osmotic potential by actively accumulating organic compounds (such as proline), in order to maintain water content, cell turgor, and associated cellular processes (Morgan, 1984; Hoekstra et al.,

2001). AMF were shown to enhance the plant cell's ability to accumulate organic compounds (reviewed by Ruiz-Lozano, 2003). Moreover, levels of ions (Mg^{2+}, K^+, and Ca^{2+}) and sugars, including soluble sugars and soluble starch, were shown to increase in mycorrhitic plants. Proline levels were reduced, suggesting osmotic adjustment of mycorrhitic plants mainly via adjustment of ion and sugar levels, rather than via proline levels (Wu and Xia, 2006). In contrast, Porcel and Ruiz-Lozano (2004) found that AM roots of soybean accumulate more proline than non-AM roots, while the opposite was observed in shoots.

Perhaps as a result of the fungus ability to enhance water uptake and transport, an increment in leaf stomatal conductance is achieved in mycorrhitic plants, accompanied by a higher rate of gas exchange (Augé et al., 1987, 1992, 2004; Augé, 1989, 2001; Ruiz-Lozano et al., 1995a,b; Duan et al., 1996; reviewed in Bethlenfalvay et al., 1987; Goicoechea et al., 1997; Cho et al., 2006). AMF were shown to affect leaf water potential and leaf osmotic potential as well, albeit to a lesser extent, perhaps also due to the enhanced water uptake and transport (Allen and Boosalis, 1983; Augé et al., 1986; Cho et al., 2006). Accordingly, enhanced cumulative transpiration was observed in mycorrhitic plant leaves (Querejeta et al., 2007).

14.3.2.2 Effect on Soil's Water-Retention Properties

Augé et al. (2001) suggested that the AM symbiosis affects the soil's water-retention properties. The fungal exudates were shown to affect soil structure (Jastrow and Miller, 1991; Oades and Waters, 1991) and promote soil aggregation (Rillig et al., 2002). This, in turn, increased the soil moisture retention (Hamblin, 1985), and reduced the changes in soil matric potential upon drought (Augé et al., 2001).

14.3.2.3 Protection against Oxidative Stress

One of the effects of drought in plants is to provoke oxidative stress, associated with degenerative reactions. Plants mediate oxidative stress by producing or activating reactive oxygen species, such as superoxide dismutases (SODs; reviewed by Blokhina et al., 2003). The AM plants possesses SOD activity (Palma et al., 1993); mycorrhizal plants demonstrate elevated transcript levels of plant-originated SOD and higher SOD activity, relative to nonmycorrhitic plants (Ruiz-Lozano et al., 1996b, 2001). Similarly, increased concentrations of antioxidant enzymes and nonenzymatic antioxidants were found in mycorrhitic citrus plants; these enzymes and compounds were suggested to protect plants against oxidative damage. In turn, such protection against oxidative stress was suggested to enhance drought tolerance (Wu et al., 2007). In addition, Porcel and Ruiz-Lozano (2004) showed lower oxidative damage to lipids in shoots of desiccated AM soybean plants relative to their non-AM counterparts. However, no correlation was found between the lower oxidative damage to lipids in AM plants and the activity of antioxidant enzymes.

In conclusion, it is evident that in most cases, AMF enhance plant protection against drought conditions. This protection may be mediated via protection against or reduction of oxidative stress, by an effect on water movement within the plants, by an alteration of soil water retention properties, or by some combinations of these two. However, how do the mycorrhizal fungi, as plant-root inhabitants, have such a pronounced effect on plant leaves? It may be that mycorrhitic roots pose the first and ultimate barrier in the protection of plant shoots from drought-associated damage: AM symbiosis may enhance osmotic adjustments in the roots that control water movement from the soil into the roots; this in turn contributes to maintaining a favorable water potential gradient in the leaves, and hence protecting the plants against oxidative stress (Porcel and Ruiz-Lozano, 2004). Moreover, soil colonization by AMF, leading to alterations in soil structure near the mycorrhitic plant, has been suggested to be more influential than root colonization on plant stomatal conductance (Augé et al., 2007).

14.3.3 High Salinity

High soil salinity may affect plants by creating both drought conditions and salt toxicity (Munns, 2002). AM symbiosis has been frequently shown to increase the resilience of host plants to salinity stress, with even greater consistency than to drought stress. Salt resistance is improved by AMF colonization in a large number of crops, including maize (Feng et al. 2002), mung bean (Jindal et al., 1993), clover (Ben Khaled et al., 2003), guayule (Pfeiffer and Bloss, 1987), *Sesbania* sp. (Giri and Mukerji, 2004), cucumber (Rosendahl and Rosendahl, 1991), lotus (Sannazzaro et al., 2007), and tomato (Al-Karaki, 2000; Al-Karaki et al., 2001). Reductions in Na^+ uptake together with an associated increase in P, N, and Mg absorption and high chlorophyll content were suggested to support the salinity-resistant phenotype of AMF-inoculated plants (Cho et al., 2006 and references therein). In addition, improved osmoregulation via polyamine and proline accumulation was demonstrated (Sannazzaro et al., 2007). In contrast, in a study on mycorrhitic lettuce responding to salt stress, mycorrhitic plants had higher water and lower proline levels (Jahromi et al., 2008). Neither P supplementation nor high nutritional condition alone could recapitulate AM-related protection against salinity stress in lettuce and alfalfa (Ruiz-Lozano et al., 1996a; Azcón and El-Atrash, 1997).

Under soil drought conditions, solutes may concentrate in the soil solution adjacent to the roots as the soil dries, leading to high-salinity conditions in close proximity to the roots (Stirzaker and Passioura, 1996). This raises the question of whether the AMF-induced drought response might be mediated by a reaction to high salinity. Not necessarily, according to Cho et al. (2006) results, they showed that addition of salt could nullify, rather than enhance, the AMF-induced drought response in plants (e.g., as reflected by stomatal conductance measurements), in some cases.

Interestingly, lower levels of response to salt stress, as reflected by a reduction in the expression of a stress-marker gene and of abscisic acid, were recorded in mycorrhitic plant roots as compared to noninoculated roots. This suggests reduced salt-stress injury as a possible mechanism for AMF enhancement of the plant's resilience to salt stress (Jahromi et al., 2008).

14.3.4 Heavy Metals

Soil contamination with heavy metals can be due to various natural activities, volcanic for example, and human activities such as metal smelting, use and drainage of chemical agroproducts (e.g., fertilizers and pesticides), mining tail dumping, and burning fossil fuels. In small quantities, some of these metals, such as Cu, Mn, nickel (Ni), and cobalt (Co) (Marschner and Romheld, 1995), are the necessary micronutrients for plant growth; others, however, such as cadmium (Cd), lead (Pb), and mercury (Hg), are not biologically functional and may be harmful to a broad variety of species including humans, other mammals, birds, and plants. Recently, exceedingly high levels of these compounds in the environment have prompted governmental and environmental organizations to dictate removal activities (such as soil remediation).

Of all the potential remediation alternatives, phytoremediation, which is the use of plants to extract, sequester, or detoxify pollutants through physical, chemical, or biological processes, can be used at relatively low cost (Raskin and Ensley, 2000). The success of phytoremediation depends on the extent of the soil contamination, bioavailability of the metal, and the ability of the plant to absorb and accumulate metals in its shoots. However, plants with exceptionally high metal-accumulating capacity often have a slow growth rate and produce limited amounts of biomass when the concentration of metal in the contaminated soil is very high and toxic. Thus, alternatives that include "fast-growing" plants have been exploited; nevertheless, their adaptation and survival in the contaminated sites has been limited (Raskin and Ensley, 2000).

To exploit the benefits of phytoremediation, the use of plant-growth-promoting microorganisms, such as AMF, has been recommended. It has been postulated that AMF enhance heavy-metal-sequestration capacity, nutrient recycling, soil-structure improvement, and chemical detoxification in the plant rhizosphere. The benefits to the plant, fungus, and the environment from mycorrhitic associations under heavy-metal contamination are discussed below.

14.3.4.1 Contribution of AMF to Plant Adaptation and Growth in Heavy-Metal-Contaminated Soils

Introduction of an AMF inoculum into heavy-metal polluted sites might be one approach to facilitate the establishment of mycorrhizal-dependent plant species there; AMF increase plant establishment and growth, despite high levels of heavy metal in the soil (Enkhtuya et al., 2002). This ability may to be a result of better plant nutrition (Taylor and Harrier, 2001; Feng et al., 2003), better water availability to the plant (Augé, 2001), and alterations via the symbiosis of soil-aggregation properties (described above; Kabir and Koide, 2000; Rillig and Steinberg, 2002). Two potential scenarios for AMF enhancement of plant growth under heavy-metal contamination can be envisioned.

1. Heavy-metal-tolerant AMF (see below) could protect plants against excessive absorption of heavy metals (Sylvia and Williams, 1992; Leyval et al., 1997) by immobilizing those metals in their own biomass (Li and Christie, 2000; Zhu et al., 2001). This fungal capacity creates a barrier for metal transport, perhaps by intracellular precipitation of the heavy metals (Joner et al., 2000; Tullio et al., 2003). Heavy metal can be precipitated onto polyphosphate granules, and as a result be compartmentalized into plastids or other membrane-rich organelles (Turnau et al., 1993; Kaldorf et al., 1999). Alternatively, adsorption of heavy metals onto plant or fungal cell walls (Joner et al., 2000) may predominate. This adsorption might be facilitated by the chelation of heavy metals by compounds such as siderophores and metallothionens, which are released by AMF or by other rhizosphere microbes, or by plant-derived compounds like the phytochelatins or phytates (Joner and Leyval, 1997).
2. Heavy-metal-tolerant AMF does not prevent plants from absorbing excessive amounts of heavy metals. Rather, due to colonization by AMF, increased uptake, transport, and subsequent accumulation of heavy metals in aboveground plant tissues occur. However, in several cases, mycorrhizal colonization leads to accumulation of heavy metals in the root (Gaur and Adholeya, 2004). Only the former (accumulation in aboveground tissues) may be suitable for the efficient phytoextraction of heavy metals; these contaminated upper plant parts can then be removed to facilitate site cleanup.

Elevation of plant tolerance to heavy-metal stress is feasible for AMF that are adapted to elevated metal concentrations in the soil, whereas different AMF have been shown to differ in their susceptibility/tolerance to heavy metals. del Val et al. (1999) reported six AMF ecotypes with consistent differences in their tolerance to the presence of heavy metals.

14.3.4.2 AMF Tolerance to Heavy Metals

Although higher mycorrhizal colonization has been demonstrated in strongly contaminated sites (Hildebrandt et al., 1999; Tonin et al., 2001), fungal spores and presymbiotic hyphae are generally sensitive to heavy metals in the absence of plants (del Val et al., 1999). However, spores from polluted soils were more tolerant to, and germinated better in, heavy-metal-polluted soil than spores from nonpolluted soils (Leyval et al., 1995; reviewed by Gaur and Adholeya, 2004). AMF tolerance varies with fungal genotype (Shalabyl, 2003). Hence, spores from polluted soils have evolved a tolerance to some of the heavy metals. Shalabyl (2003) suggested that this naturally occurring resistance is likely due to phenotypic plasticity rather than genetic changes in the spores, because tolerance is lost after one generation in the absence of heavy metals. Weissenhorn et al. (1995) suggested high tolerance of an indigenous AMF population to elevated metal concentrations in both soil and roots.

14.3.4.3 Contribution of AMF to Heavy-Metal Phytoremediation

The fact that AMF can be identified in heavy-metal-contaminated soils has encouraged the evaluation of their potential contribution to phytoremediation. Thus, it is important to use heavy-metal-tolerant isolates to guarantee the effectiveness of AM symbiosis on site. No less important is the fact that the phytoremediation potential of contaminated soils can be enhanced by inoculating hyper-accumulator plants with the mycorrhizal fungi most suited to the contaminated site. Thus, combining selected plants with specific AMF isolates that have adapted to a given heavy metal existing on site is the most promising approach in future research and application efforts for phytoremediation.

14.3.5 Extreme Soil pH

By definition, an acidic soil has relatively high concentrations of H ions, whereas pH per se is often not what restricts plant growth in acid soils. Rather, toxicities of Al and Mn and deficiencies of P, Ca, Mg, and K have been reported under low pH conditions (reviewed by Clark, 1997), and it is this that leads to plant growth retardation. Soil alkalinity is an increase of pH in water solution in the soil caused by, for example, HCO_3^- and CO_3^{2-}. Soil alkalinity leads to the formation of insoluble forms of P and micronutrients and reduced mineral uptake by the plant (Alhendawi et al., 1997).

Since on the one hand mineral depletion or toxicity under extreme pH conditions (both high and low) are restrictors of plant growth, and on the other AMF are known to alter mineral acquisition by the host plant (see earlier), mycorrhiza may enhance plant resilience to extreme soil pHs. This notion has been examined only in a few studies. Enhanced shoot acquisition of N, S, Zn, Cu, and silicon (Si) was noted for a variety of plant species grown in acidic soils and colonized with AMF. Moreover, the cationic elements so commonly deficient in acid soils (Ca, Mg, and K) were also greatly enhanced in mycorrhitic plants compared to noninoculated plants under conditions of low soil pH (Siqueira et al., 1990; Medeiros et al., 1994b; Clark and Zeto, 1996b). In addition, higher resilience to Al toxicity, in some cases via reduction in Al acquisition, was recorded in mycorrhitic plants relative to their noninoculated counterparts, resulting in overcoming growth inhibition (Clark and Zeto, 1996a,b; Medeinos et al., 1994a). Under alkaline soil conditions, AMF reduce the stress caused by HCO_3^-, leading to enhanced *Vinca* growth, increased levels of P and additional minerals, and increased levels of antioxidants in *Vinca* leaves (Cartmill et al., 2008).

Hence, under conditions of extreme pH, mycorrhization may lead to changes in mineral uptake that promote growth of the host plant. However, under conditions of no applied pH stress, reduced growth was recorded for mycorrhitic plants relative to their noninoculated counterparts (e.g., Cartmill et al., 2008).

14.3.6 Extreme Temperatures

Higher resilience to heat stress has been demonstrated for mycorrhitic plants. Heat-stressed mycorrhitic strawberry plants had a higher number of leaves and roots, increased leaf area, increased crown diameter, and increased leaf and root dry weight relative to nonmycorrhitic plants (Matsubara et al., 2004). Under moderate growth temperatures, mycorrhitic pepper plants showed higher shoot dry weights and higher leaf P levels, while root dry weight was highest for non-AM plants. At higher growth temperatures, mycorrhitic pepper growth was enhanced relative to non-AM controls, despite reduced levels of AMF colonization under these conditions; however, different AMF inocula had different effects on the growth of pepper plants at high temperatures (Martin and Stutz, 2004). Hence, mycorrhiza may enhance plant growth under high-temperature growth conditions. However, no clear mechanism for this enhancement has yet been delineated.

14.3.7 Low Light Intensity

Low light intensity may dramatically delay plant growth and development. Nevertheless, only a handful of studies have methodically examined the ability of mycorrhizal symbiosis to enhance plant growth under conditions of low light intensity. Such conditions limit plant photosynthetic ability; since mycorrhizae rely on the efflux of photosynthesized products from the plant to the fungi, their benefit to the host plant may be questionable under these conditions.

Conditions of low light intensity are especially prevalent under the tree canopy in tropical ecosystems, representing a most important growth factor there (Lee et al., 1996). Bereau et al. (2000) exposed *Dicorynia guianensis* Amshoff, an endemic Amazonian forest tree species, to different light-intensity regimes. Under low light intensity (1% of ambient sunlight), AMF-inoculated trees exhibited a higher rate of mortality and a lower number of leaflets than noninoculated trees. However, under medium light intensity (14% and 50% of ambient sunlight), the AMF-inoculated plants exhibited better growth performance than the noninoculated ones. Correspondingly, AMF development was more prominent under medium light intensity than under low light intensity (Bereau et al., 2000).

In a study by Gehring (2003), AMF enhancement of the host was less pronounced, and only significant in one out of four examined species of Australian forest trees, whereas AMF colonization in this one species was enhanced under "small gap" light conditions (i.e., medium-light-intensity regime). Together, these results suggest that, above a certain threshold of photosynthesis, mycorrhiza may benefit host growth under low-light-intensity stress. However, below this threshold, fungal development is inhibited, and rather than beneficial to the host, it may become a burden, inhibiting host growth. Under these conditions, the carbon demanded by the fungi becomes a cost that is too high for the stressed plant to bear.

14.4 Conclusions

Most of the studies examining AMF effects on their host have suggested that the mycorrhizal association enhances plant resilience to adverse abiotic stresses. Some of the associated physiological, chemical, and genetic events were identified, shedding some light on the mechanisms associated with the AM effect on plant responses. These findings allow us to raise the following issues, which may serve to further dissect this biological phenomenon and its evolution.

Enhancement of stress resilience by mycorrhizae—existence of one main mechanism?

The most pronounced contribution of AMF to their hosts' resilience to stress conditions may involve mineral and solute acquisition, distribution, and partitioning. Although this may not be the sole mechanism involved, it may explain AMF-enhanced plant stress responses to drought, salinity, mineral starvation, and extreme pH, as well as their tolerance to heavy metals. This primary effector may lead to other, indirect plant responses (such as modulations in stomatal conductivity). As such, it may serve as an alternative to the plant-originated stress-relief mechanisms. Moreover, it may be that via mycorrhization, the plant is preconditioned to withstand stress conditions, even before their occurrence. This in turn enables the plant to better withstand any potential stress exposure.

Mutualism–parasitism continuum: Cost vs. benefit under stress conditions

There is a continuum of AMF–host relationships, ranging from symbiosis to parasitism (reviewed by Johnson et al., 1997; Klironomos, 2003). Plant growth responses to different mycorrhizal isolates can range, within an ecosystem, from highly mutualistic (i.e., symbiosis) to highly parasitic (Klironomos, 2003). Various stress conditions may be one of the factors determining the coordinates of the mycorrhitic association along the mutualism–parasitism continuum. For example, under low light conditions, where carbon sources are limited and hence the cost imposed by the AMF association is greater than its benefit, the association may tend toward parasitism. On the other hand, under other stress conditions, such as drought or mineral depletion, when the carbon source is not limited, the presence of AMF may benefit the host, allowing it to withstand the stress. Under these conditions, the mycorrhizal association may be regarded as a true symbiosis, with both partners gaining mutual benefits from the association.

This raises the following question: Has the evolution of mycorrhizal symbiosis been enhanced by stress conditions? On one hand, arbuscules were discovered in an early Devonian (417–359 million years ago [MYA]) land plant, suggesting that mycorrhizae were already present in the Early Devonian, and probably much earlier (Remy et al., 1994). On the other, during the Silurian and Devonian eras, 443–359 MYA, nutrients were limited and might have constituted the primary selection pressure for plant growth (reviewed by Pierce et al., 2005). It might be that stress conditions during these eras enhanced the evolution of mycorrhizal symbiosis. Alternatively, within the evolving symbiosis, enhancement of resilience to stress conditions may have evolved for both the symbiotic partners. In either case, the ability of AMF to elevate plant's resistance to stress may have been one of the factors promoting the coevolution of plant and fungi.

Future implications for ecosystems and sustainable agriculture

AMF inoculation has been shown to have a profound effect on the structure of plant communities. For instance, AMF were shown to affect the competitive ability of salt-marsh plants such that under conditions of low nutrient supply, AMF-inoculated plants had a competitive edge over noninoculated plants (Daleo et al., 2008). In addition, the diversity of AMF in ecosystems has been shown to be an important factor enhancing plant biodiversity, and increased plant biodiversity has been shown to contribute greater ecosystem productivity (van der Heijden et al., 1998 and references therein). Hence, AMF may be considered dispersal agents (reviewed by Purin and Rillig, 2008), promoting the plants' adaptiveness to a given ecosystem, and therefore increasing ecosystem productivity.

Due to the fungi's ability to promote plant resilience to stress conditions, the mycorrhiza's contribution to the ecosystem may be sustainable, especially under prolonged stress conditions. In view of the predicted changes in the global environment, AMF symbiosis may serve as a force for the enhancement of plant adaptation to the altered, developing new ecosystems.

For future agricultural practices, it is expected that new crop varieties will be needed to withstand the developing stress conditions resulting from the impending global environmental changes. Mycorrhizal associations may have an important role in cooping with the upcoming global changes, taking two routes. One is by expanding the physiological plasticity (e.g., stress resilience), for the current variety of agricultural and horticultural crops. Two is by enhancement of development of crop with desired traits, which will be able to cope with the expected, future stress conditions via the mycorrhitic association. Either way, mycorrhiza is an expected vision as a part of a sustainable, durable agriculture.

ACKNOWLEDGMENTS

The work was funded by The Israeli Ministry of Agriculture and Rural Development.

REFERENCES

Akiyama, K. and H. Hayashi. 2006. Strigolactones: Chemical signals in fungal symbionts and parasitic weeds in plant roots. *Annals of Botany* 97:925–931.

Alhendawi, R. A., Romheld, V., Kirkby, E. A., and H. Marschner. 1997. Influence of increasing bicarbonate concentration on plant growth, organic acid accumulation in roots and iron uptake by barley, sorghum, and maize. *Journal of Plant Nutrition* 20:1731–1753.

Al-Karaki, G. N. 2000. Growth of mycorrhizal tomato and mineral acquisition under salt stress. *Mycorrhiza* 10:51–54.

Al-Karaki, G. N., Hammad, R., and M. Rusan. 2001. Response of two tomato cultivars differing in salt tolerance to inoculation with mycorrhizal fungi under salt stress. *Mycorrhiza* 11:43–47.

Allen, M. F. and M. G. Boosalis. 1983. Effects of two species of VA mycorrhizal fungi on drought tolerance of winter wheat. *New Phytologist* 93:67–76.

Augé, R. M. 1989. Do VA mycorrhiza enhance transpiration by influencing host phosphorus status? *Journal of Plant Nutrition* 12:743–753.

Augé, R. M. 2001. Water relations, drought and VA mycorrhizal symbiosis. *Mycorrhiza* 11:3–42.

Augé, R. M., Scheckel, K. A., and R. L. Wample. 1986. Greater leaf conductance of well-watered VA mycorrhizal rose plants is not related to phosphorus nutrition. *New Phytologist* 103:107–116.

Augé, R. M., Scheckel, K. A., and R. L. Wample. 1987. Leaf water and carbohydrate status of VA mycorrhizal rose exposed to water deficit stress. *Plant and Soil* 99:291–302.

Augé, R. M., Stodola, A. J. W., Brown, M. S., and G. J. Bethlenfalvay. 1992. Stomatal responses of mycorrhizal cowpea and soybean to short-term osmotic stress. *New Phytologist* 120:117–125.

Augé, R. M., Stodola, A. J. W., Tims, J. E., and A. M. Saxton. 2001. Moisture retention properties of a mycorrhizal soil. *Plant and Soil* 230:87–97.

Augé, R. M., Moore, J. L., Sylvia, D. M., and K. Cho. 2004. Mycorrhizal promotion of host stomatal conductance in relation to irradiance and temperature. *Mycorrhiza* 14:85–92.

Augé, R. M., Toler, H. D., Moore, J. L., Cho, K., and A. M. Saxton. 2007. Comparing contributions of soil versus root colonization to variations in stomatal behavior and soil drying in mycorrhizal *Sorghum bicolor* and *Cucurbita pepo*. *Journal of Plant Physiology* 164:1289–1299.

Azcón, R. and F. El-Atrash. 1997. Influence of arbuscular mycorrhizae and phosphorus fertilization on growth, nodulation and N_2 (N-15) in *Medicago sativa* at four salinity levels. *Biology and Fertility of Soils* 24:81–86.

Azcón-Aguilar, C. and J. M. Barea. 1997. Arbuscular mycorrhizas and biological control of soil-borne plant pathogens—an overview of the mechanisms involved. *Mycorrhiza* 6:457–464.

Balestrini, R. and P. Bonfante. 2005. The interface compartment in arbuscular mycorrhizae: A special type of plant cell wall? *Plant Biosystems* 139:8–15.

Balestrini, R. and L. Lanfranco. 2006. Fungal and plant gene expression in arbuscular mycorrhizal symbiosis. *Mycorrhiza* 16:509–524.

Balestrini, R., Cosgrove, D. J., and P. Bonfante. 2005. Differential location of expansin proteins during the accommodation of root cells to an arbuscular mycorrhizal fungus. *Planta* 220:889–899.

Bécard, G. and Y. Piché. 1989. Fungal growth stimulation by CO_2 and root exudates in vesicular-arbuscular mycorrhizal symbiosis. *Applied Environmental Microbiology* 55:2320–2325.

Bécard, G., Kosuta, S., Tamasloukht, M., Séjalon-Delmas, N., and C. Roux. 2004. Partner communication in the arbuscular mycorrhizal interaction. *Canadian Journal of Botany* 82:1186–1197.

Benedetto, A., Magurno, F., Bonfante, P., and L. Lanfranco. 2005. Expression profiles of a phosphate transporter gene (GmosPT) from the endomycorrhizal fungus *Glomus mosseae. Mycorrhiza* 15:620–627.

Ben Khaled, L., Gomez, A. M., Ouarraqi, E. M., and A. Oihabi. 2003. Physiological and biochemical responses to salt stress of mycorrhized and/or nodulated clover seedlings (*Trifolium alexandrinum* L.). *Agronomie* 23:571–580.

Bereau, M., Barigah, T. S, Louisanna, E., and J. Garbaye. 2000. Effects of endomycorrhizal development and light regimes on the growth of *Dicorynia guianensis* Amshoff seedlings. *Annals of Forest Science* 57:725–733.

Besserer, A., Puech-Pagès, V., Kiefer, P., Gomez-Roldan, V., Jauneau, A., Roy, S., Portais, J., Roux, C., Bécard, G., and N. Séjalon-Delmas. 2006. Strigolactones stimulate arbuscular mycorrhizal fungi by activating mitochondria. *PLoS Biology* 4:1239–1247.

Bethlenfalvay, G. J., Brown, M. S., Mihara, K. L., and A. E. Stafford. 1987. The *Glycine-Glomus-Bradyrhizobium* symbiosis. V. Effects of mycorrhiza on nodule activity and transpiration in soybean under drought stress. *Plant Physiology* 85:115–119.

Blokhina, O., Virolainen, E., and K. V. Fagerstedt. 2003. Antioxidants, oxidative damage and oxygen deprivation stress: A review. *Annals of Botany* 91:179–194.

Bolandnazar, S., Aliasgarzad, N., Neishabury, M. R., and N. Chaparzadeh. 2007. Mycorrhizal colonization improves onion (*Allium cepa* L.) yield and water use efficiency under water deficit condition. *Scientia Horticulturae* 114:11–15.

Bonfante, P., Genre, A., Faccio, A., Martini, I., Schauser, L., Stougaard, J., Webb, K. J., and M. Parniske. 2000. The *Lotus japonicus LjSym4* gene is required for the successful symbiotic infection of root epidermal cells. *Molecular Plant-Microbe Interactions* 13:1109–1120.

Brechenmacher, L., Weidmann, S., van Tuinen, D., Chatagnier, O., Gianinazzi, S., Franken, P., and V. Gianinazzi-Pearson. 2004. Expression profiling of up-regulated plant and fungal genes in early and late stages of *Medicago truncatula–Glomus mosseae* interactions. *Mycorrhiza* 14:253–262.

Brundrett, M. C. 2002. Coevolution of roots and mycorrhizas of land plants. *New Phytologist* 154:275–304.

Bucher, M. 2007. Functional biology of plant phosphate uptake at root and mycorrhiza interfaces. *New Phytologist* 173:11–26.

Buée, M., Rossignol, M., Jauneau, A., Ranjeva, R., and G. Bécard. 2000. The pre-symbiotic growth of arbuscular mycorrhizal fungi is induced by a branching factor partially purified from plant root exudates. *Molecular Plant-Microbe Interactions* 13:693–698.

Burggraaf, A. J. P. and J. E. Beringer. 1989. Absence of nuclear DNA synthesis in vesicular-arbuscular mycorrhizal fungi during in vitro development. *New Phytologist* 111:25–33

Cartmill, A. D., Valdez-Aguilar, L. A., Bryan, D. L., and A. Alarcon. 2008. Arbuscular mycorrhizal fungi enhance tolerance of vinca to high alkalinity in irrigation water. *Scientia Horticulturae* 115:275–284.

Cavagnaro, T. R., Gao, L-L., Smith, F. A., and S. E. Smith. 2001. Morphology of arbuscular mycorrhizas is influenced by fungal identity. *New Phytologist* 151:469–475.

Cho, K., Toler, H., Lee, J., Ownley, B., Stutz, J. C., Moore, J. L., and R. M. Augé. 2006. Mycorrhizal symbiosis and response of sorghum plants to combined drought and salinity stresses. *Journal of Plant Physiology* 163:517–528.

Clark, R. B. 1997. Arbuscular mycorrhizal adaptation, spore germination, root colonization, and host plant growth and mineral acquisition at low pH. *Plant and Soil* 192:15–22.

Clark, R. B. and S. K. Zeto. 1996a. Growth and root colonization of mycorrhizal maize grown on acid and alkaline soil. *Soil Biology and Biochemistry* 28:1505–1511.

Clark, R. B. and S. K. Zeto. 1996b. Mineral acquisition by mycorrhizal maize grown on acid and alkaline soil. *Soil Biology and Biochemistry* 28:1495–1503.

Daleo, P., Alberti, J., Canepuccia, Escapa, A. M., Fanjul, E., Silliman R., Bertness, M. D., and O. Iribarne. 2008. Mycorrhizal fungi determine salt-marsh plant zonation depending on nutrient supply. *Journal of Ecology* doi: 10.1111/j.1365–2745.2007.01349.x.

David-Schwartz, R., Badani, H., Wininger, S., Levy, A., Galili, G., and Y. Kapulnik. 2001. Identification of a novel genetically controlled step in mycorrhizal colonization: Plant resistance to infection by fungal spores but not extra-radical hyphae. *The Plant Journal* 27:561–569.

David-Schwartz, R., Gadkar, V., Wininger, S., Bendov, R., Galili, G., Levy, A., and Y. Kapulnik. 2003. Isolation of a *pre-my*corrhizal *i*nfection (*pmi*2) mutant of tomato, resistant to arbuscular mycorrhizal fungal colonization. *Molecular Plant-Microbe Interactions* 16:382–388.

Davies, F. T., Potter, J. R., and R. G. Linderman. 1993. Drought resistance of mycorrhizal pepper plants independent of leaf P-concentration—response in gas exchange and water relations. *Physiologia Plantarum* 87:45–53.

del Val, C., Barea, J. M., and C. Azcòn-Aguilar. 1999. Assessing the tolerance to heavy metals of arbuscular mycorrhizal fungi isolated from sewage sludge-contaminated soils. *Applied Soil Ecology* 11:261–269.

Douds, D. D. J., Pfeffer, P. E., and Y. Shachar-Hill. 2000. Carbon partitioning, cost, and metabolism of arbuscular mycorrhizas. In *Arbuscular Mycorrhizas: Physiology and Function*, Eds. Y. Kapulnik and D. D. J. Douds, pp. 107–129. Dordrecht, The Netherlands: Kluwer Academic Publishers.

Duan, X., Newman, D. S., Reiber, J. M., Green, C. D., Saxton, A. M., and R. M. Augé. 1996. Mycorrhizal influence on hydraulic and hormonal factors implicated in the control of stomatal conductance during drought. *Journal of Experimental Botany* 47:1541–1550.

Enkhtuya, B., Rydlová, J., and M. Vosátka. 2002. Effectiveness of indigenous and non-indigenous isolates of arbuscular mycorrhizal fungi in soils from degraded ecosystems and man-made habitats. *Applied Soil Ecology* 14:201–211.

Feng, G., Song, Y. C., Li, X. L., and P. Christie. 2003. Contribution of arbuscular mycorrhizal fungi to utilization of organic sources of phosphorus by red clover in a calcareous soil. *Applied Soil Ecology* 22:139–148.

Feng G., Zhang, F. S., Li, X. L., Tian, C. Y., Tang, C., and Z. Rengel. 2002. Improved tolerance of maize plants to salt stress by arbuscular mycorrhiza is related to higher accumulation of soluble sugars in roots. *Mycorrhiza* 12:185–190.

Franken, P., Donges, K., Grunwald, U., Kost, G., Rexer, K. H., Tamasloukht, M., Waschke, A., and D. Zeuske. 2007. Gene expression analysis of arbuscule development and functioning. *Phytochemistry* 68:68–74.

Gadkar, V., David-Schwartz, R., Kunik, T., and Y. Kapulnik. 2001. Arbuscular mycorrhizal fungal colonization. Factors involved in host recognition. *Plant Physiology* 127:1493–1499.

Gaur, A. and A. Adholeya. 2004. Prospects of arbuscular mycorrhizal fungi in phytoremediation of heavy metal contaminated soils. *Current Science* 86:528–534.

Gehring, C. A. 2003. Growth responses to arbuscular mycorrhizae by rain forest seedlings vary with light intensity and tree species. *Plant Ecology* 167:127–139.

Genre, A. and P. Bonfante. 2005. Building a mycorrhizal cell: How to reach compatibility between plants and arbuscular mycorrhizal fungi. *Journal of Plant Interactions* 1:3–13.

Genre, A., Chabaud, M., Timmers, T., Bonfante, P., and D. G. Barker. 2005. Arbuscular mycorrhizal fungi elicit a novel intracellular apparatus in *Medicago truncatula* root epidermal cells before infection. *Plant Cell* 17:3489–3499.

Gianinazzi-Pearson, V. and L. Brechenmacher. 2004. Functional genomics of arbuscular mycorrhiza: Decoding the symbiotic cell programme. *Canadian Journal of Botany* 82:1228–1234.

Gianinazzi-Pearson, V., Branzanti, B., and S. Gianinazzi. 1989. In vitro enhancement of spore germination and early hyphal growth of a vesicular-arbuscular mycorrhizal fungus by host root exudates and plant flavonoids. *Symbiosis* 7:243–255.

Giovannetti, M. 2000. Spore germination and pre-symbiotic mycelial growth. In *Arbuscular Mycorrhizas: Physiology and Function*, Eds. Y. Kapulnick and D. D. Jr. Douds, pp. 47–68. London: Kluwer Academic Publishers.

Giovannetti, M., Avio, L., Sbrana, C., and A. S. Citernesi. 1993a. Factors affecting appressorium development in the vesicular-arbuscular mycorrhizal fungus *Glomus mosseae* (Nicol. and Gerd.) Gerd. and Trappe. *New Phytologist* 123:115–122.

Giovannetti, M., Sbrana, C., Avio, L., Citernesi, A. S., and C. Logi. 1993b. Differential hyphal morphogenesis in arbuscular mycorrhizal fungi during pre-infection stages. *New Phytologist* 125:587–593.

Giovannetti, M., Sbrana, C., and C. Logi. 1994. Early processes involved in host recognition by arbuscular mycorrhizal fungi. *New Phytologist* 127:703–709.

Giri, B. and K. G. Mukerji. 2004. Mycorrhizal inoculant alleviates salt stress in *Sesbania aegyptiaca* and *Sesbania grandiflora* under field conditions: Evidence for reduced sodium and improved magnesium uptake. *Mycorrhiza* 14:307–312.

Goicoechea, N., Antolin, M. C., and M. Sánchez-Díaz. 1997. Gas exchange is related to the hormone balance in mycorrhizal or nitrogen-fixing alfalfa subjected to drought. *Physiologia Plantarum* 100:989–997.

Gomez-Roldan, V., Roux, C., Girard, D., Bécard, G., and V. Puech. 2007. Strigolactones: Promising plant signals. *Plant Signaling and Behavior* 2:163–164.

Güimil, S., Chang, H. S., Zhu, T., Sesma, A., Osbourn, A., Roux, C., Ioannidis, V., Oakeley, E. J., Docquier, M., Descombes, P., Briggs, S. P., and U. Paszkowski. 2005. Comparative transcriptomics of rice reveals an ancient pattern of response to microbial colonization. *Proceedings of the National Academy of Sciences USA* 102:8066–8070.

Hamblin, A. P. 1985. The influence of soil structure on water movement, crop root growth, and water uptake. *Advances in Agronomy* 38:95–158.

Harrison, M. J. 2005. Signaling in the arbuscular mycorrhizal symbiosis. *Annual Review of Microbiology* 59:19–42.

Harrison, M. J. and M. L. van Buuren. 1995. A phosphate transporter from the mycorrhizal fungus *Glomus versiforme*. *Nature* 378:626–629.

Harrison, M. J., Dewbre, G. R., and L. Liu. 2002. A phosphate transporter from *Medicago truncatula* involved in the acquisition of phosphate released by arbuscular mycorrhizal fungi. *Plant Cell* 14:2413–2429.

Hause, B. and T. Fester. 2005. Molecular and cell biology of arbuscular mycorrhizal symbiosis. *Planta* 221:184–196.

Hetrick, B. A. D., Gerschefske, K., and G. T. Wilson. 1987. Effects of drought stress on growth response in corn, sudan grass, and big bluestem to *Glomus etunicatum*. *New Phytologist* 105:403–410.

Hildebrandt, U., Kaldorf, M., and H. Bothe. 1999. The zinc violet and its colonization by arbuscular mycorrhizal fungi. *Journal of Plant Physiology* 154:709–717.

Hoekstra, F. A., Golovina, E. A., and J. Buitink. 2001. Mechanisms of plant desiccation tolerance. *Trends in Plant Science* 6:431–438.

Hohnjec, N., Vieweg, M. F., Pühler, A., Becker, A., and Küster, H. 2005. Overlaps in the transcriptional profiles of *Medicago truncatula* roots inoculated with two different Glomus fungi provide insights into the genetic program activated during arbuscular mycorrhiza. *Plant Physiology* 137:1283–1301.

Jahromi, F., Aroca, R., Porcel, R., and J. M. Ruiz-Lozano. 2008. Influence of salinity on the in vitro development of *Glomus intraradices* and on the in vivo physiological and molecular responses of mycorrhizal lettuce plants. *Microbial Ecology* 55:45–53.

Jastrow, J. D. and R. M. Miller. 1991. Methods for assessing the effects of biota on soil structure. *Agriculture, Ecosystems and Environment* 34:279–303.

Javot, H., Pumplin, N., and M. J. Harrison. 2007. Phosphate in the arbuscular mycorrhizal symbiosis: Transport properties and regulatory roles. *Plant, Cell and Environment* 30:310–322.

Jindal, V., Atwal, A., Sekhon, B. S., Rattan, S., and R. Singh. 1993. Effect of vesicular-arbuscular mycorrhizae on metabolism of moong plants under NaCl salinity. *Plant Physiology and Biochemistry* 31:475–481.

Johnson, N. C., Graham, J. H., and F. A. Smith. 1997. Functioning of mycorrhizal associations along the mutualism–parasitism continuum. *New Phytologist* 135:575–585.

Joner, E. J. and C. Leyval. 1997. Uptake of 109Cd by roots and hyphae of a *Glomus mosseae/Trifolium subterraneum* mycorrhiza from soil amended with high and low concentrations of cadmium. *New Phytologist* 135:353–360.

Joner, E. J., Leyval, C., and R. Briones. 2000. Metal binding capacity of arbuscular mycorrhizal mycelium. *Biology and Fertility of Soils* 226:227–234.

Kabir, Z. and R. T. Koide. 2000. The effect of dandelion or a cover crop on mycorrhiza inoculum potential, soil aggregation and yield of maize. *Agriculture, Ecosystems and Environment* 78:167–174.

Kaldorf, M., Kuhn, A. J., Schroder, W. H., Hildebrandt, U., and H. Bothe. 1999. Selective element deposits in maize colonized by a heavy metal tolerance conferring arbuscular mycorrhizal fungus. *Journal of Plant Physiology* 154:718–728.

Karandashov, V. and M. Bucher. 2005. Symbiotic phosphate transport in arbuscular mycorrhizas. *Trends in Plant Science* 10:22–29.

Kistner, C., Winzer, T., Pitzschke, A., Mulder, L., Sato, S., Kaneko, T., Tabata, S., Sandal, N., Stougaard, J., Webb, K. J., Szczyglowski, K., and M. Parniske. 2005. Seven *Lotus japonicus* genes required for transcriptional reprogramming of the root during fungal and bacterial symbiosis. *Plant Cell* 17:2217–2229.

Klironomos, J. N. 2003. Variation in plant response to native and exotic arbuscular mycorrhizal fungi. *Ecology* 84:2292–2301.

Krajinski, F. and A. Frenzel. 2007. Towards the elucidation of AM-specific transcription in *Medicago truncatula. Phytochemistry* 68:75–81.

Kramer, P. J. and J. S. Boyer. 1997. *Water Relations of Plants and Soils*. Academic Press, San Diego, CA.

Lee, D. W., Krishnapillay, B., Mansor, M., Mohamad, H., and S. K. Yap. 1996. Irradiance and spectral quality affect Asian tropical rain forest seedling development. *Ecology* 77:568–580.

Leyval, C., Singh, B. R., and E. J. Joner. 1995. Occurrence and infectivity of arbuscular mycorrhizal fungi in some Norwegian soils influenced by heavy metals and soil properties. *Water, Air and Soil Pollution* 84:203–216.

Leyval, C., Turnau, K., and K. Haselwandter. 1997. Effect of heavy metal pollution on mycorrhizal colonization and function: Physiological, ecological and applied aspects. *Mycorrhiza* 7:139–153.

Li, X. L. and P. Christie. 2000. Changes in soil solution Zn and pH and uptake of Zn by arbuscular mycorrhizal red clover in Zn-contaminated soil. *Chemosphere* 42:201–207.

Liu J., Blaylock, L. A., Endré, G., Cho, J., Town, C. D., VandenBosch, K. A., and J. M. Harrison. 2003. Transcript profiling coupled with spatial expression analyses reveals genes involved in distinct developmental stages of an arbuscular mycorrhizal symbiosis. *Plant Cell* 15:2106–2123.

Liu, J., Blaylock, L., and M. J. Harrison. 2004. cDNA arrays as tools to identify mycorrhiza-regulated genes: Identification of mycorrhiza induced genes that encode or generate signaling molecules implicated in the control of root growth. *Canadian Journal of Botany* 82:1177–1185.

Liu, J., Maldonado-Mendoza, I., Lopez-Meyer, M., Cheung, F., Town, C. D., and M. J. Harrison. 2007. Arbuscular mycorrhizal symbiosis is accompanied by local and systemic alterations in gene expression and an increase in disease resistance in the shoots. *The Plant Journal* 50:529–544.

Maeda, D., Ashida, K., Iguchi, K., Chechetka, S. A., Hijikata, A., Okusako, Y., Deguchi, Y., Izui, K., and S. Hata. 2006. Knockdown of an arbuscular mycorrhiza-inducible phosphate transporter gene of *Lotus japonicus* suppresses mutualistic symbiosis. *Plant Cell Physiology* 47:807–817.

Maldonado-Mendoza, I. E., Dewbre, G. R., and M. J. Harrison. 2001. A phosphate transporter gene from the extra-radical mycelium of an arbuscular mycorrhizal fungus *Glomus intraradices* is regulated in response to phosphate in the environment. *Molecular Plant-Microbe Interactions* 14:1140–1148.

Marschner, H. 1995. *Mineral Nutrition of Higher Plants*, 2nd ed. Academic Press, Cambridge.

Marschner, H. and V. Romheld. 1995. Strategies of plants for acquisition of iron. *Plant Soil* 165:262–274.

Martin, C. A. and J. C. Stutz. 2004. Interactive effects of temperature and arbuscular mycorrhizal fungi on growth, P uptake and root respiration of *Capsicum annuum* L. *Mycorrhiza* 14:241–244.

Massoumou, M., van Tuinen, D., Chatagnier, O., Arnould, C., Brechenmacher, L., Sanchez, L., Selim, S., Gianinazzi, S., and V. Gianinazzi-Pearson. 2007. *Medicago truncatula* gene responses specific to arbuscular mycorrhiza interactions with different species and genera of Glomeromycota. *Mycorrhiza* 17:223–234.

Matsubara, Y., Hirano, I., Sassa, D., and K. Koshikawa. 2004. Alleviation of high temperature stress in strawberry plants infected with arbuscular mycorrhizal fungi. *Environment Control in Biology* 42:105–111.

Matusova, R., Rani, K., Verstappen, A. W. F., Franssen, F. W. A., Beale, M. H., and H. J. Bouwmeester. 2005. The strigolactone germination stimulants of the plant-parasitic *Striga* and *Orobanche* spp. are derived from the carotenoid pathway. *Plant Physiology* 139:920–934.

Medeiros, C. A. B., Clark, R. B., and J. R. Ellis. 1994a. Effects of excess aluminum on mineral uptake in mycorrhizal sorghum. *Journal of Plant Nutrition* 17:1399–1416.

Medeiros, C. A. B., Clark, R. B., and J. R. Ellis. 1994b. Effects of excess manganese on mineral uptake in mycorrhizal sorghum. *Journal of Plant Nutrition* 17:2203–2219.

Morgan, J. M. 1984. Osmoregulation and water stress in higher plants. *Annual Review of Plant Physiology* 35:299–319.

Mosse, B. and C. Hepper. 1975. Vesicular-arbuscular mycorrhizal infections in root organ cultures. *Physiological Plant Pathology* 5:215–223.

Munns, R. 2002. Comparative physiology of salt and water stress. *Plant, Cell and Environment* 25:239–250.

Nagahashi, G. and D. D. Douds. 2000. Partial separation of root exudate components and their effects upon the growth of germinated spores of AM fungi. *Mycological Research* 104:1453–1464.

Oades, J. M. and Waters, A. G. 1991. Aggregate hierarchy in soils. *Australian Journal of Soil Research* 29:815–828.

Palma, J. M., Longa, M. A., del Rio, L. A., and J. Arines. 1993. Superoxide dismutase in vesicular-arbuscular red clover plants. *Physiologia Plantarum* 87:77–83.

Paszkowski, U. 2006. A journey through signaling in arbuscular mycorrhizal symbioses. *New Phytologist* 172:35–46.

Paszkowski, U., Jakovleva, L., and T. Boller. 2006. Maize mutants affected at distinct stages of the arbuscular mycorrhizal symbiosis. *The Plant Journal* 47:165–173.

Paszkowski, U., Kroken, S., Roux, C., and S. P. Briggs. 2002. Rice phosphate transporters include an evolutionarily divergent gene specifically activated in arbuscular mycorrhizal symbiosis. *Proceedings of the National Academy of Sciences, USA* 99:13324–13329.

Pfeiffer, C. M. and H. E. Bloss. 1987. Growth and nutrition of guayule (*Parthenium argentatum*) in a saline soil as influenced by vesicular-arbuscular mycorrhiza and phosphorus fertilization. *New Phytologist* 108:315–321.

Pfleger, F. L. and R. G. Linderman. 1994. *Mycorrhizae and Plant Health*. APS Press, St. Paul, MN.

Pierce, S., Vianelli, A., and B. Cerabolini. 2005. Essay review: From ancient genes to modern communities: The cellular stress response and the evolution of plant strategies. *Functional Ecology* 19:763–776.

Poirier, Y. and M. Bucher. 2002. Phosphate transport and homeostasis in *Arabidopsis*. In *The Arabidopsis Book*, Eds. C. R. Somerville and E. M. Meyerowitz, pp. 1–35. American Society of Plant Biologists, Rockville, MD.

Porcel, R. and J. M. Ruiz-Lozano. 2004. Arbuscular mycorrhizal influence on leaf water potential, solute accumulation, and oxidative stress in soybean plants subjected to drought stress. *Journal of Experimental Botany* 55:1743–1750.

Porcel, R., Barea, J. M., and J. M. Ruiz-Lozano. 2003. Antioxidant activities in mycorrhizal soybean plants under drought stress and their possible relationship to the process of nodule senescence. *New Phytologist* 157:135–143.

Poulin, M. J., Simard, J., Catford, J. G., Labrie, F., and Y. Piche. 1997. Response of symbiotic endomycorrhizal fungi to estrogens and antiestrogens. *Molecular Plant-Microbe Interactions* 10:481–487.

Purin, S. and M. C. Rillig. 2008. Parasitism of arbuscular mycorrhizal fungi: Reviewing the evidence. *FEMS Microbiology Letters* 279:8–14.

Querejeta, J. I., Allen, M. F., Alguacil, M. M., and A. Roldan. 2007. Plant isotopic composition provides insight into mechanisms underlying growth stimulation by AM fungi in a semiarid environment. *Functional Plant Biology* 34:683–691.

Raghothama, K. G. 2000. Phosphate transport and signaling. *Current Opinions in Plant Biology* 3:182–187.

Raskin, I. and B. D. Ensley. 2000. *Phytoremediation of Toxic Metals: Using Plants to Clean Up the Environment*. John Wiley and Sons, NY, pp. 15–31.

Rausch, C. and M. Bucher. 2002. Molecular mechanisms of phosphate transport in plants. *Planta* 216:23–37.

Rausch, C., Daram, P., Brunner, S., Jansa, J., Laloi, M., Leggewie, G., Amrhein, N., and M. Bucher. 2001. A phosphate transporter expressed in arbuscule-containing cells in potato. *Nature* 414:462–470.

Reddy, D. M. R. S., Schorderet, M., Feller, U., and D. Reinhardt. 2007. A petunia mutant affected in intracellular accommodation and morphogenesis of arbuscular mycorrhizal fungi. *The Plant Journal* 51:739–750.

Reinhardt, D. 2007. Programming good relations—development of the arbuscular mycorrhizal symbiosis. *Current Opinions in Plant Biology* 10:98–105.

Remy, W., Taylor, T. N., Hass, H., and H. Kerp. 1994. Four hundred-million-year-old vesicular arbuscular mycorrhizae. *Proceedings of the National Academy of Sciences, USA* 91:11841–11843.

Requena, N., Serrano, E., Ocon, A., and M. Breuninger. 2007. Plant signals and fungal perception during arbuscular mycorrhiza establishment. *Phytochemistry* 68:33–40.

Rillig, M. C. and P. D. Steinberg. 2002. Glomalin production by an arbuscular mycorrhizal fungus: A mechanism of habitat modification. *Soil Biology and Biochemistry* 34:1371–1374.

Rillig, M. C., Wright, S. F., and V. T. Eviner. 2002. The role of arbuscular mycorrhizal fungi and glomalin in soil aggregation: Comparing effects of five plant species. *Plant Soil* 238:325–333.

Rosendahl, C. N. and S. Rosendahl. 1991. Influence of vesicular-arbuscular mycorrhizal fungi (*Glomus* spp.) on the response of cucumber (*Cucumis sativis* L.) to salt stress. *Environmental and Experimental Botany* 31:313–318.

Ruan, Y., Kotraiah, V., and D. C. Straney. 1995. Flavonoids stimulate spore germination in *Fusarium solani* pathogenic on legumes in a manner sensitive to inhibitors of cAMP-dependent protein kinase. *Molecular Plant-Microbe Interactions* 8:929–938.

Ruiz-Lozano, J. M. 2003. Arbuscular mycorrhizal symbiosis and alleviation of osmotic stress. New perspectives for molecular studies. *Mycorrhiza* 13:309–317.

Ruiz-Lozano, J. M. and R. Azcón. 2000. Symbiotic efficiency and infectivity of an autochthonous arbuscular mycorrhizal *Glomus* sp. from saline soils and *Glomus deserticola* under salinity. *Mycorrhiza* 10:137–143.

Ruiz-Lozano, J. M., Azcón, R., and M. Gómez. 1995a. Effects of arbuscular mycorrhizal *Glomus* species on drought tolerance: Physiological and nutritional plant responses. *Applied Environmental Microbiology* 61:456–460.

Ruiz-Lozano, J. M., Gómez, M., and R. Azcón. 1995b. Influence of different *Glomus* species on the time-course of physiological plant responses of lettuce to progressive drought stress periods. *Plant Science* 110:37–44.

Ruiz-Lozano, J. M., Azcón, R., and M. Gomez. 1996a. Alleviation of salt stress by arbuscular-mycorrhizal *Glomus* species in *Lactuca sativa* plants. *Physiologia Plantarum* 98:767–772.

Ruiz-Lozano, J. M., Azcón, R., and J. M. Palma. 1996b. Superoxide dismutase activity in arbuscular-mycorrhizal *Lactuca sativa* L. plants subjected to drought stress. *New Phytologist* 134:327–333.

Ruiz-Lozano, J. M., Collados, C., Barea, J. M., and R. Azcón. 2001. Cloning of cDNAs encoding SODs from lettuce plants which show differential regulation by arbuscular mycorhizal symbiosis and by drought stress. *Journal of Experimental Botany* 52:2241–2242.

Sannazzaro, A. I., Echeverría, M., Albertó, E. O., Ruiz, O. A., and A. B. Menéndez. 2007. Modulation of polyamine balance in *Lotus glaber* by salinity and arbuscular mycorrhiza. *Plant Physiology and Biochemistry* 45:39–46.

Schüssler, A., Schwarzott, D., and C. Walker. 2001. A new fungal phylum, the Glomeromycota: Phylogeny and evolution. *Mycological Research* 105:1413–1421.

Shalabyl, A. M. 2003. Responses of arbuscular mycorrhizal fungal spores isolated from heavy metal-polluted and unpolluted soil to Zn, Cd, Pb and their interactions *in vitro. Pakistan Journal of Biological Sciences* 6:1416–1422.

Siciliano, V., Genre, A., Balestrini, R., Cappellazzo, G., deWit, P. J., and P. Bonfante. 2007. Transcriptome analysis of arbuscular mycorrhizal roots during development of the prepenetration apparatus. *Plant Physiology* 144:1455–1466.

Simpson, D. and M. J. Daft. 1991. Effects of *Glomus clarum* and water stress on growth and nitrogen fixation in 2 genotypes of groundnut. *Agriculture, Ecosystems and Environment* 35:47–54.

Siqueira, J. O., Rocha, W. F. Jr., Oliveira, E., and A. Colozzi-Filho. 1990. The relationship between vesicular-arbuscular mycorrhiza and lime: Associated effects on the growth and nutrition of brachiaria grass. *Biology and Fertility of Soils* 10:65–71.

Smith, S. E. and D. J. Read. 1997. *Mycorrhizal Symbiosis*, 2nd ed. Academic Press, San Diego, CA.

Smith, S. E., Smith, F. A., and I. Jakobsen. 2003. Mycorrhizal fungi can dominate phosphate supply to plant irrespective of growth responses. *Plant Physiology* 133:16–20.

Smith, S. E., Smith, F. A., and I. Jakobsen. 2004. Functional diversity in arbuscular mycorrhizal (AM) symbioses: the contribution of the mycorrhizal P uptake pathway is not correlated with mycorrhizal responses in growth or total P uptake. *New Phytologist* 162:511–524.

Stirzaker, R. J. and J. B. Passioura. 1996. The water relations of the root–soil interface. *Plant Cell Environment* 19:201–208.

Subramanian, K. S. and C. Charest. 1998. Arbuscular mycorrhizae and nitrogen assimilation in maize after drought and recovery. *Physiologia Plantarum* 102:285–296.

Sylvia, D. M. and S. E. Williams. 1992. Vesicular-arbuscular mycorrhizae and environmental stresses. In *Mycorrhizae in Sustainable Agriculture*, Eds. G. J. Bethlenfalvay and R. G. Linderman, pp. 101–124. ASA No. 54, Madison, WI.

Tamasloukht, M., Séjalon-Delmas, N., Kluever, A., Jauneau, A., Roux, C., Bécard, G., and P. Franken. 2003. Root factors induce mitochondrial-related gene expression and fungal respiration during the developmental switch from asymbiosis to presymbiosis in the arbuscular mycorrhizal fungus *Gigaspora rosea. Plant Physiology* 131:1468–1478.

Taylor, J. and L. A. Harrier. 2001. A comparison of development and mineral nutrition of micropropagated *Fragaria ananassa* cv. Elvira (strawberry) when colonized by nine species of arbuscular mycorrhizal fungi. *Applied Soil Ecology* 18:205–215.

Tonin, C., Vandenkoornhuyse, P., Joner, E. J., Straczek, J., and C. Leyval. 2001. Assessment of arbuscular mycorrhizal fungi diversity in the rhizosphere of *Viola calaminaria* and effect of these fungi on heavy metal uptake by clover. *Mycorrhiza* 10:161–168.

Tullio, M., Pierandrei, F., Salerno, A., and E. Rea. 2003. Tolerance to cadmium of vesicular arbuscular mycorrhizae spores isolated from a cadmium-polluted and unpolluted soil. *Biology and Fertility of Soils* 37:211–214.

Turnau, K., Kottke, I., and F. Oberwinkler. 1993. Element localization in mycorrhizal roots of *Pteridium aquilinum* L. Kuhn collected from experimental plots treated with cadmium dust. *New Phytologist* 123:313–324.

van der Heijden, M. G. A., Klironomos, J. N., Ursic, M., Moutoglis, P., Streitwolf-Engel, R., Boller, T., Wiemken, A., and I. R. Sanders. 1998. Mycorrhizal fungal diversity determines plant biodiversity, ecosystem variability and productivity. *Nature* 396:69–72.

Vierheilig, H. and Y. Piché. 2002. Signalling in arbuscular mycorrhiza: Facts and hypotheses. *Advances in Experimental Medicine and Biology* 505:23–39.

Walker, T. S., Bais, H. P., Grotewold, E., and J. M. Vivanco. 2003. Root exudation and rhizosphere biology. *Plant Physiology* 132:44–51.

Weidmann, S., Sanchez, L., Descombìn, J., Chatagnier, O., Gianinazzi, S., and V. Gianinazzi-Pearson. 2004. Fungal elicitation of signal transduction-related plant genes precedes mycorrhiza establishment and requires the *dmi*3 gene in *Medicago truncatula. Molecular Plant-Microbe Interactions* 17:1385–1393.

Weissenhorn, I., Mench, M., and C. Leyval. 1995. Bioavailability of heavy metals and arbuscular mycorrhiza in a sewage sludge amended sandy soil. *Soil Biology and Biochemistry* 27:287–296.

Wu, Q. S. and R. X. Xia. 2006. Arbuscular mycorrhizal fungi influence growth, osmotic adjustment and photosynthesis of citrus under well-watered and water stress conditions. *Journal of Plant Physiology* 163:417–425.

Wu, Q., Zou, Y., and R. Xia. 2007. Effect of *Glomus versiforme* inoculation on reactive oxygen metabolism of Citrus tangerine leaves exposed to water stress. *Frontiers of Agriculture in China* 1:438–443.

Wulf, A., Manthey, K., Doll, J., Perlick, A. M., Linke, B., Bekel, T., Meyer, F., Franken, P., Küster, H., and F. Krajinski. 2003. Transcriptional changes in response to arbuscular mycorrhiza development in the model plant *Medicago truncatula. Molecular Plant-Microbe Interactions* 16:306–314.

Yoneyama, K., Xie, X., Kusumoto, D., Sekimoto, H., Sugimoto, Y., Takeuchi, Y., and K. Yoneyama. 2007. Nitrogen deficiency as well as phosphorus deficiency in sorghum promotes the production and exudation of 5-deoxystrigol, the host recognition signal for arbuscular mycorrhizal fungi and root parasites. *Planta* 225:1031–1038.

Zhu, Y.G., Christie, P., and A. S. Laidlaw. 2001. Uptake of Zn by arbuscular mycorrhizal white clover from Zn-contaminated soil. *Chemosphere* 42:193–199.

15

Conifer Endophytes

Anna Maria Pirttilä and Piippa R. Wäli

CONTENTS

15.1 Introduction

Endophytes are fungi or bacteria living inside plants without eliciting symptoms or disease, common to a large number of plant species (Petrini, 1986; Schulz et al., 1993; Redlin and Carris, 1996). Hundreds of fungal and bacterial species can be detected in a coniferous host, living in buds, leaves, wood, and bark of trees (Petrini, 1986; Fisher and Petrini, 1990; Kowalski and Kehr, 1992; Müller and Hallaksela, 1998, 2000; Pirttilä et al., 2003). Also conifer roots are inhabited by various endophytic microorganisms (Addy et al., 2005). Most of the endophytes form local infections and are transmitted horizontally via spores, and their ecological roles in relation to host tree are suggested to vary from latent pathogens or saprophytes to neutral commensalists or sometimes even mutualists (e.g., via defensive mutualism) (Arnold, 2007; Saikkonen, 2007; Sieber, 2007).

15.2 Conifers as Host

The conifer leaves range from needle-like structures to small, compressed leaves. Majority of the conifer species are evergreen and retain leaves for several years, whereas *Larix* and *Metasequoia* are deciduous. A specific feature of the conifer leaves is the thickness of the cuticular wax layer. The cuticular wax layer consists of intracuticular and epicuticular layers where the wax is located inside or outside the cutin matrix, respectively (Jeffree, 1986). The intracuticular wax layer is likely responsible for limiting the nonstomatal water loss (Baur, 1998), whereas epicuticular waxes can contribute to transpiration barrier, protect the needle tissue against UV radiation (Reicosky and Hanover, 1978), and reduce stomatal water loss (Jeffree et al., 1971).

The epicuticular waxes form a smooth thin surface on the leaves of many plant species, but conifer needles typically have a rough thick surface consisting of tubule-shaped crystals that may be formed due to the secondary alcohol nonacosan-10-ol (Jeffree et al., 1975; Holloway et al., 1976; Jetter and Riederer,

1994; Wen et al., 2006). Other compounds found within the epicuticular wax layers are fatty acids, aldehydes, primary and secondary alcohols, ketones, alkanes, and alkyl esters (Walton, 1990; Gordon et al., 1998; Wen et al., 2006). Furthermore, triterpenoids, tocopherols, and biflavonoids can be present in the needles of some conifer species (Gordon et al., 1998; Wollenweber et al., 1998; Wen et al., 2006).

The epicuticular waxes surround the plant organs and play important ecological functions in the interaction with microbes and insects (Eigenbrode and Espelie, 1995). It is clear that the wax layer is a selective factor for infection by the horizontal endophytes. On maize, the thickness of wax layer affects the infection by bacterial endophytes (Beattie and Marcell, 2002). Wax layers can directly prevent the entry of large fungal spores into leaf tissues (Millar, 1974), whereas some fungal species are able to utilize the wax constituents as an energy source (White et al., 1991). Simulated acid rain has negative effects on epicuticular wax layers and also reduces the endophyte populations in pine (*Pinus sylvestris* L.) (Helander et al., 1993).

Another factor contributing to colonization of conifer tissues by endophytes is chemical composition. In conifers, stilbenes, lignans, flavonoids, catechins, and proanthocyanidins are found within different tissues (Pan and Lundgren, 1996). The defensive compounds are mainly terpene-based, which are produced against insect and fungal pathogens (Phillips and Croteau, 1999). When the plant tissue is injured, the volatile terpenes evaporate and semicrystalline resin acids remain in the wound, preventing the attack of insects or microbial pathogens (Phillips and Croteau, 1999). In a study by Jurc et al. (1999), growth of the endophytic fungi *Cenangium ferruginosum* and *Phialophora hoffmannii* and the facultative pathogen (*Sphaeropsis sapinea*) was studied at different terpenoid extract concentrations. *S. sapinea* and *P. hoffmannii* were not inhibited in several different concentrations of terpenoid extracts isolated from *Pinus nigra* needles. The growth of *C. ferruginosum* was stimulated in high concentration of terpene extracts of *P. nigra* ssp. *austriaca* from which it had been isolated, but not in extracts from other subspecies studied (Jurc et al., 1999). The terpenoid extract of *P. nigra* ssp. *austriaca* has high quantities of *alpha*-Pinene and Germacrene-d, which were suggested to be responsible for the stimulatory effect (Jurc et al., 1999). Other compounds that may contribute to endophyte colonization are tannins, which represent one of the most common classes of compounds found within conifers (Yu and Dahlgren, 2000). For example, in the twig bark of *Populus*, endophyte colonization frequency was negatively correlated with quantities of condensed tannins in the tissues (Bailey et al., 2005).

15.3 Diversity of Endophytes in Conifers

Conifer needles are colonized locally by a high number of endophyte species in comparison to grasses, where a systemic infection by fewer species is more common (Sieber-Canavesi and Sieber, 1993; Hata and Futai, 1996; Müller and Hallaksela, 1998; Deckert et al., 2001). The number of infections typically increases with needle age (Bernstein and Carroll, 1977; Petrini and Carroll, 1981; Hata and Futai, 1993; Sieber-Canavesi and Sieber, 1993) and with the age of the surrounding vegetation (Müller and Hallaksela, 1998; Ganley and Newcombe, 2006).

In conifers, fungal endophytes are restricted to specific areas in the needle tissue where they reside in a slow-growing state and accumulate low quantities of biomass (Suske and Acker, 1986; Deckert et al., 2001). The endophytes are thought to remain in the slow-growing state unless the status of the host changes due to injury or natural senescence. This type of restricted infection enables numerous fungal species to colonize niches in the same needle tissue (Carroll, 1994). At the microscopic level, fungal endophytes have mostly been localized intercellularly in the leaf tissue in conifers and other trees like in grasses (Johnson and Whitney, 1989; Suske and Acker, 1989; Viret and Petrini, 1993; Yang et al., 1994). However, some endophytes are found in the symplast (Stone, 1987, 1988).

Pine, spruce, and fir species have been the most studied conifers with respect to endophyte diversity. Endophyte species diversity is similar among conifer species, with some fungal endophytes being generalists and found in more than one conifer species, whereas others are specific to one host. The most well-known example of a host-specific endophyte is *Rhabdocline parkeri* on *Pseudotsuga menzienzii* (Sherwood-Pike et al., 1986). Typically, one fungal endophyte species or genus dominates a conifer species or genus. In *Pinus* and *Picea*, the members of the Rhytismataceae are the most abundant endophytic isolates (Carroll and Carroll, 1978; Hata and Futai, 1996; Müller and Hallaksela, 1998, 2000; Deckert and Peterson, 2000; Guo et al., 2003; Ortiz-Garcia et al., 2003; Arnold et al., 2007).

For example, in *Pinus sylvestris*, endophytes belonging to 86 fungal genera are found in needles, *Anthostomella* and *Lophodermium* being the most common ones (Kowalski, 1993), whereas in *P. strobus*, *Hormonema* in addition to *Lophodermium* are the main genera isolated (Deckert and Peterson, 2000). In *Pinus mugo*, the endophytes *Cenangium ferruginosum*, *Cyclaneusma minus*, and *L. pinastri* may be isolated most frequently, with the order of frequency depending on the growth site (Sieber et al., 1999). The needle endophyte *Cenangium ferruginosum* of *P. sylvestris* also responds to air pollution (Helander, 1995). In addition to these genera, *Phialocephala* spp. are detected most frequently in the needles of *P. densiflora* and *P. thunbergii* (Hata et al., 1998). The mycobiota on pine needles differ with position (Hata et al., 1998). In the buds of *Pinus sylvestris*, the yeast *Rhodotorula minuta* is found in the green photosynthetic tissues, whereas *Hormonema dematioides* inhabits the bud scales (Pirttilä et al., 2003). Furthermore, several bacterial species (*Methylobacterium*, *Mycobacterium*, and *Pseudomonas*) are found in the inner tissues of Scots pine buds (Pirttilä et al., 2000, 2005).

Most studies on pines, as well as other conifers, have been made on the needle tissues, and so far, little knowledge exists on the endophytic fungi living in the bark or wood. An exception is the spruce, which has been studied for bark and twig endophytes (Barklund and Kowalski, 1996; Müller and Hallaksela, 2000). The number of endophytic infections, as well as the species diversity in the twigs, is at the same level, but the frequency of bark and wood colonization is typically lower than in the needles (Müller and Hallaksela, 2000). One endophytic fungus that dominates in the twigs of *Picea abies* is *Tryblidiopsis pinastri*, which appears to be both host and organ-specific (Barklund and Kowalski, 1996). Root endophytes, for example, *Phialocephala fortinii*, are usually considered as possible mutualistic symbionts of trees as they are thought to provide nutritional benefits to their hosts similar to mycorrhizas. Recently, some *Phialocephala* species that were assumed to be root-specific have been found also in living and decaying wood of conifers indicating that these fungi may act also as wood saprobes (Menkis et al., 2004). Ecological role of these fungi varies from pathogenic to neutral and beneficial (Addy et al., 2005; Grünig et al., 2006).

The presence or metabolic activity of conifer endophytes is dependent on the growth season in some cases. For example *Hormonema dematioides*, living in bud scales of Scots pine, is isolated more frequently during the winter period (Hohtola, 1988; Pirttilä et al. 2003, 2005). In contrast, the bacterial endophytes of Scots pine buds were undetectable during the winter season and infrequent during the most vigorous growth of the shoot tip in midsummer, whereas the highest quantities were detected in spring and autumn prior to growth or differentiation of the bud (Pirttilä et al., 2005). Less seasonal fluctuations are found in the colonization frequency of fungal endophytes in the needles (Hata et al., 1998). The main variations in the colonization frequencies are associated with needle age (Hata et al., 1998).

15.4 Transmission of Endophytes

Endophytes are typically not abundant in young, expanding leaves of conifers, but needles become highly infected with age (Hata et al., 1993, 1998). The species composition also changes with needle age (Sieber-Canavesi and Sieber, 1993). Therefore, transmission of fungal endophytes to successive generations has been suggested to occur horizontally in conifers compared to the vertical transmission through seeds in grasses (Johnson and Whitney, 1989; Wilson, 1996). Vertical transmission is suggested to be unlikely in trees due to strong differentiated and hierarchically organized tissues of the trees (Saikkonen et al., 2004; Saikkonen, 2007). However, some candidates for vertically transmitted conifer endophytes have been suggested. Fungi belonging to *Alternaria alternata*, *Epicoccum purpurascens*, and *Ulocladium atrum* were isolated from seeds of *Pinus sylvestris* and did not exhibit pathogenicity when inoculated in to seedlings (Lilja et al., 1995). At least *Alternaria alternata* and *Epicoccum purpurascens* have earlier been isolated as needle endophytes of Scots pine (Kowalski, 1993). Gure et al. (2005) isolated fungi from seeds and female cones of *Podocarpus falcatus* and tested the pathogenicity in seedlings. One isolate, *Diaporthe* sp., resulted in increased germination of *P. falcatus* seeds and no pathogenicity to seedlings was observed. In *P. monticola*, the same strains of endophytes belonging to the genera *Hormonema* and *Cladosporium* were isolated from seeds, seedlings, and mature needles (Ganley and Newcombe, 2006). Earlier, seed-borne transmission of endophytes has been reported from Douglas fir (Bloomberg, 1966). Although the systemic nature of the seed-borne endophytes of conifers has not been confirmed, the endophytic presence in seeds probably has ecological importance if they are disseminated with their hosts, and thus have a unique opportunity to develop in young seedlings.

15.5 Recent Advances in Research of Conifer Endophyte Diversity

Although many of the conifer endophytes have been characterized according to their morphology at the genus level today, many species are still unknown. Nuclear small subunit ribosomal RNA gene regions are generally used as a molecular tool to analyze fungal taxa at a family or order level, and internal transcribed spacer (ITS) regions are mostly used to examine the phylogenetic relationships at a species or intraspecies level. Sequence-based identification has provided new insights to the diversity and ecology of endophytic fungi. Earlier, fungal diversity was thought to increase with decreasing latitude, which is typical for many organisms (Hawksworth, 1991, 2001). Higgins et al. (2007) studied the diversity of endophytic fungi in the leaf tissues of arctic and boreal plants and found that the species richness of *Picea mariana* is similar to that found in tree species of the temperate and tropical latitudes. Furthermore, the molecular sequence data has provided new information on the ecological roles of horizontally transmitted endophytic fungi in the host plant (Ganley et al., 2004).

The most recent studies based on molecular sequence data have created new knowledge on the identity and diversity of conifer endophytes (Arnold et al., 2003; Ganley et al., 2004). Estimates of the endophyte diversity have traditionally been based on the morphological similarity with known species. However, endophytes often are morphologically similar to known parasites or pathogens that colonize the same plant species, or its close relatives. Therefore it has been assumed that many of the endophytes are actually cryptic or latent, but known, parasites. Ganley et al. (2004) showed by molecular sequencing that endophytic fungi of *Pinus monticola* were not latent phases of known parasites of the host or nonhost plants. Therefore, they concluded that endophytes likely represent substantial, unknown biodiversity in woody plants. They also suggested that even the maximal estimates of the total endophyte diversity have been conservative (Ganley et al., 2004; Arnold, 2007).

15.6 Ecological Interactions

The endophytic microorganisms are heterotrophic, which means that they take the resources from the host plant. The term "endophyte" defines that they do not cause either disease symptoms or other notable harm to the host plant (Wilson, 1995). Sometimes endophytes are pathogens that are in latent stage waiting for circumstances to become more favorable for disease to develop, e.g., host defence weakening due to herbivore attack or poor growing conditions. Endophytes may also be pathogens that are endophytic in intermediary hosts waiting for access to primary hosts where they may cause disease (Arnold, 2007; Saikkonen, 2007; Sieber, 2007). Sometimes endophytes may be avirulent strains of known pathogenic species (Freeman and Rodriquez, 1993). Some endophytes are saprobes that colonize the substrate (wood or needles) early while the tree is alive and thus have a competitive advantage compared to saprobes colonizing dead plant tissues (Sieber, 2007).

Heterotrophic organisms (endophytes, pathogens, and herbivores) using the same substrate (plant consumers in case of conifers) compete with each other. As a result, many endophytes have the ability to produce compounds antagonistic to other plant consumers. When these consumers are serious pathogens or herbivores of the host tree, the bioactive compounds of endophytes may defend the host from serious damage. The possible interaction mechanisms between endophytes and conifer antagonists can be either direct or indirect, such as between pathogenic fungi and herbivorous insects (Hatcher, 1995). Direct interactions may occur through resource competition, antagonism (interference competition), or parasitism (Isaac, 1992). Indirect interactions between fungi take place through the changes in host plants, and they are affected by the growth conditions of the host. Endophytes may alter, e.g., the metabolism and the defense mechanisms of host trees. Carroll suggested in 1988 that endophytes may function as a defense system for the trees—their life cycle being several orders of magnitude shorter than the life cycle of their hosts, they could evolve faster to resist pathogens and herbivores. Furthermore, Carroll (1991a) suggested endophyte-mediated induced resistance to occur in Douglas fir. On the other hand, herbivores and endophytes may benefit each other as the damage by herbivores opens the route for endophytes to colonize conifer wood and needles that otherwise are well protected by, for example, bark and waxes, respectively. One indirect form of defensive mutualism may occur via interspecific competition (Carroll, 1991b). As some symptomless endophytic fungi of one plant may be pathogens of other plant species (Redman et al., 2001), endophytes may serve as "microbial weapons," "enemies of enemies" in interspecific competition. The well-proved defensive mutualism between grasses and their systemic *Epichloë* endophytes has raised the term endophyte synonymous to the ecological term mutualism (Saikkonen et al., 1998; Saikkonen et al., 2004). Researchers have hunted for evidence of such mutualisms in trees. For conifers, defensive mutualism between host trees and their foliar endophytes has been suggested in cases of Douglas fir, *Rhabdocline parkeri* (Carroll, 1991a), white spruce, and rugulosin-producing endophytes (Miller et al., 2002) based on experimental research *in vivo*. However, in the case of trees and horizontally transmitted endophytes, assumption of mutualistic interaction is based mainly on circumstantial evidence (Saikkonen et al., 1998; Sieber, 2007), because experiments testing for mutualism require endophyte-free control trees which may be impossible to find in natural situations (Saikkonen, 2007; Sieber, 2007).

Although some endophytes have been demonstrated to be mutualistic to their hosts in some settings, the outcome of the interaction depends on the situation: the growth conditions and the physiological state of the host, state of the life cycles, genotypes of interacting species, and other biological interactions. In recent literature, the ecological outcome of the endophyte symbiosis is often described as a labile continuum from antagonism to mutualism (Petrini, 1991; Saikkonen et al., 1998; Hirsch, 2004; Saikkonen et al., 2004; Müller and Krauss, 2005; Schulz and Boyle, 2005; Saikkonen, 2007; Sieber, 2007).

Individual conifer trees and tree populations are a diverse assemblage of interacting species. Also the endophyte community in conifer trees is diverse and co-occurs with various antagonistic species. Recent literature suggests that pathogens may colonize the tree leaf with equal probability to colonization by beneficial or harmless endophytes, and that it is the balance between various species that results in healthy growth or disease (Arnold et al., 2003; Clay, 2004; Arnold, 2007). Beneficial and harmless endophytes may serve as mutualistic guilds of species and genotypes with slightly differing roles or locations within the tree. Particular species may act beneficially in different stages of life cycle of the host, or symbionts may be required to co-occur simultaneously in order to provide beneficial effects to the host. Species composition of endophytes and prevalence of individual species as well as the ecological roles of the symbionts in one host species vary among populations and across geographic mosaics, as predicted by geographic mosaic theory of coevolution (e.g., Thompson, 2005). Since the circumstances and players are changing all the time, the ecological outcome of endophyte–conifer interaction is changing and the result is difficult to predict and generalize.

15.7 Bioactivity of Endophytes in Conifers

Production of antimicrobial compounds is more common to endophytic fungi than soil microbes (Schulz et al., 2002). The endophytic fungi of grasses protect their hosts against insects by the production of biologically active compounds (Bush et al., 1982, 1997). The needle, bark, and twig endophytes of conifers have been studied with respect to antagonism against pathogenic fungi (Noble et al., 1991; Polishook et al., 1993; Schulz et al., 1995; Pelaez et al., 2000; Pirttilä, 2001; Wang et al., 2002; Kim et al., 2004; Park et al., 2005; Dai et al., 2006), pathogenic bacteria (Yang et al., 1994; Schulz et al., 1995; Findlay et al., 1997; Stierle et al., 1998), and insect pests (Calhoun et al., 1992; Johnson and Whitney, 1994; Findlay et al., 1995a,b, 1997, 2003; Miller et al., 2002). The most studied conifer species with respect to bioactive endophytes are members of Taxaceae and *Picea*. The isolated compounds consist of a large range of various compounds, such as alkaloids, steroids, terpenoids, peptides, polyketones, flavonoids, quinols, phenols, and chlorinated compounds (reviewed in Tan and Zou, 2001; Gunatilaka, 2006).

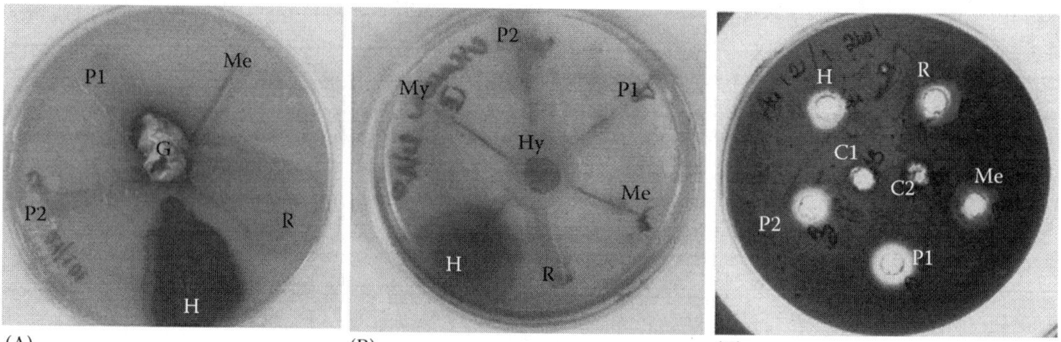

(A) (B) (C)

FIGURE 15.1 Antagonism of bud endophytes of Scots pine *Hormonema dematioides* (H), *Rhodotorula minuta* (R), *Pseudomonas synxhantha* (P1), *Pseudomonas* sp. G (P2), *Methylobacterium extorquens* (Me), and *Mycobacterium* sp. (My) against; (A) *Gremmeniella abietina* strain KR1 (G) and (B) *Hymenscypha ericae* (HY). (C) Testing antagonism of bacterial products against *Hormonema dematioides* (Pirttilä et al., 2001). C1 = Luria Bertani medium (control), C2 = D1 medium (control).

The reports of antibacterial conifer endophytes are surprisingly rare. This may reflect the low quantity of such studies rather than actual lack of antibacterial endophytes in conifers. We have discovered a high percentage of conifer endophytes with antibacterial activity (for further details, see Section 15.8).

Whereas conifers harbor both bacterial and fungal endophytes, which produce antifungal and antibacterial compounds, microbial antagonism has independently arrived in each population of symbionts. When the bioactivity of the endophytes isolated from buds of *Pinus sylvestris* was studied, *Hormonema dematioides* was found to inhibit the growth of pathogenic *Gremmeniella abietina* strains and the mycorrhizal fungus *Hymenoscypha ericae* (Figure 15.1; Pirttilä, 2001). Interestingly, the bacterial and yeast endophytes isolated from the inner tissues of the bud produced antagonistic substances *in vitro* toward *H. dematioides* that lives in the bud scales (Figure 15.1; Pirttilä, 2001). Therefore, the antagonism was well adapted to the location of the endophytes in the pine buds. Because *H. dematioides* is located in the bud scales that are the outermost parts of the bud, it may readily protect the host from pathogen attacks. *H. dematioides* was found to be pathogenic in the photosynthetic bud tissues, and therefore the inhibition of *H. dematioides* growth by the bacterial and yeast endophytes is consistent with their location inside the buds (Pirttilä, 2001).

15.8 Applications of Bioactive Conifer Endophytes

Among the most serious pests of conifers are spruce budworms (*Choristoneura*) that consist of dozens of species, subspecies, and forms. These variations comprise populations that create serious problems in timber production in the northern spruce and fir forests of the Eastern United States and Canada. The spruce budworms have periodic outbreaks that belong to the natural maturation cycle of balsam fir. The most suitable host trees for the budworm are white, red, and black spruce. Tamarack, pine, and hemlock can also become attacked by the budworms. Needle endophytes have been studied as a potential biological control agent against the spruce budworms in white spruce (*Picea glauca*) with some promising results (Miller et al., 2002; Sumarah et al., 2005). The most typical problems with applying the needle endophytes to biocontrol purposes are their horizontal transmission and the poor knowledge of the infection processes. However, Miller et al. (2002) were successful in the inoculation of spruce seedlings using the endophyte 5WS22E1 with a needle injection method, which resulted in 2%–3% infection of the seedlings detectable by culture plating. When immunodetection was used, a higher infection rate of 90% was obtained (Sumarah et al., 2005). The colonized needles contained rugulosin, which is toxic to budworms, resulting in reduced weight of the larvae (Miller et al., 2002; Sumarah et al., 2005).

Besides applications in forestry, conifer endophytes hold great promise for pharmaceutical development due to their bioactivity and high diversity. One of the most striking examples is the discovery of the endophytic fungus *Taxomyces andreanae* in the bark of Pacific yew (*Taxus brevifolia*) that produced the anticancer compound taxol (Stierle et al., 1993). After the initial discovery, a variety of endophytic fungi capable of producing taxol and/or taxane derivatives were isolated from *T. wallachiana* (Strobel et al., 1996), *T. baccata* (Caruso et al., 2000), *T. mairei* (Wang et al., 2000), *T. chinensis* (Zhou et al., 2007), *Taxodium distichum* (Li et al., 1996), *Torreya grandifolia* (Li et al., 1998), and *Wollemia nobilis* (Strobel et al., 1997). Furthermore, two novel antitumor metabolites sequoiatones A and B were isolated from an endophytic fungus *Aspergillus parasiticus* of redwood (Stierle et al., 1999). However, to date no endophytic fungi have been advanced to industrial production of taxol (Ji et al., 2006). Use of these fungi in industry is limited by low fungal biomass generated in fermentation, low yield of taxol in culture, and unknown pathway of taxol biosynthesis, regulation, and intermediates (Ji et al., 2006). A profitable yield for taxol production would be 1 mg/L whereas the top yield reported so far is a million times lower. Although taxol has become a successful, widely used natural antitumor agent and has a higher demand than production rates, biotechnological tools for large-scale production are lacking. Knowledge of taxol biosynthesis is needed for engineering the endophytic strains or generating heterologous hosts for taxol production.

In our laboratory we are screening endophytic fungi of woody plants for antibacterial activity to produce antibacterial leads for the pharmaceutical industry. We have discovered that conifers such as Scots pine (*Pinus sylvestris* L.) and Norway spruce (*Picea abies* Karst.) harbor the highest percentage (25% and 27%, respectively) of antibacterial endophytes whereas shrub herbs such as Labrador tea (*Rhododendron tomentosum* L.) and bilberry (*Vaccinium myrtillus* L.) typically contain lower colonization of antibacterial endophytes (12% and 9%, respectively) (unpublished data). The active broths of two pine isolates, identified as *Dothidea* sp., have been characterized by LC–MS and NMR, revealing an array of compounds possibly responsible for the antibacterial effect. Our studies have shown that endophytic fungi of conifers are a rich, untapped source of new antibacterials that should be utilized more efficiently. Although highly promising, the antibacterial endophytes encounter similar problems to those producing taxol: they are typically slow growing and difficult to culture on standard laboratory media. New methods and approaches are needed to effectively access the untapped biological resources of conifer endophytes for applied purposes such as the pharmaceutical development.

ACKNOWLEDGMENT

Academy of Finland is thanked for the support (Grant 1118569).

REFERENCES

Addy, H. D., Piercey, M. M., and R. S. Currah. 2005. Microbial endophytes in roots. *Canadian Journal of Botany* 83:1–13.

Arnold, A. E. 2007. Understanding the diversity of foliar endophytic fungi: Process, challenges, and frontiers. *Fungal Biology Reviews* 21:51–66.

Arnold, A. E., Henk, D. A., Eells, R. A., Lutzoni, F., and R. Vilgalys. 2007. Diversity and phylogenetic affinities of foliar fungal endophytes in loblolly pine inferred by culturing and environmental PCR. *Mycologia* 99:185–206.

Arnold, A. E., Meija, L. C., Kyllo, D., Rojas, E. I., Maynard, Z., Robbins, N., and E. A. Herre. 2003. Fungal endophytes limit pathogen damage in a tropical tree. *Proceedings of the National Academy of Sciences* 100:15649–15654.

Bailey, J. K., Deckert, R., Schweitzer, J. A., Rehill, B. J., Lindroth, R. L., Gehring, C., and T. G. Whitham. 2005. Host plant genetics affect hidden ecological players: Links among Populus, condensed tannins, and fungal endophyte infection. *Canadian Journal of Botany* 83:356–361.

Barklund, P. and T. Kowalski. 1996. Endophytic fungi in branches of Norway spruce with particular reference to *Tryblidiopsis pinastri*. *Canadian Journal of Botany* 74:673–678.

Baur, P. 1998. Mechanistic aspects of foliar penetration of agrochemicals and the effect of adjuvants. *Recent Research Development in Agricultural and Food Chemistry* 2:809–837.

Beattie, G. A. and L. M. Marcell. 2002. Comparative dynamics of adherent and nonadherent bacterial populations on maize leaves. *Phytopathology* 92:1015–1023.

Bernstein, M. E. and C. G. Carroll. 1977. Internal fungi in old-growth Douglas fir foliage. *Canadian Journal of Botany* 55:644–653.

Bloomberg, W. J. 1966. The occurrence of endophytic fungi in Douglas-fir seedlings and seeds. *Canadian Journal of Botany* 44:413–420.

Bush, L. P., Cornelius, P. L., Buckner, R. C., Varney, D. R., Chapman, R. A., Burriss II, P. B., Kennedy, C. W., Jones, T. A., and M. J. Saunders. 1982. Association of *N*-acetyl loline and *N*-formyl loline with *Epichloe typhina* in tall fescue. *Crop Science* 22:941–943.

Bush, L. P., Wilkinson, H. H., and C. L. Schardl. 1997. Bioprotective alkaloids of grass-fungal endophyte symbioses. *Plant Physiology* 114:1–7.

Calhoun, L. A., Findlay, J. A., Miller, J. D., and N. J. Whitney. 1992. Metabolites toxic to spruce budworm from balsam fir needle endophytes. *Mycological Research* 96:281–286.

Carroll, G. C. 1994. Forest endophytes: pattern and process. *Canadian Journal of Botany* 73:S1316–S1324.

Carroll, G. C. 1988. Fungal endophytes in stems and leaves: From latent pathogen to mutualistic symbiont. *Ecology* 69:2–9.

Carroll, G. C. 1991a. Fungal associates of woody plants as insect antagonists in leaves and stems. *Microbial Mediation of Plant-Herbivore Interactions*, Eds. P. Barbosa, V. A. Krischik, and C. G. Jones, 253–271. New York: John Wiley & Sons.

Carroll, G. C. 1991b. Beyond pest deterrence—alternative strategies and hidden costs of endophytic mutualisms in vascular plants. In *Microbial Ecology of Leaves*, Eds. J. H. Andrews, and S. S. Hirano, 358–375. New York: Springer-Verlag.

Carroll, G. C. and F. E. Carroll. 1978. Studies on the incidence of coniferous needle endophytes in the Pacific Northwest. *Canadian Journal of Botany* 56:3034–3043.

Caruso, M., Colombo, A. L., Fedeli, L., Pavesi, A., Quaroni, S., Saracchi, M., and Ventrella, G. 2000. Isolation of endophytic fungi and actinomycetes taxane producers. *Annals of Microbiology* 50:3–13.

Clay, K. 2004. Fungi and the food of gods. *Nature* 427:401–402.

Dai, J., Krohn, K., Florke, U., Draeger, S., Schulz, B., Kiss-Szikszai, A., et al. 2006. Metabolites from the endophytic fungus *Nodulisporium* sp from *Juniperus cedre. European Journal of Organic Chemistry* 15:3498–3506.

Deckert, R. J., Melville, L. H., and R. L. Peterson. 2001. Structural features of a *Lophodermium* endophyte during the cryptic life-cycle phase in the foliage of *Pinus strobus. Mycological Research* 105:991–997.

Deckert, R. J. and R. L. Peterson. 2000. Distribution of foliar fungal endophytes of *Pinus strobus* between and within host strees. *Canadian Journal of Forest Research* 30:1436–1442.

Eigenbrode, S. D. and K. E. Espelie. 1995. Effects of plant epicuticular lipids on insect herbivores. *Annual Review of Entomology* 40:171–194.

Findlay, J. A., Buthelezi, S., Lavoie, R., Pena-Rodriguez, L., and J. D. Miller. 1995a. Bioactive isocoumarins and related metabolites from conifer endophytes. *Journal of Natural Products* 58:1759–1766.

Findlay, J. A., Li, G., and J. A. Johnson. 1997. Bioactive compounds from an endophytic fungus from eastern larch (*Larix laricina*) needles. *Canadian Journal of Chemistry* 75:716–719.

Findlay, J. A., Li, G., Miller, J. D., and T. O. Womiloju. 2003. Insect toxins from spruce endophytes. *Canadian Journal of Chemistry* 81:284–292.

Findlay, J. A., Li, G. Q., Penner, P. E., and J. D. Miller. 1995b. Novel diterpenoid insect toxins from a conifer endophyte. *Journal of Natural Products* 58:197–200.

Fisher, P. J. and O. Petri. 1990. A comparative study of fungal endophytes in xylem and bark of Alnus species in England and Switzerland. *Mycological Research* 94:313–319.

Freeman, S. and R. J. Rodriquez. 1993. Genetic conversion of a fungal plant pathogen to a nonpathogenic, endophytic mutualist. *Science* 260:75–78.

Ganley, R. J., Brunsfeld, S. J., and G. Newcombe. 2004. A community of unknown, endophytic fungi in western white pine. *Proceedings of the National Academy of Sciences* 101:10107–10112.

Ganley, R. J. and G. Newcombe. 2006. Fungal endophytes in seeds and needles of *Pinus monticola. Mycological Research* 110:318–327.

Gordon, D. C., Percy, K. E., and R. T. Riding. 1998. Effects of UV-B radiation on epicuticular wax production and chemical composition of four *Picea* species. *New Phytologist* 138:441–449.

Grünig, C. R., Duó, A., and T. N. Sieber. 2006. Population genetic analysis of *Phialocephala fortinii* s.l. and *Acephala applanata* in two undisturbed forests in Switzerland and evidence for new cryptic species. *Fungal Genetics and Biology* 43:410–421.

Gunatilaka, L. 2006. Natural products from plant-associated microorganisms: Distribution, structural diversity, bioactivity, and implications of their occurrence. *Journal of Natural Products* 69:509–526.

Guo, L. D., Huang, G. R., Wang, Y., He, W. H., Zheng, W. H., and K. D. Hyde. 2003. Molecular identification of white morphotype strains of endophytic fungi from *Pinus tabulaeformis. Mycological Research* 107:680–688.

Gure, A., Wahlström, K., and J. Stenlid. 2005. Pathogenicity of seed-associated fungi to *Podocarpus falcatus* in vitro. *Forest Pathology* 35:23–35.

Hata, K. and K. Futai. 1993. Effect of needle aging on the total colonization rates of endophytic fungi on *Pinus thunbergii* and *Pinus densiflora* needles. *Journal of the Japanese Forestry Society* 75:338–341.

Hata, K. and K. Futai. 1996. Variation in fungal endophyte populations in needles of the genus *Pinus. Canadian Journal of Botany* 74:103–114.

Hata, K., Tsuda, M., and K. Futai. 1998. Seasonal and needle age-dependent changes of the endophytic mycobiota in *Pinus thunbergii* and *Pinus densiflora* needles. *Canadian Journal of Botany* 76:245–250.

Hatcher, P. E. 1995. Three-way interactions between plant pathogenic fungi, herbivorous insects and their host plants. *Biological Reviews* 70:639–694.

Hawksworth, D. L. 1991. The fungal dimension of biodiversity: Magnitude, significance, and conservation. *Mycological Research* 95:641–655.

Hawksworth, D. L. 2001. The magnitude of fungal diversity: The 1.5 million species estimate revisited. *Mycological Research* 105:1422–1432.

Helander, M. L. 1995. Responses of pine needle endophytes to air pollution. *New Phytologist* 131:223–229.

Helander, M. L., Neuvonen, S., Sieber, T., and O. Petrini. 1993. Simulated acid rain affects birch leaf endophyte populations. *Microbial Ecology* 26:227–234.

Higgins, K. L., Arnold, A. E., Miadlikowska, J., Sarvate, S. D., and F. Lutzoni. 2007. Phylogenetic relationships, host affinity, and geographic structure of boreal and arctic endophytes from three major plant lineages. *Molecular Phylogenetics and Evolution* 42:543–555.

Hirsch, A. M. 2004. Plant–microbe symbioses: A continuum from commensalism to parasitism. *Symbiosis* 37:345–363.

Hohtola, A. 1988. Seasonal changes in explant viability and contamination of tissue cultures from mature Scots pine. *Plant Cell Tissue and Organ Culture* 15:211–222.

Holloway, P. J., Jeffree, C. E., and E. A. Baker. 1976. Structural determination of secondary alcohols from plant epicuticular waxes. *Phytochemistry* 15:1768–1770.

Isaac, S. 1992. *Fungal-Plant Interactions*. Chapman & Hall, London.

Jeffree, C. E. 1986. The cuticle, epicuticular waxes and trichomes of plants, with reference to their structure, functions and evolution. In *Insects and the Plant Surface*, Eds. B. Juniper and R. Southwood, 23–135. London: Edward Arnold.

Jeffree, C. E., Baker, E. A., and P. J. Holloway. 1975. Ultrastructure and recrystallisation of plant epicuticular waxes. *New Phytologist* 75:539–549.

Jeffree, C. E., Johnson, R. P. C., and P. G. Jarvis. 1971. Epicuticular wax in the stomatal antechamber of sitka spruce and its effects on the diffusion of water vapour and carbon dioxide. *Planta* 98:1–10.

Jetter, R. and M. Riederer. 1994. Epicuticular crystals of nonacosan-10-ol: In vitro reconstitution and factors influencing crystal habits. *Planta* 195:257–270.

Ji, Y., Bi, J.-N., Yan, B., and X.-D Zhu. 2006. Taxol-producing fungi: A new approach to industrial production of taxol. *Chinese Journal of Biotechnology* 22:1–6.

Johnson, J. A. and N. J. Whitney. 1989. A study of fungal endophytes of needles of balsam fir (*Abies balsamea*) and red spruce (*Picea rubens*) in New Brunswick, Canada, using cultural and electron microscope techniques. *Canadian Journal of Botany* 67:3513–3516.

Johnson, J. A. and N. J. Whitney. 1994. Cytotoxicity and insecticidal activity of endophytic fungi from black spruce (*Picea mariana*) needles. *Canadian Journal of Microbiology* 40:24–27.

Jurc, D., Bojovic, S., and M. Jurc. 1999. Influence of endogenous terpenes on growth of three endophytic fungi from the needles of *Pinus nigra* Arnold. *Phyton* 39:225–229.

Kim, S., Shin, D. S., Lee, T., and K. B. Oh. 2004. Periconicins, two new fusicoccane diterpenes produced by an endophytic fungus *Periconia* sp. with antibacterial activity. *Journal of Natural Products* 67:448–450.

Kowalski, T. 1993. Fungi living in symptomless needles of *Pinus sylvestris* with respect to some observed disease processes. *Journal of Phytopathology* 139:129–145.

Kowalski, T. and R. D. Kehr. 1992. Endophytic fungal colonization of branch bases in several forest tree species. *Sydowia* 44:137–168.

Li, J. Y., Sidhu, R. S., Ford, E. J., Long, D. M., Hess, W. M., and G. A Strobel. 1998. The induction of taxol production in the endophytic fungus—*Periconia* sp. from *Torreya grandifolia*. *Journal of Industrial Microbiology and Biotechnology* 20:259–264.

Li, J. Y., Strobel, G., Sidhu, R., Hess, W. M., and E. J. Ford. 1996. Endophytic taxol-producing fungi from bald cypress, *Taxodium distichum*. *Microbiology* 142:2223–2226.

Lilja, A., Hallaksela, A. M., and R. Heinonen. 1995. Fungi colonizing Scots-pine cone scales and seeds and their pathogenicity. *Forest Pathology* 25:38–46.

Menkis, A., Allmer, J., Vasiliauskas, R., Lygis, V., Stenlid, J., and R. Finlay. 2004. Ecology and molecular characterization of dark septate fungi from roots, living stems, coarse and fine woody debris. *Mycological Research* 108:965–975.

Millar, C. S. 1974. Decomposition of coniferous leaf litter. In *Biology of Plant Litter Decomposition*. Vol. 1, Eds. C. H. Dickinson and G. J. F. Pugh, 105–128. London: Academic Press.

Miller, J. D., MacKenzie, S., Foto, M., Adams, G. W., and F. M. Findlay. 2002. Needles of white spruce inoculated with rugulosin-producing endophytes contain rugulosin reducing spruce budworm growth rate. *Mycological Research* 106:471–479.

Müller, M. M. and A.-M. Hallaksela. 1998. Diversity of Norway spruce needle endophytes in various mixed and pure Norway spruce stands. *Mycological Research* 102:1183–1189.

Müller, M. M. and A.-M. Hallaksela. 2000. Fungal diversity in Norway spruce: A case study. *Mycological Research* 104:1139–1145.

Müller, C. B. and J. Krauss. 2005. Symbiosis between grasses and asexual fungal endophytes. *Current Opinion in Plant Biology* 8:450–456.

Noble, H. M., Langley, D., Sidebottom, P. J., Lane, S. J., and P. J. Fisher. 1991. An echinocandin from an endophytic *Cryptosporiopsis* sp. and *Pezicula* sp. in *Pinus sylvestris* and *Fagus sylvatica*. *Mycological Research* 95:1439–1440.

Ortiz-Garcia, S., Gernandt, D. S., Stone, J. K., Johnston, P. R., Chapela, I. H., Salas-Lizana, R., and E. R. Alvarez-Buylla. 2003. Phylogenetics of *Lophodermium* from pine. *Mycologia* 95:846–859.

Pan, H. and L. N. Lundgren. 1996. Phenolics from inner bark of *Pinus sylvestris*. *Phytochemistry* 42:1185–1189.

Park, J. H., Choi, G. J., Lee, H. B., Kim, K. M., Jung, H. S., Lee, S. W., et al. 2005. Griseofulvin from *Xylaria* sp strain F0010, an endophytic fungus of *Abies holophylla* and its antifungal activity against plant pathogenic fungi. *Journal of Microbiology and Biotechnology* 15:112–117.

Pelaez, F., Cabello, A., Platas, G., Diez Matas, M. T., Gonzáles del Val, A., Basilio, A., et al. 2000. The discovery of enfumafungin, a novel antifungal compound produced by an endophytic *Hormonema* species biological activity and taxonomy of the producing organism. *Systematic and Applied Microbiology* 23:333–343.

Petrini, O. 1986. Taxonomy of endophytic fungi of aerial plant tissues. In *Microbiology of the Phyllosphere*, Eds. N. J. Fokkema and J. van den Heuvel, 175–187. Cambridge: Cambridge University Press.

Petrini, O. 1991. Fungal endophytes of tree leaves. In *Microbial ecology of leaves*, Eds. J. H. Andrews and S. S. Hirano, 179–197. New York: Springer-Verlag.

Petrini, O. and C. G. Carroll. 1981. Endophytic fungi in foliage of some Cupressaceae in Oregon. *Canadian Journal of Botany* 59:629–636.

Phillips, M. A. and R. B. Croteau. 1999. Resin-based defenses in conifers. *Trends in Plant Science* 4:184–190.

Pirttilä, A. M. 2001. Endophytes in the buds of Scots pine (*Pinus sylvestris* L.). Doctoral thesis. University of Oulu, Oulu, Finland.

Pirttilä, A. M., Laukkanen, H., Pospiech, H., Myllylä, R., and A. Hohtola, A. 2000. Detection of intracellular bacteria in the buds of Scotch pine (*Pinus sylvestris* L.) by *in situ* hybridization. *Applied and Environmental Microbiology* 66:3073–3077.

Pirttilä, A. M., Pospiech, H., Laukkanen, H., Myllylä, R., and A. Hohtola. 2003. Two endophytic fungi in different tissues of Scots pine buds (*Pinus sylvestris* L.). *Microbial Ecology* 45:53–62.

Pirttilä, A. M., Pospiech, H., Laukkanen, H., Myllylä, R., and A. Hohtola. 2005. Seasonal variation in location and population structure of endophytes in buds of Scots pine. *Tree Physiology* 25:289–297.

Polishook, J. D., Dombrowski, A. W., Tsou, N. N., Salituro, G. M., and J. E. Curotto. 1993. Preussomerin D from the endophyte *Hormonema dematioides*. *Mycologia* 85:62–64.

Redlin, S. C. and Carris, L. M. 1996. *Endophytic Fungi in Grasses and Woody Plants: Systematics, Ecology and Evolution*. St. Paul, MN: APS Press.

Redman, R. S., Dunigan, D. D., and R. J. Rodriguez. 2001. Fungal symbiosis from mutualism to parasitism: Who controls the outcome, host or invader? *New Phytologist* 151:705–716.

Reicosky, D. A. and J. W. Hanover. 1978. Physiological effects of surface waxes I. Light reflectance for glaucous and nonglaucous picea pungens. *Plant Physiology* 62:101–104.

Saikkonen, K. 2007. Forest structure and fungal endophytes. *Fungal Biology Review* 21:67–74.

Saikkonen, K., Faeth, S. H., Helander, M., and T. J. Sullivan. 1998. Fungal endophytes: A continuum of interactions with host plants. *Annual Review of Ecology and Systematics* 29:319–343.

Saikkonen, K., Helander, M., Ranta, H., Neuvonen, S., Virtanen, T., Suomela, J., and P. Vuorinen. 1996. Endophyte-mediated interactions between woody plants and insect herbivores? *Entomologia Experimentalis et Applicata* 80:269–271.

Saikkonen, K., Wäli, P., Helander, M., and S. H. Faeth. 2004. Evolution of endophyte–plant symbioses. *Trends in Plant Science* 9:275–280.

Schulz, B. and C. Boyle. 2005. The endophytic continuum. *Mycological Research* 109:661–686.

Schulz, B., Boyle, C., Draeger, S., Römmert, A.-K., and K. Krohn. 2002. Endophytic fungi: A source of novel biologically active secondary metabolites. *Mycological Research* 106:996–1004.

Schulz, B., Römmert, A.-K., Dammann, U., Aust, H.-J., and D. Strack. 1999. The endophyte–host interaction: A balanced antagonism? *Mycological Research* 103:1275–1283.

Schulz, B., Sucker, J., Aust, H.-J., Krohn, K., Ludewig, K., Jones, P. G., and D. Döring. 1995. Biologically active secondary metabolites of endophytic Pezicula species. *Mycological Research* 99:1007–1015.

Schulz, B., Wanke, U., Draeger, S., and H.-J. Aust. 1993. Endophytes from herbaceous plants and shrubs: Effectiveness of surface sterilization methods. *Mycological Research* 97:1447–1450.

Sherwood-Pike, M., Stone, J. K., and G. C. Carroll. 1986. *Rhabdocline parkeri*, a ubiquitous foliar endophyte of Douglas fir. *Canadian Journal of Botany* 64:1849–1855.

Sieber, T. 2007. Endophytic fungi in forest these: Are they mutualists? *Fungal Biology Review* 21:75–89

Sieber, T. N., Rys, J., and O. Holdenrieder. 1999. Mycobiota in symptomless needles of *Pinus mugo* ssp. *uncinata. Mycological Research* 103:306–310.

Sieber-Canavesi, F. and T. N. Sieber. 1993. Successional patterns of fungal communities in needles of European silver fir (*Abies alba* Mill.). *New Phytologist* 125:149–161.

Siegel, M. C., Latch, G. C. M., Bush, L. P., Fannin, F. F., Rowan, D. D., Tapper, B. A., et al. 1990. Fungal endophyte-infected grasses: Alkaloid accumulation and aphid response. *Journal of Chemical Ecology* 16:3301–3315.

Stierle, A., Stierle, D., and T. Bugni. 1999. Sequoiatones A and B: Novel antitumor metabolites isolated from a redwood endophyte. *Journal of Organic Chemistry* 64:5479–5484.

Stierle, D. B., Stierle, A. A., and A. Kunz. 1998. Dihydroramulosin from *Botrytis* sp. *Journal of Natural Products* 61:1277–1278.

Stierle, A., Strobel, G., and D. Stierle. 1993. Taxol and taxane production by *Taxomyces andreanae*, an endophytic fungus of Pacific yew. *Science* 260:214–216.

Stone, J. K. 1987. Initiation and development of latent infections by Rhabdocline parkeri on Douglas-fir. *Canadian Journal of Botany* 65:2614–2621.

Stone, J. K. 1988. Fine structure of latent infections by *Rhabdocline parkeri* on Douglas-fir, with observations on uninfected epidermal cells. *Canadian Journal of Botany* 66:45–54.

Strobel, G. A., Hess, W. M., Li, J. Y., Ford, E., Sears, J., Sidhu, R. S., and B. Summerell. 1997. *Pestalotiopsis guepinii*, a taxol-producing endophyte of the Wollemi pine, *Wollemia nobilis. Australian Journal of Botany* 45:1073–1082.

Strobel, G., Yang, X., Sears, J., Kramer, R., Sidhu, R. S., and W. M. Hess. 1996. Taxol from *Pestalotiopsis microspora*, an endophytic fungus of *Taxus wallachiana. Microbiology* 142:435–440.

Sumarah, M. W., Miller, J. D., and G. W. Adams. 2005. Measurement of a rugulosin-producing endophyte in white spruce seedlings. *Mycologia* 97:770–776.

Suske, J. and G. Acker. 1986. Internal hyphae in young, symptomless needles of *Picea abies*: Electron microscopic and cultural investigation. *Canadian Journal of Botany* 65:2098–2103.

Tan, R. X. and W. X. Zou. 2001. Endophytes: A rich source of functional metabolites. *Natural Products Reports* 18:448–459.

Thompson, J. N. 2005. The geographic mosaic of coevolution. Chicago, USA: The University of Chicago Press.

Walton, T. J. 1990. Waxes, cutin and suberin. In *Methods in Plant Biochemistry*, vol. 4, Eds. J. J. Harwood and J. Boyer, pp. 106–158.

Wang, J., Huang, Y., Fang, M., Zhang, Y., Zheng, Z., Zhao, Y., and W. Su. 2002. Brefeldin A, a cytotoxin produced by *Paecilomyces* sp. and *Aspergillus clavatus* isolated from *Taxus mairei* and *Torreya grandis. FEMS Immunological and Medical Microbiology* 34:51–57.

Wang, J., Li, G., Lu, H., Zhang, Z., Huang, Y., and W. Su. 2000. Taxol from *Tubercularia* sp. strain TF5, an endophytic fungus of *Taxus mairei. FEMS Microbiology Letters* 193:249–253.

Wen, M., Buschhaus, C., and R. Jetter. 2006. Nanotubules on plant surfaces: Chemical composition of epicuticular wax crystals on needles of *Taxus baccata* L. *Phytochemistry* 67:1808–1817.

White, J. F., Breen, J. P., and G. Morganjones. 1991. Substrate utilization in selected *Acremonium, Atkinsonella* and *Balansia* species. *Mycologia* 83:601–610.

Wilson, D. 1995. Endophyte-the evolution of a term, and clarification of its use and definition. *Oikos* 73:274–276.

Wilson, D. 1996. Manipulation of infection levels of horizontally transmitted fungal endophytes in the field. *Mycological Research* 100:827–830.

Wollenweber, E., Kraut, L., and R. Mues. 1998. External accumulation of biflavonoids on gymnosperm leaves. *Zeitschrift fur Naturforschung C* 53:946–950.

Yang, X., Strobel, G., Stierle, A., Hess, W. M., Lee, J., and J. Clardy. 1994. A fungal endophyte-tree relationship: *Phoma* sp. in *Taxus wallachiana. Plant Science* 102:1–9.

Yu, Z. and R. A. Dahlgren. 2000. Evaluation of methods for measuring polyphenols in conifer foliage. *Journal of Chemical Ecology* 26:2119–2140.

Zhou, X., Wang, Z., Jiang, K., Wei, Y., Lin, J., Sun, X., et al. 2007. Screening of taxol-producing endophytic fungi from *Taxus chinensis* var. mairei. *Applied Biochemistry and Microbiology* 43:439–443.

16

Diversity and Ecological Roles of Clavicipitaceous Endophytes of Grasses

Mariusz Tadych, Mónica S. Torres, and James F. White, Jr.

CONTENTS

16.1 Introduction

Clavicipitaceous grass endophytes compared to nonclavicipitaceous endophytes are characterized by the systemic fungal colonization of the host plant parts by only one fungal species (see Chapter 22). The result is the development of a long-term symbiosis (Stone et al., 2004). Taxonomically, these grass endophytes are included in the family Clavicipitaceae (Ascomycota). These include epiphytic and endophytic forms in the genera of *Atkinsonella* Diehl, *Balansia* Speg., *Balansiopsis* Höhn., *Dussiella* Pat., *Epichloë* (Fr.) Tul & C. Tul., *Myriogenospora* G.F. Atk., and *Parepichloë* J.F. White & P.V. Reddy. Clavicipitaceous endophytes are widespread in grasses (Poaceae) with one estimate as high as 25% of cool-season grass species harboring these endophytes (White, 1987).

The genus *Epichloë*, with anamorphs in *Neotyphodium* Glenn, C.W. Bacon & Hanlin, is the most frequently studied group of the clavicipitaceous grass endophytes. Much of the work has been done on *Neotyphodium coenophialum* (Morgan-Jones & Games) Glenn, C.W. Bacon & Hanlin (endophyte of "tall fescue," *Lolium arundinaceum* (Schreb.) Darbysh.), and *Neotyphodium lolii* (Latch, Christensen & Samuels) Glenn, C.W. Bacon & Hanlin (endophyte of "perennial ryegrass," *Lolium perenne* L.; Hoveland, 1993). Based on the relative costs and benefits to the hosts, the grass–fungal endophyte symbioses are regarded as generally mutualistic (Clay, 1988; Schardl et al., 2004; Wäli et al., 2006), although they have been found to be pathogenic in some nonagronomic or natural grassland species (Faeth, 2002; Cheplick, 2007). Endophytic species in *Epichloë* may result in partial or complete sterilization of hosts due to the production of a fungal stroma on the host flowering culms, a condition known as "choke disease" (White, 1988; White et al., 1991). However, the degree of effect on the host reproduction varies with the host–fungus combination; it may extend to the point where endophytes are entirely dependent on the survival and growth of the grass host plant for their own growth survival and dispersal. The family Clavicipitaceae also includes (1) entomopathogenic species (e.g., *Metacordyceps* G.H. Sung, J.M. Sung, Hywel-Jones & Spatafora) associated with a relatively broad range of arthropod hosts, (2) necrothrophs of scale insects and whiteflies (e.g., *Hypocrella* Sacc., *Moelleriella* Bres., *Regiocrella* Chaverri & K.T. Hodge, *Samuelsia* Chaverri & K.T. Hodge, *Torrubiella* Boud.), and (3) plant pathogenic forms such as the ergot fungus *Claviceps* Tul. and epiphytic species (e.g., *Balansia*, *Myriogrnospora*) associated with grasses, rushes, and sedges (Sung et al., 2007; Chaverri et al., 2008). Phylogenetically, the family Clavicipitaceae is derived from within the order Hypocreales (Spatafora and Blackwell, 1993; Rehner and Samuels, 1995; Spatafora et al., 2007).

Based on multigene phylogenetic analyses of the family, it has been proposed that the family diversified evolutionarily through a series of interkingdom host jumps that developed among the animal, fungal, and plant host species (Spatafora et al., 2007). Grass biotrophs of the family appear to have been derived from insect parasitic species (Koroch et al., 2004; Spatafora et al., 2007; Torres et al., 2007a). It has been hypothesized that the endophytic fungi associated with grasses (e.g., *Epichloë/Neotyphodium* and *Balansia* spp.) evolved from those insect parasites (e.g., *Metacordyceps* spp.) from which the epibiotic plant biotrophic species (e.g., *Hypocrella* spp., *Moelleriella* spp., and *Samuelsia* spp.) arose. These biotrophic fungi gained access to plant nutrients by first infecting and necrophytizing scale insects and whiteflies and then utilizing nutrients emerging on the surface of the plant through the insect's stylet or stylet wound. Finally, clavicipitaceous fungi evolved the capacity to infect grass hosts directly with forms that are epibiotic (e.g., *Myriogenospora* spp.) and others that are endophytic (e.g., *Epichloë/Neotyphodium* spp. and *Balansia* spp.). This scenario for evolution of the clavicipitaceous endophytes is an evolutionary shortcut to biotrophy as many biotrophic endophytes are believed to be derived from more virulent plant pathogenic fungi. Since clavicipitaceous endophytes were derived from insect pathogens, they did not possess enzymes or toxins for killing or degrading plant tissues; there were, thus, no defensive mechanisms in grasses that would limit their colonization of grasses. They were not adapted to cause disease or degrade plant tissues. Studies on the range of nutrients on which Clavicipitaceae grows suggest that evolution of plant biotrophy and endophytism in the family was a phenomenon of reduction of enzymatic capabilities and increasing dependence on the host plant to provide nutrients for growth (Torres, 2007). Concurrent with the reduction of enzymatic capabilities was an apparent increase in the allocation of energy to the production of particular secondary metabolites beneficial in the symbiosis (e.g., ergot alkaloids; Torres et al., 2007b, 2008). The evolutionary histories of the clavicipitaceous endophytes and their derivation from insect pathogens rather than plant pathogens may, in part, explain why they produce toxins that affect insects and other animals and have more subtle effects on plants. The insect pathogens in *Metacordyceps* related species were evolved to kill and degrade insects. It is likely that the chemical arsenal that they used in immobilizing insects is similar to that used by the endophytes now for defending the host plants.

16.2 Clavicipitaceous Endophyte Life Cycles

Species of genus *Balansia* are epibionts and endophytes of warm-season grasses. Epibiotic species survive on the surface of grasses in the crown near developing tillers, and the stroma development modifies the epidermal layer to permit leakage of nutrients from the grass to the stromal mycelium on the plant's surface. Endophytic species grow in the intercellular spaces of healthy grass plants and reproduce by forming a fungal stroma on living leaves, culms, or inflorescences of the host. For example, *Balansia nigricans* (Speg.) J.F. White, T.E. Drake & T.I. Martin develops dark pigmented stromata on internodes of the flowering culm of *Panicum* spp. (Figure 16.1a and b); sometimes the stroma completely surrounds the grass culm (Figure 16.1c).

Populations of *Balansia* spp. consist of two mating types (White et al., 1995). Ephelidial conidia (spermatia) of one mating type must be transferred to the stromata of the opposite mating type for perithecia to develop. Insects have been observed to visit stromata to feed and are hypothesized to transfer the spermatia between the mating types during the process of feeding (White and Owens, 1992). Endophytic species of *Balansia* are not widely distributed in grasses. This is likely because they tend not to be seed transmitted, or at least not commonly disseminated in seeds; this may limit their spatial dissemination in grass populations.

Species of genus *Epichloë* (anamorphs in *Neotyphodium*) are endophytes of cool-season grasses (White, 1987). Mycelium can be found in intercellular spaces of leaf sheaths, culms, rhizomes (Figure 16.2a), and seeds of grasses (Figure 16.2b). Epiphyllous mycelium may also be present sparsely on the surface of leaf blades (Figure 16.2c; White et al., 1996; Moy et al., 2000; Dugan et al., 2002; Tadych and White, 2007; Tadych et al., 2007). When the grasses flower, the fungus proliferates within and grows over the developing inflorescence and forms the fungal stroma. The inflorescence primordium remains

(a) (b) (c)

FIGURE 16.1 *Balansia nigricans*. (a) Uninfected and infected inflorescences of *Panicum* sp. (b) Ascostroma of *B. nigricans* on culm of *Panicum* sp. (c) Cross section showing ascostroma with perithecia of *B. nigricans* on infected culm of *Panicum* sp.

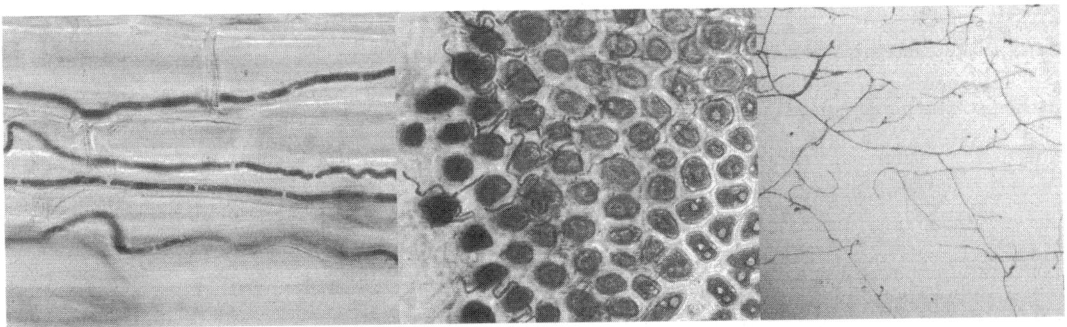

(a) (b) (c)

FIGURE 16.2 *Neotyphodium* endophytes of grasses. (a) Endophytic mycelium of the *Neotyphodium* sp. in culm of *Poa ampla* grass. (b) Convoluted hyphae of *Neotyphodium coenophialum* on and between the aleurone cells in seed-squash preparations of *Festuca arundinacea* seed. (c) Section of endophyte-infected leaf of *Poa ampla* colonized by the epiphytic stage of *Neotyphodium* showing a hyphal network, conidiogenous cells, and conidia.

alive, but in an arrested stage of development within the fungal mycelium; this prevents the development of the seed head (type-I endophytes). Stroma development is often referred to as "choke disease." Some *Epichloë* endophytes (type-II endophytes) exhibit stromata only in some of the tillers which allows for the partial seed production and, thus, vertical transmission of the fungal endophyte within seeds. As in *Balansia*, the stromata of *Epichloë* bear spermatia and fungal populations contain two mating types; the two mating types must be transferred between stromata before perithecia and ascospores develop (Bultman and White, 1987). Symbiotic flies of genus *Botanophila* Lioy (Anthomyiidae; Diptera) act as "pollinators" of stromata and vector spermatia between mating types of the fungus (Bultman and White, 1987; Rao et al., 2005). Fertilized stromata produce perithecia containing ascospores that are ejected and possibly may infect endophyte-free plants.

Stromata production and the sexual cycle of *Epichloë* occur only on grasses in the northern hemisphere. Clavicipitaceous endophytes, abundant in the southern hemisphere, are unable to form stromata and, thus, are not known to reproduce sexually. Furthermore, some species of *Epichloë* have lost capacity for the development of the sexual stage. These have been referred to as asexual or type-III endophytes and are frequently classified as species of *Neotyphodium*. The type-III endophytes are transmitted vertically through seed but many retain an epiphyllous mycelium on leaves where conidia form. In addition, the majority of type-III endophytes continue to produce conidia in culture. The presence of conidia suggests some function of these viable conidia in the life cycles of the endophytes and the potential for horizontal transmission to uninfected plant individuals. Tadych et al. (2007) demonstrated that the epiphyllous conidia on leaf surfaces (Figure 16.2c) were released from their conidiophores in water but not air currents. The conidia may spread to neighboring plants in water flowing across leaves from rain or dew. The likely, but as yet unproven, sites of infection are the tillers where the fungi likely colonize tillering meristems epiphyllously in the crowns of grass plants. A tiller colonization mechanism would result in the endophytic colonization of some of the ovules and seeds of neighboring plants, but the original mature neighboring maternal plants would not bear endophytic mycelium. This is because the clavicipitaceous endophytes are adapted for rapid growth in the rich meristem nutrients but show limited capacity to grow through mature plant tissue (Western and Cavett, 1959; White et al., 1991). Endophytic mycelium is systemic throughout the grass plant because it proliferates in the shoot meristems and is deposited in the intercellular spaces of tissues that differentiate to form aerial plant organs. Endophytic mycelium does not colonize root meristems in embryos and roots do not contain detectable quantities of mycelium of clavicipitaceous endophytes. Many aspects of the life histories of clavicipitaceous endophytes, their sexual and parasexual systems, and population dynamics are currently at best sketchy. Type-III endophytes have evolved multiple times and are distributed in both northern and southern hemispheres (White, 1988; Moon et al., 2002; Schardl and Moon, 2003; Schardl et al., 2004). These species are endophytic of leaves, culms, and rhizomes and colonize inflorescence primordia. Instead of fully enveloping the inflorescence primordium in a stroma, the endophyte grows in the tissues that differentiate to form ovules and developing seeds; within seeds, they colonize the embryo (Figure 16.2b; Philipson and Christey, 1986). There are no obvious symptoms of type-III endophyte infections observable during plant development. However, these asexual endophytes are not entirely clonal since phylogenetic analyses have demonstrated interspecific hybridization (Moon et al., 2002, 2007; Schardl et al., 2004). This interspecies "parasexual" recombination has been hypothesized to serve as a mechanism to maintain genome viability by counteracting accumulation of deleterious mutations among endophytes that cannot undergo sexual recombination and repair gene damage (Schardl and Moon, 2003).

16.3 Historical Perspective on Clavicipitaceous Endophytes

Tall fescue (*Lolium arundinaceum*) is one of the examples where grass toxicity has played a role in spurring research in endophytes (Hoveland, 1993). The grass was brought to the United States from Europe in the late 1800s. It was discovered in Kentucky in 1931, tested at the University of Kentucky, and released in 1943 as "Kentucky 31." It became popular with farmers, spreading quickly throughout the United States. It now occupies well over 16 million hectares of pasture and forage land in the United States. The problem of livestock neurotoxicosis "fescue toxicosis" became a major concern in the United States. In "fescue toxicosis," consumption of endophyte-infected tall fescue results in the reduction of forage consumption by cattle and, thus, lowers weight gains by these ungulates. Affected cattle may produce less milk, have elevated body temperatures and respiration rates, develop a rough hair coat, and demonstrate an unthrifty appearance. In 1977, it was determined that an endophyte was responsible for the toxic syndrome associated with tall fescue forage grass (Bacon et al., 1977). The endophyte *N. coenophialum* produces several alkaloids, particularly ergovaline, that are responsible for "fescue toxicosis" (Bush et al., 1997; Lane et al., 2000).

Perennial ryegrass (*Lolium perenne*) is an important forage and soil stabilization grass. In New Zealand, the toxicosis "ryegrass staggers" of sheep and cattle has been known for a long time (Fletcher et al., 1990). In 1898, the presence of fungal endophyte in the seeds of *L. perenne* was first observed (Guerin, 1898; Hanausek, 1898; Vogl, 1898). The association between the *Lolium* endophyte and "ryegrass staggers" was finally established in 1981 (Fletcher and Harvey, 1981). *Neotyphodium lolii* that infects perennial and hybrid ryegrasses was shown to synthesize several alkaloids, including lolitrem B, a tremorgenic molecule causing staggers symptoms, ergovaline causing vasoconstrictive effects and heat stress in animals, and peramine, a tripeptide that deters some insects, particularly the Argentine stem weevil, from feeding on ryegrass but is not toxic to mammals (Keogh et al., 1996). Ryegrass staggers is a very serious problem for grazing livestock (sheep, cattle, horses, and deer). However, mortality rates are low.

Sleepygrass (*Achnatherum robustum* (Vasey) Barkworth) is a perennial grass that is native to North America and locally abundant in some regions in the southwestern United States. It is frequently found to be toxic to grazing animals, i.e., horses, cattle, and sheep. After consumption of relatively small quantities of the grass, animals go to sleep for 2–3 days, and then gradually recover (White, 1987). The ergot alkaloids like lysergic and isolysergic acid amides have been identified as the sleep-inducing agents in sleepygrass. These alkaloids are produced by *N. funkii* K.D. Craven & C.L. Schardl, the endophyte isolated from this plant (Moon et al., 2007). However, reports of toxicity suggest that the level of alkaloids in native sleepygrass populations may be highly variable among populations despite the level of endophyte infection. Animal toxicity is localized in particular areas where particular strains of the endophyte may predominate (White, 1987).

Drunken horse grass (*Achnatherum inebrians* (Hance) Keng) is a perennial bunchgrass, distributed on alpine and subalpine grasslands of northwestern China and Mongolia. Horses that graze on *A. inebrians* develop staggers, and the animals walk as if they are drunk (Li et al., 2004). The symptoms persist for 6–24 h. Typically, animals completely recover after three days. Toxicity of *A. inebrians* was reported by Marco Polo and the Russian explorer Nikolai Mikhaylovich Przhevalsky. The presence of *N. gansuense* C.J. Li & Z.B. Nan and *N. gansuense* var. *inebrians* C.D. Moon & C.L. Schardl endophytes in this grass species was confirmed (Li et al., 2004; Moon et al., 2007). In a study of drunken horse grass, two major toxic alkaloids, ergonovine and lysergic acid amide, were found when the grass was infected with endophytes (Lane et al., 2000).

Dronkgras (*Melica decumbens* (L.) Weber) is a perennial grass that has a history similar to "drunken horse grass." *Melica decumbens* is endemic to Africa and has a limited distribution in South Africa, found only in the arid areas of central South Africa (White, 1987). It has also been called "staggers grass," which stems from the fact that this grass has narcotic effects on cattle, horses, donkeys, and sheep. The tremorgenic neurotoxins produced by the endophyte *N. melicicola* C.D. Moon & C.L. Schardl were found to be responsible for the narcotic effect of *M. decumbens* on grazing livestock. Consumption of the grass is usually not lethal and the animals completely recover after three days (Lane et al., 2000).

Poa huecu Parodi and several other grasses of South America (e.g., *Festuca argentina* (Speg.) Parodi, *Festuca hieronymi* Hack., and some other *Poa* spp.) are colonized by endophytic fungus *N. tembladerae* D. Cabral & J.F. White (Cabral et al., 1999; Gentile et al., 2005). The association of the endophyte with *F. hieronymi* and *P. huecu* is probably responsible for the toxic syndrome, called "tembladera" or "huecu," long known to occur in grazing animals (Rivas and Zanolli, 1909). "Tembladera" is from the Spanish word that means "tremble," and word "huecú" is from the indigenous Araucanian language of the tribes that lived in the region and means "intoxicator." Huecú toxicosis results from consumption of the endophyte-infected grass and is frequently lethal to animals. Studies of *N. tembladerae* suggest that toxicity of the infected grasses to mammals is associated with the ergot alkaloids and lolitrems produced by the fungus (Lane et al., 2000; Yue et al., 2000; Gentile et al., 2005).

16.4 Secondary Metabolites of Clavicipitaceous Fungi

Clavicipitaceous endophytes produce a variety of secondary metabolites that may accumulate in plants (Koshino et al., 1989a,b; Porter, 1994; Bacon and White, 2000; Panaccione, 2005). The most studied secondary metabolites with known biological activity include ergot alkaloids, lolines, lolitrems, and peramine (Figure 16.3; Schardl et al., 2006). These compounds are considered the primary mechanism for herbivore deterrence and antimicrobial activity (Bacon et al., 1986; Bush et al., 1997; Schardl and Philips, 1997; Bacon and Lyons, 2005). *Epichloë* has been also shown to produce auxin-like compounds (Figure 16.3) that have been hypothesized to play a role in altering the development of plant tissues (De Battista et al., 1990). Factors influencing the types and concentrations of endophyte metabolites include the strain of endophyte, plant species, plant part (e.g., leaves, leaf sheaths, culms, and seeds), age of plant, growing season, and fertilization status. Alkaloid production and concentration correlates

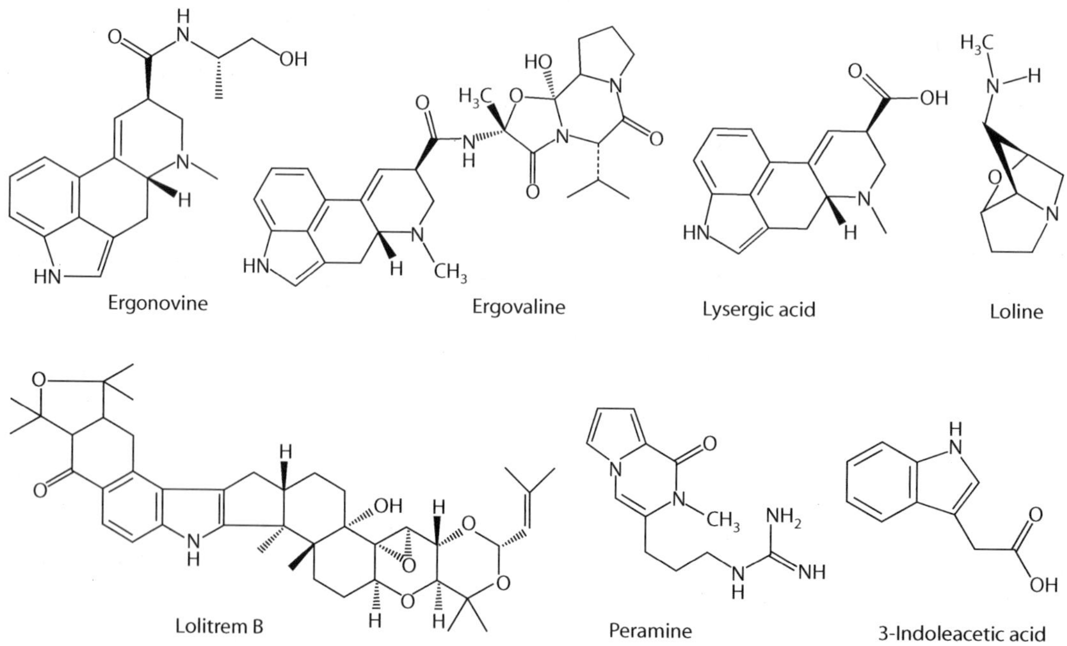

FIGURE 16.3 Selected secondary metabolites of clavicipitaceous endophytes of grasses.

directly with the hyphal density and the survival value of the plant part, with meristems and surrounding leaf sheaths and seeds exhibiting high concentrations of alkaloids (Clay, 1990).

Ergot alkaloids and lolitrems are generally limited to leaf sheaths, meristematic zones, inflorescences and seeds; lolines and peramine are also found in blades (Keogh et al., 1996; Lane et al., 2000). This greater dissemination of lolines and peramine in plants is likely due to their greater solubility in the aqueous solutions of the plant. Peramine has been shown to mobilize from within the fungus hyphae, to be taken up into the xylem and secreted by hydathodes at the leaf tip (Koulman et al., 2007). Ergot alkaloids and lolitrems are less soluble and less likely translocated for distances through aqueous plant tissues.

As endosymbiotic associations evolved, the persistence of secondary metabolites suggests that they are advantageous to the host plant or fungal endophyte. The fungal endophyte produces compounds that are involved in defense of the host plant from herbivory (Clay, 1990). Lolines are more active against insects than mammals. Peramine is less toxic to insects and mammals but is recognized to be an effective feeding deterrent to insects (Rowan and Gaynor, 1986). Ergot alkaloids and lolitrems are active against mammals, and both of these metabolites have received significant attention as they are involved in animal toxicoses such as "fescue foot" and "ryegrass staggers."

16.5 Beneficial Effects of Clavicipitaceous Endophytes on Host Fitness

Biotic and abiotic stresses due to infectious and noninfectious plant diseases, pests, and unfavorable growing conditions are major challenges for plants. Clavicipitaceous endophytes have been described as "defensive mutualists" that protect host plants against biotic and abiotic stresses (Clay, 1988, 1990, 1997; see also Chapter 2). In return, the grass–fungal endophytes have access to nutrients from the plant apoplast and also have a means of dissemination through the seed.

The host benefits frequently reported include protection from mammalian and insect herbivores, drought tolerance, resistance to nematodes, resistance to fungal pathogens, and field-improved tolerance to heavy metals (Prestidge et al., 1982; Funk et al., 1983; Arechavaleta et al., 1989; Siegel et al., 1989; West et al., 1990; Patterson et al., 1991; Riedell et al., 1991; Breen, 1993; Funk and White, 1997;

Malinowski and Belesky, 2000; Zaurov et al., 2001; Bonos et al., 2005; Panaccione et al., 2006; Wäli et al., 2006). All these characteristics result in an increased plant fitness or a greater field persistence. However, the benefits from the endophytic symbiosis may be much more variable in the natural grass populations than the introduced or agricultural grass populations (Saikkonen et al., 1999; Tintjer and Rudgers, 2006).

The actual mechanism underlining endophyte-mediated disease resistance is elusive; however, several in vitro studies have demonstrated the production of antifungal compounds of endophytic origin with the capacity to control plant pathogens (White and Cole, 1985; Siegel and Latch, 1991; Yue et al., 2000). In field conditions, it has been demonstrated that fine fescue grass colonized by the *Epichloë festucae* Leuchtm., Schardl & M.R. Siegel shows enhanced disease resistance to leaf spot diseases, e.g., dollar spot (caused by *Sclerotinia homoeocarpa* F.T. Benn.; Clarke et al., 2006) and red thread disease (caused by *Laetisaria fuciformis* (McAlpine) Burds.; Bonos et al., 2005).

Balansia species develop epiphytic and/or endophytic relationships with warm-season grasses. For example, *Balansia cyperi* Edgerton commonly causes host sterilization by suppressing inflorescence development. It has also been reported that *B. cyperi* may deter infection of the host plant (*Cyperus rotundus* L.) by other plant pathogenic fungi, although the mechanism of resistance is unknown (Stovall and Clay, 1991). In endophytic species of *Balansia* such as *Balansia henningsiana* (Møller) Diehl, endophytic mycelium can be found throughout the host tissues. Similar to *Epichloë*, some *Balansia* species also possess the capacity to produce biologically active compounds that may play a role in disease and insect resistance. For example, *Panicum agrostoides* Spreng., when infected by *B. henningsiana*, showed disease resistance to the common leaf spot fungus *Alternaria triticina* Prasada & Prabhu (Clay et al., 1989).

ACKNOWLEDGMENTS

This research was supported in part by NIH U01 TW006674 International Cooperative Biodiversity Groups (NIH Fogarty International Center) and Center for Turfgrass Science (Rutgers, The State University of New Jersey).

REFERENCES

Arechavaleta, M., Bacon, C. W., Hoveland, C. S., and D. E. Radcliffe. 1989. Effect of the tall fescue endophyte on plant response to environmental stress. *Agronomy Journal* 81:83–90.

Bacon, C. W. and P. Lyons. 2005. Ecological fitness factors for fungi within the Balansieae and Clavicipiteae. In *The Fungal Community: Its Organization and Role in the Ecosystem*, Eds. J. Dighton, J. F. White Jr., and P. Oudemans, pp. 519–531. Boca Raton, FL: CRC Taylor & Francis.

Bacon, C. W. and J. F. White Jr. 2000. Physiological adaptations in the evolution of endophytism in the Clavicipitaceae. In *Microbial Endophytes*, Eds. C. W. Bacon and J. F. White Jr., pp. 237–263. New York: Marcel Dekker.

Bacon, C. W., Porter, J. K., Robbins, J. D., and E. S. Luttrell. 1977. *Epichloë typhina* from toxic tall fescue grasses. *Applied Environmental Microbiology* 34:576–581.

Bacon, C. W., Porter, J. K., Robbins, J. D., and E. S. Luttrell. 1986. Ergot toxicity from endophyte infected weed grasses: A review. *Agronomy Journal* 78:106–116.

Bonos, S. A., Wilson, M. M., Meyer, W. A., and C. R. Funk. 2005. Suppression of red thread in fine fescues through endophyte-mediated resistance. *Applied Turfgrass Science* 10:1094.

Breen, J. P. 1993. Enhanced resistance to fall armyworm (Lepidoptera: Noctuidae) in *Acremonium* endophyte infected turfgrasses. *Journal of Economic Entomology* 86:621–629.

Bultman, T. L. and J. F. White Jr. 1987. "Pollination" of a fungus by a fly. *Oecologia* 75:317–319.

Bush, L. P., Wilkinson, H. H., and C. L. Schardl. 1997. Bioprotective alkaloids of grass–fungal endophyte symbioses. *Plant Physiology* 114:1–7.

Cabral, D., Cafaro, M. J., Saidman, B. O., Lugo, M. A., Reddy, P. V., and J. F. White Jr. 1999. Evidence supporting the occurrence of a new species of endophyte in some South American grasses. *Mycologia* 91:315–325.

Chaverri, P., Liu, M., and K. T. Hodge. 2008. A monograph of the entomopathogenic genera *Hypocrella*, *Moelleriella*, and *Samuelsia* gen. nov. (*Ascomycota*, *Hypocreales*, *Clavicipitaceae*), and their aschersonia-like anamorphs in the Neotropics. *Studies in Mycology* 60:1–66.

Cheplick, G. P. 2007. Costs of fungal endophyte infection in *Lolium perenne* genotypes from Eurasia and North Africa under extreme resource limitation. *Environmental and Experimental Botany* 60:202–210.

Clarke, B. B., White, J. F. Jr., Hurley, R. H., Torres, M. S., Sun, S., and D. R. Huff. 2006. Endophyte-mediated suppression of dollar spot disease in fine fescues. *Plant Disease* 90:994–998.

Clay, K. 1988. Fungal endophytes of grasses: A defensive mutualism between plants and fungi. *Ecology* 69:10–16.

Clay, K. 1990. Fungal endophytes of grasses. *Annual Review Ecology Systematics* 21:275–295.

Clay, K. 1997. Consequences of endophyte-infected grasses on plant biodiversity. In *Neotyphodium/Grass Interactions*, Eds. C. W. Bacon and N. S. Hill, pp. 109–124. New York: Plenum Press.

Clay, K., Cheplick, G. P., and S. Marks. 1989. Impact of the fungus *Balansia henningsiana* on *Panicum agrostoides*: Frequency of infection, plant growth and reproduction, and resistance to pests. *Oecology* 80:374–380.

De Battista, J. P., Bacon, C. W., Severson, R., Plattner, R. D., and J. H. Bouton. 1990. Indole acetic acid production by the fungal endophyte of tall fescue. *Agronomy Journal* 82:878–880.

Dugan, F. M., Sitton, J., Sullivan, R. F., and J. F. White Jr. 2002. The *Neotyphodium* endophyte of wild barley (*Hordeum brevisubulatum* subsp. *violaceum*) grows and sporulates on leaf surfaces of the host. *Symbiosis* 32:147–159.

Faeth, S. 2002. Are endophytic fungi defensive plant mutualists? *Oikos* 98:25–36.

Fletcher, L. R. and I. C. Harvey. 1981. An association of a *Lolium* endophyte with ryegrass staggers. *New Zealand Veterinary Journal* 29:185–186.

Fletcher, L. R., Hogland, J. H., and B. L. Sutherland. 1990. The impact of *Acremonium* endophytes in New Zealand, past, present and future. *Proceedings New Zealand Grassland Association* 52:227–235.

Funk, C. R. and J. F. White Jr. 1997. Use of natural and transformed endophytes for turf improvement. In *Neotyphodium/Grass Interactions*, Eds. C. W. Bacon and N. S. Hill, pp. 229–329. New York: Plenum Press.

Funk, C. R., Halisky, P. M., Johnson, M. C., Siegel, M. R., Stewart, A. V., Ahmad, S., Hurley, R. H., and I. C. Harvey. 1983. An endophytic fungus and resistance to sod webworms: Association in *Lolium perenne* L. *Biotechnology* 1:189–191.

Gentile, A., Rossi, M. S., Cabral, D., Craven, K. D., and C. L. Schardl. 2005. Origin, divergence, and phylogeny of epichloë endophytes of native Argentine grasses. *Molecular Phylogenetics and Evolution* 35:196–208.

Guerin, D. 1898. Sur la presence d'un Champignon dans l'Ivraie. *Journal Botany* 12:230–238.

Hanausek, T. F. 1898. Vorlaufige Mittheilung uber den von A Vogl in der Frucht von *Lolium temulentum* entdeckten Pilz. *Berichte der Deutsche Botanische Gesellschaf* 16:203.

Hoveland, C. S. 1993. Economic importance of *Acremonium* endophytes. *Agricultural Ecosystem Environment* 44:3–23.

Keogh, R. G., Tapper, B. A., and R. H. Fletcher. 1996. Distribution of the fungal endophyte *Acremonium lolii* and of the alkaloids lolitrem B and peramine within perennial ryegrass. *New Zealand Journal of Agricultural Research* 39:121–127.

Koroch, A., Juliani, H., Bischoff, J., Lewis, E., Bills, G., Simon, J., and J. F. White Jr. 2004. Examination of plant biotrophy in the scale insect parasitizing fungus *Dussiella tuberiformis*. *Symbiosis* 37:267–280.

Koshino, H., Yoshihara, T., Sakamura, S., Shimanuki, Y., Sato, T., and A. Tajimi. 1989a. A ring B aromatic sterol from stromata of *Epichloë typhina*. *Phytochemistry* 28:771–772.

Koshino, H., Yoshihara, T., Sakamura, S., Shimanuki, Y., Sato, T., and A. Tajimi. 1989b. Novel C-11 epoxy fatty acid from stromata of *Epichloë typhina* on *Phleum pratense*. *Agricultural Biological Chemistry* 53:2527–2528.

Koulman, A., Lane, G. A., Christensen, M. J., Fraser, K., and B. A. Tapper. 2007. Peramine and other fungal alkaloids are exuded in the guttation fluid of endophyte-infected grasses. *Phytochemistry* 68:355–360.

Lane, G. A., Christensen, M. J., and C. O. Miles. 2000. Coevolution of fungal endophytes with grasses: The significance of secondary metabolites. In *Microbial Endophytes*, Eds. C. W. Bacon and J. F. White Jr., pp. 341–388. New York: Marcel Dekker.

Li, C., Nan, Z., Paul, V. H., Dapprich, P., and Y. Liu. 2004. A new *Neotyphodium* species symbiotic with drunken horse grass (*Achnatherum inebrians*) in China. *Mycotaxon* 90:141–147.

Malinowski, D. and D. P. Belesky. 2000. Adaptations of endophyte-infected cool-season grasses to environmental stresses. *Crop Science* 40:923–940.

Moon, C. D., Miles, C. O., Jarlfors, U., and C. L. Schardl. 2002. The evolutionary origins of three new *Neotyphodium* endophyte species in grasses indigenous to the Southern Hemisphere. *Mycologia* 94:694–711.

Moon, C. D., Guillaumin, J-J., Ravel, C., Li, C., Craven, K. D., and C. L. Schardl. 2007. New *Neotyphodium* endophyte species from the grass tribes Stipeae and Meliceae. *Mycologia* 99:895–905.

Moy, M., Belanger, F., Duncan, R., Freehof, A., Leary, C., Meyer, W., Sullivan, R., and J. F. White Jr. 2000. Identification of epiphyllous mycelial nets on leaves of grasses infected by clavicipitaceous endophytes. *Symbiosis* 28:291–302.

Panaccione, D. G. 2005. Origins and significance of ergot alkaloid diversity in fungi. *FEMS Microbiology Letters* 251:9–17.

Panaccione, D. G., Cipoletti, J. R., Sedlock, A. B., Blemings, K. P., Schardl, C. L., Machado, C., and G. Seidel. 2006. Effects of ergot alkaloids on food preference and satiety in rabbits as assessed with gene-knock-out endophytes in perennial ryegrass (*Lolium perenne*). *Journal of Agricultural and Food Chemistry* 54:4582–4587.

Patterson, C. G., Potter, D. A., and F. F. Fanin. 1991. Feeding deterrence of alkaloids from endophyte-infected grasses to Japanese beetle grubs. *Entomologia Experimentalis et Applicata* 61:285–289.

Philipson, M. N. and M. C. Christey. 1986. The relationship of host and endophyte during flowering seed formation and germination of *Lolium perenne. New Zealand Journal of Botany* 24:125–134.

Porter, J. K. 1994. Chemical constituents of grass endophytes. In *Biotechnology of Endophytic Fungi of Grasses*, Eds. C. W. Bacon and J. F. White Jr., pp. 103–123. Boca Raton, FL: CRC Press.

Prestidge, R. A., Pottinger, R. P., and G. M. Barker. 1982. An association of a *Lolium* endophyte with ryegrass resistance to the Argentine stem weevil. *Proceedings New Zealand Weed Pest Control Conference* 35:199–222.

Rao, S., Alderman, S. C., Takeyasu, J., and B. Matson. 2005. The *Botanophila-Epichloë* association in cultivated *Festuca* in Oregon: Evidence of simple fungivory. *Entomologia Experimentalis et Applicata* 115:427–433.

Rehner, S. A. and G. J. Samuels. 1995. Molecular systematics of the Hypocreales: A teleomorph gene phylogeny and the status of their anamorphs. *Canadian Journal of Botany* 73:816–823.

Riedell, W. E., Kieckhefer, R. E., Petroski, R. J., and R. G. Powell. 1991. Naturally occurring and synthetic loline alkaloid derivatives: Insect feeding behavior modification and toxicity. *Journal of Entomological Science* 26:122–129.

Rivas, H. and M. Zanolli. 1909. La tembladera enfermidad propia de los animals herbívoros de las regions andinas. *Argentinian Government Report*, La Plata, Argentina, pp. 1–12.

Rowan, D. D. and D. L. Gaynor. 1986. Isolation of feeding deterrent against stem weevil from ryegrass infected with the endophyte *Acremonium loliae. Journal of Chemical Ecology* 12:647–658.

Saikkonen, K., Helander, M., Faeth, S. H., Schulthess, F., and D. Wilson. 1999. Endophyte–grass–herbivore interactions: The case of *Neotyphodium* endophytes in Arizona fescue populations. *Oecologia* 121: 411–420.

Schardl, C. L. and C. D. Moon. 2003. Processes of species evolution in *Epichloë/Neotyphodium* endophytes of grasses. In *Clavicipitalean Fungi: Evolutionary Biology, Chemistry, Biocontrol and Cultural Impact*, Eds. J. F. White Jr., C. W. Bacon, N. L. Hywel-Jones, and J. W. Spatafora, pp. 273–310. New York: Marcel Dekker.

Schardl, C. L. and T. D. Phillips. 1997. Protective grass endophytes: Where are they from and where are they going? *Plant Disease* 81:430–437.

Schardl, C. L., Leuchtmann, A., and M. J. Spiering. 2004. Symbioses of grasses with seed borne fungal endophytes. *Annual Review of Plant Biology* 55:315–340.

Schardl, C. L., Panaccione, D. G., and P. Tudzynski. 2006. Ergot alkaloids—biology and molecular biology. *The Alkaloids: Chemistry & Biology* 63:45–86.

Siegel, M. R. and G. C. M. Latch. 1991. Expression of antifungal activity in agar culture by isolates of grass endophytes. *Mycologia* 83:529–537.

Siegel, M. R., Dahlaman, D. L., and L. P. Bush. 1989. The role of endophytic fungi in grasses: New approaches to biological control of pests. In *Integrated Pest Management for Turfgrass and Ornamentals*, Eds. A. R. Leslie and R. L. Metcalf, pp. 169–186. Washington, DC: US-EPA.

Spatafora, J. W. and M. Blackwell. 1993. Molecular systematics of unitunicate perithecial ascomycetes: The Clavicipitales–Hypocreales connection. *Mycologia* 85:912–922.

Spatafora, J. W., Sung, G-H., Sung, J-M., Hywel-Jones, N. L., and J. F. White Jr. 2007. Phylogenetic evidence for an animal pathogen origin of ergot and the grass endophytes. *Molecular Ecology* 16:1701–1711.

Stone, J. K., Polishook, J. D., and J. F. White Jr. 2004. Endophytic fungi. In *Biodiversity of Fungi: Inventory and Monitoring Methods*, Eds. G. M. Mueller, G. F. Bills and M. S. Foster, pp. 241–270. Burlington, MA: Elsevier Academic Press.

Stovall, M. E. and K. Clay. 1991 Fungitoxic effects of *Balansi cyperi* (Clavicipitaceae). *Mycologia* 83:288–295.

Sung, G. H., Hywel-Jones, N. L., Sung, J. M., Luangsa-ard, J. J., Shrestha, B., and J. W. Spatafora. 2007. Phylogenetic classification of *Cordyceps* and the clavicipitaceous fungi. *Studies in Mycology* 57:5–59.

Tadych, M. and J. F. White Jr. 2007. Ecology of epiphyllous stages of endophytes and implications for horizontal dissemination. In *Proceedings of the 6th International Symposium on Fungal Endophytes of Grasses*, Eds. A. J. Popay and E. R. Thom, pp. 157–161. New Zealand: New Zealand Grassland Association.

Tadych, M., Bergen, M., Dugan, F. M., and J. F. White Jr. 2007. Evaluation of the potential role of water in the spread of conidia of the *Neotyphodium* endophyte of *Poa ampla. Mycological Research* 111:466–472.

Tintjer, T. and J. A. Rudgers. 2006. Grass–herbivore interactions altered by strains of a native endophyte. *New Phytologist* 170:513–521.

Torres, M. S. 2007. Defensive mutualism in the clavicipitaceae: An evaluation based on morphology, phylogeny and alkaloid production, PhD thesis, Rutgers University, New Brunswick.

Torres, M. S., White, J. F. Jr., and J. F. Bischoff. 2007a. *Hypocrella panamensis* sp. nov. (Clavicipitaceae, Hypocreales): A new species infecting scale insects on *Piper carrilloanum* in Panama. *Mycological Research* 111:317–323.

Torres, M. S., Singh, A. P., Vorsa, N., Gianfagna, T., and J. F. White Jr. 2007b. Were endophytes pre-adapted for defensive mutualism? In *Proceedings of the 6th International Symposium on Fungal Endophytes of Grasses*, Eds. A. J. Popay and E. R. Thom, pp. 63–67. New Zealand: New Zealand Grassland Association.

Torres, M. S., Singh, A. P., Vorsa, N., and J. F. White Jr. 2008. An analysis of ergot alkaloids in Clavicipitaceae (Hypocreales, Ascomycota) and ecological implications. *Symbiosis* 46:11–19.

Vogl, A. 1898. Mehl und die anderen Mehlprodukte der Cerealien und Leguminosen. *Zeitschrift Nahrungsmittle Untersuchung Hyg Warenkunde* 12:25–29.

Wäli, P. R., Helander, M., Nissinen, O., and K. Saikkonen. 2006. Susceptibility of endophyte-infected grasses to winter pathogens (snow molds). *Canadian Journal of Botany* 84:1043–1051.

West, C. P., Oosterhuis, D. M., and S. D. Wullschleger. 1990. Osmotic adjustment in tissues of tall fescue in response to water deficit. *Environmental Experimental Botany* 30:149–156.

Western, J. H. and J. J. Cavett. 1959. The choke disease of cocksfoot (*Dactylis glomerata*) caused by *Epichloë typhina* (Fr) Tul. *Transactions British Mycological Society* 42:298–307.

White, J. F. Jr. 1987. Widespread distribution of endophytes in the Poaceae. *Plant Disease* 71:340–342.

White, J. F. Jr. 1988. Endophyte–host associations in forage grasses. XI. A proposal concerning origin and evolution. *Mycologia* 80:442–446.

White, J. F. Jr. and G. T. Cole. 1985. Endophyte–host associations in forage grasses. III. *In vitro* inhibition of fungi by *Acremonium coenophialum. Mycologia* 77:487–489.

White, J. F. Jr. and J. R. Owens. 1992. Stromal development and mating system of *Balansia epichloë*, a leaf-colonizing endophyte of warm-season grasses. *Applied Environmental Microbiology* 58:513–519.

White, J. F. Jr., Martin, T. I., and D. Cabral. 1996. Endophyte-host associations in grasses. XXII. Conidia formation by *Acremonium* endophytes on the phylloplanes of *Agrostis hiemalis* and *Poa rigidifolia. Mycologia* 88:174–178.

White, J. F. Jr., Morrow, A. C., Morgan-Jones, G., and D. A. Chambless. 1991. Endophyte–host associations in forage grasses. XIV. Primary stromata formation and seed transmission in *Epichloë typhina*: Developmental and regulatory aspects. *Mycologia* 83:72–81.

White, J. F. Jr., Sharp, L., Martin, T. I., and A. Glenn. 1995. Endophyte–host associations in grasses. XXI. Studies on the structure and development of *Balansia obtecta. Mycologia* 87:172–181.

Yue, C., Miller, C. J., White, J. F. Jr., and M. Richardson. 2000. Isolation and characterization of fungal inhibitors from *Epichloë festucae. Journal of Agricultural and Food Chemistry* 48:4687–4692.

Zaurov, D. E., Bonos, S., Murphy, J. A., Richardson, M., and F. C. Belanger. 2001. Endophyte infection can contribute to aluminum tolerance in fine fescues. *Crop Science* 41:1981–1984.

17

Contributions of Pharmaceutical Antibiotic and Secondary Metabolite Discovery to the Understanding of Microbial Defense and Antagonism

Gerald Bills, David Overy, Olga Genilloud, and Fernando Peláez

CONTENTS

17.1 Introduction

The mention of defensive microbial mutualism immediately brings to mind the popular visions of antibiosis and intermicrobial warfare. Antibiotics were originally defined as microbially produced secondary metabolites that inhibit the growth (static) or kill (cidal) other microorganisms. In the health care sciences, this definition has been expanded to also include semisynthetic and synthetic compounds. However, the scope of this chapter encompasses the fungi and bacteria and their small molecules that perturb, modify, or inhibit the growth, behavior, or physiology of other organisms. It has been hypothesized that during the early biochemical evolution, precursors of these secondary metabolites had a metabolic role in modulating macromolecular functions in the producer (Davies, 1990). All small molecules made by microorganisms, including antibiotics, are thought to have biological functions, even though in most cases, the function is unknown. Furthermore, we now know that organisms have dedicated a significant portion of their genomes to producing such compounds because of the selective advantages conferred upon them as a result of the interactions of the compounds with other organisms (Vining, 1990; Stone and Williams, 1992; Demain and Fang, 2000). Different variations of these hypotheses have been termed as the "competition" or "function–evolution" hypothesis (Wiener, 1996; Wink, 2003). However, counter viewpoints, termed as the "screening hypothesis" or the "diversity hypothesis" (Firn and Jones, 2003; Fischbach and Clardy, 2007) are based on the assumption that molecules with potent biological activity evolve infrequently and that selection occurs for organisms that synthesize multiple molecules from a given pathway because the probabilities of making an advantageous one with potent biological activity are increased.

The arguments favoring the function–evolution hypothesis have been reviewed extensively and fall into several broad categories.

1. Secondary metabolites originate from complex and energetically expensive pathways that are unlikely to have developed unless the products had advantageous functions (Wink, 2003). In general, only those organisms that lack an immune system are prolific producers of secondary metabolites, where these are thought to act as an alternative defense system. Conversely, organisms that are affected by secondary metabolites have often developed elaborate resistance mechanisms to avoid, detoxify, or excrete antibiotics and toxins produced by organisms. For example, natural bacterial populations demonstrate a broad range of antibiotic resistance mechanisms that appear to have evolved among many bacterial lineages, even before the selection pressures exerted by modern use of antibiotics (Walsh, 2003a; Aminov and Mackie, 2007; Dantas et al., 2008).
2. Most purified natural products and crude natural product mixtures are known to affect cells and whole organisms, even at concentrations below that of the primary, affected biochemical target. For example, we know from our own work and that of others that antibiosis is observed among 10%–50% of the crude extracts derived from fungi and bacteria screened against a narrow range of human pathogenic bacteria and fungi. Extending the panel of screening organisms by including other model or environmental target organisms detects more antibiotics from a wider population of source organisms, thus increasing the percentages of organisms yielding antibiotic activity. Subinhibitory concentrations of antibiotics and other secondary metabolites secreted in the environment are potent modulators of transcription and other cellular processes in bacterial and yeast cells (Davies et al., 2006; Perlstein et al., 2006). Such an antibiotic binding to bacterial ribosomes may exhibit hormetic characteristics; in other words, transcription modulation or protein synthesis inhibition is concentration dependent (Yim et al., 2007).
3. Secondary metabolites are produced in biologically relevant quantities in nature and are not artifacts of laboratory fermentation conditions. Many natural products have been experimentally proven to mediate interactions between various types of microorganisms, plants, and animals, indicating that they indeed have antagonistic capacities in nature (Janzen, 1977; Wicklow, 1981, 1988; Demain and Fang, 2000; Wink, 2003). Targeted disruption of regulating

genes, pathway promoters, or key early pathway enzymes can eliminate metabolites and alter the outcome of interorganism interactions (Dekkers et al., 2007; Rohlfs et al., 2007; Sugui et al., 2007; Potter et al., 2008).

4. Cloning and sequencing of biosynthetic pathways and genomic sequencing have elucidated the genetics underlying secondary metabolite biosynthesis, regulation, and resistance and have left no doubt about the evolution of functional natural products. Microorganisms regulate their metabolite synthesis, probably to ensure that the energy and precursors used in their synthesis are expended only in the environments where the metabolites are advantageous. In other cases, synthesis may be regulated globally or tightly coordinated with growth and developmental stages (Hoffmeister and Keller, 2007). It is also evident from the filamentous fungal (Pezizomycotina) and filamentous bacterial (Actinomycetes) genomes studied to date that they all have the pathways to produce small molecules that can affect the behavior or growth of another organism. In every case, the number of presumed secondary metabolites gene clusters exceeds the number of metabolites detectable in a given species (Scherlach and Hertweck, 2006; Arvas et al., 2007; Bergmann et al., 2007; Gross, 2007; Hornung et al., 2007), suggesting the presence of a significant number of unexpressed pathways.

These unexpressed gene clusters may encode for not only the known secondary metabolite biosynthetic pathways (not expressed by the organism under previously studied laboratory conditions) but also, and more importantly, the unknown biosynthetic pathways related to "novel" secondary metabolites. Previously described as "cryptic" or "silent" gene clusters, they have been aptly named "orphan" gene clusters because they are not truly silent but rather anticipated, at least under responsive environmental conditions (Gross, 2007). Orphan pathways have been defined as a "biosynthetic loci for which the corresponding metabolite is (currently) unknown." Putative functional gene clusters encoding for orphan biosynthetic pathways have been found not only in well-studied *Streptomyces* species, such as *S. coelicolor* A3(2) or *S. avermitilis* ATCC31267, but also in cyanobacteria (Kaneko et al., 2008), myxobacteria (Schneiker et al., 2007), and fungi (Arvas et al., 2007). The challenge in studying the natural function of these potentially novel small molecules from orphan pathways is inducing their production in the laboratory and identifying their effects on other organisms. Detection and manipulation of the same pathway products in nature represent the ultimate challenge.

Bioactive microbial metabolites and their derivatives have historically afforded an unequaled source of therapeutic agents used to treat infectious diseases. These successes exploited the innate properties of microbially synthesized small molecules, namely, their vast chemical diversity, intrinsic cell permeability, and biomolecular specificity compared with the corresponding spectrum available in synthetic or combinatorial chemistry libraries. Microbial metabolites can be regarded as a reservoir of privileged chemical scaffolds selected by evolutionary pressures to interact with a diversity of proteins, nucleic acids, and other cellular targets that only started in the last century to have been commandeered for drug development (Yim et al., 2006). Despite the seemingly infinite expanse of microbial diversity, most of the antibiotics used for the treatment of human and animal infectious diseases are ultimately derived from a limited range of easily cultivated actinomycetes and fungi. The most important antibacterials include the β-lactams (penicillins, cephalosporins, carbapenems, and their synthetic derivatives) that are nonribosomal peptides produced by both soil bacteria and fungi, the macrolide erythromycin (e.g., their synthetic derivatives azithromycin and clarithromycin) produced by *Saccharopolyspora erythraea*, and the glycopeptide vancomycin produced by *Amycolatopsis orientalis*, which are all products of actinomycetes. The most important naturally produced antifungals are the polyene amphotericin B, obtained from *Streptomyces nodosus* and other *Streptomyces* species, and fungal-derived lipopeptides known as the echinocandins (e.g., their semisynthetic derivatives caspofungin and micafungin).

Since the beginning of the golden era of antibiotics, industrial discovery has focused its attention on actinomycetes and fungi because they can be easily obtained from soils and other organic materials, they are easy to cultivate, and through process optimization and strain selection, factory-scale production can generally be achieved. However, in the past two decades, experts have worried that the approach has reached the limit of its usefulness because the discovery rates for new structural classes of antibiotics have plummeted (Walsh, 2003b; Clardy et al., 2006; Luzhetskyy et al., 2007), and they have argued for new discovery paradigms that penetrate deeper into the more inaccessible regions of the microbial metabolite reservoir.

We believe that established approaches for microbial metabolite discovery will continue to be the source of the majority of the new natural products. However, it must be acknowledged that much of the antibiotic landscape is overtrodden and continued screening of familiar microorganisms with the same tools is unlikely to yield new antibiotics. Rapid DNA sequencing methods are now in hand to predictably map the origins of known antibiotics in phylogenetic space, and the diversity of newly acquired strains can now be assessed against the background of phylogenies of the archaea, bacteria, and eukaryotes. At the same time, several international collaborative efforts are being made to generate DNA barcodes of fungi and bacteria (http://barcoding.si.edu; http://www.barcodeoflife.org) that can yield species-level identifications. Such frameworks can steer metabolomic and phylogenetic strain dereplication toward less-explored microorganisms. In addition, chemical isolation techniques and instrumentation have vastly improved in recent years, and compounds can now be isolated and characterized on the basis of minute quantities of material leading to identification and isolation of minor metabolites that were inaccessible one or two decades ago (Harris, 2005; Kohn, 2008). These improvements alone may be insufficient to reverse declining discovery rates, and emerging technologies have the potential to attain the eventual goal of identifying all antibiotics from all organisms.

The genetic and cellular bases for the manner in which model microorganisms and plants manufacture, store, activate, transport, or allocate secondary metabolite classes are now better understood. Understanding the rules according to which organisms interact and defend themselves chemically and isolating and identifying chemical defenses relevant to microbial survival could lead to novel approaches for chemical therapies. Below, we outline some technologies that can be applied in the immediate future while others which are beyond the scope of this chapter, e.g., genomic mining and pathway engineering, may need long-term investments. At the same time, we recognize that drug discovery-motivated research has led to an enormous knowledge of the biology and biosynthesis of many microbial metabolites, especially those that have reached the market. We finish by analyzing how lessons learned from the drug discovery and development efforts have contributed in understanding the potential roles of some of these well-known metabolites in nature and suggest several candidate metabolite families produced by fungi and actinomycetes as model systems for elucidating ecological functions.

17.2 Eliciting Chemical Defensive Metabolites by Interorganism Interactions

Most secondary metabolites are the products from the conditional pathways that are expressed in a particular context or situation. If that context-dependent stimulus can be artificially replicated, organisms having orphan gene clusters may be stimulated to produce their metabolites, giving rise to a range of diverse and novel chemistry, often with bioactive potential (Gross, 2007). Numerous isolated examples suggest that the expression of orphan biosynthetic pathways can be enticed by the coculturing of two or more organisms (Perry et al., 1984; Tokimoto et al., 1987; Burgess et al., 1999; Lu and Shen, 2004; Maldonado et al., 2004a,b; Oh et al., 2005; Zhu and Lin, 2006). A popular hypothesis is that microorganisms in nature need external chemical stimuli to produce their full spectrum of secondary metabolites, often perceiving specific signals originating from nearby organisms (microbes, plants, and animals) to discriminate the presence of a noncompetitive or a competitive environment. Alternatively, competitive chemical interactions can lead to a reaction followed by the production of biologically active metabolites (which may not be qualitatively observed in axenic culture extracts).

Many examples exist on how the interactions of fungi are manifested phenotypically and how those interactions are accompanied by responses at the chemical level. Within proximal interaction zones between two adjacent fungal cultures, chemical communication is visually most evident in the formation of barrage zones manifested as colorful interaction zones in culture or darkened stromatic zones in wood (Florey et al., 1949; Rayner et al., 1994; Boddy, 2000). In vitro confrontation of strains may stimulate or alter patterns of the volatile terpenoids emitted from the mycelia (Boddy, 2000; Hynes et al., 2007). When *Heterobasidion annosum* was challenged with different antagonistic fungi in Petri dish culture, crystals were secreted on its aerial hyphae that were determined to consist of 7,8-dihydro-9-hydroxy-5,7,7-trimethylcyclopenta-(g)-2-benzopyran-1(6H)-one (Sonnenbichler et al., 1983). Further studies on paired cultures of *H. annosum* and *Gloeophyllum abietinum* demonstrated that a series of benzofurane derivatives from *H. annosum*, and isocoumarine derivatives (e.g., oosponol) by *G. abietinum*, were produced exclusively in dual agar culture (Sonnenbichler et al., 1989, 1993). Dual Petri dish cultures of an *Acremonium* sp. with the mycoparasite *Mycogone rosea* resulted in the new lipoaminopeptide metabolites, acremostatins A, B, and C (Degenkolb et al., 2002). *Lambertella corni-maris*, a mycoparasite of the apple pathogen *Monilinia fructigena*, produces high titers of the antifungal lambertellols A, B, and C and lambertellin in acid environments, but not when grown on near-neutral potato–sucrose medium (Murakami et al., 2007). When the two organisms were cocultured on potato–sucrose agar or in liquid media, acidification of the medium by *M. fructigena* induced *L. corni-maris* to produce lambertellols A, B, and C, causing hyphal deterioration in the *M. fructigena*. By coculturing two unidentified endophytic strains, isolated from an estuarine mangrove, two antibacterial metabolites showing activity against *Escherichia coli*, *Pseudomonas pyocyanea*, and *Staphylococcus aureus* (marinamide, a 1-isoquinolone analog, and its methyl ester) were isolated (Zhu and Lin, 2006). Neither metabolite was produced in axenic cultures. Cofermentation of an *Aspergillus flavus* and an *A. nidulans* strain impaired in penicillin production stimulated the production of aspergillic acids, not previously observed in axenic culture extracts even after intensive concentration; only cocultured extracts had antibiotic activity against *E. coli* and *S. aureus* (Perry et al., 1984).

Within bacterial communities, quorum sensing is integrated within global regulatory networks and plays a key role in developing and maintaining the structure of these communities (Waters and Bassler, 2005). Competition between bacteria or bacterial biofilms occurs through the formation of quorums and the excretion of antibacterial metabolites. Within a quorum, bacteria of the same kind manifest and produce species-specific chemical signals, and different classes of sensing signals have been shown to coexist. Numerous gram-negative bacteria, e.g., *Vibrio* spp., *Erwinia carotovora*, *Pseudomonas aeruginosa*, and *Agrobacterium tumefaciens* among others, produce *N*-acyl-homoserine lactones (AHL) which play a key role in the LuxI/LuxR quorum sensing system (Lerat and Moran, 2004), while Actinomycetes, on the other hand, synthesize δ-butyrolactones, structural analogs of AHLs, to control morphological differentiation and secondary metabolite production. The A-factor was the first of these molecules to be described in *Streptomyces* spp. several decades ago (Khokhlov et al., 1967). These global regulators are produced in micromolar concentrations and accumulate in a growth-dependent manner. Once a threshold concentration has been exceeded, the transcription of regulatory factors activates a gene expression cascade, resulting in morphological differentiation and the production and secretion of enzymes, virulence factors, and secondary metabolites, many of which represent potential niche-defense mechanisms (Horinouchi, 2007). δ-butyrolactones regulate antibiotic production, are strictly species-specific, and are suspected to be widespread across all actinomycete taxa. As many as 14 different δ-butyrolactones have been found so far in seven *Streptomyces* species (Takano, 2006) and other non-*Streptomyces* actinomycetes, e.g., *Amycolatopsis mediterranei*, *Actinoplanes teichomyceticus*, *Micromonospora echinospora*, or *Kitasatospora setae* (Choi et al., 2003). Moreover, 33 δ-butyrolactones receptor homologs have been identified recently of which 22 were associated with antibiotic biosynthetic genes (Takano, 2006).

In the context of screening bacterial extracts for the discovery of novel antimicrobial compounds, promoting quorum sensing associations might result in the production of new natural bacterial products. Burgess et al. (1999) studied the effects of exposing nonantibiotic producing epibiotic bacteria (in conventional culture systems) isolated from the surface of seaweeds, nudibranchs, and starfish to the cells or culture supernatants from marine bacteria or laboratory bacteria, e.g., *Bacillus subtilis*, *E. coli*, and *P. aeruginosa*. The induction of the production of antimicrobial metabolites from those epibiotic bacteria clearly suggested the occurrence of a chemically induced defense response when confronted with a competing organism (Burgess et al., 1999). In 12 cases of 53 different species of marine bacteria added to established cultures of *Streptomyces tenjimariensis*, production of the antibiotic istamaycin was significantly stimulated up to twice the levels observed in axenic cultures (Slattery et al., 2001). Bacteriocins are antimicrobial peptides that inhibit the growth of related species of bacteria, particularly species with the same nutritional requirements. A total of 41 out of 82 bacterial strains belonging to the genera *Bacillus*, *Enterococcus*, *Lactobacillus*, *Lactococcus*, *Leuconostoc*, *Listeria*, *Pediococcus*, *Staphylococcus*, and *Streptococcus* induced the production of the bacteriocin plantaricin from a strain of

Lactobacillus plantarum in coculture, where none of the strains studied produced plantaricin in axenic culture (Maldonado et al., 2004b). Evidence suggested that plantaricin production was environmentally mediated via a quorum-sensing mechanism (Maldonado et al., 2004b). More recently, the production of the antibiotic pyrrolnitrin by *Serratia plymuthica*, a plant biocontrol bacterium used to suppress *Verticillium* spp. infections, was shown to be directly under the control of AHLs produced by the strain (Liu et al., 2001). Cocultivation experiments employing different species of *Streptomyces* have shown that other diffusible substances not related with γ-butyrolactones exist which can stimulate sporulation and antibiotic production in a large number of species (Ueda et al., 2000). These findings clearly suggest the existence of a widespread interspecies signaling systems in actinomycetes involved in the regulation of antibiotic production.

Interactions between eukaryotes and bacteria are frequently mediated by small pharmacologically active molecules (Dudler and Eberl, 2006). Although the signaling mechanisms that govern inter kingdom communication are still unknown in most cases, many recent reports indicate that prokaryotic symbionts are often the source of bioactive small molecules attributed to the eukaryote hosts. The polyketide pederin, a potent antitumor compound produced as a defense against predators by terrestrial *Paederus fuscipes* (Coleoptera, Staphylinidae), is in fact produced by an uncultured bacterial symbiont related to *Pseudomonas* and vertically transmitted to the female offspring via eggs (Piel, 2002). The patellamides are heterocyclic peptides found in ascidians (*Didemnidae*) harboring the unculturable cyanobacterium *Prochloron* and suspected to be produced by the symbiont. The proof of the direct involvement of a cyanobacterium in the synthesis of these compounds was only obtained from the genome sequence of *Prochloron didemnin*, a symbiont of *Lissocladium patella* containing patellamides A and C, which showed the presence of a nonribosomal peptide synthase gene cluster encoding the precursor of patellamides octapeptides (Schmidt et al., 2000). The antitumor compounds bryostatins are produced by the bryozoan *Bugula neritina* to protect larvae against fish predators. The structure of the molecule and the cloning of a DNA fragment containing part of the gene responsible for the biosynthesis of bryostatin suggested that this was produced by a bacterial symbiont, the proteobacterium *Candidatus Endobugula sertula*. Many molecules structurally related to bacterial metabolites have been isolated from sponges and tunicates, many of them shown to harbor complex assemblages of symbiotic bacteria; in many cases, consisting of more than 160 different bacteria species (Hentschel et al., 2001). The production of theopalauamide has been associated with a filamentous proteobacterium symbiont *Candidatus Entotheonella palauensis*, also related to bacterial symbionts detected in *Theonella swinhoei* specimens producing structurally related compounds. When cultivated in a mixed culture on agar plates containing sponge extracts, no metabolites could be detected from this strain, suggesting a role of theopalauamide compounds in bacterial symbiosis (Schmidt et al., 2000). Whereas the sources of the bacterial metabolites in sponges have been identified in many cases, the unculturability and complexity of many of these microbial consortia still represent an exceptional challenge not only to identify and isolate the producing organisms but also to provide the unknown necessary signals to emulate the symbiotic relationship and elicit the production of the desired metabolite.

In vitro investigations involving coculturing of organisms from different kingdoms have resulted in the discovery of novel antimicrobial metabolites whose production was dependent upon cocultivation. For example, using mixed culture methods employing more than 50 combinations of marine-derived fungi and marine actinomycetes, Fenical and collaborators induced the production of two novel emericellamides (A and B) having modest antibacterial activities against MRSA, from an *Emericella* sp. cocultured with a *Salinispora arenicola* strain (Oh et al., 2007); the biosynthesis of four new diterpenoids (libertellenones A–D showing cytotoxicity against a human colon carcinoma cell line), from a mixed fermentation of a *Libertella* sp. and an unidentified protobacterium (Oh et al., 2005); and the production of a chlorinated antibiotic benzophenone from a *Pestalotia* sp. cultured with the same unidentified protobacterium (Cueto et al., 2001). When a strain, tentatively identified as *Arthrinium saccharicola*, was paired in cocultivation with 14 different marine bacteria (Miao et al., 2006), the fungus grew faster when the spent culture media of six bacterial species were added into the fungal cultures relative to the controls while the fungus grew slower when the cells of these bacterial species were added. Furthermore, some spent bacterial culture media enhanced the antibacterial metabolite production of the fungus (Miao et al., 2006).

Using a different approach to cross-kingdom coculturing, Kobayashi and collaborators (1995) carried out coculturing experiments using plant callus challenged with fungal spores. In two occasions, this form of coculturing yielded antimicrobial extracts; a novel gram-positive antibiotic tetramic acid, trichosetin, was induced in dual culture with *Trichoderma harzianum* grown on *Catharanthus roseus* callus (Marfori et al., 2002), and the antifungal compound phytolaccoside B from the dual culture of *Botrytis fabae* grown on *Phytolacca americana* callus (Kobayashi et al., 1995). In each documented case, the induced metabolites were only produced when the participating organisms were cocultured in the same environment.

Many of the authors from the previously mentioned examples suggested that coculturing organisms should be adapted to natural product screening programs and can be employed as a novel approach for isolating pharmaceutically interesting metabolites. The body of evidence of the isolation of novel, biologically active metabolites from cocultured organisms is conclusive, albeit highly fragmented. So far, a cohesive strategy for exploiting these phenomena has not emerged. Many factors have to be considered before the approach can be applied within an industrial screening process. A major difficulty is how to select the organisms that should be cocultured, because the number of potential organism combinations will grow exponentially with the numbers of microorganisms examined. Crossing all combinations will quickly result in an unmanageable number of fermentations. An additional legitimate concern that will need to be addressed is the inherent irreproducibility of mixed culture experiments, considering that even in conventional axenic fermentations used in natural products discovery, production of a given metabolite may be highly variable (including lost production) among a series of theoretically identical experiments. Large-scale screening experiments with appropriate controls are needed to determine if frequency of stimulation of new metabolites is sufficiently profitable to warrant the increased effort in preparing many permutations of randomly paired cofermentations. At least in the early stages of developing such a screening system, evaluation of dual cultures would require comparisons with the corollary axenic cultures, further increasing the total number of individual fermentations. More importantly, in the event of the discovery of a potential lead metabolite, both physical and biological issues must be solved regarding the scale-up and coculturing of the organisms in large-scale fermentors.

One possible approach that could overcome some of the problems anticipated in controlling mixed cultures would be to grow organism pairs in compartmentalized vessels with each being separated by diffusible membrane that permits small-molecule communication (Figure 17.1). Some of the earliest experiments documenting the microbial antibiosis employed the partitioned coculture techniques using

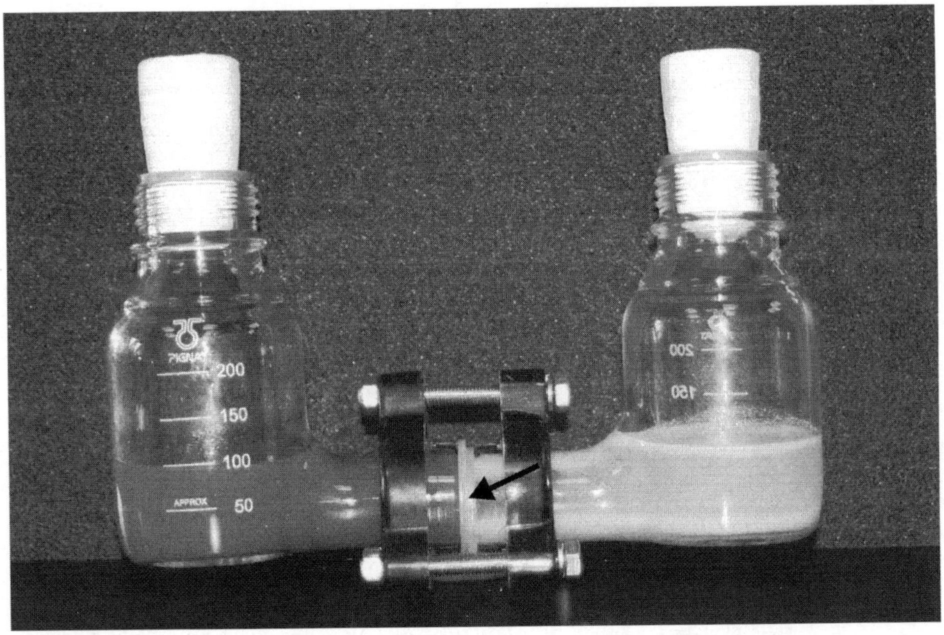

FIGURE 17.1 (See color insert following page 206.) Prototype for a shaker-mounted dual fermentation vessel. The design was inspired by dual culture systems for *Symbiobacterium thermophilum* (Ohno et al., 1999). Each fermentation bottle is separated by a dialysis membrane mounted at the junction between the bottles arrow.

porcelain cylinders of one bacterial culture suspended inside a jar with a second culture or a bacterial culture in a colloidal bag suspended in the flask of the second bacterium (Florey et al., 1949). Mearns-Spragg et al. (1998) improvised dual culture fermentations by sealing living or killed bacterial cell suspensions in dialysis tubing and adding the dialysis bags to growing cultures of marine bacteria. We have tried this technique on a small scale, but aseptic filling and sealing of dialysis tubes with a live culture is difficult, and therefore the technique is very prone to contamination. Dual culture vessels and large-scale fermentors partitioned by dialysis membranes were custom built to grow *Symbiobacterium thermophilum* in one half of a vessel isolated from its obligate symbiont, *Bacillus* sp. (Ohno et al., 1999; Ueda et al., 2002). To our knowledge, dual culture vessels are not commercially available, but development of economical, robust, and reusable systems would find multiple applications in microbial ecology and facilitate the incorporation of coculture systems as a routine discovery tool.

An alternative to the approach of coculturing two organisms in the same vessel that could alleviate the complexity described above would be to supplement the fermentation of a single organism with culture extracts, filtered culture supernatants, or autoclaved cellular material to the fermentation media. Several cases have been described that demonstrate an effect upon metabolite expression when extracts or filtrates of one biological organism have been added to the growth medium of a second organism suggesting that orphan pathways were unlocked and novel chemistry was expressed that otherwise was not observed when the organism was grown free of extraneous culture extracts. For example, culture extracts from a *Trichoderma* sp. induced the production of two antifungal substances by *Lentinus edodes* (Tokimoto et al., 1987). Marine bacterial cell-free supernatants supplemented into growth media were also found to induce antibiotic production from a marine bacterial strain not previously determined to produce antibiotics (Burgess et al., 1999). In the case of plantaricin production from *Lactobacillus plantarum*, cell-free supernatants from the inducing strains did not induce the production of plantaricin (Maldonado et al., 2004b); rather, induction of plantaricin resulted from the addition of autoclaved and heat-treated cell suspensions to the cultivation media (Maldonado et al., 2003). The use of autoclaved host plant tissues for the cultivation of pathogenic species of Penicillia from the series *Corymbifera* have been beneficial to stimulate the production of low titer, antioxidant metabolites; expression was observed from necrosis *in planta* (Overy et al., 2006). It should be noted here that although evidence exists to prove that the addition of organismal supplements to the media can have a stimulatory effect upon the producing organism, this is not always the case. For example, stimulation of libertellenone production by a *Libertella sp.* was not observed when the fungal culture broth was supplemented with dead bacterial cells, cell-free supernatant or a reconstituted ethyl acetate extract (Oh et al., 2005).

17.3 Stimulation of Metabolite Production by Natural Product Metabolites and Metabolic Inhibitors

Continuing to extend and simplify the preceding concept of exploiting interorganism interactions, sublethal doses of synthetic metabolic inhibitors or bioactive metabolites can be used to elicit *de novo* secondary metabolite biosynthesis or modulate the secondary metabolite profile of an organism, with the goal of exploiting its physiological and morphological responses to a range of chemical factors, i.e., elicitors. Some of the responses might be defense reactions elicited in one organism in response to molecules from another in order to ensure their survival, persistence, and competitiveness. Other responses might be caused by natural ligands that bind to receptors initiating signal transduction pathways. In other cases, addition of small bioactive molecules may cause perturbations or shunting in biosynthetic pathway or induce stress responses in the challenged organism.

For example, additions of a pipecolic acid-containing cyclopeptide (cyclopeptide 90–215) during the initiation of fermentations of different strains of *Metarhizium anisopliae* increased the production of destruxins A, B, and E up to 1.3 to 12.5 times while suppressing helvolic acid production (Espada and Dreyfuss, 1997). A marine-derived isolate of *Phomopsis asparagi* produced three new chaetoglobosin analogues when cultured in the presence of a sponge-derived F-actin inhibitor, jasplakinolide (Christian et al., 2005). Production of penicillin in *Penicillium chrysogenum* cultures was enhanced when supplemented with alginate and locust-bean (*Ceratonia siliqua*) gum oligosaccharides (Ariyo et al., 1997, 1998).

When mannuronate (OM) and guluronate (OG) oligosaccharides, obtained through the acid hydrolysis of alginate, were added to shake flask cultures of two *Penicillium chrysogenum* strains (a high penicillin G producer and a wild-type strain), biomass-based penicillin G yields in OG- and OM-supplemented cultures increased as much as 50%–150% (Ariyo et al., 1997). Differences in cell dry weight between the control and the elicited cultures were negligible, but changes were observed in the extracellular concentrations of the penicillin production pathway intermediate D-(L-α-aminoadipyl)-L-cysteinyl-D-valine (ACV). Additional stimulation of penicillin G yields and efficient conversion of ACV were reported using mannan oligosaccharides from enzymatic hydrolysis of locust-bean gum (Ariyo et al., 1998). Other *Penicillium* strains, including a low penicillin producer containing a single copy of the penicillin gene cluster and a high penicillin producer containing multiple copies of the penicillin gene cluster, were used to investigate elicitation of penicillin biosynthesis in alginate- and OM-supplemented cultures (Liu et al., 2001). Penicillin G was overproduced in all elicited cultures either in defined or complex media. Elicitor-induced transcription of the three penicillin biosynthetic genes, pcbAB, pcbC, and penDE, was suggested to be the cause (Liu et al., 2001).

The *Streptomycete* metabolite, dioctatin A (DotA), which inhibits human dipeptidyl aminopeptidase II, strongly inhibited aflatoxin production by *Aspergillus parasiticus* (IC_{50} 4.0 μM) (Yoshinari et al., 2007). Levels of norsolorinic acid, an early precursor of aflatoxin, were strongly reduced as well as the mRNA levels of genes encoding aflatoxin biosynthetic enzymes. Addition of DotA at a concentration of 50 μM did not affect mycelial growth but inhibited conidiation. The mRNA levels of *brlA*, a gene which encodes a conidiation-specific transcription factor, were also reduced by DotA. At the same time, DotA significantly stimulated kojic acid production. The results indicated that DotA has pleiotropic effects on regulatory mechanisms of *A. flavus* secondary metabolite production and differentiation, leading to the inhibition of aflatoxin production.

Hyphal morphology and timing of mycelial emergence were determining factors in the outcome of coculture, fungal/fungal interactions varied not only for pairings on different media but also for replicate pairings made under identical conditions (Rayner et al., 1994; Boddy, 2000). Emergent mycelial phases, occurring both at interspecies interfaces and in self-combinations, were often accompanied by the accumulation of dark pigments in barrage zones. Hyphal-wall fractions, degraded by autoclaving or photo-oxidation, were more active in stimulating morphological changes than protoplasmic fractions from the same fungus; however, morphological differentiation was most stimulated by *B. subtilis* extracts that strongly promoted mycelial cord formation and rhythmic mycelial advancement. Addition of sublethal concentrations of a respiration decoupler, 2,4-dinitrophenol, to the medium induced effects most closely simulating the emergence patterns at interaction interfaces and changes in secondary metabolite profiles (Griffith et al., 1994a,b; Rayner et al., 1994).

When bacteria, including natural and recombinant strains of *Streptomyces* and *Bacillus*, were fermented in the presence of low concentrations (up to 3%) of dimethyl sulfoxide (DMSO), patterns of the production of secondary metabolites exhibited significant qualitative and quantitative alterations (Chen et al., 2000). The effect was observed with a variety of metabolite families, including chloramphenicol (chorismate), thiostrepton (peptide), and tetracenomycin (polyketide). The increased antibiotic production was not caused by changes in the growth rate; biomass yields were similar in media with and without DMSO. Although the mode of action of DMSO was not elucidated, DMSO perturbation was suggested to have affected cells at the level of translation (Chen et al., 2000). The naturally occurring auxin indole-acetic acid, a phytohormone found in the rhizosphere and produced by *Streptomyces* spp., is also involved in regulating cellular differentiation and antibiotic production in *Streptomyces* (Matsukawa et al., 2007). After addition of exogenous indole-acetic acid, induction of aerial mycelium formation and antibiotic production was observed in more than half of the antibiotic producing *Streptomyces* spp. tested. Molecular evidence of this upregulation was shown in the case of the rhodomycin biosynthetic pathway in *S. purpurascens*.

Implementation of primary screening strategies employing arbitrary combinations of microorganisms with different elicitor compounds, metabolic inhibitors, or precursor molecules will face the same challenges as random screening of cocultured organisms. Without prior information on specific pathways and their regulation, the choices of organisms and challenge compounds will likely need to be empirical; therefore, screen design will be accompanied by issues regarding the numbers of compounds, selection and timing of effective dosages, and the number of permutations to be tested relative to untreated controls. The complexity of the problem was illustrated by a study challenging various *Aspergillus* species with the signal transduction inhibitors that interfere with calcineurin, i.e., FK506 and cyclosporine A (Hanlon, 2006). Metabolite perturbation was detected based on changes in pigmentation of cultures rather than differentials in bioactivity. Media supplemented with these compounds showed an increased yellow pigmentation in *A. terreus*, and the pigment was determined to be a series of aspulvinones and terriquinones. Additional pigment upregulating compounds were sought by high-throughput screening of conidial suspensions of *A. terreus* dispersed in 384-well microtiter plates across concentration gradients of compounds from a large collection of metabolic inhibitors and natural products. After incubation, plates were scanned by spectrophotometry on a plate reader to identify wells with pigmentation changes relative to untreated control cultures. Additional pigment upregulators were found, e.g., heat-shock protein 90 inhibitors (geldanamycin, radicicol) and the histone deacetylase inhibitor (trichostatin A), but these compounds decreased the growth relative to controls. Increased pigmentation associated with the aspulvinones/terriquinones was hypothesized to be a result of stress response, and induction of stress response was suggested as a general method for releasing cryptic biosynthetic pathways (Hanlon, 2006). The next step will be to determine whether chemically induced stress response is generally applicable to large sets of *Aspergillus* species and other fungi.

A promising approach for chemical manipulation of secondary metabolism comes from the study of the regulatory protein *LaeA*, identified as a master transcriptional regulator of diverse secondary metabolite gene clusters in *Aspergillus nidulans* (Bok et al., 2006) and *A. fumigatus* (Bok and Keller, 2004; Perrin et al., 2007; Shwab et al., 2007). *LaeA* appears to be a protein methyltransferase with limited homology to histone methyltransferases (Bok and Keller, 2004). The influence of *LaeA* on secondary metabolite clusters may be explained by the fact that many clusters are located in highly divergent telomere-proximal regions characterized by frequent chromosomal rearrangements which may give rise to subsequent alterations in secondary metabolite expression. Frequent telomere-proximal rearrangements may contribute to biosynthetic pathway divergence and niche adaptation between different species of fungi or strains of the same species. Local gene regulation by *LaeA* exhibits a chromosome positional preference because movement of genes into or out of a secondary metabolite cluster leads to respective gain or loss of transcriptional regulation; the mechanism is thought to occur via regulation of nucleosome positioning and heterochromatin formation (Bok et al., 2006). Therefore, manipulation of *LaeA* could be a key target for comprehensive changes in the complement of secondary metabolites. Deletion of the gene *hdaA* encoding *A. nidulans* histone deacetylase (HDAC) activated two telomere-proximal gene clusters correlated with increased levels of the corresponding secondary metabolites (sterigmatocystin and penicillin), but a telomere-distal cluster (terraquinone) was unaffected. Treatment of two unrelated fungi, *Alternaria alternata* and *Penicillium chrysogenum*, with HDAC inhibitors caused an overproduction of several metabolites and a production of new metabolites relative to controls, suggesting a widely conserved mechanism of HDAC repression of some secondary-metabolite gene clusters among filamentous fungi (Shwab et al., 2007). The idea of chemical interference with heterochromatin formation to allow transcription of metabolite pathways was validated by applying different HDAC inhibitors, e.g., suberoylanilide hydroxamic acid and DNA methyltransferase inhibitors, e.g., 5-azacytidine, to a broad sample of filamentous fungi (Williams et al., 2008). Chemical profiling of treated fermentations demonstrated that application of epigenetic effectors often resulted in an enhanced metabolite profiles, including the production of a compound absent in control fermentations and sometimes the production of novel compounds.

17.4 OSMAC Approach: Empirical Manipulation of Primary and Secondary Metabolism

Manipulation of nutritional and environmental factors promoting secondary metabolite biosynthesis is critical to success in microbial natural products discovery. Slight variations in environmental or nutritional conditions have the potential to impact the quantity and diversity of metabolic products. The deliberate elaboration of cultural parameters in order to augment the metabolic diversity of a microorganism has been called the OSMAC (one strain, many compounds) approach (Höfs et al., 2000; Bode et al., 2002; Scherlach and Hertweck, 2006). Implicit in the OSMAC approach is an empirical strategy that addresses the uncertainties of how to grow and modulate the metabolism of uncharacterized strains. The likelihood that one or more growth regimes would be adequate for growth and product formation increases as an organism is empirically tested among multiple conditions. In the absence of information on gene regulation among the seemingly infinite numbers of microorganisms, empirical manipulations are the most direct option for inducing changes in secondary metabolite expression during primary screening and have been used to find the end products of orphan biosynthetic pathways predicted by bioinformatic analyses of homologous biosynthetic pathway (Gross, 2007; Schroeder et al., 2007). Application of OSMAC in screening regimes using individual shake flasks or tubes can be labor intensive. As a result, when screening large numbers of unknown organisms, uncertainty as to the relative effort to devote to screening more phylogenetic diversities or more physiological parameters usually leads to a trade-off where organisms are grown in no more than one to four conditions (Yarbrough et al., 1993; Wildman, 1997).

Recently, our laboratory adapted a miniaturized fermentation system for bacteria to grow on filamentous fungi. Miniaturized parallel fermentations allowed systematic variation of nutritional parameters among large sets of screening strains, and thus, the probability of successful discovery of bioactivity from a given strain was increased (Bills et al., 2008). Through automation, we were able to simultaneously replicate and extract large sets of fungi (80–236 strains) across 8–12 media conditions with minimal effort. The effects of increased numbers of media on detection of antifungal signals from extracts were modeled by interpolation; results from screening models supported the idea that OSMAC increased the probability of detecting antifungal and antibacterial phenotypes. This conclusion can be extended to the actinomycetes and other bacteria which are even more amenable to miniaturized fermentations (unpublished results). A consequence of this approach is that often the majority of strains will be active in only one or a few media among all the media tested (Yarbrough et al., 1993; Wildman, 1997; Tormo et al., 2003; Scherlach and Hertweck, 2006; Schroeder et al., 2007; Bills et al., 2008). The consistent pattern of microorganisms yielding bioactivity in one or a few media when many were tested exemplifies the importance of microbial nutrition to the outcomes of antibiotic and secondary metabolite screening.

The use of miniaturized parallel fermentations in deepwell 96- or 24-well microplates can facilitate the survey of many fermentation parameters, including aeration, light, temperature, osmolarity, pH, and nutrition. The long history of fermentation sciences indicates that growing microorganisms with complex plant-, microbe-, and animal-derived substrata, e.g., fruit and vegetable juices, oils, seed meals, brans, flours, peptones, protein hydrolysates, and yeast extracts is a successful strategy for varying secondary metabolite profiles. Complex microbe-, plant-, or animal-derived components not only provide a wide range of complex carbohydrates and peptide nitrogen, vitamins, and minerals but may also contain natural plant- or animal-derived effectors that may stimulate metabolite synthesis (Ariyo et al., 1998; Aldred et al., 1999; Overy et al., 2006). Microorganisms can be grown on natural solid surfaces, e.g., plant seeds and grains (Barrios-Gonzales and Mejia, 1996) or media dispersed within inert solid supports, e.g., agar, perlite, vermiculite (Singh et al., 2002), expanded clay aggregates (Nielsen et al., 2004), or synthetic membranes (Bigelis et al., 2006) in order to promote cell differentiation, production of aerial mycelium, and sporulation, which are often correlated with increased secondary metabolite diversity (Demain and Fang, 2000; Coyle et al., 2007; Komon-Zelazowska et al., 2007). Addition of halogen salts may promote the biosynthesis of organohalogens metabolites or cause substitutions of one halogen for another, i.e., bromine for chlorine (Spinnler et al., 1994; Stadler et al., 1995; Peters and Spiteller, 2006; Clark et al., 2007). Even varying the water source for media, whether distilled water or tap water containing metal cations, can influence metabolite profiles (Paranagama et al., 2007). Temporal succession in metabolite profiles and appearance of ephemeral metabolites could be readily exploited by replicating and extracting them at different times to generate temporal arrays.

17.5 Natural Functions of Medically and Agriculturally Significant Microbial Metabolites Remain Unknown

Have observations of microbial defensive mutualisms provided useful direction for selection of pharmacologically relevant metabolites or have such observations simply guided organic chemists toward the most metabolically relevant organisms to include in natural products screening collections? Fungi from competitive habitats, e.g., animal dung, wood from freshwater streams, and mycoparasites have been emphasized as important sources for searching new metabolites (Wicklow, 1981; Degenkolb et al., 2003, 2007; Gloer, 2007). But what are competitive habitats? The strategy seems self-evident because most microorganisms exist in competitive environments. The exceptions might be extremophiles that rely on specialized physiological adaptations to escape competition or some specialized biotrophic parasites that reside protected within their hosts.

On one hand, a seemingly unlimited pool of microorganisms producing biologically active molecules exists in nature. At the same time, it has been hypothesized that for all important molecular targets in an organism, e.g., *Homo sapiens*, naturally molecules exist in other species that affect those targets (Tulip and Bohlin, 2002). If both assumptions are true, then why are the intersections among small natural molecules and their targets so difficult to identify and why has the knowledge on microbe–organism interactions not made a greater direct impact on discovery of pharmacological lead molecules? Is it because way few systems have been adequately studied or is it that the starting points for lead discovery fail to take the basic ecological knowledge into account?

One obvious reason for this gap is that the lead discovery strategies in pharmaceutical research treat the natural compounds as screening currency and the process usually circumvents the exploration of a metabolite's natural function. Pharmaceutical lead discovery focuses on the identification of low molecular weight compounds with a predefined mechanism of action or molecular interaction. The process starts with recognition of medical need and the identification of molecular targets that are hypothesized to be relevant for a given pathology. Based on these targets, in vitro screens are developed that allow the testing of vast collections of synthetic or natural molecules made available for drug discovery. Screening of collections is usually carried out in specialized laboratories dedicated to high throughput screening, a process heavily supported by automation technologies. The screening of those collections leads to the identification of primary hits, which are usually triaged by means of a panel of secondary assays aiming to provide additional information related to the potency and specificity and other biological and pharmacological properties of the compound set defined by biological activity in the primary screens. Compounds surviving the elimination process and considered by medicinal chemists to have the desirable structural features (and lacking undesirable features) are nominated as lead compounds and subsequently subjected to a process of successive rounds of chemical derivatization to improve their in vitro and in vivo properties. The goal of the lead optimization process is to deliver a few optimized compounds that, after passing the test of rigorous preclinical safety models, will become candidates to enter into the clinical trials.

The major exceptions to this modern discovery paradigm, which has dominated the pharmaceutical industry since the beginning of the 1990's, have been the searches for antibiotic activity against human pathogens where crude biological activity was first detected in whole-organism assays and the molecular mechanism of antibiosis was elucidated only later. Other whole-cell or whole-organism assays include mammalian cell cultures to find anticancer leads and multicellular organisms, e.g., insects, helminthes, and plants to find insecticides, antiparasitic agents, and herbicides. This kind of search identifies functional bioactive molecules upfront but has limited possibilities to discern ecological function because the screening process channels the extracts from many source organisms through a narrow window defined by single parameter, in vitro growth inhibition of human pathogens or agricultural pests; therefore, the ecological relevance between the extract source and the target organism will be tenuous at best. Historically, besides antibiotics, insecticides, and other cytotoxic agents, a limited number of therapeutic areas have benefited from microbial natural products discovery. In these therapeutic areas, the relationships of natural interactions are readily envisioned, or at least easily imaginable, because many of the targets are widely conserved; they include lipid and steroid biosynthesis, immunological agents, neurotransmitters, and ion channel antagonists (Samuelson, 1999).

The potential economic value of microbial-derived pharmaceuticals and agricultural products has motivated the elucidation of molecular targets of many microbially derived antibiotics, antiparasitics, cytotoxins, and other bioactive molecules. Why then have chemical ecologists not taken greater advantage of molecular mechanisms as a starting point for experimentally determining as to why these molecules are produced in nature and their relevance to the producing organism's fitness and survival? Due to their wide recognition and the extensive accumulated knowledge, we believe that the organisms producing these important compounds merit the chemical ecological studies aimed at dissecting their function in nature.

Most of the organisms discussed in the following section are easily manipulated in the laboratory and can be readily obtained from nature, and if not, then from culture collections. For most, reference compounds are commercially available for simulation experiments. In most cases, the biosynthetic pathways are characterized or can be predicted; gene disruptions or mutations could be used to generate the producing and nonproducing strains for controls and confirm hypothesized functions (Coyle et al., 2007; Rohlfs et al., 2007; Potter et al., 2008). The greatest challenge remains in the design of predictive bioassays to study their effects on other organisms in simulated and natural environments (Wiener, 1996; Thines et al., 2004).

We think the proposed roles of a selection of commercially successful microbial products in the natural history of the organisms that produce them deserve reflection. Many of these molecules were discovered by empirical screening of microbial extracts in pathogen-based whole-cell screens, or in the case of lovastatin, in a biochemical assay targeting a specific step in cholesterol biosynthesis using liver microsome preparations. Only in the case of ergotamine was there a long ethnopharmacological history that eventually culminated in the isolation and characterization of ergotamine and ergometrine and recognition that their vasoconstrictive properties could have pharmacological applications (van Dongen and de Groot, 1995; Giger and Engel, 2006).

17.5.1 Fungal Metabolites

17.5.1.1 Ergotamine

Ergotamine (Figure 17.2), the main ergopeptine of the ergot fungus *Claviceps purpurea*, was the cause of mass poisoning in Europe in the Middle Ages caused by eating bread from rye contaminated with sclerotia of the fungus (van Dongen and de Groot, 1995; Tfelt-Hansen et al., 2000). Due to its strong vasoconstrictor effects, it was used in traditional medicine and later introduced into clinical practice (as Gynergen, Methergine, Hydergine, Cafergot, Ergostat, and other brand names) to precipitate childbirth, control postpartum hemorrhage, and for the treatment of hypertension and migraines. At therapeutically relevant doses, ergotamine acts mainly as an agonist of α-adrenoceptors, $5\text{-HT}_{1B/1D}$, and dopamine D_2 receptors (Tfelt-Hansen et al., 2000). Because ergotamine is predominantly deposited in the reproductive sclerotia of *C. purpurea*, one of the more obvious ecological roles, along with other coproduced ergot alkaloids, would be to protect the sclerotium and symbiotic grasses from mycophagy and herbivory. Ergotamine concentrates in the outer rind rather than the center of the sclerotia suggesting that the first mouthful encountered by "would be" mycophagists may discourage consumption of the sclerotia (Wicklow, 1988; Panaccione, 2005).

17.5.1.2 β-Lactams

Antibiosis against bacteria was the first described biological activity of fungal metabolites that resulted in a therapeutically useful drug. The discovery and industrial development of penicillin (Figure 17.2) from *Penicillium chrysogenum* stands out as a hallmark in the history of medicine. Cephalosporin (Figure 17.2), like penicillin, is a β-lactam inhibitor of bacterial cell wall synthesis. Originally discovered from the species *Acremonium chrysogenum* in the 1940s, cephalosporin has been at the origin of several generations of semisynthetic derivatives (cefuroxime, cephalexin, cefaclor, cefotaxime, ceftriaxone, ceftazidime, and cefixime among others), some of which are still among the best selling antibiotics in the market today (Strohl, 1997). Surprisingly, in light of their historical and economic significance, little has been done outside of Petri dish interactions to elucidate the biological role of the β-lactams in nature and their significance to the organisms that produce them. Although their function is generally assumed to be antibiosis against bacteria, ecological studies of the β-lactams have largely consisted of studying their effects on microbial communities in vivo at therapeutic concentrations. Conceivably, β-lactams may have functions other than bacterial antibiosis (Brakhage, 1998). Penicillin G's precursor, δ-(L-α-aminoadipyl)-L-cysteinyl-D-valine, may have a function in amino acid transport or could be involved in cysteine fixation, which is toxic at high concentrations. Another possibility is based on the observation that penicillin biosynthesis occurs in microbodies and aromatic molecules in the environment, e.g., lignin that could be covalently bound to isopenicillin N. Such reactions could

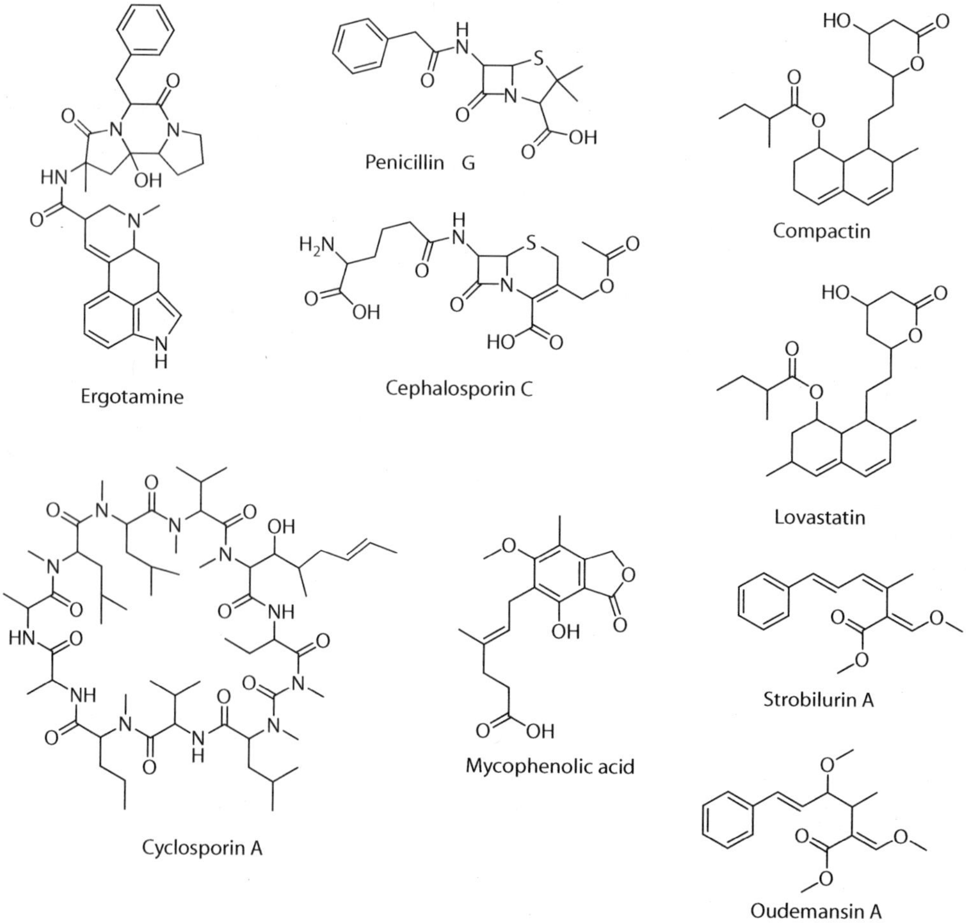

FIGURE 17.2 Structures of fungal metabolites with significant medical and agricultural applications.

be a mechanism of detoxifying these compounds and excreting them to the environment. Because the biosynthetic pathways in *Penicillium* and *Acremonium* have been elucidated (Brakhage, 1998), experimental genetic tools to test the relative fitness of wild-type β-lactam-producing strains, over-producing mutants, and different pathway knockouts are available to carry out such experiments.

17.5.1.3 Statins

The statins are the most widely used cholesterol-lowering drugs on the market today. They are inhibitors of 3-hydroxymethyl-glutaryl-CoA reductase (HMGR), a rate-limiting enzyme responsible for sterol biosynthesis. Compactin (ML-236B) (Figure 17.2) was the first statin structure elucidated, coincidentally as a cholesterologenesis inhibitor from the fermentation of *Penicillium citrinum* (Endo et al., 1976; Endo, 1985) and as an antifungal metabolite from a fermentation of *P. brevicompactum* (Brown et al., 1976). In the late 1970s, another statin, lovastatin (Figure 17.2, also named mevinolin) was discovered from an *Aspergillus terreus* strain (Alberts et al., 1980; Shu, 1998; Patchett, 2002) and *Monascus ruber* (Endo, 1979). Lovastatin was the first natural statin introduced in the market and sold under the brand name Mevacor. Investments in their large-scale production have motivated the full characterization of the gene clusters for the biosynthesis of lovastatin and compactin (Hutchinson et al., 2005).

The statins have profound pleiotropic effects on a variety of cellular functions involving sterol metabolism, terpene biosynthesis, and isoprenylation and bring into question as to whether the sole function of the statins in nature is the modulation of sterol synthesis through the inhibition of HMGR. In plants, compactin and lovastatin have been important and useful in elucidating the contribution of HMGR activity to the production of particular isoprenoids. Exogenous application of the statins stunts plant growth and development, presumably by limiting the availability of mevalonate for isoprenoid metabolism (Bach and Lichtenthaler, 1983; Hashizume et al., 1983; Hata et al., 1987; Narita and Gruissem, 1989). Inhibited incorporation of radioactive mevalonate into sesquiterpenoids has been demonstrated to be dose dependent in elicitor-treated tobacco cell-suspension cultures and in vitro measurements of sesquiterpene cyclase activity (Chappell et al., 1995). The modulating effect of compactin upon terpene biosynthesis was exemplified using cell cultures of *Cupressus lusitanica* (Mexican cypress) as the production of β-thujaplicin, an antimicrobial terpenoid tropolone (derived from geranyl pyrophosphate and a monoterpene intermediate), was significantly suppressed in elicited cell cultures following the addition of compactin (Yamaguchi et al., 1997).

HMGR is crucial to insect development and reproduction. Insect juvenile hormone, which regulates embryonic development, represses metamorphosis, and induces the synthesis of vitellogenin in insects, is synthesized from mevalonic acid. Its synthesis is inhibited by compactin, lovastatin, and other synthetic statins (Monger et al., 1982; Hiruma et al., 1983; Feyereisen and Farnsworth, 1987; Guerrero and Rosell, 2005). Via inhibition of HMGR, the statins affect insect dolichols that behave as donors of oligosaccharide residues in the glycosylation of proteins (Casals et al., 1996, 1997; Zapata et al., 2002) and aggregation pheromones that are acyclic isoprenoids (Ivarsson et al., 1993; Ivarsson and Birgersson, 1995; Ozawa et al., 1995)

Both naturally produced and synthetic statins have been shown to have an antifungal effect through the inhibition of HMGR and the subsequent suppression of ergosterol synthesis (Trenin, 1998; Macreadie et al., 2006). Aside from acting upon cell membrane targets, the statins have also been demonstrated to influence sexual reproduction and spore germination in fungi. Compactin effectively blocked tremerogen A-10 (MW 1,480), an S-polyisoprenyl peptide-mating pheromone secreted by the heterobasidiomycete *Tremella mesenterica*, causing a large precursor polypeptide (MW 28,000) to accumulate in the cell membrane. The result suggested that the addition of a lipid residue to a polypeptide precursor is important for the production of tremerogen A-10, especially in the intracellular transport and processing of the precursor molecules (Miyakawa et al., 1985). Natural and synthetic statins (lovastatin, simvastatin, rosuvastatin, and atorvastatin) efficiently inhibited sporangiospore germination in zygomycetes in the absence or presence of a constant concentration of antifungal protein (PAF). PAF and lovastatin acted synergistically on sporangiospore germination of *Mycotypha africana*, and the combinations PAF–rosuvastatin and PAF–atorvastatin had similar effects on *Syncephalastrum racemosum* (Galgoczy et al., 2007).

17.5.1.4 Cyclosporines and Mycophenolic Acid

The discovery of cyclosporine A (Figure 17.2) (Sandimmune, Neoral) and its application to renal transplantation was an important medical milestone, and cyclosporine soon became the most widely used immunosuppressive drugs (Plosker and Barradell, 1996). Numerous natural members of the cyclosporine family have been reported (Traber, 1997). Cyclosporines, synthesized by a nonribosomal peptide synthetase, are produced by a wide range of fungi in the the Hypocreales, many of which are insect and plant parasites (Traber and Dreyfuss, 1996). Cyclosporine A was originally discovered because of its antifungal activity, and although not clinically useful, its antifungal properties may be ecologically relevant (Rodriguez et al., 2006). Research on cyclosporine A and the actinomycete metabolite FK506 (see below) has been critical in revealing the important role of cyclophilins and FK506-binding proteins in inhibiting the calcium/calmodulin-dependent phosphatase calcineurin in immune regulation and other widespread calcium-activated signaling pathways, ranging from the recovery from cell cycle arrest by a mating factor in yeast and stimulation of humoral immune response and inhibition of phagocytosis in insects (Vilcinskas et al., 1999) to the opening of an inward potassium channel involved in stomatal opening in plant guard cells. Therefore, the potential roles of cyclosporines in nature are likely to be more complex than simple antibiosis.

Some years after the development of cyclosporine A, mycophenolic acid (Figure 17.2), another immunosuppressant fungal metabolite, reached the market (Bentley, 2000). This compound, produced by *Penicillium* spp. and other fungi (Buckingham, 2007), was discovered as an antibiotic in the early twentieth century (it was crystallized before penicillin), but its immunosuppressive properties were recognized much later. The compound shows antibiotic, antifungal, antiviral, and antitumor activities in animal studies as well as activity against *Trichomonas*, *Eimeria*, and other parasitic protozoa. Although not therapeutically useful in any of these areas, a mycophenolic acid derivative was developed as a clinically useful immunosuppressant and marketed under the name CellCept (mycophenolate mophetil) and Myfortic (sodium mycophenolate). In vitro, predominant excretion of mycophenolic acid into the growth media by the fungus *P. brevicompactum* has been demonstrated to coincide with the development and deposition of the metabolite in aerial mycelium (Bird and Campbell, 1982), and deposition of mycophenolic acid by *P. brevicompactum* into infected plant tissues was confirmed to occur *in planta* in infected *Zingiber officinale* rhizomes (Overy and Frisvad, 2005). Aside from its antimicrobial properties, it has been hypothesized that in nature, mycophenolic acid may act synergistically with other fungal secondary metabolites to deter mycophagy and herbivory through secondary mycotoxicosis (Wicklow, 1988; Overy and Frisvad, 2005).

17.5.1.5 Strobilurins

The methoxyacrylate antibiotics, strobilurins and oudemansins (Figure 17.2) (Anke and Steglich, 1999), are fungal metabolites that have provided the molecular template for several marketed agricultural fungicides (azoxystrobin, kresoxim-methyl, trifloxystrobin, and others). This is an important market; sales of azoxystrobin (Heritage) alone amounted to $415 million in 1999. These compounds are nanomolar-level inhibitors of the respiratory chain that act by binding to the mitochondrial cytochrome *b* complex. The producing fungi have amino acid substitutions in their cytochrome *b* conferring them resistance. Although the methoxyacrylates are inhibitory to fungi, the target is not fungal-specific, and it has been reported that these compounds may also be phytotoxic, insecticidal, or cytotoxic to mammalian cell lines, depending on the derivative (Anke, 1995; Anke and Steglich, 1999). However, despite their mode of action against a highly conserved target, oral toxicity in mammals is low, for reasons that are unclear as yet. This toxicity differential has allowed their successful development as fungicides.

The strobilurins and the related oudemansins are produced by many species of basidiomycetes belonging to the genera *Strobilurus*, *Oudemansiella*, *Mycena*, *Crepidotus*, and others as well as by the ascomycete *Bolina lutea* (Anke and Steglich, 1999). Their role in nature has been hypothesized to be antifungal-mediated resource defense. In the case of *Oudemansiella mucida* grown on sterilized beech wood, in situ production and excretion into host tissues have been confirmed (Anke, 1995). In situ concentrations of strobilurin D from *Mycena tintinnabulum* were sufficient to inhibit the growth from competing fungi (Engler et al., 1998). In vitro cocultures of *O. mucida* and *Xerula melanotricha* with *Penicillium notatum* or *P. turbatum* produced elevated levels of strobilurins while culture filtrates or autoclaved cells of these Penicillia did not induce strobilurin production (Kettering et al., 2004).

17.5.2 Actinomycete Metabolites

Unfortunately, no clear role in the natural environment has been demonstrated so far for any of the highly successful antibiotics and bioactive metabolites produced by actinomycetes discussed below. The traditional hypothesis that they may confer a survival advantage on the producing organism (the "competitive hypothesis") would be consistent with the metabolic energy required for the production of many of these compounds and the extensive and conserved genetic information to program their biosynthesis. Another argument favoring this view would be the fact that *Streptomyces,* being sessile organisms, cannot escape from stress as can other motile bacterial species. In the complex soil environment, with spatially and temporally varying stresses, having the capability to counteract stress would justify the enormous investment in genes encoding regulators, transport proteins, degrading and other nutritional enzymes, and antibiotics biosynthesis identified in the *Streptomyces* species (Challis and Hopwood, 2003). *Streptomyces* antibiotics are typically produced during colonial development at the transition phase when vegetative mycelial growth is slowing due to nutrient exhaustion, and aerial mycelium emerges at the expense of nutrients released by the breakdown of vegetative hyphae, in an apparently highly structured process (Manteca et al., 2005). One possibility would be that antibiotics would be secreted to defend the mycelial nutrient reservoir from other soil microbes. The fact that other microorganisms with complex life cycle differentiation also produce antibiotics, e.g., fungi, myxobacteria, and cyanobacteria, supports this concept. An extreme "attraction" model has been proposed, hypothesizing that competitors would be attracted by the nutrients released from degraded vegetative mycelium just to be killed by the antibiotics produced, and therefore being recycled for the benefit of the developing *Streptomyces* colony. This "fatal attraction" concept has been discussed in depth in relation with *Myxococcus* life cycle (Shi and Zusman, 1993). Other lines of evidence supporting the "competition hypothesis" come from studies in laboratory and in the field showing that established populations of actinomycetes can prevent colonization by other microorganisms such as other bacteria or phytopathogenic fungi (Wiener, 1996; Demain and Fang, 2000).

In contrast with this view, several groups have emphasized the role of antibiotics in nature as mediators of chemical communication and intra- and inter-species signaling (Davies et al., 2006; Hoffman et al., 2007). This hypothesis is essentially based on the observations of many pleiotropic effects of a number of antibiotics at sublethal concentrations that mediate transcriptional regulation (Davies et al., 2006; Yim et al., 2007). The concept is that the natural role of antibiotics at naturally achieved concentrations is likely to be that of signaling molecules, and only at much higher concentrations, these molecules would be antibiotic in action. It is true that sometimes the distinction between antibiotic and signaling molecule can be subtle. For instance, *Rhizobium leguminosarum* produces an AHL autoinducer involved in the symbiotic relationship with the plant host that was originally reported as a bacteriocin because of its effects on inhibiting bacterial growth. Thus, although this autoinducer is a well-characterized signaling molecule, it shares some properties with canonical antibiotics (Hoffman et al., 2007). In other words, the functional definition of antibiotics, based on growth inhibition effect in laboratory conditions, can lead to assumptions about the ecological role of those molecules that may not be necessarily correct. Both hypotheses likely may be true, so that the natural role of antibiotics may include both possibilities (Hoffman et al., 2007).

17.5.2.1 Actinomycin D (Dactynomycin)

Actinomycin was the first antibiotic isolated by Waksman in 1940, and it is one of the oldest drugs used against a variety of cancers. The actinomycins are a family of bicyclic chromopeptide lactones that share the chromophoric phenoxazinone dicarboxylic acid actinocin to which there are attached two pentapeptide lactones of nonribosomal origin. Actinomycin D (Figure 17.3), also known as dactinomycin, has been used as an antitumor agent since the 1960s. The molecule binds to DNA duplexes and acts as a transcription inhibitor binding DNA at the transcription initiation complex and preventing RNA polymerase elongation (Goldberg et al., 1962). The conformation of the molecule determined by specific intramolecular hydrogen bonds is well adapted to be intercalated into a right-handed DNA helix, favoring the establishment of hydrophobic interactions that stabilize the DNA/antibiotic complex (Williams et al., 1989). This property of intercalation is also observed in other molecules, such as echinomycin or trisotin A, that establish a highly stable drug–DNA complex through their selective binding to CpG dinucleotide sequences in DNA double helices.

Because of its DNA binding mechanism, actinomycin D and other actinomycins are toxic to a many organisms and they are potent antibiotics with activity against gram-positive and gram-negative bacteria. Not surprisingly, *Streptomyces* species are resistant to actinomycin. Actinomycin was first isolated from *Streptomyces antibioticus* and is produced by many *Streptomyces* strains (Waksman and Woodruff, 1941). In our own laboratory, we have observed that actinomycin is the most frequently detected antibiotic across many communities of *Streptomyces* species. More than 10% of the *Streptomyces* strains isolated from diverse geographic origins and substrata may produce this compound when cultivated under appropriate conditions (unpublished results). In contrast, actinomycins are relatively uncommon among other actinomycetes families. The widespread occurrence of actinomycin raises obvious questions about its ecological role. Obviously, such a common metabolite may have a role in antibiosis in natural environments. Unfortunately, as it is the case with the rest of actinomycete metabolites described in this chapter, little is known about the regulation of actinomycin production in natural conditions. *Streptomyces antibioticus* mutants that are disrupted in the *relA* gene are deficient in actinomycin production (Hoyt and Jones, 1999). The *relA* gene encodes for a ppGpp synthetase involved in the stringent response to carbon and energy starvation energy in bacteria, suggesting that ppGpp plays a role in controlling antibiotic production in *S. antibioticus* when the conditions in culture reach nutrient starvation level above some threshold (Hoyt and Jones, 1999). How to extrapolate this type of *in vitro* studies to the natural environment remains to be clarified.

17.5.2.2 Streptomycin and Other Aminoglycosides

The aminoglycosides are one of the most frequent families of antibiotics produced by actinomycetes. Their history started with the discovery of the broad-spectrum antibiotic streptomycin (Figure 17.3) in 1944, the first useful drug against tuberculosis (Schatz et al., 1944). This was followed by the successive

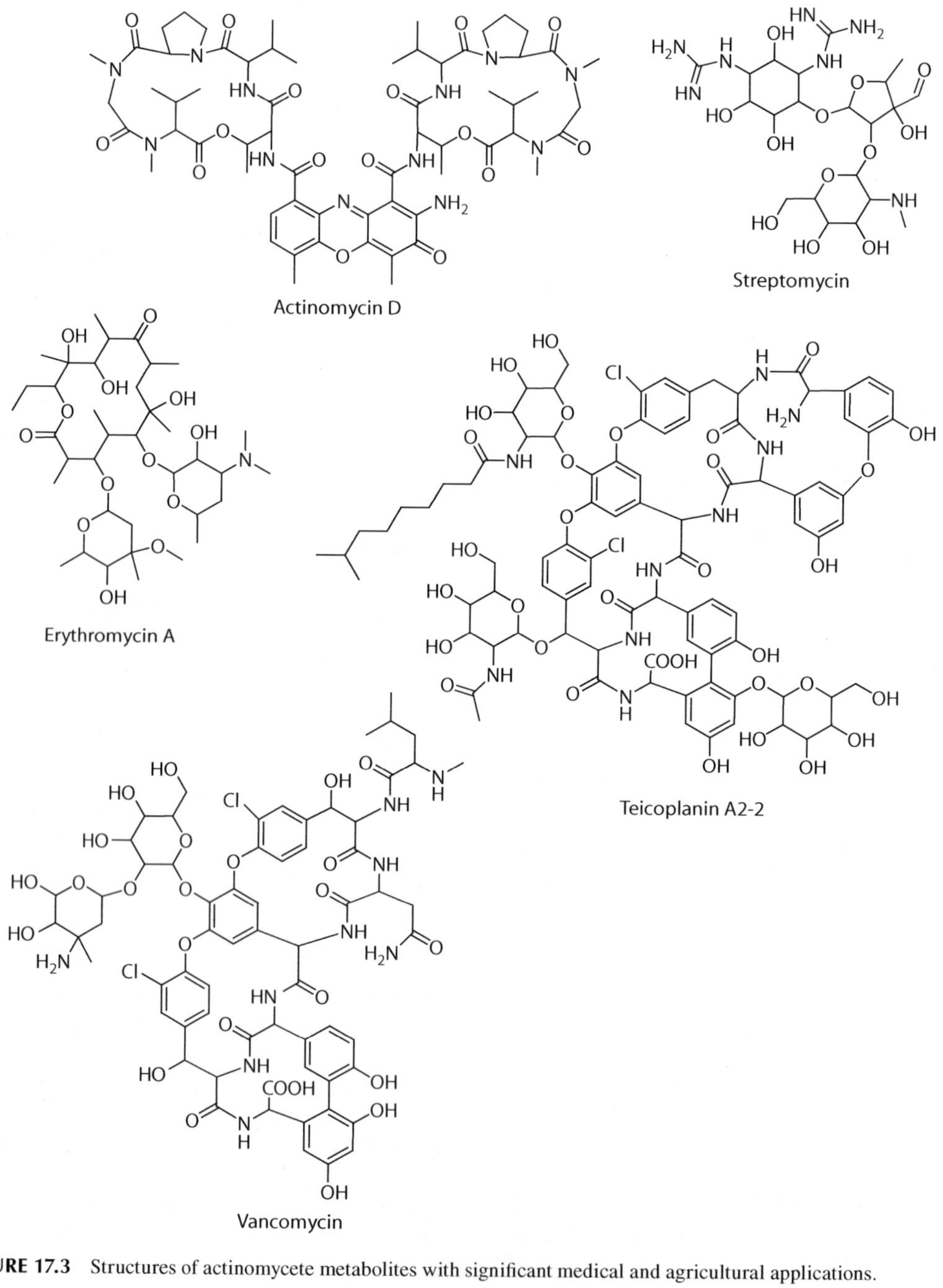

FIGURE 17.3 Structures of actinomycete metabolites with significant medical and agricultural applications.

discovery of new aminoglycosides for the treatment of gram-negative infections, such as neomycin, kanamycin, and gentamicin (Waksman and Lechevalier, 1949; Umezawa, 1958; Weinstein et al., 1963), and the development of many related and semisynthetic compounds in the 1970s. All aminoglycosides present a similar mode of action, inhibiting protein synthesis by binding to specific sites of the 30S subunit of the bacterial ribosome (Mingeot-Leclercq et al., 1999). Transport across the cytoplasmic membrane is dependent upon electron transport and inhibited by divalent cations, hyperosmolarity, or anaerobiosis. Once in the cytoplasm, aminoglycosides bind to the ribosomes, perturbing the elongation of the nascent chain. The aberrant proteins that are produced as a result of the impaired proof-reading translation process can be inserted in cell membrane where an altered permeability would facilitate aminoglycoside transport.

Aminoglycosides are also known to produce a wide range of pleiotropic phenotypic responses at subinhibitory concentrations, which would be caused by transcription modulation. This includes suppression of nonsense mutations, stress responses, and increased mutation frequencies in several bacterial species (Davies et al., 2006). These responses are very specific, not general for all translation inhibitors, including other antibiotics acting through binding to the 30S subunit. Another well-described effect of aminoglycosides at subinhibitory concentrations is the induction of gram-negative bacteria (*P. aeruginosa* and *E. coli*) to grow as biofilms, which are markedly more resistant to many antibiotics than free living cells (Hoffman et al., 2005). This induction effect, which is observed within a narrow range of concentrations and not shown by other antibiotics, is mediated in *P. aeruginosa* by a gene named *arr* (aminoglycoside response regulator) that regulates the levels of the second messenger cyclic-di-GMP. These functional interactions with other cellular processes might be related with an alternative role of these compounds in the environment beyond antibiosis (Davies et al., 2006).

17.5.2.3 Erythromycin

Erythromycin A (Figure 17.3) was the first drug from a new class of antimicrobial agents, the macrolides, to be introduced in the clinic (McGuire et al., 1952). It was originally used as an alternative therapy to β-lactam agents to treat infections with gram-positive pathogens, e.g., *Staphylococcus* and *Streptococcus* species. The compound is a 14-membered macrocyclic lactone produced by *Saccharopolyspora erythraea* and prevents bacterial protein biosynthesis by binding to the 50S ribosomal unit, interfering with the elongation process of nascent polypeptide chains. It binds specifically to the 23S rRNA and blocks translation, stimulating the dissociation of peptidyl-tRNA during translocation of the nascent chain (Walsh, 2003a). Oleandomycin is a natural analog of erythromycin A produced by a strain of *Streptomyces antibioticus* sharing the same mode of action but exhibiting weak inhibitory activity. Resistances to macrolides are mediated by methylation of the target 23S rRNA and efflux mechanisms. Macrolides have been successful antibiotics in clinical practice, having originated a number of semisynthetic analogs, such as azithromycin (Zythromax) and clarithromycin (Biaxin, Klacid), that reached blockbuster status in the 1990s.

In spite of the intensive research in molecular genetics and biochemistry carried out on this family of compounds as model biosynthetic systems, little is known about the potential role of these macrocyclic lactones in the life cycle of the producing organisms. The macrolide producers devote more than 60 Kb of DNA to the synthesis of any of these compounds using a modular polyketide synthase assembly line involving at least 28 active sites, further hydroxylation and glycosylation steps, and self-protection by ribosomal RNA methylation (Reeves et al., 1999; Staunton and Weissman, 2001; Challis and Hopwood, 2003). The complete genome sequence of *Saccharopolyspora erythraea* showed that the strain contains 25 gene clusters of known secondary metabolites and up to 72 genes known to confer resistance to major classes of antibiotics (Oliynyk et al., 2007). It is intriguing that none of the hypothetical polyketide products predicted from the sequence have been ever detected in experiments that were carried out by testing more than 50 production conditions (Boakes et al., 2004). It has been proposed that the energy cost of producing these molecules would be compensated by a selective advantage for securing nutrient sources in a highly competitive environment.

Like aminoglycosides, macrolides have been reported to induce a number of pleiotropic effects at subinhibitory concentrations in diverse bacterial species. These include prevention of flagella formation and motility in *P. aeruginosa* and *Proteus mirabilis*, decreased biofilm formation in *Mycobacterium avium*, inhibition of alginic acid production, biofilm formation and virulence factor production in *P. aeruginosa*, and diverse transcriptional modulation effects on a wide range of genes in *B. subtilis*, *Streptococcus pneumoniae*, and *Enterococcus faecalis* (Davies et al., 2006; Yim et al., 2007). Again, these findings have argued to support the hypothesis of the role of antibiotics in bacterial communication rather than chemical warfare as traditionally envisioned.

17.5.2.4 Vancomycin and Teicoplanin

Vancomycin and teicoplanin (Figure 17.3) are glycopeptide antibiotics produced by actinomycetes that are used in the clinic to treat serious gram-positive infections resistant to other antibiotics, including methicillin resistant *Staphylococcus aureus* (MRSA). Both metabolites inhibit bacterial cell wall biosynthesis by blocking the final steps of the synthesis of the peptidoglycan, inducing the lysis of the cell. Vancomycin was the first compound of this class, isolated in 1955 from *Amycolatopsis orientalis*. Initially, its use was reserved for patients with β-lactams allergies or for infections with organisms resistant to other agents. Since the late 1970s, however, its widespread use favored the development of new multiple resistant *S. aureus* strains (VRSA) while resistance in strains of *Enterococcus faecium* (VRE) was first described in 1986. As other members of this family, the molecule recognizes the dipeptide motif D-Ala-D-Ala on peptidoglycan precursors, inhibiting both the transglycosylation and transpeptidation steps (Walsh, 2003a). Receptor-binding studies have revealed that there is a remarkable complementarity between vancomycin and the cell wall peptide (Williams et al., 1989).

Teicoplanin is a related lipoglycopeptide produced by *Actinoplanes teichomyceticus*, discovered in 1984. Its structure is related to vancomycin because it shares a glycosylated heptapeptide scaffold but presents a hydrophobic acid chain that alters its physical properties, and this factor determines its differential activity (Zhu, 1999; Kahne et al., 2005). Numerous analogs have been isolated from actinomycetes in the vancomycin (balhimycin, chloroeremomycin) and the teicoplanin families (A47934 and A40926); some have been used as starting points for the development of the semisynthetic analogs dalvabancin, oritavancin, and telavancin, currently in advanced clinical trials (Decousser et al., 2007; Laohavaleeson et al., 2007; Poulakou and Giamarellou, 2008). In spite of the clinical importance of these compounds, and the numerous studies directed to understand the biosynthesis of these and other related compounds (van Wageningen et al., 1998; Li et al., 2004), the potential roles of these compounds in the environment remain unknown. The presence of γ-butyrolactones regulators and receptors in crude cell extracts of *A. orientalis* and *A. teichomyceticus* suggest that the synthesis of these compounds could be controlled by microbial quorum sensing signals, as is the case with other actinomycete antibiotics (Choi et al., 2003).

17.5.2.5 Amphotericin B and Other Polyenes

Polyene macrolides are potent antifungals commonly produced by *Streptomyces* species. Amphotericin B (Figure 17.4) is an amphipathic polyene macrolide produced by *S. nodosus* and has been the drug of choice in the treatment of serious systemic fungal infections for more than 45 years. Like other polyenes, such as nystatin, pimaricin, or natamycin, amphotericin B produces pores in fungal cell membranes by binding with sterols, and especially with ergosterol for which it presents a high affinity (Bolard, 1986). Disrupted membrane integrity leads to the loss of ions, small molecules, and oxidative enzymes, driving to cell death. The interaction of amphotericin B with phospholipids to form nonbilayer lipid phases has been also associated with pore formation (Hartsel and Bolard, 1996). In spite of its toxicity, especially its high nephrotoxicity, amphotericin B presents a broad spectrum of action against human fungal pathogens that renders it useful in the treatment of many fungal infections (candidiasis, cryptococcosis, histoplasmosis, coccidioidomycosis, and aspergillosis among others). Surprisingly, it has little utility against phytopathogens, suggesting that its natural role may not be related to antibiosis against fungi in the same environment. On the contrary, other polyenes, e.g., nystatin, candicidin, or pimaricin, show activity against infections of *Botrytis tulipae* on tulips, *Fulvia fulva* on tomatoes, or *Ascochyta pisi* in pea seeds, respectively (Worthington, 1988). Amphotericin

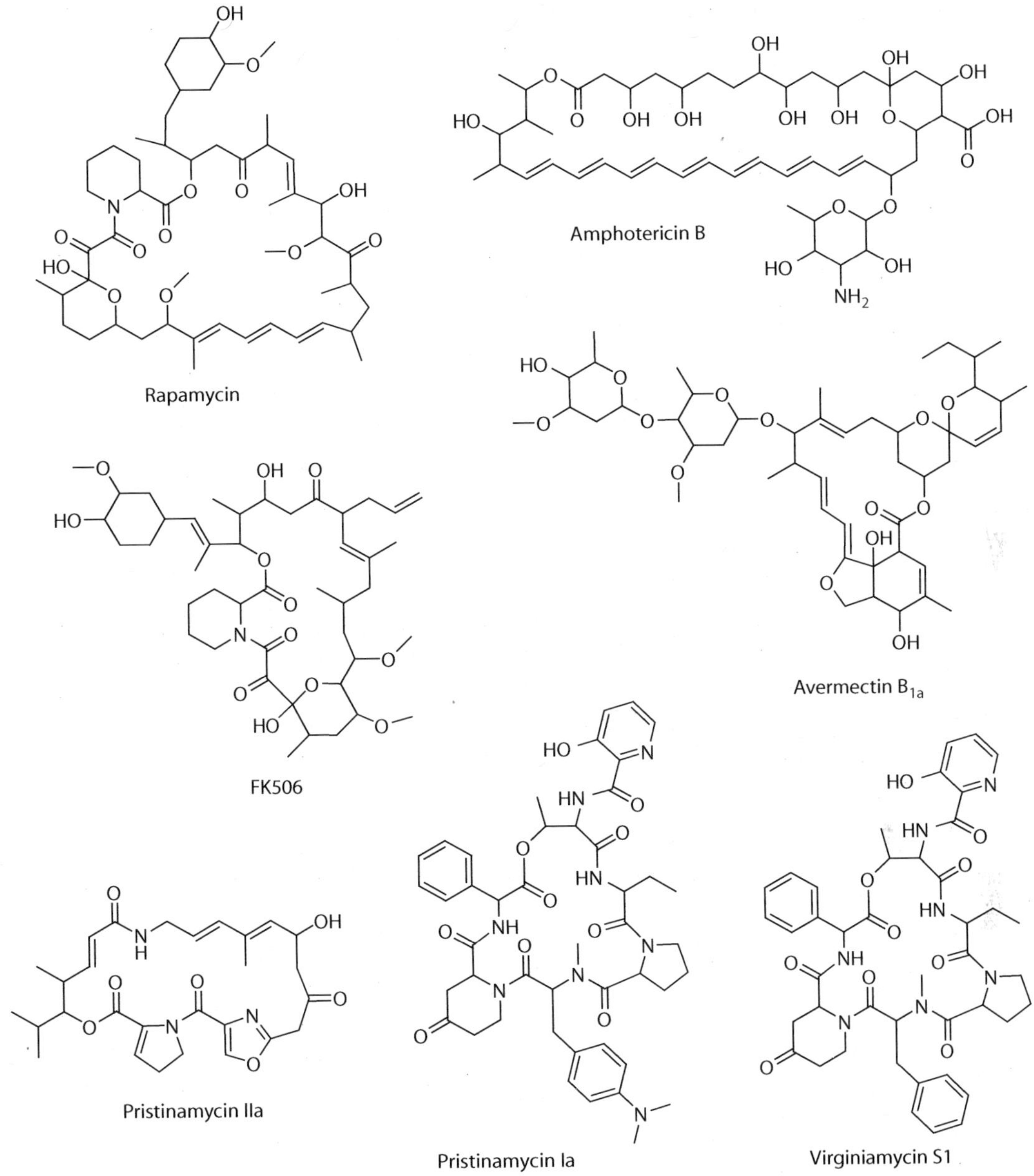

FIGURE 17.4 Structures of actinomycetes metabolites with significant medical and agricultural applications.

B also exhibited activity against some viruses, inhibiting cellular infection by HIV in vitro and protozoans, e.g., *Leishmania brasiliensis* and *L. mexicana*, that are rich in ergosterol precursors in their membranes (Ellis, 2002).

Regarding the regulation of the synthesis of these polyenes, pathway-specific regulator genes have been recently identified in the biosynthetic pathways of two antifungal polyenes sharing a high-sequence homology with the gene *sanG* that has been described to control sporulation as well as the synthesis of another antifungal compound, nikkomycin. It has been suggested that this could be a mechanism by which *Streptomyces* strains could respond to fungal competition by producing antifungal antibiotics and sporulating efficiently (Liu et al., 2005).

17.5.2.6 Rapamycin and FK506

Rapamycin (Figure 17.4) (also known as sirolimus, marketed as Rapamune) is a potent immunosuppressant produced by *Streptomyces hygroscopicus* that also possesses antifungal and antineoplasic activity. This macrocyclic lactone inhibits lymphocyte T activation and proliferation that occurs in response to antigenic and cytokine stimulation (Dumont et al., 1990; Dumont and Su, 1995). It binds to immunophilin, the cytosolic FK-binding protein 12 (FKBP-12), shown to be a peptidyl–prolyl *cis–trans* isomerase, to form the FKBP-12 complex. This complex binds a key regulatory kinase, mTOR, suppressing cytokine driven T-cell proliferation and halting progression from G1 to S phase of cell cycle.

FK506 (Figure 17.4) (tacrolimus, Fujimycin, Prograf) is a 23-membered macrocyclic lactone isolated from *S. tsukubaensis*, structurally related to rapamycin and also showing antifungal and immunosuppressive activities. Rapamycin and FK506 share the same common cellular receptor FKBP but present a different mechanism of action. Similar to cyclosporine A, FK506 suppresses T-cell activation at the level of lymphokine production and prevents IL-2R expression (Tocci et al., 1989; Dumont et al., 1990). In contrast, rapamycin has little effect on lymphokine production but markedly suppresses IL-2 T-cell proliferation. The fact that two structurally related molecules binding to the same target protein can elicit similar cell responses, but mediated by different mechanisms, is one of the best characterized examples of the subtle factors ruling the interaction between natural small molecules and the cell machinery.

FKBPs are present in abundance in a variety of eukaryotic cells, including fungi (*Saccharomyces cerevisiae*, *Neurospora crassa*), where they are supposed to mediate the cytotoxic effect of FK506. Calcineurin inhibition by FK506 acts synergistically with antifungal agents that affect fungal cell wall biosynthesis, e.g., echinocandins and nikkomycin (Steinbach et al., 2007). Proteins analogous to FKBP have been isolated from various streptomycetes, particularly in strains of *S. hygroscopicus* that produce these compounds. Bacterial FKBP analogs are inhibited by all of these immunosuppressive agents at the nanomolar level, suggesting that any protective mechanism in bacteria is based on effective secretion systems (Pahl and Keller, 1992).

17.5.2.7 Avermectin

The avermectins are a complex of chemically related 16-membered macrocyclic lactones produced by *Streptomyces avermitilis* that exhibit potent anthelmintic activity without significant antibacterial or antifungal activity (Burg et al., 1979). The avermectins are produced as a mixture of eight components from the fermentation of *S. avermitilis*, avermectin B_{1a} being (Figure 17.4) the most abundant component. Avermectins interact with high affinity and stereoselectively with nematode-specific glutamate-gated chloride channels, hyperpolarizing neuron membranes and causing paralysis and death in nematode and arthropod species (Shoop et al., 1995). Ivermectin is the 22,23-dihydro derivative of avermectin B1 and the first avermectin that was commercialized (Campbell et al., 1983). It is active at extremely low dosage against a wide variety of nematode and arthropod parasites, but has greater safety than avermectin B_1. Since then, ivermectin has been the treatment of choice for nematode and arthropod infections in cattle and pets (Ivomec), as well as to treat filariasis infections (heartworm) in dogs (Heartgard), and human onchocerciasis or river blindness (Mectizan). The complete genome sequence of *S. avermitilis* has revealed the complexity and richness of genes involved in the regulation and production of secondary metabolites. From the 25 secondary metabolite clusters that were identified from genomic analysis, many of the compounds still remain uncharacterized (Omura et al., 2001). We could expect that the production of avermectin would not escape this global regulation network that ensures the survival of the producer strain.

17.5.2.8 Streptogramins

Streptogramins represent a unique class of antibacterials that occur as pairs of structurally unrelated molecules that include pristinamycins, virginiamycins, oestreomycins, and mikamycins. Group A (or M) (M-virginiamycins and II-pristinamycins) are polyunsaturated macrolactones whereas group B (or S) (S-virginiamycins and I-pristinamycins) are cyclic hexadepsipeptides. Pristinamycin II_A (Figure 17.4) and virginiamycin M_1 are coproduced with pristinamycin I_A (Figure 17.4) by *Streptomyces pristinaespiralis* and with virginiamycin S_1 (Figure 17.4) by *Streptomyces virginiae*. They are produced as natural mixtures by different *Streptomyces* species, but none of these compounds have been found to be produced alone. Both groups of molecules alone are bacteriostatic, as they inhibit bacterial protein synthesis at the peptidyl transfer step, but their combination is bactericidal. They act synergistically in vivo against many bacterial pathogens, reducing the possibility of emergence of resistant strains. Synergy between type A and B components originates from the initial conformational changes imposed on the peptidyl transferase by type A streptogramins, which not only inhibit different stages of protein synthesis but also increase the ribosomal affinity up to 40-fold for type B streptogramins (Di Giambattista et al., 1989; Cocito et al., 1997). The synthesis of water-soluble semisynthetic derivatives of pristinamycin I_A and II_B has allowed the development of injectable formulations (RP-59500, Synercid) containing the quinupristin/dalfopristin combination used for the treatment of multidrug-resistant infections (Bonfiglio and Furneri, 2001). The clustering of the biosynthetic genes of pristinamycins I and II within a 200 Kb region of *S. pristinaespiralis* has been used as an evolutionary argument in support of the idea of a strong selective pressure in the producing strains to maintain the coproduction of streptogramins (Bamas-Jacques et al., 1999). An alternative proposal is a horizontal transfer acquisition of the second pathway and subsequent chromosomal rearrangement to facilitate coregulated production (Challis and Hopwood, 2003). In any case, the synergistic effects described above could represent a selective advantage for the producer organism against competing organisms that would ensure the maintenance of both pathways.

A similar situation where evolutionary pressure has been proposed as the driver for the coproduction of two metabolites by the same strain is the production of clavulanic acid and cefamycin C by a number of *Streptomyces* species. Interestingly, all the producers of clavulanic acid also produce cefamycin C whereas the contrary is not always true. Considering that clavulanic acid does not possess any antibacterial activity of its own, being an inhibitor of β-lactamases that potentiates the antibacterial activity of β-lactam antibiotics (such as cefamycin), it makes perfect sense from an evolutionary perspective that the production of clavulanic acid needs to be accompanied by that by a β-lactam antibiotic to provide any advantage for the producing organism (Challis and Hopwood, 2003).

17.6 Horizontally Transferred Fungal Endophytes: Exemplary Opportunities to Understand the Natural Roles of Antibiotics

Pharmaceutical lead discovery has contributed significantly to the discovery of potent biologically active molecules from fungal endophytes and the biochemistry underlying their activities. The contributions of scientists at the Swiss company Sandoz (now Novartis) to chemical basis of ergotamine and ergot alkaloid activities are well known (Giger and Engel, 2006). Starting in at least the early 1980s, horizontally transmitted fungal endophytes (Arnold, 2007) were regularly investigated as sources of secondary metabolites in pharmaceutical screening laboratories at Sandoz (Petrini and Dreyfuss, 1981; Dreyfuss and Petrini, 1984; Dreyfuss, 1986). With the incorporation of dedicated mycological expertise in other companies, e.g., Merck and Glaxo, in the late 1980s (Bills and Polishook, 1991; Wildman and Jones, 1991), plant endophytes became one of the dominant microorganisms exploited for natural products discovery during the 1990s.

The chemical protective attributes of vertically transmitted grass endophytes (Clavicipitaceae) are well studied at the organismal, ecophysiological, and genomic levels (Chapter 10, this volume). In contrast, horizontally transmitted endophytes are a polyphyletic assemblage spanning the greater part of the Ascomycetes; they are capable of producing much of the known spectrum of bioactive fungal metabolites (Gunatilaka, 2006). In spite of the sustained awareness of their chemical diversity, attempts to develop a parallel hypothetical framework explaining their chemical protective benefits to their plant hosts have been less than satisfying because it has been difficult to assess the *in planta* growth and metabolism of cryptic endophyte infections. Many arguments are based on correlative inferences centered on the fact that in the laboratory, these fungi can make a range of antagonistic, toxic, or exquisite mode-of-action-specific metabolites, and therefore it is argued that these metabolites must exert a protective effect against herbivores and pathogens of the host. Experiments demonstrating the production of the substances in the living or senescent plant are rare, and that fungal metabolites are responsible for herbivore or pathogen exclusion *in planta* are even less frequent (Miller et al., 2002; Weber et al., 2004; Wicklow et al., 2005; Lardner et al., 2006). We are aware of at least two cases where endophytes have been manipulated to provide chemical deterrents to herbivores or pathogens of the host, thus demonstrating that these kinds of experiments are possible (Sumarah et al., 2005; Wicklow et al., 2005). At some stage in their life history, these fungi may assume a pathogenic or necrotrophic mode of nutrition, so in many cases, these toxic metabolites are likely to be under the same adaptive influences as in their plant pathogenic or saprobic counterparts. However, more satisfying explanations of the protective potential of horizontally transmitted endophytes have been based on experiments demonstrating that endophyte infections induce or increase the expression of intrinsic host defense mechanisms (Arnold, 2007).

Below, we point out several outstanding endophyte metabolites with potent mode-of-action-based activities that may contribute significantly to the endophyte–host relationship. Some of these fungi are quite widespread among many plants. Like the situations described above for medically and historically significant metabolites, e.g., the statins, β-lactams, and actinomycin, the tools and experimental systems are available to investigate the roles of these significant endophyte metabolites in simulated and natural systems.

17.6.1 Echinocandins and Pneumocandins

One of the main targets for antifungal therapy under investigation in the last few decades has been the fungal cell wall. β-glucan, a polysaccharide composed of glucose monomers linked by (1,3)-β or (1,6)-β bonds, is widespread among fungi and forms an essential amorphous, insoluble component of the cell wall, guaranteeing many of its physical properties (Latge, 2007). Antifungal metabolites within the echinocandin family target the cell wall, specifically inhibiting β-glucan synthesis, and have inherently selective activity in vitro as well as in vivo in different animal models. Much of their basic biology and early history of the development of echinocandins and pneumocandins as clinical candidates have been extensively reviewed (Bartizal et al., 1993; Kurtz and Douglas, 1997). These molecules have been successfully used to develop several antifungal agents currently approved to treat serious fungal infections by *Candida* and *Aspergillus* species: CANCIDAS (caspofungin acetate), MYCAMINE (micafungin), and ERAXIS (anidulafungin) (Barrett, 2002; Denning, 2003; Vicente et al., 2003; de la Torre and Reboli, 2007).

The echinocandins are cyclic hexapeptides with a nonpolar amide-linked acyl side chain. The echinocandin A (Figure 17.5) and other echinocandins were originally discovered from the strains of *Aspergillus* species in the early 1970s by several pharmaceutical discovery laboratories (Turner and Current, 1997). Other members of the class, showing different substitution patterns on the hexapeptide ring or in the side chain, were subsequently discovered from other fungi during the following decades (Turner and Current, 1997; Hino et al., 2001; Vicente et al., 2003). At present, the number of known naturally produced members of the echinocandin class is close to 30.

Pneumocandin A_0 (Figure 17.5) was originally discovered from *Glarea lozoyensis*, being the most abundant compound among the several pneumocandins produced by this fungus (Schwartz et al., 1989, 1992). A minor component from *G. lozoyensis* fermentations, pneumocandin B_0, was used as the starting point for the medicinal chemistry efforts leading to the development of CANCIDAS (Bartizal et al., 1993; Conners and Pollard, 2005). The original pneumocandin-producing strain was isolated from a water sample from Spain (Bills et al., 1999), and the ecological role of this species remained uncertain. We have recently found additional pneumocandin A_0-producing isolates of *G. lozoyensis* from plant litter collected in southern Argentina and New Hampshire (unpublished results).

Pneumocandin A_0 and other pneumocandins are commonly found among the endophytic isolates from woody plants because they are consistently produced by certain species of *Pezicula* and their anamorphs, *Cryptosporiopsis* (Dreyfuss, 1986; Noble et al., 1991). In our own laboratory, we have observed that the production of pneumocandins is widespread across many *Pezicula* species and their anamorphic *Cryptosporiopsis* states (unpublished results). Cryptocandin, a similar member of the echinocandin class, was described from a strain of *Cryptosporiopsis* cf. *quercina*, isolated as an endophyte of *Tripterigeum wilfordii* stems (Strobel et al., 1999). Other related compounds, e.g., WF11899A (Figure 17.5), that have been reported from fungal endophytes belong to a subfamily of echinocandins characterized by the

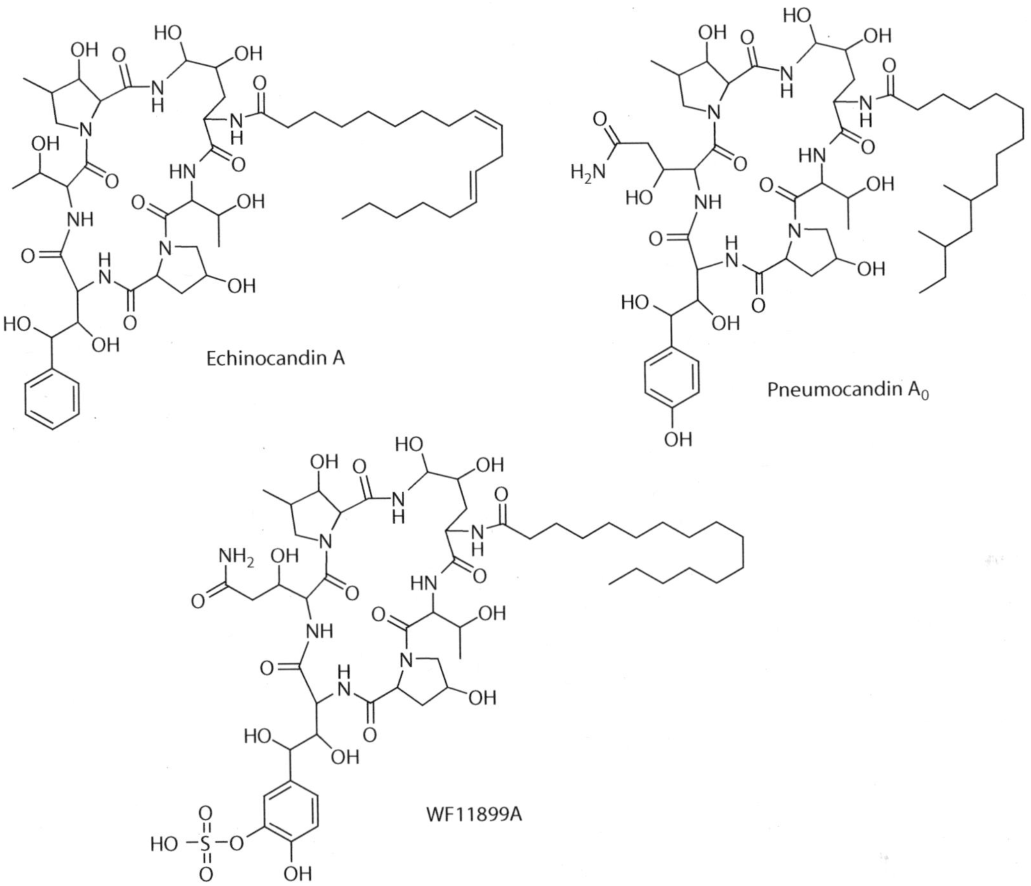

FIGURE 17.5 Structures of echinocandin A and pneumocandins.

presence of a sulphate group at the *para* or *meta* position of the homotyrosine in the hexapeptide ring. The sulphated echinocandins have been investigated extensively by researchers at Fujisawa Pharmaceuticals (now Astellas), who used WF11899A as the starting point for the commercial antifungal MYCAMINE. Several compounds in this category have been reported from the strains of *Coleophoma empetri* and *C. crateriformis* isolated from plant material (Hino et al., 2001; Kanasaki et al., 2006a,b). The specificity of the pneumocandins and echinocandins for fungal 1,3-β-D-glucan synthase along with their widespread distribution among many types of filamentous ascomycetes points to their natural role in antifungal-mediated antagonism of other fungi and yeasts. However, experimental tests of the hypothesis remain to be carried out.

17.6.2 Enfumafungin, Arundifungin, and Other Acidic Triterpenes

The success of the lipopeptide class of β-glucan synthesis inhibitors has prompted interest in the search for other structural types with improved features over the echinocandins (especially because of their lack of oral absorption). However, to date, besides cyclic lipopeptides (echinocandins and other molecules), only two other types of fungal-derived β-glucan synthesis inhibitors are known essentially: the papulacandins and the acidic triterpenes. The papulacandins were discovered in the late 1970s and since then, a series of structurally related glycolipids have been discovered over the years; however, neither papulacandins nor any of their relatives have been developed as drugs (Schwartz, 2001; Barrett, 2002). The most recently discovered class of β-glucan synthesis inhibitors are triterpenes, containing a polar moiety (Onishi et al., 2000). This polar moiety can be a glycoside as in enfumafungin and ascosteroside, a succinate as in arundifungin, or a sulfate-derivative amino acid as in ergokonin A. Although these compounds were inactive or only weakly active in vivo in the mouse models (Vicente et al., 2003), they could represent a new paradigm in the search for antifungal drugs that act on the fungal cell wall.

Some of these triterpenic compounds are commonly found among the endophytic fungi. A good example is enfumafungin (Figure 17.6), originally isolated from a strain identified as *Hormonema* sp. (Peláez et al., 2000). Subsequently, the fungus was recovered from various substrata, including living and decaying leaves of *Juniperus* spp. and other plants, and even from rock surfaces, and it was named

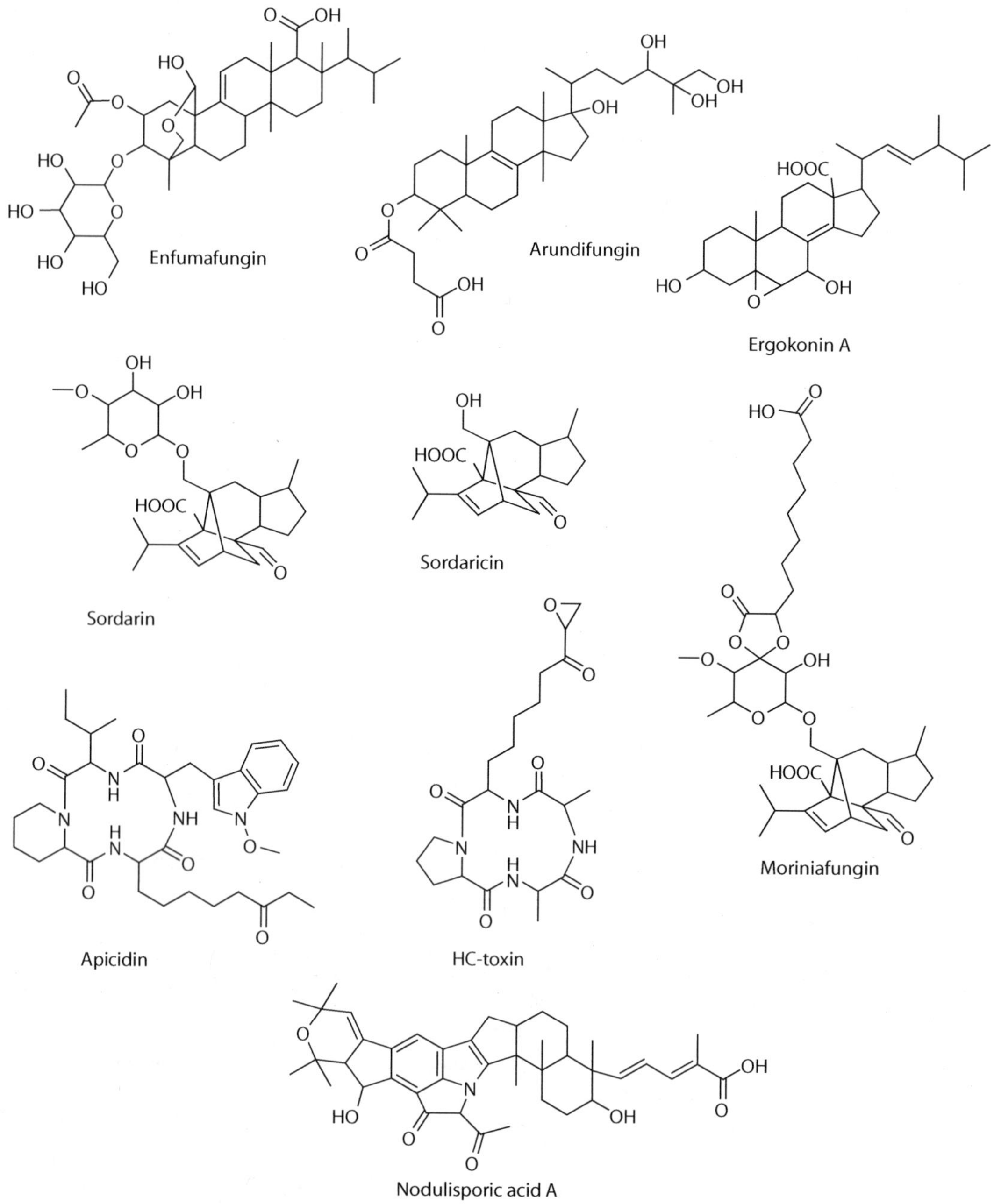

FIGURE 17.6 Structures of bioactive metabolites from fungal endophytes.

Hormonema carpetanum (Bills et al., 2004). Despite intensive screening efforts across tens of thousands of strains, we have yet to find enfumafungin produced by any other fungal species. In contrast, another member of this triterpene class, arundifungin (Figure 17.6), is distributed widely; it is not only found in the common endophyte, *Arthrinium arundis* and other *Arthrinium* species, but also in different fungi around the globe. For example, we found arundifungin produced by a *Selenophoma*-like coelomycete recovered from the twigs of *Olea europaea* and a putative member of the *Sclerotiniaceae* from the leaves of *Quercus ilex*, as well as from fungi isolated from soil and plant litter in the tropical, temperate, and extremely cold environments (Cabello et al., 2001; Weber et al., 2007).

17.6.3 Sordarin and Sordarin Analogs

Sordarin (Figure 17.6), a diterpene glycoside, and its aglycon, sordaricin (Figure 17.6), were originally isolated from the terrestrial ascomycete *Sordaria araneosa* as antifungal compounds lacking antibacterial activity. Their discovery and biology have been reviewed extensively (Odds, 2001; Domínguez and Martín, 2005). Fungal protein biosynthesis has been identified as the target of sordarins where the compounds block the interaction of elongation factor 2 (EF2) at the ribosomal P-protein stalk (the region of the large ribosomal subunit where soluble factors interact during translation) (Gomez-Lorenzo and García-Bustos, 1998; Justice et al., 1998). Although the protein EF2 is conserved across all the eukaryotic organisms, sodarins exclusively inhibit fungal protein synthesis and do not affect mammalian or plant cells.

Sordarins are produced by fungi from among the Xylariales, Microascales, and Sordariales from different habitats, including endophytic isolates. Extensive work on the biology of the sordarins by Glaxo Wellcome was originally focused on the production of sordarins in fermentations of *Graphium putredinis* (Kennedy et al., 1998). At Merck, work of sordarin production originally focused on *Rosellinia subiculata*, a fungus isolated from decayed wood and that can cause endophytic infections of plants (Bills et al., 2002). We have also observed sordarin from *Zopfiella marina* CBS 155.77 and *Eutypa tetragona* CBS 284.87 (unpublished data). Sordarin and the related derivative, xylarin, have been isolated from *Xylaria longipes* (Schneider et al., 1995). There is at least one report of a sordarin-like metabolite, BE-31405, from a *Penicillium* species (Okada et al., 1999). Zofimarin, from the marine ascomycete *Zopfiella marina*, is a related analog (Tanaka et al., 2002). Hypoxysordarin from the mangrove ascomycete, *Hypoxylon croceum*, differs from sordarin in the substitution with an unusual side chain (Daferner et al., 1999). Unlike sordarin, which has little efficacy against filamentous fungi, hypoxysordarin exhibits more potent antifungal activities toward several filamentous fungi, e.g., *Absidia glauca*, *Mucor miehei*, *Paecilomyces variotii*, *Penicillium notatum*, and *Penicillium islandicum* in vivo. Another recently discovered member of the family is moriniafungin (Figure 17.6) from an endophytic isolate of *Morinia pestalozzioides* (Amphisphaeriaceae). Moriniafungin is one of the most potent natural sordarins found to date and shows a much broader antifungal spectrum (Basilio et al., 2006; Collado et al., 2006). It is unique because of its 2-hydroxysebacic acid residue linked to C-30 of the sordarose residue of sordarin through a 1,3-dioxolan-4-one ring and is the first example of a fungal natural product containing a 1,3-dioxolan-4-one ring. The plant triterpenoid glycosides spilacleosides A and B isolated from *Ruscus aculeatus* are the only other natural products known to contain this ring system.

Some limited experimentation with a sordarin-producing coprophilous fungus, *Podospora pleiospora*, demonstrated that sordarin, its aglycone sordaricin, and sordarin analogs were produced in a simulated natural environment of autoclaved rabbit dung (Weber et al., 2005). Consistent with its observed in vitro activity, sordarins produced in simulated dung habitats exhibited activity toward the yeasts, *Nematospora coryli* and *Sporobolomyces roseus*, but not against the filamentous fungus, *Penicillium claviforme*. Sordarin production by *P. pleiospora* was unaffected by the presence of other fungi when grown in liquid medium cocultures.

17.6.4 Apicidin, HC-Toxin, and Other Tetrapeptide HDAC Inhibitors

The cyclic tetrapeptide apicidin (Figure 17.6), produced by endophytic *Fusarium* species, was identified as a broad-spectrum in vitro and in vivo agent effective against a range of apicomplexan parasites, including *Plasmodium berghei* malaria in mice (Darkin-Rattray et al., 1996; Singh et al., 1996; Meinke and Liberator, 2001; Singh et al., 2002). Apicidin and related fungal tetrapetides, e.g., trapatoxin, HC-toxin, and chlamydocin, are also potent inhibitors of other protozoans, e.g., the poultry parasite, *Eimeria tenella*, and affect the HDACs of many eukaryotic organisms (Darkin-Rattray et al., 1996). Unlike apicidin, similar cyclic peptides such as trapoxins (Kijima et al., 1993; Taunton et al., 1996), chlamydocin, HC-toxin (Walton et al., 1985; Walton, 2006), and others contain an epoxyketone moiety. These compounds produce the irreversible inhibition of HDAC at nanomolar concentrations by covalently binding to the enzyme through the epoxide group (Kijima et al., 1993; Masuoka et al., 2000). The apicidins produced by plant-inhabiting Fusaria may function as a phytotoxin in their host plants in a manner similar to the host-specific plant toxin, HC-toxin (Figure 17.6), produced by *Cochliobolus carbonum* (Brosch et al., 1995; Walton, 2006).

Besides its potential as a target for an antiparasitic agent, inhibition of HDAC has emerged as a potential strategy in human cancer therapy because reversible histone acetylation–deacetylation plays a fundamental role in regulating transcription and chromatin assembly. Inhibitors of HDAC are known to have detransforming activity, i.e., they are able to induce morphological reversion from oncogen-transformed to normal cells (Kim et al., 2004; Gallinari et al., 2007; Li et al., 2007). A plethora of cultured tumor cell lines have been shown to be susceptible to HDAC inhibitors, and studies using rodent models for cancer have shown that these compounds reduce the growth of tumors and metastases in vivo; in some cases, without noticeable side effects (Krämer et al., 2001). Inhibition of HDAC has been shown to result in upregulation of some genes and downregulation of others. The synthesis of the gene p21^{WAF1} (an inhibitor of cyclin-dependent kinases) is the effect most consistently observed in cells treated with HDAC, and its expression could explain the G1 arrest and G2/M block frequently observed in treated cells, though other factors may be involved in the mode-of-action of these compounds (Krämer et al., 2001). In addition to their antitumor effects, tetrapeptide inhibitors of HDAC are inhibitors of angiogenesis (the formation of new blood vessels) in carcinoma mouse models, upregulating p53 and other tumor suppressor genes and reducing the expression of angiogenic factors such as VEGF. Their antiangiogenic properties have led to their investigation as anticancer agents (Kim et al., 2004). The first HDAC inhibitor reaching the market for the treatment of cutaneous T-cell lymphoma has been vorinostat (Zolinza), also known as SAHA (suberoyl anilide hydroxamic acid), a synthetic analog of the *Streptomyces hygroscopicus* metabolite trichostatin A.

17.6.5 Nodulisporic Acids

Nodulisporic acids were first isolated from fermentation extracts of an endophytic fungus, *Nodulisporium* sp., MF 5954, ATCC 74245, and recovered from a dried specimen of *Bontia daphnoides* collected in Hawaii. Nodulisporic acids are novel indole diterpenes that exhibit potent insecticidal properties and were discovered by screening for activity against the larvae of the blowfly *Lucilia sericata* (Ondeyka et al., 1997). Nodulisporic acid not only kills blowflies but also mosquitoes (*Aedes aegypti*) and fruit flies (*Drosophila melanogaster*) (Ostlind et al., 1997). In the blowfly and mosquito assays, nodulisporic acid was more potent than paraherquamide, malathion, and DDT but less potent than ivermectin (Ostlind et al., 1997). Nodulisporic acid A (Figure 17.6) was about 10-fold more potent than ivermectin in a flea model. Electrophysiological recordings of grasshopper neurones and radioligand-binding studies on membrane preparations from *D. melanogaster* demonstrated that nodulisporic acids modulate insect glutamate-gated chloride channels in a similar fashion to ivermectin (Kane et al., 2000; Smith et al., 2000; Meinke et al., 2002). In vivo studies in dogs have shown that nodulisporic acid A exhibited potent systemic efficacy with no adverse effects. Derivatives of these natural products with improved properties have been also reported (Meinke et al., 2002).

Nodulisporic acids were produced by multiple isolates of a species of the genus *Nodulisporium*, associated with different types of plants and plant-derived materials, and morphologically similar to the anamorphs of some *Hypoxylon* species. Therefore, it was thought that fungus has a widespread pantropical distribution (Polishook et al., 2000). Recently, this undescribed *Nodulisporium* sp. was hypothesized to represent a unique phylogenetic and chemotypical lineage within the Hypoxyloideae (Stadler and Hellwig, 2005). Nodulisporic acids were the first indole diterpenes, compounds generally associated with the Eurotiales and the Clavicipitaceae, to be found in a fungus belonging to the Xylariaceae.

An insecticidal metabolite that is produced by a widespread tropical endophyte with these extraordinary potency levels and a specific mode-of-action biased toward insects would provide an ideal situation for testing the popular idea that horizontally transmitted endophytes are capable of chemically mediated protection of their hosts from herbivory.

ACKNOWLEDGMENTS

We thank various colleagues for making comments on the earlier versions of this paper. The cocultivation prototype was built in collaboration with Pignat, S.A., Genas, France.

REFERENCES

Alberts, A. W., Chen, J., Kuron, G., Hunt, V., Huff, J., Hoffman, C., Rothrock, J., et al. 1980. Mevinolin—a highly potent competitive inhibitor of hydroxymethylglutaryl-coenzyme-A reductase and a cholesterol-lowering agent. *Proceedings of the National Academy of Sciences, USA* 77:3957–3961.

Aldred, D., Magan, N., and B. S. Lane. 1999. Influence of water activity and nutrients on growth and production of squalestatin S1 by a *Phoma* sp. *Journal of Applied Microbiology* 87:842–848.

Aminov, R. I. and R. I. Mackie. 2007. Evolution and ecology of antibiotic resistance genes. *FEMS Microbiology Letters* 271:147–161.

Anke, T. 1995. The antifungal strobilurins and their possible ecological role. *Canadian Journal of Botany* 73:S940–S945.

Anke, T. and W. Steglich. 1999. Strobilurins and oudemansins. In *Drug Discovery from Nature*, Eds. S. Grabley and R. Thiericke, pp. 320–334. Berlin, Germany: Springer-Verlag.

Ariyo, B., Tamerler, C., Bucke, C., and T. Keshavarz. 1998. Enhanced penicillin production by oligosaccharides from batch cultures of *Penicillium chrysogenum* in stirred-tank reactors. *FEMS Microbiology Letters* 166:165–170.

Ariyo, B. T., Bucke, C., and T. Keshavarz. 1997. Alginate oligosaccharides as enhancers of penicillin production in cultures of *Penicillium chrysogenum*. *Biotechnology and Bioengineering* 53:17–20.

Arnold, A. E. 2007. Understanding the diversity of foliar endophytic fungi: Progress, challenges, and frontiers. *Fungal Biology Reviews* 21:51–66.

Arvas, M., Kivioja, T., Mitchell, A., Saloheimo, M., Ussery, D., Penttila, M., and S. Oliver. 2007. Comparison of protein coding gene contents of the fungal phyla Pezizomycotina and Saccharomycotina. *BMC Genomics* 8, DOI: 10.1186/1471-2164-1188-1325.

Bach, T. J. and H. K. Lichtenthaler. 1983. Inhibition by mevinolin of plant growth, sterol formation and pigment accumulation. *Physiologia Plantarum* 59:50–60.

Bamas-Jacques, N., Lorenzon, S., Lacroix, P., de Swetschin, C., and J. Crouzet. 1999. Cluster organization of the genes of *Streptomyces pristinaespiralis* involved in pristinamycin biosynthesis and resistance elucidated by pulsed-field gel electrophoresis. *Journal of Applied Microbiology* 87:939–948.

Barrett, D. 2002. From natural products to clinically useful antifungals. *Biochimica et Biophysica Acta-Molecular Basis of Disease* 1587:224–233.

Barrios-Gonzales, J. and A. Mejia. 1996. Production of secondary metabolites by solid state fermentation. *Biotechnology Annual Review* 2:85–121.

Bartizal, K., Abruzzo, G. K., and D. M. Schmatz. 1993. Biological activity of the pneumocandins. In *Cutaneous Antifungal Agents: Selected Compounds in Clinical Practice*, Eds. J. W. Rippon and R. A. Fromtling, pp. 421–458. New York: Marcel Dekker.

Basilio, A., Justice, M., Harris, G., Bills, G., Collado, J., Cruz, M.d.l., Díez, M. T., et al. 2006. The discovery of moriniafungin, a novel sordarin derivative produced by *Morinia pestalozzioides*. *Biorganic & Medicinal Chemistry* 14:560–566.

Bentley, R. 2000. Mycophenolic acid: A one hundred year odyssey from antibiotic to immunosuppressant. *Chemical Reviews* 100:3801–3825.

Bergmann, S., Schumann, J., Scherlach, K., Lange, C., Brakhage, A. A., and C. Hertweck. 2007. Genomics-driven discovery of PKS-NRPS hybrid metabolites from *Aspergillus nidulans*. *Nature Chemical Biology* 3:213–217.

Bigelis, R., He, H., Yang, H. Y., Chang, L.-P., and M. Greenstein. 2006. Production of antifungal antibiotics using polymeric solid supports in solid-state and liquid fermentation. *Journal of Industrial Microbiology and Biotechnology* 33:815–826.

Bills, G., Platas, G., Fillola, A., Jiménez, M. R., Collado, J., Vicente, F., Martín, J., et al. 2008. Enhancement of antibiotic and secondary metabolite detection from filamentous fungi by growth on nutritional arrays. *Journal of Applied Microbiology* 104:1644–1658.

Bills, G. F. and J. D. Polishook. 1991. Microfungi from *Carpinus caroliniana. Canadian Journal of Botany* 69:1477–1148.

Bills, G. F., Platas, G., Peláez, F., and P. Masurekar. 1999. Reclassification of a pneumocandin-producing anamorph, *Glarea lozoyensis* gen. et sp. nov., previously identified as *Zalerion arboricola. Mycological Research* 103:179–192.

Bills, G. F., Dombrowski, A. W., Horn, W. S., Jansson, R. K., Rattray, M., Schmatz, D., and R. E. Schwartz. 2002. *Rosellinia subiculata* ATCC 74386 and fungus ATCC 74387 for producing sordarin compounds for fungi control. U.S. Patent 6,436,395.

Bills, G. F., Collado, J., Ruibal, C., Peláez, F., and G. Platas. 2004. *Hormonema carpetanum* sp. nov., a new lineage of dothideaceous black yeasts from Spain. *Studies in Mycology* 50:149–157.

Bird, B. A. and I. M. Campbell. 1982. Disposition of mycophenolic acid, brevianamide A, asperphenamate, and ergosterol in solid cultures of *Penicillium brevicompactum. Applied and Environmental Microbiology* 43:345–348.

Boakes, S., Oliynyk, M., Cortés, J., Böhm, I., Rudd, B. A. M., Revill, W. P., Staunton, J., and P. F. Leadlay. 2004. A new modular polyketide synthase in the erythromycin producer *Saccharopolyspora erythraea. Journal of Molecular Microbiology and Biotechnology* 8:73–80.

Boddy, L. 2000. Interspecific combative interactions between wood-decaying basidiomycetes. *FEMS Microbiology Ecology* 31:185–194.

Bode, H. B., Bethe, B., Höfs, R., and A. Zeek. 2002. Big effects from small changes: Possible ways to explore nature's chemical diversity. *ChemBioChem* 3:619–627.

Bok, J. W. and N. P. Keller. 2004. LaeA, a regulator of secondary metabolism in *Aspergillus* spp. *Eukaryotic Cell* 3:527–535.

Bok, J. W., Noordermeer, D., Kale, S. P., and N. P. Keller. 2006. Secondary metabolic gene cluster silencing in *Aspergillus nidulans. Molecular Microbiology* 61:1636–1645.

Bolard, J. 1986. How do polyene macrolide antibiotics affect cellular membrane properties. *Biochimica et Biophysica Acta* 864:257–304.

Bonfiglio, G. and P. M. Furneri. 2001. Novel streptogramin antibiotics. *Expert Opinion on Investigational Drugs* 10:185–198.

Brakhage, A. A. 1998. Molecular regulation of β-lactam biosynthesis in filamentous fungi. *Microbiology and Molecular Biology Reviews* 62:547–585.

Brosch, G., Ramsom, R., Lechner, T., Walton, J. D., and P. Loidl. 1995. Inhibition of maize histone deacetylases by HC-toxin, the host-selective toxin of *Cochliobolus carbonum. Plant Cell* 7:1941–1950.

Brown, A. G., Smale, T. C., King, T. J., Hasenkamp, R., and R. H. Thompson. 1976. Crystal and molecular structure of compactin, a new antifungal metabolite from *Penicillium brevicompactum. Journal of the Chemical Society, Perkin Transactions* 1:1165–1170.

Buckingham, J. 2007. *Dictionary of Natural Products on CD-Rom.* London, U.K.: Chapman & Hall/CRC Press.

Burg, R. W., Miller, B. M., Baker, E. E., Birnbaum, J., Currie, S. A., Hartman, R., Kong, Y.-L., et al. 1979. Avermectins, new family of potent anthelmintic agents: Producing organism and fermentation. *Antimicrobial Agents and Chemotherapy* 15:361–367.

Burgess, J. G., Jordan, E. M., Bregu, M., Mearns-Spragg, A., and K. G. Boyd. 1999. Microbial antagonism: A neglected avenue of natural products research. *Journal of Biotechnology* 70:27–32.

Cabello, M. A., Platas, G., Collado, J., Díez, M. T., Martín, I., Vicente, F., Meinz, M., et al. 2001. Arundifungin, a novel antifungal compound produced by fungi: Biological activity and taxonomy of the producing organisms. *International Microbiology* 4:93–102.

Campbell, W. C., Fisher, M. H., Stapley, E. O., Albers-Schonberg, G., and T. A. Jacob. 1983. Ivermectin: A potent new antiparasitic agent. *Science* 221:823–828.

Casals, N., Buesa, C., Piulachs, M. D., Cabañó, J., Marrero, P. F., Bellés, X., and F. G. Hegardt. 1996. Coordinated expression and activity of 3-hydroxy-3-methylglutaryl coenzyme A synthase and reductase in the fat body of *Blattella germanica* L. during vitellogenesis. *Insect Biochemistry and Molecular Biology* 26:837–843.

Casals, N., Martín, D., Buesa, C., Piulachs, M. D., Hegardt, F., and X. Bellés. 1997. Expression and activity of 3-hydroxy-3-methylglutaryl-CoA synthase and reductase in the fat body of ovariectomized and allatectomized *Blattella germanica. Physiological Entomology* 22:6–12.

Challis, G. L. and D. A. Hopwood. 2003. Synergy and contingency as driving forces for the evolution of multiple secondary metabolite production by *Streptomyces* species. *Proceedings of the National Academy of Sciences, USA* 100:14555–14561.

Chappell, J., Wolf, F., Proulx, J., Cuellar, R., and C. Saunders. 1995. Is the reaction catalyzed by 3-hydroxy-3-methylglutaryl coenzyme-a reductase a rate-limiting step for isoprenoid biosynthesis in plants. *Plant Physiology* 109:1337–1343.

Chen, G. H., Wang, G. Y. S., Li, X., Waters, B., and J. Davies. 2000. Enhanced production of microbial metabolites in the presence of dimethyl sulfoxide. *Journal of Antibiotics* 53:1145–1153.

Choi, S.-U., Lee, C.-K., Hwang, Y.-I., Kinosita, H., Lee, C.-K., Hwang, Y.-I., Kinosita, H., and T. Nihira. 2003. Gamma-butyrolactone autoregulators and receptor proteins in non-*Streptomyces* actinomycetes producing commercially important secondary metabolites. *Archives of Microbiology* 180:303–307.

Christian, O. E., Compton, J., Christian, K. R., Mooberry, S. L., Valeriote, F. A. and P. Crews. 2005. Using Jasplakinolide to turn on pathways that enable the isolation of new chaetoglobosins from phomospis asparagi. *Journal of Natural Products* 68:1592–1597.

Clardy, J., Fischbach, M. A., and C. T. Walsh. 2006. New antibiotics from bacterial natural products. *Nature Biotechnology* 24:1541–1550.

Clark, B. R., Lacey, E., Gill, J. H., and R. J. Capon. 2007. The effect of halide salts on the production of *Gymnoascus reessii* polyenylpyrroles. *Journal of Natural Products* 70:665–667.

Cocito, C., Di Giambattista, M., Nyssen, E., and P. Vannuffel. 1997. Inhibition of protein synthesis by streptogramins and related antibiotics. *Journal Antimicrobial Chemotherapy* 39(Suppl.1):7–13.

Collado, J., Platas, G., Bills, G. F., Basilio, Á., Vicente, F., Tormo, J. R., Hernández, P., Díez, M. T., and F. Peláez. 2006. Studies on *Morinia*. Recognition of *Morinia longiappendiculata* sp. nov. as a new endophytic fungus, and a new circumscription of *Morinia pestalozzioides*. *Mycologia* 98:616–627.

Conners, N. and D. Pollard. 2005. Pneumocandin B0 produced by fermentation of the fungus *Glarea lozoyensis*: Physiological and engineering factors affecting titer and structural analouge formation. In *Handbook of Industrial Mycology*, Ed. Z. An, pp. 515–538. New York: Marcel Dekker.

Coyle, C. M., Kenaley, S. C., Rittenour, W. R., and D. G. Panaccione. 2007. Association of ergot alkaloids with conidiation with *Aspergillus fumigatus*. *Mycologia* 99:804–811.

Cueto, M., Jensen, P. R., Kauffman, C., Fenical, W., Lobkovsky, E., and J. Clardy. 2001. Pestalone, a new antibiotic produced by a marine fungus in response to bacterial challenge. *Journal of Natural Products* 64:1444–1446.

Daferner, M., Mensch, S., Anke, T., and O. Sterner. 1999. Hypoxysordarin, a new sordarin derivative from *Hypoxylon croceum*. *Zeitschrift für Naturforschung* 54c:474–480.

Dantas, G., Sommer, M. O. A., Oluwasegun, R. D., and G. M. Church. 2008. Bacteria subsisting on antibiotics. *Science* 320:100–103.

Darkin-Rattray, S. J., Gurnett, A. M., Myers, R. M., Dulski, P. M., Crumley, T. M., Alloco, J. J., Cannova, C., et al. 1996. Apicidin: A novel antiprotozoal agent that inhibits parasite histone deacetylase. *Proceedings of the National Academy of Sciences, USA* 93:13143–13147.

Davies, J. 1990. What are antibiotics? Archaic functions for modern activities. *Molecular Microbiology* 4:1227–1231.

Davies, J., Spiegelman, G. B., and G. Yim. 2006. The world of subinhibitory antibiotic concentrations. *Current Opinion in Microbiology* 9:445–453.

de la Torre, P. and A. C. Reboli. 2007. Anidulafungin: A new echinocandin for candidal infections. *Expert Review of Anti-infective Therapy* 5:45–52.

Decousser, J. W., Bourgeois-Nicoloos, N., and F. Doucet-Populaire. 2007. Dalbavancin, a long-acting lipoglycopeptide for the treatment of multidrug-resistant Gram-positive bacteria. *Expert Review of Anti-Infective Therapy* 5:557–571.

Degenkolb, T., Heinze, S., Schlegel, B., Strobel, G., and U. Grafe. 2002. Formation of new lipoaminopeptides, acremostatins A, B, and C, by co-cultivation of *Acremonium* sp Tbp-5 and *Mycogone rosea* DSM 12973. *Bioscience Biotechnology and Biochemistry* 66:883–886.

Degenkolb, T., Berg, A., Gams, W., Schlegel, B., and U. Grafe. 2003. The occurrence of peptaibols and structurally related peptaibiotics in fungi and their mass spectrometric identification via diagnostic fragment ions. *Journal of Peptide Science* 9:666–678.

Degenkolb, T., Kirschbaum, J., and H. Bruckner. 2007. New sequences, constituents, and producers of peptaibiotics: An updated review. *Chemistry & Biodiversity* 4:1052–1067.

Dekkers, K. L., You, B.-J., Gowda, V. S., Liao, H.-L., Lee, M.-H., Bau, H.-H., Ueng, P. P., and K. R. Chung. 2007. The *Cercospora nicotianae* gene encoding dual *O*-methyltransferase and FAD-dependent monooxygenase domains mediates cercosporin toxin biosynthesis. *Fungal Genetics and Biology* 44:444–454.

Demain, A. L. and A. Fang. 2000. The natural functions of secondary metabolites. *Advances in Biochemical Engineering/Biotechnology* 69:1–39.

Denning, D. W. 2003. Echinocandin antifungal drugs. *Lancet* 362:1142–1151.

Di Giambattista, M., Chinali, G., and C. Cocito. 1989. The molecular basis of the inhibitory activities of type A and B synergimycins and related antibiotics on ribosomes. *Journal Antimicrobial Chemotherapy* 24:485–507.

Domínguez, J. M. and J. J. Martín. 2005. Sordarins: Inhibitors of fungal elongation factor-2. In *Handbook of Industrial Mycology*, Ed. Z. An, pp. 335–353. New York: Marcel Dekker.

Dreyfuss, M. M. 1986. Neue Erkenntnisse aus einem pharmakologischen Pilz-screening. *Sydowia* 39:22–36.

Dreyfuss, M. M. and O. Petrini. 1984. Further investigations on the occurrence and distribution of endophytic fungi in tropical plants. *Botanica Helvetica* 94:33–40.

Dudler, R. and L. Eberl. 2006. Interactions between bacteria and eukaryotes via small molecules. *Current Opinion in Biotechnology* 17:268–273.

Dumont, F. J. and Q. Su. 1995. Mechanism of action of the immunosuppressant rapamycin. *Life Sciences* 58:373–395.

Dumont, F. J., Melino, M. R., Staruch, M. J., Koprak, S. L., Fischer, P. A., and N. H. Sigal. 1990. The immunosuppressive macrolides FK-506 and rapamycin act as reciprocal antagonists in murine T cells. *Journal of Immunology* 144:1418–1424.

Ellis, D. 2002. Amphotericin B: Spectrum and resistance. *Journal of Antimicrobial Chemotherapy* 49:7–10.

Endo, A. 1979. Monacolin K, a new hypocholesterolemic agent produced by a *Monascus* species. *Journal of Antibiotics* 32:852–854.

Endo, A. 1985. Compactin ML-236B and related compounds as potential cholesterol-lowering agents that inhibit HMG-CoA reductase. *Journal of Medicinal Chemistry* 28:401–405.

Endo, A., Kuroda, M., and Y. Tsujita. 1976. ML-236A, ML-236B, and ML-236C, new inhibitors of cholesterogenesis produced by *Penicillium citrinum. Journal of Antibiotics* 29:1346–1348.

Engler, M., Anke, T., and O. Sterner. 1998. Production of antibiotics by *Collybia nivalis*, *Omphalotus olearius*, a *Favolaschia* and a *Pterula* species on natural substrates. *Zeitschrift für Naturforschung* 53:318–324.

Espada, A. and M. M. Dreyfuss. 1997. Effect of the cyclopeptolide 90–215 on the production of destruxins and helvolic acid by *Metarhizium anisopliae. Journal of Industrial Microbiology and Biotechnology* 19:7–11.

Feyereisen, R. and D. E. Farnsworth. 1987. Characterization and regulation of HMG-CoA reductase during a cycle of juvenile hormone synthesis. *Molecular and Cellular Endocrinology* 53:227–238.

Firn, R. D. and C. G. Jones. 2003. Natural products—a simple model to explain chemical diversity. *Natural Product Reports* 20:382–391.

Fischbach, M. A. and J. Clardy. 2007. One pathway, many products. *Nature Chemical Biology* 3:353–355.

Florey, H. W., Chain, E., Heatley, N. G., Jennings, M. A., Sanders, A. G., Abraham, E. P., and M. E. Florey. 1949. *Antibiotics. A Survey of Penicillin, Streptomycin and Other Antimicrobial Substances from Fungi, Actinomycetes, Bacteria, and Plants. Vol. I.* London: Oxford University Press.

Galgoczy, L., Papp, T., Lukacs, G., Leiter, E., Pocsi, I., and C. Vagvolgyi. 2007. Interactions between statins and *Penicillium chrysogenum* antifungal protein PAF. To inhibit the germination of sporangiospores of different sensitive Zygomycetes. *FEMS Microbiology Letters* 270:109–115.

Gallinari, P., Di Marco, S., Jones, P., Pallaoro, M., and C. Steinkuhler. 2007. HDACs, histone deacetylation and gene transcription: From molecular biology to cancer therapeutics. *Cell Research* 17:195–211.

Giger, R. K. A. and G. Engel. 2006. Albert Hofmann's pioneering work on ergot alkaloids and its impact on the search of novel drugs at Sandoz, a predecessor company of Novartis—Dedicated to Dr. Albert Hofmann on the occasion of his 100th birthday. *Chimia* 60:83–87.

Gloer, J. B. 2007. Applications of fungal ecology in the search for new bioactive natural products. In *Environmental and Microbial Relationships,* 2nd ed. *The Mycota IV*, Eds. C. P. Kubicek and I. S. Druzhinina, pp. 257–283. Berlin Heidelberg: Springer-Verlag.

Goldberg, I. H., Rabinowitz, M., and E. Reich. 1962. Basis of actinomycin action. I. DNA binding and inhibition of RNA-polymerase synthetic reactions by actinomycin. *Proceedings of the National Academy of Sciences, USA* 48:2094–2101.

Gomez-Lorenzo, M. G. and J. F. Garcia-Bustos. 1998. Ribosomal P-protein stalk function is targeted by sordarin antifungals. *Journal of Biological Chemistry* 273:25041–25044.

Griffith, G. S., Rayner, A. D. M., and H. G. Wildman. 1994a. Interspecific interactions and mycelial morphogenesis of *Hypholoma fasciculare* Agaricaceae. *Nova Hedwigia* 59:47–75.

Griffith, G. S., Rayner, A. D. M., and H. G. Wildman. 1994b. Interspecific interactions, mycelial morphogenesis and extracellular metabolite production in *Phlebia radiata* Aphyllophorales. *Nova Hedwigia* 59:331–344.

Gross, H. 2007. Strategies to unravel the function of orphan biosynthesis pathways: Recent examples and future prospects. *Applied Microbiology and Biotechnology* 75:267–277.

Guerrero, A. and G. Rosell. 2005. Biorational approaches for insect control by enzymatic inhibition. *Current Medicinal Chemistry* 12:461–469.

Gunatilaka, A. A. L. 2006. Natural products from plant-associated microorganisms: Distribution, structural diversity, bioactivity, and implications of their occurrence. *Journal of Natural Products* 69:509–526.

Hanlon, A. 2006. Chemical modulation of secondary metabolite production in *Aspergillus sp.* MS Thesis. Cambridge: Massachusetts Institute of Technology. 18 p.

Harris, G. 2005. The isolation and structure elucidation of fungal metabolites. In *Handbook of Industrial Mycology*, Ed. An, Z., pp. 187–268. New York: Marcel Dekker.

Hartsel, S. and J. Bolard. 1996. Amphotericin B: New life for an old drug. *Trends in Pharmacological Sciences* 17:445–449.

Hashizume, T., Matsubara, S., and A. Endo. 1983. Compactin (Ml-236b) as a new growth inhibitor of plant callus. *Agricultural and Biological Chemistry* 47:1401–1403.

Hata, S., Takagishi, H., and H. Kouchi. 1987. Variation in the content and composition of sterols in alfalfa seedlings treated with compactin (Ml-236b) and mevalonic acid. *Plant and Cell Physiology* 28:709–714.

Hentschel, U., Schmid, M., Wagner, M., Fieseler, L., Gernert, C., and J. Hacker. 2001. Isolation and phylogenetic analysis of bacteria with antimicrobial activities from the Mediterranean sponges *Aplysina aerophoba* and *Aplysina cavernicola. FEMS Microbiology Ecology* 35:305–312.

Hino, M., Fujie, A., Iwamoto, T., Hori, Y., Hashimoto, M., Tsurumi, Y., Sakamoto, K., Takase, S., and S. Hashimoto. 2001. Chemical diversity in lipopeptide antifungal antibiotics. *Journal of Industrial Microbiology and Biotechnology* 27:157–162.

Hiruma, K., Yagi, S., and A. Endo. 1983. Ml-236b (compactin) as an inhibitor of juvenile-hormone biosynthesis. *Applied Entomology and Zoology* 18:111–115.

Hoffman, L. R., D'Argenio, D. A., MacCoss, M. J., Zhang, Z., Jones, R. A., and S. I. Miller. 2005. Aminoglycosides antibiotics induce bacterial biofilm formation. *Nature* 436:1171–1175.

Hoffman, L., D'Argenio, D., Bader, M., and S. Miller. 2007. Microbial recognition of antibiotics: Ecological, physiological, and therapeutic implications. *Microbe* 2:175–182.

Hoffmeister, D. and N. P. Keller. 2007. Natural products of filamentous fungi: Enzymes, genes, and their regulation. *Natural Product Reports* 24:393–516.

Höfs, R., Walker, M., and A. Zeeck. 2000. Hexacyclinic acid, a polyketide from *Streptomyces* with a novel carbon skeleton. *Angewandte Chemie* 39:3258–3261.

Horinouchi, S. 2007. Mining and polishing of the treasure trove in the bacterial genus *Streptomyces. Bioscience Biotechnology and Biochemistry* 71:283–299.

Hornung, A., Bertazzo, M., Dziarnowski, A., Schneider, K., Welzel, K., Wohlert, S. E., Holzenkampfer, M., et al. 2007. A genomic screening approach to the structure-guided identification of drug candidates from natural sources. *Chem Bio Chem* 8:757–766.

Hoyt, S. and G. H. Jones. 1999. *relA* is required for actinomycin production in *Streptomyces antibioticus. Journal of Bacteriology* 181:3824–3829.

Hutchinson, C. R., Kennedy, J., Park, C., Auclair, K., Kendrew, S. G., and J. Vederas. 2005. Molecular genetics of lovastatin and compactin biosynthesis. In *Handbook of Industrial Mycology*, Ed. An, Z., pp. 479–491. New York: Marcel Dekker.

Hynes, J., Muller, C. T., Jones, T. H., and L. Boddy. 2007. Changes in volatile production during the course of fungal mycelial interactions between *Hypholoma fasciculare* and *Resinicium bicolor. Journal of Chemical Ecology* 33:43–57.

Ivarsson, P. and G. Birgersson. 1995. Regulation and biosynthesis of pheromone components in the double spined bark beetle *Ips duplicatus* Coleoptera: Scolytidae. *Journal of Insect Physiology* 41:843–849.

Ivarsson, P., Schlyter, F., and G. Birgersson. 1993. Demonstration of de-novo pheromone biosynthesis in *Ips duplicatus* (Coleoptera, Scolytidae): Inhibition of ipsdienol and e-myrcenol production by compactin. *Insect Biochemistry and Molecular Biology* 23:655–662.

Janzen, D. H. 1977. Why fruits rot, seeds mold, and meat spoils. *The American Naturalists* 111:691–713.

Justice, M. C., Hsu, M. J., Tse, B., Ku, T., Balkovec, J., Schmatz, D. and J. Nielson. 1998. Elongation factor 2 as a novel target for selective inhibition of fungal protein synthesis. *Journal of Biological Chemistry* 273:3148–3151.

Kahne, D., Leimkuhler, C., Lu, W., and C. Walsh. 2005. Glycopeptide and lipoglycopeptide antibiotics. *Chemical Reviews* 105:425–448.

Kanasaki, R., Abe, F., Kobayashi, M., Katsuoka, M., Hashimoto, M., Takase, S., Tsurumi, Y., et al. 2006a. FR220897 and FR220899, novel antifungal lipopeptides from *Coleophoma empetri* no. 14573. *Journal of Antibiotics* 59:149–157.

Kanasaki, R., Sakamoto, K., Hashimoto, M., Takase, S., Tsurumi, Y., Fujie, A., Hino, M., Hashimoto, S., and Y. Hori. 2006b. FR209602 and related compounds, novel antifungal lipopeptides from *Coleophoma crateriformis* no. 738. I. Taxonomy, fermentation, isolation and physico-chemical properties. *Journal of Antibiotics* 59:137–144.

Kane, N. S., Hirschberg, B., Qian, S., Hunt, D., Thomas, B., Brochu, R., Ludmerer, S. W., et al. 2000. Drug-resistant *Drosophila* indicate glutamate-gated chloride channels are targets for the antiparasitics nodulisporic acid and ivermectin. *Proceedings of the National Academy of Sciences, USA* 97:3949–13954.

Kaneko, T., Nakajima, N., Okamoto, S., Suzuki, I., Tanabe, Y., Tamaoki, M., Nakamura, Y., et al. 2008. Complete genomic structure of the bloom-forming toxic cyanobacterium *Microcystis aeruginosa* NIES-843. *DNA Research* 14:247–256.

Kennedy, T. C., Webb, G., Cannell, R. J. P., Kinsman, O. S., Middleton, R. F., Sidebottom, P. J., Taylor, N. L., Dawson, M. J., and A. D. Buss. 1998. Novel inhibitors of fungal protein synthesis produced by a strain of *Graphium putredinis*—Isolation, characterisation and biological properties. *Journal of Antibiotics* 51:1012–1018.

Kettering, M., Sterner, O., and T. Anke. 2004. Antibiotics in the chemical communication of fungi. *Zeitschrift für Naturforschung* 59c:816–823.

Khokhlov, A. S., Tovarova, I. I., Borisova, L. N., Pliner, S. A., Shevchenko, L. A., Komitskaya, E. Y., Ivkina, N. S., and I. A. Rapoport. 1967. The A-factor, responsible for streptomycin biosynthesis by mutant strains of *Actinomyces streptomycini. Doklady Akademii Nauk SSSR* 177:232–235.

Kijima, M., Yoshida, M., Sugita, K., Horinouchi, S., and T. Beppu. 1993. Trapoxin, an antitumor cyclic tetrapeptide, is an irreversible inhibitor of mammalian histone deacetylase. *Journal of Biological Chemistry* 268:22429–22435.

Kim, S. H., Ahn, S., Han, J. W., Lee, H. W., Lee, H. Y., Lee, Y. W., Kim, M. R., Kim, K. W., Kim, W. B., and S. Hong. 2004. Apicidin is a histone deacetylase inhibitor with anti-invasive and anti-angiogenic potentials. *Biochemical and Biophysical Research Communications* 315:964–970.

Kobayashi, A., Hagihara, K., Kajiyama, S., Kanzaki, H., and K. Kawazu. 1995. Antifungal compounds induced in the dual culture with *Phytolacca americana* callus and *Botrytis fabae. Zeitschrift für Naturforschung* 50c:398–402.

Kohn, F. E. 2008. High impact technologies for natural products screening. In *Natural Compounds as Drugs. Vol. 1*, Eds. F. Petersen and R. Amstutz, pp. 176–210. Basel: Birkhäuser Verlag AG.

Komon-Zelazowska, M., Neuhof, T., Dieckmann, R., von Dohren, H., Herrera-Estrella, A., Kubicek, C. P., and I. S. Druzhinina. 2007. Formation of atroviridin by *Hypocrea atroviridis* is conidiation associated and positively regulated by blue light and the G protein GNA3. *Eukaryotic Cell* 6:2332–2342.

Krämer, O. H., Gottlicher, M., and T. Heinzel. 2001. Histone deacetylase as a therapeutic target. *Trends in Endocrinology and Metabolism* 12:294–300.

Kurtz, M. B. and C. M. Douglas. 1997. Lipopetide inhibitors of fungal glucan sythase. *Medical Mycology* 35:79–86.

Laohavaleeson, S., Kuti, J. L., and D. P. Nicolau. 2007. Telavancin: A novel lipoglycopeptide for serious Gram-positive infections. *Expert Opinion on Investigational Drugs* 16:347–357.

Lardner, R., Mlahoney, N., Zanker, T. P., Molyneux, R. J., and E. S. Scott. 2006. Secondary metabolite production by the fungal pathogen *Eutypa lata*: Analysis of extracts from grapevine cultures and detection of those metabolites in planta. Australian Journal of Grape and Wine Research 12:107–114.

Latge, J. P. 2007. The cell wall: A carbohydrate armour for the fungal cell. *Molecular Microbiology* 66:279–290.

Lerat, E. and N. A. Moran. 2004. The evolutionary history of quorum-sensing systems in bacteria. *Molecular Biology and Evolution* 21:903–913.

Li, T.-L., Huang, F., Haydock, S. F., Mironenko, T., Leadlay, P. F., and J. B. Spencer. 2004. Biosynthetic gene cluster of the glycopeptide antibiotic teicoplanin: Characterization of two glycosyltransferases and the key acyltransferase. *Chemistry & Biology* 11:107–119.

Li, X. H., Li, J. X., Li, S. R., Xiu, Z. L., and N. Nishino. 2007. Cyclic peptide histone deacetylase inhibitors. *Progress in Chemistry* 19:762–768.

Liu, G., Casqueiro, J., Gutierrez, S., Kosalkova, K., Castillo, N. I., and J. F. Martin. 2001. Elicitation of penicillin biosynthesis by alginate in *Penicillium chrysogenum*, exerted on pcbAB, pcbC, and penDE genes at the transcriptional level. *Journal of Microbiology and Biotechnology* 11:812–818.

Liu, G., Tian, Y., Yang, H., and H. Tan. 2005. A pathway-specific transcriptional regulatory gene for nikkomycin biosynthesis in *Streptomyces ansochromogenes* that also influences colony development. *Molecular Microbiology* 55:1855–1866.

Lu, C. and Y. Shen. 2004. Harnessing the potential of chemical defense from antimicrobial activities. *BioEssays* 26:808–813.

Luzhetskyy, A., Pelzer, S., and A. Bechthold. 2007. The future of natural products as a source of new antibiotics. *Current Opinion in Investigational Drugs* 8:608–613.

Macreadie, I. G., Johnson, G., Schlosser, T., and P. I. Macreadie. 2006. Growth inhibition of *Candida* species and *Aspergillus fumigatus* by statins. *FEMS Microbiology Letters* 262:9–13.

Maldonado, A., Ruiz-Barba, J. L., and R. Jiménez-Díaz. 2003. Purification and genetic characterization of plantaricin NC8, a novel coculture-inducible two-peptide bacteriocin from *Lactobacillus plantarum* NC8. *Applied and Environmental Microbiology* 69:383–389.

Maldonado, A., Jiménez-Díaz, R., and J. L. Ruiz-Barba. 2004a. Induction of plantaricin production in *Lactobacillus plantarum* NC8 after coculture with specific Gram-positive bacteria is mediated by an autoinduction mechanism. *Journal of Bacteriology* 186:1556–1564.

Maldonado, A., Jiménez-Díaz, R., and J. L. Ruiz-Barba. 2004b. Production of plantaricin NC8 by *Lactobacillus plantarum* NC8 is induced in the presence of different types of Gram-positive bacteria. *Archives of Microbiology* 181:8–16.

Manteca, A., Fernández, M., and J. Sánchez. 2005. A death round affecting a young compartimentalized mycelium precedes aerial mycelium dismantling in confluent surface cultures of *Streptomyces antibioticus. Microbiology* 151:3689–3697.

Marfori, E. C., Kajiyama, S., Fukusaki, E., and A. Kobayashi. 2002. Trichosetin, a novel tetramic acid antibiotic produced in dual culture of *Trichoderma harzianum* and *Catharanthus roseus* callus. *Zeitschrift für Naturforschung* 57c:465–470.

Masuoka, Y., Shin-Ya, K., Kim, Y. B., Yoshida, M., Nagai, K., Suzuki, K., Hayakawa, Y., and H. Seto. 2000. Diheteropeptin, a new substance with TGF-beta like activity, produced by a fungus, *Diheterospora chlamydosporia*. I. Production, isolation and biological activities. *Journal of Antibiotics* 53:788–792.

Matsukawa, E., Nakagawa, Y., Iimura, Y., and M. Hayakawa. 2007. Stimulatory effect of indole-3-acetic acid on aerial mycelium formation and antibiotic production in *Streptomyces* spp. *Actinomycetologica* 21:32–39.

McGuire, J. M., Bunch, R. L., Anderson, R. C., Boaz, H. E., Flynn, H. E., Powell, H. M., and J. W. Smith. 1952. Ilotycin, a new antibiotic. *Antibiotics and Chemotherapy* 2:281–283.

Mearns-Spragg, A., Bregu, M., Boyd, K. G., and J. G. Burgess. 1998. Cross-species induction and enhancement of antimicrobial activity produced by epibiotic bacteria from marine algae and invertebrates, after exposure to terrestrial bacteria. *Letters in Applied Microbiology* 27:142–146.

Meinke, P. T. and P. Liberator. 2001. Histone deacetylase: A target for antiproliferative and antiprotozoal agents. *Current Medicinal Chemistry* 8:211–235.

Meinke, P. T., Smith, M. M., and W. L. Shoop. 2002. Nodulisporic acid: Its chemistry and biology. *Current Topics in Medicinal Chemistry* 2:655–674.

Miao, L., Kwong, T. F. N., and P. Y. Qian. 2006. Effect of culture conditions on mycelial growth, antibacterial activity, and metabolite profiles of the marine-derived fungus *Arthrinium* c.f. *saccharicola. Applied Microbiology and Biotechnology* 72:1063–1073.

Miller, J. D., Mackenzie, S., Foto, M., Adams, G. W., and J. A. Findlay. 2002. Needles of white spruce inoculated with rugulosin-producing endophytes contain rugulosin reducing spruce growth rate. *Mycological Research* 106:471–479.

Mingeot-Leclercq, M.-P., Glupczynski, Y., and P. M. Tulkens. 1999. Aminoglycosides: activity and resistance. *Antimicrobial Agents and Chemotherapy* 43:727–737.

Miyakawa, T., Miyama, R., Tabata, M., Tsuchiya, E., and S. Fukui. 1985. A study on the biosynthesis of tremerogen a-10, a polyisoprenyl peptide mating pheromone of *Tremella mesenterica*, using an inhibitor of mevalonate synthesis. *Agricultural and Biological Chemistry* 49:1343–1347.

Monger, D. J., Lim, W. A., Kezdy, F. J., and J. H. Law. 1982. Compactin inhibits insect HMG-CoA reductase and juvenile-hormone biosynthesis. *Biochemical and Biophysical Research Communications* 105:1374–1380.

Murakami, T., Takada, N., Harada, Y., Okuno, T., and M. Hashimoto. 2007. Stimulation of the biosynthesis of the antibiotics lambertellols by the mycoparasitic fungus *Lambertella corni-maris* under the acidic conditions produced by its host fungus in vitro. *Bioscience Biotechnology and Biochemistry* 71:1230–1235.

Narita, J. O. and W. Gruissem. 1989. Tomato hydroxymethylglutaryl CoA reductase is required early in fruit development but not during ripening. *Plant Cell* 1:181–190.

Nielsen, K. F., Larsen, T. O., and J. C. Frisvad. 2004. Lightweight expanded clay aggregates LECA, a new upscaleable matrix for production of microfungal metabolites. *Journal of Antibiotics* 57:29–36.

Noble, H. M., Langley, D., Sidebottom, P. J., Lane, S. J., and P. J. Fisher. 1991. An echinocandin from and endophytic *Cryptosporiopsis* sp. and *Pezicula* sp. in *Pinus sylvestris* and *Fagus sylvatica. Mycological Research* 95:1439–1440.

Odds, F. C. 2001. Sordarin antifungal agents. *Expert Opinion on Therapeutic Patents* 11:283–294.

Oh, D. C., Jensen, P. R., Kauffman, C. A., and W. Fenical. 2005. Libertellenones A-D: Induction of cytotoxic diterpenoid biosynthesis by marine microbial competition. *Bioorganic & Medicinal Chemistry* 13:5267–5273.

Oh, D. C., Kauffman, C. A., Jensen, P. R., and W. Fenical. 2007. Induced production of emericellamides A and B from the marine-derived fungus *Emericella* sp. in competing co-culture. *Journal of Natural Products* 70:515–520.

Ohno, M., Okano, I., Watsuji, T. O., Kakinuma, T., Ueda, K., and T. Beppu. 1999. Establishing the independent culture of a strictly symbiotic bacterium *Symbiobacterium thermophilum* from its supporting *Bacillus* strain. *Bioscience Biotechnology and Biochemistry* 63:1083–1090.

Okada, H., Nagashima, M., Kamiya, S., Kojiri, K., and H. Suda. 1999. Antifungal composition. U.S. Patent 5,922,709.

Oliynyk, M., Samborskyy, M., Lester, J. B., Mironenko, T., Scott, N., Dickens, S., Haydock, S. F., and P. F. Leadlay. 2007. Complete genome sequence of the erythromycin-producing bacterium *Saccharopolyspora erythraea* NRRL23338. *Nature Biotechnology* 25:447–453.

Omura, S., Ikeda, H., Ishikawa, J., Hanamoto, A., Takahashi, C., Shinose, M., Takahashi, Y., et al. 2001. Genome sequence of an industrial microorganism *Streptomyces avermitilis*: Deducing the ability of producing secondary metabolites. *Proceedings of the National Academy of Sciences, USA* 98:12215–12220.

Ondeyka, J. G., Helms, G. L., Hensens, O. D., Goetz, M. A., Zink, D. L., Tsipouras, A., Shoop, W., et al. 1997. Nodulisporic acid A, a novel and potent insecticide from a *Nodulisporium* sp., isolation, structure determination and chemical transformations. *Journal of the American Chemical Society* 119:8809–8816.

Onishi, J., Meinz, M., Thompson, J., Curotto, J., Dreikorn, S., Rosenbach, M., Douglas, C., et al. 2000. Discovery of novel β1,3-glucan synthase inhibitors. *Antimicrobial Agents and Chemotherapy* 44:368–377.

Ostlind, D. A., Felcetto, T., Misura, A., Ondeyka, J., Smith, S., Goetz, M., Shoop, W., and W. Mickle. 1997. Discovery of a novel indole diterpene insecticide using first instars of *Lucilia sericata. Medical and Veterinary Entomology* 11:407–408.

Overy, D. P. and J. C. Frisvad. 2005. Mycotoxin production and postharvest storage rot of ginger (*Zingiber officinale*) by *Penicillium brevicompactum. Journal of Food Protection* 68:607–609.

Overy, D. P., Smegsgaard, J., Frisvad, J. C., Phipps, R. K., and U. Thrane. 2006. Host-derived media used as a predictor for low abundant, in planta metabolite production from necrotrophic fungi. *Journal of Applied Microbiology* 101:1292–1300.

Ozawa, R., Matsumoto, S., Kim, G. H., Uchiumi, K., Kurihara, M., Shono, T., and T. Mitsui. 1995. Intracellular signal transduction of Pban action in lepidopteran insects—Inhibition of sex-pheromone production by compactin, an HMG CoA reductase inhibitor. *Regulatory Peptides* 57:319–327.

Pahl, A. and U. Keller. 1992. FK-506-binding proteins from streptomycetes producing immunosuppressive macrolactones of the FK-506 type. *Journal of Bacteriology* 174:5888–5894.

Panaccione, D. G. 2005. Origins and significance of ergot alkaloid diversity in fungi. *FEMS Microbiology Letters* 251:9–17.

Paranagama, P. A., Wijeratne, E. M. K., and A. A. L. Gunatilaka. 2007. Uncovering biosynthetic potential of plant-associated fungi: Effect of culture conditions on metabolite production by *Paraphaeosphaeria quadriseptata* and *Chaetomium chiversii*. *Journal of Natural Products* 70:1939–1945.

Patchett, A. A. 2002. Alfred Burger Award address in medicinal chemistry. Natural products and design: Interrelated approaches in drug discovery. *Journal of Medicinal Chemistry* 45:5609–5616.

Peláez, F., Cabello, A., Platas, G., Díez, M. T., Gonzalez, A., Basilio, A., Martín, I., et al. 2000. The discovery of enfumafungin, a novel antifungal compound produced by endophytic *Hormonema* species, biological activity, and taxonomy of the producing organisms. *Systematic and Applied Microbiology* 23:333–343.

Perlstein, E. O., Ruderfer, D. M., Ramachandran, G., Haggarty, S. J., Kruglyak, L., and S. L. Schreiber. 2006. Revealing complex traits with small molecules and naturally recombinant yeast strains. *Chemistry & Biology* 13:319–327.

Perrin, R. M., Fedorova, N. D., Bok, J. W., Cramer, R. A., Wortman, J. R., Kim, H. S., Nierman, W. C., and N. P. Keller. 2007. Transcriptional regulation of chemical diversity in *Aspergillus fumigatus* by LaeA. *PLoS Pathogens* 3:e50.

Perry, M. J., Makins, J. F., Adlard, M. W., and G. Holt. 1984. Aspergillic acids produced by mixed cultures of *Aspergillus flavus* and *Aspergillus nidulans*. *Journal of General Microbiology* 130:319–323.

Peters, S. and P. Spiteller. 2006. Chloro- and bromophenols from culture of *Mycena alcalina*. *Journal of Natural Products* 69:1809–1812.

Petrini, O. and M. Dreyfuss. 1981. Endophytische Pilze in epiphytischen Araceae, Bromiliaceae und Orchidaceae. *Sydowia* 34:135–148.

Piel, J. 2002. A polyketide synthase-peptide synthetase gene cluster from an uncultured bacterial symbiont of *Paederus* beetles. *Proceedings of the National Academy of Sciences, USA* 99:14002–14007.

Plosker, G. L. and L. B. Barradell. 1996. Cyclosporin—A review of its pharmacological properties and role in the management of graft versus host disease. *Clinical Immunotherapeutics* 5:59–90.

Polishook, J. D., Ondeyka, J. G., Dombrowski, A. W., Peláez, F., Platas, G., and A. M. Terán. 2000. Biogeography and relatedness of *Nodulisporium* strains producing nodulisporic acid. *Mycologia* 92:1125–1137.

Potter, D. A., Stokes, J. T., Redmond, C. T., Schardl, C. L., and D. G. Panniccione. 2008. Contribution of ergot alkaloids to suppression of grass-feeding caterpillar assessed with gene knockout endophytes in perrennial ryegrass. *Entomologia Experimentalis et Applicata* 126:138–147.

Poulakou, G. and H. Giamarellou. 2008. Oritavancin: A new promising agent in the treatment of infections due to Gram-positive pathogens. *Expert Opinion on Investigational Drugs* 17:225–243.

Rayner, A. D. M., Griffith, G. S., and H. G. Wildman. 1994. Induction of metabolic and morphogenetic changes during mycelial interactions among species of higher fungi. *Biochemical Society Transactions* 22:389–394.

Reeves, A. R., English, R. S., Lampel, J. S., Post, D. A., and T. J. Vanden Boom. 1999. Transcriptional organization of the erythromycin biosynthetic gene cluster of *Saccharopolyspora erythraea*. *Journal of Antibiotics* 181:7098–7106.

Rodriguez, M. A., Cabrera, G., and A. Godeas. 2006. Cyclosporine A from a nonpathogenic *Fusarium oxysporum* suppressing *Sclerotinia sclerotiorum*. *Journal of Applied Microbiology* 100:575–586.

Rohlfs, M., Albert, M., Keller, N. P., and F. Kempken. 2007. Secondary chemicals protect mould from fungivory. *Biology Letters* 5:523–525.

Samuelson, G. 1999. *Drugs of Natural Origin—A Textbook of Pharmacognosy*, 4th ed., Stockholm: Pharmaceutical Press.

Schatz, A., Bugie, E., and S. Waksman. 1944. Streptomycin, a substance exhibiting antibiotic activity against Gram-positive and Gram-negative bacteria. *Proceedings of the Society for Experimental Biology and Medicine* 55:66–69.

Scherlach, K. and C. Hertweck. 2006. Discovery of aspoquinolones A-D, prenylated quinoline-2-one alkaloids from *Aspergillus nidulans*, motivated by genome mining. *Organic & Biomolecular Chemistry* 4:3517–3520.

Schmidt, E. W., Obraztsova, A. Y., Davidson, S. K., Faulkner, D. J., and M. G. Haygood. 2000. Identification of the antifungal peptide-containing symbiont of the marine sponge *Theonella swinhoei* as a novel deltaproteobacterium, *Candidatus Entotheonella palauensis*. *Marine Biology* 136:969–977.

Schneider, G., Anke, H., and O. Sterner. 1995. Xylarin, an antifungal *Xylaria* metabolite with an unusual tricyclic uronic acid moiety. *Natural Products Letters* 7:309–316.

Schneiker, S., Perlova, O., Kaiser, O., Gerth, K., Alici, A., Altmeyer, M. O., Bartels, D., et al. 2007. Complete genome sequence of the myxobacterium *Sorangium cellulosum. Nature Biotechnology* 25:1281–1289.

Schroeder, F. C., Gibson, D. M., Churchill, A. C. L., Sojikul, P., Worsthorn, E. J., Krasnoff, S. B., and J. Clardy. 2007. Differential analysis of 2D NMR spectra: New natural products from a pilot-scale fungal extract library. *Angewandte Chemie* 119:919–922.

Schwartz, R. E. 2001. Cell wall active antifungal agents. *Expert Opinion on Therapeutic Patents* 11:1761–1772.

Schwartz, R. E., Giacobbe, R. A., Bland, J. A., and R. L. Monaghan. 1989. L-671,329, a new antifungal agent. 1. Fermentation and isolation. *Journal of Antibiotics* 42:163–167.

Schwartz, R. E., Sesin, D. F., Joshua, H., Wilson, K. E., Kempf, A. J., Goklen, K. A., Kuehner, D., et al. 1992. Pneumocandins from *Zalerion arboricola*. 1. Discovery and isolation. *Journal of Antibiotics* 45:1853–1866.

Shi, W. and D. R. Zusman. 1993. Fatal attraction. *Nature Biotechnology* 366:414–415.

Shoop, W. L., Mrozik, H., and M. H. Fisher. 1995. Structure and activity of avermectins and milbemycins in animal health. *Veterinary Parasitology* 59:139–156.

Shu, Y.-Z. 1998. Recent natural products based drug development: A pharmaceutical industry perspective. *Journal of Natural Products* 61:1053–1071.

Shwab, E. K., Bok, J., Tribus, M., Galehr, J., Graessle, S., and N. P. Keller. 2007. Histone deacetylase activity regulates chemical diversity in *Aspergillus. Eukaryotic Cell* 6:1656–1664.

Singh, S. B., Zink, D. L., Polishook, J. D., Dombrowski, A. W., DarkinRattray, S. J., Schmatz, D. M., and M. A. Goetz. 1996. Apicidins: Novel cyclic tetrapeptides as coccidiostats and antimalarial agents from *Fusarium pallidoroseum. Tetrahedron Letters* 37:8077–8080.

Singh, S. B., Zink, D. L., Liesch, J. M., Mosley, R. T., Dombrowski, A. W., Bills, G. F., Darkin-Rattray, S. J., Schmatz, D. M., and M. A. Goetz. 2002. Structure and chemistry of apicidins, a class of novel cyclic tetrapeptides without a terminal alpha-keto epoxide as inhibitors of histone deacetylase with potent antiprotozoal activities. *Journal of Organic Chemistry* 67:815–825.

Slattery, M., Rajbhandari, I., and K. Wesson. 2001. Competition-mediated antibiotic induction in the marine bacterium *Streptomyces tenjimariensis. Microbial Ecology* 41:90–96.

Smith, M. M., Warren, V. A., Thomas, B. S., Brochu, R. M., Ertel, E. A., Rohrer, S., Schaeffer, J., et al. 2000. Nodulisporic acid opens insect glutamate-gated chloride channels: Identification of a new high-affinity modulator. *Biochemistry* 39:5543–5554.

Sonnenbichler, J., Lamm, V., Gieren, A., Holdenrieder, O., and H. Lotter. 1983. A cyclopentabenzopyranone produced by the fungus *Heterobasidion annosum* in dual cultures. *Phytochemistry* 22:1489–1491.

Sonnenbichler, J., Bliestle, I. M., Peipp, H., and O. Holdenrieder. 1989. Secondary fungal metabolites and their biological activities. 1. Isolation of antibiotic compounds from cultures of *Heterobasidion annosum* synthesized in the presence of antagonistic fungi or host plant cells. *Biological Chemistry Hoppe-Seyler* 370:1295–1303.

Sonnenbichler, J., Peipp, H., and J. Dietrich. 1993. Secondary fungal metabolites and their biological activities. 3. Further metabolites from dual cultures of the antagonistic basidiomycetes *Heterobasidion annosum* and *Gloeophyllum abietinum. Biological Chemistry Hoppe-Seyler* 374:467–473.

Spinnler, H. E., de Jong, E., Mauvais, G., Semon, E., and J. L. le Quere. 1994. Production of halogenated compound by *Bjerkander adusta. Applied Microbiology and Biotechnology* 42:212–221.

Stadler, M. and V. Hellwig. 2005. Chemotaxonomy of the Xylariaceae and remarkable bioactive compounds from Xylariales and their associated asexual stages. *Recent Research Developments in Phytochemistry* 9:41–93.

Stadler, M., Anke, H., and O. Sterner. 1995. Metabolites with nematocidal and antimicrobial activities from the ascomycete *Lachnum papyraceum* (Karst.). Karst. 3. Production of novel isocoumarin derivatives, isolation, and biological activities. *Journal of Antibiotics* 48:261–266.

Staunton, J. and K. J. Weissman. 2001. Polyketide biosynthesis: A millennium review. *Natural Product Reports* 18:380–416.

Steinbach, W. J., Cramer, R. A., Perfect, B. Z., Henn, C., Nielsen, M., Heitman, J., and J. R. Perfect. 2007. Calcineurin inhibition or mutation enhances cell wall inhibitors against *Aspergillus fumigatus. Antimicrobial Agents and Chemotherapy* 51:2979–2981.

Stone, M. J. and D. H. Williams. 1992. On the evolution of functional secondary metabolites natural products. *Molecular Microbiology* 6:29–34.

Strobel, G. A., Miller, R. V., Martinez-Miller, C., Teplow, M. M. C. D. B., and W. M. Hess. 1999. Cryptocandin, a potent antimycotic from the endophytic fungus *Cryptosporiopsis* cf. *quercina. Microbiology* 145:1919–1926.

Strohl, W. R. 1997. Industrial antibiotics: Today and the future. In *Biotechnology of Antibiotics. Drugs and the Pharmaceutical Sciences. Vol 82*, Ed. W. R. Strohl, pp. 1–47. New York: Marcel Dekker.

Sugui, J. A., Pardo, J., Chang, Y. C., Zarember, K. A., Nardone, G., Galvez, E. M., Muellbacher, A., Gallin, J. I., Simon, M. M., and K. J. Kwon-Chung. 2007. Gliotoxin is a virulence *Aspergillus* factor of *fumigatus*: gliP deletion attenuates virulence in mice immunosuppressed with hydrocortisone. *Eukaryotic Cell* 6:1562–1569.

Sumarah, M. W., Miller, J. D., and G. W. Adams. 2005. Measurement of a rugulosin-producing endophyte in white spruce seedlings. *Mycologia* 97:770–776.

Takano, E. 2006. Gamma-butyrolactones: *Streptomyces* signaling molecules regulating antibiotic production and differentiation. *Current Opinion in Microbiology* 9:287–294.

Tanaka, M., Moriguchi, T., Kizuka, M., Ono, Y., Miyakoshi, S. I., and T. Ogita. 2002. Microbial hydroxylation of zofimarin, a sordarin-related antibiotic. *Journal of Antibiotics* 55:437–441.

Taunton, J., Collins, J. L., and S. L. Schreiber. 1996. Synthesis of natural and modified trapoxins, useful reagents for exploring histone deacetylase function. *Journal of the American Chemical Society* 118:10412–10422.

Tfelt-Hansen, P., Saxena, P. R., Dahlöf, C., Pascual, J., Láinez, M., Henry, P., Diener, H.-C., Schoenen, J., Ferrari, M. D., and P. J. Goadsby. 2000. Ergotamine in the acute treatment of migraine. A review and European consensus. *Brain* 123:9–18.

Thines, E., Anke, H., and R. W. S. Weber. 2004. Fungal secondary metabolites as inhibitors of infection-related morphogenesis in phytopathogenic fungi. *Mycological Research* 108:14–25.

Tocci, M. J., Matkovich, D. A., Collier, K. A., Kwok, P., Dumont, F., Lin, S., Degudicibus, S., Siekierka, J. J., Chin, J., and N. I. Hutchinson. 1989. The immunosuppressant FK506 selectively inhibits expression of early T cell activation genes. *Journal of Immunology* 143:718–726.

Tokimoto, K., Fujita, T., Takeda, Y., and Y. Takaishi. 1987. Increased or induced formation of antifungal substances in cultures of *Lentinus edodes* by the attack of *Trichoderma* spp. *Proceedings of the Japan Academy Series B, Physical and Biological Sciences* 63:277–280.

Tormo, J. R., García, J. B., DeAntonio, M., Feliz, J., Mira, A., Díez, M. T., Hernández, P., and F. Peláez. 2003. A method for the selection of production media for actinomycete strains based on their metabolite HPLC profiles. *Journal of Industrial Microbiology and Biotechnology* 30:582–588.

Traber, R. 1997. Biosynthesis of cyclosporins. In *Biotechnology of Antibiotics. Drugs and the Pharmaceutical Sciences. Vol 82*, Ed. W. R. Strohl, pp. 279–314. New York: Marcel Dekker.

Traber, R. and M. M. Dreyfuss. 1996. Occurrence of cyclosporins and cyclosporin-like peptolides in fungi. *Journal of Industrial Microbiology and Biotechnology* 17:397–401.

Trenin, A. S. 1998. Fungal secondary metabolites that inhibit sterol biosynthesis. *Applied Biochemistry and Microbiology* 34:117–123.

Tulip, M. and L. Bohlin. 2002. Functional versus chemical diversity: Is biodiversity important for drug discovery? *Trends in Pharmacological Sciences* 22:225–231.

Turner, W. W. and W. Current. 1997. Echinocandin antifungal agents. In *Biotechnology of Antibiotics. Drugs and the Pharmaceutical Sciences. Vol 82*, Ed. W. R. Strohl, pp. 315–334. New York: Marcel Dekker.

Ueda, K., Kawai, S., Ogawa, H.-O., Kiyama, A., Kubota, T., Kawanobe, H., and T. Beppu. 2000. Wide distribution of interspecific stimulatory events on antibiotic production and sporulation among *Streptomyces* species. *Journal of Antibiotics* 53:979–982.

Ueda, K., Saka, H., Ishikawa, Y., Kato, T., Takeshita, Y., Shiratori, H., Ohno, M., Hosono, K., Wada, M., and T. Beppu. 2002. Development of a membrane dialysis bioreactor and its application to a large-scale culture of a symbiotic bacterium, *Symbiobacterium thermophilum. Applied Microbiology and Biotechnology* 60:300–305.

Umezawa, H. 1958. Kanamycin: Its discovery. *Annals of the New York Academy of Sciences* 76:20–26.

van Dongen, P. W. J. and A. N. J. A. de Groot. 1995. History of ergot alkaloids from ergotism to ergometrine. *European Journal of Obstetrics and Gynecology and Reproductive Biology* 60:109–116.

van Wageningen, A. M. A., Kirkpatrick, P. N., Williams, D. H., Harris, B. R., Kershaw, J. K., Lennard, N. J., Jones, M., Jones, S. J. M., and P. J. Solenberg. 1998. Sequencing and analysis of genes involved in the biosynthesis of a vancomycin group antibiotic. *Chemistry & Biology* 5:155–162.

Vicente, M. F., Basilio, A., Cabello, A., and F. Pelaez. 2003. Microbial natural products as a source of antifungals. *Clinical Microbiology and Infection* 9:15–32.

Vilcinskas, A., Jegorov, A., Landa, Z., Götz, P., and V. Matha. 1999. Effects of beauverolide L and cyclosporin A on humoral and cellular immune response of the greater wax moth, *Galleria mellonella. Comparative Biochemistry and Physiology Part C: Pharmacology, Toxicology and Endocrinology* 122:83–92.

Vining, L. C. 1990. Function of secondary metabolites. *Annual Review of Microbiology* 44:395–427.

Waksman, S. A. and H. A. Lechevalier. 1949. Neomycin, a new antibiotic active against streptomycin-resistant bacteria, including tuberculosis organisms. *Science* 109:305–307.

Waksman, S. A. and H. B. Woodruff. 1941. *Actinomyces antibioticus*, a new soil organism antagonistic to pathogenic and non-pathogenic bacteria. *Journal of Bacteriology* 42:231–249.

Walsh, C. 2003a. *Antibiotics: Actions, Origins, Resistance*. Washington, DC: ASM Press.

Walsh, C. 2003b. Where will new antibiotics come from? *Nature Reviews Microbiology* 1:65–70.

Walton, J. D. 2006. HC-toxin. *Phytochemistry* 67:1406–1413.

Walton, J. D., Earle, E. D., Stahelin, H., Grieder, A., Hirota, A., and A. Suzuki. 1985. Reciprocal biological activities of the cyclic tetrapeptides chlamydocin and HC-toxin. *Experientia* 41:348–350.

Waters, C. M. and B. L. Bassler. 2005. Quorum sensing: Cell-to-cell communication in bacteria. *Annual Review of Cell and Developmental Biology* 21:319–346.

Weber, R. W. S., Stenger, E., Meffert, A., and M. Hahn. 2004. Brefeldin A production by *Phoma medicaginis* in dead pre-colonized plant tissue: A strategy for habitat conquest? *Mycological Research* 108:662–671.

Weber, R. W. S., Meffert, A., Anke, H., and O. Sterner. 2005. Production of sordarin and related metabolites by the coprophilous fungus *Podospora pleiospora* in submerged culture and in its natural substrate. *Mycological Research* 109:619–626.

Weber, R. W. S., Kappe, R., Paululat, T., Mosker, E., and H. Anke. 2007. Anti-*Candida* metabolites from endophytic fungi. *Phytochemistry* 68:886–892.

Weinstein, M. J., Luedemann, G. M., Oden, E. M., Wagman, G. H., Rosselet, J. P., Marquez, J. A., Coniglio, C. T., Charney, W., Herzog, H. L., and J. Black. 1963. Gentamicin, a new antibiotic complex from *Micromonospora. Journal of Medicinal Chemistry* 6:463–464.

Wicklow, D. T. 1981. Interference competition and the organization of fungal communities. In *The Fungal Community, Its Organization and Role in the Ecosystem*, Eds. D. T. Wicklow and G. C. Carroll, pp. 351–375. New York: Marcel Dekker, Inc.

Wicklow, D. T. 1988. Metabolites in the coevolution of fungal chemical defence systems. In *Coevolution of Fungi with Plants and Animals*, Eds. K. A. Pirozynski and D. L. Hawksworth, pp. 174–201. New York: Academic Press.

Wicklow, D. T., Roth, S., Deyrup, S. T., and J. B. Gloer. 2005. A protective endophyte of maize: *Acremonium zeae* antibiotics inhibitory to *Aspergillus flavus* and *Fusarium verticillioides. Mycological Research* 109:610–618.

Wiener, P. 1996. Experimental studies on the ecological role of antibiotic production in bacteria. *Evolutionary Ecology* 10:405–421.

Wildman, H. G. 1997. Potential of tropical microfungi within the pharmaceutical industry. In *Biodiversity of Tropical Microfungi*, Ed. K. D. Hyde, pp. 29–46. Hong Kong: Hong Kong University Press.

Wildman, H. G. and R. J. Jones. 1991. Isolation of fungal endophytes from root samples of trees blown over at The Royal Botanic Gardens, Kew, during the 1987 hurricane. *The Mycologist* 5:180–182.

Williams, D. H., Stone, M. J., Hauck, P. R., and S. K. Rahman. 1989. Why are secondary metabolites (natural products) biosynthesized? *Journal of Natural Products* 52:1189–1208.

Williams, R. B., Henrikson, J. C., Hoover, A. R., Lee, A. E., and R. H. Cichewicz. 2008. Epigenetic remodeling of the fungal secondary metabolome. *Organic & Biomolecular Chemistry* 6:1895–1897.

Wink, M. 2003. Evolution of secondary metabolites from an ecological and molecular perspective. *Phytochemistry* 64:3–19.

Worthington, P. A. 1988. Antibiotics with antifungal and antibacterial activity against plant diseases. *Natural Product Reports* 5:47–66.

Yamaguchi, T., Itose, R., and K. Sakai. 1997. Biosynthesis of heartwood tropolones I. Incorporation of mevalonate and acetate into beta-thujaplicin (hinokitiol) of *Cupressus lusitanica* cell cultures. *Journal of the Faculty of Agriculture Kyushu University* 42:131–138.

Yarbrough, G. C., Taylor, D. P., Rowlands, R. T., Crawford, M. S., and L. L. Lasure. 1993. Screening microbial metabolites for new drugs—theoretical and practical issues. *Journal of Antibiotics* 46:535–544.

Yim, G., Huimi, H., and J. Davies. 2006. The truth about antibiotics. *International Journal of Medical Microbiology* 296:163–170.

Yim, G., Wang, H. H., and J. Davies. 2007. Antibiotics as signalling molecules. *Philosophical Transactions of the Royal Society B: Biological Sciences* 362:1195–1200.

Yoshinari, T., Akiyama, T., Nakamura, K., Kondo, T., Takahashi, Y., Muraoka, Y., Nonomura, Y., Nagasawa, H., and S. Sakuda. 2007. Dioctatin A is a strong inhibitor of aflatoxin production by *Aspergillus parasiticus*. *Microbiology* 153:2774–2780.

Zapata, R., Martín, D., Piulachs, M.-D., and X. Bellés. 2002. Effects of hypocholesterolaemic agents on the expression and activity of 3-hydroxy-3-methylglutaryl-CoA reductase in the fat body of the German cockroach. *Archives of Insect Biochemistry and Physiology* 49:177–186.

Zhu, F. and Y. Lin. 2006. Marinamide, a novel alkaloid and its methyl ester produced by the application of mixed fermentation technique to two mangrove endophytic fungi from the South China Sea. *Chinese Science Bulletin* 51:1426–1430.

Zhu, J. 1999. Recent developments in reversing glycopeptide-resistant pathogens. *Expert Opinion on Therapeutic Patents* 9:1005–1019.

Part IV

Fungal Endophyte as Model System to Understand Defensive Mutualism

18

Extensions to and Modulation of Defensive Mutualism in Grass Endophytes

Thomas L. Bultman, Terrence J. Sullivan, Michael H. Cortez, Timothy J. Pennings, and Janet L. Andersen

CONTENTS

18.1 Introduction

The hypothesis that fungal endophytes (Ascomycota: Clavicipitaceae) are defensive mutualists with their grass hosts was first formally proposed by Clay (1988) and Cheplick and Clay (1988). They postulated that the mutualism was based on the reciprocal exchange of resources and services between the symbionts. Grasses provide energy—carbohydrates from photosynthesis and mineral nutrients—and a stable place to live. In exchange, the fungi provide defense from herbivores and pathogens of the grasses. The mechanism of this defense is the production of toxic alkaloids like lolines, peramines, ergots, and lolitrems (Bush et al., 1997). In addition, other benefits may be provided, like drought tolerance, but the primary benefit postulated to operate is herbivore defense.

Defense against herbivory is one of the primary selective pressures on plants (Cyr and Pace, 1993), and the defensive mutualism hypothesis has been useful in guiding and evaluating research on grass endophyte–herbivore interactions over the past 20 years. Data have generally supported the hypothesis when investigators have worked with agronomic, nonnative species like tall fescue (*Lolium arundinacea*), meadow fescue (*Lolium pratense*), and perennial ryegrass (*Lolium perenne*) (Clay, 1991; Latch, 1993; Breen, 1994; Clement et al., 1994; Bush et al., 1997; Lehtonen et al., 2005; Wali et al., 2008). Yet, support often has been weak or even lacking when native, nonagronomic species have been studied

(Saikkonen et al., 1998; Faeth and Bultman, 2002). Work on nonagronomic species led Faeth (2002) to question the assumption that all endophytic fungi are defensive mutualists. As research has accumulated, it has become clear that in addition to domestication, several variables influence the degree of mutualism including plant and fungal genotype, fungal mode of transmission, and environmental conditions (Saikkonen et al., 1998; Müller and Krauss, 2005). The degree to which the defensive mutualism hypothesis explains the interaction between fungal endophytes and grasses is difficult to ascertain due to the limited phylogenetic scope of published work. However, new work is continually shedding light on our understanding. Here we review three areas of recent and active research on defensive mutualism in grass endophytes. First, we discuss an extension of the defensive mutualism hypothesis that endophytes provide not only constitutive resistance to their hosts, but also wound-inducible resistance. Second, we discuss the modulation of the mutualism by one environmental variable, available nutrients (particularly nitrogen). And third, we review how accounting for natural enemies of herbivores might modify the defense provided to the grass by its fungal symbiont. Our aim is not only to highlight this new work, but also suggest where gaps in our knowledge remain and where new contributions are needed.

18.2 Inducible Responses

18.2.1 Background

It is well documented that many plants have the ability to be plastic in their response to herbivory, and only produce defenses when the likelihood for herbivore pressure is high (Karban and Baldwin, 1997). The genetic mechanisms behind active induced resistance are beginning to be understood. For example, in *Brassica oleracea*, expression of the gene BoLOX, a gene closely related to the JA biosynthesis mediating AtLOX2 in *Arabidopsis thaliana*, increases 10-fold after damage (Zheng et al., 2007). In *A. thaliana*, several stress-related genes are induced by herbivores (Kusnierczyk et al., 2007).

Endophytes can have community-level effects due to how they restructure herbivore communities (Clay and Holah, 1999). Most research on *Neotyphodium*-provided defense has treated it as constitutive, and while they can provide constitutive defenses in many host species, recent work has found that the defenses provided by *Neotyphodium* endophytes can change after herbivore damage. Since induced responses can also have community-level effects (Van Zandt and Agrawal, 2004), induced variation in endophyte defenses are likely to as well. Below we review current knowledge on the inducibility of endophyte-produced defenses in pooid grasses.

18.2.2 Induced Resistance in Tall Fescue/*N. coenophialum* Symbiosis

The antiherbivore traits of tall fescue/*N. coenophialum* are well known (Clay, 1991; Bush et al., 1997). While these defenses occur constitutively, they can also be modified by herbivory. Chemically, the production of *N*-formyl and *N*-acetyl loline increases in *N. coenophialum*-infected tall fescue after clipping (Eichenseer et al., 1991). These changes also affect higher trophic levels. Bultman et al. (2004) artificially damaged tall fescue plants, both with and without *N. coenophialum* and followed the population growth of *Rhopalosiphum padi* (bird cherry-oat aphid) over 4 days and found that damaged, infected plants inhibited population growth relative to undamaged, uninfected plants. Conversely, *R. padi* population growth rates were higher on infected plants that were damaged compared to undamaged suggesting that herbivory makes uninfected tall fescue more palatable to herbivores (Bultman et al., 2004).

In a subsequent experiment to assess if herbivory could also elicit a response, damage by fall armyworm (*Spodoptera frugiperda*) was included. Eight-week old tall fescue, both with and without *N. coenophialum*, was damaged either mechanically by cutting all above ground tissue 3 cm above the soil or by fall armyworm herbivory, or left undamaged. Toxicity was tested using an aphid bioassay. Four *R. padi* were placed in mesh bags on the youngest, fully expanded leaf, and the total *R. padi* population size was assessed 1 week later. Damage, both artificial and by fall armyworm, increased toxicity of *N. coenophialum*-infected plants relative to undamaged controls, while damage to *N. coenophialum*-free hosts increased the *R. padi* population indicating damage made them more susceptible (Sullivan et al., 2007b).

The molecular mechanism behind these changes appears to be due, in part, to changes in gene expression in the endophyte, not changes in relative biomass of the endophyte. *LolC*, putatively a pyridoxal phosphate-containing enzyme, is part of the gene complex that produces loline and its expression is correlated with loline production (Spiering et al., 2005). The amount of *lolC* transcript can be quantified with RT-PCR and normalized using the transcript for beta-tubulin (*tubB*) to account for variation in the overall transcription rate and/or the endophyte concentration in the sample.

We found mechanically damaging tall fescue/*N. coenophialum* symbiota upregulated the production of *lolC* transcript. Samples were taken at 10 and 24 days and a significant interaction between damage type and time from damage was found. At 10 days, damaged symbiota had higher *lolC* production than the undamaged controls 10 days after damage (Sullivan et al., 2007b). No difference was found 24 days postdamage. Induced resistance by *N. coenophialum*, then, is due to metabolic changes in the endophytes in response to damage.

18.2.3 Taxonomic Variation in Induced Resistance

There is considerable variation in the interaction between *Neotyphodium* endophytes and their host grasses. Some of this is likely due to the evolutionary history of the group. *Neotyphodium* endophytes are often the product of interspecific hybridization events (Moon et al., 2004). These hybrid lineages can be identified by the presence of duplicate copies of many genes, each copy from a different ancestral species (Schardl et al., 1994; Tsai et al., 1994; Schardl and Craven, 2003; Moon et al., 2004). Hypotheses regarding any selective advantage are not well tested (but see Sullivan and Faeth, 2008), but the presence of multiple copies of alkaloid-producing genes led to the hypothesis that hybrids are favored due to increased and/or varied toxin production (Clay and Schardl, 2002). We would also expect that hybridization could create interspecific variation in the plasticity of antiherbivore resistance. *Neotyphodium uncinatum*, symbiotic with meadow fescue (*Lolium pratense*), is of hybrid origin (Craven et al., 2001). In an experiment in which we tested for induced resistance in the *N. uncinatum/L. pratense* symbiota, neither artificial nor herbivore damage caused the symbiota to become more toxic to subsequent fall armyworm herbivores (Sullivan et al., 2007a). Additionally, molecular evidence supports the finding that damage does not increase the defensive response, and may actually slightly depress it. Being the product of interspecific hybridization, *N. uncinatum* carries multiple copies of *lolC* (designated *lolC-1* and *lolC-2*). Following herbivory, damaged symbiota downregulated production of *lolC-1* relative to the undamaged controls, but there was no difference in *lolC-2* expression. The reduction of *lolC-1* expression may have minimal ecological consequences, though, as expression of both *lolC-1* and *lolC-2* was approximately 15-fold greater than the expression of *lolC* in *N. coenophialum*, suggesting that even after being downregulated, the potential production of lolines was still very high. Several hypotheses explaining plant defense predict a trade-off between growth and defense (Stamp, 2003). If the constitutive production of lolines is high enough, not only may there be no advantage to increasing production postdamage, but also reducing production to reallocate resources for regrowth may be the optimal strategy if the associated reduction in toxicity is negligible.

18.2.4 Evolutionary Origins of Inducible Resistance

Lolines are generally associated with *Neotyphodium* spp., although they are also produced by some lineages of *E. festucae*. The most common hypothesis explaining this variation is related to life history differences. *Neotyphodium* endophytes are asexual, maternally inherited obligate symbionts, so their fitness is directly linked to their hosts. Conversely, many *Epichloë* species reproduce sexually via stromata that choke their host's florescences. These choking endophytes do not rely completely upon their hosts for reproduction and would be expected to be less mutualistic as a result (Ewald, 1994). Recent phylogenetic analyses, however, show that loline production predates the *Neotyphodium/Epichloë* clade (Torres et al., 2007) and the lack of loline production in many *Epichloë* species is due to loss of the trait, that is, defensive mutualisms in clavicipitacean fungi appear to have evolved before the parasitisms of *Epichloë* (Kutil et al., 2007).

A recent study by Gonthier et al. (2008) found that induced resistance may also be an ancestral trait. *Epichloë glyceriae*, symbiotic with the grass *Glyceria striata*, had not previously been known to produce

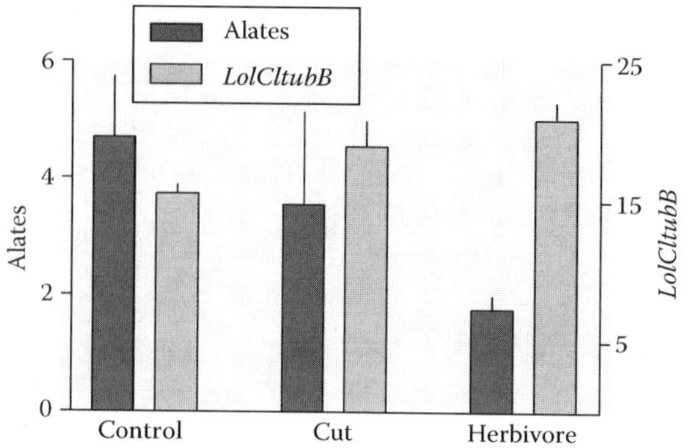

FIGURE 18.1 Effects of damage type on relative *lolC* expression (dark bars, left y-axis) and *N*-acetylnorloline concentration (light bars, right y-axis) in *Epichloë glyceriae* infected *Glyceria striata*. The difference in *lolC/tubB* is significant between control and cut treatments (Tukey's HSD, $p = 0.026$). *N*-acetylnorloline was significantly higher in the cut treatment relative to the control and herbivore treatments (Kruskal–Wallis H, $p = 0.002$).

constitutive defenses for its hosts (Leuchtmann et al., 2000). It was recently found, however, that after mechanical damage, *E. glyceriae* will produce lolines detectable by both an aphid bioassay and by quantifying *lolC* transcript (Figure 18.1). While *Epichloë* spp. are not considered defensive mutualists because they seldom provide constitutive defenses to their hosts, these results suggest they still contain the genetic mechanisms to provide antiherbivore compounds.

In summary, even with data from just a few endophyte species, it is clear that there is considerable variation in the plasticity of the defense response, likely due to the history of hybridization in the group, although varying selection pressures due to variation in herbivory and resource availability probably also play a role, and caution should be exercised about making any generalizations. Also, while the impacts on specific herbivores are clear, it is not known how induced responses in endophytes affect the rest of the community.

18.3 Nutrient Availability

18.3.1 Introduction

An environmental factor that could modulate the defensive mutualism is the availability of nutrients to plants. In particular, soil nitrogen (N) is likely a nutrient of prime importance because the ring structure of alkaloids contains N (Bush et al., 1993; Bush et al., 1997; Schardl et al., 2007). Hence, their production should depend upon N availability. Furthermore, herbivores often prefer and perform better when feeding on plant tissues high in N (Mattson, 1980; Slansky and Rodriquez, 1982); therefore, N should be a key nutrient influencing the interaction between grasses, their endophytic fungi, and herbivores.

18.3.2 Graphical Model

We developed a graphical model that predicts how N availability should influence N content in plant tissues, alkaloid production by the fungus, and herbivore performance. Our intent is to provide a framework for future studies on N and the grass–endophyte interaction. We make the following assumptions: (1) plant absorption of N increases linearly with soil N concentration, (2) endophytes use N from the plant for their production of alkaloids, and (3) herbivores perform better on N-rich plant tissues. Our model has similarities to the model presented by Faeth and Fagan (2002) except that they were concerned with uninfected grasses as well as low and high alkaloid-producing endophytes. They also were ultimately interested in N flux in plants, while we are primarily concerned with herbivore performance.

We assume that N uptake by roots should increase in a more or less linear fashion with N concentration in the soil (Figure 18.2a); as has been found experimentally in some plants (Gray and Schlensinger, 1983).

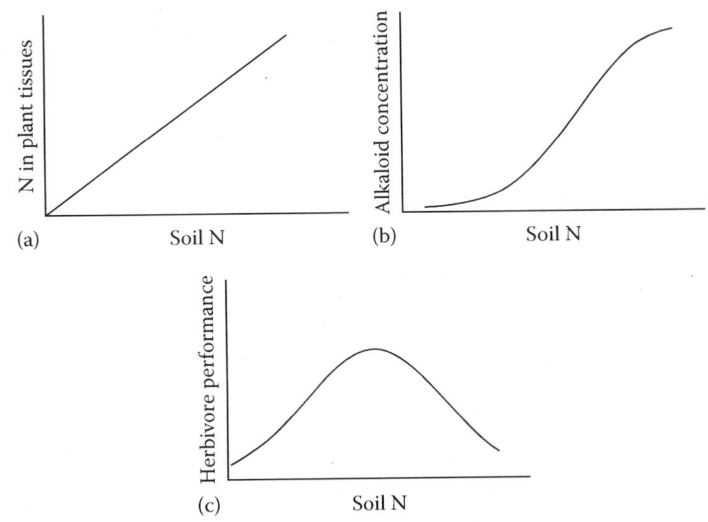

FIGURE 18.2 Graphical model of hypothetical impacts of changes in soil nitrogen (N) on: (a) nitrogen content in grass tissues, (b) concentration of endophyte-produced alkaloids, and (c) herbivore performance. Our model predicts that the protection the plant/endophyte symbiotum provides against herbivores will be greatest at intermediate soil N levels. At high soil N levels, alkaloids will counter benefits to herbivores of high N in plant tissues and at low soil N levels, protection is not needed due to low suitability of plant tissues (low N content).

At very high concentrations, likely beyond that which would occur even in fertilized soils, the curve should level off, but for our purposes we assume a linear response. While it is possible endophytes may increase nutrient uptake by altering fine root structure and modifying chemical environments near root zones (Malinowski et al., 1999). Evidence for this is limited to tall fescue and may not be a general phenomenon, so we have not included that possibility.

The dose response by endophytes to N availability in terms of alkaloid production is not known. The only published paper on nutrient manipulation that included more than two treatment levels is that by Cheplick et al. (1989) in which they used no nutrients, 10% Peters solution and 100% Peters solution (1.5 g/L 20–20–20). While they did not measure alkaloid production, they did find a somewhat geometric increase in plant productivity. We suggest alkaloid production would follow a similar pattern and would level off at very high levels of soil N; thus, we suggest a sigmoidal response in alkaloid production to increases in N availability to the plant (Figure 18.2b).

Combining Figures 18.2a and b should define conditions for herbivore performance. That is, increasing N in plant tissue generally leads to elevated herbivore performance (McNeill and Southwood, 1978; Mattson, 1980), while higher levels of alkaloids should reduce herbivore performance. Accounting for both of these effects predicts that herbivore performance should be highest at intermediate soil N levels (Figure 18.2c). Therefore, we predict plants will benefit most from a defensive mutualism with endophytes at high levels of soil N; plants at intermediate N levels will suffer greater herbivory while those at very low N levels will not benefit from defense due to the low suitability of their tissues for herbivores. There is some evidence for these predictions.

18.3.3 Empirical Evidence

Several researchers have found that alkaloid production by endophytes increases with N availability. For example, ergot alkaloids were produced in higher concentrations with increasing N availability to tall fescue (Lyons et al., 1986; Belesky et al., 1988; Arechavaleta et al., 1992). Similarly, nutrient augmentation of perennial ryegrass resulted in elevated peramine (Latch, 1993; Krauss et al., 2007) and lolitrem B levels (Latch, 1993). Unfortunately, studies in which N availability has been manipulated over a wide gradient have not been performed, so the exact shape of the curve of alkaloid response to changes in soil N is not known. Moreover, soil fertility is not always positively correlated with alkaloid concentration (see Faeth et al., 2002; Rasmussen et al., 2006). Future studies that determine the shape of the curve for individual alkaloids in multiple grass/endophyte associations would help us to better understand the conditions influencing the defensive mutualism.

Based on the defensive mutualism hypothesis one can predict that plant growth would be greater with greater N availability and might be less than that for uninfected grasses under conditions of very low N availability (due to the physiological cost of harboring the endophyte when nutrients are limiting). This prediction was met in a greenhouse experiment with tall fescue (Cheplick et al., 1989). Infected seedlings receiving high nutrient input obtained greater biomass, while those receiving no nutrient addition had lower biomass compared to uninfected plants. Cheplick (2007) also demonstrated a cost of infection in perennial ryegrass genotypes from Eurasia and North Africa. Root:shoot ratio and the proportion of shoots that was alive and photosynthetic was reduced in infected compared to uninfected grasses under extremely poor nutrient conditions. Ahlholm et al. (2002) conducted greenhouse experiments with the grasses *L. pratense* and *Festuca rubra*. They manipulated NPK fertilizer (low and high treatments) and found *L. pratense* infected with *Neotyphodium uncinatum* showed decreased performance in nutrient-poor soil. Results for *F. rubra* infected with *Epichloë festucae* showed plants performed better when fertilized, but did not show a cost of infection at low nutrient levels. In contrast, Lewis (2004) found infected perennial ryegrass at both low and high levels of N fertilizer produced more biomass than uninfected plants. However, low N-treated plants did receive weekly liquid fertilizer, so the low fertilizer treatment may not have been low enough to uncover a cost of infection. Faeth and Sullivan (2003) also failed to uncover a cost of infection. They transplanted Arizona fescue (*Festuca arizonica*) plants into the field. Some plants received augmented nutrients while others received none. They found no advantage of infection (and even found some disadvantages of infection) for various measures of plant growth and reproduction and no interaction between infection and nutrients. Yet, their plants were in native (albeit low nutrient) soil, where there may have been enough nutrients to negate any cost of infection.

If N availability and alkaloid production are generally linked, one would expect endophyte-infected grasses to be more resistant to herbivores when growing in fertile soil. Davidson and Potter (1995) found fall armyworm weight gain and developmental rate were not greater when feeding on fertilized compared to unfertilized tall fescue infected with *N. coenophialum*. Preference by aphids (*Schizaphis graminum* and *R. padi*) showed a similar pattern (Davidson and Potter, 1995). In like manner, Lehtonen et al. (2005) found that bird-cherry oat aphid performance decreased on endophyte-infected meadow fescue with increasing availability of nutrients. The generality of these findings for other grass/endophyte associations awaits future research.

There are very few studies that have explored herbivore defense across varied nutrient conditions in nonagronomic grasses. In a study in subarctic Finland, Wali et al. (2008) studied meadow fescue (*L. pratense*), a forage grass used widely in Finland, in experimental fields. Seeds were transplanted into the soil at experimental plots which differed in soil fertility and the incidence of snow mold pathogens. Two different grass cultivars were used (Kasper and Salten). They found that endophyte infection enhanced plant performance at the high-nutrient site, but that the results were most pronounced for one of the two cultivars. Further, they detected a cost of infection at the low nutrient site. Their results show that the benefits of endophyte infection can depend upon cultivar and abiotic environmental conditions.

We have recently conducted an experiment with tall fescue. Seeds of Kentucky-31 infected with the endophyte *N. coenophialum* were planted into pots of a mixture of Sunshine LC1 potting soil, sand, vermiculite in a 1:1:1 ratio. All plants were fertilized with a 1.5 g/L solution of 15% N: 16% P: 17% K once at 5 weeks after germination. Following this, plants were randomly assorted into one of four treatment groups: 0, 300, 600, or 1000 ppm N in the form of NH_4NO_3. All plants received other macro- and micro-nutrients via Hoaglands solution (100 mL every week), with pH adjusted to 6.0–6.3. At 12 weeks a bioassay was conducted using bird-cherry oat aphid preference. Nitrogenous compounds are a limiting resource for aphids (Weibull, 1987; Terra, 1988), and generalist aphids use amino acids as a cue in selecting their host (Tosh et al., 2003). Thus, we decided to assess aphid preference among our treatment plants. One pot of each nutrient treatment was placed into the corners of screen cages (44.1 cm × 44.1 cm × 41.1 cm). A heavily aphid-infested pot of barley was placed into the center of each cage. There were 12 replicated cages within each treatment. After 5 days the number of winged aphids present on each experimental

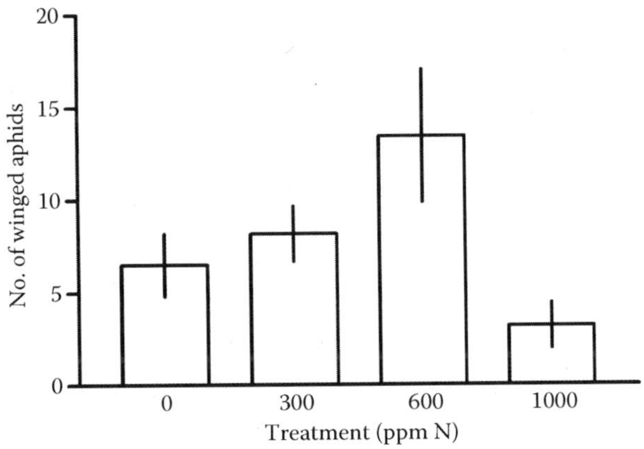

FIGURE 18.3 Number of winged *R. padi* aphids colonizing tall fescue plants grown under varied N availability. There was a significant difference among treatments ($F_{3,30} = 2.89$, $p < 0.05$).

plant was counted. We found that the number of winged aphids (standardized for plant dry mass) colonizing treatment plants was nonrandom with aphids colonization showing a peak at intermediate levels of N application (Figure 18.3). These results agree with the prediction from our graphical model (Figure 18.2c) that plants will benefit most from a defensive mutualism with endophytes at high levels of soil N and will experience the highest levels of herbivory at intermediate N availability.

18.4 Multitrophic Interactions

18.4.1 Empirical Evidence

Given the sometimes strong negative effects of endophytes on primary consumers, one might expect these fungal symbionts, and the alkaloids they produce, to have influences that reach to secondary consumers and beyond. Ecologists have increasingly recognized the importance of trophic cascades in community structuring (Oksanen et al., 1981; Polis, 1999). Bottom-up cascades occur when levels of resources and productivity at the base of the food web regulate productivity at higher levels, while top-down cascades occur when higher-level consumers are important in regulating levels below them. Endophytes may play roles in both types of cascades. The role of endophytes in multitrophic systems was reviewed by Faeth and Bultman (2002) and more recently by Chaneton and Omacini (2006). Thus, we focus our review on the most recent contributions and interested readers should consult these earlier reviews for a more complete picture of what is known about grass endophytes in multitrophic systems. We end with a discussion of a theoretical model aimed at predicting how endophytes might alter a grass–herbivore–parasitoid tritrophic interaction.

The large sheep industry of New Zealand rests primarily on perennial ryegrass as a forage plant and much of the grass is infected with *N. lolii* (Easton and Fletcher, 2006). Lolitrem B alkaloid produced by the fungus causes a neurotoxic syndrome called ryegrass staggers in sheep (Fletcher and Harvey, 1981). An additional challenge in this pastoral system is Argentine stem weevil, an important pest of the grass. While the endophyte provides protection against the weevil (through peramine alkaloid production) (Popay et al., 1995), additional biocontrol has been sought through the introduction of a parasitoid (*Microctonus hyperodae*) from South America where it is a native, natural enemy of the weevil (Goldson et al., 1990, 1993). The use of two biocontrol methods, one bottom-up (endophyte-infected grass) and one top-down (parasitoids), sets the stage for questions about how these two processes interact. Are they compatible with one another or does one interfere with the other?

Goldson et al. (2002) found that parasitism decreased with increasing peramine alkaloid concentration in ryegrass but because peramine concentration did not correlate with weevil density, the investigators felt that lower parasitism resulted from reduced parasitoid searching/attack efficiency in highly infected paddocks (and not from a reduction in parasitoid performance). Nonetheless, endophyte infection did decrease parasitoid development and survival in laboratory experiments (Barker and Addison, 1996; Bultman et al., 2003). This result is qualitatively similar to that found for a tritrophic system based on endophyte-infected tall fescue in North America (Bultman et al., 1997). Furthermore, *M. hyperodae* survival and development varied when reared from weevil hosts fed different isolates of the fungus (Bultman et al., 2003). From this preliminary work it appears that bottom-up control and top-down control may be, to some extent, at odds with one another. Endophyte infection provides resistance to the weevil, but may diminish the effectiveness of top-down control through reduced parasitoid preference or performance. Yet, in a well-conceived study, Urrutia et al. (2007) found no effect of *N. lolii* in the diet of Argentine stem weevil on parasitoid development, adult size (tibia length) or egg complement. Much work remains to be done in this intriguing agroecosystem which provides an opportunity to better understand the interaction of opposing trophic cascades.

The University of Zurich laboratory of the late Christine Müller also actively studied endophytes within multitrophic systems. One study had a multifactorial design in which fertilizer, endophyte infection, and plant cultivar were manipulated (Krauss et al., 2007). Perennial ryegrass with or without infection by *N. lolii* was established in trays placed in soil in an experimental field site. Aphids and their parasitoids were allowed to naturally colonize the grasses. They found that fertilizer increased plant biomass and abundance of both aphids and parasitoids; there was a strong bottom-up trophic cascade. They also found that fertilizer resulted in elevated levels of peramine alkaloids in infected plants. Oddly, endophyte infection did not reduce aphid or parasitoid abundances—endophytes did not cause a bottom-up cascade, in contrast to similar studies (for example, Omacini et al., 2001). The authors suggested that relatively low levels of peramine found in their infected plants may explain the lack of aphid resistance (*N. lolii* does not produce lolines).

Harri et al. (2008) have also recently studied trophic cascades in perennial ryegrass-endophyte systems. They used the aphid *Metopolophium festucae* and its primary parasitoid *Aphidius ervi* in experiments conducted on potted plants covered with inverted PET bottles. They found the endophyte had no effect on aphid performance. Similarly, infection did not influence parasitism rate or parasitoid survival, but did strongly reduce parasitoid reproduction. Hence, their work shows a delayed effect of endophytes that would have been missed if they had not followed parasitoids through to reproduction. Moreover, the study also shows that the defense endophytes provide their host can be circumvented if herbivores can withstand the endophyte-produced alkaloids and transfer toxic the effects to natural enemies.

In a laboratory-based study, de Sassi et al. (2006) found that ladybird beetles (*Coccinella septempunctata*) preying on *R. padi* aphids feeding on endophyte-infected perennial ryegrass showed reduced fecundity and impaired reproductive performance compared to those preying on aphids fed uninfected grass. As with the New Zealand perennial ryegrass and North American tall fescue systems, the work of Harri et al. (2008) and de Sassi et al. (2006) suggests the defensive mutualism endophytes provide may be weakened when natural enemies are considered.

Studies on entomopathogenic nematodes mirror the results found for insect parasitoids and predators. Black cutworm (*Agrotis ipsilon*) was less susceptible to the *Steinernema carpocapse* nematode when it consumed endophyte-infected perennial ryegrass (Kunkel et al., 2004). The mechanism appears to be sequesterization of endophyte-produced alkaloids, some of which suppress nematode symbiotic bacteria (which normally aid in killing the host). In like manner, fall armyworm fed endophyte-infected perennial ryegrass experienced lower mortality from *S. carpocapsae* nematodes (Richmond et al., 2004). Thus, as with parasitoids, endophytes weaken the top-down cascades initiated by predacious nematodes.

Another group of predators that may be affected by endophytes is the spiders. Finkes et al. (2006) used replicated, successional fields in which endophyte infection (*N. coenophialum*) in tall fescue was manipulated. They found the number of spider families and morphospecies was greater in the absence of endophyte. They also found a shift in spider species composition showing that the invasive endophyte-infected tall fescue can alter the structure of these grassland food webs. They attributed the effects on spiders to reductions in abundance of spider prey by endophyte-infected grass, a trend that would follow what Bernard et al. (1997) found in tall fescue pastures. Interestingly, endophyte infection in perennial ryegrass resulted in more (not less) predatory invertebrates in paddocks in New Zealand (Prestidge and Marshall, 1997). Since many variables (agricultural vs. nonagricultural setting, climate, geography, etc.) differed between these studies it is difficult to draw conclusions. Nonetheless, the difference between endophyte-infected perennial ryegrass and tall fescue may be a major contributing factor, the latter of which may have stronger negative effects on insect herbivores than most if not all other grass/endophyte associations.

An intriguing study by Lehtonen et al. (2005) shows how sequesterization of endophyte mycotoxins is not limited to some herbivores. They found that meadow fescue infected with *N. uncinatum* can be parasitized by the root hemiparasitic plant *Rhinanthus serotinus*. The hemiparasite acquires mycotoxins from the endophyte which in turn enhance its resistance to an aphid herbivore. Consequently, the hemiparasite realizes increased performance while its host, meadow fescue, suffers. As with weevils, aphids, cutworm, and fall armyworm, the protection endophytes may provide can be altered when other species, like hemiparasites, are considered.

While the relative impact of beneficial (protection from herbivores) and detrimental (negative effect on natural enemies and concomitant release of herbivore populations) effects on plants have not been thoroughly quantified, it is likely benefits outweigh costs in most asexual agronomic grasses. The net effect of endophytes on other grasses is far less clear.

18.4.2 Mathematical Model

A theoretical modeling approach to the multitrophic level interaction is attractive for several reasons. First, mathematical models can serve as a guide to empirical work; they will reveal if the parameters we choose to include in the model are in fact the key parameters in the system. If our model poorly describes nature, we can begin to explore other parameters. In this way they can suggest possibilities for experiments or observational protocols (Roughgarden et al., 1989). Second, while tritrophic level interactions are generally accepted as appropriate frameworks within which to study plant–herbivore interactions (Price et al., 1980), very little theoretical work has been developed to describe them. Furthermore, a theoretical approach here will reveal, in a predictive way, how individual factors interact to influence plants; hence, the experiments suggested from theoretical predictions are particularly key to determining the mutualistic benefits endophytes provide grasses.

In a first attempt at a predictive model, we have chosen to use traditional Lotka–Volterra differential equations of population growth, one for each species:

$$\mathrm{d}G/\mathrm{d}t = aG[1 - (G/k)] - bH[G/(d + G)] \tag{18.1}$$

$$\mathrm{d}H/\mathrm{d}t = \alpha bH[G/(d + G)] - mPH - qH \tag{18.2}$$

$$\mathrm{d}P/\mathrm{d}t = BmPH - sP \tag{18.3}$$

where Equations 18.1 through 18.3 are equations for the growth of grass, herbivores, and parasitoids, respectively. G is the mass of grass, H the mass of herbivores, and P the number of parasitoids. The maximum potential growth rate of the grass is $a * b$ while that for the herbivore is α. The carrying capacity of the grass is k and b is the rate at which herbivores consume grass. The death rate of herbivores dying from causes other than parasitization is q. The number of herbivores being parasitized is a function of the number of herbivores, the number of parasitoids, and the attack rate of parasitoids once they encounter herbivores (m). The maximum potential growth rate of parasitoids is $B * m$ and their death rate is s. We assume that each species experiences density-dependent growth (or loss). We also assume that growth in the grass is limited by the herbivore and that growth in the herbivore is limited by its consumption of grass. Likewise, we assume growth of the parasitoid depends upon how many herbivores it parasitizes. Finally, we assume that each species grows in a continuous rather than discrete fashion. All of these assumptions are over-simplifications, but our intent is not to develop a model of exact realism, but rather to attempt to capture the important aspects of the interaction.

To begin studying the model, we analyzed the coexistence equilibria of the system. These are points where the three species coexist and their densities do not change. Of interest is the stability of the equilibrium points. An equilibrium point is stable if species densities perturbed from coexistence equilibrium densities return to those equilibrium values, it is unstable otherwise. Stable equilibria tell us that under some set of conditions the species will tend to coexist at constant densities, while unstable equilibria imply population cycling or species extinction. We found one coexistence equilibrium point in our model that was stable or unstable depending on parameter values.

TABLE 18.1

Values of Model Parameters Used to Generate Stabile Equilibrium Points

Parameter	E+ Value	E− Value	Source[a]
a	5.1%	4.6%	Literature
b	628.6%	239.6%	Laboratory
α	14.5%	15.7%	Laboratory
m	0.129 wasps/m^2	0.173 wasps/m^2	Laboratory
q	1%	1%	Laboratory
s	15%	15%	Laboratory
B	160 eggs/g[b]	160 eggs/g[b]	Estimate
k	863.5 g/m^2	781.3 g/m^2	Literature
d	100 g/m^2	100 g/m^2	Estimate

Note: Data are for Tall Fescue infected (E+) or uninfected (E–) with *Neotyphodium coenophialum.*

[a] values determined empirically in the laboratory at Hope College, from the literature or mathematically estimated.

[b] eggs/gram of caterpillar that survive to adulthood.

Using empirical data for uninfected or infected tall fescue that students either collected in the laboratory or obtained from the literature (Table 18.1), we inserted values for each parameter into our model and obtained a predicted trajectory of density change for each species (tall fescue—with and without *N. coenophialum* infection, fall armyworm, and *Euplectrus comstockii* parasitoids, see Bultman et al., 1997) (Figures 18.4 and 18.5). Two outcomes of Figures 18.4 and 18.5 are noteworthy. First, our model predicts greater cycling among the densities of the three species when tall fescue is infected with fungal endophyte. Both the density of caterpillars and the density of wasps reach greater maximums than they would in a system with uninfected grass. The interactions are stronger and more asymmetric. Intuitively, this makes sense in that endophyte-infected tall fescue has negative (ranging from mild to moderate) impacts on fall armyworm performance (Clay et al., 1985) and fairly strong negative influences on parasitoid performance (Bultman et al., 1997). These effects should intensify fluctuations in densities of

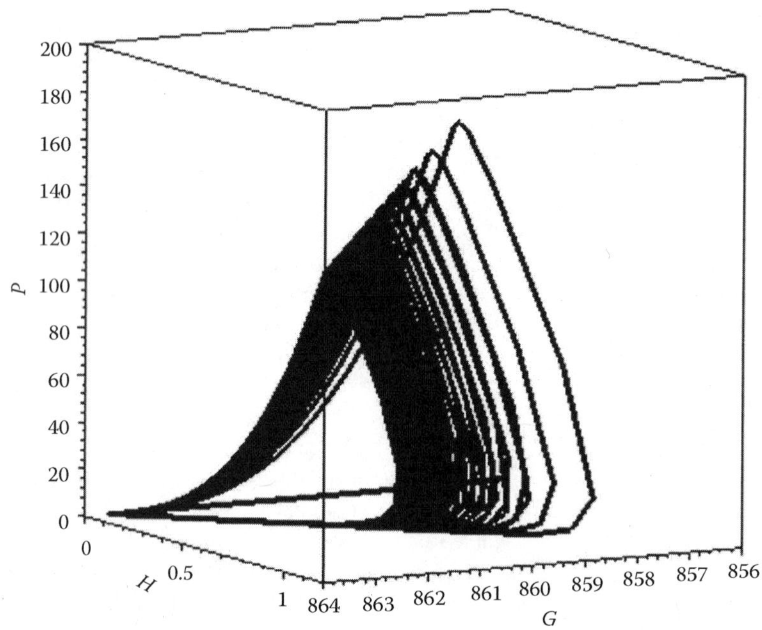

FIGURE 18.4 Changes in densities of uninfected plant biomass (*G*), herbivore biomass (*H*), and parasitoid numbers (*P*) over time as predicted by our nonlinear model of differential equations (see text).

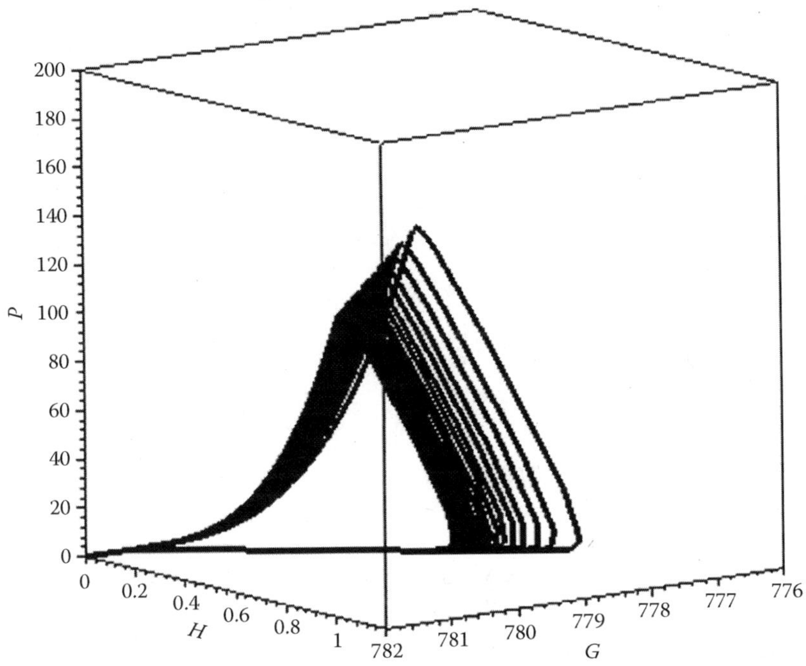

FIGURE 18.5 Changes in densities of plant (infected with *N. coenophialum*) biomass (*G*), herbivore biomass (*H*), and parasitoid numbers (*P*) over time as predicted by our nonlinear model of differential equations (see text).

the three species over time. Second, from the viewpoint of the grass, the presence of the endophyte is advantageous—grass biomass is greater at the equilibrium point than it is when the grass lacks fungal endophyte. The advantage of the mathematical model is that it simultaneously accounts for all three species in the interaction; trying to make a prediction about grass biomass production based on individual experiments on herbivore feeding and others on herbivore–parasitoid interactions would be very difficult if not impossible. Only when one puts all the species together, as the model does, can one begin to see the overall impact of fungal infection on the grass. The next step in this process is to conduct mesocosm or field experiments in which all three species are together and collect data that evaluate predictions of the model-like density fluctuations and total biomass accumulation, for example.

18.5 Are Grass Endophytes Defensive Mutualists?

The defensive mutualism hypothesis, which was primarily based on data from tall fescue/*N. coenophialum* (Cheplick and Clay, 1988; Clay, 1988), can be a useful framework by which to study grass–endophyte interactions. However, many investigators have assumed that all grass–endophyte interactions are mutualistic (reviewed in Saikkonen et al., 2004, 2006). Accumulating evidence suggests this assumption is often false and that variation in plant and fungal genotype, environmental conditions, mode of fungal transmission, and affects of higher order consumers can all modify the relationship between grass host and its fungal symbiont (Saikkonen et al., 1998; Müller and Krauss, 2005).

Faeth (2002) questioned if endophytes, particularly those in nonagronomic, native grasses, act as defensive mutualists. We assert that we lack the data to adequately answer this question. For domesticated agronomic grasses, like tall fescue and perennial ryegrass, the answer appears to be "yes" with little qualification. However, data for nonagronomic plants are limited. The best studied is Arizona fescue (*Festuca arizonica*), which Stan Faeth's group has been working with in the southwestern United States. The grass can be infected with a *Neotyphodium* endophyte that is strictly asexual. Thus, one might expect it to protect its host from herbivores (Faeth and Sullivan, 2003). Despite this expectation, there has been no evidence that the endophyte provides herbivore defense (Lopez et al., 1995; Saikkonen et al., 1999; Tibbets and Faeth,

1999). Most questions about the mutualistic benefits grass endophytes provide have stemmed from the now extensive studies on this grass. Unfortunately, beyond that there are few studies of nonagronomic grasses.

In contrast to Arizona fescue, *Brachypodium sylvanticum* (native to Europe) infected with *E. sylvantica* shows strong resistance to fall armyworm, compared to uninfected plants (Brem and Leuchtmann, 2001). The fungus can, but rarely does, form choke. So, it generally acts like an asexual endophyte. Herbivore resistance was also found for a native fescue in subarctic Yukon. Koh and Hik (2007) studied *Festuca altaica*, a perennial bunchgrass that is circumboreal in range (Gleason and Cronquist, 1963). They studied plants occurring along a grazing gradient from within a boulder field (where grazing by marmots and pikas has historically been high) to 80 m outside the boulder field into a meadow (where grazing pressure was lower). They found endophyte infection of grasses from within the boulder field deterred the vertebrate herbivores, but those from outside the boulder field did not. Furthermore, frequency of endophyte infection and hyphal density within plants both were greatest in areas where grazing pressure was highest. The basis for this grazing deterrence is as yet unknown. These results suggest intraspecific variation in the plant/fungus symbiotum that is finely tuned to herbivore grazing pressure.

Another native North American perennial grass, *Achnatherum robustum* (sleepy grass), has long been reported to be toxic and narcotic to livestock. The grass can be infected with a *Neotyphodium* endophyte (an as yet unidentified species). However, careful sampling from populations of the grass revealed that only one population produces high levels of alkaloids, with alkaloid concentration declining with increasing distance from the focal population (Faeth et al., 2006). Hence, protection the endophyte provides appears to vary spatially and is often low. As with *F. altaica*, sleepy grass and its endophyte may show intraspecific variation that is finely tuned to herbivore grazing pressure (although this has not been tested). A congener of sleepy grass, *Achnatherum inebrians* (drunken horse grass), is native to parts of China. Li et al. (2006) found that naturally occurring plants infected with *Neotyphodium gansuense* had lower numbers of mites (*Tetranychus cinnabarinus*) and aphids (*R. padi*).

Lastly, we found that *G. striata*, a native grass in eastern North America and infected with *E. glyceriae*, provides wound-inducible resistance to aphids (Gonthier et al., 2008, see above). Beyond these examples, there are virtually no data that test the defensive mutualism hypothesis in nonagronomic, native grasses. Yet, based on this small sample, it appears that nonagronomic, native grasses can provide some herbivore resistance, but the protection is quite variable.

18.6 Future Directions in Grass/Endophyte Research

As we stated above, one clear need in grass/endophyte research continues to be the study of nonagronomic, native grass species. Research to date has been dominated by agronomic grasses, which can skew our interpretation of the interaction in general (Saikkonen et al., 2006). We need data on a wide range, both in transmission mode and phylogeny, of grass endophytes.

A second area of need is field experiments. Studies by Rudgers et al. (2004), Faeth and Sullivan (2003), Omacini et al. (2001), and Krauss et al. (2006) all show the power of carefully designed field experiments. Studies in which natural colonization of herbivores and their natural enemies is allowed and which are conducted for a long term show great promise in shedding light on the grass/endophyte system. These studies have the advantage of allowing all possible interactions to occur simultaneously, rather than focusing on two-way interactions as in most laboratory-based studies. Coupling field experiments with predictions from theoretical models (like those developed above, for example) should provide insight into the important mechanisms operating to produce the patterns observed in these experiments.

Because N should be a key nutrient in the grass/endophyte interaction (see above), we feel more studies are needed to understand its affects. Most of these studies are best conducted in the laboratory/greenhouse where N concentration can be carefully controlled. Ultimately, the study of N and other nutrients as well as possible plant–endophyte signaling compounds will shed light on how plant and fungus respond to one another, to consumers, and to changes in soil fertility.

Finally, recent work on inducible resistance shows that this can be an important aspect of the protection endophytes offer their hosts (see above). We need to test for this in a wider range of plant species, particularly nonagronomic, native species. We need more work on *Epichloë* species to see if the trait of inducibility is ancestral, as recent work suggests it is (Gonthier et al., 2008). Further, we need to expand the initial investigations (Sullivan et al., 2007b) of the molecular mechanism of inducibility. Beyond upregulation of loline biosynthesis genes, what other fungal and plant genes are up- and down-regulated following damage to the plant? And, what role are these genes playing in the inducible response? Microarray analyses should provide avenues by which these questions can begin to be probed.

ACKNOWLEDGMENTS

Undergraduate researchers T. Bowman, K. Harrison, L. Jackson, L., B. Jager, R. Johnson, B. McMahon, J. Molenhouse, M. Nelsen, C. Rose, J. Skoug, D. Visser, and A. Zambrano all contributed to the development of the tritrophic mathematical modeling and gathering empirical data for model parameters. Funding was provided by NSF-CRUI (DBI-03300840) to TLB and NSF-REU (DBI-0139035) to the Biology Department at Hope College.

REFERENCES

Ahlholm, J.U., Helander, M., Lehtimäki, M., Wäli, P., and Saikkonen, K. (2002). Vertically transmitted fungal endophytes: Different responses of host-parasite systems to environmental conditions. *Oikos 99*: 173–183.

Arechavaleta, M., Bacon, C.W., Plattner, R.D., Hoveland, C.S., and D.E. Radcliffe. (1992). Accumulation of ergopeptide alkaloids in symbiotic tall fescue grown under deficits of soil water and nitrogen fertilizer. *Appl. Enviorn. Micro. 58*: 857–861.

Barker, G.M. and Addison, P.J. (1996). Influence of clavicipitaceous endophyte infection in ryegrass on development of the parasitoid *Microctonus hyperodae* Loan (Hymenoptera: Braconidae) in *Listronotus bonariensis* (Kuschel) (Coleoptera: Curculionidae). *Biol. Control 7*: 281–287.

Belesky, D.P., Stuedemann, J.A., Plattner, R.P., and Wilkinson, S.R. (1988). Ergopeptine alkaloids in grazed tall fescue. *Agron. J. 80*: 209–212.

Bernard, E.C., Gwinn, K.D., Pless, C.D., and Williver, C.D. (1997). Soil invertebrate species diversity and abundance in endophyte-infected tall fescue pastures. In *Neotyphodium/Grass Interactions*, C.W. Bacon and N.S. Hill (Eds.). Plenum Press, New York, pp. 124–135.

Breen, J.P. (1994). Acremonium endophyte interactions with enhanced plant-resistance to insects. *Annual Review of Entomology 39*: 401–423.

Brem, D. and A. Leuchtmann. (2001). *Epichloe* grass endophytes increase herbivore resistance in the wood land grass *Brachypodiumsylvaticum. Oecologia 126*: 522–530.

Bush, L.P., Gray, S., and Burhan, W. (1993). Accumulation of alkaloids during growth of tall fescue. In *Proc. XVII Int. Grassland Congr.* Palmerston North, New Zealand, pp. 1379–1381.

Bush, L.P., Wilkinson, H.W., and Schardl, C.L. (1997). Bioprotective alkaloids of grass-fungal endophyte symbioses. *Plant Physiol. 114*: 1–7.

Bultman, T.L., Bell, G., and Martin, W.D. (2004). A fungal endophyte mediates reversal of wound-induced resistance and constrains tolerance in a grass. *Ecology 85*: 679–685.

Bultman, T.L., Borowicz, K.L., Schneble, R.M., Coudron, T.A., and Bush, L.P. (1997). Effect of a fungal endophyte on the growth and survival of two *Euplectrus* parasitoids. *Oikos 78*: 170–176.

Bush, L.P., Wilkinson, H.H., and Schardl, C.L. (1997). Bioprotective alkaloids of grass-fungal endophyte symbioses. *Plant Physiol. 114*: 1–7.

Chaneton, E.J. and M. Omacini. (2006). Bottom-up cascades induced by fungal endophytes in multitrophic systems. In *Indirect Interaction Webs: Nontrophic Linkages through Induced Plant Traits*, Eds. Ohgushi, T., Craig, T.P., and P.W. Price, Cambridge University Press.

Cheplick, G.P. (2007). Costs of fungal endophyte infection in *Lolium perenne* genotypes from Eurasia and North Africa under extreme resource limitation. *Environ. Exp. Bot. 69*: 202–210.

Cheplick, G.P. and Clay, K. (1988). Acquired chemical defenses in grasses: The role of fungal endophytes. *Oikos 52*: 309–318.

Cheplick, G.P., Clay, K., and Marks, S. (1989). Interactions between infection by endophytic fungi and nutrient limitation in the grasses *Lolium perenne* and *Festuca arundinacea. New Phytol. 11*: 89–97.

Clay, K. (1991). Fungal endophytes, grasses, and herbivores. In *Microbial Mediation of Plant-Herbivore Interactions*, P. Barbosa, V.A. Krischik, and C.G. Jones (Eds.). Wiley, New York, pp. 199–226.

Clay, K. (1988). Fungal endophytes of grasses: A defensive mutualism between plants and fungi. *Ecology 69*: 10–16.

Clay, K., Hardy, T.N., and Hammond, A.M., Jr. (1985). Fungal endophytes of grasses and their effects on an insect herbivore. *Oecologia 66*: 1–6.

Clay, K. and Holah, J. (1999). Fungal endophypte symbiosis and plant diversity in successional fields. *Science 285*: 1742–1744.

Clay, K. and Schardl, C.L. (2002). Evolutionary origins and ecological consequences of endophyte symbiosis with grasses. *Am. Nat. 160*: S99–S127.

Clement, S.L., Kaiser, W.J., and Eichenseer, H. (1994). *Acremonium* endophytes in germ plasms of major grasses and their utilization for insect resistance. In *Biotechnology of Endophytic Fungi of Grasses*, C.W. Bacon, and J.F. White, Jr. (Eds.). CRC Press, Boca Raton, FL, pp. 185–199.

Craven, K.D., Blankenship, J.D., Leuchtmann, A., Hignight, K., and Schardl, C.L. (2001). Hybrid fungal endophytes symbiotic with the grass *Lolium pratense. Sydowia 53*: 44–73.

Cyr, H. and Pace, M.L. (1993). Magnitude and patterns of herbivory in aquatic and terrestrial ecosystems. *Nature 361*: 148–150.

Davidson, A.W. and Potter, D.A. (1995). Response of plant-feeding, predatory, and soil-inhabiting invertebrates to *Acremonium* endophytes and nitrogen fertilization in tall fescue. *J. Econ. Entomol. 88*: 367–379.

de Sassi, C., Müller, C.M., and Krauss, J. (2006). Fungal plant endosymbionts alter life history and reproductive success of aphid predators. *Proc. R. Soc. B. 273*: 1301–1306.

Easton, H.S. and Fletcher, L.R. (2006). The importance of endophyte in agricultural systems—changing plant and animal productivity. In *Fungal Endophytes of Grasses*, A.J. Popay and E.R. Thom (Eds.). New Zealand Grassland Association, Dunedin, New Zealand, pp. 11–18.

Eichenseer, H., Dahlman, D.L., and Bush, L.P. (1991). Influence of endophyte infection, plant age and harvest interval on *Rhopalosipum padi* survival and its relation to quantity of N-formyl and N-acetyl loline in tall fescue. *Ent. Exp. Appl. 60*: 29–38.

Ewald, P.W. (1994). *Evolution of Infectious Disease*. Oxford University Press, New York.

Faeth, S.H. (2002). Are endophytic fungi defensive mutualists? *Oikos 98*: 25–36.

Faeth, S.H. and Bultman, T.L. (2002). Endophytic fungi and interactions among host plants, herbivores, and natural enemies. In *Multitrophic Level Interactions*, T. Tscharntke and B.A. Hawkins (Eds.). Cambridge University Press, Cambridge, U.K., pp. 89–123.

Faeth, S.H., Bush, L.P., and Sullivan, T.J. (2002). Peramine alkaloid variation in *Neotyphodium*-infected Arizona fescue: Effects of endophyte and host genotype and environment. *J. Chem. Ecol. 28*: 1511–1526.

Faeth, S.H. and Fagan, W.F. (2002). Fungal endophytes: Common hosts plant symbionts but uncommon mutualists. *Intergr. Compar. Biol. 42*: 360–368.

Faeth, S.H., Gardner, D.R., Haues, C.J., Jani, A., Wittlinger, S.K., and Jones, T.A. (2006). Temporal and spatial variation in alkaloid levels in *Achantherum robustum*, a native grass infected with the endophyte *Neotyphodium. J. Chem. Ecol. 32*: 307–324.

Faeth, S.H. and Sullivan, T.J. (2003). Mutualistic asexual endophytes in a native grass are usually parasitic. *Am. Nat. 161*: 316–325.

Finkes, L.K., Cady, A.B., Mulroy, J.C., Clay, K., and Rudgers, J.A. (2006). Plant-fungus mutualism affects spider composition in successional fields. *Ecol. Lett. 9*: 347–356.

Fletcher, L.R. and Harvey, I.C. (1981). An association of a *Lolium* endophyte with ryegrass staggers. *NZ Veterinary J. 29*: 185–186.

Gleason, H.A. and A. Cronquist. (1963). *Manual of Vascular Plants of Northeastern United States and adjacent Canada*, D. Van Nostrand, New York.

Goldson, S.L., McNeill, M.R., Profitt, J.R., Barker, G.M., Addison, P.J., Barratt, B.I.P., and Ferguson, C.M. (1993). Systematic mass rearing and release of *Microctonus hyperodae* (Hym.: Braconidae, Euphorinae), a parasitoid of the Argentine stem weevil *Listronotus bonariensis* (Kuschel) (Col.: Curculionidae) and records of its establishment in New Zealand. *Entomophaga 38*: 527–536.

Goldson, S.L., McNeill, M.R., Stufkens, M.W., Profitt, J.R., Pottinger, R.P., and Farrell, J.A. (1990). Importation and quarantine of *Microctonus hyperodae*, a South American parasitoid of Argentine stem weevil. *Proc. 43rd New Zealand Weed and Pest Control Conf.*, pp. 334–338.

Goldson, S.L., Profitt, J.R., Fletcher, L.R., and Baird, D.B. (2002). Multitrophic interaction between the ryegrass *Lolium perenne*, its endophyte *Neotyphodium lolii*, the weevil pest *Listronotus bonariensis*, and its parasitoid *Microctonus hyperodae. NZ J. Agri. Res. 34*: 227–233.

Gonthier, D.J., Sullivan, T.J., Brown, K.L., Wurtzel, B., Lawal, R., VandenOever, K., Buchan, Z., and Bultman, T.L. (2008). Stroma-forming endophyte *Epichloë glyceriae* provides wound-inducible herbivore resistance to its grass host. *Oikos 117*: 629–633.

Gray, J.T. and Schlensinger, W.H. (1983). Nutrient use by evergreen and deciduous shrubs in southern California. II. Experimental investigations of the relationship between growth, nitrogen uptake and nitrogen availability. *J. Ecol. 71*: 43–56.

Harri, S.A., Krauss, J., and Müller, C.B. (2008). Trophic cascades initiated by fungal plant endosymbionts impair reproductive performance of parasitoids in the second generation. *Oecologia 157*: 399–407.

Karban, R. and Baldwin, I.T. (1997). *Induced Responses to Herbivory*. University of Chicago Press, Chicago, Illinois.

Koh, S. and Hik, D.S. (2007). Herbivory mediates grass-endophyte relationships. *Ecology 88*: 2752–2757.

Krauss, J., Harri, S.A., Bush, L., Husi, R., Bigler, L., Power, S.A., and Müller, C.B. (2007). Effects of fertilizer, fungal endophytes, and plant cultivar on the performance of insect herbivores and their natural enemies. *Funct. Ecol. 211*: 107–116.

Kunkel, B.A., Grewal, P.A., and Quigley, M.F. (2004). A mechanism of acquired resistance against an entomopathogenic nematode by *Agrotis ipsilon* feeding on perennial ryegrass harboring a fungal endophyte. *Biol. Control 29*: 100–108.

Kusnierczyk, A., Winge, P., Midelfart, H., Armbruster, W.S., Rossiter, J.T., and Bones, A.M. (2007). Transcriptional responses of Arabidopsis thaliana ecotypes with different glucosinolate profiles after attack by polyphagous Myzus persicae and oligophagous Brevicoryne brassicae. *J. Exp. Bot. 58*: 2537–2552.

Kutil, B.L., Greenwald, C., Liu, G., Spiering, M.J., Schardl, C.L., and Wilkinson, H.H. (2007). Comparison of loline alkaloid gene clusters across fungal endophytes: Predicting the co-regulatory sequence motifs and the evolutionary history. *Fung. Gen. Biol. 44*: 1002–1010.

Latch, G.C.M. (1993). Physiological interactions of endophytic fungi and their hosts: biotic stress tolerance imparted to grasses by endophytes. *Agri. Ecosys. Environ. 44*: 143–156.

Lehtonen, P., Helander, M., and Saikkonen, K. (2005). Are endophyte-mediated effects on herbivores conditional on soil nutrients? *Oecologia 142*: 38–45.

Leuchtmann, A., Schmidt, D., and Bush, L.P. (2000). Different levels of protective alkaloids in grasses with stroma-forming and seed-transmitted *Epichloë/Neotyphodium* endophytes. *J. Chem. Ecol. 26*: 1025–1036.

Lewis, G.C. (2004). Effects of biotic and abiotic stress on the growth of three genotypes of *Lolium perenne* with and without infection by the fungal endophyte *Neotyphodium lolii*. *Ann. Appl. Biol. 144*: 53–63.

Li, C., Zhang, X., Li, F., Nan, Z., and Schardl, C.L. (2006). Disease and pest resistance of endophyte infected and non-infected drunken horse grass. In *Fungal Endophytes of Grasses*, A.J. Popay and E.R. Thom (Eds.). New Zealand Grassland Association, Dunedin, New Zealand, pp. 111–114.

Lopez J.E., Faeth, S.H., and M. Miller. (1995). Effect of endophytic fungi or herbivory by redlegged grasshoppers (Orthoptera: Acrididae) on Arizona fescue. *Environmental Entomology 24*: 1576–1580.

Lyons, P.C., Plattner, R.D., and Bacon, C.W. (1986). Occurrence of peptide and clavine ergot alkaloids in tall fescue grass. *Science 232*: 487–489.

Malinowski, D.P., Brauer, D.K., and Belesky, D.P. (1999). *Neotyphodium coenophialum*-endophyte affects root morphology of tall fescue grown under phosphorus deficiency. *J. Agron. Crop Sci. 183*: 53–60.

Mattson, J.M., Jr. (1980). Herbivory in relation to plant nitrogen content. *Ann. Rev. Ecol. Syst. 11*: 119–161.

McNeill, S. and Southwood, T.R.E. (1978). The role of nitrogen in the development of insect/plant relationships. In *Biochemical Aspects of Plant and Animal Coevolution*, J.B. Harborne (Ed). Academic Press, London, pp. 77–98.

Moon, C.D., Craven, K.D., Leuchtmann, A., Clement, S.L., and Schardl, C.L. (2004). Prevalence of interspecific hybrids amongst asexual fungal endophytes of grasses. *Mol. Ecol. 13*: 1455–1467.

Müller, C.B. and Krauss, J. (2005). Symbiosis between grasses and asexual fungal endophytes. *Curr. Opin. Plant Bio. 8*: 450–456.

Oksanen, L., Fretwell, S.D., Arruda, J., and Niemela, P. (1981). Exploitation ecosystems in gradients of primary productivity. *Am. Nat. 118*: 240–261.

Omacini, M., Chaneton, E.J., Ghersa, C.M., and Müller, C.B. (2001). Symbiotic fungal endophytes control insect host-parasite interaction webs. *Nature 409*: 78–81.

Polis, G.A. (1999). Why are parts of the world green? Multiple factors control productivity and the distribution of biomass. *Oikos 86*: 3–15.

Popay, A.J., Hume, D.E., Mainland, R.A., and Saunders, C.J. (1995). Field resistance to Argentine stem weevil (*Listonotus bonariensis*) in different ryegrass cultivars infected with an endophyte deficient in lolitrem B. *NZ J. Agri. Res. 38*: 519–528.

Prestidge, R.A. and Marshall, S.L. (1997). The effects of *Neotyphodium*-infected perennial ryegrass on the abundance of invertebrate predators. In *Neotyphodium/Grass Interactions*, C.W. Bacon and N.S. Hill (Eds.). Plenum Press, New York, pp. 195–198.

Price, P.W., Bouton, C.E., Gross, P., McPherson, A.A., Thomspon, J.N., and Weis, A.E. (1980). Interactions among three trophic levels: Influence of plants on interactions between insect herbivores and natural enemies. *Ann. Rev. Ecol. Syst. 11*: 41–65.

Rasmussen, S., Parsons, A.J., Basset, S., Christensen, M.J., Hume, D.E., Johnson, L.J., Johnson, R.D., Simpson, W.R., Stacke, C., Voisey, C.R., Xue, H., and Newman, J.A. (2006). High nitrogen supply and carbohydrate content reduce fungal endophyte and alkaloid concentration in *Lolium perenne. New Phytol. 173*: 787–797.

Richmond, D.S., Kunkel, B.A., Somasekhar, N., and Grewal, P.S. (2004). Top-down and bottom-up regulation of herbivores: Spodoptera frugiperda turns tables on endophyte-mediated plant defense and virulence of an entomopathogenic nematode. *Ecol. Entomol. 29*: 353–360.

Roughgarden, J., May, R.M., and Levin, S.A. (1989). Introduction. In *Perspectives in Ecological Theory*, J. Roughgarden, R.M. May, and S.A. Levin, (Eds.). Princeton University Press, Princeton, NJ, pp. 3–10.

Rudgers, J.A., Koslow, J.M., and K. Clay. (2004). Endophytic fungi alter relationships between diversity and ecosystem properties. *Ecology Letters 7*: 42–51.

Saikkonen, K., Faeth, S.H., Helander, M., and Sullivan, T.J. (1998). Fungal endophytes: A continuum of interactions with host plants. *Annu. Rev. Ecol. Syst. 29*: 319–343.

Saikkonen, K., Helander, M., Faeth, S.H., Schulthess, F., and D. Wilson. (1999). Endophyte-grass-herbivore interactions: The case of *Neotyphodium* endophytes in Arizona fescue populations. *Oecologia 121*: 411–420.

Saikkonen, K., Lehtonen, P., Helander, M., Koricheva, J., and Faeth, S.H. (2006). Model systems in ecology: Dissection the endophyte-grass literature. *Trends Plant Sci. 11*: 428–433.

Saikkonen, K., Wali, P., Helander, M., and Faeth, S.H. (2004). Evolution of endophyte-plant symbioses. *Trends Plant Sci. 9*: 275–280.

Schardl, C.L. and Craven, K.D. (2003). Interspecific hybridization in plant-associated fungi and oomycetes: A review. *Mol. Ecol. 12*: 2861–2873.

Schardl, C.L., Grossman, R.B., Nagabhyru, P., Faulkner, J.R., and Mallik, U.P. (2007). Loline alkaloids: Currencies of mutualism. *Phytochemistry 68*: 980–996.

Schardl, C.L., Leuchtmann, A., Tsai, H.F., Collett, M.A., Watt, D.M., and Scott, D.B. (1994). Origin of a fungal symbiont of perennial ryegrass by interspecific hybridization of a mutualist with the ryegrass choke pathogen, *Epichloë typhina. Genetics 136*: 1307–1317.

Slansky, F.E. and Rodriquez, J.G. (1982). *Nutritional Ecology of Insects, Mites, Spiders and Related Invertebrates*. Wiley, New York.

Spiering, M.J., Moon, C.D., Wilkinson, H.H., and Schardl, C.L. 2005. Gene clusters for insecticidal loline alkaloids in the grass-endophytic fungus *Neotyphodium uncinatum. Genetics 169*: 1403–1414.

Stamp, N. 2003. Out of the quagmire of plant defense hypotheses. *Quart. Rev. Biol. 78*: 23–55.

Sullivan, T.J., Bultman, T.L., Rodstrom, J., Vandop, J., Librizzi, J., Graham, C., Sielaff, A., and Fernandez, L. (2007a). Inducible defenses provided by *Neotyphodium* to *Lolium arundinacea* and *Lolium pratense*: An ecological and molecular approach. In *Proc. 6th Int. Symp. Fungal Endophytes Grasses*, A. Popay and E.R. Thom (Eds.). New Zealand Grassland Association, Dunedin, New Zealand, pp. 59–62.

Sullivan, T.J. and Faeth, S.F. (2008). Local adaptation in *Festuca arizonica* infected with hybrid and nonhybrid *Neotyphodium* endophytes. *Microb. Ecol. 55*: 687–704.

Sullivan, T.J., Rodstrom, J., Vandop, J., Librizzi, J., Graham, C., Schardl, C.L., and Bultman, T.L. (2007b). Symbiont-mediated changes in Lolium arundinaceum inducible defenses: Evidence from changes in gene expression and leaf composition. *New Phytol. 176*: 673–679.

Terra, W. (1988). Physiology and biochemistry of insect digestion: An evolutionary perspective. *Brasil. J. Med. Biol. Res. 21*: 675–734.

Tibbets, T.M. and S.H. Faeth. (1999). *Neotyphodium* endophytes in grasses: Deterrents or promoters of herbivory by Leaf-cutting ants? *Oecologia 118*: 297–305.

Torres, M.S., Singh, A.P., Vorsa, N., Gianfagna, T., and White, J.F. (2007). Were endophytes pre-adapted for defensive mutualism? In *Proc. 6th Int. Symp. Fungal Endophytes Grasses*, A. Popay and E.R. Thom (Eds.). New Zealand Grassland Association, Dunedin, New Zealand, pp. 62–68.

Tosh, C.R., Powell, G., Holmes, N.D., and Hardie, J. (2003). Reproductive response of generalist and specialist aphids with the same genotype to plant secondary compounds and amino acids. *J. Insect Physiol. 49*: 1173–1182.

Tsai, H.F., Liu, J.S., Staben, C., Christensen, M.J., Latch, G.C.M., Siegel, M.R., and Schardl, C.L. (1994). Evolutionary diversification of fungal endophytes of tall fescue grass by hybridization with *Epichloë* species. *Proc. Natl. Acad. Sci. USA 91*: 2542–2546.

Urrutia, M.A., Wade, M.R., Phillips, C.B., and Wratten, S.D. (2007). Influence of host diet on parasitoid fitness: Unraveling the complexity of a temperate pastoral agroecosystem. *Entomol. Exper. Appl. 123*: 63–71.

Van Zandt, P.A. and Agrawal, A.A. (2004). Community-wide impacts of herbivore-induced plant responses in milkweed (*Asclepias syriaca*). *Ecology 85*: 2616–2629.

Wali, P., Helander, M., Nissinen, O., Lehtonen, P., and Saikkonen, K. (2008). Endopohyte infection and environmental affect performance of meadow fescue in subarctic conditions. *Grass Forage Sci. 63*: 324–330.

Weibull, J. (1987). Seasonal changes in the free amino acids of oat and barley phloem sap in relation to plant growth stage and growth of *Rhopalosiphum padi*. *Ann. Appl. Biol. 111*: 729–737.

Zheng, S.J., van Dijk, J.P., Bruinsma, M., and Dicke, M. (2007). Sensitivity and speed of induced defense of cabbage (*Brassica oleracea* L.): Dynamics of BoLOX expression patterns during insect and pathogen attack. *Mol. Plant Microbe Interact. 20*: 1332–1345.

19

Conceptual Model for the Analysis of Plant–Endophyte Symbiosis in Relation to Abiotic Stress

Gregory P. Cheplick

CONTENTS

19.1 Introduction

Probably no plant in nature is ever completely devoid of microbial symbionts living both in and on it. The term *endophyte* literally means "within plant" and is used to refer to organisms such as bacteria and fungi that live inside a host (Wilson, 1995). The term implies nothing as to whether or not the endosymbiont grows within or between host cells. Typically, the endosymbiont is obligatorily dependent on its host, while the host can survive without its endosymbiont. Although in the past, use of the term *endophyte* had sometimes been restricted to mutualistic associations (Wilson, 1995; Stone et al., 2000), it is now mostly presumed not to imply anything about the nature of the endosymbiotic relationship. In Wilson's (1995) definition, endophytes were also presumed to be "unapparent and asymptomatic infections" that did not cause disease symptoms. However, because there are examples of fungal endophytes that can behave like plant pathogens at least during some portion of their life cycle, it is unnecessarily restrictive to limit the use of the term in this way.

This chapter will focus on symbioses between fungal endophytes and plants in relation to abiotic stress. Fungal endophytes can be found in any plant organ: roots, stems, leaves, flowers, and seeds. Here the emphasis will be on root endophytes (arbuscular mycorrhizal [AM] fungi) and leaf endophytes, especially those common in the grass family (Cheplick and Faeth, 2009). Although there are differences between the two groups and they are taxonomically unrelated (Table 19.1), both are obligate biotrophs that obtain their nutrition from carbohydrates made by the host. These nutritional symbioses will be explored using a cost–benefit perspective. The model and empirical patterns presented will permit examination of the widespread assertion that fungal endophytes act as defensive mutualists by providing their hosts with some level of protection from, or tolerance to, various abiotic stresses (Bacon, 1993; Belesky and Malinowski, 2000; Malinowski and Belesky, 2000; Brundrett, 2002; Entry et al., 2002; Malinowski et al., 2005; Rodriguez et al., 2005).

TABLE 19.1

Comparison and Contrast of AM and Leaf Endophytic Fungi

Feature	Mycorrhizal Fungi	Leaf Endophytic Fungi
Classification	Glomeromycota[a]	Ascomycota
Fungal nutrition	Obligate biotrophs	Obligate biotrophs
Haustoria	Present, finely branched	Absent
Host specificity	Low	High
Interaction	Predominantly mutualistic	Parasitic to mutualistic
Location in host	Roots	Leaves, stems
	Inter and intracellular	Intercellular
Number of species	<200	>500
Reproduction	Asexual	Asexual and/or sexual

[a] Previously Zygomycota, order Glomales; Glomeromycota is now recognized as a fungal phylum. Schüßler, A., Schwarzott, D., and Walker, C., *Mycol. Res.*, 105, 1413, 2001.

19.2 Continuum of Interactions

Plant associations with fungal endophytes have an extraordinarily long evolutionary history, going back at least 400 million years (Brundrett, 2002; Krings et al., 2007). At present, these associations are highly variable with a continuum of life history strategies for endophytic fungi (Schulz and Boyle, 2005). Regarding the nature of the association, there is also a continuum of possible types of interaction. Although the endophyte mostly benefits from the association, from the plant's perspective the interactions range along a continuum from parasitic to mutualistic (Johnson et al., 1997; Saikkonen et al., 1998). Plant–endophyte symbioses are also remarkably labile, varying in relation to environmental conditions, the intensity of infection, and the genotypes of the interacting partners (Johnson et al., 1997; Saikkonen et al., 1998; Cheplick and Faeth, 2009). Even for AM, long considered a textbook example of a mutualistic relationship, the fungal endophytes do not invariably benefit the host plants under all conditions (Buwalda and Goh, 1982; Smith and Read, 1997; Reynolds et al., 2005).

To quantitatively depict where along the interaction continuum a particular plant–endophyte association lies, it is necessary to compare the performance of the host with and without its endosymbiont. Mycorrhizal researchers have expressed "mycorrhizal dependency" as a percentage difference in the biomass of plants with and without mycorrhizal fungi growing in the same soil under identical conditions (Smith and Read, 1997). However, it is probably more precise to label such a comparison as plant "responsiveness" to mycorrhiza infection rather than dependency (for further discussion see Janos, 2007). Of course, analogous calculations can be made using data from plants infected and uninfected by leaf endophytes.

One broadly applicable index that can be employed to quantify various ecological interactions involving plants is the index of relative interaction intensity (RII) championed by Armas et al. (2004), for its simplicity and statistical properties. It can be used for any type of interaction, including competitive and symbiotic interactions. It was formulated as

$$\mathrm{RII} = \frac{B_{\mathrm{w}} - B_{\mathrm{wo}}}{B_{\mathrm{w}} + B_{\mathrm{wo}}}$$

where B is individual plant biomass when growing with (w) or without (wo) its interacting partner (Armas et al., 2004). The authors recognized that other metrics of plant growth and reproduction could also be used. The RII is at or near zero when the interaction is commensalistic, negative when antagonistic (e.g., competitive or parasitic interactions), and positive when mutualistic. Thus the RII can be employed to place specific plant–endophyte associations along the continuum of possible interactions. Cheplick (2007) illustrates application of the RII index to depict the variable nature of host genotype–endophyte relationships in *Lolium perenne* infected by *Neotyphodium lolii,* when grown in abiotically stressful conditions. Similarly, in this chapter the index will be used to illuminate a conceptual model that relates position along the parasitism–mutualism continuum to levels of abiotic stress.

19.2.1 Nature of Abiotic Stress

What is stress to a plant? Although entire volumes have been devoted to the physiology and molecular biology of plants under stress (e.g., Nilsen and Orcutt, 1996; Orcutt and Nilsen, 2000; Hirt and Shinozaki, 2003; Jenks and Hasegawa, 2005), from an ecological perspective most resources can be limiting to a plant (Harper, 1977) and the definitions are somewhat problematic. Grime (2001) described stresses as phenomena that restricted photosynthetic production, such as shortage of light, water, or soil minerals. In their classic text *Plant Physiology* (4th ed.), Salisbury and Ross (1992, p. 595) maintained "any change in environmental conditions that results in plant response that is less than the optimum might be considered stressful." Cheplick (1991) developed the concept of a "fundamental phenotype" to describe the optimum phenotype that could occur for a given genotype in an ideal nonlimiting environment. Of course, this almost never occurs in nature, and the given environment of a plant will always constrain the degree to which a fundamental phenotype is realized (Cheplick, 1991). Clearly all plants in nature are likely to be under some level of "stress" to the extent that they are displaced from achieving their genotypically based (inherent) growth and reproductive potential. Due to their lengthy evolutionary history in many types of more or less stressful environments, both tolerance and avoidance mechanisms for dealing with stress can be expected in plants of natural ecosystems (Orcutt and Nilsen, 2000).

Three primary categories of environmental stressors that affect plant physiology and growth are (1) physical, which includes natural abiotic factors such as drought, temperature, and light; (2) chemical, which includes air pollutants, heavy metals, and toxins; and (3) biotic, which includes competition, herbivory, and disease (Nilsen and Orcutt, 1996). Endosymbionts have the potential to alter plant responses to all types of stresses (see Chapters 4, 12, 13, 16 through 19, and 21); however, here the focus will be on abiotic stresses as a general feature of the environment of most plants. In the model that follows, particular attention will be paid to how the relative costs and benefits of endophyte infection might change as abiotic stress levels increase.

19.3 Conceptual Model

A variety of scenarios can be developed for the hypothetical relation of plant–endophyte interactions to increasing levels of abiotic stress (Figure 19.1). The three traditional interactions of mutualism, commensalism, and parasitism can each be relatively consistent and stable across the range of abiotic stress levels for at least some types of endophytic associations. Although the trends visualized in Figure 19.1a are shown as straight lines, curvilinear patterns are also possible. The increase in the strength of the mutualistic interaction, as rendered by an increase in the positive RII values with increasing stress is in accord with the premise that endophytic fungi act as defensive mutualists. Some mycorrhizal fungi may show this pattern of interaction with their host plants: as soil minerals such as phosphorus become more limiting, mycorrhizal infection is likely to be increasingly beneficial to the host. However, the behavior of the interaction may be quite different under conditions that are more benign and nonlimiting (moving from right to left along the x-axis in Figure 19.1b), perhaps becoming commensalistic (curve I) or parasitic (curve II) as costs of harboring the endosymbiont exceed the benefits (Johnson et al., 1997; Morgan et al., 2005; Janos, 2007).

When the costs and benefits of endophyte infection are approximately balanced and do not change greatly along the abiotic stress gradient, the RII values will remain at or near zero (Figure 19.1a). However, a threshold level of stress could cause a change in the relative balance in costs and benefits, tipping the scale toward mutualism (if benefits begin to exceed costs—curve I, Figure 19.1b) or toward parasitism (if costs begin to exceed benefits—curve IV). This model clearly underscores the importance of cost–benefit considerations to theoretical understanding of the evolution of mutualistic (and other) interactions (Holland et al., 2002; Neuhauser and Fargione, 2004; Foster and Wenseleers, 2006).

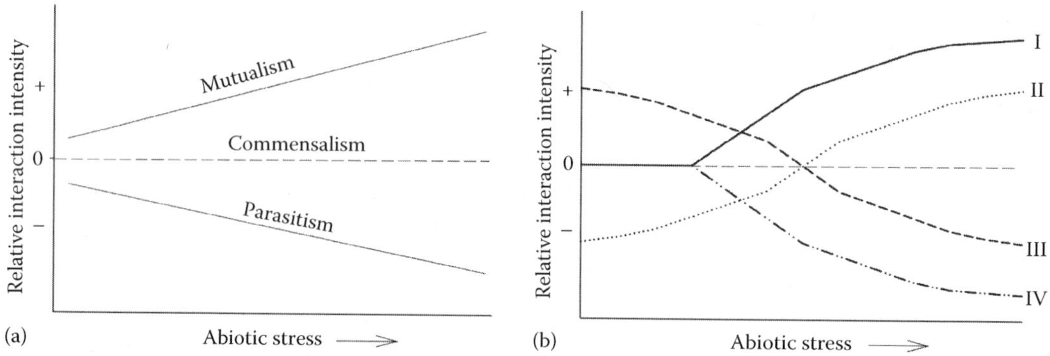

FIGURE 19.1 Hypothetical patterns of change in the relationship of plants and their endophytic symbionts along a gradient of increasing abiotic stress. The RII (Armas et al., 2004) expresses the extent to which a relationship is mutualistic (+), commensalistic (0), or parasitic (−) based on the performance of the host with and without the endophyte. (a) Relatively stable mutualism, commensalism, and parasitism across the entire stress gradient. (b) Four curves depicting variable patterns of change in the nature of the plant–endophyte relationship along the stress gradient.

In endophytic fungi which are predominantly parasitic in nature, the RII is continuously below zero (i.e., negative) and is predicted to become increasingly negative as abiotic stress levels increase (Figure 19.1a). If they cause disease symptoms, such fungi could be considered to be plant pathogens. Burdon (1987), in his review of *Diseases and Plant Population Biology* (p. 81), stated: "diseased plants in general appear to be less able to withstand unfavourable environmental conditions…than do healthy ones." In his discussion on the effects of mineral nutrients, he noted that as the availability of major nutrients such as phosphorus or potassium increases, the level of disease symptoms may decline. He went on to state: "deficiencies of most minor nutrients generally result in plants that are more susceptible to a variety of diseases" (Burdon, 1987, p. 82).

Biotic stress such as competition can also render hosts less tolerant to pathogen infection (Gilbert, 2002). However, disease tolerance can evolve in plant communities often infected by disease-causing fungi (e.g., Roy et al., 2000), and the precise relationship between abiotic stress levels and position along the parasitic–mutualistic continuum is liable to vary greatly across plant–pathogen systems (Rodriguez et al., 2004; Rodriguez et al., 2005; Kogel et al., 2006).

Some systemic endophytic fungi (*Epichloë* spp.) that cause choke disease effectively render the grasses they infect sexually sterile. The choke-inducing stromata produced by these endophytes as part of their sexual cycle cause complete abortion of the host inflorescences (Meijer and Leuchtmann, 2000; Schardl and Leuchtmann, 2005). However, vegetative vigor can be improved in plants that show parasitic castration by fungi (Clay, 1991). Thus, RII values based solely on vegetative biomass could be positive, while RII values based on seed production would be exceedingly negative. These considerations underscore the complexities inherent in the use of descriptive indices to place plant–endophyte associations along the parasitic–mutualistic continuum.

19.3.1 Empirical Patterns

One of the primary difficulties of attempting to determine the pattern of plant–endophyte interactions that will occur as abiotic stress levels change is that many experimental studies do not provide multiple stress treatments. Often these studies compare the growth of plants with and without the endosymbiont in an unstressed control to a single stress treatment. Thus, in a drought stress study there may simply be two experimental conditions: plants are subjected to a well-watered control and a drought stress treatment (e.g., Al-Karaki, 1998; Cheplick et al., 2000; and many others). Clearly, the more treatment levels the better the likelihood of detecting a particular pattern in the RII for a specific plant–endophyte symbiosis as abiotic stress is increased. A few examples in which there are at least three stress levels applied to both uninfected and endophyte-infected hosts follow. They illustrate that while the nature of plant–endophyte interactions often change in relation to abiotic conditions, they neither necessarily function in a consistent manner as defensive mutualisms nor change in predictable ways.

Undoubtedly, the most thoroughly investigated grass–endophyte symbiosis that has been examined in relation to abiotic stress is that of tall fescue (*Lolium arundinaceum* [Schreb.] Darbyshire, formerly *Festuca arundinacea*) and its endophyte *Neotyphodium coenophialum* (Bacon, 1993; Belesky and Malinowski, 2000). Because tall fescue in many studies has showed improved growth and vigor under abiotic stresses such as drought or low soil fertility (West, 1994; Malinowski and Belesky, 2000), it is sometimes characterized as an exemplar of a mutualistic plant–fungus symbiosis. However, as pointed out by Cheplick and Faeth (2009), there are several counterexamples of experimental conditions in which the benefits of infection to tall fescue have not been detected. For example, endophyte-infected (E+) seedlings had significantly greater dry mass than uninfected (E–) seedlings at high concentrations of mineral fertilizer, but when no fertilizer was applied and seedlings were presumably under nutrient stress, E– seedlings had greater dry mass (Cheplick et al., 1989). Thus, the RII curve for seedlings becomes significantly negative under the most stressful condition (Figure 19.2a). Cheplick et al. (1989) speculated that this result was due to a metabolic cost (in terms of available photosynthate) to the host of endophyte infection. For adult tall fescue plants, the relative advantage of endophyte infection was increasingly greater with increasing nutrient levels and all RII values were positive (Figure 19.2a).

In the two experiments that explored the effect of *N. coenophialum* on the responses of tall fescue to variable nitrogen addition rates and water levels (Arachevaleta et al., 1989), endophyte effects on host dry mass were positive. In the first experiment, N was applied at low (11 mg pot^{-1}), medium (73 mg pot^{-1}), or high rates (220 mg pot^{-1}). Plotted RII values based on dry mass after 160 days are close to +0.25 across the entire N supply gradient (Figure 19.2b), indicating a consistent mutualistic relationship between the host and endosymbiont. In the second experiment, N levels were varied in the same way, but three water

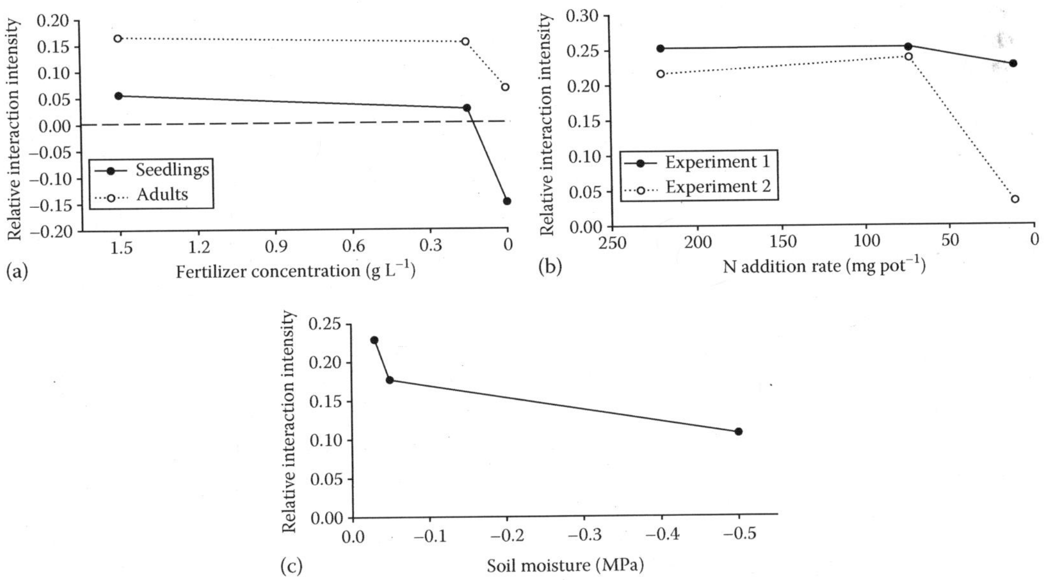

FIGURE 19.2 RII for the tall fescue-endophyte symbiosis along abiotic stress gradients. (a) RII at three concentrations of applied mineral fertilizer for seedlings and adults. Calculated from dry mass data in Figures 19.1 and 19.3 of Cheplick et al. (1989). (b) RII at three nitrogen addition rates in two greenhouse experiments. Calculated from dry mass data in Tables 19.1 and 19.2 of Arachevaleta et al. (1989). (c) RII at three levels of soil moisture stress. Calculated from dry mass data in Figure 19.4 of Arachevaleta et al. (1989).

stress treatments were also imposed (Arachevaleta et al., 1989). The authors first presented data that were averaged over the two least stressful soil moisture levels (−0.03 and −0.05 MPa) for the three N treatments. The lower curve in Figure 19.2b shows RII values based on these data. Again, mutualism is evident at the high and medium levels of nitrogen addition; however, at the lowest nitrogen level, the RII was near zero, suggesting a commensalistic symbiosis (Figure 19.2b, experiment 2). When the data were averaged over the three nitrogen addition rates and examined in relation to drought stress only, all RII values remained positive, but the strength of the mutualistic relationship appeared to diminish at the most stressful moisture level (−0.50 MPa; Figure 19.2c). Indeed, ANOVA results in the original paper reported a significant ($P < 0.01$) interaction of infection status with soil moisture level (Arachevaleta et al., 1989).

The ecologically important forage grass *L. perenne* (perennial ryegrass) and its endophytic fungus *N. lolii* have also been examined in relation to abiotic stress (Cheplick et al., 1989, 2000; Cheplick, 1997, 2004, 2007; Ravel et al., 1997; Hesse et al., 2003; Lewis, 2004). Although there is evidence that endophyte infection can improve host survival and growth under soil nutrient or drought stresses (Ravel et al., 1997; Lewis, 2004), there is much genotypic variability among hosts in their phenotypic responses to endophyte infection (Cheplick, 1997, 2004; Cheplick and Cho, 2003; Hesse et al., 2003, 2004). Because endophyte genotypes (haplotypes) are typically not characterized, each host genotype–endophyte combination is probably best considered a unique entity (West, 2007).

As an illustration of the complex nature of the plant–endophyte relationship, RII values were calculated for vegetative and reproductive traits of the three genotypes of *L. perenne* examined in the drought experiment of Hesse et al. (2003). The three host genotypes with their endophytes were originally collected from different sites in central Germany, that varied in yearly soil moisture conditions. Tillers were cloned from these plants to propagate the genotypes and some were treated with a systemic fungicide to eliminate the endophyte. To minimize possible aftereffects of the fungicide, E− and E+ replicates were grown in the field for 1 year before beginning the experiment (Hesse et al., 2003). After further propagation, five clones per genotype per infection status were subjected to a drought stress in which substrate water content was 30%–40% of the maximum water-holding capacity; the other five clones were maintained at 80%–85% under an unstressed control. Shoots were harvested at 11 weeks and allowed to regrow for an additional 10 weeks. Data were recorded on the shoot dry mass, the number of vegetative and reproductive tillers, and the number and mass of mature seeds produced.

RII values based on two vegetative traits (shoot mass and tiller number) recorded at 11 and 21 weeks were mostly negative or near zero in both watered and drought groups. Only the regrown shoot mass of genotype DRY (collected from the driest site) was greater in E+ plants, with RII = +0.30. However, in the control the RII values ranged from −0.11 to −0.33 in this genotype. Genotype FLOOD-DRY (collected from a site dry in summer, but periodically flooded) had very negative RII values in both watered and drought groups, ranging from 0.29 to −0.57 for three vegetative traits and time periods for which means were significantly reduced by endophyte infection. Genotype WET (collected from a wet floodplain) showed a significant reduction in tiller number at 21 weeks for E+ plants in both watered (RII = −0.18) and drought conditions (RII = −0.38). In short, there was little evidence for a mutualistic grass–endophyte relationship based on vegetative traits, with the possible exception of regrowth ability in the DRY genotype following drought (Hesse et al., 2003). For the other two genotypes, endophytes were commensalists or antagonists regardless of soil moisture conditions.

RII values based on three reproductive traits revealed a somewhat different pattern (Table 19.2). Note that this table indicates whether or not the differences between E− and E+ plants were significant ($P < 0.05$) for each genotype and treatment. Despite major differences among genotypes, there was no consistent pattern for RII values in relation to drought stress. Genotype FLOOD-DRY showed consistent positive effects of endophytes on all reproductive traits (Table 19.2). Genotype WET showed consistent negative effects of endophyte infection, with an especially low RII of −0.60 under drought stress. Thus, the plant–endophyte symbiosis appears to be mutualistic for genotype FLOOD-DRY, but antagonistic for genotype WET. Genotypic variation in endophyte-mediated effects on morphological traits of perennial ryegrass are common and have been demonstrated in other studies (Cheplick et al., 2000; Cheplick and Cho, 2003; Hesse et al., 2004).

TABLE 19.2

RIIs Based on Reproductive Traits in the Perennial Ryegrass–Endophyte Symbiosis in a Watered Control and a Drought Treatment. Three Host Genotype–Endophyte Combinations, Collected from Three Habitats that Varied in Seasonal Supply of Soil Water, Were Studied

Trait	Genotype	Watered	Drought
Number of reproductive tillers	DRY	−0.1084*	+0.0153
	FLOOD-DRY	+0.2944*	+0.1706*
	WET	−0.0323	−0.0880*
Number of seeds per spike	DRY	+0.0238	−0.0725
	FLOOD-DRY	+0.1897	+0.2126*
	WET	−0.2491*	−0.3809*
Seed mass per plant	DRY	−0.1461	−0.0749
	FLOOD-DRY	+0.5210*	+0.4490*
	WET	−0.2699*	−0.6044*

Source: Hesse, U., Schöberlein, W., Wittenmayer, L., Förster, K., Warnstorff, K., Diepenbrock, W., and Merbach, W., *Grass Forage Sci.*, 58, 407, 2003.

*, denote that differences between E+ and E− plants were significant ($P < 0.05$).

Considerable variation in plant–endophyte relationships can also be found in arbuscular mycorrhiza, although in comparison to leaf endophytes, mutualism may be more prevalent under abiotically stressful conditions (Johnson et al., 1997; Smith and Read, 1997; Herman, 2000; Augé, 2001; Entry et al., 2002). A few examples in which at least three stress levels were imposed on plants infected by AM fungi and on uninfected control plants will be described. RII values, based on biomass or seed yield, were calculated from means presented in the original sources: a study of maize (*Zea mays*) in relation to soil moisture (Table 5 of Sylvia et al., 1993) and a study of subterranean clover (*Trifolium subterraneum*) in relation to soil phosphorus (Table 4.1 of Smith and Read, 1997).

In the study of maize, seeds were sown into field plots where one side of each plot was inoculated with AM while the other side remained uninoculated (Sylvia et al., 1993). Three water-management treatments were applied: fully irrigated (no stress), irrigated after 2–3 days of leaf wilting (moderate stress), and watered by natural rainfall (severe stress). Biomass and seed yield were recorded for 3 years. In the first year, effects of AM and moisture stress were significant ($P < 0.01$) and the RII values show an increase with increasing stress for both biomass (Figure 19.3a) and seed yield (Figure 19.3b). Like the upper line in Figure 19.1a, the symbiotic relationship became increasingly mutualistic as abiotic conditions became more stressful. However, in the second year, AM inoculation did not significantly affect biomass or seed yield (Sylvia et al., 1993). The curve for biomass RII is close to zero regardless of moisture stress (Figure 19.3a). In the third year, AM was again a significant factor ($P < 0.01$), but RII values were lowest when stress was moderate (Figure 19.3a and b). Thus, despite year to year variation, from the host's perspective AM fungi were at best mutualistic, and at worst commensalistic, in terms of their symbiotic relationship.

In the second study, the dry mass of roots and shoots of subterranean clover were obtained after uncolonized and AM-colonized plants were grown at three levels of added phosphorus for 31–35 days (data of Oliver et al., 1983, reported in Smith and Read, 1997). The curve of RII, based on total biomass (Figure 19.3c), reveals a classic case of AM having beneficial effects when plants are in P-deficient conditions, but showing detrimental effects in P-rich conditions (Johnson et al., 1997; Morgan et al., 2005). That is, the plant–endophyte symbiosis is initially parasitic when soil nutrients are plentiful but then becomes increasingly mutualistic as P becomes limiting. This approximates curve II in the conceptual model (Figure 19.1b) and suggests that the costs can exceed the benefits of AM infection when soil conditions are not stressful. Neuhauser and Fargione (2004) predicted such a shift along the mutualism–parasitism continuum in a modified Lotka–Volterra style, predator–prey model for plant–mycorrhizal fungi interactions.

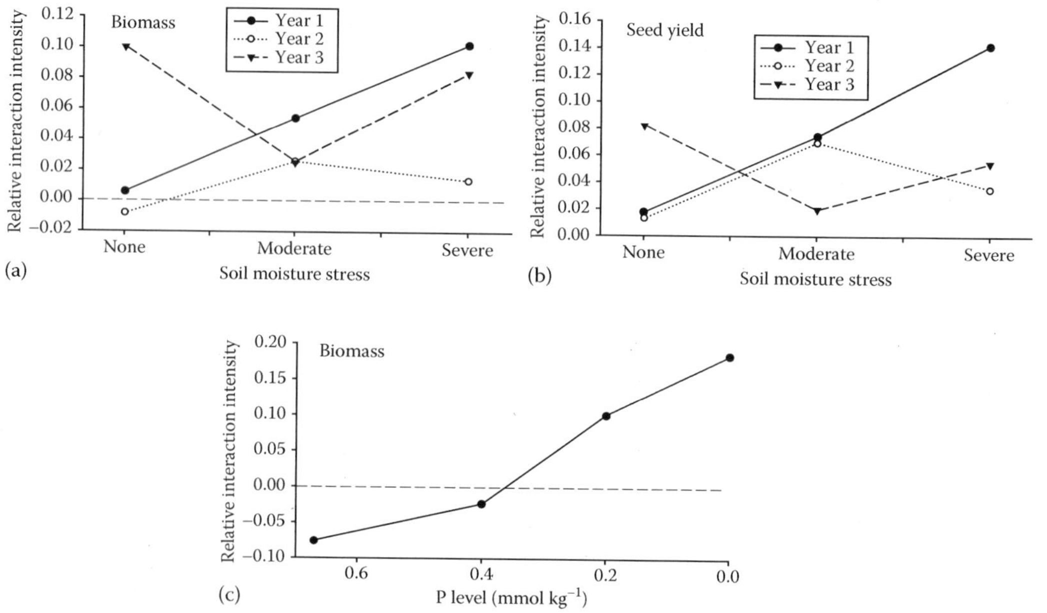

FIGURE 19.3 RII for two plant–AM fungal associations along abiotic stress gradients. (a) RII for *Z. mays* at three levels of soil moisture stress under field conditions. Calculated from dry mass data for three years in Table 5 of Sylvia et al. (1993). (b) Same as (a), except RIIs were calculated from seed yield data. (c) RII for *T. subterraneum* at four levels of added phosphorus. Calculated from dry mass data in Table 4.1 of Smith and Read (1997).

19.4 Endophyte Infection Intensity

Another important factor sometimes overlooked in the studies of endosymbionts and their plant hosts in relation to abiotic stress is the extent to which roots and leaves are infected by mycorrhizal fungi or leaf endophytes, respectively. The degree of fungal infection will here be denoted endophyte infection intensity, defined as the number of hyphae (or any other measure correlated with fungal abundance) per unit of host tissue (Cheplick and Faeth, 2009). In the cost–benefit models of Holland et al. (2002), this is delineated as the population size or density of the endosymbiont. They describe how the net effect of one mutualistic partner on the second partner can increase with increasing density, at least up to a point. However, in some associations where costs exceed benefits at high densities (i.e., at a high intensity of infection) of one partner, the net effect may level off or eventually decline (Holland et al., 2002). For endosymbiotic infection of host plants, this possibility certainly exists, as symbiotic fungi depend on photosynthates produced by the host for their own metabolic needs.

A simple graphical model can be employed to show the relationship of the benefits and costs of endosymbiotic infection to their host plants as the intensity of infection increases (Figure 19.4). For simplicity, the benefit of infection is considered to progressively increase with infection intensity under some hypothetical set of homogeneous conditions, while acknowledging that the net beneficial effect could easily decline at the highest levels of infection intensity as the endosymbionts use more of the host's available photosynthate. Thus, costs are expected to rise as infection intensity increases, but the relation of the cost function to the benefit function will vary among different plant–endophyte symbioses. This cost–benefit relation will determine at what levels of infection intensity the host should experience a net benefit.

The first cost function increases more slowly than the benefit function (cost 1 in Figure 19.4) and therefore, the net benefit of the symbiosis (and the RII) increases as infection intensity increases. Such an increasingly mutualistic scenario might exist for AM if mycorrhizal fungi improve mineral nutrient

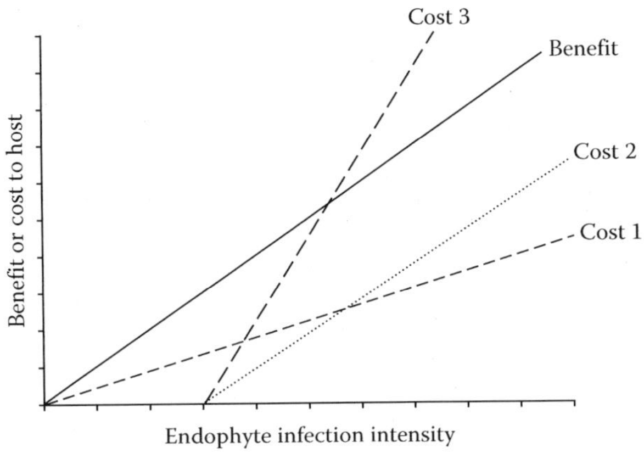

FIGURE 19.4 Graphical model showing the benefit and costs of endophytes to their host in relation to the intensity of infection (i.e., fungal abundance or density).

acquisition by the host when soil fertility is low, for example, provided that carbohydrate costs are moderately low at most intensities of infection. Whenever costs increase in a parallel manner to the benefit (cost 2 in Figure 19.4), a consistent mutualistic relationship is observed regardless of endophyte infection intensity. Note in this second scenario that it is assumed that a specific intensity of infection must be realized before costs become detectable. A third possibility is that the cost function increases more rapidly than the benefit function (cost 3 in Figure 19.4) with increasing infection intensity. In this scenario, the plant–endophyte symbiosis is mutualistic (positive RII) only at the lowest levels of infection intensity. At the infection intensity where the cost and benefit functions intersect, the symbiosis is effectively commensalistic (RII = 0). With greater intensity of infection beyond this point the symbiosis become parasitic (negative RII).

Distinguishing among different cost–benefit functions in relation to the intensity of endophyte infection is important to analyzing plant–endophyte symbioses across abiotic stress gradients. For example, if a positive growth effect of infection on a host mostly occurs under benign conditions when soil resources are prevalent, is this net benefit due to a higher intensity of infection under these conditions? And might the carbohydrate cost of supporting a high density of endosymbionts be relatively lower under benign conditions, thereby rendering a greater net benefit of infection to the host? As abiotic stress increases, does endophyte infection intensity decline, resulting in reduced benefit–cost ratio and a drop in RII below zero (curves III and IV in Figure 19.1)? It is clearly important to quantify whenever possible the intensity of endophyte infection in grass–endophyte experiments (Groppe et al., 1999; Rasmussen et al., 2007). This is especially true when environmental conditions are variable in the extent to which they act as abiotic stresses to the growth of the host plant.

19.4.1 More Empirical Patterns

For the leaf endophytes of grasses, infection intensity has been simply estimated by counting the number of stained hyphae traversing the field of view under a light microscope at 400× (ca. 800 μm diameter). This is a straightforward but relatively crude technique, and false negatives can occur whenever the endophytic hyphae are sparsely or unevenly distributed within the plant (Cheplick and Faeth, 2009). Nevertheless, when a few leaf sheath segments are sampled from the same individual, and at least three counts in different fields of view are made, fairly consistent mean infection frequencies can be obtained for different host genotypes over multiple years (Cheplick, 2008).

Prior to initiation of a drought experiment with 13 genotypes of perennial ryegrass (*L. perenne*), endophyte infection intensity was recorded by counting hyphae under a light microscope at 400× (Cheplick et al., 2000). Each host genotype was represented by 10 infected (E+) and 10 uninfected (E−) individuals from which endophytes had been eliminated by a systemic fungicide. The number of tillers produced by E+ and E− hosts in a watered control and in a drought stress treatment (2 weeks of no added water followed by 3 weeks of recovery) was used to calculate the RII values. The RII values for each genotype are plotted against endophyte infection intensity in Figure 19.5a and b. In the analyses reported in Cheplick et al. (2000), endophyte infection, host genotype, and their interaction significantly affected tiller number. Although the treatment also affected this variable, there was no interaction of infection with treatment. In the plants that had been drought stressed, RII values were near or below zero (Figure 19.5b), while in the unstressed control both positive and negative values were found, depending on genotype (Figure 19.5a). There was no strong relationship between RII and the intensity of infection, although genotypes with negative values all had more than five hyphae per microscopic field of view, in both drought stressed and control groups.

In another drought experiment with *L. perenne*, eight E+ and eight E− plants per each of 10 genotypes were subjected to three sequential water stress periods of 11–14 days and permitted to recover in 7 weeks (Cheplick, 2004). These plants were compared to the same number of individuals of the same genotypes in a watered control. Total dry mass (shoots plus roots) following the recovery period was used to calculate the RII values to plot against infection intensity (Figure 19.5c and d). Analysis of covariance

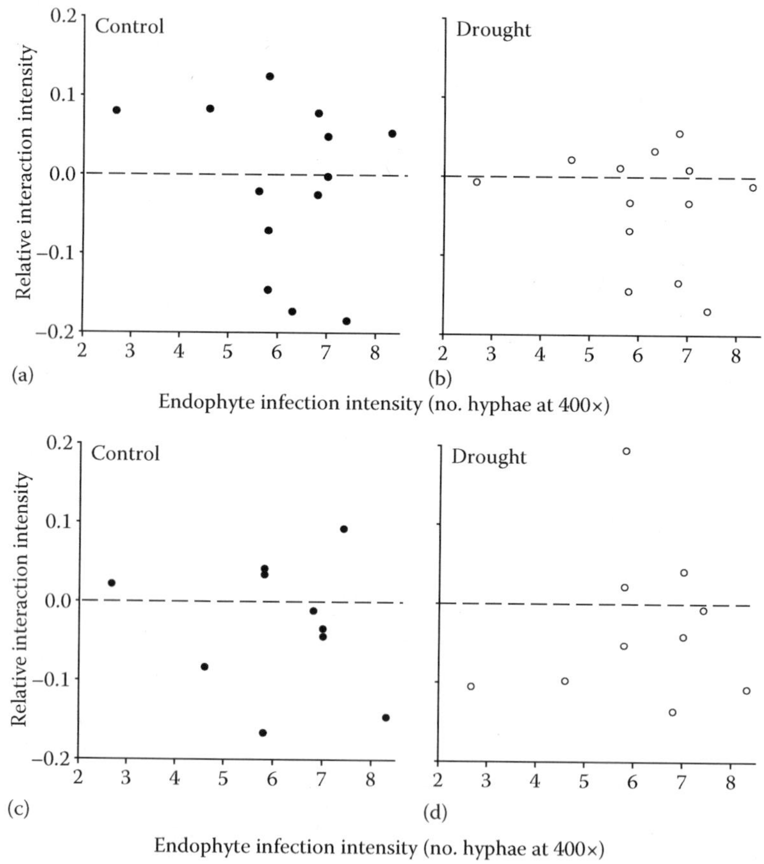

FIGURE 19.5 RII for the perennial ryegrass–endophyte symbiosis in unstressed control and previously drought-stressed plants versus endophyte infection intensity. (a) Means for 13 host genotypes not subjected to stress and (b) means for the same genotypes subjected to drought and following 3 weeks of recovery. Both calculated from data on the number of live tillers in Figure 19.2 of Cheplick et al. (2000). (c) Means for 10 host genotypes not subjected to stress and (d) means for the same genotypes subjected to drought and following 7 weeks of recovery. Calculated from previously unreported dry mass data in Cheplick (2004).

(using plant size prior to experimental treatment as the covariate) revealed significant effects of host genotype, endophyte infection, and drought on total mass (Table 1 in Cheplick, 2004). In addition, both genotype and endophyte showed a significant interaction with drought stress. However, although genotype-specific infection intensity varied greatly, RII was not correlated with infection intensity (Figure 19.5c and d).

In a study of the perennial grass *Bromus erectus* and its endophyte *Epichloë bromicola*, Gropp et al. (1999) employed a quantitative technique based on the polymerase chain reaction (PCR) to estimate fungal DNA amounts as a marker for endophyte "concentration" within host tissues. RII values were calculated from their data on the number of tillers and vegetative dry mass. The resulting plot reveals that RII becomes increasingly positive with higher intensities of endophyte infection (Figure 19.6). This suggests that the relative benefits of infection to host growth (and reproduction—see Groppe et al., 1999) become increasingly greater with higher levels of endophyte infection. However, it is not known how the infection intensity might change along an abiotic stress gradient and alter this relationship in this grass–endophyte association.

PCR-based techniques were also used to quantify the concentration of three strains of endophytic fungi (*N. lolii*) that infected two cultivars of perennial ryegrass (Rasmussen et al., 2007). Plants were grown under high or low nitrogen conditions and the alkaloid concentration was analyzed. The concentration of endophytes within host leaves varied with ryegrass cultivar and endophyte strain. Because endophytes are the source of the alkaloids measured, there was a positive linear relationship between the concentration of three of the four alkaloids and endophyte concentration (Rasmussen et al., 2007). Interestingly, endophyte levels were reduced by 40% under high nitrogen conditions. This result leads to the intriguing speculation that the absence of mutualistic effects of endophytes on some host species reared under benign conditions might be due to reduced infection intensity. If this reduction results in reduced concentrations of bioprotective alkaloids, then a usually defensive mutualism may breakdown over time due to unfavorable cost–benefit ratios (Sachs and Simms, 2006).

For plant–AM fungal associations, the greatest net benefit to the host may occur under specific abiotic stresses such as low water and mineral availability (Sylvia et al., 1993; Smith and Read, 1997; Al-Karaki, 1998; Augé, 2001; Entry et al., 2002). It is often not clear to what extent the greater benefit to infected hosts might be due to increased colonization or greater infection intensity under abiotic stress. In some field situations, drought may promote greater colonization by AM fungi (Augé, 2001); however, in a study of two wheat cultivars, AM fungi colonization was lower under water stress, although plants showed enhanced biomass relative to uninfected controls (Al-Karaki et al., 2004). Also, in the arid grasslands of the Namib Desert in southwestern Africa, the percent mycorrhizal colonization of grass

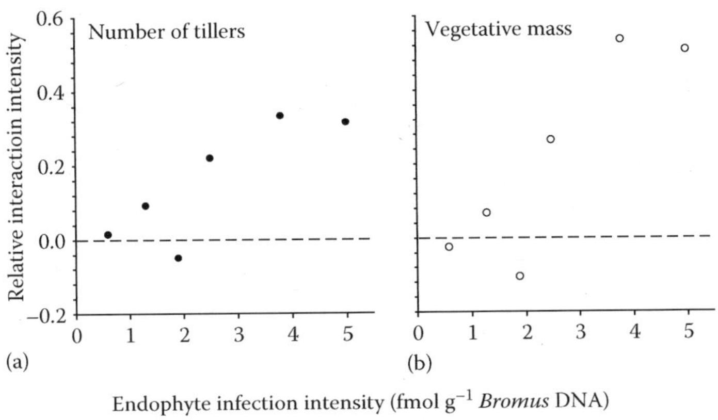

FIGURE 19.6 RII for the *B. erectus–E. bromicola* symbiosis versus endophyte infection intensity assessed by PCR-based techniques. (a) Means based on the number of tillers (calculated from Figure 19.1A of Groppe et al., 1999). (b) Means based on vegetative dry mass (calculated from Figure 19.1C of Groppe et al., 1999).

species was positively associated with soil moisture (Jacobson, 1997). In the 3-year drought experiment with maize described earlier (Sylvia et al., 1993), colonization by AM fungi was not significantly different among treatments in the first and third years, although AM fungi significantly increased biomass and seed yield (Figure 19.3a and b). Ironically, in the second year in which fungal colonization in the severe drought treatment was about twice as great as in the other two irrigated treatments (Sylvia et al., 1993), AM fungi had *no* significant effect on biomass or seed yield.

Host responsiveness to mycorrhizas increases along a gradient of phosphorus availability up to a point, but then diminishes at the highest levels (Janos, 2007; Figure 19.3c). It is presently unclear the extent to which changes in colonization and infection intensity along such abiotic gradients determines the relative benefits or costs of AM fungi to their hosts. However, it is known that the relative abundance and density of AM fungi can decline under fertile soil conditions (Clapperton and Reid, 1992; Johnson, 1993).

Gange and Ayres (1999) developed a simple graphical model to depict the relationship between the "density" of AM colonization of roots and the benefit derived by the host. The "benefit" was expressed as the percentage change in a measured variable of an infected plant relative to the mean value for the variable of uninfected plants. When expressed in this relative manner, the benefit reflects the proportional growth improvement attributable to AM fungi (i.e., responsiveness; Janos, 2007). Their proposed curvilinear relation between AM colonization and benefit (Gange and Ayres, 1999) is shown in Figure 19.7a, without presenting quantitative information. At low colonization densities, the AM levels are not great enough to produce a maximum benefit (point A in Figure 19.7a). At some intermediate level of colonization (point B), maximum benefit is achieved and positive RII values would reveal a mutualistic symbiosis. Continued colonization beyond this optimum results in increasing carbohydrate drain by the AM fungi and a drop in host benefit (Gange and Ayres, 1999). At the colonization level where no net host benefit occurs (RII = 0, point D in Figure 19.7a), any benefit is exactly balanced by the cost (Janos, 2007). Still, higher levels of colonization can result in carbohydrate costs that exceed the benefit (Clapperton and Reid, 1992) and generate negative RII values.

Four examples of empirical tests on their model were presented by Gange and Ayres (1999), one of which is illustrated in Figure 19.7b. Colonization density is the percent of the root length of *Cirsium arvense* colonized by AM fungi. The host benefit was based on dry mass and was calculated as described earlier. Although the scattering of data points is considerable (Figure 19.7b), both second- and third-order polynomial equations explained 65% of the variation. Much of this scatter probably reflects genotypic variation in both plants and endosymbionts, and the functioning of particular host–symbiont combinations.

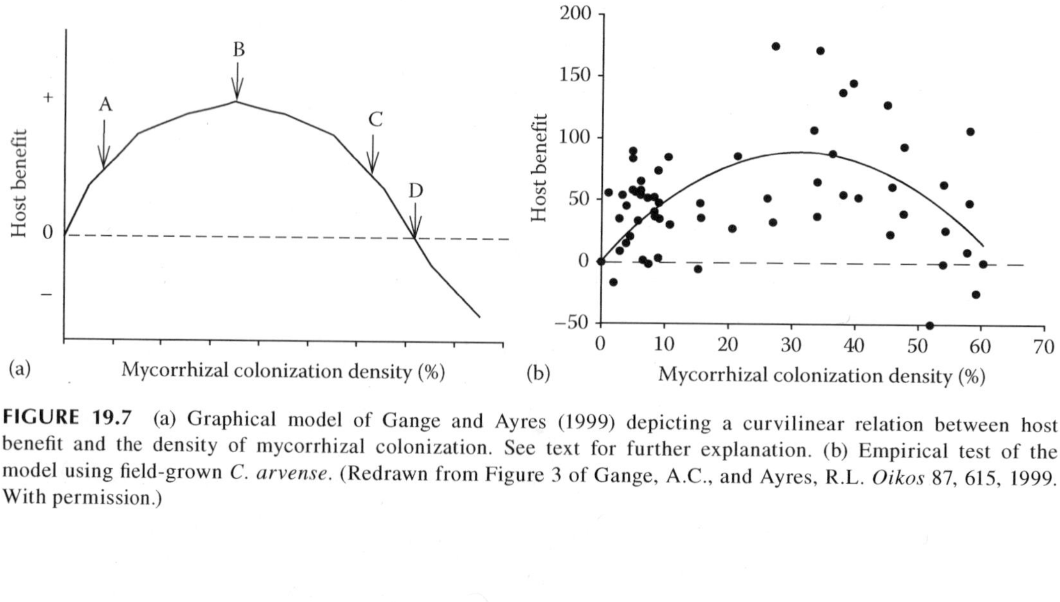

FIGURE 19.7 (a) Graphical model of Gange and Ayres (1999) depicting a curvilinear relation between host benefit and the density of mycorrhizal colonization. See text for further explanation. (b) Empirical test of the model using field-grown *C. arvense*. (Redrawn from Figure 3 of Gange, A.C., and Ayres, R.L. *Oikos* 87, 615, 1999. With permission.)

Nonetheless, this approach emphasizes the necessity of accounting for the intensity of endosymbiont infection when trying to model the environmental and genotypic parameters that account for variation in host responses to mycorrhizal fungi (Johnson et al., 1997).

19.5 Conclusions

Like biotic factors such as competitors and herbivores, the abiotic stresses explored in this chapter can act as potent, evolutionarily significant agents of natural selection in plant populations. Because microbial endosymbiosis with plants appears to be ubiquitous in nature, a host phenotype is the result of a complex ecological interaction that entails multiple genotypes (host plus all endosymbiont species) acting in combination with prevailing environmental conditions. Along abiotic stress gradients, the Darwinian fitness of a host population infected by endosymbionts may be assessed by a cost–benefit analysis of host phenotypes that quantify the relative intensity of microbial infection. Such an analysis must, of course, incorporate the responses of control plants devoid of microbial endosymbionts along the same gradients.

Variability in the nature of the symbiotic relationship between a plant and an endophyte across abiotic gradients reflects changes in the cost–benefit function and endophyte infection intensity. Upon closer inspection of these relationships, the question invariably arises: how broadly applicable is the "defensive mutualism" hypothesis in plant–endophyte symbioses? It is clear for plants infected by AM fungi (Augé, 2001) and a few grass–endophyte associations that have been examined in detail (mostly tall fescue and perennial ryegrass; West, 1994; Cheplick and Faeth, 2009) that infected hosts can exhibit greater tolerance to drought conditions. In addition, many plants infected by AM fungi (Smith and Read, 1997) and several agronomic grasses infected by leaf endophytes (Cheplick and Faeth, 2009) show improved growth when soil nutrients are low relative to uninfected counterparts. However, cost–benefit analysis of plant–endophyte symbioses along abiotic stress gradients reveals enough variation to question the broad applicability of the defensive mutualism concept. Additional research on a wider array of plant–fungal interactions exposed to variable abiotic conditions in the field will be necessary to improve the understanding of the effects of abiotic stress on plant populations and the communities in which they occur.

REFERENCES

Al-Karaki, G. N. 1998. Benefit, cost and water-use efficiency of arbuscular mycorrhizal durum wheat grown under drought stress. *Mycorrhiza* 8:41–45.

Al-Karaki, G., McMichael, B., and J. Zak. 2004. Field response of wheat to arbuscular mycorrhizal fungi and drought stress. *Mycorrhiza* 14:263–269.

Arachevaleta, M., Bacon, C. W., Hoveland, C. S., and D. E. Radcliffe. 1989. Effect of the tall fescue endophyte on plant response to environmental stress. *Agronomy Journal* 81:83–90.

Armas C., Ordiales, R., and F. I. Pugnaire. 2004. Measuring plant interactions: A new comparative index. *Ecology* 85:2682–2686.

Augé, R. M. 2001. Water relations, drought and vesicular-arbuscular mycorrhizal symbiosis. *Mycorrhiza* 11:11–42.

Bacon, C. W. 1993. Abiotic stress tolerances (moisture, nutrients) and photosynthesis in endophyte-infected tall fescue. *Agriculture, Ecosystems and Environment* 44:123–141.

Belesky, D. P. and D. P. Malinowski. 2000. Abiotic stresses and morphological plasticity and chemical adaptations of *Neotyphodium*-infected tall fescue plants. In *Microbial Endophytes*, Eds. C. W. Bacon and J. F. White Jr., pp. 455–485. Marcel Dekker, New York.

Brundrett, M. C. 2002. Coevolution of roots and mycorrhizas of land plants. *New Phytologist* 154:275–304.

Burdon, J. J. 1987. *Diseases and Plant Population Biology*. Cambridge University Press, Cambridge.

Buwalda, J. G. and K. M. Goh. 1982. Host-fungus competition for carbon as a cause of growth depressions in vesicular-arbuscular perennial ryegrass. *Soil Biology and Biochemistry* 14:103–106.

Cheplick, G. P. 1991. A conceptual framework for the analysis of phenotypic plasticity and genetic constraints in plants. *Oikos* 62:283–291.

Cheplick, G. P. 1997. Effects of endophytic fungi on the phenotypic plasticity of *Lolium perenne* (Poaceae). *American Journal of Botany* 84:34–40.

Cheplick, G. P. 2004. Recovery from drought stress in *Lolium perenne* (Poaceae): Are fungal endophytes detrimental? *American Journal of Botany* 91:1960–1968.

Cheplick, G. P. 2007. Costs of fungal endophyte infection in *Lolium perenne* genotypes from Eurasia and North Africa under extreme resource limitation. *Environmental and Experimental Botany* 60:202–210.

Cheplick, G. P. 2008. Host genotype overrides fungal endophyte infection in influencing tiller and spike production of *Lolium perenne* (Poaceae) in a comman garden experiment. *American Journal of Botany* 95:1063–1071.

Cheplick, G. P. and R. Cho. 2003. Interactive effects of fungal endophyte infection and host genotype on growth and storage in *Lolium perenne. New Phytologist* 158:183–191.

Cheplick, G. P. and S. H. Faeth. 2009. *Ecology and Evolution of the Grass-Endophyte Symbiosis*. Oxford University Press, Oxford.

Cheplick, G. P., Clay, K., and S. Marks. 1989. Interactions between infection by endophytic fungi and nutrient limitation in the grasses *Lolium perenne* and *Festuca arundinacea. New Phytologist* 111:89–97.

Cheplick, G. P., Perera, A., and K. Koulouris. 2000. Effect of drought on the growth of *Lolium perenne* genotypes with and without fungal endophytes. *Functional Ecology* 14:657–667.

Clapperton, M. J. and D. M. Reid. 1992. A relationship between plant growth and increasing VA mycorrhizal inoculum density. *New Phytologist* 120:227–234.

Clay, K. 1991. Parasitic castration of plants by fungi. *Trends in Ecology and Evolution* 6:162–166.

Entry, J. A., Rygiewicz, P. T., Watrud, L. S., and P. K. Donnelly. 2002. Influence of adverse soil conditions on the formation and function of arbuscular mycorrhizas. *Advances in Environmental Research* 7:123–138.

Foster, K. R. and T. Wenseleers. 2006. A general model for the evolution of mutualisms. *Journal of Evolutionary Biology* 19:1283–1293.

Gange, A. C. and R. L. Ayres. 1999. On the relation between arbuscular mycorrhizal colonization and plant 'benefit'. *Oikos* 87:615–621.

Gilbert, G. S. 2002. Evolutionary ecology of plant diseases in natural ecosystems. *Annual Review of Phytopathology* 40:13–43.

Grime, J. P. 2001. *Plant Strategies, Vegetation Processes, and Ecosystem Properties*, 2nd edn. John Wiley & Sons, New York.

Groppe, K., Steinger, T., Sanders, I., Schmid, B., Wiemken, A., and T. Boller. 1999. Interaction between the endophytic fungus *Epichloë bromicola* and the grass *Bromus erectus*: Effects of endophyte infection, fungal concentration and environment on grass growth and flowering. *Molecular Ecology* 8:1827–1835.

Harper, J. L. 1977. *Population Biology of Plants*. Academic Press, San Diego, CA.

Herman, P. 2000. Biodiversity and evolution in mycorrhizae of the desert. In *Microbial Endophytes*, Eds. C. W. Bacon and J. F. White Jr., pp. 141–160. Marcel Dekker, New York.

Hesse U., Schöberlein, W., Wittenmayer, L., Förster, K., Warnstorff, K., Diepenbrock, W., and W. Merbach. 2003. Effects of *Neotyphodium* endophytes on growth, reproduction and drought-stress tolerance of three *Lolium perenne* L. genotypes. *Grass and Forage Science* 58:407–415.

Hesse, U., Hahn, H., Andreeva, K., Förster, K., Warnstorff, K., Schöberlein, W., and W. Diepenbrock. 2004. Investigations on the influence of *Neotyphodium* endophytes on plant growth and seed yield of *Lolium perenne* genotypes. *Crop Science* 44:1689–1695.

Hirt, H. and K. Shinozaki, Eds. 2003. *Plant Responses to Abiotic Stress* (*Topics in Current Genetics*). Springer-Verlag, Berlin.

Holland, J. N., DeAngelis, D. L., and J. L. Bronstein. 2002. Population dynamics and mutualism: Functional responses of benefits and costs. *American Naturalist* 159:231–244.

Jacobson, K. M. 1997. Moisture and substrate salinity determine VA-mycorrhizal fungal community distribution and structure in an arid grassland. *Journal of Arid Environments* 35:59–75.

Janos, D. P. 2007. Plant responsiveness to mycorrhizas differs from dependence upon mycorrhizas. *Mycorrhiza* 17:75–91.

Jenks, M. A. and P. M. Hasegawa, Eds. 2005. *Plant Abiotic Stress*. Blackwell Publishers, Oxford.

Johnson, N. C. 1993. Can fertilization of soil select less mutualistic mycorrhizae? *Ecological Applications* 3:749–757.

Johnson, N. C., Graham, J. H., and F. A. Smith. 1997. Functioning of mycorrhizal associations along the mutualism-parasitism continuum. *New Phytologist* 135:575–585.

Kogel, K.-H., Franken, P., and R. Hückelhoven. 2006. Endophyte or parasite—what decides? *Current Opinion in Plant Biology* 9:358–363.

Krings, M., Taylor, T. N., Hass, H., Kerp, H., Dotzler, N., and E. J. Hermsen. 2007. Fungal endophytes in a 400-million-yr-old land plant: Infection pathways, spatial distribution, and host responses. *New Phytologist* 174:648–657.

Lewis, G. C. 2004. Effects of biotic and abiotic stress on the growth of three genotypes of *Lolium perenne* with and without infection by the fungal endophyte *Neotyphodium lolii*. *Annals of Applied Biology* 144:53–63.

Malinowski, D. P. and D. P. Belesky. 2000. Adaptations of endophyte-infected cool-season grasses to environmental stresses: Mechanisms of drought and mineral stress tolerance. *Crop Science* 40:923–940.

Malinowski, D. P., Belesky, D. P., and G. C. Lewis. 2005. Abiotic stresses in endophytic grasses. In *Neotyphodium in Cool-Season Grasses*, Eds. C. A. Roberts, C. P. West, and D. E. Spiers, pp. 187–199. Blackwell Publications, Iowa.

Meijer, G. and A. Leuchtmann. 2000. The effects of genetic and environmental factors on disease expression (stroma formation) and plant growth in *Brachypodium sylvaticum* infected by *Epichloë sylvatica*. *Oikos* 91:446–458.

Morgan, J. A. W., Bending, G. D., and P. J. White. 2005. Biological costs and benefits to plant-microbe interactions in the rhizosphere. *Journal of Experimental Botany* 56:1729–1739.

Neuhauser, C. and J. E. Fargione. 2004. A mutualism-parasitism continuum model and its application to plant-mycorrhizae interactions. *Ecological Modeling* 177:337–352.

Nilsen, E. T. and D. M. Orcutt. 1996. *The Physiology of Plants Under Stress: Abiotic Factors*. John Wiley & Sons, New York.

Orcutt, D. M. and E. T. Nilsen. 2000. *The Physiology of Plants Under Stress: Soil and Biotic Factors*. John Wiley & Sons, New York.

Rasmussen, S., Parsons, A. J., Bassett, S., Christensen, M. J., Hume, D. E., Johnson, L. J., Johnson, R. D., Simpson, W. R., Stacke, C., Voisey, C. R., Xue, H., and J. A. Newman. 2007. High nitrogen supply and carbohydrate content reduce fungal endophyte and alkaloid concentration in *Lolium perenne*. *New Phytologist* 173:787–797.

Ravel, C., Courty, C., Coudret, A., and G. Charmet. 1997. Beneficial effects of *Neotyphodium lolii* on the growth and the water status in perennial ryegrass cultivated under nitrogen deficiency or drought stress. *Agronomie* 17:173–181.

Reynolds, H. L., Hartley, A. E., Vogelsang, K. M., Bever, J. D., and P. A. Schultz. 2005. Arbuscular mycorrhizal fungi do not enhance nitrogen acquisition and growth of old-field perennials under low nitrogen supply in glasshouse culture. *New Phytologist* 167: 869–880.

Rodriguez, R. J., Redman, R. S., and J. M. Henson. 2004. The role of fungal symbionts in the adaptation of plants to high stress environments. *Mitigation and Adaptation Strategies for Global Change* 9:261–272.

Rodriguez, R. J., Redman, R. S., and J. M. Henson. 2005. Symbiotic lifestyle expression by fungal endophytes and the adaptation of plants to stress: Unraveling the complexities of intimacy. In *The Fungal Community: Its Organization and Role in the Ecosystem*, 3rd edn., Eds. J. Dighton, J. F. White, and P. Oudemans, pp. 683–695. Taylor & Francis, Boca Raton, FL.

Roy, B. A., Kirchner, J. W., Christian, C. E., and L. E. Rose. 2000. High disease incidence and apparent disease tolerance in a North American Great Basin community. *Evolutionary Ecology* 14:421–438.

Sachs, J. L. and E. L. Simms. 2006. Pathways to mutualism breakdown. *Trends in Ecology and Evolution* 21:585–592.

Saikkonen, K., Faeth, S. H., Helander, M., and T. J. Sullivan. 1998. Fungal endophytes: A continuum of interactions with host plants. *Annual Review of Ecology and Systematics* 29:319–343.

Salisbury, F. B. and C. W. Ross. 1992. *Plant Physiology*, 4th edn. Wadsworth, Belmont, CA.

Schardl, C. L. and A. Leuchtmann. 2005. The *Epichloë* endophytes of grasses and the symbiotic continuum. In *The Fungal Community: Its Organization and Role in the Ecosystem*, 3rd edn., Eds. J. Dighton, J. F. White Jr., and P. Oudemans, pp. 475–503. Taylor & Francis, Boca Raton, FL.

Schüßler, A., Schwarzott, D., and C. Walker. 2001. A new fungal phylum, the Glomeromycota: Phylogeny and evolution. *Mycological Research* 105:1413–1421.

Schulz, B. and C. Boyle. 2005. The endophytic continuum. *Mycological Research* 109:661–686.

Smith, S. E. and D. J. Read. 1997. *Mycorrhizal Symbiosis*, 2nd edn. Academic Press, San Diego, CA.

Stone, J. K., Bacon, C. W., and J. F. White Jr. 2000. An overview of endophytic microbes: Endophytism defined. In *Microbial Endophytes*, Eds. C. W. Bacon and J. F. White Jr., pp. 3–29. Marcel Dekker, New York.

Sylvia, D. M., Hammond, L. C., Bennett, J. M., Haas, J. H., and S. B. Linda. 1993. Field response of maize to a VAM fungus and water management. *Agronomy Journal* 85:193–198.

West, C. P. 1994. Physiology and drought tolerance of endophyte-infected grasses. In *Biotechnology of Endophytic Fungi of Grasses*, Eds. C. W. Bacon and J. F. White Jr., pp. 87–99. CRC Press, Boca Raton, FL.

West, C. P. 2007. Plant influence on endophyte expression. In *Proceedings of the 6th International Symposium on Fungal Endophytes of Grasses*, Eds. A. J. Popay and E. R. Thom, pp. 117–121. New Zealand Grassland Association, Christchurch, New Zealand.

Wilson, D. 1995. Endophyte—the evolution of a term, and clarification of its use and definition. *Oikos* 73:274–276.

20

Habitat-Adapted Symbiosis as a Defense against Abiotic and Biotic Stresses

Rusty J. Rodriguez, Claire Woodward, Yong-Ok Kim, and Regina S. Redman

CONTENTS

20.1 Introduction

Since their establishment on land between 400 and 450 mya (Cleal and Thomas, 1999), plants have been confronted with a variety of biotic and abiotic stresses. The sessile nature of plants forced them to develop stress tolerance mechanisms which allowed plants to colonize a wide diversity of habitats from the Arctic to Antarctica. There has been a tremendous amount of research undertaken to elucidate how plants respond to and tolerate stress and the contributions of stress tolerance to plant distribution and biogeography (Brown et al., 1996; Gaston, 1996; Lowry and Lester, 2006). However, the nature of plant adaptation and subsequent establishment in high stress habitats remains unresolved (Leone et al., 2003; Maggio et al., 2003; Tuberosa et al., 2003). Although genetic mechanisms of plant stress tolerance are complex and not fully elucidated, they are thought to be based on adaptive changes in the plant genome (Robe and Griffiths, 2000; Givnish, 2002; Pan et al., 2006; Schurr et al., 2006). Interestingly, all plants have the ability to perceive abiotic stress and transduce signals at the cellular, tissue, and whole plant level to mitigate the impacts of stress, yet relatively few species thrive in high stress habitats (Bohnert et al., 1995; Iba, 2002; Bartels and Sunkar, 2005).

The majority of plant physiological studies do not consider a ubiquitous aspect of plant biology, fungal symbiosis. All plants in natural ecosystems are thought to have symbiotic associations with mycorrhizal and/or endophytic fungi which can have profound impact on plant fitness (Petrini, 1996; Brundrett, 2006). For example, there are numerous reports of symbiotic fungi conferring tolerance to abiotic (drought, metals, temperature, and salinity) and biotic (disease, herbivory) stresses (Read, 1999; Stone et al., 2000; Rodriguez et al., 2004, 2008; Selosse et al., 2004; Schulz, 2006). Although the mechanisms of symbiotically conferred stress tolerance are not yet elucidated, it is clear that symbiotic fungi can provide a mechanism for stress tolerance in plants. In this chapter, we discuss recent studies indicating that one group of fungal symbionts (class 2 endophytes) have the ability to adapt to stress in a habitat-specific manner and confer stress tolerance to host plants, a phenomenon we define as habitat-adapted symbiosis (Rodriguez et al., 2008). The proceeding discussion will focus on stress tolerance conferred by class 2 endophytes, potential physiological mechanisms, and ecological significance of this phenomenon.

20.2 Fungal Endophytes

Fungal endophytes can be subdivided into four classes based on host range, colonization pattern, transmission, and conferred fitness benefits (Table 20.1; RJR et al., in review). These symbionts are ubiquitous in nature and fossil records suggest that plant–endophyte associations have been in existence for more than 400 million years (Krings et al., 2007). Since the first description of a fungal endophyte in the early 1900s (Rayner, 1915), there have been more than 1000 reports on fungal endophytes. Collectively, these studies demonstrate that endophytes represent diverse taxa and are ubiquitous in nature. However, few studies have focused on the ecological or evolutionary significance of endophytes or performed comparative studies with symbiotic and nonsymbiotic plants. Recently, such comparative studies have been performed with class 2 endophytes revealing that they are required for the survival of at least some plants in high stress habitats (Rodriguez et al., 2008).

20.3 Endophyte-Conferred Abiotic Stress Tolerance

20.3.1 Temperature

In several locations around the world, geothermal activity has resulted in the establishment of sustained geothermal soils. Temperatures of geothermal soils fluctuate seasonally and are regulated by the intensity of geothermal activity and precipitation. For example, in Yellowstone National Park (YNP), geothermal soils range from 20°C in the moist winter to 60°C or higher in the dry summer months (Rodriguez et al., 2004). The geothermal soils of YNP are dispersed throughout the park and support a small community of nine vascular plants and some bryophytes (Stout and Al-Niemi, 2002). The most thermotolerant plant species (*Dichanthelium lanuginosum*) is able to tolerate root zone temperatures of 57°C and is common throughout YNP. Although the mechanism of heat tolerance was not known in *D. lanuginosum*, one study indicated that there was a correlation between root zone temperature and the accumulation of heat-shock proteins (Stout and Al-Niemi, 2002).

A symbiotic analysis of *D. lanuginosum* revealed that all of the plants sampled ($N = 100$) were colonized with one class 2 endophyte that represented >95% of all fungi isolated from surface-sterilized

TABLE 20.1

Fungal Endophytes

Fungal Group	Clavicipitaceous	Nonclavicipitaceous		
Class	**1**	**2**	**3**	**4**
Host range	Limited	Broad	Broad	Broad
Colonization	Shoot	Shoot & root	Shoot	Root
Transmission	V&H	V&H	H	H
Fitness benefits	NHA	NHA&HA	NHA	NHA

Note: V = vertical transmission; H = horizontal transmission; NHA = Non-habitat-adapted benefits are common among endophytes regardless of the habitat of origin; HA = Habitat-adapted benefits result from habitat-specific selective pressures such as pH, temperature, salinity, etc.

plant tissues (Redman et al., 2002). The endophyte was characterized as *Curvularia protuberata* and isolated from leaf, stem, roots, and seed coats but not from seeds. A combination of seed coat removal and surface sterilization resulted in the generation of *C. protuberata*-free plants allowing for comparative studies of endophyte-colonized and endophyte-free plants to determine the ecological role of the fungus (Redman et al., 2002). When grown nonsymbiotically, *D. lanuginosum* and *C. protuberata* tolerated temperatures up to 40°C and 38°C, respectively. However, experiments in a geothermal soil simulator indicated that *C. protuberata* (isolate Cp4666D)-colonized *D. lanuginosum* plants could tolerate continuous root zone temperatures of 50°C and temperature regimes of 65°C for 10 h/day followed by 37°C for 14 h/day. These studies were proceeded by a number of experiments to determine if Cp4666D was necessary for the survival of *D. lanuginosum* under field conditions. Geothermal soils representing a range of temperatures from 30°C to 45°C were identified in YNP, soil removed and pasteurized to eliminate any resident fungi and replaced into the holes from which they were derived (Redman et al., 2002). Cp4666D-colonized and endophyte-free *D. lanuginosum* plants were then transplanted into the pasteurized soils in spring. After 1 year of growth, symbiotic plants had significantly more biomass than nonsymbiotic plants at all soil temperatures. In fact, biomass differences between symbiotic and nonsymbiotic plants increased with soil temperature (6.5, 6.8, and 13.4 g differences at 30°C, 35°C, and 40°C, respectively) and nonsymbiotic plants did not survive at soil temperatures of 45°C.

Collectively, these studies indicate that *C. protuberata* (isolate Cp4666D) confers heat tolerance to *D. lanuginosum* and, in doing so, is also protected against elevated soil temperatures. In fact, the survival of *D. lanuginosum* in geothermal soils of YNP appears to be dependent on its association with *C. protuberata*. This is one of the most clear examples of a plant/fungal endophyte association resulting in a mutualism. While the ability of endophytes to confer fitness benefits to plants is known, defining the symbiosis can be difficult. Positive fitness benefits are easily observed in plants but it is challenging to observe them in endophytes.

To assess the specificity of the association between Cp4666D and *D. lanuginosum*, a host range study was performed with genetically divergent plant species (Rodriguez et al., 2008). Remarkably, Cp4666D is capable of asymptomatically colonizing a diversity of plants including both monocots (rice, wheat, and dunegrass) and eudicots (tomato, watermelon, and squash). Thermal stress studies with tomato and watermelon revealed that Cp4666D confers heat tolerance to both of these species (Rodriguez et al., 2008) indicating that Cp4666D is not functionally restricted to *D. lanuginosum*. Therefore, we hypothesize that the symbiotic communication responsible for heat tolerance is conserved and predates the divergence of monocots and eudicots between 140–235 mya (Wolfe et al., 1989; Yang et al., 1999; Chaw et al., 2004).

Although Cp4666D colonized genetically divergent hosts and conferred heat tolerance, it was not clear if endophyte-conferred heat tolerance resulted from habitat-specific adaptation or was a general trait of *C. protuberata*. Therefore, the ability to confer heat tolerance to plants was determined for a *C. protuberata* isolate (CpMH206) from the grass *Deschampsia flexuosa* growing in nongeothermal soils in Scotland. A study with tomato and *D. lanuginosum* revealed that CpMH206 and Cp4666D asymptomatically colonized these plants to the same extent; however, only Cp4666D conferred heat tolerance (Rodriguez et al., 2008). This suggests that the ability to confer heat tolerance is a habitat-adapted phenomenon.

Recently, Cp4666D was found to contain a double stranded RNA (designated CThTV) virus that is involved with endophyte-conferred heat tolerance (Márquez et al., 2007). Exposure of Cp4666D to freeze/thaw cycles resulted in the generation of a virus-free isolate designated as Cp4666DVF. Remarkably, Cp4666DVF did not confer heat tolerance to any host plants although it asymptomatically colonized plants to the same extent as Cp4666D. Reintroduction of CThTV back into Cp4666DVF via anastomosis reestablished the ability of this isolate to confer heat tolerance. While the role that CThTV plays in endophyte-conferred heat tolerance is not yet known, it is clear that all partners in this three-way symbiosis are required for survival in the geothermal habitats of YNP. In the coming years, researchers will determine if three-way symbioses are common in plant species and begin to decipher the biochemical contribution of each partner in stress tolerance and survival.

20.3.2 Salinity

Coastal habitats are known to impose salt stress on plants either by inundation during high tides and storms or from wave-induced spray. The proximity of plants to the tidal zone dictates the severity of salt stress. Although coastal plants can tolerate higher levels of salt than plants from nonsaline habitats, the majority of coastal species are not halophytes (Kearney, 1904; Barbour and Dejong, 1977). Current understanding of how plants tolerate salt stress is based on studies of halophytes and genetic model systems (Bartels and Sunkar, 2005). Although gene expression patterns associated with salt tolerance have been defined, it is not known how those patterns generate or how plants adapt to salt stress. Regardless, the ability of plants to tolerate salt stress is considered to be genetically complex and the result of adaptive changes in the plant genome. However, a recent study indicates that at least in one coastal plant, a class 2 fungal endophyte is responsible for salt tolerance and survival in coastal habitats (Rodriguez et al., 2008).

Leymus mollis is a dunegrass that commonly occurs on beaches throughout the San Juan Island archipelago in Puget Sound off the coast of Washington state. In many locations, the roots of *L. mollis* get inundated with salt water during high tides. A study of *L. mollis* on beaches located in the University of Washington's Cedar Rocks Biological Preserve on Shaw Island revealed that all of the plants analyzed ($N = 100$) were symbiotic with one class 2 fungal endophyte (Rodriguez et al., 2008). The endophyte was identified as *Fusarium culmorum* and isolated from all parts of the plant (rhizome, root, stem, leaves, and seed coats) but not from seeds. This made it possible to remove seed coats, surface sterilize seed, and propagate plants free of the endophyte. *F. culmorum* has the developmental genetics necessary to colonize plants via appressoria so that endophyte-free plants can be recolonized by exposing tissues to fungal conidia (Redman et al., 2001). Based on studies with *D. lanuginosum* in geothermal soils, we hypothesized that *F. culmorum* was conferring salt tolerance to *L. mollis* on Puget Sound beaches. Laboratory and field studies were performed to test the hypothesis and determine the ecological significance of this symbiosis. In laboratory studies, nonsymbiotic plants and plants colonized with *F. culmorum* (isolate FcRed1) from Shaw Island were continuously exposed to a concentration range (0–500 mM) of NaCl. Nonsymbiotic plants wilted and died at NaCl concentrations of 100 mM and above while symbiotic plants remained healthy at all NaCl concentrations indicating that FcRed1 conferred salt tolerance to *L. mollis* (Rodriguez et al., 2008). The ecological significance of this symbiosis was assessed by transplanting nonsymbiotic ($N = 20$) and symbiotic (FcRed1 colonized; $N = 20$) plants on a Shaw Island beach capable of supporting this plant species. Unlike the YNP study, the beach substrates were not sterilized prior to transplanting. Experimental plants were harvested after 3 months of growth to assess health, biomass, and fungal colonization. There were significant differences in plant health and total biomass with 100% of the FcRed1-colonized plants surviving and achieving an average biomass of 19.16 g while only 40% of the nonsymbiotic plants survived obtaining an average of 17.58 g (±9.23 g). Microbiological analysis indicated that all of the symbiotic plants were still colonized with FcRed1. More interesting was the fact that the surviving nonsymbiotic plants were also colonized with *F. culmorum*. Fungal community structure analysis indicated that FcRed1 was present in the beach substrates but at very low abundance. We surmised that the survival and final biomass of initially nonsymbiotic plants was dependent on the timing of in situ colonization by FcRed1.

To assess the specificity of the symbiosis between FcRed1 and *L. mollis*, a host range study was performed with genetically divergent plant species (Rodriguez et al., 2008). FcRed1 was able to asymptomatically colonize both monocots (wheat, rice, and turf) and eudicots (tomato, watermelon, and squash). Additional experiments indicated that FcRed1 conferred salt tolerance (300–500 mM) to tomato, watermelon, and rice plants. Clearly, FcRed1 is not limited to establishing symbiosis with *L. mollis* and is able to confer salt tolerance to genetically distant species. Therefore, we hypothesize that the symbiotic communication responsible for salt tolerance is conserved and predates the divergence of monocots and eudicots to 140–235 mya (Wolfe et al., 1989; Yang et al., 1999; Chaw et al., 2004).

To determine if FcRed1-conferred salt tolerance involved habitat-specific adaptations, a comparative study was performed with an isolate of *F. culmorum* (Fc18) from an agricultural field in the Netherlands that did not impose salt stress (Rodriguez et al., 2008). Both fungi asymptomatically colonized *L. mollis*, tomato, and rice plants to equivalent levels but only FcRed1 conferred salt tolerance. The levels of salt tolerance observed in Fc18-colonized plants were not significantly different from nonsymbiotic controls which begin wilting at 100 mM NaCl. Depending on the plant species, FcRed1 conferred tolerance to NaCl levels between 300 and 500 mM. These results indicate that the ability of FcRed1 to confer salt tolerance likely requires habitat-specific adaptations.

20.3.3 Drought

One aspect of class 2 endophytes that does not appear to involve habitat-specific adaptations is the ability to confer drought tolerance to both monocot and eudicot plants (Rodriguez et al., 2008). All five of the fungal endophytes described above (Cp4666D, CpMH206, Cp4666DVF, FcRed1, and Fc18) confer drought tolerance regardless of the habitat of origin or the plant host colonized. This indicates that although CpMH206 and Fc18 do not confer habitat-specific stress tolerance (heat and salt, respectively), they do communicate with plants to confer drought tolerance. It is tempting to speculate that endophyte-conferred drought tolerance reflects an evolutionary legacy. This is based on fossil records that suggest that fungal endophytes have been associated with plants for >400 mya and these associations may have predated the movement of plants onto land (Krings et al., 2007). In fact, it has been suggested that fungi may have perpetuated the movement of plants onto land by providing them with a mechanism to mitigate periods of desiccation (Pirozynski and Malloch, 1975). Regardless, the fact that the endophytes described above all confer drought tolerance to both monocots and eudicots suggests an evolutionary dynamic similar to that described above for symbiotically conferred heat and salt tolerance (the symbiotic communication for drought tolerance likely predates the divergence of monocots and eudicots).

20.4 Endophyte-Conferred Biotic Stress Tolerance

There are several biotic stresses that plants are confronted with, including attack by fungal pathogens. The interaction between plant hosts and fungal pathogens is thought to be either compatible or incompatible. During compatible fungal pathogen attack, the fungus will disseminate through the host tissue faster than host defenses can respond, resulting in the development of disease symptoms such as necrosis, wilting, and plant death (Goodman et al., 1986; Morel and Dangl, 1997). If the interaction is incompatible, a hypersensitive response will ensue and the ingress of the pathogen is thwarted by induction of oxidative stress and subsequent death of plant cells surrounding the infection site (Govrin and Levin, 2000).

In general, pathogenic fungi are considered to be very different from mutualistic fungi and may be an evolutionary predecessor to mutualists (Schardl and Leuchtmann, 2005). However, there are several reports indicating that different fungal species within a genus and different isolates of a species may express different symbiotic lifestyles ranging from mutualism to parasitism. This variation in lifestyle expression within a genus or species may occur due to the host physiology or edaphic conditions (Francis and Read, 1995; Graham et al., 1996; Graham and Eissenstat, 1998) and is defined as the Symbiotic Continuum (Johnson et al., 1997; Schardl and Leuchtmann, 2005; Schulz and Boyle, 2005). Recently, a host range study of pathogenic *Colletotrichum* species revealed that individual isolates could express pathogenic or mutualistic symbiotic lifestyles depending on the host species they colonized (Redman et al., 2001). When expressing a mutualistic lifestyle, the fungi asymptomatically colonize host plants without activating host defense systems (Redman et al., 1999a). For example, individual isolates of *Colletotrichum magna* and *Colletotrichum orbiculare* that are virulent pathogens of cucurbits can asymptomatically colonize tomato plants and confer disease resistance to virulent species of *Colletotrichum*, *Fusarium*, and *Phytophtora*. Similarly, the banana pathogen *C. musae* can asymptomatically colonize certain pepper plants and confer disease resistance. *Colletotrichum*-conferred disease resistance appears to be based on the very rapid and strong activation of host defense systems when plants that are colonized with an isolate expressing a mutualism are challenged with virulent pathogens (Redman et al., 1999a). Interestingly, when these fungi are expressing mutualistic lifestyles, they also confer growth enhancement and drought tolerance to plants. In contrast, *Colletotrichum coccodes*, a virulent tomato pathogen, only expressed a pathogenic lifestyle indicating that it is unable to switch lifestyles or the number of host species tested was too limited. Regardless, fungi that were previously viewed solely as pathogens have the potential to confer biotic stress tolerance to plants and individual isolates can span the Symbiotic Continuum (Redman et al., 2001).

Mutagenesis studies revealed that the ability of *C. magna* to switch between pathogenic and mutualistic lifestyles involves a single genetic locus (Freeman and Rodriguez, 1993). Subsequent restriction enzyme-mediated integration mutation studies revealed that the integration of transformation vectors into different genomic locations caused pathogenic isolate of *C. magna* to express one of the three nonpathogenic lifestyles (mutualism, intermediate mutualism, or commensalism-based disease resistance; (Redman et al., 1999b). The genomic region (genetic locus designated fungal symbiotic lifestyle; FSL) responsible for the conversion of pathogenic to mutualistic lifestyles has been isolated and sequenced, and it does not align with any known sequence present in public databases (Redman et al. 2009. in prep.). The potential functional role of the FSL locus was assessed by using the FSL sequences to construct a gene disruption vector. Transformation of three phylogenetically distinct *Colletotrichum* species (*C. coccodes*, *C. lindemuthianum*, and *C. orbiculare*) resulted in the generation of nonpathogenic transformants that asymptomatically colonized and conferred disease resistance to wildtype isolates in their respective hosts (Redman et al., in prep.). These results not only demonstrate functionality of the FSL locus but also reiterate the observation that there can be as less as one genetic locus that differentiates a pathogen from a mutualist.

Symbiotic lifestyle switching has interesting evolutionary implications. The ability of pathogenic fungi to express nonpathogenic lifestyles suggests that they may have evolved as flexible symbionts to expand host ranges for increased dissemination or stress avoidance. If this is the case, then in one location, flexible symbionts may be virulent in one host species and mutualistic in another. Studies by Freeman et al. (2001) indicate that pathogens capable of expressing nonpathogenic lifestyles can asymptomatically colonize weed species that act as refugia for producing virulent inoculum capable of inducing disease in nearby susceptible agricultural crops (Freeman et al., 2001). It is tempting to speculate that the ability to express alternate symbiotic lifestyles is not unique to *Colletotrichum* isolates and, in fact, may be common among fungal pathogens. If that is the case, current paradigms in plant ecology, disease, and evolution will need to be readdressed.

20.5 Biochemical Mechanisms of Symbiotically Conferred Stress Tolerance

Plants perceive abiotic stresses through receptors on the cell membrane and subsequently elicit the production of secondary messengers such as reactive oxygen species (ROS), Ca^{2+}, and inositol phosphates to transduce stress-related signals within cells. The messengers act to upregulate stress responsive genes and this results in the adaptive responses by the plant necessary for stress tolerance and survival. The reaction to stress at the cellular level is coordinated into a whole plant response by the change in the level of hormones such as abscisic acid (ABA), ethylene, and jasmonates (Christmann et al., 2006). The responses to the abiotic stresses like drought, heat, and salinity, discussed below, are intricately linked through the "cross-talk" of the stress response pathways to ellicit responses that are common to the stresses or uniquely tailored for a specific stress.

Soils with high salinity cause both osmotic and ionic stress in the plant. Low soil water potential makes it hard for roots to absorb both water and nutrients. In addition, NaCl competes with other ions, resulting in a decrease in cellular levels of Ca^{2+}, K^+, Mg^{2+}, and NO_3^- (Niu et al., 1995) that can induce nutrient deficiencies. As the salt load in the cell exceeds the capacity of the compartmentalization into the vacuole, the salts rapidly accumulate in the cytoplasm. Increases in cellular Na^+ and Cl^- ions result in a disruption of ionic and osmotic equilibrium, and at concentrations above 100 mM, most enzymes are inhibited strongly, including those essential for the cellular metabolism. Endomembrane permeability is changed as cations are displaced, grana swell and become disorganized, and RUBISCO efficiency declines. These cellular effects cause an overall reduction in plant growth through the inhibition of both cell division and expansion along with a reduction in photosynthesis and eventual tissue death as the ions reach toxic levels.

Drought conditions also cause a water deficit in plants. Initially, the changes in soil water potential cause signals to be emitted from the root to the shoot to reduce transpiration and leaf expansion. Nutrient uptake by the roots becomes increasingly difficult; the xylem pH increases as the nitrate uptake decreases. The apoplastic pH increases, and this may mediate the release of ABA from mesophyll cells and cause the closure of stomata (Felle et al., 2005). The desiccation of plant cells causes an increase in the electrolyte concentration (Bahrun et al., 2002) to levels that are toxic to enzyme function, resulting in a breakdown of cellular metabolism. Plant membranes become porous, and this loss of membrane integrity leads to a disruption of mitochondria and chloroplast function (Flexas and Medrano, 2002). Collectively, the decrease in membrane integrity and loss of organelle function cause a decrease in plant growth, particularly of the leaf tissue (Salah and Tardieu, 1997) and senescence of older leaves (Rivero et al., 2007).

Heat stress is deleterious to photosynthesis; initially, stomatal closure causes a decrease in internal CO_2 concentration and this causes an increase in the oxygenation reaction and energy production via the energy wasteful photorespiration process. As drought stress increases in severity, the thylakoid and PSII complexes are disrupted (Vani and Saradhi, 2001; Kim and Portis, 2005; Tang et al., 2007). Carbon metabolism is impeded as rubisco activase and the associated proteins become inhibited (Salvucci and Crafts-Brandner, 2004) and ATP synthesis in inhibited. Heat-shock proteins are found to increase in concentration; these act as chaperones, protecting the integrity of essential proteins (Kotak et al., 2007) as well as maintaining membrane fluidity and composition.

Physiological responses that are common to all salt, drought, and heat stress are based on the fact that all three stresses affect water relations and result in the generation of ROS. To prevent the loss of intracellular water, the plant lowers the water potential inside the cell by producing osmolytes. Osmolytes are molecules that can be accumulated to high concentrations without disturbing the cell metabolism and include sugars and sugar alcohols (raffinose, sucrose, trehalose, mannitol, and sorbitol), amino acids (proline) (Yoshiba et al., 1997), and amines (glycine betaine) (Waditee et al., 2005). Not only the osmolytes lower the water potential of the cell but also act to stabilize the structure of membranes and enzymes and scavenge ROS. ROS are normally produced at a very low concentration in plants. However, when plant metabolism is disrupted by abiotic stress, ROS concentrations significantly increase, resulting in cellular damage through chemical oxidation, deesterification of membrane lipids, protein denaturation, and mutation of nucleic acids (Apel and Hirt, 2004). The ROS themselves act as secondary messengers (Neill et al., 2002), leading to hormone-mediated responses such as the further closure of leaf stomata, production of heat-shock proteins to chaperone proteins, and initiating an increase in the level of antioxidant scavengers such as catalase, guaicol peroxidase, and ascorbate peroxidase (Kotak et al., 2007). Plants limit the leaf dehydration and subsequent osmotic imbalance of the cells by closing the leaf stomates. This can occur through direct evaporation from the guard cells or by ABA promoting K^+ efflux and subsequent loss of turgor pressure and stomatal closure. The resulting reduction in transpiration simultaneously decreases the influx of CO_2, decreasing photosynthesis and upsetting the delicate balance of ROS production and antioxidant quenching.

Recent understanding of plant physiological responses to abiotic stresses does not include potential influences of fungal endophytes on plant responses. For example, when exposed to heat, salt, or drought stress, nonsymbiotic panic grass and tomato plants increase osmolyte concentrations (Rodriguez et al., 2008). However, plants that are symbiotic with an endophyte that confers tolerance to each stress (heat and drought tolerance by Cp4666D; salt and drought tolerance by FcRed1; and only drought tolerance by CpMH206 & Fc18) either did not increase in osmolyte levels or had lower osmolyte concentration compared to nonstressed controls (Márquez et al., 2007; Rodriguez et al., 2007, 2008). Therefore, increased osmolyte concentrations is not the method of heat, salt, or drought tolerance in symbiotic plants.

Although the mechanisms associated with endophyte-conferred stress tolerance are not yet known, there are three physiological aspects of symbiotic plants that provide insight into this phenomenon: Symbiotic plants consume less water, have greater biomass, and produce less ROS compared to nonsymbiotic plants (Rodriguez et al., 2008). A concomitant decrease in water consumption and increase in biomass suggest that symbiotic plants have greater water-use efficiency than nonsymbiotic plants. This may be especially useful under restricted water conditions where it would be advantageous for the plants to require less water to survive and fulfill their life cycle. The herbicide Paraquat can be used to test tissue sensitivity to ROS (Vaughn and Duke, 1983). The involvement of ROS in endophyte-conferred stress tolerance was determined with Paraquat (Rodriguez et al., 2008) which is reduced by photosystem I and oxidized by molecular oxygen, forming superoxide molecules that cause photobleaching. In the absence of stress, leaf tissues of symbiotic and nonsymbiotic plants (panic grass, dunegrass, and tomato) exposed to Paraquat were not susceptible to photobleaching, showing that no ROS was produced. However, when the symbiotic and nonsymbiotic plants were exposed to stress (panic grass with heat stress, tomato with heat and salt stress, and dunegrass with salt stress), leaf tissues of nonsymbiotic plants bleached white and symbiotic plants remained green. This suggests that the symbiotic plants had a greater capacity to either scavenge or prevent the production of ROS.

20.6 Evolutionary and Ecological Significance of Habitat-Adapted Symbiosis

In order to appreciate the potential significance of symbiotically conferred stress tolerance, it is important to consider the evolutionary history and the geographic distribution of plants. Two historical events that required significant adaptive responses in plants were the oxygenation of earth's atmosphere and the colonization of terrestrial habitats. It is estimated that between 0.5 and 1 billion years ago, atmospheric oxygen levels increased to levels conducive for aerobic respiration and photosynthesis (Canfield and Teske, 1996). During photosynthesis, plants generate metabolic energy (ATP) by transferring electrons from an electron donor (H_2O) through an electron transport chain to a terminal electron acceptor (CO_2). During this process, the electron donor (H_2O) is oxidized to produce O_2 and the terminal electron acceptor (CO_2) is reduced to generate organic compounds such as carbohydrates. A byproduct of both aerobic respiration and photosynthesis is the production of high-energy ROS such as superoxide radicals and peroxide ions (Apel and Hirt, 2004). Although there are several different ROS that can be formed, all of them cause oxidative damage to proteins, DNA, and lipids. Under normal physiological conditions, cells are metabolically balanced and ROS are eliminated by chemical and/or enzymatic antioxidation systems (Apel and Hirt, 2004; Changbin and Dickman, 2005). However, exposure to abiotic stress results in metabolic imbalances causing an increase in ROS to levels beyond the capacity of antioxidant systems. The ability of fungal endophytes to protect plants against detrimental effects of ROS may represent an ancient aspect of these symbiotic associations, one that likely predates the divergence of monocots and eudicots (est. 200 mya) and may reflect the occurrence of these associations very early in plant evolution (Rodriguez and Redman, 2005).

Fossil records indicate that plants colonized terrestrial habitats approximately 400 mya (Cleal and Thomas, 1999). When plants moved from aquatic habitats onto land, they were confronted with periods of desiccation and the need to transport sufficient amounts of water from below the ground tissues to prevent the desiccation of aerial tissues. Plants developed root and vascular systems to efficiently absorb and transport water. However, plants have been confronted with periods of desiccation throughout their evolutionary history. Class 2 fungal endophytes increase the efficiency of water usage in plants which may provide an important mechanism for mitigating potential impacts of desiccation and allow for more efficient vascular transportation of water and nutrients.

The geographical distribution of vascular plants has been studied intensively for more than 275 years (Martyn, 1729). Plant distribution patterns across complex habitats are well documented (Crisci, 2001), but little is known about potential biological processes contributing to these patterns. In general, there is an inverse relationship between the biological diversity of plants and abiotic stress (Brown et al., 1996; Gaston, 1996; Lowry and Lester, 2006). Stress tolerances in plants involve genetically complex processes (Bohnert et al., 1995; Alpert, 2000; Iba, 2002; Tester and Davenport, 2003; Wang et al., 2003; Bartels and Sunkar, 2005) that have evolved several times throughout history (Oliver et al., 2000). The complex genetic nature of stress tolerances suggests that these processes evolve by slow adaptive changes in accordance with Darwinian expectations. Plant stress tolerance conferred by class 2 endophytes appears to represent a deviation from Darwinian evolution for several reasons: (1) this is an intergenomic epigenetic process; (2) stress tolerance is achieved rapidly after colonization by a stress tolerance conferring endophyte; (3) endophyte-conferred stress tolerance is transferable between genetically unrelated plant species; (4) at least some plants from high stress habitats do not survive in those habitats without the appropriate endophytes. It seems that in at least some species, class 2 endophytes must play a significant role in distribution patterns by virtue of conferring the stress tolerance required for survival in specific habitats (Redman et al., 2002; Márquez et al., 2007; Rodriguez et al., 2008).

The observations presented here raise some interesting questions in plant biology. Are class 2 endophytes required for many, most, or all plants to colonize habitats imposing abiotic stresses? After 400 million years of evolution, why plants have not, in high stress habitats, circumvented the need for endophyte symbioses? Do endophytes provide a mechanism to explain plant distribution patterns? Based on symbiotic communication, is it possible to predict species distribution patterns? What role do endophytes have in plant biogeography? Clearly, there are confounding ecological factors that prevent high frequency transfer of endophytes between unrelated plant species. Otherwise, we would expect to observe a higher diversity of plants in habitats imposing high levels of abiotic stresses, especially if they are adjacent to high biodiversity habitats devoid of those stresses. For example, the low biodiversity geothermal soils of YNP are surrounded by high biodiversity cool soil habitats and the endophyte (Cp4666D) from a geothermal plant (*D. lanuginosum*) can colonize plants from the surrounding cool soils and confer heat tolerance. Yet, plants from the cool soil habitats are not expanding into the geothermal soil habitats. As we begin to decipher the biochemical/genetic communication between plants and fungal endophytes, we will begin to address the questions listed above and better understand the ecological factors that regulate plant distribution patterns.

ACKNOWLEDGMENTS

This work was supported by the U.S. Geological Survey, and support funding from NSF (0414463) and US/IS BARD (3260-01C).

REFERENCES

Alpert, P. 2000. The discovery, scope, and puzzle of desiccation tolerance in plants. *Plant Ecology* 151:5–17.

Apel, K. and H. Hirt. 2004. Reactive oxygen species: Metabolism, oxidative stress, and signal transduction. *Annual Review of Plant Biology* 55:373–399.

Bahrun, A., Jensen, C., Asch, F., and V. Mogensen. 2002. Drought-induced changes in xylem pH, ionic composition, and ABA concentration act as early signals in field-grown maize (*Zea mays* L.). *Journal of Experimental Botany* 53:251–263.

Barbour, M. G. and T. M. Dejong. 1977. Response of west coast beach taxa to salt spray, seawater inundation, and soil-salinity. *Bulletin of the Torrey Botanical Club* 104:9–34.

Bartels, D. and R. Sunkar. 2005. Drought and salt tolerance in plants. *Critical Reviews in Plant Science* 24:23–58.

Bohnert, H. J., Nelson, D. E., and R. G. Jensen. 1995. Adaptations to environmental stresses. *The Plant Cell* 7:1099–1111.

Brown, J. H. C., Stevens, G. C., and D. M. Kaufman. 1996. The geographic range: Size, shape, boundaries, and internal structure. *Annual Review of Ecology and Systematics* 27:597–623.

Brundrett, M. C. 2006. Understanding the roles of multifunctional mycorrhizal and endophytic fungi. In *Microbial Root Endophytes*, Eds. B. J. E. Schulz, C. J. C. Boyle, and T. N. Sieber, pp. 281–293. Berlin: Springer-Verlag.

Canfield, D. E. and A. Teske. 1996. Late proterozoic rise in atmospheric oxygen concentration inferred from phylogenetic and sulphur-isotope studies. *Nature* 382:127–132.

Changbin, C. and M. B. Dickman. 2005. Proline suppresses apoptosis in the fungal pathogen *Colletotrichum trifolii*. *Proceedings of the National Academy of Sciences* 102:3459–3464.

Chaw, S., Chang, C., Chen, H., and W. Li. 2004. Dating the monocot–dicot divergence and the origin of core eudicots using whole chloroplast genomes. *Journal of Molecular Evolution* 58:424–441.

Christmann, A., Moes, D., Himmelbach, A., Yang. Y., Tang, Y., and E. Grill. 2006. Integration of abscisic acid signaling into plant responses. *Plant Biology* 8:314–325.

Cleal, C. J. and B. A. Thomas. 1999. *Plant Fossils: The History of Land Vegetation*, pp. 316. The Woodbridge, U. K.: Boydell Press.

Crisci, J. V. 2001. The voice of historical biogeography. *Journal of Biogeography* 28:157–168.

Felle, H., Herrmann, A., and R. K. H. Hückelhoven. 2005. Root-to-shoot signaling: Apoplastic alkalinization, a general stress response and defence factor in barley (*Horeum vulgare*). *Protoplasma* 227:17–24.

Flexas, J. and H. Medrano. 2002. Drought-inhibition of photosynthesis in C3 plants: Stomatal and non-stomatal limitations revisited. *Annals of Botany* 89:183–189.

Francis, R. and D. J. Read. 1995. Mutualism and antagonism in the mycorrhizal symbiosis, with special reference to impacts on plant community structure. *Canadian Journal of Botany-Revue Canadienne De Botanique* 73:S1301–S1309.

Freeman, S., Horowitz, S., and A. Sharon. 2001. Pathogenic and nonpathogenic lifestyles in *Colletotrichum acutatum* from strawberry and other plants. *Phytopathology* 91:986–992.

Freeman, S., and R. J. Rodriguez. 1993. Genetic conversion of a fungal plant pathogen to a nonpathogenic, endophytic mutualist. *Science* 260:75–78.

Gaston, K. 1996. Species-range-size distributions: Patterns, mechanisms and implications. *Trends in Ecology and Evolution* 11:197–201.

Givnish, T. J. 2002. Ecological constraints on the evolution of plasticity in plants. *Evolutionary Ecology* 16:213–242.

Goodman, R. N., Kiraly, Z., and K. R. Wood. 1986. *The Biochemistry and Physiology of Plant Infectious Disease*, 433 pp. Colombia, MO: University of Missouri Press.

Govrin, E. M. and A. Levin. 2000. The hypersensitive response facilitates plant infection by the necroptrophic pathogen *Botrytis cinerea. Current Biology* 13:751–757.

Graham, J. H., Drouillard, D. L., and N. C. Hodge. 1996. Carbon economy of sour orange in response to different *Glomus* spp. *Tree Physiology* 16:1023–1029.

Graham, J. H. and D. M. Eissenstat. 1998. Field evidence for the carbon cost of citrus mycorrhizas. *New Phytologist* 140:103–110.

Iba, K. 2002. Acclimative response to temperature stress in higher plants: Approaches of gene engineering for termperature tolerance. *Annual Review of Plant Biology* 53:225–245.

Johnson, N. C., Graham, J. H., and F. A. Smith. 1997. Functioning of mycorrhizal associations along the mutualism–parasitism continuum. *New Phytologist* 135:575–586.

Kearney, T. H. 1904. Are plants of sea beaches and dunes true halophytes? *Botanical Gazette* 37:424–436.

Kim, K. and A. R. J. Portis. 2005. Temperature dependence of photosynthesis in Arabidopsis plants with modifications in Rubisco activase and membrane fluidity. *Plant Cell Physiology* 46:522–530.

Kotak, S., Larkindale, J., Lee, U., von Koskull-Döring, P., Vierling, E., and K. D. Scharf. 2007. Complexity of the heat stress response in plants. *Current Opinion in Plant Biology* 10:310–316.

Krings, M., Taylor, T. N., Hass, H., Kerp, H., Dotzler, N., and E. J. Hermsen. 2007. Fungal endophytes in a 400-million-yr-old land plant: Infection pathways, spatial distribution, and host responses. *New Phytologist* 174:648–657.

Leone, A., Perrotta, C., and B. Maresca. 2003. Plant tolerance to heat stress: Current strategies and new emergent insight. In *Abiotic Stresses in Plants*, Eds. L. S. di Toppi and B. Pawlik-Skowronska, pp. 1–22. London: Kluwer Academic.

Lowry, E. and S. E. Lester. 2006. The biogeography of plant reproduction: Potential determinants of species' range sizes. *Journal of Biogeography* 33:1975–1982.

Maggio, A., Bressan, R. A., Ruggiero, C., Xiong, L., and S. Grillo. 2003. Salt tolerance: Placing advances in molecular genetics into a physiological and agronomic context. In *Abiotic Stresses in Plants*, Eds. L. S. di Toppi and B. Pawlik-Skowronska, pp. 53–70. London: Kluwer Academic.

Márquez, L. M., Redman, R. S., Rodriguez, R. J., and M. J. Roossinck. 2007. A virus in a fungus in a plant—three way symbiosis required for thermal tolerance. *Science* 315:513–515.

Martyn, J. 1729. An account of some observations relating to natural history, made in a journey to the peak in Derbyshire. *Philosophical Transactions* 36:22–32.

Morel, J. B. and J. L. Dangl. 1997. The hypersensitive response and the induction of cell death in plants. *Cell Death and Differentiation* 8:671–683.

Neill, S., Desikan, R., and J. Hancock. 2002. Hydrogen peroxide signaling. *Current Opinion in Plant Biology* 5:388–395.

Niu, X., Bressan, R. A., Hasegawa, P. M., and J. M. Pardo. 1995. Ion homeostasis in NaCl stress environments. *Plant Physiology* 109:735–742.

Oliver, M. J., Tuba, Z., and B. D. Mishler. 2000. The evolution of vegetative dessication tolerance in land plants. *Plant Ecology* 151:85–100.

Pan, X. Y., Geng, Y. P., Zhang, W. J., Li, B., and J. K. Chen. 2006. The influence of abiotic stress and phenotypic plasticity on the distribution of invasive *Alternanthera philoxeroides* along a riparian zone. *Acta Oecologica—International Journal of Ecology* 30:333–341.

Petrini, O. 1996. Ecological and physiological aspects of host-specificity in endophytic fungi. In *Endopytic Fungi in Grasses and Woody Plants*, Eds. S. C. Redlin and L. M. Carris, pp. 87–100. St. Paul, MN: APS Press.

Pirozynski, K. A., and D. W. Malloch. 1975. The origin of land plants a matter of mycotrophism. *Biosystems* 6:153–164.

Rayner, M. C. 1915. Obligate symbiosis in *Calluna vulgaris. Annals of Botany* 29:97–133.

Read, D. J. 1999. Mycorrhiza—the state of the art. In *Mycorrhiza*, Eds. A. Varma and B. Hock, pp. 3–34. Berlin: Springer-Verlag.

Redman, R. S., Beckwith, F., and R. J. Rodriguez. 2009. Generation of Nonpathogenic endophytic mutants by gene disruption in the genus. *Colletotrichum*. In preparation.

Redman, R. S., Dunigan, D. D., and R. J. Rodriguez. 2001. Fungal symbiosis: From mutualism to parasitism, who controls the outcome, host or invader? *New Phytologist* 151:705–716.

Redman, R. S., Freeman, S., Clifton, D. R., Morrel, J., Brown, G., and R. J. Rodriguez. 1999a. Biochemical analysis of plant protection afforded by a nonpathogenic endophytic mutant of *Colletotrichum magna. Plant Physiology* 119:795–804.

Redman, R. S., Ranson, J., and R. J. Rodriguez. 1999b. Conversion of the pathogenic fungus *Colletotrichum magna* to a nonpathogenic endophytic mutualist by gene disruption. *Molecular Plant Microbe Interactions* 12:969–975.

Redman, R. S., Sheehan, K. B., Stout, R. G., Rodriguez, R. J., and J. M. Henson. 2002. Thermotolerance conferred to plant host and fungal endophyte during mutualistic symbiosis. *Science* 298:1581.

Rivero, R. M., Kojima, M., Gepstein, A., Sakakibara, H., Mittler, R., and S. Gepstein. 2007. Delayed leaf senescence induces extreme drought tolerance in a flowering plant. *Proceedings of the National Academy of Sciences* 104:19631–19636.

Robe, W. E. and H. Griffiths. 2000. Physiological and photosynthetic plasticity in the amphibious, freshwater plant, *Littorella uniflora*, during the transition from aquatic to dry terrestrial environments. *Plant Cell and Environment* 23:1041–1054.

Rodriguez, R. and R. Redman. 2005. Balancing the generation and eliminaton of reactive oxygen species. *Proceedings of the National Academy of Sciences* 102:3175–3176.

Rodriguez, R. J., Henson, J., Van Volkenburgh, E., Hoy, M., Wright, L., Beckwith, F. et al. 2008. Stress tolerance in plants via habitat-adapted symbiosis. *International Society of Microbial Ecology* 2:404–416.

Rodriguez, R. J., Redman, R. S., and J. M. Henson. 2004. The role of fungal symbioses in the adaptation of plants to high stress environments. *Mitigation and Adaptation Strategies for Global Change* 9:261–272.

Salah, H. and F. Tardieu. 1997. Control of leaf expansion rate of droughted maize plants under fluctuating evaporative demand (a superposition of hydraulic and chemical messages?). *Plant Physiology* 114:893–900.

Salvucci, M. E. and S. J. Crafts-Brandner. 2004. Relationship between the heat tolerance of photosynthesis and the thermal stability of rubisco activase in plants from contrasting thermal environments. *Plant Physiology* 134:1460–1470.

Schardl, C. and A. Leuchtmann. 2005. The *Epichloe* endophytes of grasses and the symbiotic continuum. In *The Fungal Community: Its organization and Role in the Ecosystem*, eds. J. Dighton, J. F. White Jr., and P. Oudemans, pp. 475–503. Boca Raton, FL: Taylor & Francis.

Schulz, B. and C. Boyle, C. 2005. The endophytic continuum. *Mycological Research* 109:661–686.

Schulz, B. J. E. 2006. Mutualistic interactions with fungal root endophytes. In *Microbial Root Endophytes*, Eds. B. J. E. Schulz, C. J. C. Boyle, and T. N. Sieber, pp. 261–280. Berlin: Springer-Verlag.

Schurr, U., Walter, A., and U. Rascher. 2006. Functional dynamics of plant growth and photosynthesis—from steady-state to dynamics—from homogeneity to heterogeneity. *Plant Cell and Environment* 29:340–352.

Selosse, M. A., Baudoin, E., and P. Vandenkoornhuyse. 2004. Symbiotic microorganisms, a key for ecological success and protection of plants. *Comptes Rendus Biologies* 327:639–648.

Stone, J. K., Bacon, C. W., and J. F. White Jr. 2000. An overview of endophytic microbes: Endophytism defined. In *Microbial Endophytes*, Eds. C. W. Bacon and J. F. White Jr., pp. 3–30. New York: Marcel Dekker.

Stout, R. G. and T. S. Al-Niemi. 2002. Heat-tolerance flowering plants of active geothermal areas in Yellowstone National Park. *Annals of Botany* 90:259–267.

Tang, Y., Wen, X., Lu, Q., Yang, Z., Cheng, Z., and C. Lu. 2007. Heat stress induces an aggregation of the light-harvesting complex of photosystem II in spinach plants. *Plant Physiology* 143:629–638.

Tester, M. and R. Davenport. 2003. Na^+ tolerance and Na^+ transport in higher plants. *Annals of Botany* 91:503–527.

Tuberosa, R., Grillo, S., and R. P. Ellis. 2003. Unravelling the genetic basis of drought tolerance in crops. In *Abiotic Stresses in Plants*, Eds. L. S. di Toppi and B. Pawlik-Skowronska, pp. 71–122. London: Kluwer Academic.

Vani, B. and P. P. Saradhi. 2001. Characterization of high temperature induced stress impairments in thylakoids of rice seedlings. *Indian Journal of Biochemistry and Biophysics* 38:220–229.

Vaughn, K. C. and S. O. Duke. 1983. In situ localization of the sites of paraquat action. *Plant Cell and the Environment* 6:13–20.

Waditee, R., Bhuiyan, M. N., Rai, V., Aoki, K., Tanaka, Y., and T. Hibino. 2005. Genes for direct methylation of glycine provide high levels of glycinebetaine and abiotic-stress tolerance in *Synechococcus* and *Arabidopsis. Proceedings of the National Academy of Sciences* 102:1318–1323.

Wang, W., Vincur, B., and A. Altman. 2003. Plant responses to drought, salinity and extreme temperatures: Towards genetic engineering for stress tolerance. *Planta* 218:1–14.

Wolfe, K. H., Gouy, M., Yang, Y., Sharp, P. M., and W. H. Li. 1989. Date of the monocot–dicot divergence estimated from chloroplast DNA sequence data. *Proceedings of the National Academy of Sciences* 86:6201–6205.

Yang, Y. W., Lai, K. N., Tai, P. Y., and W. H. Li. 1999. Rates of nucleotide substitution in angiosperm mitochondrial DNA sequences and dates of divergence between *Brassica* and other angiosperm lineages. *Journal of Molecular Evolution* 48:597–604.

Yoshiba, Y., Kiyosue, T., Nakashima, K., Yamaguchi-Shinozaki, K., and K. Shinozaki. 1997. Regulation of levels of proline as an osmolyte in plants under water stress. *Plant Cell Physiology* 38:1095–10102.

21

Insect Herbivory and Defensive Mutualisms between Plants and Fungi

Alison J. Popay

CONTENTS

21.1 Introduction

Fungal endophytes are ubiquitous in the plant kingdom. Some are pathogenic but many form a symbiotic relationship with their host which spans the continuum between parasitism and mutualism. Many of those interactions between plant and fungus will alter the chemical composition of the plant which, positively or negatively, has an impact on insect herbivores. If the primary function of a mutualistic relationship is to provide protection to both the partners, then the mutualism can be described as defensive. The Pooid grasses and their associations with fungal endosymbionts are a very well-documented and fascinating example of a mutualism considered to be defensive in its role. Despite the widespread occurrence of other fungal endophytes, however, further examples of defensive mutualisms are rare.

The endophytes in grasses (Clavicipitaceae, Ascomycota) belong to the tribe Balansiae which is distinct from other members of the Clavicipitaceae because their infections are generally both perennial and systemic in their hosts. Residing within this tribe, *Neotyphodium* spp. Glenn, Bacon, and Hanlin (= *Acremonium* sect. *Albanosa* Morgan-Jones and Gams), are obligate biotrophic endosymbionts which live asymptomatically in the aboveground tissues of their hosts and are vertically transmitted in seeds. *Neotyphodium* spp., because of their economic importance, have been extensively studied in the last 25 years, and their association with their hosts is more comprehensively understood than any other endophyte–host relationship. There are several other genera within this tribe, all of which are capable of a teleomorphic state resulting in the production of external stromata on leaves or inflorescences of host plants, which often result in abortion of inflorescences. *Epichloë* is the most widely studied of these genera. Both *Neotyphodium* and *Epichloë*, fungal endophytes, share a commonality of features which indicates a phylogenetic relationship between the two genera (e.g., Schardl and Tsai, 1997; Wilkinson and Schardl, 1997).

The defensive mutualism hypothesis as originally proposed by Clay (1988) is based on the premise that a reduction in herbivory due to the production of fungal metabolites was the raison d'être for the evolution of clavicipitaceous fungi from localized parasitic infections to systemic symbionts with host grasses. Clay (1988) uses the example of *Claviceps* producing ergot alkaloids that could be toxic to grazing animals if accidentally ingested. Such localized infections could be easily avoided by insect herbivores whereas systemic infections provide host plants with a more complete defense system, allowing the plant to "sequester" the alkaloids produced by the fungus (Clay, 1988).

In this chapter five features that define the symbiotic relationship between *Neotyphodium* and underpin the defensive mutualism hypothesis will be examined:

1. Endophyte defends its host from herbivory.
2. A range of secondary metabolites are produced by the fungus which are deterrent and/or toxic to herbivores.
3. Host protection from herbivory is allocated according to the value of particular plant parts.
4. Plant and the fungus benefit from the defense against herbivory.
5. Coevolution has resulted in adaptive responses of endophyte associations to herbivory and of herbivores to endophyte infection of hosts.

21.2 Defense against Herbivory

Much evidence accumulated in the late 1980s and early 1990s demonstrating the effect of endophyte on insect herbivores and giving credence to the defensive mutualism hypothesis. During this period, over 40 species of insects in a range of insect orders were found to be sensitive to the presence of endophyte in plants or to alkaloids produced by them (Breen, 1994; Clement et al., 1994; Popay and Rowan, 1994). The majority of those studies reported effects of endophytes in cultivated tall fescue and ryegrass on herbivores, and there were relatively few studies carried out in native environments. While this is still the case, the knowledge of endophyte–plant interactions in native environments and their effects on herbivory is gradually accumulating. These studies have shown undoubtedly that both sexual and asexual forms of the endophyte are capable of anti-insect activity, but, as might be predicted, there are a diverse range of responses related to host, endophyte, and the insect herbivore.

There are parallels in the insect responses elicited by the sexual *Epichloë* infections and the asexual *Neotyphodium* spp. in their native hosts as well as in the cultivated grasses. In particular, insects such as fall armyworm (*Spodoptera frugiperda*) and various aphid species have been the focus of considerable research. *Epichloë* infections in a range of native grass species in the United States and South America reduce feeding, growth, development, and survival of fall armyworm which is native to these regions (Cheplick and Clay, 1988; White et al., 1993; Gonthier et al., 2008). Leaves of naturally and artificially infected *Brachypodium sylvaticum* plants, which are not known to produce any of the common alkaloids associated with endophyte infection, also reduce growth, development, and survival of fall armyworm but with little evidence of feeding deterrency (Brem and Leuchtmann, 2001). On the other hand, fall armyworm damage was not affected by *Epichloë elymi* inoculated into its natural wild cereal host, *Elymus hystrix*, and exposed to natural herbivory outdoors (Tintjer and Rudgers, 2006). In the only study on fall armyworm involving an asexual *Neotyphodium* endophyte in a native grass, the larvae showed a preference for endophyte-free *Bromus setifolius* over plants infected with *N. temblederae* (White et al., 2001). Among the cultivated grasses, Ball et al. (2006), in an extensive study in which a range of endophyte isolates in tall fescue, meadow fescue, and perennial ryegrass were tested against fall armyworm, found that responses were highly dependent on host species as well as on endophyte strain. Interhost comparisons where the same endophyte was present in all the three species showed that endophyte activity against fall armyworm was stronger in meadow fescue and perennial ryegrass than in tall fescue. Braman et al. (2002) also reported similar results. Other studies have consistently demonstrated that the common strain of endophyte in ryegrass, *N. lolii*, adversely affects this insect (Clay et al., 1985, 1993; Hardy et al., 1986; Bultman and Ganey, 1995). In contrast to this, *N. coenophialum*-infected tall fescue has been reported to have negative effects on fall armyworm (Clay et al., 1985, 1993; Hardy et al., 1986), no effects (Breen, 1993a; Davidson and Potter, 1995), and even positive effects (Bultman and Conard, 1997; Bultman and Bell, 2003). Other Lepidopteran grazers responsive to endophyte infection in cultivated ryegrass include the bluegrass webworm (*Parapediasia teterella*) (Koga et al., 1997; Richmond and Shetlar, 1999) and sod webworms (*Crambus* spp.) (Funk et al., 1983; Murphy et al., 1993). Some Lepidopteran insects such as black cutworm (*Agrotis ipsilon*) (Williamson and Potter, 1997; Richmond and Shetlar, 2001) and common cutworm (*A. infusa*) (McDonald et al., 1993) are not affected or only mildly affected by endophyte in perennial ryegrass. However, strong negative effects of perennial ryegrass infected with a natural *N. lolii* × *E. typhina* hybrid (Lp1) on weight gain and survival of neonate black cutworm larvae have recently been reported (Potter et al., 2008). Infected tall fescue appears to have no effect on black cutworm (Williamson and Potter, 1997).

Different aphid species also display variable responses to endophyte infection which may relate to their specialization on certain hosts. Among eight *E. typhina* associations with *Festuca* hosts, five significantly reduced the survival of the greenbug (*Schizaphis graminum*) but none affected the bird cherry oat aphid (*Rhopalosiphum padi*) (Siegel et al., 1990). Interestingly, this apparent species-specific effect of *E. typhina* in *Festuca* hosts on aphids is analogous to *N. lolii* infection of ryegrass which also appears to be more active against *S. graminum* than against *R. padi* (Siegel et al., 1990; Breen, 1993b). However, not all effects on aphids become apparent in short term experiments. Effects of endophyte infection on populations of *R. padi* were only discernible after four generations of reduced lifespan and fecundity (Meister et al., 2006). Numerous studies have shown that both *R. padi* and *S. graminum* clearly prefer endophyte-free tall fescue to that infected with the common strain of endophyte in choice tests while their survival and reproduction is reduced when they are confined to endophyte-infected plants (Johnson et al., 1985; Latch et al., 1985; Eichenseer et al., 1991; Eichenseer and Dahlman, 1992; Breen, 1993b). However, genetic differences in both host and fungus will influence the degree of resistance in both domestic tall fescue (Hunt and Newman, 2005; Bultman et al., 2006) and wild tall fescue (Clement et al., 2001, 2007a). In the latter study, accessions of wild tall fescue from Morocco and Sardinia elicited responses from *R. padi* similar to those in the cultivar "Kentucky 31," whereas accessions from Tunisia showed little or no resistance.

As with *S. graminum*, Russian wheat aphid (*Diuraphis noxia*) is affected by endophyte infection in both tall fescue (Clement et al., 1990; Springer and Kindler, 1990) and perennial ryegrass (Clement et al., 1992). Some accessions of *Neotyphodium* endophyte associations with wild barley (*Hordeum* spp.) also reduce populations of this aphid (Kindler and Springer, 1991) and another, *Metopolophium dirhodum*, but have no effect on *R. padi* (Clement et al., 1997, 2005). Like *R. padi*, populations of *M. dirhodum* on endophyte-infected perennial ryegrass gradually decline over several generations (Meister et al., 2006). The common strain of *N. lolii* in ryegrass also has relatively mild effects on populations of a root aphid (Popay and Gerard, 2007), whereas endophyte infection of tall fescue virtually eliminates this aphid (Jensen and Popay, 2007). Infection of meadow fescue with its natural endophyte *N. unicinatum* also has a potent effect on *A. lentisci* (Schmidt and Guy, 1997).

The anti-insect properties of endophyte were first described when it was discovered that perennial ryegrass infected with *N. lolii* reduced populations of Argentine stem weevil (*Listronotus bonariensis*) in New Zealand (Prestidge et al., 1982). Since then several different endophyte strains sourced from Europe and inoculated into New Zealand cultivars have been found to be just as effective as the common strain in reducing damage to perennial ryegrass (Popay and Wyatt, 1995; Popay et al., 1999) and tall fescue (Popay et al., 2005). Other grass species hosting natural endophytes, such as *Bromus anomalous* infected with *N. starrii* and an Australasian native grass (*Echinopogon ovatus*) infected with an endophytic fungus serologically related to *N. lolii*, also affect Argentine stem weevil (Bell and Prestidge, 1991, 1999; Miles et al., 1998). An insect species very similar to Argentine stem weevil, the bluegrass billbug (*Sphenophorus parvulus*), responds very similarly to endophyte infection of perennial ryegrass (Johnson-Cicalese and White, 1990; Murphy et al., 1993; Richmond et al., 2000).

Other wild grass–endophyte symbioses have effects on such diverse insect herbivores as Hessian fly (*Mayetiola destructor*) (Clement et al., 2005), locusts (*Locusta migratoria*) (Bazely et al., 1997), and leaf cutting ants (White et al., 2001). Cultivated ryegrass and tall fescue infected with their respective common strains of endophytes can also adversely influence a wide range of other insects. Included amongst those are leafhoppers which show species-specific responses to endophyte infection (Kirfman et al., 1986; Muegge et al., 1991); hemipteran plant bugs (Saha et al., 1987; Mathias et al.,

1990; Carriére et al., 1998; Yue et al., 2000); pasture mealybug (*Balanococcus poae*) (Pennell and Ball, 1999; Sabzalian et al., 2004; Pennell et al., 2005); and African black beetle (*Heteronychus arator*) (Ball et al., 1994). Highly variable effects of endophyte have also been recorded for the white grubs (Scarabaeidae) which graze on roots of grasses. There is no conclusive evidence for endophytes in perennial ryegrass affecting white grubs such as the New Zealand native grass grub (*Costelytra zealandica*) (Prestidge and Ball, 1993), but in various fescue species *Neotyphodium* infection has been shown to affect feeding damage, survival, and growth of various white grub species (Potter et al., 1992; Crutchfield and Potter, 1995; Koppenhoffer et al., 2003). However, clear evidence for the effects of endophyte-infected tall fescue on scarab larvae in the field has been difficult to obtain (Potter et al., 1992; Murphy et al., 1993; Davidson and Potter, 1995; Koppenhoffer et al., 2003). On the other hand, a strong effect of meadow fescue infected with its natural endophyte *N. uncinatum* on the New Zealand native, *C. zealandica*, has been found in both field and laboratory trials (Fletcher et al., 2000; Popay et al., 2003).

Not surpisingly, there are insect herbivores which have been found not to be affected by endophyte infection. *Neotyphodium* infection of *Festuca arizonica*, a native of Arizona, had no effect on feeding on clipped leaves by the redlegged grasshopper (*Melanoplus femurrubrum*) (Lopez et al., 1995) or by another native grasshopper, *Xanthippus corallipes* (Saikkonen et al., 1999), and had no effect on leaf cutting ant (*Acromyrmex versicolor*) queen survival, worker production, or size of fungal gardens (Tibbets and Faeth, 1999). The common endophyte strain in perennial ryegrass has no effect on frit fly (*Oscinella frit*) and leatherjackets (*Tipula* spp.) (Lewis and Clements, 1986; Lewis and Vaughan, 1997) and cereal leaf beetles (*Oulema melanopus*) (Clement et al., 2007b).

In something of a paradox, endophyte may enhance the fitness of some insects (Lopez et al., 1995; Siakkonen et al., 1999; Tibbets and Faeth, 1999; Bultman and Bell, 2003; Popay et al., 2004). The reasons are unknown but are likely to be related to altered plant chemistry due to the presence of the endophyte (Rasmussen et al., 2008) that improves individual insect fitness, where there is either no effect of the alkaloids on the insects concerned or effects are negated by increased host plant quality.

The studies so far carried out on native grasses and their endophytes do not provide conclusive evidence of defensive mutualism. Effects range from reduced herbivory to no apparent effect, to cases where infection may increase herbivory. Those that show reduced herbivory at least indicate the potential of these native associations to defend their hosts in their natural environments, but more in-depth research is required to determine if that defense equates with plant benefit. Evidence for reduced herbivory in the economically important grasses, tall fescue and perennial ryegrass, have really underpinned the defensive mutualism hypothesis. It has been argued that in cultivating these plants, plant breeding has selected for robust associations which strongly protect against herbivores (Faeth and Sullivan, 2003). That may well be true but it is also useful to consider the defensive role of new strains of endophytes in tall fescue and ryegrass that have been sourced mainly from natural grasslands in Eurasia. New Zealand has utilized the rich diversity of endophytes found in this region to introduce new endophyte strains into agricultural environments that exploit their anti-insect properties but eliminate or at least reduce the risk of toxicity to grazing animals. These new endophytes have been inoculated into existing cultivars but have not undergone the plant breeding selection process that the associations between the common strain and their respective tall fescue and ryegrass hosts have been subjected to. The biological activity shown by these new endophytes in domestic cultivars will surely reflect the spectrum of activity they potentially have in the environments from which they originated.

For example, two endophytes with identical characteristics determined by SSR (M. Faville, personal communication) and found at the same locality in a European ryegrass seed collection (i.e., as far as is known these are the same strain of endophyte) were tested for their effects on Argentine stem weevil. One endophyte (AR40) in this collection was tested in its parent plant; the other (AR37) had been inoculated into the cultivar 'Grasslands Nui' (Popay and Wyatt, 1995). Despite the differences in host plant, these endophytes had very similar effects on Argentine stem weevil with very little or no effect on adult feeding and oviposition but marked decreases in larval damage compared with endophyte-free counterparts (Table 21.1). Further testing has shown that AR37 has activity against a broad spectrum of insects, including black beetle (*Heteronychus arator*) (Ball et al., 1994), porina caterpillars (*Wiseana cervinata*) (Jensen and Popay, 2004), pasture mealybug (*Balanococcus poea*) (Pennell et al., 2005), and root aphids

TABLE 21.1

Effect of a Unique Strain of Endophyte in Its Original Perennial Ryegrass Host and in a Selected Cultivar on Argentine Stem Weevil Adult Feeding, Oviposition, and Larval Damage (see Popay and Wyatt, 1995 for Full Results)

Host	Infection Status	Adult Feeding Score/ Tiller	No. of Eggs/10 Tillers	% Tillers with Larval Damage
Original	E+	3.5	4.0	1
	E–	3.1	4.2	27
Grasslands Nui	E+	3.7	3.5	9
	E–	3.8	1.8	44

(*Aploneura lentisci*) (Popay and Gerard, 2007). With the possible exception of *A. lentisci*, none of the insects for which AR37 provides resistance are present in the locality where this endophyte originates. Nevertheless it seems unlikely that the capability this endophyte has of defending its host from a wide range of insects would have arisen by chance in its native habitat. The same arguments apply to the introduction of other novel endophytes that originate from natural grasslands in Europe into agricultural environments. In New Zealand, AR1 has been inoculated into local ryegrass cultivars primarily to provide resistance to Argentine stem weevil without harmful effects on grazing livestock and also provides protection from pasture mealybug (Pennell et al., 2005). Similarly, AR542 (Max P or Max Q) has been inoculated into tall fescue cultivars used in agriculture in the United States, Australia, and New Zealand. In New Zealand, infestations of black beetle (Popay et al., 2005), pasture mealybug (Pennell and Ball, 1999), and root aphid (Jensen and Popay, 2007) are almost eliminated from tall fescue hosting this endophyte. AR542 also confers resistance to the bird cherry oat aphid, although the level of resistance is slightly less than that from the common strain (Hunt and Newman, 2005; Bultman et al., 2006). Although insect pressure is undoubtedly higher in intensive agriculture with minimal plant diversity, the existence of these endophytes in their native habitats is evidence in itself of a role for the fungus in plant defense in these environments.

21.3 Secondary Metabolites

The production of secondary metabolites by the fungus is fundamental to the endophyte–host interactions that influence herbivory. The type of alkaloid produced is a function of the strain of endophyte, but concentrations within plants are determined by a variety of factors. It is the type of alkaloid and its concentration and distribution within the plant which will determine the effect of an endophyte on any particular insect.

At least 20 secondary metabolites are known to be produced by *Neotyphodium* species in ryegrass and tall fescue with another 17 produced by *Balansia* symbiota (Bacon and White, 2000). Firstly, of this diverse array, only four classes of compounds have been the focus of much research (reviewed by Lane et al., 2000). Second, the ergopeptine alkaloids and a water soluble guanidinium alkaloid, called peramine, are common to many fungal endophytes in tall fescue and ryegrass and are commonly synthesized within the *Epichloë* taxonomic group (Clay and Schardl, 2002). Tremorgenic indole diterpenoids (lolitrems) are produced only by certain endophytes in perennial ryegrass. The pyrrolizidine alkaloids (lolines) are a characteristic of associations of tall fescue, meadow fescue, and *L. multiflorum* harbouring respectively *N. coenophialum*, *N. uncinatum*, and *N. occultans*. *Epichloe festucae* produces lolines in its host *L. giganteum* as do several other wild endophyte–grass associations (see review by Schardl et al., 2007).

The four main groups of alkaloids common to many endophyte strains have anti-insect activity. Peramine is a potent feeding deterrent to Argentine stem weevil (Rowan et al., 1990) and the primary alkaloid responsible for resistance to this insect in endophyte-infected ryegrass. Peramine has been implicated in observed adverse effects of ryegrass infected with *N. lolii* on the aphid *S. graminum* (Siegel et al., 1990), but other insects such as black beetle (Ball et al., 1997a) and fall armyworm show no sensitivity to this alkaloid (Ball et al., 2006). The ergopeptine alkaloids are deterrent and/or toxic to a range of insects including Argentine stem weevil adults (Dymock et al., 1988), black beetle adults (Ball et al., 1997a), fall armyworm (Clay and Cheplick, 1989), the large milkweed bug (*Oncopeltus fasciatus*) (Yates et al., 1989), and Japanese beetle larvae (*Popillia japonica*) (Patterson et al., 1991). Lolitrem B reduces growth and development of Argentine stem weevil larvae (Prestidge and Gallagher, 1985) but has no effect on adults (Dymock et al., 1988) or on black beetle (Ball et al., 1997a). The loline alkaloids have a broad spectrum of activity against insects, including fall armyworm and European corn borer (*Ostrinia nubilalis*) (Riedell et al., 1991), porina caterpillars (*Wiseana* spp.) and grass grub (*C. zealandica*) larvae (Popay and Lane, 2000), Japanese beetle larvae (Patterson et al., 1991), aphids (Seigel et al., 1990; Wilkinson et al., 2000), and the large milkweed bug (Yates et al., 1989). Several other insects which do not feed directly on plants are killed by contact or oral activity of *N*-formyl loline (Dahlmann et al., 1997). The wide ranging biological activity of this group of compounds prompted Schardl et al. (2007) to term their role in the symbiosis as "currencies of mutualism."

While research into the chemistry of the endophytes has focused mainly on these four groups of compounds, there are many other intermediaries in the biosynthetic pathways which may also have biological activity. Genetic manipulation of endophytes to eliminate or disable genes within pathways can assist in determining whether or not certain alkaloids are solely responsible for observed biological effects. The effect of peramine was confirmed in this way when a mutant of *E. festucae* unable to produce peramine due to the deletion of the *perA* cosmid had no effect on Argentine stem weevil adult feeding (Tanaka et al., 2005). Endophyte gene knockouts that eliminate complex ergot alkaloids or all ergot alkaloids reduced resistance compared to the wild-type Lp1 indicating that these alkaloids were responsible for some of the effects of the endophyte on neonate black cutworm (Potter et al., 2008). The results also indicate that other secondary metabolites are involved. The possibility that alkaloids act in concert, either additively or synergistically, to confer resistance to insects has seldom been addressed in endophyte research. Certainly it is not always easy to identify a single alkaloid responsible for observed effects on an insect as Ball et al. (2006) found when they tested a wide range of endophyte–plant associations and carried out alkaloid testing against fall armyworm. Salminen et al. (2005) found that four alkaloids in tall fescue and three in perennial ryegrass accounted for 47% and 70%, respectively, of the variation in weight change in fall armyworm that fed on these plants.

21.4 Allocation of Defense

Neotyphodium species are typical hyphal fungi that grow within plants. Hyphae are concentrated in the stem apex region, infecting the axillary buds from which new tillers develop. Mycelia are usually confined to the leaf sheath area but can extend to the leaf lamina in some plant/endophyte associations (Christensen et al., 1997; Moy et al., 2000). Colonisation of the leaf by hyphae continues as long as the leaf is growing and ceases when leaf growth ceases although they remain metabolically active throughout the life of the leaf (Schmid et al., 2000; Christensen et al., 2002). Hyphae come to reside in the mature seed in reproductive tillers after first invading the inflorescence primordium and floral apices and then the ovaries and developing ovules, and in this way are maternally transmitted to the next generation (Philipson and Christey, 1986). The endophyte does not occur in the roots, but it can be transferred vegetatively to new plants via stolons or rhizomes (Hinton and Bacon, 1985).

Distribution of alkaloids within the plant is similar to the distribution of the hyphae but not always directly related to it (Spiering et al., 2005). Both ergovaline (Lane et al., 1997a) and lolitrem B (Ball et al., 1997b) generally occur in the highest concentrations in the leaf sheaths and developing inflorescences in ryegrass whereas concentrations of peramine tend to be higher in leaf blades than in leaf sheaths (Ball et al., 1997c). Only trace amounts of these three alkaloids have been reported in the roots. The highest concentrations of loline alkaloids in meadow fescue are allocated to young leaves in spring (Justus et al., 1997) which are likely to be highly vulnerable to herbivory. Pseudostems also have high concentrations during the vegetative growth stage in late summer and autumn following seed dispersal. A similar pattern of distribution of loline alkaloids is found in tall fescue. Unlike the other alkaloids, lolines can also be translocated to roots in biologically active concentrations.

With alkaloids concentrated in the crown and basal area of tillers of the plant, it is not surprising that insect herbivory in this region is often markedly reduced by endophyte infection. For *R. padi* aphids whose survival is severely affected by endophyte in tall fescue, a preference for inhabiting stems rather than leaf blades may have consequences for its survival on endophyte-infected plants. Hunt and Newman (2005) when comparing the effects of the common and AR542 strains of endophyte in tall fescue on *R. padi* used both, a clip cage experiment in which aphids were confined to the leaf sheath and another in which the aphids were enclosed on whole plants. The endophyte killed those aphids confined to the stems by clip cages whereas the enclosed plants allowed survival of a small population. The presence of endophyte may also shift the distribution of insects on plants. Psudostems and the crown of the plant are the favored locale of the chinch bug but Mathias et al. (1990) observed that after 7 days exposure to endophyte-infected or endophyte-free perennial ryegrass, 72% of chinch bugs were located on the leaf sheath on endophyte-free plants compared with only 29% on endophyte-infected plants. Endophytes are highly active against billbugs and Argentine stem weevil which destroy meristems causing tiller death. Similarly adult African black beetle feeding at the base of tillers are highly sensitive to a wide range of endophytes as is pasture mealybug which sucks the plant sap on and near the crowns of plants. Endophyte infection of perennial ryegrass affects bluegrass webworm, but that activity appears to be confined to the leaf sheaths (Kanda et al., 1994).

Less herbivory on or near the crown and on the pseuodstems of grasses protects not only the plant but also the endophyte. On the other hand, adaptation to grazing means that defoliation may not only have no adverse effects on many grasses but may also benefit the plant by stimulating growth. Under these circumstances there would be little advantage for the grass to invest in defense of leaf blades. Thus studies using only leaf material may be of limited value in determining the endophyte's ability to protect its host from the detrimental effécts of herbivory. A trial carried out to investigate feeding damage by black field cricket on endophyte-infected tall fescue (Popay and Evans, unpublished data) illustrates this point. On average leaf blades were removed by the crickets on 96% of all tillers on every plant regardless of endophyte infection status. However, the severity of damage to tiller pseudostems was less on the endophyte-infected plants and consequently survival and recovery of endophyte-infected plants was significantly better than for endophyte-free. If this study focused only on the leaf blades, the conclusion would have been altogether different.

Protection of the seed as the sole means of propagation and dispersal for both the plant and the systemic fungus is also vital for this mutualistic relationship. Nevertheless this is one of the least studied aspects of the endophyte symbiosis. Cheplick and Clay (1988) were the first to demonstrate that ground seed of *N. lolii*-infected perennial ryegrass and particularly of *N. coenophialum*-infected tall fescue reduced the population growth and survival of the common flour beetle (*Tribolium castaneum*). Infected tall fescue seed also reduces the survival and population growth of two stored-grain pests, the confused flour beetle (*T. confusum*) (Yoshimatsu et al., 1998) and the sawtoothed grain beetle (*Oryzaephilus surinamensis*) (Yoshimatsu et al., 1999). In a field test, endophyte-free ryegrass seed disappeared at a faster rate than infected seeds although the seed predator was not identified (Popay et al., 2000).

The reservoir of alkaloids in the seed may also serve to protect the vulnerable seedling stage during the period between germination and when the endophyte growth begins in the new plant. Whether endophyte in the seed was viable or not, germinating seedlings of endophyte-infected perennial ryegrass deterred Argentine stem weevil adult feeding for atleast 10 days after germination (Stewart, 1985). Consumption of germinating seedlings by *C. zealandica* was reduced by *N. uncinatum* in meadow fescue and by several different isolates in tall fescue (Popay and Tapper, 2007).

21.5 Costs and Benefits to Plant Host and Fungus

Evolutionary theory predicts that for those organisms that are obligate biotrophic symbionts such as *Neotyphodium* spp., benefits to the host must accrue from the relationship in order that it be maintained. In an evolutionary context, however, plant fitness is not defined by the plant's ability to produce vegetative growth as it is in an agricultural context but rather by its ability to propagate. Furthermore, although survival is a key attribute for success of plants in both the contexts, in the agricultural sense survival of individuals is important whereas in the evolutionary sense survival of the species may be more so. Within an ecological framework, temporal and spatial scales apply to the concepts of costs and benefits, both of which may shape mutualisms (Bronstein, 2001). The functional significance of each individual's relationship within a diverse community of associations will be determined by the selection pressures present in a particular locality at a particular time and will be a dynamic process.

Undoubtedly fungal endophyte infection reduces insect herbivory. Even so, the numerous examples given above and a meta-analysis of the literature (Saikkonen et al., 2006) are insufficient to prove that the grass–endophyte symbiosis has been shaped by defensive mutualism. The key is to demonstrate that reduced herbivory is beneficial to the plant. Thus, in the presence of herbivores (invertebrate and/or vertebrate) which compromise growth, survival, and/or reproductive potential of endophyte-free plants, the benefits of hosting the endophyte must outweigh the costs. The corollary is that in the absence of herbivory (or other benefits that the endophyte may confer such as tolerance of abiotic stress) the endophyte may be parasitic on the plant, but it does not impact on the growth or reproduction of the plant to the extent that survival is decreased. The native North American grass, *Festuca arizonica*, is a good example of this case (Faeth and Hamilton, 2006).

Reduced herbivory by a rodent has been shown to be a major factor in giving endophyte-infected tall fescue a selective advantage in the field (Clay, 1996; Clay and Holah, 1999). In New Zealand where perennial ryegrass is a major component of pastures, increased yields of ryegrass have been directly attributed to reduced insect herbivory as a result of endophyte infection (Popay et al., 1999; Pennell et al., 2005; Hume et al., 2007). Although native grass–endophyte associations are known to be capable of defense against insect herbivory, the advantage this provides in the environments where these associations naturally occur still remains unknown. The endophyte may reduce or enhance plant growth (e.g., Groppe et al., 1999; Hesse et al., 2004; Iannone and Cabral, 2006; Olejniczak and Lembicz, 2007). Certainly, *Neotyphodium* infection has been found to generally decrease the performance of its native Arizona fescue host (Faeth and Sullivan, 2003) although infection does not appear to compromise long term survival of endophyte-infected plants (Faeth and Hamilton, 2006). In another study, species-specific interactions in growth response of endophyte-infected *F. rubra* and *F. pratensis* plants to different nutrient and watering regimes were recorded, but the authors concluded for both species that the cost of endophyte infection outweighed the benefits in a resource-limiting environment (Ahlholm et al., 2002a). In their comprehensive analysis of the literature, Saikkonen et al. (2006) concluded that endophyte did not enhance plant performance or competitive ability. Many of the studies investigating the effect of endophyte on plant growth, however, have removed plants from their natural environment and tested them in the apparent absence of biotic stress. The obvious question to ask is would the results of these studies have been the same if the plants had been exposed to herbivory?

Intraspecific communities of endophyte-infected and uninfected grasses subjected to biotic and abiotic pressures represent part of a dynamic plant community in which insect herbivory can contribute to population fluctuations (Verkaar, 1987; Maron, 1998; Van der Putten, 2003). Shifts in infection frequency of individuals can tell us much about the benefits or otherwise of endophyte infection in such communities since they are assumed to be indicative of the fitness of infected individuals relative to their uninfected counterparts (e.g., Clay and Schardl, 2002; Jensen and Rolund, 2004). However, conclusions are often drawn as to the role of endophyte within these communities based on the measurements of infection levels at one point in time when it is the fluctuations in endophyte infection over temporal and spatial dimensions which may be the most informative. Population dynamics of many herbivorous insects are characterized by short periods of high density resulting in severe damage to their hosts followed by long periods of low density (e.g., Varley et al., 1973; Berryman, 1987; Abbott and Dwyer, 2007). Outbreaks may be localized or widespread. Under these circumstances, maintaining heterogeneity of infected and endophyte-free individuals will have considerable advantages to the species, particularly if the endophyte imposes some cost to the plant in the absence of abiotic or biotic stress. During periods of low herbivore density the proportion of infected plants in the population may decline but such a trend would be dramatically reversed during an insect outbreak which, at least in part may be triggered by the availability of endophyte-free hosts (Figure 21.1). The outbreak will greatly reduce the total plant population as endophyte-free plants are removed with a concomitant increase initially in the proportion of infected individuals, followed by high levels of recruitment of infected plants into the population from reproducing mother plants. Herbivore density will decline to low levels due to low resource availability,

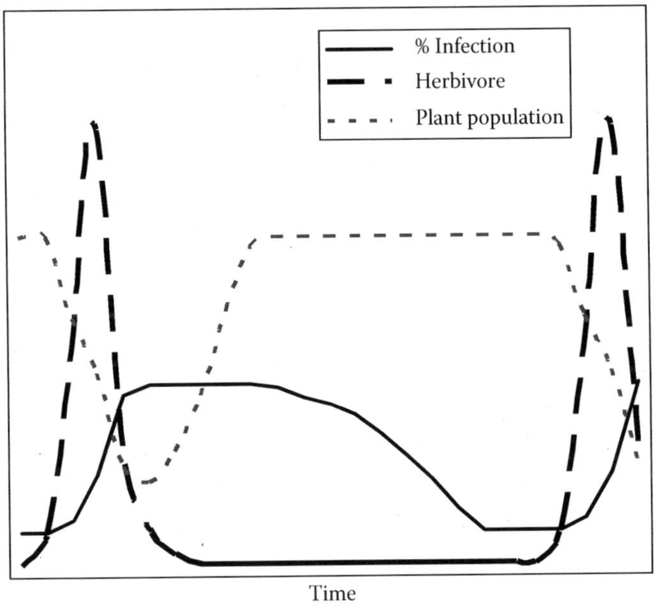

FIGURE 21.1 Hypothetical changes in endophyte infection frequency and plant populations with time in response to herbivore selection pressures.

although this will also depend on the availability of alternative hosts. Eventually the decrease in selection pressure on endophyte-free plants and a small competitive edge over their infected counterparts may lead to a decline in infection frequency in that locality.

There is very little information on temporal changes in endophyte frequency in natural grassland communities, but increase in endophyte infection over time is a well-documented phenomenon in pasture (e.g., Hume and Barker, 2005). In New Zealand increases in frequency of infection can occur rapidly and have been attributed to preferential herbivory of endophyte-free plants by insects such as Argentine stem weevil (Prestidge et al., 1984, 1985; Popay et al., 1999), black beetle (Popay et al., 1999), and pasture mealybug (Pennell et al., 2005). The lack of plant diversity in agricultural plant communities combined with the constant pressure from invertebrate and vertebrate herbivores is likely to mean that effects on infection frequency are exaggerated and the time scales over which they occur often compressed. The constant grazing pressure also means that a high infection frequency is the equilibrium state.

Although herbivory can naturally select for resistance traits in plants, there is often considerable variation in the strength of that resistance within a population (Simms and Rausher, 1987). For the endophyte symbiosis there are likely to be advantages to both the endophyte and the host in maintaining that heterogeneity within populations governed by having, firstly, a natural diversity of endophyte strains within a population and, secondly, a strong influence of host plant genotype on endophyte expression. In losing the ability to sexually reproduce, the systemic *Neotyphodium* endophytes have become reliant on their hosts for adaptation to environmental conditions. Both endophyte strain and host plant genotype which affects expression of the attributes pertaining to each specific strain are major factors determining the outcome of the symbiosis (Lane et al., 2000; Clay and Schardl, 2002). A natural community made up of a diversity of endophyte–host plant genotype associations is likely to be resilient in its responses to abiotic or biotic perturbations. Within such communities, where the endophyte mutualism varies from strongly mutualistic to parasitic, a high degree or interplant variability in antiherbivore defenses is inherent, allowing plasticity in plant response to sporadic selection pressures. On the other hand where there is less diversity and more constant herbivore pressure as in agricultural environments, the strength of the mutualism in terms of plant benefit may be inversely related to the variability in host genotype/endophyte associations.

21.6 Coevolution—Host, Endophyte, and Insect

Coevolution between plants and their natural enemies is widely believed to be the process that has generated the extraordinary biological diversity that exists in nature. In this cyclic and reciprocal interaction (Mauricio, 2001; Rausher, 2001), natural enemies, such as herbivores, select for resistance traits in plants, which subsequently leads to the natural enemies evolving responses to those traits in a way that allows them to consume the plant. There is abundant evidence for the latter but the former is still a subject of debate (Mauricio, 2001; Rausher, 2001).

For the fungus itself there is evidence that it evolved from pathogens (Clay and Schardl, 2002) of animal origin via an interkingdom host jump from animals to plants (Spatofora et al., 2007). Questions still remain as to whether defense against herbivory has been a driving force for the development of the systemic asexual state which has arisen mainly through interspecific hybridization of two and sometimes three ancestors (Schardl et al., 2004). The abundance of the asexual form would seem to be evidence in itself of a fitness advantage for maternally transmitted symbionts (Clay and Schardl, 2002).

So have *Neotyphodium* endophytes relinquished sexual reproduction in order to enhance their ability to defend their hosts or are these relationships a case of "don't bite the hand that feeds you?" Both sexual and asexual endophytes produce secondary metabolites which reduce insect herbivory, and there is no particular evidence that defense is stronger in the asexual symbionts. So what if there are any differences? Increased herbivory may occur on *Epichloë* isolates exhibiting greater sexual reproduction relative to other isolates and on stroma-bearing tillers compared to non-stroma-bearing tillers (Tintjer and Rudgers, 2006). Similarly, Brem and Leuchtmann (2001), reported tillers with stromata showed more leaf damage by unspecified microherbivores than vegetative or flowering tillers in wild populations of *Brachypodium sylvatica*, almost all of which were infected with the endophyte *E. sylvatica*. Furthermore, known alkaloids were absent or at very low concentrations in plants associated with stroma-forming *Epichloë* (Leuchtmann et al., 2000). The increased herbivory and lack of protection of stroma-bearing plants and tillers may have implications for the transmission of the fungus. Thus increased transmission efficiency plus the addition of seed and seedling protection to the plant survival kit, would provide considerable advantages to the asexual *Neotyphodium* symbionts.

If insect herbivory has selected for mutualism between plant and endophyte then there are likely to have been corresponding effects on the insect herbivore that have allowed it to adapt to the resistance traits created by endophyte infection. Detoxification of alkaloids is one option for the insect, but this is an unlikely scenario given the diverse array of metabolites present. The availability of alternative hosts (uninfected conspecifics or different species) also works against an insect's capacity to evolve mechanisms to overcome the harmful effects of these alkaloids. Under these circumstances, the evolution of behaviors that enhance the ability of an insect to select favorable hosts is likely to be critical to its survival. Thus for those insect species whose fitness is challenged by endophyte infection, an ability to detect and avoid infection would be highly advantageous and of considerable adaptive value to insects whose life cycle is completed entirely on its host. We can speculate that this is the case for Argentine stem weevil which originates in South America where endophytes are abundant in many native grasses. This adult weevil feeds on the leaf blades of grass hosts, oviposits in the leaf sheath and the larvae feed within the central tiller. In a good example of the "mother knows best" principle, feeding and oviposition preferences for endophyte-free plants of the highly mobile adult weevil greatly influence the fitness of the less mobile and vulnerable offspring. Because alkaloids produced by the fungus can be toxic to the larvae, avoidance of infected plants is achieved by the detection of the alkaloid peramine by the adult, resulting in both reduced feeding and oviposition on these plants. Peramine itself appears to have no toxicity to the weevil but is the main alkaloid present in the leaf blades which the adult feeds. Has the weevil evolved a coadaptive response to endophyte through an ability to detect peramine and therefore avoid the toxicity of other alkaloids to its offspring? Contrast this with the response of this insect to another endophyte, AR37, which originates in Europe. This endophyte also appears to be toxic to the larvae (Table 21.1) but does not produce peramine. Adult weevils feed and oviposit freely on ryegrass infected with this endophyte resulting in a great waste of reproductive effort. Investigation of the interactions between Argentine stem weevil and the many species of native grasses that harbor endophyte infection where this insect originates in South America may provide some interesting insights into this hypothesis. Endophyte-infected plants are often repellant as well as toxic to other insects, also suggesting that the insects have evolved mechanisms to detect infection in order to avoid these plants.

21.7 Factors Affecting the Mutualism

Concentrations of alkaloids *in planta* vary considerably as a result of various factors, and this will impact on the herbivore. As already mentioned, host plant genotype, in particular, has a major influence on the quantities of alkaloids that are produced. A four- to ten-fold difference in concentrations of different alkaloids has been recorded between individual ryegrass plants taken from the field and then grown under the same conditions (Latch, 1994; Ball, 1995a,b). Similarly Easton et al. (2002) found that concentrations of peramine and ergovaline consistently varied across two perennial ryegrass families. Corresponding host plant influences on alkaloid expression have been recorded in fescue associations (e.g. Agee and Hill, 1994; Adcock et al., 1997; Hiatt and Hill 1997; Faeth et al., 2002).

Environmental factors also influence the expression of the symbiosis. Seasonal changes in alkaloid content of plants occur in concert with seasonal changes in the concentration of endophyte (di Menna et al., 1992; Ball et al., 1995a,b; Justus et al., 1997) and may alter the strength of resistance to herbivory (Popay and Mainland, 1991). In addition, environmental stresses often alter alkaloid content in plants. Water deficit, for instance, elevates ergovaline concentrations in ryegrass (Barker et al., 1993; Lane et al., 1997b) and tall fescue (Arechavaleta et al., 1992), although peramine and lolitrem B levels are less consistently affected by such conditions (Barker et al., 1993; Lane et al., 2000). Drought stress increased the adverse effects of *N. coenophialum* in tall fescue on fall armyworm performance (Bultman and Bell, 2003).

It has been argued that bioactivity of endophytes is more apparent in highly fertile situations that occur in agriculture than in the relatively impoverished soils of natural grasslands (Saikkonen et al., 2006). Lehtoneen et al. (2005) found that the magnitude of the effect of *N. unicnatum* infection of meadow fescue on performance of *R. padi* increased with increasing availability of soil nutrients. Evidence is mixed, however. Fertilization may increase *in planta* alkaloid concentrations (Lyons et al., 1986; Rottinghaus et al., 1991; Arachevaleta et al., 1992), but high nitrogen fertilization can also decrease it (Rasmussen et al., 2007). Furthermore, high nitrogen conditions lessened the adverse effects of *N. coenophialum* in tall fescue on fall armyworm (Bultman and Conard, 1997).

21.8 What of Other Defensive Mutualisms?

Other defensive mutualisms exist among the diverse asymptomatic fungal endophytes that inhabit plants (Carroll, 1988). Unlike the grasses, most are transmitted horizontally outside the host tissues via spores and, although they may provide resistance to herbivores, the resistance is localized rather than systemic. A well known example is the elm tree which often hosts the fungus *Phomopsis oblonga* under its bark without any apparent detriment to the plant. The fungus produces chemical compounds that disrupt breeding of elm bark beetles (Claydon et al., 1985) leading to a decline in beetle populations. The beetles show a clear preference for logs free of fungal infection. In another example, Douglas fir needles are frequently infected with an endophyte *Rhabdocline parkeri* which has an asexual form on which conidia are produced and which then invade galls produced by gall midges. Presence of the conidia clearly antagonizes the midges causing high mortality of the larvae (Carroll, 1988). Similar correlations between fungal infection and insect mortality in galls occur on white spruce and California live oak. A few of the many fungal endophytes that have been isolated from conifer needles produce anti-insect compounds. An extract from *R. parkeri* and another from a fungal endophyte in needles of white spruce reduced the growth and survival of spruce budworm (Miller et al., 2002). While some of these mutualistic associations do benefit the plant, others may be of no consequence. In their study of birch trees, Ahlholm et al. (2002b) concluded that the presence of microfungi had no effect on the performance of insect herbivores.

21.9 Conclusions

There is a broad spectrum of anti-insect activity among the grasses and their fungal endophytic symbionts engendered by the production of alkaloids by the fungal partner. Distribution of the fungus and the alkaloids ensures that protection is strongest at the base of plants where herbivory is most likely to threaten the survival of both plant and fungus. In addition, a reservoir of alkaloid in the seed is likely to reduce predation of both seed and seedlings. Defense against herbivory is advantageous to the plant, although there is still much to be learnt about those advantages in native grassland environments. Selection for strong associations that have robust insect protection and plant populations with a high frequency of infection as occurs in agriculture may intuitively be a beneficial path for the evolution of endosymbiosis in native environments. However, there are also advantages to both the endophyte and its host in maintaining heterogeneity within populations. Where infection frequencies are low to moderately high, the overall cost of the symbiosis for the population will be reduced, but there will still be resilience for the species as a whole in the face of sporadic insect outbreaks. Heterogeneity within populations governed by a strong influence of host plant on endophyte expression will have similar advantages.

According to evolutionary principles, those organisms that are obligate biotrophic symbionts, such as *Neotyphodium* spp., should be beneficial to their hosts in order that the relationship be maintained. The endophyte itself is not altruistic and the benefits that it confers will also ensure its own survival in perpetuity. There is good evidence that there is defensive mutualism between the fungal endophytes and the Pooid grasses, and indeed between other fungal endophytes and their hosts. This does not imply that all associations among the huge diversity that exists confer that benefit. Evolution takes many paths and undoubtedly there are some associations that may provide no advantages to their host or may benefit them in other ways.

REFERENCES

Abbott, K. C. and G. Dwyer. 2007. Food limitation and insect outbreaks: Complex dynamics in plant-herbivore models. *Journal of Animal Ecology* 76:1004–1014.

Adcock, R. A., Hill, N. S., Bouton, J. H., Boerma, H. R., and G. O. Ware. 1997. Symbiont regulation and reducing ergot alkaloid concentration by breeding endophyte-infected tall fescue. *Journal of Chemical Ecology* 2:691–704.

Agee, C. S. and N. S. Hill. 1994. Ergovaline variability in *Acremonium*-infected tall fescue due to environment and plant genotype. *Crop Science* 34:221–226.

Ahlholm, J. U., Helander, M., Lehtimaki, S., Wali, P., and K. Saikkonen. 2002a. Vertically transmitted fungal endophytes: Different responses of host-parasite systems to environmental conditions. *Oikos* 99:173–183.

Ahlholm, J. U., Helander, M., Elamo, P., Saloniemi, I., Neuvonen, S., Hanhimaki, S., and K. Saikkonen. 2002b. Micro-fungi and invertebrate herbivores on birch trees: Fungal mediated plant-herbivore interactions or responses to host quality? *Ecology Letters* 5:648–655.

Arechavaleta, M., Bacon, C. W., Plattner, R. D., Hoveland, C. S., and D. E. Radcliffe. 1992. Accumulation of ergopeptide alkaloids in symbiotic tall fescue grown under deficits of soil water and nitrogen fertilizer. *Applied and Environmental Microbiology* 58:857–861.

Bacon, C. W. and J. F. White Jr. 2000. Physiological adaptations in the evolution of endophytism in Clavicipitaceae. In *Microbial Endophytes*, Eds. C. W. Bacon and J. F. White Jr., pp. 237–261. New York: Marcel Dekker.

Ball, O. J. P., Christensen, M. J., and R. A. Prestidge. 1994. Effect of selected isolates of *Acremonium* endophyte on adult black beetle (*Heteronychus arator*) feeding. *Proceedings of the 47th New Zealand Plant Protection Conference*, pp. 227–231. Waitangi, New Zealand.

Ball, O. J. P., Lane, G. A., and R. A. Prestidge. 1995a. *Acremonium lolii*, ergovaline and peramine production in endophyte-infected perennial ryegrass. *Proceedings of the 48th New Zealand Plant Protection Conference*, pp. 224–228. Hastings, New Zealand.

Ball, O. J. P., Prestidge, R. A., and J. M. Sprosen. 1995b. Interrelationships between *Acremonium lolii*, peramine, and lolitrem B in perennial ryegrass. *Applied and Environmental Microbiology* 61:1527–1533.

Ball, O. J. P., Miles, C. O., and R. A. Prestidge. 1997a. Ergopeptine alkaloids and *Neotyphodium lolii*-mediated resistance in perennial ryegrass against adult *Heteronychus arator* (Coleoptera: Scarabaeidae). *Journal of Economic Entomology* 90:1382–1391.

Ball, O. J. P., Barker, G. M., Prestidge, R. A., and J. M. Sprosen. 1997b. Distribution and accumulation of the mycotoxin lolitrem B in *Neotyphodium lolii*-infected perennial ryegrass. *Journal of Chemical Ecology* 23:1435–1449.

Ball, O. J. P., Barker, G. M., Prestidge, R. A., and D. R. Lauren. 1997c. Distribution and accumulation of the alkaloid peramine in *Neotyphodium lolii*-infected perennial ryegrass. *Journal of Chemical Ecology* 23:1419–1434.

Ball, O. J. P., Coudron, T. A., Tapper, B. A., Davies, E., Trently, D., Bush, L. P., Gwinn, K. D., and A. J. Popay. 2006. Importance of host plant species, *Neotyphodium* endophyte isolate, and alkaloids on feeding by *Spodoptera frugiperda* (Lepidoptera: Noctuidae) larvae. *Journal of Economic Entomology* 99:1462–1473.

Barker, D. J., Davies, E., Lane, G. A., Latch, G. C. M, Nott, H. M., and B. A. Tapper. 1993. Effect of water deficit on alkaloid concentrations in perennial ryegrass endophyte associations. *Proceedings of the 2nd International Symposium on Acremonium/Grass Interactions*, Eds. D. E. Hume, G. C. M. Latch, and H. S. Easton, pp. 67–71. Palmerston North, New Zealand.

Bazely, D. R., Vicari, M., Emmerich, S., Filip, L., Lin, D., and A. Inman. 1997. Interactions between herbivores and endophyte-infected *Festuca rubra* from the Scottish islands of St Kilda, Benbecula and Rum. *Journal of Applied Ecology* 34:847–860.

Bell, N. L. and R. A. Prestidge. 1991. The effects of the endophytic fungus *Acremonium starii* on feeding and oviposition of the Argentine stem weevil. *Proceedings of the 44th New Zealand Weed and Pest Control Conference*, pp. 181–184, Tauranga, New Zealand.

Bell, N. L. and R. A. Prestidge. 1991. The effects of the endophytic fungus *Neotyphodium starii* on larval development of *Listronotus bonariensis* (Coleoptera: Curculionidae). *Proceedings of the 7th Australasian Conference on Grassland Invertebrate Ecology*, pp. 279–285. Perth, Australia.

Berryman, A. A. 1987. The theory and classification of outbreaks. In *Insect Outbreaks*, Eds. P. Barbosa, J. C. Schultz. New York: Academic Press.

Braman, S. K., Duncan, R. R., Engelke, M. C., Hanna, W. W., Hignight, K., and D. Rush. 2002. Grass species and endophyte effects on survival and development of fall armyworm (Lepidoptera: Noctuidae). *Journal of Economic Entomology* 95:487–492.

Breen, J. P. 1993a. Enhanced resistance to fall armyworm (Lepidoptera: Noctuidae) in *Acremonium* endophyte-infected turfgrasses. *Journal of Economic Entomology* 86:621–629.

Breen, J. P. 1993b. Enhanced resistance to three species of aphids (Homoptera: Aphididae) in *Acremonium* endophyte-infected turfgrasses. *Journal of Economic Entomology* 86:1279–1286.

Breen, J. P. 1994. *Acremonium* endophyte interactions with enhanced plant resistance to insects. *Annual Review of Entomology* 39:401–423.

Brem, D. and A. Leuchtmann. 2001. Epichloe grass endophytes increase herbivore resistance in the woodland grass *Brachypodium sylvaticum. Oecologia* 126:522–530.

Bronstein, J. L. 2001. The costs of mutualism. *American Zoologist* 41:825–839.

Bultman, T. L. and D. T. Ganey. 1995. Induced resistance to fall armyworm (Lepidoptera: Noctuidae) mediated by a fungal endophyte. *Environmental Entomology* 24:1196–1200.

Bultmann, T. L. and N. J. Conard. 1997. Effects of endophytic fungus, nitrogen, and plant damage on performance of fall armyworm. In *Neotyphodium/Grass Interactions*, Eds. C. W. Bacon and N. S. Hill, pp. 145–148. New York: Plenum Press.

Bultman, T. L. and G. D. Bell. 2003. Interaction between fungal endophytes and environmental stressors influences plant resistance to insects. *Oikos* 103:182–190.

Bultman, T. L., Pulas, C., Grant, L., Bell, G. D., and T. J. Sullivan. 2006. Effects of fungal endophyte isolate on performance and preference of bird cherry oat aphid. *Environmental Entomology* 35:1690–1695.

Carriere, Y., Bouchard, A., Bourassa, S., and J. Brodeur. 1998. Effect of endophyte incidence in perennial ryegrass on distribution, host-choice, and performance of the hairy chinch bag (Hemiptera: Lygaeidae). *Journal of Economic Entomology* 91:324–328.

Carroll, G. 1988. Fungal endophytes in stems and leaves: From latent pathogen to mutualistic symbiont. *Ecology* 69:2–9.

Cheplick, G. P. and K. Clay. 1988. Acquired chemical defences in grasses: The role of fungal endophytes. *Oikos* 52:309–318.

Christensen, M. J., Ball, O. J. P., Bennett, R. J., and C. L. Schardl. 1997. Fungal and host genotype effects on compatibility and vascular colonisation by *Epichloë festucae. Mycological Research* 101:493–501.

Christensen, M. J., Bennett, R. J., and J. Schmid. 2002. Growth of *Epichloe/Neotyphodium* and p-endophytes in leaves of *Lolium* and *Festuca* grasses. *Mycological Research* 106:93–106.

Clay, K. 1988. Clavicipitaceous fungal endophytes of grasses: Coevolution and the change from parasitism to mutualism. In *Coevolution of Fungi with Plants and Animals*, Eds. K. A. Pirozynski and D. L. Hawksworth, pp. 79–105. London: Academic Press.

Clay, K. 1996. Interactions among fungal endophytes, grasses and herbivores. *Researches on Population Ecology* 38:191–201.

Clay, K. and G. P. Cheplick. 1989. Effect of ergot alkaloids from fungal endophyte-infected grasses on fall armyworm (*Spodoptera frugiperda*). *Journal of Chemical Ecology* 15:169–182.

Clay, K. and J. Holah. 1999. Fungal endophyte symbiosis and plant diversity in successional fields. *Science (Washington)* 285:1742–1744.

Clay, K. and C. L. Schardl. 2002. Evolutionary origins and ecological consequences of endophyte symbiosis with grasses. *Special issue Consequences of Infection for Ecology and Evolution of Hosts, Symposium of the American Society of Naturalists, Knoxville, Tennessee, Summer 2001. American Naturalist* 160:S99–S127.

Clay, K., Hardy, T. N., and A. M. J. Hammond. 1985. Fungal endophytes of grasses and their effects on an insect herbivore. *Oecologia* 66:1–6.

Clay, K., Marks, S., and G. P. Cheplick. 1993. Effects of insect herbivory and fungal endophyte infection on competitive interaction. *Ecology* 74:1767–1777.

Claydon, N., Grove, J. F., and M. Pople. 1985. Elm bark beetle boring and feeding deterrents from *Phomopsis oblonga. Phytochemistry* 24:937–943.

Clement, S. L., Elberson, L. R., Waldson, B. L., and S. S. Quisenberry. 2007a. Variable performance of bind cherry-oat aphid on *Neotyphodium*-infected wild tall fescue from Tunisia. *Proceedings of the 6th International Symposium on Fungal Endophytes of Grasses. Grassland Research and Practice Series No. 13*, Eds. A. J. Popay and E. R. Thom, pp. 337–340, Christchurch, New Zealand.

Clement, S. L., Elberson, L. R., Miller, T., Kynaston, M., and T. D. Phillips, 2007b. Cereal leaf bettle: Is performance affected by grass endophytes? *Proceedings of the 6th International Symposium on Fungal Endophytes of Grasses. Grassland Research and Practice Series No. 13*, Eds., A. J. Popay and E. R. Thom, p. 345, Christchurch, New Zealand.

Clement, S. L., Pike, K. S., Kaiser, W. J., and A. D. Wilson. 1990. Resistance of endophyte-infected plants of tall fescue and perennial ryegrass to the Russian wheat aphid (Homoptera: Aphididae). *Journal of the Kansas Entomological Society* 63:646–648.

Clement, S. L., Lester, D. G., Wilson, A. D., and K. S. Pike. 1992. Behavior and performance of *Diuraphis noxia* (Homoptera:Aphididae) on fungal endophyte-infected and uninfected perennial ryegrass. *Journal of Economic Entomology* 85:583–588.

Clement, S. L., Kaiser, W. J., and H. Eichenseer. 1994. *Acremonium* endophytes in germplasms of major grasses and their utilisation for insect resistance. In *Biotechnology of Endophytic Fungi of Grasses*, Eds. C. W. Bacon and J. F. White Jr., pp. 185–200. Boca Raton, FL: CRC Press.

Clement, S. L., Wilson, A. D., Lester, D. G., and C. M. Davitt. 1997. Fungal endophytes of wild barley and their effects on *Diuraphis noxia* population development. *Entomologia Experimentalis et Applicata* 82:275–281.

Clement, S. L., Elberson, L. R., Youssef, N. N., Davitt, C. M., and R. P. Doss. 2001. Incidence and diversity of *Neotyphodium* fungal endophytes in tall fescue from Morocco, Tunisia, and Sardinia. *Crop Science* 41:570–576.

Clement, S. L., Elberson, L. R., Bosque-Perez, N. A., and D. J. Schotzko. 2005. Detrimental and neutral effects of wild barley–*Neotyphodium* fungal endophyte associations on insect survival. *Entomologia Experimentalis et Applicata* 114:119–125.

Crutchfield, B. A. and D. A. Potter. 1995. Damage relationships of Japanese beetle and southern masked chafer (Coleoptera: Scarabaeidae) grubs in cool-season turfgrasses. *Journal of Economic Entomology* 88:1049–1056.

Dahlman, D. L., Siegel, M. R., and L. P. Bush. 1997. Insecticidal activity of N-formylloline, *XVIII International Grasslands Conference*, Canada, pp. 13–15.

Davidson, A. W. and D. A. Potter. 1995. Response of plant-feeding, predatory, and soil-inhabiting invertebrates to *Acremonium* endophyte and nitrogen fertilisation in tall fescue turf. *Journal of Economic Entomology* 88:367–379.

di Menna, M. E., Mortimer, P. H., Prestidge, R. A., Hawkes, A. D., Sprosen, J. M., and M. E. Di Menna. 1992. Lolitrem B concentrations, counts of *Acremonium lolii* hyphae, and the incidence of ryegrass staggers in lambs on plots of *A. lolii*-infected perennial ryegrass. *New Zealand Journal of Agricultural Research* 35:211–217.

Dymock, J. J., Rowan, D. D., and I. R. McGee. 1988. Effects of endophyte-produced mycotoxins on Argentine stem weevil and the cutworm *Graphania mutans. Proceedings of the 5th Australasian Grassland Invertebrate Ecology Conference*, pp. 35–43, Melbourne, Australia.

Easton, H. S., Latch, G. C. M., Tapper, B. A., and O. J. P. Ball. 2002. Ryegrass host genetic control of concentrations of endophyte-derived alkaloids. *Crop Science* 42:51–57.

Eichenseer, H. and D. L. Dahlman. 1992. Antibiotic and deterrent qualities of endophyte-infected tall fescue to two aphid species (Homopera: Aphididae). *Environmental Entomology* 21:1046–1051.

Eichenseer, H., Dahlman, D. L., and L. P. Bush. 1991. Influence of endophyte infection, plant age and harvest interval on *Rhopalosiphum padi* survival and its relation to quantity of N-formyl and N-acetyl loline in tall fescue. *Entomologia Experimentalis et Applicata* 60:29–38.

Faeth, S. H. and T. J. Sullivan. 2003. Mutualistic asexual endophytes in a native grass are usually parasitic. *American Naturalist* 161:310–325.

Faeth, S. H. and C. E. Hamilton. 2006. Does an asexual endophyte symbiont alter life stage and long-term survival in a perennial host grass? *Microbial Ecology* 52:748–755.

Faeth, S. H., Bush, L. P., and T. J. Sullivan. 2002. Peramine alkaloid variation in *Neotyphodium*-infected Arizona tall fescue: Effects of endophyte and and host genotype and environment. *Journal of Chemical Ecology* 28:1511–1526.

Fletcher, L. R., Popay, A. J., Stewart, A. V., and B. A. Tapper. 2000. Herbage and sheep production from meadow fescue with and without the endophyte *Neotyphodium uncinatum. Proceedings of the 4th International Neotyphodium/Grass Interactions Symposium*, Eds. H. P. Volker, and P. D. Dapprich, pp. 447–453. Soest, Germany.

Funk, C. R., Halisky, P. M., Johnson, M. C., Siegel, M. R., Stewart, A. V., Ahmad, S., Hurley, R. H., and I. C. Harvey. 1983. An endophytic fungus and resistance to sod webworms: Association in *Lolium perenne* L. *Biotechnology* 1:189–191.

Gonthier, D. J., Sullivan, T. J., Brown, K. L., Wurtzel, B., Lawal, R., VandenOever, K., Buchan, Z., and T. L. Bultman. 2008. Stroma-forming endophyte *Epichloe glyceria* provides wound inducible herbivore resistance to its grass host. *Oikos* 117:629–633.

Groppe, K., Steinger, T., Sanders, I., Schmid, B., Wiemken, A., and T. Boller. 1999. Interaction between the endophytic fungus *Epichloe bromicola* and the grass *Bromus erectus*: Effects of endophyte infection, fungal concentration and environment on grass growth and flowering. *Molecular Ecology* 8:1827–1835.

Hardy, T. N., Clay, K., and A. M. J. Hammond. 1986. Leaf age and related factors affecting endophyte-mediated resistance to fall armyworm (Lepidoptera: Noctuidae) in tall fescue. *Environmental Entomology* 15:1083–1089.

Hesse, U., Hahn, H., Andreeva, K., Forster, K., Warnstorff, K., Schoberlein, W., and W. Diepenbrock. 2004. Investigations on the influence of *Neotyphodium* endophytes on plant growth and seed yield of *Lolium perenne* genotypes. *Crop Science* 44:1689–1695.

Hiatt, E. E. I. and N. S. Hill. 1997. *Neotyphodium coenophialum* mycelial protein and herbage mass effects on ergot alkaloid concentration in tall fescue. *Journal of Chemical Ecology* 23:2721–2736.

Hinton, D. M. and C. W. Bacon. 1985. The distribution and ultrastructure of the endophyte of toxic tall fescue. *Canadian Journal of Botany* 63:35–42.

Hume, D. E. and D. J. Barker. 2005. Growth and management of endophytic grasses in pastoral agriculture. In *Neotyphodium in Cool-Season Grasses*, Eds. C. A. Roberts, C. P. West, and D. E. Spiers, pp. 201–226. Iowa, IA: Blackwell Publishing.

Hume, D. E., Ryan, D. L., Cooper, B. M., and A. J. Popay. 2007. Agronomic performance of AR37-infected ryegrass in northern New Zealand. *Proceedings of the New Zealand Grassland Association* 69:201–205.

Hunt, M. G. and J. A. Newman. 2005. Reduced herbivore resistance from a novel grass-endophyte association. *Journal of Applied Ecology* 42:762–769.

Iannone, I. J. and D. Cabral. 2006. Effects of the *Neotyphodium* endophyte status on plant performance of *Bromus auleticus*, a wild native grass from South America. *Symbiosis* 41:61–69.

Jensen, A. M. D. and N. Rolund. 2004. Occurrence of *Neotyphodium* endophytes in permanent grassland with perennial ryegrass (*Lolium perenne*) in *Denmark. Agriculture, Ecosystems and Environment* 104:419–427.

Jensen, J. G. and A. J. Popay. 2004. Perennial ryegrass infected with AR37 endophyte reduces survival of porina larvae. *New Zealand Plant Protection* 57:323–328.

Jensen, J. G. and A. J. Popay. 2007. Reductions in root aphid populations by non-toxic endophyte strains in tall fescue. *Proceedings of the 6th International Symposium on Fungal Endophytes of Grasses. Grassland Research and Practice Series No. 13*, Eds. A. J. Popay and E. R. Thom, pp. 341–344, Christchurch, New Zealand.

Johnson, M. C., Dahlman, D. L., Siegel, M. R, Bush, L. P., Latch, G. C. M., Potter, D. A., and D. R. Varney. 1985. Insect feeding deterrents in endophyte-infected tall fescue. *Applied and Environmental Microbiology* 49:568–571.

Johnson-Cicalese, J. M. and R. H. White. 1990. Effect of *Acremonium* endophytes on four species of billbug found on New Jersey turfgrasses. *Journal of American Society of Horticultural Science* 115:602–604.

Justus, M., Witte, L., and T. Hartmann. 1997. Levels and tissue distribution of loline alkaloids in endophyte-infected *Festuca pratensis. Phytochemistry* 44:51–57.

Kanda, K., Hirai, H., Koga, H., and K. Hasegawa. 1994. Endophyte-enhanced resistance in perennial ryegrass and tall fescue to bluegrass webworm, *Parapediasia teterrella. Japanese Journal of Applied Entomology and Zoology* 38:141–145.

Kindler, S. D. and T. L. Springer. 1991. Resistance to Russian wheat aphid in wild *Hordeum* species. *Crop Science* 31:94–97.

Kirfman, G. W., Brandenburg, R. L., and G. B. Garner. 1986. Relationship between insect abundance and endophyte infestation level in tall fescue in Missouri. *Journal of the Kansas Entomological Society* 59:552–554.

Koga, H., Hirai, Y., Kanda, K. I., Tsukiboshi. T., and T. Uematsu. 1997. Successive transmission of resistance to bluegrass webworm to perennial ryegrass and tall fescue plants by artificial inoculation with Acremonium endophytes. *JARQ, Japan Agricultural Research Quarterly* 31:109–115.

Koppenhofer, A. M., Cowles, R. S., and E. M. Fuzy. 2003. Effects of turfgrass endophytes (Clavicipitaceae: Ascomycetes) on white grub (Coleoptera: Scarabaeidae) larval development and field populations. *Environmental Entomology* 32:895–906.

Lane, G. A., Ball, O. J. P., Davies, E., and C. Davidson. 1997a. Ergovaline distribution in perennial ryegrass naturally infected with endophyte. In *Third International Symposium on Neotyphodium/Grass Interactions*, Eds. C. W. Bacon and N. S. Hill, pp. 65–67. New York, Athens, GA: Plenum Press.

Lane, G. A., Tapper, B. A., Davies, E., Christensen, M. J., and G. C. M. Latch. 1997b. Occurrence of extreme alkaloid levels in endophyte-infected perennial ryegrass, tall fescue and meadow fescue. In *Neotyphodium/Grass Interactions*, Eds. C. W. Bacon and N. S. Hill, pp. 433–436. New York, Athens, GA: Plenum Press.

Lane, G. A., Christensen, M. J., and C. O. Miles. 2000. Coevolution of fungal endophytes with grasses: The significance of secondary metabolites. In *Microbial Endophytes*, Eds. C. W. Bacon and J. F. White Jr., pp. 341–388. New York: Marcel Dekker.

Latch, G. C. M. 1994. Influence of *Acremonium* endophytes on perennial grass improvement. *New Zealand Journal of Agricultural Research* 37:311–318.

Latch, G. C. M., Christensen, M. J., and D. L. Gaynor. 1985. Aphid detection of endophyte infection in tall fescue. *New Zealand Journal of Agricultural Research* 28:129–132.

Lehtonen, P., Helander, M., and K. Saikkonen. 2005. Are endophyte-mediated effects on herbivores conditional on soil nutrients. *Oecologia* 142:38–45.

Leuchtmann, A., Schmidt, D., and L. P. Bush. 2000. Different levels of protective alkaloids in grasses with stroma-forming and seed-transmitted *Epichloe/Neotyphodium* endophytes. *Journal of Chemical Ecology* 26:1025–1036.

Lewis, G. C. and R. O. Clements. 1986. A survey of ryegrass endophyte (*Acremonium loliae)* in the U.K. and its apparent ineffectuality on a seedling pest. *Journal of Agricultural Science Cambridge* 107:633–638.

Lewis, G. C. and B. Vaughan. 1997. Evaluation of a fungal endophyte (*Neotyphodium lolii*) for control of leatherjackets (*Tipula* spp.) in perennial ryegrass. *Tests of Agrochemicals and Cultivars* 18:34–35.

Lopez, J. E., Faeth, S. H., and M. Miller. 1995. Effect of endophytic fungi on herbivory by redlegged grasshoppers (Orthoptera: Acrididae) on Arizona fescue. *Environmental Entomology* 24:1576–1580.

Lyons, P. C., Plattner, R. D., and C. W. Bacon. 1986. Occurrence of peptide and clavine ergot alkaloids in tall fescue grass. *Science, USA* 232:487–489.

Maron, J. L. 1998. Insect herbivory above- and below-ground: Individual and joint effects on plant fitness. *Ecology* 79:1281–1293.

Mathias, J. K., Hellman, J. L., and R. H. Ratcliffe. 1990. Association of an endophytic fungus in perennial ryegrass and resistance to hairy chinch bug (Hemiptera: Lygaeidae). *Journal of Economic Entomology* 83:1640–1646.

Mauricio, R. 2001. An ecological genetic approach to the study of coevolution. *American Zoologist* 41:916–927.

McDonald, G., Noske, A., van Heeswijck, R., and W. E. Frost. 1993. The role of perennial ryegrass endophyte in the management of pasture pests in south eastern Australia. *Proceedings of the 6th Australasian Grassland Invertebrate Ecology Conference* pp. 122–128, Hamilton, New Zealand.

Meister, B., Krauss, J., Harri, S. A., Schneider, M. V., and C. B. Muller. 2006. Fungal endosymbionts affect aphid population size by reduction of adult lifespan and fecundity. *Basic and Applied Ecology* 7:244–252.

Miles, C. O., di Menna, M. E., Jacobs, S. W. L., Garthwaite, I., Lane, G. A., Prestidge, R. A., Marshall, S. L., Wilkinson, H. H., Schardl, C. L., Ball, O. J. P., and G. C. M. Latch. 1998. Endophytic fungi in indigenous Australasian grasses associated with toxicity to livestock. *Applied and Environmental Microbiology* 64:601–606.

Miller, J. D., Mackenzie, S., Foto, M., Adams, G. W., and J. A. Findlay. 2002. Needles of white spruce inoculated with rugulosin-producing endophytes contain rugulosin reducing spruce budworm growth rate. *Mycological Research* 106:471–479.

Moy, M., Belanger, F., Duncan, R., Freehoff, A., Leary, C., Meyer, W., Sullivan, R., and J. F. White Jr. 2000. Identification of epiphyllous mycelial nets on leaves of grasses infected by clavicipitaceous endophytes. *Symbiosis* 28:291–302.

Muegge, M. A., Quisenberry, S. S., Bates, G. E., and R. E. Joost. 1991. Influence of *Acremonium* infection and pesticide use on seasonal abundance of leafhoppers and froghoppers (Homoptera:Cicadellidae:Cercopidae) in tall fescue. *Environmental Entomology* 20:1531–1536.

Murphy, J. A., Sun, S., and L. L. Betts. 1993. Endophyte-enhanced resistance to billbug (Coleoptera: Curculionidae), sod webworm (Lepidoptera: Pyralidae) and white grub (Coleoptera: Scarabaeidae) in tall fescue. *Environmental Entomology* 22:699–703.

Olejniczak, P. and M. Lembicz. 2007. Age-specific response of the grass *Puccinellia distans* to the presence of a fungal endophyte. *Oecologia* 152:485–494.

Patterson, C. G., Potter, D. A., and F. F. Fannin. 1991. Feeding deterrency of alkaloids from endophyte-infected grasses to Japanese beetle larvae. *Entomologia Experimentalis et Applicata* 61:285–289.

Pennell, C. and O. J. P. Ball. 1999. The effects of *Neotyphodium* endophytes in tall fescue on pasture mealy bug (*Balanococcus poae*). *Proceedings of the 52nd New Zealand Plant Protection Conference*, pp. 259–263, Auckland, New Zealand.

Pennell, C., Popay, A. J., Ball, O. J. P., Hume, D. E., and D. B. Baird. 2005. Occurrence and impact of pasture mealybug (*Balanococcus poae*) and root aphid (*Aploneura lentisci*) on ryegrass (*Lolium* spp.) with and without infection by *Neotyphodium* fungal endophytes. *New Zealand Journal of Agricultural Research* 48:329–337.

Philipson, M. N. and M. C. Christy. 1986. The relationship of host and endophyte during flowering, seed formation, and germination of *Lolium perenne*. *New Zealand Journal of Botany* 24:125–134.

Popay, A. J. and R. A. Mainland. 1991. Seasonal damage by Argentine stem weevil to perennial ryegrass pastures with different levels of *Acremonium lolii*. *Proceedings of the 44th New Zealand Weed and Pest Control Conference*, pp. 171–175, Tauranga, New Zealand.

Popay, A. J. and D. D. Rowan. 1994. Endophytic fungi as mediators of plant-insect interactions. In *Insect–Plant Interactions*, Ed. E. A. Bernays, pp. 84–103. Boca Raton, FL: CRC Press.

Popay, A. J. and R. T. Wyatt. 1995. Resistance to Argentine stem weevil in perennial ryegrass infected with endophytes producing different alkaloids. *Proceedings of the 48th New Zealand Plant Protection Conference*, Hastings, pp. 229–236.

Popay, A. J. and G. A. Lane. 2000. The effect of crude extracts containing loline alkaloids on two New Zealand insect pests. In *4th International Neotyphodium/Grass Interactions Symposium*, eds. V. H. Paul and P. D. Dapprich. Germany: Soest.

Popay, A. J. and P. J. Gerard. 2007. Cultivar and endophyte effects on a root aphid, *Aploneura lentisci*, in perennial ryegrass. *New Zealand Plant Protection* 60:223–227.

Popay, A. J. and B. A. Tapper. 2007. Endophyte effects on consumption of seed and germinated seedlings of ryegrass and fescue by grass grub (*Costelytra zealandica*). *Proceedings of the 6th International Symposium on Fungal Endophytes of Grasses, Grasslands Research and Practice Series No. 13*, Eds. A. J. Popay and E. R. Thom, pp. 353–355, Christchurch, New Zealand.

Popay, A. J., Hume, D. E., Baltus, J. G., Latch, G. C. M., Tapper, B. A., Lyons, T. B., Cooper, B. M., Pennell, C. G., Eerens, J. P. J., and S. L. Marshall. 1999. Field performance of perennial ryegrass (*Lolium perenne*) infected with toxin-free fungal endophytes (*Neotyphodium* spp.). In *Ryegrass endophyte: An Essential New Zealand Symbiosis Grassland Research and Practice Series 7*, Eds. D. R. Woodfield and C. Matthew, pp. 113–122.

Popay, A., Marshall, S., and J. Baltus. 2000. Endophyte infection influences disappearance of perennial ryegrass seed. *New Zealand Plant Protection* 53:8–10.

Popay, A. J., Townsend, R. J., and L. R. Fletcher. 2003. The effect of endophyte (*Neotyphodium uncinatum*) in meadow fescue on grass grub larvae. *New Zealand Plant Protection* 56:123–128.

Popay, A. J., Silvester, W. B., and P. J. Gerard. 2004. New endophyte isolate suppresses root aphid, *Aploneura lentisci*, in perennial ryegrass. In *5th International Symposium on Neotyphodium/Grass Interactions*, Eds. R. Kallenbach, C. J. Rosenkrans, and T. R. Lock, p. 317, Fayetteville, Arkansas.

Popay, A. J., Jensen, J. G., and B. M. Cooper. 2005. The effect of non-toxic endophytes in tall fescue on two major insect pests. *Proceedings of the New Zealand Grassland Association* 67:169–173.

Potter, D. A., Patterson, C. G., and C. T. Redmond. 1992. Influence of turfgrass species and tall fescue endophyte on feeding ecology of Japanese beetle and southern masked chafer grubs (Coleoptera: Scarabaeidae). *Journal of Economic Entomology* 85:900–909.

Potter, D. A., Stokes, J. T., Redmond, C. T., Schardl, C. L., and D. G. Panaccione. 2008. Contribution of ergot alkaloids to suppression of a grass-feeding caterpillar assessed with gene knockout endophytes in perennial ryegrass. *Entomologia Experimentalis et Applicata* 126:138–147.

Prestidge, R. A. and R. T. Gallagher. 1985. Lolitrem B—a stem weevil toxin isolated from *Acremonium*-infected ryegrass. *Proceedings, New Zealand Weed and Pest Control Conference* 38:38–40.

Prestidge, R. A. and O. J. P. Ball. 1993. The role of endophytes in alleviating plant biotic stress in New Zealand. In *2nd International Symposium on Acremonium Grass Interactions*, Eds. D. E. Hume, G. C. M. Latch and H. S. Easton, pp. 141–151. AgResearch, Palmerston North, New Zealand.

Prestidge, R. A., Pottinger, R. P., and G. M. Barker. 1982. An association of *Lolium* endophyte with ryegrass resistance to Argentine stem weevil. *Proceedings of the 35th New Zealand Weed and Pest Control Conference*, pp. 119–122, Hamilton, New Zealand.

Prestidge, R. A., van der Zijpp, S., and D. Badan. 1984. Effects of Argentine stem weevil on pastures in the Central Volcanic Plateau. *New Zealand Journal of Experimental Agriculture* 12:323–331.

Prestidge, R. A., di Menna, M. E., van der Zijpp, S., and D. Badan. 1985. Ryegrass content, *Acremonium* endophyte and Argentine stem weevil in pastures in the volcanic plateau, *Proceedings of the 38th New Zealand Weed and Pest Control Conference*, pp. 41–44, Rotorua, New Zealand.

Rasmussen, S., Parsons, A. J., Bassett, S., Christensen, M. J., Hume, D. E., Johnson, L. J., Johnson, R. D., Simpson, W. R., Stacke, C., Voisey, C. R., Xue, H., and J. A. Newman. 2007. High nitrogen supply and carbohydrate content reduce fungal endophyte and alkaloid concentration in *Lolium perenne. New Phytologist* 173:787–797.

Rasmussen, S., Parsons, A. J., Fraser, K., Xue, H., and J. A. Newman. 2008. Metabolic profiles of *Lolium perenne* are differentially affected by nitrogen supply, carbohydrate content, and fungal endophyte infection. *Plant Physiology* 146:1440–1453.

Rausher, M. 2001. Co-evolution and plant resistance to natural enemies. *Nature* 411:857–864.

Richmond, D. S. and D. J. Shetlar. 1999. Larval survival and movement of bluegrass webworm in mixed stands of endophytic perennial ryegrass and Kentucky bluegrass. *Journal of Economic Entomology, December* 92:1329–1334.

Richmond, D. S. and D. J. Shetlar. 2001. Black cutworm (Lepidoptera: Noctuidae) larval emigration and biomass in mixtures of endophytic perennial ryegrass and Kentucky bluegrass. *Journal of Economic Entomology* 94:1183–1186.

Richmond, D. S., Neimczyk, H., and D. J. Shetlar. 2000. Overseeding endophytic perennial ryegrass into stands of Kentucky bluegrass to manage bluegrass billbug (Coleoptera: Curculionidae). *Journal of Economic Entomology* 93:1662–1668.

Riedell, W. E., Kiechefer, R. E., Petroski, R. J., and R. G. Powell. 1991. Naturally occurring and synthetic loline alkaloid derivatives: Insect feeding behavior modification and toxicity. *Journal of Entomological Science* 26:122–129.

Rowan, D. D., Dymock, J. J., and M. A. Brimble. 1990. Effect of fungal metabolite peramine and analogs on feeding development of Argentine stem weevil (Listronotus bonariensis). *Journal of Chemical Ecology* 16:1683–1695.

Sabzalian, M. R. Hatami, B., and A. Mirlohi. 2004. Mealybug, *Phenococcus solani*, and barley aphid, *Sipha maydis*, response to endophyte-infected tall and meadow fescue. *Entomologia Experimentalis et Applicata* 113:205–209.

Saha, D. C., Johnson-Cicalese, J. M., Halisky, P. M., van Heemstra, M. I., and C. R. Funk. 1987. Occurrence and significance of endophytic fungi in the fine fescues. *Plant Disease* 71:1021–1024.

Saikkonen, K., Helander, M., Faeth, S. H., Schulthess, F., and D. Wilson. 1999. Endophyte–grass–herbivore interactions: The case of *Neotyphodium* endophytes in Arizona fescue populations. *Oecologia* 121: 411–420.

Saikkonen, K., Lehtonen, P., Helander, M., Koricheva, J., and S. H. Faeth. 2006. Model systems in ecology: Dissecting the endophyte–grass literature. *Trends in Plant Science* 11:428–433.

Salminen, S. O., Richmond, D. S., Grewal, P. S., and P. S. Grewal. 2005. Influence of temperature on alkaloid levels and fall armyworm performance in endophytic tall fescue and perennial ryegrass. *Entomologia Experimentalis et Applicata* 115:417–426.

Schardl, C. L. and H.-F. Tsai. 1997. Molecular biology and evolution of grass endophytes. *Natural Toxins* 1:171–184.

Schardl, C., Leuchtmann, A., and M. J. Spiering. 2004. Symbioses of grasses with seedborne fungal endophytes. *Annual Review of Plant Biology* 55:315–340.

Schardl, C. L., Grossman, R. B., Nagabhyru, P., Faulkner, J. R., and U. P. Mallik. 2007. Loline alkaloids: Currencies of mutualism. *Phytochemistry* 68:980–996.

Schmid, J., Spiering, M. J., and M. J. Christensen. 2000. Metabolic activity, distribution and propogation of grass endophytes *in planta*: Investigations using the gus reporter gene system. In *Microbial Endophytes*, Eds. C. W. Bacon and J. F. White Jr., pp. 295–322. New York: Marcel Dekker.

Schmidt, D. and P. L. Guy. 1997. Effects of the presence of the endophyte *Acremonium uncinatum* and of an insecticide treatment on seed production of meadow fescue. *Revue Suisse d'Agriculture* 29:97–99.

Siegel, M. R., Latch, G. C. M., Bush, L. P., Fannin, F. F., Rowan, D. D., Tapper, B. A., Bacon, C. W., and M. C. Johnson. 1990. Fungal endophyte-infected grasses: Alkaloid accumulation and aphid response. *Journal of Chemical Ecology* 16:3301–3315.

Simms, E. L. and M. D. Rausher. 1987. Costs and benefits of plant resistance to herbivory. *American Naturalist* 130:570–581.

Spatafora, J. W., Sung, G. H., Sung, J. M., Hywel-Jones, N. L., and J. F. White Jr. 2007. Phylogenetic evidence for an animal pathogen origin of ergot and the grass endophytes. *Molecular Ecology* 16:1701–1711.

Spiering, M. J., Lane, G. A., Christensen, M. J., and J. Schmid. 2005. Distribution of the fungal endophyte *Neotyphodium lolii* is not a major determinant of the distribution of fungal alkaloids in *Lolium perenne* plants. *Phytochemistry* 66:195–202.

Springer, T. L. and S. D. Kindler. 1990. Endophyte-enhanced resistance to Russian wheat aphid and the incidence of endophytes in fescue species. *Proceedings of International Symposium on Acremonium/Grass Interactions*, Eds. S. S. Quisenberry and R. E. Joost, pp. 194–195. Baton Rouge, LA: Louisiana Agriculture Experimental Station.

Stewart, A. V. 1985. Perennial ryegrass seedling resistance to Argentine stem weevil. *New Zealand Journal of Agricultural Research* 28:403–407.

Tanaka, A., Tapper, B. A., Popay, A., Parker, E. J., and B. Scott. 2005. A symbiosis expressed non-ribosomal peptide synthetase from a mutualistic fungal endophyte of perennial ryegrass confers protection to the symbiotum from insect herbivory. *Molecular Microbiology* 57:1036–1050.

Tibbets, T. M. and S. H. Faeth. 1999. *Neotyphodium* endophytes in grasses: Deterrents or promoters of herbivory by leaf-cutting ants. *Oecologia* 118:287–305.

Tintjer, T. and J. A. Rudgers. 2006. Grass–herbivore interactions altered by strains of a native endophyte. *New Phytologist* 170:513–521.

Van der Putten, W. H. 2003. Plant defense belowground and spatiotemporal processes in natural vegetation. *Ecology* 84:2269–2280.

Varley, G. C., Gradwell, G. R., and M. P. Hassell. 1973. *Insect Population Ecology*. Oxford, U.K.: Blackwell Scientific.

Verkaar, H. J. 1987. Population dynamics—the influence of herbivory. *New Phytologist* 106:49–60.

White, J. F. Jr., Glenn, A. E., and K. F. Chandler. 1993. Endophyte–host associations in grasses. XVII. Moisture relations and insect herbivory of the emergent stromal leaf of *Epichloe*. *Mycologia* 85:195–202.

White, J. F. Jr., Sullivan, R. F., Balady, G. A., Gianfagna, T. J., Yue, Q., Meyer, W. A., and D. Cabrnal. 2001. A fungal endosymbiont of the grass Bromus setifolius: Distribution in some Andean populations, identification, and examination of beneficial properties. *Symbiosis* 31:241–257.

Wilkinson, H. H. and C. L. Schardl. 1997. The evolution of mutualism in grass–endophyte associations. In *Neotyphodium/Grass Interactions*, Eds. C. W. Bacon and N. S. Hill, pp. 13–25. New York, Athens, GA: Plenum Press.

Wilkinson, H. H., Siegel, M. R., Blankenship, J. D., Mallory, A. C., Bush, L. P., and C. L. Schardl. 2000. Contribution of fungal loline alkaloids to protection from aphids in a grass-endophyte mutualism. *Molecular Plant Microbe Interactions* 13:1027–1033.

Williamson, R. C. and D. A. Potter. 1997. Turfgrass species and endophyte effects on survival, development, and feeding preference of black cutworms (Lepidoptera: Noctuidae). *Journal of Economic Entomology Oct* 90:1290–1299.

Yates, S. G., Fenster, J. C., and R. J. Bartelt. 1989. Assay of tall fescue seed extracts, fractions, and alkaloids using the large milkweed bug. *Journal of Agricultural and Food Chemistry* 37:354–357.

Yoshimatsu, S., Arimura, K., and T. Shimanuki. 1998. Comparison of population growth rates of confused flour beetle, *Tribolium confusum* Jaquelin (Coleoptera: Tenebrionidae), on endophyte-infected or endophyte-uninfected seeds of ground tall fescue and perennial ryegrass. *Journal of Applied Entomology and Zoology* 42:227–229.

Yoshimatsu, S., Shimanuki, T., and K. Arimura. 1999. Influence of endophyte-infected tall fescue, *Festuca arundinacea* Shreb., seeds on the adult survival and reproduction of the grain beetle, *Oryzaephilus surinamensis* (L.) (Coleoptera: Silvanidae). *Japanese Journal of Entomology* 2:51–56.

Yue, Q., Johnson-Cicalese, J., Gianfagna, T. J., and W. A. Meyer. 2000. Alkaloid production and chinch bug resistance in endophyte-inoculated chewings and strong creeping red fescues. *Journal of Chemical Ecology* 26:279–292.

22

Fungal Endophytes: Defensive Characteristics and Implications for Agricultural Applications

Luis C. Mejía, Edward Allen Herre, Ajay P. Singh, Vartika Singh, Nicholi Vorsa, and James F. White, Jr.

CONTENTS

22.1 Introduction

Endophytes are the subject of intensive research, in part because of the potential they hold in agriculture as a source of beneficial effects to their host plants, such as increased vigor and tolerance to a range of abiotic and biotic stresses (Backman and Sikora, 2008; Kuldau and Bacon, 2008). They are defined as organisms that asymptomatically infect the internal tissues of plants during at least part of their life cycle (see Petrini, 1991; Wilson, 1995; Saikkonen et al., 1998; Stone et al., 2000). In particular, fungal endophytes have been reported from all plant species surveyed, including representatives from all ecosystems. In addition, these fungi can be isolated from different plant organs and tissues, including roots, stems, branches, leaves, flowers, and fruits (Saikkonen et al., 1998; Stone et al., 2000, 2004; Rodriguez et al., 2004; Arnold, 2007). Commonly found fungal endophytes (excluding mycorrhizal associations) belong to diverse classes of Ascomycota, mostly Dothidiomycetes, Leotiomycetes, and Sordariomycetes, although Basidiomycota endophytes have been observed to be common in some hosts (Stone et al., 2004; Crozier et al., 2006; Sieber, 2007; Thomas et al., 2008).

One type of fungal endophyte–host association that has been frequently examined is that found in fungal endophyte–grass associations, specifically referring to endophytic fungi from the family Clavicipitaceae (tribe Balansiae) in cool-season grasses (Pooideae) (White et al., 2000), Two well known examples of this type of association are between *Neotyphodium coenophialum* and tall fescue (*Festuca arundinaceae)* and *Neotyphodium lolii* and perennial ryegrass (*Lolium perenne).* In this association, the fungal partner can confer some protection to their host against herbivores, or an enhanced ability to overcome abiotic stresses (e.g., drought, heavy metals) through production of mycotoxins or other fungal derived molecules, while the host provides nutrients and a stable environment (Clay, 1988; Arechavaleta et al., 1989; Clay and Schardl, 2002). The symbiotic relationship between clavicipitalean endophytes and their grass hosts has been exploited commercially in the turf and forage grass industry, specifically by the production of endophyte-infected grass varieties with enhanced tolerance to abiotic and biotic stresses (e.g., MaxQ, AR37 endophyte).

A second type of host–endophyte interaction that also often exhibits similar mutualistic benefits with agricultural potential is that between non-Clavicipitaceous endophytic fungi (see Schulz and Boyle, 2005) and woody plants (see Table 22.1). In this group, different studies have shown that fungal endophytes can make substantial contributions to their host's capacity to tolerate or avoid adverse abiotic and biotic factors, ranging from drought stress to herbivory to pathogens (Alvarez et al., 2008; Carroll, 1988; Saikkonen et al., 1996; Faeth and Hammon, 1997; Arnold et al., 2003; Campanile et al., 2007; Herre et al., 2007; Miller et al., 2002 see Table 22.1). In particular, some field trials have shown that inoculation with particular endophyte strains can benefit their hosts by limiting the damage by pests or reducing the dispersal capabilities of their pathogens (Narisawa et al., 2000; Mejía et al., 2008b; Miller et al., 2008).

The host–endophyte properties of agriculturally important crops (e.g., vegetables, cereals, fruits, and ornamental flowers) as well as other plants that do not fit neatly into the two major host categories previously mentioned (i.e., cool-season grasses and woody plants) require intensive study to determine their relevance for crop protection and production. Studies have been conducted on these plants and some fungal endophytes have been determined to increase plant productivity and resistance to diseases (see Sieber et al., 1988; Narisawa et al., 2002; Waller et al., 2005; Rodriguez et al., 2005; D'Amico et al., 2008; Sutton et al., 2008). In some cases, the beneficial effects of fungal endophytes on their hosts may involve additional partners. This is the case with the tripartite interaction involving a plant, a virus, and a fungal endophyte. In this case, the fungus *Curvularia protuberata* provides thermotolerance to its wild host *Dichanthelium lanuginosum* only when infected with a specific virus (Redman et al., 2002; Márquez et al., 2007). In some cases, the beneficial effects have been reproduced in hosts other than the original source of the endophytic isolate. However, in these studies, the effect of endophytes appears to be greater on their wild hosts (Márquez et al., 2007).

Currently, some general attributes of fungal endophytes such as their transmission mode, colonization pattern, and species diversity have been determined for both the grass and the woody plant–fungal endophyte associations. However, specific characteristics that likely vary for particular plant–fungal endophyte species interactions need to be considered for successful application to agriculture. For instance, target plant tissues and organs, plant life cycle (annual vs. perennial), crop production condition (greenhouse vs. open field system), and fungal endophyte life cycles are likely to be crucial for determining suitable matches between endophyte species and their hosts.

In this chapter, we highlight the importance of some general attributes of plant–fungal endophyte symbioses and their implications for practical uses in agriculture. Additionally, we will consider a case study of *Cryptosporella* (synonym *Ophiovalsa*, Mejía et al., 2008a), a fungal genus with dominant species in assemblages of endophytes from several hosts in hardwood forests of the Northern Hemisphere (Stone et al., 2004; Sieber, 2007). In particular, we will focus on the growth inhibitory activity of *Cryptosporella wehmeyeriana* on the plant pathogenic bacterium *Xanthomonas campestris* pv. *campestris*, and provide a preliminary identification of the compounds responsible for this activity.

22.2 Fungal Endophytes: Diversity, Transmission Mode, and Dominance

22.2.1 Diversity and Transmission Mode

Some common features have been found between the association of cool-season grasses and species of *Neotyphodium* and the associations between non-Clavicipitaceous endophytes and woody plants. However, marked differences have been observed between these two types of associations, which hold important implications for practical application of endophytes in agriculture. Species of *Neotyphodium* are transmitted vertically from mother to offspring through seeds, except when producing the sexual stage or when conidia are produced epiphytically (see Tadych et al., 2007). In these last two cases, the transmission is horizontal (i.e., laterally from plant to plant). Additionally, *Neotyphodium* spp. and other clavicipitaceous endophytes associated with cool-season grasses establish systemic and nonorgan

TABLE 22.1

Representative Studies Conducted *in Planta* on Host of Agricultural Importance Showing that Fungal Endophytes Can Enhance Plant Productivity and Resistance against Pathogens

Host	Host Family	Endophyte	Benefit Conferred by Endophyte	Reference
Phragmites	Poaceae	*Stagonospora* spp.	Improved host vigor	Ernst et al. (2003)
Brassica campestris	Brassicaceae	*Heteroconium chaetospira*	Suppresion of clubroot and *Verticillium* yellows	Narisawa et al. (2000)
Cucumis sativus	Cucurbitaceae	*Clonostachys rosea*	Growth enhancement and productivity	Sutton et al. (2008)
Quercus cerris, *Q. pubescens*	Fagaceae	*Fusarium tricinctum and Alternaria alternata*	Reduce seedling mortality due to *Diplodia corticola* pathogen	Campanile et al. (2007)
Geranium sp.	Geraniaceae	*Clonostachys rosea*	Growth enhancement and productivity	Sutton et al. (2008)
Theobroma cacao	Malvaceae	Mix of different fungal species	Limit leaf damage due to *Phytophthora palmivora*	Arnold et al. (2003)
Theobroma cacao	Malvaceae	*Clonostachys rosea*	Limit reproduction of the fungal pathogen *Moniliophthora roreri*	Mejía et al. (2008b)
Theobroma cacao	Malvaceae	*Colletotrichum gloeosporioides*	Control incidence of *Phytopthora* spp.	Mejía et al. (2008b)
Theobroma cacao	Malvaceae	*Gliocladium catenulatum*	Reduce incidence of *Crinipellis perniciosa* (Witche's Broom disease of cacao)	Rubini et al. (2005)
Musa	Musaceae	*Fusarium* spp.	Reduce number of *Rhadopolus similis/g* of roots	Pocasangre et al. (2001)
Musa AAA	Musaceae	*Fusarium oxysporum* and *Trichoderma atroviride*	Control of the nematode *Rhadopolus similis*	Zum Felde et al. (2006)
Brachiaria brizantha	Poaceae	*Acremonium implicatum*	Reduce number and size of lesions by *Dreschlera* sp.	Kelemu et al. (2001)
Festuca arundinacea	Poaceae	*Acremonium coenophialum*	Drougth tolerance	Arechavaleta et al. (1989)
Festuca arundinacea	Poaceae	*Acremonium coenophialum*	Control nematode	West et al. (1988)
Festuca arundinacea	Poaceae	*Acremonium coenophialum*	Limit nematode reproduction	Kimmons et al. (1990)
Festuca arundinaceae	Poaceae	*Acremonium coenophialum*	Reduce seedling loss due to *Rhizoctonia zeae* seedling disease	Gwinn and Gavin (1992)
Festuca spp.	Poaceae	*Epichloe festucae*	Suppression of Red threat (*Laetisaria fusiformis*)	Bonos et al. (2005)
Festuca spp.	Poaceae	*Epichloe festucae*	Suppression of dollar spot disease by *Sclerotinia homoeocarpa*	Clarke et al. (2006)
Lolium perenne	Poaceae	*Acremonium lolli*	Reduction of galls caused by *Meloidogyne naasi*	Stewart et al. (1993)

(continued)

TABLE 22.1 (continued)

Representative Studies Conducted *in Planta* on Host of Agricultural Importance Showing that Fungal Endophytes Can Enhance Plant Productivity and Resistance against Pathogens

Host	Host Family	Endophyte	Benefit Conferred by Endophyte	Reference
Hordeum vulgare	Poaceae	*Piriformospora indica*	Tolerance to salt stress, increase in yield and resistance to pathogens	Waller et al. (2005)
Oryza sativa	Poaceae	*Fusarium*	Reduce galling severity by *Meloidogyne graminicola*	Sikora et al. (2008)
Triticum aestivum	Poaceae	*Chaetomium* spp. and *Phoma* sp.	Reduce density of pustules of the rust pathogen *Puccinia recondita*	Dingle and Mcgee (2003)
Zea mays	Poaceae	*Acremonium zeae*	Interfere with *Aspergillus flavus* infection	Wicklow et al. (2005)
Rosa sp.	Rosaceae	*Clonostachys rosea*	Growth enhancement and productivity	Sutton et al. (2008)
Lycopersicon esculentum	Solanaceae	*Fusarium oxysporum*	Reduce infection by the nematode *Meloidogyne incognita*	Hallman and Sikora (1995)
Lycopersicon esculentum	Solanaceae	*Fusarium oxysporum*	Induce resistance toward *Meloidogyne incognita*	Dababat and Sikora (2007)
Solanum melongena	Solanaceae	*Heteroconium chaetospira, Phialocephala fortinii*	Suppresion of *Verticillium* wilt	Narisawa et al. (2002)

Note: For fungal endophyte effects on insect deterrence or control see Breen (1994), Azevedo et al. (2000), Kuldau and Bacon (2008), and Rowan and Latch (1994).

specific colonization of aboveground tissues, and are generally considered host specific and to have a long evolutionary history of relationships with their hosts (Schardl et al., 1997; Clay and Schardl, 2002). In some grasses, the diversity of endophyte species can be high (see Sánchez Márquez et al., 2007). However, in cool-season grasses, the norm is the occurrence of one or a few fungal endophyte species per host individual, usually with one species, able to establish systemic colonization of aboveground tissues of their hosts (e.g., *N. coenophialum* on tall fescue *F. arundinaceae*). The long-term evolutionary relationships between clavicipitalean endophytes and cool-season grasses, their low fungal endophyte species diversity per host individuals, systemic colonization pattern, and vertical transmission mode, are the factors that may contribute to the persistence of desirable effects from the *Neotyphodium* species when this association is artificially manipulated. For instance, high levels of tissue colonization can be maintained for long periods in grasses after plants are artificially inoculated with Clavicipitaceous endophyte species and placed under either greenhouse or field conditions (see Clay and Holah, 1999). Additionally, while the frequencies of infection of these fungi vary depending on abiotic and biotic factors, these frequencies are expected to be high in nature (see Shelby and Dalrymple, 1993; Wäli et al., 2007).

By contrast, the associations between fungal endophytes and woody plants are characterized by a high diversity of fungal species, usually exceeding more than 30 species per host (see Stone et al., 2004). These endophytes are transmitted horizontally, and little information exists regarding their evolutionary history or symbiotic interactions with their hosts (Saikkonen et al., 1998; but see Sieber, 2007). In addition, fungal endophytes from leaves of woody plants tend to form localized infections (Stone, 1986; Wilson and Carroll, 1994). Thus, maintaining high densities of a particular fungal endophyte strain or species in a given host seems more challenging in woody plants and vegetable crops than in the Clavicipitaceous endophytes–grass system. This high diversity of endophyte species represents a challenge to answer the major question of what are the roles of fungal endophytes in general and those in woody plants and other nongrass host in particular, for example, that specifically determine which endophyte strains or species have a positive effect on their hosts, which are latent pathogens, and which species just remain there. Recent studies have been conducted with the aim of identifying the role of fungal endophytes in woody plants and other nongrass hosts and to specifically test defensive mutualism hypotheses, (see Clay, 1988; Carroll, 1988) specifically addressing the questions of whether these fungi help their hosts to tolerate herbivore and pathogen damage. Because of the high diversity of endophyte species in woody plants and their horizontal transmission mode, it is not simple to reconcile this association with current mutualism theory. It has been hard to simplify this complex system of multiple species interactions and to clearly determine the effects that these symbionts have on their hosts. It is poorly understood whether an assemblage of multiple fungal endophyte species work synergistically in a given woody plant host, or whether different species or strains perform different roles (see Arnold et al., 2003; Herre et al., 2007). For example, based on the high diversity of fungal taxa associated with woody plants and vegetable crops, the number of fungal-derived molecules with direct or indirect effects on their hosts in this association could be expected to be much more diverse than that observed in the Clavicipitaceous endophyte–grass systems (see Schulz and Boyle, 2005). Nonetheless, some woody hosts and their associated endophytic mycoflora have been experimentally manipulated to address questions on the roles of these fungi, and beneficial effects have been observed to be provided by fungal endophytes (see Wilson, 1996; Wilson and Faeth, 2001; Arnold et al., 2003; Sumarah et al., 2008).

Horizontal transmission of endophytes as observed in woody plants and vegetables crops increases the likelihood that many different species colonize plants. With many encounters some of the fungi are able to get into the plants and persist for multiple generations. When the host plant encounters stresses that are a threat to its survival (plant diseases, insect herbivores, or environmental factors), those endophytes that enable hosts to overcome the stresses will increase in frequency during the stress and persist within plants, with varying frequencies in the future. Those that do not improve host fitness in the face of stress will move to a new host or become extinct. It is also important to notice that for plants, keeping multiple fungal endophytes species may be a faster way of evolving extrinsic defense mechanisms than what their host could do, because these symbionts have shorter life cycles or because they have similar rates of evolution compared to plant pest, and pathogens (see Carroll, 1988; Herre et al., 2007). While perennial plants may not generate new sources of defense as quickly as their pest and pathogens, their symbionts may keep track of the enemies of their host.

22.2.2 Dominance of Endophyte Species in Particular Host Assemblages

Besides the diversity of endophyte species associated with woody hosts, in these plants there is usually a set of species that dominate the assemblage in a given host. While the endophyte species determined to be dominant in a particular host or host organ can be an artifact of the method used for isolating or detecting them (e.g., endophyte isolates grow differentially on different media and uncultivable endophyte species may occur), it is also likely that dominant species are good at colonizing their hosts. It has been shown experimentally that fungal endophytes from woody plants determined to be dominant in a given host based on surveys using culturing methods are good colonizers of the organs they were isolated from.

The pattern of dominance of one or a few endophyte species over an assemblage of species in a given host has been reproduced in a simplified form experimentally (see Wille et al., 2002; Mejía et al., 2008b). Additionally, these dominant species are better colonizers than rarely found or singleton species for a given host (Wilson and Carroll, 1994; Wilson, 1996; Mejía et al., 2008b). Mechanistically, this dominance of one or few endophyte species can be explained by the specialization of some species at degrading specific compounds produced by their hosts (see Saunders and Kohn, 2008). The other possibility is that dominant endophyte species produce toxic compounds to other endophytes occupying the same niche. Alternatively, dominant species bring an advantage to the host under some stresses, so they have been naturally selected and have become dominant over time. Moreover, evidence suggests that some mutualistic symbioses between fungal endophytes and their hosts are the result of specific adaptations to stresses following a habitat-specific manner (Rodriguez et al., 2004).

Knowing what endophyte species are dominant in a particular host is of great relevance when the host is intended to be inoculated with a selected endophyte strain that has shown some promise at benefiting the host. In some hosts, good endophyte colonizers are not necessarily the ones with more toxic or antibiotic potential on host pathogens and pests. It has been observed that in woody plants there is apparently a trade-off between the production of compounds with antibiotic properties and mycelial growth in fungal endophytes (Mejía et al., 2008), so endophyte species with antibiotic or toxic capabilities on plant pathogens and pests in a given host are not necessarily good *in planta* colonizers.

22.3 Fungal Endophytes: General Life Cycle

Observation of fungal endophyte life cycles may help in the search for good fungal endophyte candidates for agricultural applications (e.g., for biocontrol, growth improvement, etc.). Studies conducted on temperate broad leaf and particularly in evergreen tropical rain forests, suggest that a particular host is constantly receiving the fungal spores from the environment, and a subset of these spores are able to germinate in a given host, infect, and colonize. Some fungal endophyte species will infect one or a few hosts (endophytes with specific or limited host range) while others will infect several hosts (generalist endophytes). Most of these endophytes will establish localized infections (Stone, 1986; Petrini, 1991). On these hosts, fungal endophyte colonization goes usually from undetectable or low levels in very young tissues such as recently emerged leaves and shoots to high levels in mature tissues and to full colonization in older ones. Jointly with this pattern of colonization there is generally an increase in the diversity of endophyte species in host tissues through time. In some cases, this diversity reaches a peak in mature tissues that is later followed by a decrease in the diversity in older tissues, but with a group of few species that tend to be preferentially associated, or specific to a particular host (Wilson and Carroll, 1994; Faeth and Hammon, 1997; Wilson et al., 1997; Herre et al., 2007). Some of these endophytes will sporulate on the dead tissue and the cycle will begin again (see Herre et al., 2007; Promputtha et al., 2007).

22.4 Relevance of Ecological Studies

To appropriately address questions on the roles of fungal endophytes, we consider it important to conduct ecological studies designed to determine the identities of fungal endophyte species associated with a particular host under different growth conditions and environments (e.g., sampling of the host in its

TABLE 22.2
Summary of Recommendations for Selection and Application of Fungal Endophytes to Agricultural Plants

1. Survey and collect fungal endophytes from the target host species and organs, including sampling of organs at different ages. Sampling is recommended under both cultivated and wild conditions, in particular at the center of origin of the host. Mature organs close to senescence, are likely to harbor dominant or better adapted fungal endophyte strains. Dominant endophyte species are likely to outcompete rare ones, and thus be more easily administered to the host and remain within its tissues.
2. Classify of fungal endophyte strains by morphology and molecular methods. Strains accurately determined to belong to species known as pathogenic on target and co-occurring plant species should be avoided for application in the field. Test pathogenicity of selected endophyte strains on nontarget hosts. Co-occurring crops should be tested.
3. *In vitro* screening for bioactivity on target pathogens and pests should be combined with comparison of endophyte colonized (E+) and noncolonized (E–) plants to determine likely mechanism of beneficial function. This work also evaluates Koch's postulates and determines the pathogenicity of the endophytes to be used. Small scale testing should be conducted under a range of nursery or field conditions.

natural distribution area, exotic environments, and agricultural systems), seasons, and tissues (see Table 22.2). Studies on the chocolate tree *Theobroma cacao* and associated endophyte mycoflora may help to illustrate the complexity of endophyte–woody plant interactions, determining their roles and the challenges of making a practical application of the effects that these fungi have in their hosts. *Theobroma cacao* and some congeneric species as well as co-occurring plant species have been recently surveyed for their endophytic mycoflora. The surveys have been conducted in or near the center of origin of *T. cacao*, in exotic environments, and under a wide range of conditions and seasons. While the number of fungal endophyte species and morphospecies reported in this host is extremely high, in the order of 1000, studies suggest that there are a group of species preferentially associated with it and that a subset of species or genera tend to be localized in particular tissues (e.g., leaf vs. trunk; Arnold et al., 2003; Evans et al., 2003; Van Bael et al., 2005; Crozier et al., 2006; Samuels et al., 2006; Herre et al., 2007; Rojas et al., 2008). A major goal is to find endophyte species in this host, that can help the trees to better tolerate or resist damage by pests and pathogens. Similar surveys and approaches to find endophytes antagonistic to plant pests have been conducted on coffee plants (Santamaría and Bayman, 2005; Posada et al., 2007; Vega et al., 2008). These crops can be manipulated under laboratory and greenhouse conditions, so that plants with (E+) and without (E–) fungal endophytes can be compared under different conditions and against particular pests and pathogens. Importantly it has been shown that fungal endophytes can limit pathogen damage to hosts (Table 22.1). For example, strains of *Clonostachys* and *Trichoderma* show promise for their antagonistic activity to important cacao pathogens (Arnold et al., 2003; Evans et al., 2003; Rubini et al., 2005; Posada and Vega, 2006; Samuels et al., 2006; Tondje et al., 2006; Mejia et al., 2008b). However, a major challenge is ensuring that the chosen endophyte strains remain viable and active within host plants for extended periods. Towards that end, information on the colonization ability, persistence, and activity of a given endophyte within a given host is critical.

Importantly, when attempting to use a particular endophyte to directly antagonize a pest or a pathogen, it is not only important to find evidence for antagonistic activity in vitro. What is probably more important is to identify endophytes that are good colonizers of the host and remain active within it. This is especially important because woody plants can accumulate many endophyte species that will compete for the same habitat, over time. Dominant endophyte species in a given host may outcompete selected endophyte species under field conditions. Extensive endophyte surveys on particular host to determine patterns of species dominance and their capacity to colonize target organs or tissues can be tedious but is likely to be fruitful, if not essential, in the long term.

22.5 Fungal Endophytes: Defensive Characteristics

Fungi, including fungal endophytes, are known for their ability to produce a diverse range of molecules (e.g., antibiotics, toxins, peptides) that positively or negatively affect other organisms (Petrini et al., 1992;

Tan and Zou, 2001; Gunatilaka, 2006; Zhang et al., 2006). Further, many of these molecules are believed to play an integral role in fungal development and survival in specific environments. To better appreciate the effects of these molecules in relation to fungal endophyte niches, it is important to understand the nature of the endophytic habitat and of the particular plant–fungal endophyte association. The endophytic habitat (i.e., the internal tissues of the host) can be rich in nutrients and endophyte species (Arnold et al., 2000; White et al., 2000; Kuldau and Bacon, 2008). Endophytic fungi tend to be localized in extracellular spaces, however, some are located intracellularly and multiple species can occur within a small area (Stone, 1986; Stone et al., 2000; Lodge et al., 1996; Herre et al., 2005, 2007).

Independent of the endophyte–plant association, it is likely in the interest of a particular fungal endophyte species not to be displaced by other species, and to efficiently exploit the available resources provided by the host (see Herre et al., 2007). It is plausible to think that general ecological tenets of species interactions would apply to endophytic communities as they do to the host species. This would imply competition for resources among endophytes with particular endophyte species being better at exploiting specific resources and at colonizing particular plant tissues and organs. Fungal endophyte species that produce compounds antagonistic to other endophytes, pathogens, and pests that occupy or depend on the same habitat (the inner plant) will have an advantage at colonizing particular host tissues. Identifying fungal endophyte species that can help their host to neutralize specific pathogens or pests (directly via inhibitory or toxic compounds produced *in planta* or indirectly via induction of host defense mechanisms) is the quest for the "holy grail" of applied fungal endophyte research on crop protection and relevant to developing clean technologies for pest management in agriculture.

There is ample evidence from *in vitro* studies showing that fungal endophytes inhibit the growth of plant pests and pathogens (White and Cole, 1985; Tunali and Marshal, 2000; Evans et al., 2003; Holmes et al., 2004; data presented here). The inhibitory compounds have been identified in multiple cases (Calhoun et al., 1992; Schulz et al., 1995; Strobel et al., 2001; Daisy et al., 2002; Schwarz et al., 2004; Aneja et al., 2005; Wicklow et al., 2005; Wang et al., 2007). These compounds can be defensive in nature for the fungus in the endophytic habitat. However, with exception of the alkaloids produced by clavicipitaceous endophytes (see Siegel et al., 1990; Kuldau and Bacon, 2008), there is little evidence for *in planta* production of inhibitory compounds. When the toxic or inhibitory compounds have been detected *in planta*, their levels have been too low to actually stop the progress of a pathogen (see Schulz and Boyle, 2005). Nevertheless, a recent study under open nursery conditions found that levels of Rugulosin in the needles of *Picea glauca* infected with a Rugulosin-producing fungal endophyte were at the concentration necessary to reduce the weight of the budworm *Choristoneura fumiferana* (Miller et al., 2008).

Often biological activities by fungal endophytes have been tested to known standard organisms *in vitro*. However, it is not a prerequisite that inhibitory compounds need to be produced *in planta* to be useful for practical agricultural applications. Research on treatments of some crops with fungal endophyte-derived compounds has been shown to have beneficial effects compared to controls in terms of protection against pests and pathogens (Daisy et al., 2002; Lacey and Neven, 2006). Other studies have shown that inoculation of plants with fungal endophytes helps protect the treated plants against a wide variety of plant enemies (Table 22.1). While a general mechanism of action has been proposed in most of these cases, the details of the mechanisms are largely unknown (see Herre et al., 2007).

22.6 Antagonistic Activity of *Cryptosporella wehmeyeriana* on *Xanthomonas campestris* p.v. *campestris*

In temperate broadleaf forests, spore production of some fungal endophyte species has a marked seasonality. For instance, fungal endophyte species from the family Gnomoniaceae (Diaporthales) have marked seasonalities. While conidia of these fungi can be found through the growing season, ascospores of most fungi from this family can be found in dead and over-wintered leaves and twigs (Sogonov et al. 2008; Wilson and Carroll, 1994; Wilson et al., 1997) more likely during spring. Furthermore, genera and species from this family are dominant in assemblages of fungal endophytes associated with particular hosts in temperate broadleaf forests. (Stone et al., 2004; Sieber, 2007). For example, species

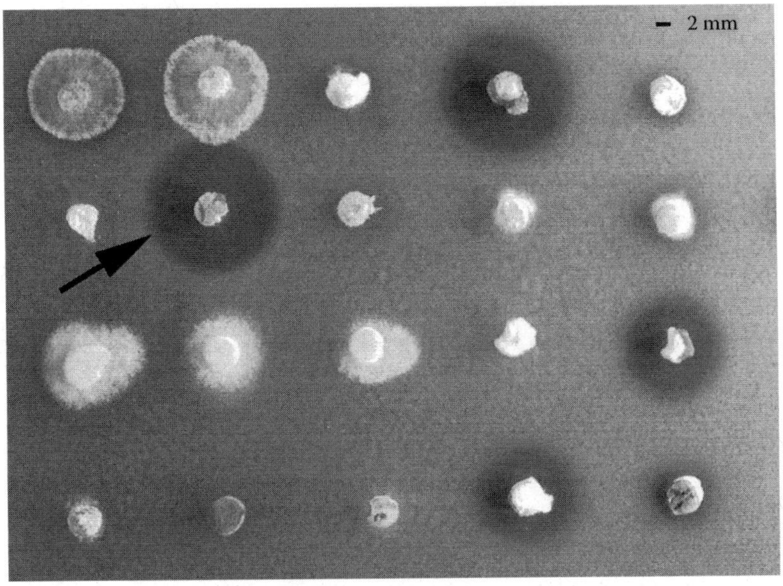

FIGURE 22.1 *In vitro* assay testing fungal endophytes activity on the growth of *Xanthomonas campestris campestris*. Plugs of agar with mycelia of fungal endophytes were plated almost at the same time with the bacterium on Potato Dextrose Agar (see methodology in the text). The arrow shows inhibition on the bacterium growth due to a diffusible compound coming from the agar plug with mycelium of *Cryptosporella wehmeyeriana*.

of *Cryptosporella* and their anamorphs are dominant endophytes on twigs from trees of the family Betulaceae. *Apiognomonia* spp. are dominant on leaves of Fagaceae in temperate broadleaf forests (see Fisher and Petrini, 1990; Sieber, 2007), and we have observed a species of *Ophiognomonia* to be dominant on leaves of *Alnus acuminata* (Betulaceae) in tropical cloud mountain forests in Central America (Mejía and white, unpublished). Moreover, when fungi such as *Ophiognomonia* sporulate, they cover large tissue area. In species of *Cryptosporella,* perithecia are found covering big patches of several cm^2 on dead twigs. We have observed that some of these species also produce inhibitory compounds to plant pathogenic fungi and bacteria (see Figure 22.1). As has been suggested by Fisher et al. (1984a,b), the primary function of these inhibitory compounds may be competition against antagonists. We suggest that the dominance of the *Cryptosporella* species in extensive areas of their hosts is facilitated by the production of compounds antagonistic to other species that occupy the same habitat. Furthermore, there is usually no evidence of growth of other fungi co-occurring with fruiting bodies of *Cryptosporella*. Based on these observations, we hypothesized that relatively "pure microstands" of *Cryptosporella* could be due to the production of antagonistic compounds to the growth of potentially co-occurring species (e.g., bacteria or fungi). To test this hypothesis and evaluate the antimicrobial potential of a group of dominant fungal endophytes, including *C. wehmeyeriana*, we have done *in vitro* assays of these fungi to evaluate their capacity to inhibit the growth of common plant pathogenic bacteria, including *Xanthomonas campestris* p.v. *campestris.*

These *in vitro* assays have been conducted following a methodology similar to that of Peláez et al., 1998. From these assays we have found that *C. wehmeyeriana* (isolated from *Tilia americana*) has strong inhibitory activity on the growth of *X. campestris* p.v. *campestris* (Figure 22.1). Our preliminary chemical analyses using HPLC equiped with a Photodiode Array Detector (PDA) indicate that major types of bioactive compounds from *C. wehmeyeriana* extracts are phenolic acids and their derivatives, flavonoids. Some representative compounds identified in this study are shown in Figure 22.2. Based on HPLC retention time and UV spectra, the compounds were characterized as phenolic acid derivatives, quercetin derivatives. Phenolic compounds are well known to occur in plant tissues and they

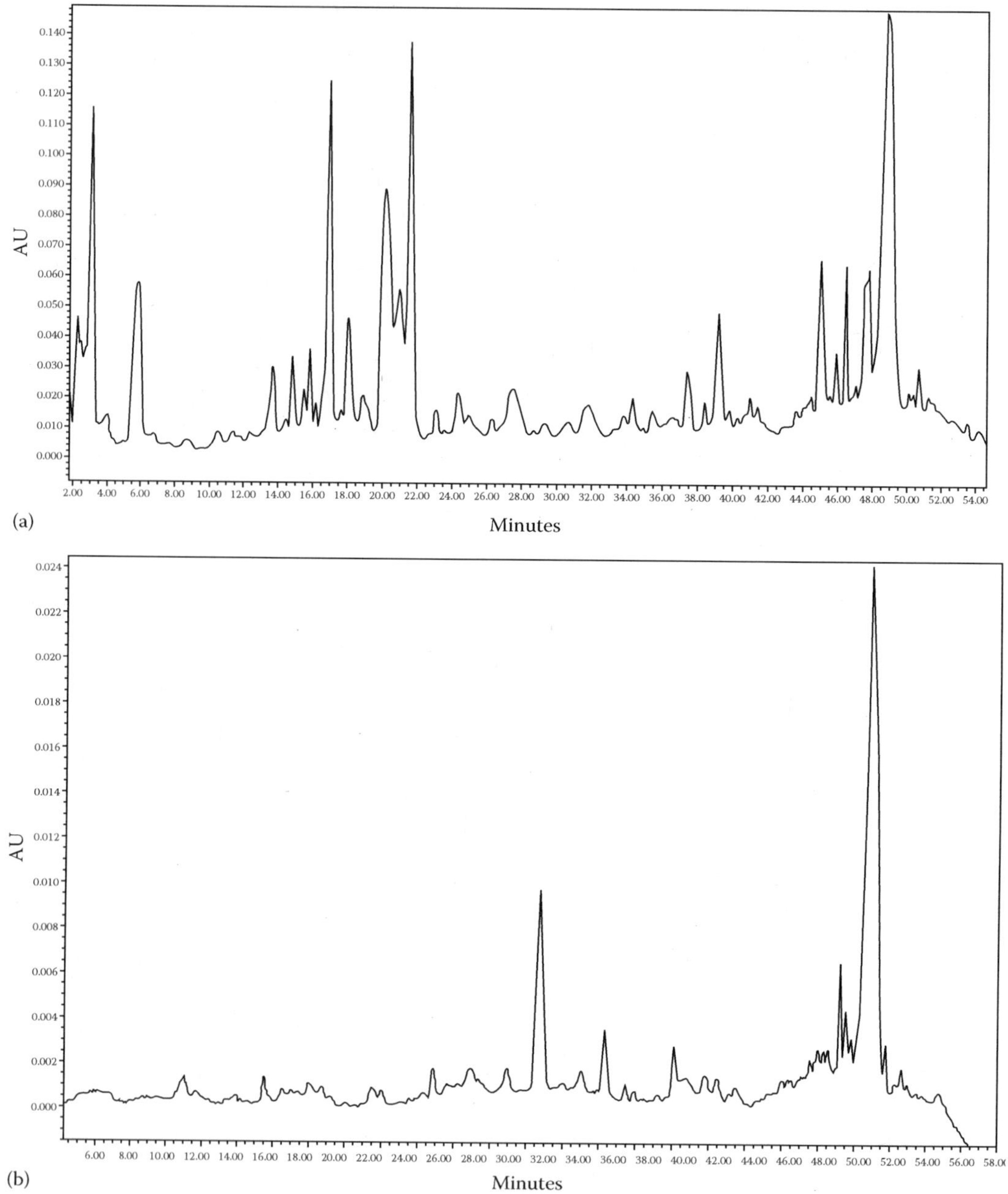

FIGURE 22.2 Representative bioactive compounds produced by *Cryptosporella wehmeyeriana*. Phenolic acids (a), and quercetin derivatives (b), as detected by HPLC-PDA at 280 and 366 nm, respectively.

have been implicated in plant disease resistance. Specifically, some flavonoids have been reported as phytoalexins that help in the defense response against insects, fungi, and bacteria (Nicholson and Hammerschmidt, 1992; Xu and Lee, 2001; McNally et al., 2003; Pereira et al., 2007). Furthermore, it has been observed that flavonoids are involved in the defense and hypersensitive reaction of cotton against *Xanthomonas campestris* pv. *malvacerum* (Dai et al., 1996). Here we report that *C. wehmeyeriana* produces compounds that could benefit its host, specifically by inhibiting plant pathogenic bacteria. These results emphasize previous observations that fungal endophytes produce compounds with antibacterial activity including phenolic compounds and that this activity may be significant for host protection from natural enemies (see Yang et al., 1994). The extent to which are the antibacterial compounds produced by *C. wehmeyeriana* are synthesized *in planta* and beneficial for its host remain to be determined. Certainly, *T. americana* twigs with these compounds will be less hospitable for pathogenic bacteria.

22.7 Conclusions

Endophytes are symbiotic organisms that spend part or their entire life asymptomatically within plants. Some fungal endophytes have been observed to benefit their hosts by helping them to tolerate stressful abiotic and biotic conditions. Several lines of evidence support a defensive mutualism for some plant–endophytic fungi interactions. In these interactions, the fungal partner protects its host in two major ways: (1) directly via the production of compounds with antagonistic activity on host natural enemies or (2) indirectly by activating host defense responses to plant pests and pathogens.

Most benefits provided by fungal endophytes seem agriculturally exploitable. Some of these benefits have been exploited agriculturally only in the association between few fungi from the family Clavicipitaceae and cool-season grasses. Crops cultivated in nurseries or greenhouses could be easier systems to control than open field systems. For example, focusing protection efforts on fruits only during the relatively short period that they develop might prove to be a tractable and effecient strategy for using fungal endophytes to prevent production losses.

Ecological studies aimed at assessing the diversity and species composition of endophyte assemblages for a given host are important when selecting fungal endophytes with potential agricultural applications. Attributes of particular plant–fungal endophyte interactions such as transmission mode, species diversity, and endophytes life cycles are important factors to be considered when practical applications of the effects endophytes have on their hosts are intended.

We have found that a fungus (*C. wehmeyeriana*) that lives asymptomatically inside plant branches inhibits the growth of a common plant pathogenic bacterium (*Xanthomonas*) *in vitro*. This fungus produces phenolic compounds, quercetin derivative flavonoids, which may be responsible for this activity. These compounds may represent an extrinsic source of defense against bacterial infection for the host of *C. wehmeyeriana*.

The observed cases of defensive mutualism between endophytic fungi and their host present the question of why plants get damaged by pests and pathogens if they are already infected by endophytes in nature? While pests and pathogens are common in natural plant ecosystems, the norm is a balance whereby plant populations usually do not get completely eliminated by the effects of pests and pathogens. Plant infection by multiple endophyte species that may perform different roles as have been discussed here, could be good for plant species in the long term. Frequencies of infection by endophyte species that could bring an advantage to their hosts do not need to be high. Endophyte species found at low frequencies in contemporary time may have performed better under stressful conditions in the past. For plants that harbor multiple species, it is like keeping an extra arsenal of weapons ready when the stressful conditions arrive.

It is important to note that fungal endophytes can also negatively affect the physiology of the host, and it is important to determine to what extent this occurs and to know the costs and benefits of fungal endophyte infections under stress and nonstress conditions (see Santos Rodriguez et al., 2000; Arnold and Engelbrecht, 2007; E.A. Herre, unpublished). Based on the evidence provided (Table 22.1 and text), overall it is plausible to think that fungal endophytes jointly with other endophytic organisms boost the capacity of the plant to tolerate adverse abiotic and biotic factors. Additionally, it is tempting to think that in some cases they may perform functions similar to what the microflora of mammals do for the immune system of their hosts.

A review of the research literature on fungal endophytes by Saikkonen et al., (2006) suggests that it is likely that mutualistic effects of fungal endophytes occur frequently in agroecosystems. In agroecosystems, crops may get depauperate in their symbiotic endophytes especially if the crops have been moved far away from its center of origin (i.e., where their potentially coevolved symbionts and natural enemies are more likely to be found, see Evans, 1999) or if their genotypes have undergone extensive artificial selection (breeding). While agroecosytems may not be good promoters of fungal endophyte diversity, inoculating and keeping good endophytes inside target tissues may be readily accomplished.

Whether or not the major role of the majority of fungal endophyte species is to protect their hosts against natural enemies, multiple studies in a range of plant lineages indicate that they can confer several advantages including survival or tolerance to specific adverse factors spanning a wide range of abiotic and biotic stressful conditions. The studies reviewed here encourage deepening research on prospection of these fungi for applicability in agricultural systems. The promise of practical use of endophytes for crop protection and production is starting to be realized.

ACKNOWLEDGMENTS

We thank Gerald Bills for providing advice on *in vitro* assays for detecting fungal bioactivity, Donald Kobayashi for providing the strain of *Xanthomonas campestris* used in the *in vitro* assays reported here, and Gary Samuels for providing helpful comments that helped to improve an early version of this chapter. We also thank Fogarty International Center for ICBG grant NIH U01 TW006674. Finally, we thank The Smithsonian Tropical Research Institute and its Short Term Research Fellowship Program for support to LCM on endophyte research. Trade brands or commercial products mentioned in this chapter are only for providing information and does not imply endorsement or recommendation by the authors or their respective institutions.

REFERENCES

Alvarez, P., White, J. F., Jr., Gil, N., Svenning, J. C., Balslev, H., and T. Kristiansen. 2008. Light converts endosymbiotic fungus to pathogen, influencing seedling survival and host tree recruitment. Available from Nature Precedings (http://hdl.handle.net/10101/npre.2008.1908.1).

Aneja, M., Gianfagna, T., and P. Hebbar. 2005. *Trichoderma harzianum* produces nonanoic acid, an inhibitor of spore germination and mycelial growth of two cacao pathogens. *Physiological and Molecular Plant Pathology* 67:304–307.

Arechavaleta, M., Bacon, C. W., Hoveland, C. S., and D. Radcliffe. 1989. Effect of the tall fescue endophyte on plant response to environmental stress. *Agronomy Journal* 81:83–90.

Arnold, A. E. 2007. Understanding the diversity of foliar endophytic fungi: Progress, challenges and frontiers. *Fungal Biology Reviews* 21:51–66.

Arnold, A. E., Maynard, Z., Gilbert, G. S., Coley, P. D., and T. A., Kursar. 2000. Are tropical fungal endophytes hyperdiverse? *Ecology Letters* 3:267–274.

Arnold, A. E., Mejía, L. C., Kyllo, D., Rojas, E., Maynard, Z., Robbins, N., and E. A. Herre. 2003. Fungal endophytes limit pathogen damage in a tropical tree. *Proceedings of the National Academy of Sciences of the USA* 100:15649–15654.

Arnold, A. E. and B. M. J. Engelbrecht. 2007. Fungal endophytes nearly double minimum leaf conductance in seedlings of a neotropical tree species. *Journal of Tropical Ecology* 23:369–372.

Azevedo, J. L., Maccheroni, W. Jr., Pereira, J. O., and W. L. de Araújo. 2000. Endophytic microorganisms: A review on insect control and recent advances on tropical plants. *Electronic Journal of Biotechnology* 3:41–65.

Backman, P. A. and R. A. Sikora. 2008. Endophytes: An emerging tool for biological control. *Biological Control* 46:1–3.

Bonos, S. A., Wilson, M. M., Meyer, W. A., and C. R. Funk. 2005. Suppression of red thread in fine fescues through endophyte-mediated resistance. *Applied Turfgrass Science*, doi:10.1094/ATS-2005-0725-01-RS.

Breen, J. P. 1994. *Acremonium* endophyte interactions with enhanced plant resistance to insects. *Annual Review of Entomology* 39:401–423.

Calhoun, L. A., Findlay, J. A., Miller, J. D., and N. J. Whitney. 1992. Metabolites toxic to spruce budworm from balsam fir needles endophytes. *Mycological Research* 96:281–286.

Campanile, G., Ruscelli, A., and N. Luisi. 2007. Antagonistic activity of endophytic fungi towards *Diplodia corticola* assessed by *in vitro* and *in planta* tests. *European Journal of Plant Pathology* 117:237–246.

Carroll, G. 1988. Fungal endophytes in stems and leaves: From latent pathogen to mutualistic symbiont. *Ecology* 69(1):2–9.

Clarke, B. B., White, J. F., Hurley, R. H., Torres, M. S., Sun, S., and D. R. Huff. 2006. Endophyte-mediated suppression of dollar spot disease in fine fescues. *Plant Disease* 90:994–998.

Clay, K. 1988. Fungal endophytes of grasses: A defensive mutualism between plants and fungi. *Ecology* 69:10–16.

Clay, K. and J. Holah. 1999. Fungal endophyte symbiosis and plant diversity in successional fields. *Science* 285:1742–1744.

Clay, K. and C. Schardl. 2002. Evolutionary origins and ecological consequences of endophyte symbiosis with grasses. *American Naturalist* 160:99–127.

Crozier, J., Thomas, S. E., Aime, M. C., Evans, H. C., and K. A. Holmes. 2006. Molecular characterization of fungal endophytic morphospecies isolated from stems and pods of *Theobroma cacao. Plant Pathology* 55:783–791.

Daisy, B. H., Strobel, G. A., Castillo, U., Ezra, D., Sears, J., Weaver, D. K., and J. B. Runyon. 2002. Napthalene, an insect repellent, is produced by *Muscodor vitigenus*, a novel endophytic fungus. *Microbiology* 148:3737–3741.

Dababat, A. A. and R. Sikora 2007. Induced resistance by the mutualistic endophyte *Fusarium oxysporum* 162 toward *Meloidogyne incognita* on tomato. *Biocontrol Science and Technology* 17:969–975.

Dai, G. H., Nicole, M. Andary, C., Martinez, C., Bresson, E., Boher, B., and J. Daniel. 1996. Flavonoids accumulate in cell walls, middle lamellae and callose-rich papillae during an incompatible interaction between *Xanthomonas campestris* pv. *malvacearum* and cotton. *Physiological and Molecular Plant Pathology* 49:285–306.

D' Amico, M., Frisullo, S., and M. Cirulli. 2008. Endophytic fungi occurring in fennel, lettuce, chicory, and celery—commercial crops in southern Italy. *Mycological Research* 112:100–107.

Dingle, J. and P. A. Mcgee. 2003. Some endophytic fungi reduce the density of pustules of *Puccinia recondita* f.sp. tritici in wheat. *Mycological Research* 107(3):310–316.

Ernst, M., Mendgen, K. W., and S. G. R. Wirsel. 2003. Endophytic fungal mutualists: Seed-borne *Stagonospora* spp. enhance reed biomass production in axenic microcosms. *Molecular Plant Microbe Interactions* 16:580–587.

Evans, H. C., Holmes, K. A., and S. E. Thomas. 2003. Endophytes and mycoparasites associated with an indigenous forest tree, *Theobroma gileri*, in Ecuador and a preliminary assessment of their potential as biocontrol agents of cocoa diseases. *Mycological Progress* 2:149–160.

Evans, H. C. 1999. Classical biological control. In *Research Methodology in Biocontrol of Plant Diseases, with Special Reference to Fungal Diseases of Cocoa*, Eds. U. Krauss and K. P. Hebbar, pp. 29–37. CATIE: Turrialba, Costa Rica.

Faeth, S. H. and K. E. Hammon. 1997. Fungal endophytes in oak trees: Experimental analyses of interactions with leafminers. *Ecology* 78(3):820–827.

Fisher, P. J. and O. Petrini. 1990. A comparative study of fungal endophytes in xylem and bark of Alnus species in England and Switzerland. *Mycological Research* 94:313–319.

Fisher, P. J., Anson, A. E., and O. Petrini. 1984a. Antibiotic activity of some endophytic fungi from ericaceous plants. *Bot. Helvetica* 94:249–253.

Fisher, P. J., Anson, A. E., and O. Petrini. 1984b. Novel antibiotic activity of an endophyte *Cryptosporiopsis* sp.isolated from *Vaccinum myrtillus. Transactions of the British Mycological Society* 83:145–148.

Gunatilaka, A. A. L. 2006. Natural products from plant-associated microorganisms: Distribution, structural diversity, bioactivity, and implications of their occurrence. *Journal of Natural Products* 69:509–526.

Gwinn, K. D. and A. M. Gavin. 1992. Relationship between endophyte infestation level of tall fescue seed lots and *Rhizoctonia zeae* seedling disease. *Plant Disease* 76:911–914.

Hallman, J. and R. Sikora. 1995. Influence of *Fusarium oxysporum*, a mutualistic fungal endophyte, on *Meloidogyne incognita* infection of tomato. *Journal of Plant Disease Protection* 101:475–481.

Herre, E. A., Van Bael, S. A., Maynard, Z., Robbins, N., Bischoff, J., Arnold, A. E. Rojas, E., Mejía, L. C., Cordero, R. A., Woodward, C., and D. A. Kyllo. 2005. Tropical plants as chimera: Some implications of foliar endophytic fungi for the study of host plant defense, physiology, and genetics. In *Biotic Interactions in the Tropics: Their Role in the Maintenance of Species Diversity*, Eds. D. F. R. P. Burslem, M. A. Pinard, and S. E. Hartley, pp. 226–237. Cambridge University Press: Cambridge, U.K.

Herre, E. A., Mejia, L. C., Kyllo, D. A., Rojas, E., Maynard, Z., Butler, A., and S. A. Van Bael, 2007. Ecological implications of anti-pathogen effects of tropical fungal endophytes and mycorrhizae. *Ecology* 88:550–558.

Holmes, K. A., Schroers, H. J., Thomas, S. E., Evans, H. C., and G. J. Samuels. 2004. Taxonomy and biocontrol potential of a new species of *Trichoderma* from the Amazon basin of South America. *Mycological Progress* 3:199–210.

Kelemu, S., White, J. F., Muñoz, F., and Y. Takayama. 2001. An endophyte of tropical forage grass *Brachiaria brizantha*. Isolating, identifying and characterizing the fungus and determining its antimycotic properties. *Canadian Journal of Microbiology* 47:55–62.

Kimmons, C. A., Gwinn, K. D., and E. C. Bernard. 1990. Nematode reproduction on endophyte-infected and endophyte-free tall fescue. *Plant Disease* 74:757–761.

Krauss, J., Härri, S. A., Bush, L., Husi, R., Bigler, L., Power, S. A., and C. B. Müller. 2007. Effects of fertilizer, fungal endophytes and plant cultivar on the performance of insect herbivores and their natural enemies. *Functional Ecology* 21:107–116.

Kuldau, G. and C. Bacon. 2008. Clavicipitaceous endophytes: Their ability to enhance resistance of grasses to multiple stresses. *Biological Control* 46:57–71.

Lacey, L. A. and L. G. Neven. 2006. The potential of the fungus, *Muscodor albus*, as microbial control agent of potato tuber moth (Lepidoptera: Gelechiidae) in stored potatoes. *Journal of Invertebrate Pathology* 91:195–198.

Lodge, D. J., P. J. Fisher, and B. C. Sutton. 1996. Endophytic fungi of *Manilkara bidentata* leaves in Puerto Rico. *Mycologia* 88:733–738.

Márquez, L. M., Redman, R. S., Rodriguez, R. J., and M. J. Roosink. 2007. A virus in a fungus in a plant: Three-way symbiosis required for thermal tolerance. *Science* 315:513–515.

McNally, D. J., Wurms, K. V., Labbe, C., and R. R. Belanger. 2003. Synthesis of C-glycosil flavonoid phytoalexins as site-specific response to fungal penetration in cucumber. *Physiological and molecular Plant Pathology* 63:293–303.

Mejía, L. C., Castlebury, L. A., Rossman, A. Y., Sogonov, M. V., and J. F. White Jr. 2008a. Phylogenetic placement and taxonomic review of the genus *Cryptosporella* and its synonyms *Ophiovalsa* and *Winterella*. *Mycological Research* 112:23–35.

Mejía, L. C., Rojas, E. I., Maynard, Z., Van Bael, S., Arnold, A. E., Hebbar, P. H., Samuels, G. J., Robbin, N., and E. A. Herre. 2008b. Endophytic fungi as biocontrol agents of *Theobroma cacao* pathogens. *Biological Control* 46:4–14.

Miller, J. D., Mackenzie, S., Foto, M., Adams, G. W., and J. A Findlay. 2002. Needles of white spruce inoculated with rugulosin producing endophytes contain regulosin reducing spruce budworm growth rate. *Mycological Research* 106:471–479.

Miller, J. D., Sumarah, M. W., and G. W. Adams. 2008. Effect of Rugulosin-producing endophyte in *Picea glauca* on *Choristoneura fumiferana*. *Journal of Chemical Ecology* 34:362–368.

Narisawa, K., Ohki, K. T., and T. Hashiba. 2000. Supression of clubroot and *Verticillium* yellows in Chinese cabbage in the field by the root endophytic fungus, *Heteroconium chaetospira*. *Plant Pathology* 49:141–146.

Narisawa, K., Kawamata, H., Currah, R. S., and T. Hashiba. 2002. Suppression of *Verticillium* wilt in eggplant by some fungal root endophytes. *European Journal of Plant Pathology* 108:103–109.

Nicholson, R. L. and R. Hammerschmidt. 1992. Phenolic compounds and their role in disease resistance. *Annual Review of Phytopathology* 30:369–389.

Peláez, F., Collado, J., Arenal, F., Basilio, A. Cabello, A., Díez Matas, M. T., García, J. B., Gonzáles Del Val, A., Gonzáles, V., Gorrochategui, J., Hernández, P., Martín, I., Platas, G., and F. Vicente. 1998. Endophytic fungi from plants living on gypsum soils as a source of secondary metabolites with antimicrobial activity. *Mycological Research* 102:755–761.

Pereira, J. A., Oliveira, I., Souza, A., Valentão, P., Andrade, P. B., Ferreira, I. C. F. R., Ferreres, F., Bento, A., Seabra, R., and L. Estevinho. 2007. Walnut (*Juglans regia* L.) leaves: Phenolic compounds, antibacterial activity and antioxidant potential of different cultivars. *Food and Chemical Toxicology* 45:2287–2295.

Petrini, O. 1991. Fungal endophytes of tree leaves. In *Microbial Ecology of Leaves*, Eds. J. H. Andrews and S. S. Hirano, pp. 179–197. Springer Verlag: New York.

Petrini, O., Sieber, T. N., Toti, L., and O. Viret. 1992. Ecology, metabolite production and substrate utilization in endophytic fungi. *Natural Toxins* 1:185–196.

Pocasangre, L., Sikora, R. A., Vilich, V., and R. P. Schuster. 2001. Survey of banana endophytic fungi from Central America and screening for biological control of burrowing nematode (*Rhadopolus similis*). *Acta Horticulturae* 531:283–290.

Posada, F. and F. E. Vega. 2006. Inoculation and colonization of coffee seedlings (*Coffea arabica* L.) with the fungal entomopathogen *Beauveria bassiana* (Ascomycota: Hypocreales). *Mycoscience* 47:284–289.

Posada, F., Aime, M. C., Peterson, S., Rehner, S. A., and F. E. Vega. 2007. Inoculation of coffee plants with the fungal entomopathogen *Beauveria bassiana* (Ascomycota: Hypocreales). *Mycological Research* 111:748–757.

Promputtha, I., Lumyong, S., Dhanasekaran, V., McKenzie, E. H. C., Hyde, K. D., and R. Jeewon. 2007. A phylogenetic evaluation of whether endophytes become saprotrophs at host senescence. *Microbial Ecology* 53:579–590.

Redman, R. S., Sheehan, K. B., Stout, R. G., Rodriguez, R. J., and J. H. Henson. 2002. Thermotolerance generated by plant/fungus simbiosis. Science 298:1581.

Rodriguez, R. J., Redman, R. S., and J. M. Henson. 2004. The role of fungal symbioses in the adaptation of plants to high stress environments. *Mitigation and Adaptation Strategies for Global Change* 9:261–272.

Rodriguez, R. J., Redman, R. S., and J. M. Henson. 2005. Symbiotic lifestyle expression by fungal endophytes and the adaptation of plants to stress: Unraveling the complexities of intimacy. In *The Fungal Community, Its Organization and Role in the Ecosystem*, Eds. J. Dighton, J. F. White, Jr., and P. Oudemans, pp. 683–695. Taylor & Francis: Boca Raton, FL.

Rojas, E. I., Herre, E. A., Mejía, L. C., Arnold, E. A., Chaverri, P., and G. J. Samuels. 2008. *Endomelanconium endophyticum*, a new *Botryosphaeria* leaf endophyte from Panama. *Mycologia* 100(5):760–775.

Rowan, D. D. and G. C. M. Latch. 1994. Utilization of endophyte-infected perennial ryegrasses for increased insect resistance. In *Biotechnology of Endophytic Fungi of Grasses*, Eds. C. W. Bacon and J. F. White Jr., pp. 169–183. CRC Press: Boca Raton, FL.

Rubini, M. R., Silva-Ribeiro, R. T., Pomella, A. W. V., Maki, C. S., Araújo, W. L., dos Santos, D. R., and J. L. Azevedo. 2005. Diversity of endophytic fungal community of cacao (*Theobroma cacao* L.) and biological control of *Crinipellis perniciosa*, causal agent of Witches' Broom Disease. *International Journal of Biological Sciences* 1:24–33.

Saikkonen, K., Helander, M., Ranta, H., Neuvonen, S., Virtanen, T., Suomela, J., and P. Vuorinen. 1996. Endophyte-mediated interactions between woody plants and insect herbivores. *Entomologia Experimentalis et Applicata* 80:269–271.

Saikkonen, K., Faeth, S. H., Helander, M., and T. J. Sullivan. 1998. Fungal endophytes: A continuum of interactions with host plants. *Annual Review of Ecology and Systematics* 29:319–343.

Saikkonen, K., P. Lehtonen, M. Helander, J. Koricheva, and S. H. Faeth. 2006. Model systems in ecology: Dissecting the endophyte-grass literature. *Trends in Plant Science* 11(9):428–433.

Samuels, G. J., Suarez, C., Solis, K., Holmes, K. A., Thomas, S., Ismaiel, A., and H. C. Evans. 2006. *Trichoderma theobromicola* and *T. paucisporum*: Two new species isolated from cacao in South America. *Mycological Research* 110:381–392.

Sánchez Márquez, S., Bills, G. F., and I. Zabalgogeazcoa. 2007. The endophytic mycobiota of the grass *Dactylis glomerata. Fungal Diversity* 27:171–195.

Santamaría, J. and P. Bayman. 2005. Fungal endophytes and epiphytes of coffee leaves (*Coffea arabica*). *Microbial Ecology* 50:1–8.

Santos Rodriguez Costa Pinto, L., Azevedo, J. L., Pereira, J. O., Carneiro Vieira, M. L., and C. A. Labate. 2000. Symptomless infection of banana and maize by endophytic fungi impairs photosynthetic efficiency. *New Phytologist* 147:609–615.

Saunders, M. and L. M. Kohn. 2008. Host-synthesized secondary compounds influence the in vitro interactions between fungal endophytes of maize. *Applied and Environmental Microbiology* 74(1):136–142.

Schardl, C. L., Leuchtmann, A., Chung, K.-R., Penny, D., and M. R. Siegel. 1997. Coevolution by common descent of fungal symbionts (*Epichloë* spp.) and grass hosts. *Molecular Biology and Evolution* 14:133–143.

Schulz, B. and C. Boyle. 2005. The endophytic continuum. *Mycological Research* 109(6):661–686.

Schulz, B., Sucker, J., Aust, H. J., Krohn, K., Ludewig, K., Jones, P. G., and D. Doring. 1995. Biologically active secondary metabolites of endophytic *Pezicula* species. *Mycological Research* 99:1007–1015.

Schwarz, M., Köpcke, B., Weber, R. W. S., Sterner, O., and H. Anke. 2004. 3-Hydroxypropionic acid as a nematicidal principle in endophytic fungi. *Phytochemistry* 65:2239–2245.

Shelby, R. A. and L.W. Dalrymple. 1993. Long-term changes of endophyte infection in tall fescue stands. *Grass and Forage Science* 48:356–361.

Sieber, T. N. 2007. Endophytic fungi in forest trees: Are they mutualists? *Fungal Biology Reviews* 21:75–89.

Sieber, T. N., Riesen, T. K., Müller, E., and P. M. Fried. 1988. Endophytic fungi in four winter wheat cultivars (*Triticum aestivum L.*) differing in resistance against Stagonospora nodorum (Berk.) Cast. & Germ.=Septoria nodorum (Berk.) Berk, *Journal of Phytopathology* 122:289–306.

Sogonov, M. V., Castlebury, L., Rossman, A. Y., Mejía, L. C., and J. F. White Jr. Leaf –inhabiting Gnomoniaceae. *In review, Studies in Mycology* 62:1–79.

Stewart, T. M., Mercer, C. F., and J. L. Grant. 1993. Development of *Meloidogyne naasi* on endophyte-infected and endophyte-free perennial ryegrass. *Australassian Plant Pathology* 22:40–41.

Stone, J. K. 1986. Initiation and development of latent infections by *Rhabdocline parkeri* on Douglas-fir. *Canadian Journal of Botany* 65:2614–2621.

Stone, J. K., Bacon, C. W., and J. F. White Jr. 2000. An overview of endophytic microbes: Endophytism defined. In *Microbial Endophytes*, Eds. C. W. Bacon and J. F. White Jr., pp. 3–29. Marcel Dekker, Inc.: New York.

Stone, J. K., Polishook, J. D., and J. F. White Jr. 2004. Endophytic Fungi. In *Biodiversity of Fungi, Inventory and Monitoring Methods*, Eds. G. M. Mueller, G. F. Bills, and M. S. Foster, pp. 241–270. Elsevier Academic Press: Burlington, MA.

Strobel, G. A., Dirkse, E., Sears, J., and C. Markworth. 2001. Volatile antimicrobials from *Muscodor albus*, a novel endophytic fungus. *Microbiology* 147:2943–2950.

Sutton, J. C., Liu, W., Ma, J., Brown, W. G., Stewart, J. F., and G. D. Walker. 2008. Evaluation of the fungal endophyte *Clonostachys rosea* as an inoculant to enhance growth, fitness and productivity of crop plants. *Acta Horticulturae (ISHS)* 782:279–286.

Sumarah, M. W., Adams, G. W., Berghout, J., Slack, G. J., Wilson, A. M., and J. D. Miller. 2008. Spread and persistence of a rugulosin-producing endophyte in *Picea glauca* seedlings. *Mycological Research* 112:731–736.

Tadych, M., Bergen, M., Dugan, F. M., and J. F. White. 2007. Evaluation of the potential role of water in spread of conidia of the *Neotyphodium* endophyte of *Poa ampla. Mycological Research* 111:4666–4672.

Tan, R. X. and W. X. Zou. 2001. Endophytes: A rich source of functional metabolites. *Natural Product Reports* 18:448–459.

Thomas, S. E., Crozier, J., Aime, M. C., Evans, H. C., and K. A. Holmes. 2008. Molecular characterization of fungal endophytic morphospecies associated with the indigenous forest tree, *Theobroma gileri*, in Ecuador. *Mycological Research* 112:852–860.

Tondje, P. R., Hebbar, K. P., Samuels, G., Bowers, J. H., Weise, S., Nyemb, E., Begoude, D., Foko, J., and D. Fontem. 2006. Bioassay of *Genicolosporium* species for *Phytophthora megakarya* biological control on cacao pod husk pieces. *African Journal of Biotechnology* 5(8):648–652.

Tunali, B. and D. Marshall. 2000. Antagonistic effects of endophytes against several root-rot pathogens of wheat. In *Durum Wheat Improvement in the Mediterranean Region: New Challenges*, Eds. C. Rojo, M. M. Nachit, N. Di Fonzo, and J. L. Arauz, pp. 381–386. CIHEAM-IAMZ: Zaragoza, Spain.

Van Bael, S. A., Maynard, Z., Robbins, N., Bischoff, J., Arnold, A. E., Rojas, E., Mejia, L. C., Kyllo, D. A., and E. A. Herre. 2005. Emerging perspectives on the ecological roles of endophytic fungi in tropical plants. In *The Fungal Community: Its Organization and Role in the Ecosystem*, Eds. J. Dighton, P. Oudemans, and J. F. White Jr., pp. 181–193. 3rd edition, CRC Press: Boca Raton, FL.

Vega, F., Posada, F., Aime, M. C., Pava-Ripoll, M., Infante, F., and S. A. Rehner. 2008. Entomopathogenic fungal endophytes. *Biological Control* 46:72–82.

Waller, F., Achatz, B., Baltruschat, H., Fodor, J., Becker, K., Fischer, M., Heier, T., Hückelhoven, R., Neumann, C., von Wettstein, D., Franken, P., and K. H. Kogel. 2005. The endophytic fungus *Piriformospora indica* reprograms barley to salt-stress tolerance, disease resistance, and higher yield. *Proceedings of the National Academy of Sciences of the USA* 102(38):13386–13391.

Wäli, P. R., Ahlholm, J. U., Helander, M., and K. Saikkonen.2007. Occurrence and genetic structure of the systemic grass endophyte *Epichloe festucaë* in fine fescue populations. *Microbial Ecology* 53:20–29.

Wang, F. W., Jiao, R. H., Cheng, A. B., Tan, S. H., and Y. C. Song. 2007. Antimicrobial potentials of endophytic fungi residing in *Quercus variabilis* and brefeldin A obtained from *Cladosporium* sp. *World Journal of Microbiology and Biotechnology* 23:79–83.

West, C. P., Izekor, E., Oosterhuis, D. M., and R. T. Robbins. 1988. The effect of *Acremonium coenophialum* on the growth and nematode infestation of tall fescue. *Plant Soil* 112:3–6.

White, J. F., Jr. and G. T. Cole. 1985. Endophyte–host associations in forage grasses III. *In vitro* inhibition of fungi by *Acremonium coenophialum. Mycologia* 77(3):487–489.

White, J. F., Jr., Reddy, P. V., and C. W. Bacon. 2000. Biotrophic endophytes of grasses: A systemic appraisal. In *Microbial Endophytes*, Eds. C. W. Bacon and J. F. White Jr., pp. 49–62. Marcel Dekker, Inc.: New York.

Wicklow, D. T., Roth, S., Deyrup, S. T., and J. B. Gloer. 2005. A protective endophyte of maize: *Acremonium zea* antibiotics inhibitory to *Aspergillus flavus* and *Fusarium verticillioides. Mycological Research* 109(5):610–618.

Wilson, D. 1995. Endophyte—The evolution of a term and clarification of its use and definition. *Oikos* 73:274–276.

Wilson, D. 1996. Manipulation of infection levels of horizontally transmitted fungal endophytes in the field. *Mycological Research* 100:827–830.

Wilson, D. and G. C. Carroll. 1994. Infection studies of *Discula quercina*, an endophyte of *Quercus garryana. Mycologia* 86:635–647.

Wilson, D. and S. H. Faeth. 2001. Do fungal endophytes result in selection for leafminer ovipositional preference? *Ecology* 82(4):1097–1111.

Wilson, D., Barr, M. E., and S. H. Faeth. 1997. Ecology and description of a new species of *Ophiognomonia* endophytic in the leaves of *Quercus emoryi. Mycologia* 89:537–546.

Wille, P., Boller, T., and O. Kaltz. 2002. Mixed inoculation alters infection success of strains of the endophyte *Epichloë bromicola* on its grass host *Bromus erectus. Proceedings of the Royal Society of London B.* 269:397–402.

Xu, H.-X. and S. F. Lee. 2001. Activity of plant flavonoids against antibiotic-resistant bacteria. *Phytoteraphy Research* 15:39–43.

Yang, X., Strobel, G., Stierle, A., Hess, W. M., Lee, J., and J. Clardy. 1994. A fungal endophyte-tree relationship: *Phoma* sp. in *Taxus wallachiana. Plant Sciences* 102:1–9.

Zhang, H. W., Song, Y. C., and R. X. Tan. 2006. Biology and chemistry of endophytes. *Natural Product Report* 23:753–771.

Zum Felde, A., Pocasangre, L. E., Carñizares Monteros, C. A., Sikora, R. A., Rosales, F. E., and A. S. Riveros. 2006. Effect of combined inoculations of endophytic fungi on the biocontrol of *Radopholus similis. Info Musa* 15:12–18.

23

Endophytic Niche and Grass Defense

Charles W. Bacon, Dorothy M. Hinton, and Anthony E. Glenn

CONTENTS

23.1 Introduction

Recent advances have extended our understanding of the basic biology, genetics, and molecular biology of many fungal endophytes. Molecular biology in particular has made it possible for identification and phylogeny of the clavicipitalean fungi (Schardl and Siegel, 1992; Glenn et al., 1996; Schardl et al., 1997; Wilkinson and Schardl, 1997; Leuchtmann, 1999; Moon et al., 1999; Clay and Schardl, 2002; Schardl and Schardl and Moon, 2003; Haarmann et al., 2005; Spiering et al., 2005). Additionally, recent advances have extended our knowledge of host–fungal interactions that result in both quantitative and qualitative measures of mutualistic responses to most clavicipitalean-infected grasses and sedges (Clay, 1988; Wilkinson and Schardl, 1997; Schardl, 2001; Clay and Schardl, 2002; Spiering et al., 2005). Fungal endophytes do have common sites of colonization within grass plants, although the distribution of the endophytic habit may not be entirely distributed intercellularly throughout the plant axis. For example, some endophytes form reproductive structures on the surface of the plant, i.e., epibiotic, although part of the thallus is either subcuticular or endophytic, and their perennial nature is due to being localized to the plant's apical meristematic tissues. These epibiotic endophytes also occupy areas that are components of the apoplasm, indicating a need to better define the endophytic niche beyond the boundaries of the "intercellular spaces" of plants as it is commonly defined by others (Yoshihara et al., 1988; Prat et al., 1997; Jarvis, 1998). The apoplast is the plant's internal milieu from which internal homeostasis is maintained (Sakurai, 1998) and through which environmental stimuli are received including growth responses to auxins and other functions. This underlines the importance of apoplasm as the habitat for endophytic fungi. In this review we will use the term "apoplast" as a synonym for "intercellular spaces." However, there are distinctions and the pertinent ones are mentioned in the discussions that follow.

Intercellular spaces are very prominent components of plant tissues and apparently are essential to the survival of plants since they appear as well-developed structures in tissue of plants found living in the lower Devonian period, approximately 400 million years ago (Edwards et al., 1998). Further, all major groups of plants have intercellular spaces, and most of these are similar in structure and perhaps function. Function is not well understood, but we will utilize the interpretations of current plant physiologists to underline their importance to plants. We refer readers to the excellent reviews of Jarvis et al. (2003) and Sattelmacher (2001) dealing with all the aspects of the intercellular space which are only briefly described here for the discussion relative to endophytes.

The future technological use and modification of endophyte-infected plants or the endopytes themselves depend not only on our understanding of the genetics of the fungus, but also on our understanding of the intercellular spaces, the apoplasm and its contents, and the interactions between fungus and plant that occur within the apoplast. We do not imply that these fungi are found only in the intercellular spaces since considerable amount of the hyphae of these fungi is also found within cell walls. We refer to both the intercellular spaces and the wall spaces where hyphae may be embedded as the "intercellular matrix." As will be explained below, our expanded definition of the apoplasm does however include the cell walls and cell wall contact sites. Very little research has been conducted on the nature of the intercellular habitat within which the symbiosis is accomplished and the mutualistic responses are expressed. The doubtless dynamic interactions between endophyte and plant host represent unexplored areas for students of endophytes and their ecologies. Because this area of study represents a new frontier for biologists, we will present an analysis of the endophytic niche based on the current viewpoints and research in this area.

Most important is the fact that clavicipitalean fungi are biotrophic, and there are no penetrations of hyphae, i.e., haustoria, into the cells of the mesophyll, phloem, or xylem. The finding of a very limited host defense reactions relative to pathogenesis-related proteins in endophyte-infected grasses as opposed to high levels in defective symbiota suggests salient differences in the endophyte associations with their grass hosts as opposed to classical pathogens. Our early studies on nutrient flow to the endophyte theorized that this was accomplished by modification of host cell membrane, e.g., directed flow. However, current research on the nature of the intercellular spaces of plants and the continuity of the apoplast with the symplast indicate that this need not be the case.

We will review the intercellular fluid-filled matrix and its contents, including the apoplast and what is structurally referred to as the intercellular space. We will restrict our discussion to the clavicipitalean fungi as inhabitants of this structure and the interaction of nutrients supplied by the apoplast in relation to the accumulation of defensive compounds produced by this group of biotrophic parasites. We will also review the development and morphological nature of the intercellular spaces in plants with as much emphasis on grasses as possible, and also review the chemical makeup of nutrients contained within, the physiological nature of nutrient flow and its transfer to the fungus, and briefly discuss some benefits derived from the association. The fungal endophyte within the intercellular spaces is physically separated from the host but there must be communication between them. Thus, we will also briefly review recent information that might suggest that signaling pathways have relevance to the physiologies and metabolisms of both the grass and the fungus relative to the final response of the symbiota to biotic and abiotic stresses and crosstalk that might be operative, resulting in the mutualistic responses characteristic of these associations.

23.2 Origin of the Endophytic Habit and the Apoplastic–Symplastic Connection

The initial observations of fungi dwelling within intercellular spaces implied that these spaces were created primarily by activities of the fungus. As we shall see, these spaces are for the most part natural cavities, which may be altered due to the activities of the fungi or hosts. As early as 1937 the intercellular spaces of plants were defined and their origins studied and grouped into three broad types to include intercellular splits, triangular gaps, and quadrangular lacuna (Martens, 1937). Intercellular spaces of most plants consist of a series of nonliving connecting components referred to as the apoplasm that is distinct from the cellular to cellular contacts, which are referred to as the symplasm. Thus, the apoplast is broadly defined as being comprised of the compartments beyond the plasmalemma that consists of the interfibrillar and intermicellar spaces of the cell walls, the xylem, and its gas- and water-filled intercellular space (Sattelmacher, 2001). We find that clavicipitalean endophytes colonize most of these locations, with the exception of the xylem. Indeed, microbial endophytes that colonize the xylem are not as harmless to the association since endophytically colonized xylem become quickly functionless. The apoplast extends the entire length of the plant axis, from the cuticle of the leaves, petioles, and stems, downwards to the roots. The results of the studies using current techniques such as microcasting and specific labeling experiments have established that the apoplast plays fundamental roles in several plant processes, including providing the site for atmospheric interaction and atmospheric exchange, intercellular signaling, water and nutrient transport, and as a habitat for diverse microbes. Indeed, the functional components derived from this intercellular system (apoplasm and intercellular network) are so important that some plant physiologists have concluded that it should be considered a third vascular system in plants (Pyke, 1991; Prat et al., 1997).

The intercellular spaces are located in the cortical tissue of the root and in the parenchyma tissue of stems and leaves. They are formed by the juxtapositions of three to four cells followed by the dissolution of the middle lamellae. The intercellular spaces in the leaves and stems are morphologically different from those in the root. The volume occupied by intercellular spaces consists of a significant portion of the plant axis, and in leaves as much as 6% of the leaf tissue consists of intercellular spaces (Altus and Canny, 1985; Tetlow and Farrar, 1993). Much of the intercellular spaces and resulting apoplasm is the result of cells separating along the middle lamella, which naturally suggests that these cells adhere or are glued to each other by the middle lamella. However, according to Knox (1990) this is not as simple as it appears. While it is true that this line is important in distinguishing one cell from another, these cells are never separated by the middle lamella, and therefore there is no need for them to be glued to each other (Knox, 1990). These cells are formed in an adherent state during cell plate formation and remain together for life (Knox, 1990). Further, it is considered that the stability and contact of cell walls to cell walls is maintained in grasses by covalent cross-linking of hydroxycinnamoyl esters with arabinoxylans, which are cell wall polymers specific for grasses (Ng et al., 1997). When cells separate such as during the formation of intercellular spaces, these polymers are locally dismantled along a very precise line, presumably by enzymes specific for this process, as well as the initial process of covalent cross-linking (Roland, 1978; Kolloffel and Linssen, 1984; Jeffree et al., 1986).

Turgor pressure is another contributor to the process of intercellular space formation (Jarvis, 1998). Cytokinins can also reduce intercellular adhesion producing more and larger intercellular spaces (Faure et al., 1998). And endophytic organisms may also contribute to this process directly or indirectly by producing exogenous levels of cytokinins or production of cellulolytic enzymes specific for this process. In general most individual intercellular spaces that are produced by three cells (tricellular junctions) are formed by turgor pressure. Others are formed by constriction in the cell walls resulting in cell walls being pulled away randomly (Apostolakos et al., 1991). The very large intercellular spaces such as that occurring in aerenchyma are formed by one or more cells undergoing lysis, rather than by schizogenous cell separation. Such large intercellular spaces occur in leaf sheaths of tall fescue and are formed 30 days earlier in endophyte-infected plants regardless of the level of nitrogen fertilization (Arechavaleta et al., 1989), which suggests that the effect is due to endophyte-stimulated development, i.e., maturity, of leaves.

A single intercellular space requires coordination between at least three adjacent cells with targeted extracellular metabolism at specific surface locations of each (Jarvis et al., 2003). Further, a network of intercellular spaces would therefore require the coordination between large numbers of cells. Both processes require signaling mechanisms, which are presently unknown. Thus, there are several options through which primary cell walls of plants separate producing a complex multidimensional network of intercellular spaces in which clavicipitalean endophytes reside, interacting with the grass host by providing protective molecules and absorbing nutrients from the host. In summary, the processes describe herein produce schizogenous intercellular spaces that are formed by cell walls of three or four adjacent cells. Additionally, intercellular spaces are formed by the breakdown of specific cell types producing larger lysigenous lacunae that are restricted to various locations along the plant axis.

Thus, as currently described, the plant body consists of two major interconnected but continuous systems. The basic and well-known system consists of plasmalemma bound layer of protoplasmic cells, organized into layers of plant tissue types. The other system consists of the nonmembrane bound micro spaces within cell walls, most intercellular spaces, and the lumina of dead cells. It is this system that is referred to as the apoplast. In simple terms it is the space within the plant body that is not membrane-limited but connected to the symplastic system.

23.3 Morphology of Intercellular Spaces

The major tissue types in plants consist of parenchyma, sclerenchyma, and collenchyma, and intercellular spaces are most abundant in the parenchyma or ground tissue. The intercellular spaces of roots in tall fescue and most temperate grasses are typically schizogenous (Figure 23.1A–I), although in tropical and older stems of temperature grasses there may also be larger lysigenous lacunae resulting in two basic types, empty and fluid-filled (Figure 23.1A and B). The composition of these spaces varies with some consisting only of air or dry flaky deposits (Figure 23.1B), but the majority of spaces consists of fluids,

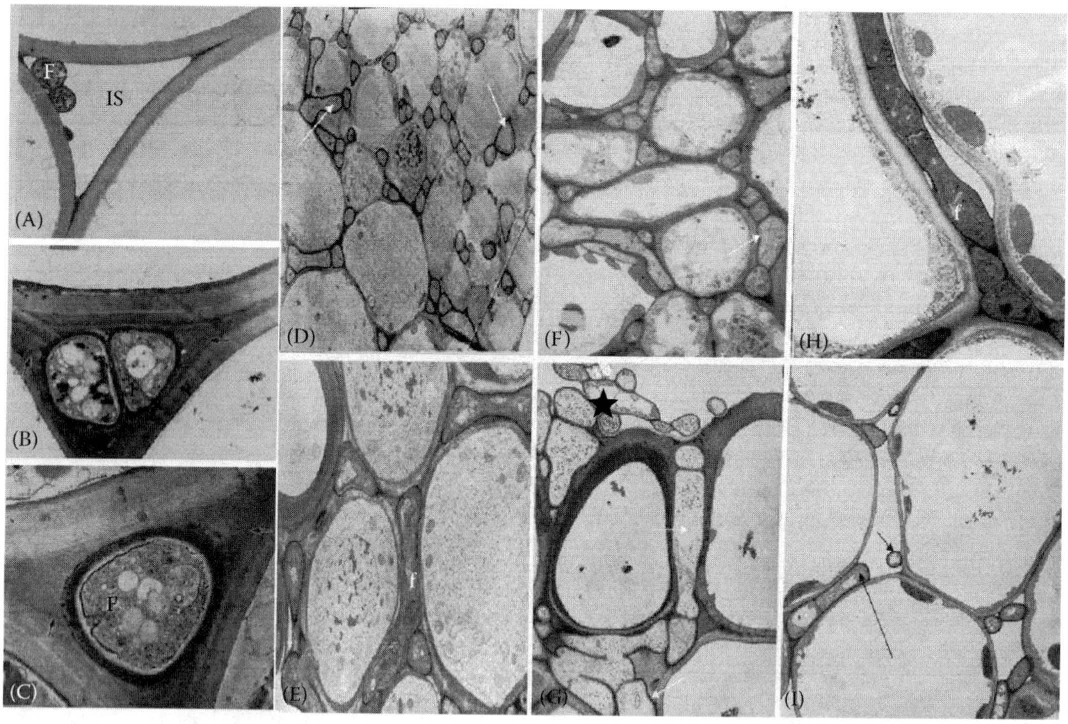

FIGURE 23.1 (A) Section through flower stalk of tall fescue (*Festuca arundinacea*) infected with the endophyte *N. coenophialum* showing a liquid-filled schizogenous intercellular space found in old tissue; F, fungus, IS, intercellular space, ×8,000. (B) Section through a matrix-containing schizogenous intercellular space showing two hyphae made through the leaf sheath of tall fescue, ×28,000. (C) Another section through an intercellular matrix-filled space in tall fescue showing one hypha, ×28,000. (D) Cross section of the ligule of prairie wedgegrass (*Sphenopholis obtusata*) showing several endophytic hyphae of *Epichloë amarillans*, ×8,500. (E) Cross section through leaf sheath of autumn bentgrass showing hyphae, f, among schizogenous cells of enlarged host cells that have been further separated by hyphae, ×16,500. (F) Cross section of the sheath of prairie wedgegrass, showing an extensive amount of host cell separations by hyphae (arrow), producing an extensive network of fungi and host cells, ×16,500. (G) Cross section through conidial stroma showing enlarged cells of host, the prairie wedgegrass, with endophytic fungi completely separating host cells (arrows) emerging through the epidermis to produce stromatic hyphae, *, ×16,500. (H) Longitudinal section through inflorescence stem of *Agrostis hiemalis* infected with *E. amarillans* showing endophytic hyphae between two host cells, ×16,500. (I) Cross section of the pith cells of tall fescue showing the presence of hyphae within large intercellular spaces of stem tissue undergoing senescence which apparently produces empty cells, ×8,500.

which in grasses are shown to contain organic and inorganic compounds (see below), and the origin of most of these are directly derived from the host plant's metabolism or from the soil via roots.

The intercellular spaces of the green portions of grasses also consist of gases and fluids. However, the location of the intercellular space may become altered, producing exaggerated structures. For example, the internodes of grasses consist of larger lysigenous intercellular spaces that have been stretched to produce a large central cavity referred to as rexigenous space (Romberge et al., 1993), similar to Figure 23.1A. As described above, the term apoplasm refers to the spaces that are aqueous-filled, indicating a requirement for habitation by microorganisms, but rexigenous cavities in tall fescue are dry and gas-filled and void of endophytic hyphae indicating that they are not rexigenous (see Figure 23.1A). However, the surrounding tissues of this cavity consist of intercellular spaces that contain the highest density of endophytic hyphae than any other location along the plant axis. Rexigenous spaces also occur in the primary surface layer of roots and are discussed here to distinguish these from endophyte-inhabiting spaces. These spaces are also found in the leaf sheaths of tall fescue (Arechavaleta et al., 1989), a location where endophytic hyphae are found in high density but within normal intercellular spaces, i.e., schizogenous.

The intercellular spaces in leaves are predominantly located in the extrafascicular plant parenchymatous tissues and consist of oxygen and air-saturated water and fluid. The movement of fluids through the apoplastic route occurs, but the connection to the symplasm, if any, is complex and uncertain. Nevertheless, since the apoplasm is envisioned as one large interconnected system (Prat et al., 1997), and while some fungal endophytes do not live within the blades of leaves, i.e., the species of *Neotyphodium*, this parenchymatous area is very rich in photosynthates, which are transported throughout the plant axis within the apoplasm in route to the symplasm (phloem loading), and as a result supplies nutrients to inhabitants of the intercellular spaces along the axis, i.e., leaf sheaths, stems, and roots. Another area not colonized by clavicipitalean endophytes is the roots, and apparently this absence is for all clavicipitalean endophytes, although roots are rich in intercellular spaces. This suggests that there are certain physiological requirements for the fungus, e.g., high oxygen concentrations, although there may also be physiological or structural limitations as well.

Symptomless expressions of endophyte-infected grasses while restricted to some clavicipitalean fungi do occur and are not necessarily characterized as disease but rather the production of reproductive structures of the fungus. Further, the production of these structures is not an annual event. In some instances of infection, primarily those involving the *Balansia*-infected warm season grasses, the seed production is considerably reduced or entirely absent. Several ultrastructural studies (Rykard, 1983; Hinton and Bacon, 1985; Rykard et al., 1985; White and Owens, 1992; White et al., 1997) reveal that host tissues associated with stromatic development in the species of *Epichloë*, *Balansia*, and *Myriogenospora* are altered. Most of the changes observed are in alterations in the epidermal layers of leaves and inflorescences immediately below the stromata. Host cell walls proximal to stromata are also altered and appeared thicker (Figure 23.1G). These were the only alterations observed in the host cell morphology, although Archevaleta et al. (1989) reported the early developmental changes in tissues of leaf blades of *Neotyphodium*-infected tall fescue, producing large spaces, resulting possibly from the dissolution and cell separations of large areas. These areas are either a modified aerenchyma-type resulting from schizogenous cell separations or rexigenous spaces. The associations of endophytes within these spaces have not been observed. These observations strongly suggest that fungal endophytes primarily occupy natural intercellular spaces, although we have evidence that they also cause cell separations (Figure 23.1F and H), and this is done primarily during the production of stromata for those that produce these structures on either the abaxial or adaxial leaf surface (Figure 23.1G).

23.4 Nutrients within Intercellular Spaces

23.4.1 Nutrients and Concentrations

Mauseth and Fujii (1994) used microcasting techniques coupled with microscopy to present the first visualization of the interconnected nature of the intercellular spaces. Current research indicates that nutrient transport within grass tissues is considered to occur primarily through an apoplastic route via the cell wall continuum (Madore and Webb, 1981; Dong et al., 1994; Canny, 1995). The water-filled interconnecting intercellular spaces provide pathways of low hydraulic resistance for apoplastic flow of materials from the xylem (Van der Weele, 1996), resulting in nutrients that are derived from roots via the soil and from leaves via photosynthesis. Thus, the concentrations of solutes in both the apoplasm and symplasm are interactive with the phloem and xylem, dispelling an earlier notion that the apoplasm is relatively free of nutrients. However, there is a large variation in inorganic and organic nutrients within the apoplasm indicating the mixed origin of nutrients from tissues of the roots and leaves (Table 23.1). The apoplastic concentrations of sugar are dependent on the dynamics and nature of phloem loading, which is not only plant species related but in grasses it varies within species. Further, the total concentration of nutrients available is expected to be altered dramatically by the presence of endophytic organisms.

The concentration of nutrients contained within the apoplasm might not be regulated by active uptake mechanisms commonly observed in membranes of cells (Madore and Webb, 1981; Canny and Huang, 1993; Tetlow and Farrar, 1993; Canny, 1995). However, an analysis of the contents of the apoplasm indicates that it consists of sucrose, glucose, and fructose in ratios close to those characteristic of surrounding cells, indicating a direct exchange with specific sugars compartmentalized within membrane-bound cells and equilibrations of these with nonmembrane-bound cells. Further, there is an immediate replacement of sugars into the apoplasm when sugars are removed from the intercellular spaces, and they are replaced within 60 min (Kursanov and Brovchenko, 1970). The composition of the apoplast is variable and within a species it depends on the photosynthetic activity at a particular time of a day, charge of specific ions, age of the plant, soil nutritional status, soil type, and nutrient cycling (Sattelmacher, 2001).

Detailed analysis of the apoplasm and symplasm of sugar cane indicates similarity to other compounds isolated from other grasses and plants (Tejera et al., 2006), suggesting a commonality of apoplastic nutrients in plants, at least in grasses. However, most of the nutrients occur at a twofold higher concentration in the symplasm than the apoplasm. Further, the dynamic nature of materials within the apoplast

TABLE 23.1

Nutrients and Growth Promoters Reported in the Apoplasm of Leaves and Roots of Plants[a]

Sugars/Sugar Alcohols	Organic Acids	Amino Acids	Inorganic/Organic Compounds
Galactinol	γ-Aminobutyric acid	Arginine	Ammonium
Galactose	Malic	Aspartic acid	Potassium
Fructose	Oxalic acid	Asparagine	Calcium
Inositol	Tartaric acid	Alanine	Sulphur
Raffinose	Fumaric	Cystine	Phosphorus
Stachyose	Citric acid	Glycine	Chloride
Sucrose		Histidine	Nitrite
Verbascose		Isoleucine	Nitrate
		Leucine	Soluble protein
		Lysine	IAA
		Methionine	ABA
		Serine	Cytokinins
		Tyrosine	
		Threonine	
		Cysteine	
		Glutamic acid	
		Proline	
		Phenylalanine	
		Valine	

Source: Modified from Kuldau, G. and Bacon, C.W., *Biol. Control*, 2008 (in press).

[a] Canny and Huang (1993); Canny and McCully (1988); Kursanov and Brovchenko (1970); Madore and Webb (1981); Tetlow and Farrar (1993); Tejera et al. (2006); and Hartung et al. (1992).

of grasses is suggested from studies of pathogenic rust fungi examined early during pathogenesis, i.e., its endophytic stage. For example, in the brown rust of barley the concentration of nutrients within the apoplasm is dynamic as it is cycled between the host and the fungus. An analysis of nutrients in the intercellular space indicates that it is rich in substances necessary to support the growth of fungal endophytes, which is reinforced by the large number of apparently healthy hyphae observed within the intercellular spaces (Figure 23.1D). Further, the concentrations of apoplastic nutrients reported to occur in fungal endophyte-infected apoplasm is higher than nonendophytic apoplasm, but in both the instances the concentrations of each are sufficient to support fungal growth (Huber and Moreland, 1980; Kneale and Farrar, 1985; Farrar and Farrar, 1986) and, a priori, for the synthesis of various secondary metabolites necessary for defense.

23.4.2 Nutrient Uptake and Exchange within the Apoplasm

We have presented evidence to indicate the free flow of nutrients between the apoplasm and symplasm. The basic question is the manner by which nutrients are taken up by endophytic hyphae and reexchanged with the host in other forms or converted to substances useful for the synthesis of endophyte-specific compounds, some of which are useful secondary defensive compounds.

Within the apoplasm signals may be produced by the endophyte that are recognized by the host to provoke transcription of specific endophytic recognition genes, or the converse. There is a direct correlation between phloem loading of low concentrations of sugars and pH, which require a proton gradient across the plasma membrane of the sieve elements and companion elements (Giaquinta, 1976, 1977; Delrot and Bonnemain, 1981). Therefore, altering the pH in the apoplast can affect sugar concentration, which has been documented to occur during the biotrophic phase of the brown rust fungus on barley (Tetlow and Farrar, 1993). It was demonstrated that in host tissue infected by this rust fungi, the pH was increased from 6.6 to 7.3, resulting in a 35% to 40% decrease in total soluble carbohydrate concentration. While this work concentrated on diseased tissues, it nevertheless indicates the dynamic nature of the interaction between an endophytic system and a host resulting in the alteration of sugar concentrations. It is our opinion that this concentration gradient would favor the flow of continued carbohydrates to an endophyte. By altering the pH of the apoplasm, fungal endophytes can alter indirectly the activity of specific enzymes and the kinetics of sugar uptake of host cells, thus increasing the concentration of sugars in the apoplasm.

Based on our model of the endophytic stage of the pathogenic brown rust, there was an increase in the volume of the intercellular spaces in infected hosts (Tetlow and Farrar, 1993). Therefore, the extent of an endophytic microorganism's alteration on the host occurs both at the physiological and morphological levels of expression. Without doubt, the occurrence of protein, ammonium, nitrate, nitrite, several amino acids, organic acids, and sugars in the apoplasm serve to regulate the biological activity of endophytic fungi and contributes to the diversity of secondary metabolites produced by specific genotypes of each endophyte species or strain. Finally and perhaps the major point to remember is the fact that the apoplasm reflects the concentrations of nutrients within the symplasm and the reverse, which suggests that there is no real need for an active process removing nutrients from the host.

23.5 In Planta Synthesis of Defense Metabolite Production

The discovery that *Neotyphodium coenophialum* can produce ergot alkaloids that were associated with cattle and other livestock toxicity (Bacon et al., 1977; Lyons et al., 1986) provided the first evidence that endophytic associations were responsible for antiherbivory in mammals. It was subsequently discovered that this class of secondary metabolites is characteristic of the Clavicipitaceae, which was historically attributed only to species of the genus *Claviceps*. Since this discovery, a primary focus for most studies of endophytic fungi is based on a fundamental principle of defensive mutualisms (Clay, 1988) that has been established as a major benefit afforded by the association. This principle of "defensive mutualism" has since been extended to include several other beneficial effects on the fitness of endophyte-infected grasses to environmental and biological stresses.

Several alkaloids comprise the major classes of secondary metabolites found with symbiotic tall fescue and perennial rye grasses. These include specific ergot alkaloids such as ergovaline (Lyons et al., 1986), the aminopyrrolizidine alkaloids, the pyrolopyrazine alkaloids, the indole diterpenoid alkaloids, and recently the 11,12-epoxy-janthitrems. These chemical classes are found variously in species of grasses infected with endophytes, but not all infected grasses will contain all classes of alkaloids. Indeed, their occurrence in grasses indicates that there are chemical analogs reflective of strains of endophytes. The final quantitative amounts of these compounds are related in part to specific host genotypes (Agee, 1992; Agee and Hill, 1994; Adcock et al., 1997). In surveys, it was reported that the pyrolopyrazine alkaloids, represented by peramine, were present in the majority of endophyte-infected hosts (Siegel et al., 1990; Dahlman et al., 1991). The ergot alkaloids followed next, which were followed by the lolines, four aminopyrrolizidine alkaloids, and lastly by lolitrem, the indole diterpenoid alkaloid.

Most of these alkaloids have been established as being synthesized by the fungus without chemical modifications by the host grass, which was established first in cultures of the fungi (Bacon et al., 1979, 1981; Porter et al., 1979; Bacon, 1985, 1988; Blankenship et al., 2001) and later demonstrated to be absent in grasses freed of their endophyte (Bush et al., 1982; Lyons et al., 1986). The direct synthesis of several other alkaloids, primarily those specific to symbiotic perennial ryegrasses, has not been determined. However, molecular analyses suggest that it is the fungus that produces most if not all the defensive alkaloids in perennial ryegrass. For example, Young et al. (2005) and Christopher and Mantle (1987) reported the molecular cloning of a gene cluster for lolitrem biosynthesis, which is the tremorgenic toxin found in perennial ryegrass that is toxic to livestock.

Several experiments have demonstrated the effects of soil nutrients on the contents and final expression of ergot alkaloids in symbiotic tall fescue (Arechavaleta et al., 1992; Malinowski et al., 1998, 2000; Malinowski and Belesky, 1999a). In addition to the soil nutrients there is a genotype interaction that affects the final expression of most secondary metabolites in symbiotic grasses (Hill et al., 1990, 1991; Agee and Hill, 1994; Roylance et al., 1994).

23.5.1 Invertebrate Pest Toxicities

Lists of animal pests deterred and not deterred by *Balansia*-, *Neotyphodium*-, and *Epichloë*-infected grasses include insect, mites, and nematodes (Clay, 1990; Breen, 1994; Popay and Bonos, 2005; Kuldau and Bacon, 2008). Insect toxicities from endophyte-infected grasses include approximately 45 species belonging to the following families: Aphididae, Chrysomelidae, Cichadllidae, Curculionidae, Gryllidae, Lygaeidae, Noctulidae, Pyralidae, Scarabaeidae, and Tenebrionidae. There is considerable evidence that specific alkaloids, especially the pyrrolopyrazine alkaloid peramine, are responsible for the toxicity observed (Rowan et al., 1986; Ball et al., 1995), although other studies suggest either synergistic or potentiating effects of the ergot alkaloid ergovaline with the pyrrolizidine loline alkaloids (Siegel et al., 1990; Wilkinson et al., 2000). There is the possibility that both explanations are possible since the production of the major insect toxin, peramine, is not found in all *Neotyphodium/Epichloë* species, but nonperamine-producing endophytic species are also toxic to insects. In contrast, feeding of *Spodoptera frugiperda* and *Agrostis ipsilon* caterpillars on endophyte-infected perennial ryegrass provides protection against the parasitic nematode *Steinernema carpocapsae* possibly through synergism involving ergot alkaloids (Kunkel et al., 2004).

Recently there is a suggestion that the janthitrems also contribute to insect toxicity, which is found in some strains of *Neotyphodium* species found in perennial ryegrass (Tapper and Lane, 2004). This class of compounds has a varied effect on insects; it is not specific to an insect order; and the toxicities may affect specific life stages of an insect species. Toxic effects range from mild to acute and evidence of toxicity include feeding deterrence, reduced survival, high mortality, reduced development, lower weight gain, and altered feeding behavior (Tapper and Lane, 2004). These and other specific insect toxins, as well as a list of insects unaffected by *Epichloë/Neotyphodium*-infected grasses are contained in the comprehensive reviews by Clement et al. (1994, 2005), Breen (1994), Lewis and Clements (1986), Popay and Rowan (1994), and Rowan and Latch (1994).

23.5.2 Plant Hormones and Plant Growth

Increased plant growth is another widespread effect observed in endophyte-infected grasses, although not necessarily in response to stresses. However, an increase in growth will prevent a variety of abiotic and biotic stresses, increasing plant vigor or persistence, which are considered here as an essential component to stress resistance. An increase in the rate of growth and herbage yield may be due to physiological response of the grass from an increase in endogenous levels of plant hormones, which may be an additive effect either from the fungal endophyte-produced auxins or from an increase in the contents of nutrients and water in apoplasts. Enhanced plant growth observed in endophyte-infected grasses may be attributed to either or both production of growth hormones and phytohormones, such as IAA, which has been demonstrated to accumulate in vitro in cultures of *N. coenophialum* (Porter et al., 1985; De Battista et al., 1990) and related species (Porter et al., 1977, 1985).

The occurrence of small molecular weight indole compounds such as 3-indoleacetic acid, 3-indole ethanol, and several indole glycerols suggest that they may serve as growth hormones (Porter et al., 1977, 1985; Porter, 1995), although most of these have also been isolated only from endophytes grown in vitro. Further, the interactive nature of plant and endophyte-produced growth hormones must be considered in the final analysis. In addition to effects on the growth of grasses, an overproduction of certain growth hormones are also considered to reduce the frequency of flowering observed on most *Balansia*-infected grasses (Rykard et al., 1985). Contrastingly, in other endophytes demonstrated to produce plant growth substances (De Battista et al., 1990), there is an almost doubling in the yield of seed (Clay, 1987).

It has been suggested that the loline alkaloids, in addition to a role in drought tolerance as discussed in Section 23.6.1, serve as allelochemicals responsible for the phenomenon of allelopathy observed in plants, particularly rosaceous species, grown in soils planted previously with endophyte-infected tall fescue (Petroski et al., 1990). The results of such a phenomenon produce a competitive edge for infected grasses, resulting in an increase in the population density.

23.5.3 Tolerance to Fungal Diseases

Biological control of diseases is perhaps the major effect observed in grasses infected by endophytic fungi. In vitro suppression of plant pathogens by endophytic fungi is demonstrated in vitro (White and Cole, 1985; Siegel and Latch, 1991), which may not be the mechanism expressed under natural conditions, but it is this observation that created interest in the potential use of endophytic fungi for disease control. Field data support the suppression of various diseases of grasses, which correlates with most of the in vitro studies that drive the interests in fungal endophytes for disease control. The association of disease resistance with specific chemical components has not been established in planta, and fungal disease suppression in endophyte-infected grasses is not always clear. There is resistance to *Sclerotinia homeocarpa* (dollar spot) in chewing fescue, hard fescue, and strong creeping red fescue, but there was an increase in disease incidence to *Pythium* blight in *N. coenophialum*-infected tall fescue (Blank, 1992; Clarke et al., 2006). A similar increase in symptoms from *Drechslera* spp. was observed in *N. uncinatum*-infected meadow fescue (Panka et al., 2004). On the other hand, using the identical grass genotypes Panka et al. (2004) also demonstrated that there was a decrease in disease occurrence from *Puccinia coronata* in endophyte-infected plants.

Specific chemicals, such as several indole compounds, a sesquiterpene and a diacetamide, as well as some unidentified volatile compounds (Yue et al., 2000, 2001) have been associated with *Epichloë* species and proposed to enhance resistance to *Cladosporium* leaf spot (*Cladosporium phlei*) and stem rust (*Puccinia graminis*) on endophyte-infected *Phleum pretense* (Yoshihara et al., 1985; Koshino et al., 1988, 1989). Later studies confirmed mitigation of these two diseases by endophytes under field conditions (Welty et al., 1986; Welty and Barker, 1993; Greulich et al., 1999).

A proposed mechanism of disease control by microbial endophytes in plants is induced systemic resistance (Kloepper and Beauchamp, 1992; Clement et al., 1994; Chen et al., 1995), which has only been partially examined in endophyte-infected grasses (Roberts et al., 1992; Kunkel et al., 2004). The reduction in the numbers of *Alternaria*, *Cladosporium*, and *Fusarium* species on leaves of endophyte-infected *Agropyron cristatum*, *Elymus cylindricus*, and *F. rubra* compared to noninfected grasses (Chen et al.,

1995; Raymond et al., 2003) is suggestive of induced systemic resistance. Another idea suggests that epiphyllous mycelial nets observed in some endophyte–grass associations play a key role in the defense against pathogens by niche exclusion (Moy et al., 2000). Induced systemic resistance as a plausible mechanism is strengthened by the demonstration of this mechanism in the control of plant diseases by bacteria (Kloepper and Beauchamp, 1992).

Regardless of the mechanism of action, there are several fungal pathogens that are controlled by endophyte infection: *Alternaria triticina, Cercospora, Cryphonectria parasitica* (in vitro), *Cladosporium phlei, Laetisaria fuciformis, Sclerotinia homeocarpa, Puccinia coronata*, and *Rhizotonia* (Yoshihara et al., 1985; Koshino et al., 1989; Gwinn and Gavin, 1992). Additionally, barley yellow dwarf virus is significantly reduced which is an effect due to the reduction of virus spread by controlling the aphid vector (Mahmood et al., 1993).

23.5.4 Unidentified and Natural Plant Components

The biological spectrum of activity expressed by the endophytes or their four classes of alkaloids, singularly or in combination, varies from antifeeding behavior in vertebrates and invertebrates to drought tolerance by the plant. The basic chemical structures of these alkaloids and their effects on ruminants and other animals have been presented and reviewed (Bush et al., 1979, 1993, 1997; Fletcher and Harvey, 1981; Porter, 1994; Ball et al., 1995; Cross et al., 1995; Jackson et al., 1996; Oliver et al., 1998; Schuenemann et al., 2005). It is essential, however, that we stress the importance for additional searches for defensive roles derived from chemically identified and unknown compounds in symbiotic grasses. Some compounds identified but with no known defensive roles include the indole glycerols, gamma aminobutyric acid, harman, norharman, and halostachine (Bond et al., 1984; Yates et al., 1987; Riedell et al., 1991; Powell and Petroski, 1992; Bush et al., 1993; TePaske et al., 1993).

Finally, there is a need to examine the role of basic plant metabolites that are present in higher levels in endophyte-infected grasses, which may play a role in the defense of vertebrate and invertebrate pests. Symbiotic plants accumulate more phosphorus and increased exudation of phenolic-like compounds in the rhizosphere (Malinowski and Belesky, 2000; Malinowski et al., 2004; Bacon et al., 2005; Bacetty et al., 2007). Similarly, increases in the level of specific plant phenolic compounds in symbiotic grasses are suspected of acting synergistically with ergot alkaloids to deter predation of the nematode *Pratylenchus scribneri* (Bacetty et al., 2007).

23.6 Defense to Abiotic Stresses

23.6.1 Tolerance to Drought and Poor Soil Nutrition

Tolerance to drought and the ability to withstand poor soil nutrition and quality are discussed as defense mechanisms because these relate to persistence and survival. Enhanced drought tolerance has been documented in several studies of endophyte-infected species (Arechavaleta et al., 1992; Lewis and Vaughan, 1997; Malinowski et al., 1998; Lewis, 2004; Malinowski et al., 2005). However, endophyte-infected grasses also respond to poor soil nutrition as evidenced by an increased growth rate (Rice et al., 1990; Arechavaleta et al., 1992; Lewis et al., 1996; Malinowski and Belesky, 1999b; Malinowski, 2000; Lewis, 2004; Malinowski et al., 2005). More recently, specific fungal secondary metabolites have been implicated in growth response and drought-tolerance mechanisms, such as the production of loline alkaloids that affects osmotic potential, which reduces the effects of drought stress (Bush et al., 1997; Hahn et al., 2007). This suggests that the loline alkaloids have a dual role in grass protection, insect deterrence, and drought tolerance. At the cellular level, there is an association of endophyte status with dehydrins, a group of intrinsically unstructured proteins formed abundantly during late embryogenesis (Carson et al., 2004) and is associated with protection from drought and temperature stresses in several grasses, including tall fescue. Finally, endophyte-infected grasses also show an increase in the rate and length of root growth (Richardson et al., 1990), which is certainly expected to play a role in drought tolerance and nutrient acquisition. Perhaps one or all of the above are related to the final phenotypic expression within specific plant genotypes for drought tolerance.

In addition to the above-mentioned endophyte-mediated stress protection that is associated with secondary metabolites, resistances to stresses in endophyte-infected grasses that have no known metabolic association are also known. These include tolerance to low soil pH and low levels of soil phosphorus, favorable growth in soils under high and low mineral stresses, low levels of soil phosphorus, and relief from soil aluminum toxicity (Belesky and Fedders, 1995; Malinowski et al., 2000; Malinowski et al., 2004) as well other abiotic soil stresses (Malinowski and Belesky, 1999a,b; Malinowski et al., 2000; Malinowski et al., 2004). It is our contention that the tolerance to these abiotic soil stresses might be related to chemical modification and as discussed below may well be a remediating property of endophytic fungi.

23.7 Crosstalk Between Endophyte and Plant Intercellular Spaces

23.7.1 Plant-Endophyte Signaling

The pathways leading from the perception of herbivory, disease or abiotic stresses to the response leading to the production of specific classes of secondary metabolites require complex but regulatory interplay between the host and the fungus. This implies the existence of highly specific networks of interacting pathways, if indeed there are responses to biotic and abiotic environmental stimuli in symbiotic grasses. That is, it is tacitly assumed that toxins and others substances attributed to pest and stress defenses were present in the intercellular spaces as happenstances and not initiated or increased by a stimulus. However, specific observations and data have documented that specific classes of compounds within symbiotic grasses do in fact accumulate in response to environmental stimuli (Arechavaleta et al., 1989, 1992; Hill et al., 1990; Malinowski et al., 1998; Malinowski and Belesky, 2000; Rasmussen et al., 2007). Therefore, channels of communications between a host and its endophyte within the intercellular spaces are suggested by this data. The term "crosstalk" is used for this communication and is commonly used to explain complex interactions in defense signaling to produce positive, negative, or neutral results (Bostock, 2005; Mundy et al., 2006). Viewed from this perspective, all pathways relating to biotic and abiotic stimuli may crosstalk at the biochemical level of cell signaling pathways (Bostock, 2005). Crosstalk between signaling pathways can explain how a singular pathway is regulated or how multiple unrelated pathways are regulated to initiate the desired solution to an environmental stimulus. The coordinated production of similar compounds within the symbiotum suggests that signaling pathways might also provide the mechanisms for a consistent, amplified reaction to a large number of environmental stimuli. Examples of signaling pathways and crosstalk are rapidly being analyzed in pathogenic fungi (Leon et al., 2001; Kunkel and Brooks, 2002; Taylor and McAinsh, 2004; Bostock, 2005; Mundy et al., 2006), but these may not be as complex as crosstalk within mutualistic relationships.

The term "crosstalk" generally denotes a common biochemical component of a cell, such as the mitogen-activated protein (MAP). MAP is a key signal-transducing enzyme that is used for a variety of cascading signaling stimulatory responses in cells to the environment. Crosstalk is used by two or more signal transduction pathways that have different cellular outcomes, but however, the term is also being used to express ecological interactions at the organismal level (Taylor et al., 2004). For example, crosstalk could be used when considering how the interaction of a plant with one organism (grass endophyte) influences the plant's defense responses when it is challenged by a different organism (herbivory or pathogenic attacks). Thus the term can be applied to molecular genetic and biochemical analyses of cell signaling pathways and networks as well as ecological analyses of interactions between a plant and its environmental stimuli (biotic and abiotic). Recognition, signaling pathways, and gene expression would be the focus of crosstalk at the cellular and molecular level, while metabolic and physiological changes, effects on attackers, and effects on the host plant are the focus of crosstalk at the ecological and agronomic level (Taylor et al., 2004). The endophytic associations of clavicipitalean fungi with grass hosts are unique biological systems allowing for both cellular and ecological aspects of crosstalk to be thoroughly evaluated together.

Responses to external stimuli are initiated at the cell surfaces (fungal and plant), and resulting biochemical signals are relayed to the target genes, producing a physiological or molecular change due to differential gene expression. In order to maintain a compatible association as observed among clavicipitalean fungi and their hosts, two basic outcomes from crosstalk must include regulated suppression of the grass's natural disease defense responses and control of regulated hyphal growth (Tanaka et al., 2006). The degree to which crosstalk between endophytes and their hosts contribute to compatible symbiotic unions is not known. Below is a brief review of the few examples of signaling within symbiotic grasses with some early and preliminary attempts at understanding potential crosstalk.

23.7.2 Examples of Crosstalk

Growth of endophytic hyphae, as well as hyphae in culture, is nonbranching, although in most instances there is tight and characteristic coiling along the long axis of cells. Coiling is especially evident in cultures of freshly isolated endophytes, i.e., emerging from stubble on agar media. In planta hyphae of most *Neotyphodium* and *Epichloë* infections grow parallel to the long axis of the plant tissue. Similar nonbranching hyphae are also observed with the *Balansia* species. Christensen et al. (2002) concluded that in planta growth is tightly regulated and synchronized with the growth of the host leaf and stem. More recently, Christensen et al. (2008) have proposed a totally new model of fungal growth involving intercalary cell division and expansion for the clavicipitaceous endophytes that challenges the traditional model of growth by hyphal tip extension. This intercalary model helps explain how fungal growth along the length of the hyphal filaments enables the fungus to grow at the same rate as the host without shearing the hyphae.

The intercellular growth of *Epichloë* and *Neotyphodium* endophytes is vigorous yet does not elicit typical host responses, even during the growth of the *Epichloë* species that produce perithecial stromata typical of the choke disease. As shown in Figure 23.1, hyphae vertically enlarge the preformed schizogenous intercellular spaces presumably due to the activities of many fungal extracellular enzymes, which occasionally produce separation of cell walls a considerable distance along the long axis of the plant (Figure 23.1E and H). Contrastingly, when these endophytes are introduced artificially into nonnatural, incompatible hosts, there is a negative reaction, which may take a few months to produce visible symptoms, including a gradual decline of the plant followed by death (Christensen, 1995; Christensen et al., 1997). The basic interpretation of this reaction is highly suggestive of signaling mechanisms between the host and the fungus.

Most of the research concerning molecular components of crosstalk and in planta growth of grass endophytes is being done by B. Scott and his research laboratory (Takemoto et al., 2006; Tanaka et al., 2006, 2007). The foundation of crosstalk signaling is being strengthened by analyses of specific endophytic genes. For example, in planta differential expression of genes to environmental stimuli involved in the production of the lolines, diterpenes (Spiering et al., 2005; Young et al., 2005), peramine (Tanaka et al., 2005), and ergot alkaloids (Panaccione, 2001) suggest that host signaling is required for the expression of some genes (Tanaka et al., 2006). Recently, there has been some research to suggest that while fungal endophytes share in some of the signaling mechanisms reported in other systems, there are apparently some mechanisms that are very specific to the endophytic habit.

Some of the earliest information on signaling and crosstalk developed from studies on reactive oxygen species (ROS), which contribute to the early defense reactions of plants to numerous pathogens. ROS is produced by NADPH oxidase and is now known to regulate hyphal growth of *Epichloë festucae* during colonization of its host *L. perenne* (Tanaka et al., 2006). Infection of plants with a mutant strain of *E. festucae* having a deletion of the NADPH oxidase gene *noxA* resulted in increased fungal biomass in planta, loss of a mutualistic fungus–plant interaction, and eventual death of *L. perenne*. ROS production was essential for the symbiotic regulation of growth and in planta interaction (Tanaka et al., 2006). Further, ROS accumulation was detected cytochemically in the endophyte extracellular matrix and at the interface between the matrix and host cell walls in wild-type infected plants but not in plants infected with the mutant endophyte (Tanaka et al., 2006). Moreover, ROS has been documented in *C. purpurea* during its specialized interactions (highly orientated, nonbranching growth into the host) and suppression of host defense (Tudzynski and Scheffer, 2004; Scheffer et al., 2005), which is strikingly similar to the observations made with grass endophytes. Finally, the production of superoxide dismutase by the ryegrass endophyte *N. lolii* is interpreted as a component of the mechanism needed for maintaining the level of superoxide formed during the association to ensure proper and sufficient colonization of host tissue (Zhang et al., 2007).

While much of the endophyte–host association data is so far restricted only to a few host grasses, we believe that the general principles will apply to all endophytic systems involving the clavicipitalean fungi that have highly regulated temporal and spatial in planta growth. For example, the involvement of ROS in fungal infections may be a general feature for plant pathogenic fungi but is likely a conserved, critical aspect of in planta growth for clavicipitalean fungi. Additional examples of signaling may include increased or decreased accumulation of some apoplastic compounds under drought conditions, suppression or activation of the host defense reaction (Hahn et al., 2007), dehydrin expression in drought tolerance (Guerber et al., 2007), and siderophore production and maintenance of iron for the integrity of the symbiosis (Johnson et al., 2007). Continued research with further examples of signaling and crosstalk should demonstrate the basis for the apparent smooth and intricate interaction between fungal endophytes and grass hosts, which during evolution have merged into an apparent singular "organism" with specific genetic modifications for survival under a diversity of environmental conditions and stimuli. Thus, both cellular and ecological aspects of crosstalk can be assessed together using the grass–endophyte association.

REFERENCES

Adcock, R. A., Hill, N. S., Boerma, H. R., and G. O. Ware. 1997. Sample variation and resource allocation for ergot alkaloid characterization in endophyte-infected tall fescue. *Crop Science* 37:31–35.

Agee, C. S. 1992. Environmental and genotypic effects on ergovaline content in tall fescue. PhD thesis, The University of Georgia.

Agee, C. S. and N. S. Hill. 1994. Ergovaline variability in *Acremonium*-infected tall fescue due to environment and plant genotype. *Crop Science* 34:21–226.

Altus, D. P. and M. J. Canny. 1985. Loading of assimilates in wheat leaves. *Plant Cell and Environment* 8:275–285.

Apostolakos, P., Galatis, B., and E. Panteris. 1991. Microtubules in cell morphogenesis and intercellular space formation in *Zea mays* leaf mesophyll and *Pilea cadierei* epithem. *Journal of Plant Physiology* 148:591–601.

Arechavaleta, M., Bacon, C. W., Hoveland, C. S., and D. E. Radcliffe. 1989. Effect of the tall fescue endophyte on plant response to environmental stress. *Agronomy Journal* 81:83–90.

Arechavaleta, M., Bacon, C. W., Plattner, R. D., Hoveland, C. S., and D. E. Radcliffe. 1992. Accumulation of ergopeptide alkaloids in symbiotic tall fescue grown under deficits of soil water and nitrogen fertilizer. *Applied and Environmental microbiology* 58:857–861.

Bacetty, A. A., Snook, M. E., Glenn, A. E., Bacon, C. W., Nagabhyru, P. N., and C. L. Schardl. 2007. Nematotoxic effects of endophyte-infected tall fescue toxins and extracts in an in vitro bioassay using the nematode *Pratylenchus scribneri*. In *Proceedings of the 6th International Symposium on Fungal Endophytes of Grasses*, eds. A. J. Popay and E. R. Thom, pp. 357–361. Christchurch: New Zealand Grassland Association.

Bacon, C. W. 1985. A chemically defined medium for the growth and synthesis of ergot alkaloids by the species of *Balansia. Mycologia* 77:418–423.

Bacon, C. W. 1988. Procedure for isolating the endophyte from tall fescue and screening isolates for ergot alkaloids. *Applied and Environmental Microbiology* 54:2615–2618.

Bacon, C. W., Bacetty, A. A., Snook, M. E., Glenn, A. E., Noe, J., Hill, N., Bouton, J., and T. Stratton. 2005. Determination of nematode toxins in endophyte-infected novel and native tall fescues. In *Proceedings of Annual Progress Report SERA-IEG 8 Tall Fescue Toxicosis/Endophyte Workshop*, Chapel Hill, TN, October 17–18, 2005.

Bacon, C. W., Porter, J. K., and J. D. Robbins. 1979. Laboratory production of ergot alkaloids by species of *Balansia. Journal General Microbiology* 113:119–126.

Bacon, C. W., Porter, J. K., and J. D. Robbins. 1981. Ergot alkaloid biosynthesis by isolates of *Balansia epichloë* and *B. henningsiana. Canadian Journal Botany* 59:2534–2538.

Bacon, C. W., Porter, J. K., Robbins, J. D., and E. S. Luttrell. 1977. *Epichloë typhina* from toxic tall fescue grasses. *Applied and Environmental Microbiology* 34:576–581.

Ball, O. J. P., Prestidge, R. A., and J. M. Sprosen. 1995. Interrelationships between *Acremonium lolii*, peramine, and lolitrem B in perennial ryegrass. *Applied and Environmental Microbiology* 61:1527–1533.

Belesky, D. P. and J. M. Fedders. 1995. Tall fescue development in response to *Acremonium coenophialum* and soil acidity. *Crop Science* 35:529–533.

Blank, C. A. 1992. Interactions of tall fescue seedlings infected with *Acremonium coenophialum* with soilborne pathogens. MS thesis, University of Tennessee.

Blankenship, J. D., Spiering, M. J., Wilkinson, H. H., Fannin, F. F., Bush, L. P., and C. L. Schardl. 2001. Production of loline alkaloids by the grass endophyte, *Neotyphodium uncinatum*, in defined media. *Phytochemistry* 58:395–401.

Bond, J., Powell, J. B., Undersander, D. J., Moe, P. W., Tyrell, H. F., and R. R. Oltjen. 1984. Forage composition and growth and physiological characteristics of cattle grazing several varieties of tall fescue during summer conditions. *Journal Animal Science* 59:584–593.

Bostock, R. M. 2005. Signal crosstalk and induced resistance: Straddling the line between cost and benefit. *Annual Review of Phytopathology* 43:545–580.

Breen, J. P. 1994. *Acremonium* endophyte interactions with enhanced plant resistance to insects. *Annual Review of Entomology* 39:401–423.

Bush, L. P., Bolling, J., and S. G. Yates. 1979. Animal disorders. In *Tall Fescue*, eds. R. C. Buckner and L. P. Bush, pp. 247–292. Madison, WI: American Society of Agronomy.

Bush, L. P., Cornelius, P. C., Buckner, R. C., Varney, D. R., Chapman, R. A., Burrus, P. B., Kennedy, C. W., Jones, T. A., and M. J. Saunders. 1982. Association of *N*-acetyl loline and *N*-formyl loline with *Epichloë typhina* in tall fescue. *Crop Science* 22:941–943.

Bush, L. P., Fannin, F. F., Siegel, M. R., Dahlman, D. L., and H. R. Burton. 1993. Chemistry, occurrence and biological effects of saturated pyrrolizidine alkaloids associated with endophyte–grass interactions. *Agriculture Ecosystem and Environment* 44:81–102.

Bush, L. P., Wilkinson, H. H., and C. L. Schardl. 1997. Bioprotective alkaloids of grass–fungal endophyte symbioses. *Plant Physiology* 114:1–7.

Canny, M. J. 1995. Apoplastic water and solute movement: new rules for an old space. *Annual Review of Plant Physiology and Molecular Biology* 46:215–236.

Canny, M. J. and C. X. Huang. 1993. What is in the intercellular spaces of roots? Evidence from the cryo-analytical-scanning electron microscope. *Physiologia Plantarum* 87:561–568.

Canny, M. J. and M. E. McCully. 1988. The xylem sap of maize roots: its collection, composition, and formation. *Australian Journal of Plant Physiology* 15:557–566.

Carson, R. D., West, C. P., Reyes, B. D. L., Rajguru, S., and C. A. Guerber. 2004. Endophyte effects on dehydrin protein expression and membrane leakage in tall fescue. In *5th International Symposium on Neotyphodium/Grass Interactions,* May 23–26, eds. R. Kallenbach, C. T. Rosenkrans, and L. Ryan, p. 202. Fayeteville: University of Arkansas.

Chen, C., Bauske, E. M., Musson, G., Rodriguez-Kabana, R., and J. W. Kloepper. 1995. Biological control of *Fusarium* wilt of cotton by use of endophytic bacteria. *Biological Control* 6:83–91.

Christensen, M. J. 1995. Variation in the ability of *Acremonium* endophytes of *Lolium perenne Festuca arundinacea* and *F. pratensis* to form compatible associations in the three grasses. *Mycoogical Research* 99:466–470.

Christensen, M. J., Ball, O. J. P., Bennett, R., and C. L. Schardl. 1997. Fungal and host genotype effects on compatibility and vascular colonisation by *Epichloë festucae. Mycological Research* 101:493–501.

Christensen, M. J., Bennett, R. J., Ansari, H. A., Koga, H., Johnson, R. D., Bryan, G. T., Simpson, W. R., Koolaard, J. P., Nickless, E. M., and C. R. Voisey. 2002. *Epichloë* endophytes grow by intercalary hyphal extension in elongating grass leaves. *Fungal Genetics and Biology* 45:84–93.

Christensen, M. J., Bennett, R. J., and J. Schmid. 2002. Growth of *Epichloë/Neotyphodium* and p-endophytes in leaves of *Lolium* and *Festuca* grasses. *Mycological Research* 106:93–106.

Christopher, W. M. and P. G. Mantle. 1987. Paxilline biosynthesis by *Acremonium loliae*, a step toward defining the origin of lolitrem neurotoxins. *Phytopathology* 26:969–971.

Clarke, B. B., White, J. F., Jr., Hurley, R. H., Torres, M. S., Sun, S., and D. F. Huff. 2006. Endophyte mediated suppression of dollar spot disease in fine fescues. *Plant Disease* 90:994–998.

Clay, K. 1987. Effects of fungal endophytes on the seed and seedling biology of *Lolium perenne* and *Festuca arundinacea. Oecologia* 73:358–362.

Clay, K. 1988. Fungal endophytes of grasses a defensive mutualism between plants and fungi. *Ecology* 69:10–16.

Clay, K. 1990. Insects, endophytic fungi and plants. In *Pests Pathogens and Plant Communities*, eds. J. Burdon and S. R. Leather. Oxford: Blackwell Scientific Publications.

Clay, K. and C. L. Schardl. 2002. Evolutionary origins and ecological consequences of endophyte symbiosis with grasses. *American Naturalist* 160:S99–S127.

Clement, S. L., Elberson, L. R., Bosque-Pérez, N. A., and D. J. Schotzko. 2005. Detrimental and neutral effects of wild barley—*Neotyphodium* fungal endophyte associations on insect survival. *Entomologia Experimentalis et Applicata* 114:119–125.

Clement, S. L., Kaiser, W. J., and H. Eichenseer. 1994. *Acremonium* endophytes in germplasms of major grasses and their utilization for insect resistance. In *Biotechnology of Endophytic Fungi of Grasses*, eds. C. W. Bacon and J. F. White Jr., pp.185–199. Boca Raton: CRC Press.

Cross, D. L., Redmond, L. M., and J. R. Strickland. 1995. Equine fescue toxicosis: Signs and solutions. *Journal of Animal Science* 73:899–908.

Dahlman, D. L., Eichenseer, H., and M. R. Siegel. 1991. Chemical perspectives on endophytes–grass interactions and their implications to insect herbivory. In *Microbial Mediation of Plant Herbivore Interactions*, eds. P. Barbosa, V. A. Krischik, and C. G. Jones. New York: John Wiley & Sons.

De Battista, J. P., Bacon, C. W., Severson, R. F., Plattner, R. D., and J. H. Bouton. 1990. Indole acetic acid production by the fungal endophyte of tall fescue. *Agronomy Journal* 82:878–880.

Delrot, S. and J. L. Bonnemain. 1981. Involvement of protons as a substrate for the sucrose carrier during phloem loading in *Vicia fava* leaves. *Plant Physiology* 67:560–564.

Dong, Z., Canny, M. J., McCully, M. E., Roboredo, M. R., Cabadilla, C. F., Ortega, E., and R. Rodés. 1994. A nitrogen-fixing endophyte of sugarcane stems. A new role for the apoplast. *Plant Physiology* 105:1139–1147.

Edwards, D., Kerp, H., and H. Hass. 1998. Stromata in early land plants an anatomical and ecophysiological approach. *Journal of Experimental Botany* 49:255–278.

Farrar, S. C. and J. F. Farrar. 1986. Compartmentation and fluxes of sucrose in intact leaf blades of barley. *New Phytologist* 103:645–647.

Faure, J. D., Vittorioso, P., Santoni, V., Fraisier, V., Prinsen, E., Barlier, I., Van Onckelen, H., Caboche, M., and C. Bellini. 1998. The pasticcino gene of *Arabidopsis thaliana* are involved in the control of cell division and differentiation. *Journal of Experimental Botany* 49:255–278.

Fletcher, L. R. and I. C. Harvey. 1981. An association of a *Lolium* endophyte with ryegrass staggers. *New Zealand Veterinary Journal* 29:185–186.

Giaquinta, R. 1976. Evidence for phloem loading from the apoplast chemical modification of membrane sulfhydryl groups. *Plant Physiology* 57:872–875.

Giaquinta, R. T. 1977. Phloem loading of sucrose, pH dependence and selectivity. *Plant Physiology* 59:750–753.

Glenn, A. E., Bacon, C. W., Price, R., and R. T. Hanlin. 1996. Molecular phylogeny of *Acremonium* and its taxonomic implications. *Mycologia* 88:369–383.

Greulich, F., Horio, E., Shimanuki, T., and T. Yoshihaa. 1999. Field results confirm natural plant protection by the endophytic fungus *Epichloë typhina* against the pathogenic fungus *Cladosporium phlei* on timothy leaves. *Annal of the Phytopathological Society of Japan* 65:454–459.

Guerber, C. A., West, C. P., Carson, R. D., and A. M. Haveley. 2007. Dehydrin expression in drought-stressed tall fescue. In *Proceedings of the 6th International Symposium on Fungal Endophytes of Grasses*, eds. A. J. Popay and E. R. Thom, pp. 225–227. Dunedin: New Zealand Grassland Association.

Gwinn, K. D. and A. M. Gavin. 1992. Relationship between endophyte infestation level of tall fescue seed lots and *Rhizoctonia zeae* seedling disease. *Plant Disease* 76:911–914.

Haarmann, T., Machado, C., Lübbe, Y., Correia, T., Schardl, C. L., Panaccione, D. G., and P. Tudzynski. 2005. The ergot alkaloid gene cluster in *Claviceps purpurea* extension of the cluster sequence and intra species evolution. *Phytochemistry* 66:1312–1320.

Hahn, D., Fiehn, O., McManus, M. A., and D. B. Scott. 2007. Metabolic profiling of endophyte-infected and endophyte-free ryegrass grown under sufficient water supply and drought. In *Proceedings of the 6th International Symposium on Fungal Endophytes of Grasses*, eds. A. J. Popay and E. R. Thom, p. 189. Dunedin: New Zealand Grassland Association.

Hartung, W., Weiler, E. W., and J. W. Radin. 1992. Auxin and cytokinins in the apoplastic solution of dehydrated cotton leaves. *Journal of Plant Physiology* 140:324–327.

Hill, N. S., Parrott, W. A., and D. D. Pope. 1991. Ergopeptine alkaloid production by endophytes in a common tall fescue genotype. *Crop Science* 31:1545–1547.

Hill, N. S., Stringer, W. C., Rottinghaus, G. E., Belesky, D. P., Parrott, W. A., and D. D. Pope. 1990. Growth, morphological, and chemical component responses of tall fescue to *Acremonium coenophialum*. *Crop Science* 30:156–161.

Hinton, D. M. and C. W. Bacon. 1985. The distribution and ultrastructure of the endophyte of toxic tall fescue. *Canadian Journal Botany* 63:36–42.

Huber, S. C. and D. E. Moreland. 1980. Translocation efflux of sugars across the plasmalemma of mesophyll protoplasts. *Plant Physiology* 65:560–562.

Jackson, J. A., Varney, D. R., Petroski, R. J., Powell, R. G., Bush, L. P., Siegel, M. R., Hemken, R. W., and P. M. Zavos. 1996. Physiological responses of rats fed loline and ergot alkaloids from endophyte-infected tall fescue. *Drug and Chemical Toxicology* 19:85–96.

Jarvis, M. C. 1998. Intercellular separation forces generated by intracellular pressure. *Plant Cell and Environment* 21:1308–1310.

Jarvis, M. C., Briggs, S. P. H., and J. P. Knox. 2003. Intercellular adhesion and cell separation in plants. *Plant Cell and Environment* 26:977–989.

Jeffree, C. E., Dale, J. E., and S. C. Fry. 1986. The genesis of intercellular spaces in developing leaves of *Phaseolus vulgaris* L. *Protoplasma* 132:90–98.

Johnson, L. J., Steringa, M., Koulman, M., Christensen, C. R., and R. J. Johnson. 2007. Biosynthesis of an extracellular siderophore is essential for maintenance of mutualistic endophyte–grass symbioses. In *Proceedings of the 6th International Symposium on Fungal Endophytes of Grasses*, eds. A. J. Popay and E. R. Thom, pp. 177–179. Dunedin: New Zealand Grassland Association.

Kloepper, J. W. and C. J. Beauchamp. 1992. A review of issues related to measuring colonization of plant roots by bacteria. *Canadian Journal of Microbiology* 38:1219–1232.

Kneale, J. and J. F. Farrar. 1985. The localization and frequency of haustoria in colonies of brown rust on barley leaves. *New Phytologist* 101:495–505.

Knox, J. P. 1990. Cell-adhesion, cell-separation and plant morphogenesis. *Plant Journal* 2:137–141.

Kolloffel, C. and P. W. T. Linssen. 1984. The formation of intercellular spaces in the cotyledons of developing and germinating pea seeds. *Protoplasma* 120:12–19.

Koshino, H., Terada, S., Yoshihara, T., Sakamura, S., Shimanuki, T., Sato, T., and A. Tajimi. 1988. Three phenolic acid derivatives from stromata of *Epichloë typhina* on *Phleum pratense*. *Phytochemistry* 27:1333–1338.

Koshino, H., Yoshihara, T., Sakamura, S., Shimanuki, T., Sato, T., and A. Tajimi. 1989. A ring B aromatic sterol from stromata of *Epichloë typhina*. *Phytochemistry* 28:771–772.

Kuldau, G. and C. W. Bacon. 2008. Clavicipitaceous endophytes their ability to enhance grass resistance to multiple stresses. *Biological Control* 46:57–71.

Kunkel, B. N. and D. M. Brooks. 2002. Cross talk between signaling pathways in pathogen defense. *Current Opinions in Plant Biology* 5:325–331.

Kunkel, B. A., Grewal, P. S., and M. F. Quigley. 2004. A mechanism of acquired resistance against an entomopathogenic nematode by *Agrostis ipsilon* feeding on perennial ryegrass harboring a fungal endophyte. *Biological Control* 29:100–108.

Kursanov, A. L. and M. I. Brovchenko. 1970. Sugars in the free space of leaf plates their origin and possible involvement in transport. *Canadian Journal Botany* 48:1243–1250.

Leon, J., Rojo, E., and J. Sanchez-Serrano. 2001. Wound signalling in plants. *Journal of Experimental Botany* 52:1–9.

Lewis, G. C. 2004. Effects of biotic and abiotic stress on the growth of three genotypes of *Lolium perenne* with and without infection by the fungal endophyte *Neotyphodium lolii*. *Annals of Applied Biology* 144:53–63.

Lewis, G. C., Bakken, A. K., MacDuff, J. H., and N. Raistrick. 1996. Effect of infection by the endophytic fungus *Acremonium lolii* on growth and nitrogen uptake by perennial ryegrass (*Lolium perenne*) in flowing solution culture. *Annals Applied Biology* 129:451–460.

Lewis, G. C. and R. O. Clements. 1986. A survey of ryegrass endophyte (*Acremonium loliae)* in the U.K. and its apparent ineffectuality on a seedling pest. *Journal of Agricultural Science Cambridge* 107:633–638.

Lewis, G. C. and B. Vaughan. 1997. Evaluation of a fungal endophyte (*Neotyphodium lolii*) for control of leatherjackets (*Tipula* spp.) in perennial ryegrass. *Annals Applied Biology Supplement* 130:34–35.

Lyons, P. C., Plattner, R. D., and C. W. Bacon. 1986. Occurrence of peptide and clavine ergot alkaloids in tall fescue. *Science* 232:487–489.

Madore, M. and J. A. Webb. 1981. Lead free space analysis and veins loading in *Curcurbita pepo*. *Canadian Journal Botany* 59:2550–2557.

Mahmood, T., Gergerich, R. C., Milius, E. A., West, C. P., and C. J. D'Arcy. 1993. Barley yellow dwarf viruses in wheat, endophyte-infected and endophyte-free tall fescue and other hosts in Arkansas. *Plant Disease* 77:225–228.

Malinowski, D. P., Alloush, G. A., and D. P. Belesky. 2000. Leaf endophyte *Neotyphodium coenophialum* modifies mineral uptake in tall fescue. *Plant and Soil* 227:115–126.

Malinowski, D. P. and D. P. Belesky. 1999a. *Neotyphodium coenophialum*-endophyte infection affects the ability of tall fescue to use sparingly available phosphorus. *Journal of Plant Nutrition* 22:835–853.

Malinowski, D. P. and D. P. Belesky. 1999b. Tall fescue aluminum tolerance is affected by *Neotyphodium coenophialum* endophyte. *Journal of Plant Nutrition* 22:1335–1349.

Malinowski, D. P. and D. P. Belesky. 2000. Adaptations of endophyte-infected cool-season grasses to environmental stresses mechanisms of drought and mineral stress tolerance. *Crop Science* 40:923–940.

Malinowski, D. P., Belesky, D. P., Hill, N. S., Baligar, V. C., and J. M. Fedders. 1998. Influence of phosphorus on the growth and ergot alkaloid content of *Neotyphodium coenophialum*-infected tall fescue (*Festuca arundinacea* Schreb.). *Plant Soil* 198:53–61.

Malinowski, D. P., Belesky, D. P., and G. C. Lewis. 2005. Abiotic stresses in endophytic grasses. In *Neotyphodium in Cool-Season Grasses*, eds. C. A. Roberts, C. P. West, and D. E. Spiers, pp. 187–199. Iowa: Blackwell Publishing.

Malinowski, D., Leuchtmann, A., Schmidt, D., and J. Nosberger. 1998. Growth and water status in meadow fescue (*Festuca pratensis*) is affected by *Neotyphodium* and *Phialophora* endophytes. *Agronomy Journal* 89:673–678.

Malinowski, D. P., Zuo, H., Belesky, D. P., and G. A. Alloush. 2004. Evidence for copper binding by extracellular root exudates of tall fescue but not perennial ryegrass infected with *Neotyphodium* spp. endophytes. *Plant and Soil* 267:1–12.

Martens, P. 1937. L'origine des espaces intercellulaires. *Cellule* 46:355–388.

Mauseth, J. D. and T. Fujii. 1994. Resin-casting—A method for investigating apoplastic spaces. *American Journal of Botany* 81:104–110.

Moon, C. D., Tapper, B. A., and B. Scott. 1999. Identification of *Epichloë* endophytes in planta by a microsatellite-based PCR fingerprinting assay with automated analysis. *Applied and Environmental Microbiology* 65:1268–1279.

Moy, M., Belanger, F. C., Duncan, R. R., Freehoff, A., Leary, C., Sullivan, R., and J. F. White Jr. 2000. Identification of epiphyllous mycelial nets on leaves of grasses infected by clavicipitaceous endophytes. *Symbiosis* 28:291–302.

Mundy, J., Nielsen, H. B., and P. Brodersen. 2006. Crosstalk. *Trends in Plant Science* 11:63–64.

Ng, A., Greenshields, R. N., and K. W. Waldron. 1997. Oxidative cross-linking of corn bran hemicellulose formation of ferulic acid dehydrodimers. *Carbohydrate Research* 303:459–462.

Oliver, J. W., Strickland, J. R., Waller, J. C., Fribourg, H. A., Linnabary, R. D., and L. K. Abney. 1998. Endophytic fungal toxin effect on adrenergic receptors in lateral saphenous veins (cranial branch) of cattle grazing tall fescue. *Journal Animal Science* 76:2853–2856.

Panaccione, D. G., Johnson, R. D., Wang, J. H., Young, C. A., Damrongkool, P., Scott, B., and C. L. Schardl. 2001. Elimination of ergovaline from a grass-*Neotyphodium* endophyte symbiosis by genetic modification of the endophyte. *Proceedings of the National Academy of Sciences of the USA* 98:12820–12825.

Panka, D., Podkowka, L., and R. Lamparski. 2004. Preliminary observations on the resistance of meadow fescue (*Festuca pratensis* Huds.) infected by *Neotyphodium uncinatum* to disease and pest and nutritive value. In *Proceeding 5th International Symposium on Neotyphodium/Grass Interactions*, May 23–26, eds. R. Kallenbach, C. F. Rosenkrans, and T. R. Lock, p. 401. Fayetteville: University of Arkansas.

Petroski, R. J., Dornbos, D. L. Jr., and R. G. Powell. 1990. Germination and growth inhibition of annual ryegrass (*Lolium multiflorum* L.) and Alfalfa (*Medicago sativa*) by loline alkaloids and synthetic *N*-acylloline derivatives. *Journal of Agriculture and Food Chemistry* 38:1716–1718.

Popay, A. J. and S. A. Bonos. 2005. Biotic responses in endophytic grasses. In *Neotyphodium in Cool-Season Grasses*, eds. C. A. Roberts, C. P. West, and D. E. Spiers, pp. 163–185. New York: Blackwell Publishing.

Popya, A. J. and D. D. Rowan. 1994. Endophytic fungi as mediators of plant–insect interactions. In *Insect–Plant Interactions*, ed. E.A. Bernays, pp. 84–103. Boca Raton: CRC Press.

Porter, J. K. 1994. Chemical constituents of grass endophytes. In *Biotechnology of Endophytic Fungi of Grasses*, eds. C.W. Bacon and J. F. White Jr., pp. 103–123. Boca Raton: CRC Press.

Porter, J. K. 1995. Analysis of endophyte toxins *Fescue* and other grasses toxic to livestock. *Journal Animal Science* 73:871–880.

Porter, J. K., Bacon, C. W., Cutler, H. G., Arrendale, R. F., and J. D. Robbins. 1985. In vitro auxin production by *Balansia epichloë*. *Phytochemistry* 24:1429–1431.

Porter, J. K., Bacon, C. W., and J. D. Robbins. 1979. Ergosine, ergosinine, chanoclavine I from *Epichloë typhina*. *Journal of Agriculture and Food Chemistry* 27:595–598.

Porter, J. K., Bacon, C. W., Robbins, J. D., Himmelsbach, D. S., and H. C. Higman. 1977. Indole alkaloids from *Balansia epichloë* (Weese). *Journal of Agriculture and Food Chemistry* 25:88–93.

Powell, R. G. and R. J. Petroski. 1992. Alkaloid toxins in endophyte-infected grasses. *Natural Toxins* 23:185–193.

Prat, R., Andre, J. P., Mutsftschiev, S., and A. M. Catesson. 1997. Three-dimensional study of the intercellular gas space in a complex network of intercellularigna radiata. *Protoplasma* 196:69–77.

Pyke, K. A. 1991. Temporal and spatial development of the cells of the expanding 1st leaf of *Arabidopsis thaliana* (L.) Heynh. *Journal of Experimental Botany* 42:1407–1416.

Rasmussen, S., Parsons, A. J., Liu, Q., Xue, H., and J. A. Newman. 2007. High nutrient supply and carbohydrate content reduce endophyte and alkaloid concentration. In *Proceedings of the 16th International Symposium on Fungal Endophytes of Grasses*, eds. A. J. Popay, and E. R. Thom, pp. 135–138. Dunedin: New Zealand Grassland Research Association.

Raymond, S. L., Smith, T. K., and H. V. L. N. Swamy. 2003. Effects of feeding a blend of grains naturally contaminated with *Fusarium* mycotoxins on feed intake, serum chemistry, and hematology of horses, and the efficacy of a polymeric glucomannan mycotoxin adsorbent. *Journal of Animal Science* 81:2123–2130.

Rice, J. S., Pinkerton, B. W., Stringer, W. C., and D. J. Undersander. 1990. Seed production in tall fescue as affected by fungal endophyte. *Crop Science* 30:1303–1305.

Richardson, M. D., Hill, N. S., and C. S. Hoveland. 1990. Rooting patterns of endophyte infected tall fescue grown under drought stress. *Agronomy Abstracts* 81:129.

Riedell, W. E., Chapman, G. W., Petroski, R. J., and R. G. Powell. 1991. Naturally occurring and synthetic loline alkaloid derivatives insect feeding behavior modification and toxicity. *Journal of Entomological Science* 26:122–129.

Roberts, C. A., Marek, S. M., Niblack, T. L., and A. L. Karr. 1992. Parasitic *Meloidogyne* and *Acremonium* increase chitinase in tall fescue. *Journal of Chemical Ecology* 18:1007–1116.

Roland, J.-C. 1978. Cell wall differentiation and stages involved with intercellular gas space opening. *Journal of Cell Science* 32:325–336.

Romberge, J. A., Hejnowicz, Z., and J. F. Hill. 1993. *Plant Structure, Function and Development*. Springer-Verlag, New York.

Rowan, D. D., Hunt, M. B., and D. L. Gaynor. 1986. Peramine, a novel insect feeding deterrent from ryegrass infected with the endophyte *Acremonium loliae*. *Journal Chemical Society Chemical Communication* 935–936.

Rowan, D. D. and G. C. M. Latch. 1994. Utilization of endophyte-infected perennial ryegrasses for increased insect resistance. In *Biotechnology of Endophytic Fungi of Grasses*, eds. C.W. Bacon and J. F. White, Jr., pp. 169–183. Boca Raton: CRC Press.

Roylance, J. T., Hill, N. S., and C. S. Agee. 1994. Ergovaline and peramine production in endophyte-infected tall fescue independent regulation and effects of plant and endophyte genotype. *Journal of Chemical Ecology* 20:2171–2183.

Rykard, D. M. 1983. Comparative morphology of the conidial states and host–parasite relationship in members of Balansiae (Clavicipitaceae). PhD thesis. University of Georgia.

Rykard, D. M., Bacon, C. W., and E. S. Luttrell. 1985. Host relations of *Myriogenospora atramentosa* and *Balansia epichloë* (Clavicipitaceae). *Phytopathology* 75:950–956.

Sakurai, N. 1998. Dynamic function and regulation of apoplast in the plant body. *Journal of Plant Research* 111:133–148.

Sattelmacher, B. 2001. The apoplast and its significance for plant mineral nutrition. *New Phytologist* 149:167–192.

Schardl, C. L. 2001. *Epichloë festucae* and related mutualistic symbionts of grasses. *Fungal Genetics and Biology* 33:69–82.

Schardl, C. L. and A. Leuchtmann. 1999. Three new species of *Epichloë* symbiotic with North American grasses. *Mycologia* 91:95–107.

Schardl, C. L., Leuchtmann, A., Chung, K.-R., Penny, D., and M. R. Siegel. 1997. Co-evolution by common descent of fungal symbionts (*Epichloë* spp.) and grass hosts. *Molecular Biology and Evolution* 14:133–143.

Schardl, C. L. and C. D. Moon. 2003. Processes of species evolution in *Epichloë /Neotyphodium* endophytes of grasses. In *Clavicipitalean Fungi*, eds. J. F. White Jr., C. W. Bacon, N. L. Hywel-Jones, and J. W. Spatafora, pp. 273–310. New York: Marcel Dekker.

Schardl, C. L. and M. R. Siegel. 1992. Genetics of *Epichloë typhina* and *Acremonium coenophialum*. In *Acremonium/Grass Interactions*, eds. S. S. Quisenberry and R. E. Joost, pp. 169–185. Amsterdam: Elsevier Science Publishers.

Scheffer, J., Ziv, C., Yarden, O., and P. Tudzynski. 2005. The COT1 homologue CPCOT1 regulates polar growth and branching and is essential for pathogenicity in *Claviceps purpurea. Fungal Genetics and Biology* 42:107–118.

Schuenemann, G. M., Edwards, J. L., Hopkins, F. M., Rohrbach, N. R., Adair, H. S., Scenna, F. N., Waller, J. C., Oliver, J. W., Saxton, A. M., and F. N. Schrick. 2005. Fertility aspects in yearling beef bulls grazing endophyte-infected tall fescue pastures. *Reproduction, Fertility and Development* 17:479–486.

Siegel, M. R. and G. C. M. Latch. 1991. Expression of antifungal activity in agar culture by isolates of grass endophytes. *Mycologia* 83:529–537.

Siegel, M. R., Latch, G. C. M., Bush, L. P., Fannin, F. F., Rowan, D. D., Tapper, B. A., Bacon, C. W., and M. C. Johnson. 1990. Fungal endophyte-infected grasses: Alkaloid accumulation and aphid response. *Journal of Chemical Ecology* 16:3301–3315.

Spiering, M. J., Moon, C. D., Wilkinson, H. H., and C. L. Schardl. 2005. Gene clusters for insecticidal loline alkaloids in the grass-endophytic fungus *Neotyphodium uncinatum. Genetics* 169:1403–1414.

Takemoto, D., Tanaka, A., and B. Scott. 2006. Ap67Phox-like regulator is recruited to control hyphal branching in a fungal–grass mutualistic symbiosis. *The Plant Cell* 18:2807–2821.

Tanaka, A., Christensen, M. J., Takemoto, D., Park, P., and B. Scott. 2006. Reactive oxygen species play a role in regulating a fungus–perennial ryegrass mutualistic interaction. *The Plant Cell* 18:1052–1066.

Tanaka, A., Christensen, M. J., Takemoto, D., and B. Scott. 2007. Endophyte production of reactive oxygen species is critical for maintaining the mutualistic symbiotic interaction between *Epichloë festucae* and Pooid grasses. In *Proceedings of the 6th International Symposium on Fungal Endophytes of Grasses*, eds. A. J. Popay and E. R. Thom, pp. 185–188. Dunedin: New Zealand Grassland Association.

Tanaka, A., Tapper, B., Popay, A. J., Parker, L. J., and B. Scott. 2005. A symbiosis expressed non-ribosomal peptide synthetase from a mutualistic fungal endophyte of perennial ryegrass confers protection to the symbiotum from insect herbivory. *Molecular Microbiology* 57:1036–1050.

Tapper, B. A. and G. A. Lane. 2004. Janthitrems found in a *Neotyphodium* endophyte of perennial ryegrass. In *Proceedings of the 6th International Symposium on Fungal Endophytes of Grasses*, eds. A. J. Popay and E. R. Thom, pp. 167–170. Dunedin: New Zealand Grassland Association.

Taylor, J. E., Hatcher, P. E., and N. D. Paul. 2004. Crosstalk between plant responses to pathogens and herbivores: A view from the outside in. *Journal of Experimental Botany* 55:159–168.

Taylor, J. E. and M. R. McAinsh. 2004. Signalling crosstalk in plants: Emerging issues. *Journal of Experimental Botany* 55:147–149.

Tejera, N., Ortega, E., Rodes, R., and C. Lluch. 2006. Nitrogen compounds in the apoplastic sap of sugarcane stem: Some implications in the association with endophytes. *Journal of Plant Physiology* 163:80–85.

TePaske, M. R., Powell, R. G., and S. L. Clement. 1993. Analyses of selected endophyte-infected grasses for the presence of loline-type and ergot-type alkaloids. *Journal Agricultural Food Chemistry* 41:2299–2303.

Tetlow, I. J. and J. F. Farrar. 1993. Apoplastic sugar concentration and pH in Barley leaves infected with brown rust. *Journal of Experimental Biology* 44:929–936.

Tudzynski, P. and J. Scheffer. 2004. *Claviceps purpurea*: Molecular aspects of a unique pathogenic lifestyle. *Molecular Plant Pathology* 5:377–388.

Van der Weele, C. M. 1996. Water in aerenchyma spaces in roots: A fast diffusion path for solutes. *Plant and Soil* 184:131–141.

Welty, R. E., Azevedo, M. D., and K. L. Cook. 1986. Detecting viable *Acremonium* endophytes in leaf sheaths and meristems of tall fescue and perennial grasses. *Plant Disease* 70:431–435.

Welty, R. E. and R. E. Barker. 1993. Reaction of twenty cultivars of tall fescue to stem rust in controlled and field environments. *Crop Science* 33:963–967.

White, J. F., Jr., Bacon, C. W., and D. M. Hinton. 1997. Modifications of host cells and tissues by the biotrophic endophyte *Epichloë amarillans* (Clavicipitaceae; Ascomycotina). *Canadian Journal of Botany* 75:1061–1069.

White, J. F., Jr. and G. T. Cole. 1985. Endophyte–host associations in forage grasses. III. In vitro inhibition of fungi by *Acremonium coenophialum. Mycologia* 77:487–489.

White, J. F., Jr. and J. R. Owens. 1992. Stromal development and mating system of *Balansia epichloë*, a leaf-colonizing endophyte of warm-season grasses. *Applied and Environmental Microbiology* 58:513–519.

Wilkinson, H. H. and C. L. Schardl. 1997. The evolution of mutualism in grass–endophyte associations. In *Biotechnology of Endophytic Fungi of Grasses*, eds. C. W. Bacon and N. S. Hill, pp. 13–25. New York: Plenum Press.

Wilkinson, H. H., Siegel, M. R., Blankenship, J. D., Mallory, A. C., Bush, L. P., and C. L. Schardl. 2000. Contribution of fungal loline alkaloids to protection from aphids in a grass–endophyte mutualism. *Molecular Plant-Microbe Interactions* 13:1027–1033.

Yates, S. G., Fenster, J. C., Bartelt, R. J., and R. G. Powell. 1987. Toxicity assay of tall fescue extracts, fractions and alkaloids using the large milkweed bug. In *Proceedings of the Tall Fescue Mtg.: Southern Region Information Exchange Group 37*, November. 17–18, Memphis, TN.

Yoshihara, T., Koshino, H., Togiya, S., Terada, T., Tsukada, S., Sakamura, S., Shimanuki, T., Sato, T., and A. Tajimi. 1988. Fungitoxic compounds from stromata of *Epichloë typhina*. Abstracts of Papers, *5th International Congress of Plant Pathology*, August 20–27, p. 331.

Yoshihara, T., Togiya, S., Koshino, H., Sakamura, S., Shimanuki, T., Sato, T., and A. Tajimi. 1985. Three fungitoxic sequiterpenes from stromata of *Epichloë typhina. Tetrahedron Letters* 26:5551–5554.

Young, C. A., Bryant, M. K., Christensen, M. J., Tapper, B. A., Bryan, G. T., and B. Scott. 2005. Molecular cloning and genetic analysis of a symbiosis-expressed gene cluster for lolitrem biosynthesis from a mutualistic endophyte of perennial ryegrass. *Molecular Genetics and Genomics* 274:13–29.

Yue, Q., Miller, C. J., White, J. F., Jr., and M. D. Richardson. 2000. Isolation and characterization of fungal inhibitors from *Epichloë festucae. Journal of Agricultural and Food Chemistry* 48:4687–4692.

Yue, Q., Wang, C. L., Gianfagna, T. J., and W. A. Meyer. 2001. Volatile compounds of endophyte-free and infected tall fescue (*Festuca arundinacea* Schreb.). *Phytochemistry* 58:935–941.

Zhang, N., Raftery, M., Richardson, K., Christensen, M. J., and J. Schmid. 2007. *Neotyphodium lolii* induces a limited host defense response by *Lolium perenne*. In *Proceedings of the 6th International Symposium on Fungal Endophytes of Grasses*, eds. A. J. Popay and E. R. Thom, pp. 199–202. Dunedin: New Zealand Grassland Association.

Index

R

S

T

V

X

Z